Cacti of the Trans-Pecos & Adjacent Areas

Cacti of the Trans-Pecos & Adjacent Areas

 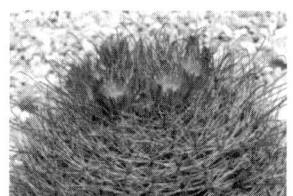 

A. Michael Powell and
James F. Weedin

Texas Tech University Press

Copyright © 2004 Texas Tech University Press

All rights reserved. No portion of this book may be reproduced in any form or by any means, including electronic storage and retrieval systems, except by explicit prior written permission of the publisher except for brief passages excerpted for review and critical purposes.

This book is typeset in Sabon. The paper used in this book meets the minimum requirements of ANSI/NISO Z39.48–1992 (R1997). ∞

Publication of this book was made possible by the contributors acknowledged on p. 510.

Library of Congress Cataloging-in-Publication Data
Powell, A. Michael.
   Cacti of the Trans-Pecos and adjacent areas / A. Michael Powell and James F. Weedin.
     p. cm.
Includes bibliographical references (p. 479–95).
  ISBN 0-89672-531-6 (cloth : alk. paper)
  1. Cactus—Trans-Pecos (Tex. and N.M.)—Identification.
I. Weedin, James F. II. Title.
QK495.C11P69 2004
583'.56'09764—dc22

                                                  2003024641

Designed by Barbara Werden

Printed in the United States of America

04 05 06 07 08 09 10 11 12 / 9 8 7 6 5 4 3 2 1

Texas Tech University Press
Box 41037
Lubbock, Texas 79409-1037 USA
800.832.4042
www.ttup.ttu.edu

*To Sharon C. Yarborough and
Keith A. Yarborough:
friends, colleagues, authors, supporters, and
benefactors.*

*To Shirley A. Powell and Teresa Johnson Weedin,
partners in exploring.*

Contents

Preface ix

Acknowledgments xii

Introduction 1

Biology of Cacti 4

Morphology and Anatomy 4

Physiology (= Primary Compound Chemistry) 20

Secondary Compound Chemistry 26

Ecology and Biogeography 30

Evolution 36

Pollination Biology 39

Uses and Other Ethnobotany 44

Horticulture 57

Conservation 61

Classification of Cacti 65

Key to the Genera 68

Descriptive Cactus Flora 73

Selected Glossary 465

Literature Cited 479

Index 497

Preface

THE Chihuahuan Desert Region (Wauer and Riskind, 1977) holds more species of cacti than any other comparable area in North America (Hernández and Bárcenas, 1995). This center of cactus diversity lies mostly in northern Mexico; it extends into the United States only in the Trans-Pecos region of Texas and some adjacent parcels in New Mexico and Arizona. This area west of the Pecos River involves about 32,000 square miles, or 20.5 million acres, equivalent in size to the state of Maine. The Trans-Pecos shares some species and vegetation types with the Edwards Plateau ("Hill Country") of Central Texas, the Tamaulipan thorn-scrub of the Rio Grande valley, the Great Plains of the Texas Panhandle and eastern New Mexico, and the Apachean floristic region of southeastern Arizona. However, by far the greatest part of the Trans-Pecos vegetation and flora pertains to the Chihuahuan Desert. Consequently, the Trans-Pecos is one of the major centers of cactus distribution in the United States.

In Trans-Pecos Texas, the southern Big Bend area is best known as cactus country. The lowest elevation in the entire Chihuahuan Desert Region lies in the Boquillas Basin near the Rio Grande, just downslope from the southeastern foothills of the Chisos Mountains. Endemic taxa and subfossil remains of desert plants indicate that the Boquillas Basin served as a major biological refugium during the last Ice Age and thus must have been one of the major dispersal centers for Chihuahuan Desert species.

The present work is the first that emphasizes the cacti of Trans-Pecos Texas. Several publications with wider geographic coverage that have included the Trans-Pecos cactus flora are cited in the Introduction. Most of the sources are out of print. The cactus books most widely used to identify Texas cacti were those by Del Weniger and Lyman Benson. The extensive nomenclatural differences between the Weniger and Benson publications were disconcerting to many who wanted to use the scientific names of southwestern cacti. One objective of our cactus treatment was to reconcile the differences between Weniger and Benson, where possible, and to provide alternative nomenclature that reflects contemporary taxonomic research.

Two advantages of regional treatments such as *Cacti of the Trans-Pecos and Adjacent Areas* are that (1) more detailed information can be included about the biology of cacti in general and about each species and (2) identification is simpler without the need to "wade through" extraneous species. One of our objectives in preparing the current *Cacti* was to allow preliminary identification by matching unidentified cacti with color photographs of plants in flower and in fruit. Distribution maps, keys, and descriptions provide a second level of information for systematic identification that supersedes the first impression gained from the pictures. In addition to the means for field, greenhouse, herbarium, and laboratory identifications, *Cacti* includes explanatory sections on morphology and anatomy, physiology, secondary compound chemistry, ecology and biogeography, chromosome numbers, evolution, pollination biology, uses and other ethnobotany, horticulture, conservation, diagnostic vegetative and floral characters, floral and fruit phenology, technical features of sterile specimens that allow identification of dried specimens and fragments, and anything else of potential relevance for confidence in identification of populations or individual species. For each species we include basionym citation, nomenclatural synonyms, common names that have come to our attention, and a translation of the Latinized international name.

Cacti of the Trans-Pecos and Adjacent Areas should allow confident identification of cacti throughout Texas, excepting varieties of *Echinocereus reichenbachii* and several eastern species of prickly pears. We have prepared full treatments for all cactus taxa known to occur in the Trans-Pecos except for certain chromosome races of *Opuntia leptocaulis* and certain prickly pears recognized by David Ferguson. There is at least brief mention of essentially all Texas taxa of Cactaceae. The Cis-Pecos Texas cactus species are not keyed but are diagnosed either in the context of discussing Trans-Pecos cactus species or in abridged treatments inserted as appendixes at the ends of genera sections. Photographs representing most of these taxa are included, except for the opuntias. In treating the relatively large and taxonomically complex genus *Opuntia,* we at least mention all of the Cis-Pecos species recognized in standard reference works such as Benson (1969b, 1982) and Britton and Rose ([1937] 1963).

The book was designed for use by dedicated nonprofessionals, self-taught hobbyists and naturalists, and serious students of cacti. Visitors to the national parks, state parks, and other natural areas in regions adjacent to the Trans-Pecos will use the book. The manual will be important to professionals in national and state park resource interpretation, wildlife biology, ecology, range management, and environmental consulting.

Many aspects of format follow the companion books *Trees and Shrubs of the Trans-Pecos and Adjacent Areas* (Powell, 1998a) and *Grasses of the Trans-Pecos and Adjacent Areas* (Powell, 2000). Metric system measurements are used in keys, descriptions, and discussions about the species. The English system is used for elevations and distances. Abbreviations for states are two letters in caps, following the U.S. Postal Service and current trends in scientific writing. Directions are in caps (e.g., N, E, S, W, or NW). Only one direction, NE, has the same abbreviation as a state, Nebraska. Author citations for plant names mostly follow those in the Missouri Botanical Garden, TROPICOS, and in Brummitt and Powell (1992).

Most of the morphological data were taken from specimens housed in the Sul Ross State University (SRSC) herbarium and from living specimens at Sul Ross. Some measurements also were taken from pertinent literature.

The generalized distribution maps reflect our current information about the ranges of relevant taxa in North America. Although we prefer to use geographically explicit dot maps for indicating taxon distribution, in this case we elected to use a more generalized mapping approach for several reasons. Dot maps are most effective when localities documented by herbarium specimens are numerous, showing patterns that reflect actual distributions. Relatively few herbarium specimens are available for many of the cacti. To fashion the range maps, we have used a combination of documented localities, remembered localities from our collective field experiences, and our understanding (sometimes insufficient) of the habitat requirements of each species. Distributional information recorded in various published sources has been incorporated or not, on a species-by-species basis, because mistakes are rampant in the literature. The generalized distributions portrayed in the present work are most accurate when dealing with taxa of restricted ranges. The mapped distributions are less precise when portraying wider-ranging taxa. The astute observer will notice where certain localized landforms (marginally conspicuous at this map scale, in most cases) have been ignored because they are predictably islands of unsuitable habitat (e.g., mountains within the ranges of the deep-soil low plains cacti). The high Chisos Mountains are among the smallest such habitat islands mapped as holes in species ranges at this scale.

The small inset maps show approximate ranges of widely distributed taxa. Spots on the inset map denote regional occurrence of species outside the Trans-Pecos and not necessarily specifically documented localities. Taxa with limited ranges outside the Trans-Pecos (only one spot, at this scale) are the minority, because most of the species included in the present work extend into two or more other states of the United States and/or Mexico. Distributional information for each taxon is summarized verbally in the text.

The term "cultivated" in figure legends denotes photography of specimens grown in containers, in the experimental *Opuntia* garden, or in the formal Cactus Garden at Sul Ross State University. The photographs were taken by the first author, except for those otherwise specifically acknowledged.

Acknowledgments

WE are especially grateful to Allan D. Zimmerman for making recommendations about the original book outline, for contributing much information about the Trans-Pecos cacti, and for reviewing and editing the manuscript, which in its current form does not necessarily reflect his opinions. A generous portion of Allan's special knowledge about cacti is infused in the current work. Allan also refined the mapped distributions of many species. Allan, Dale, and Marian Zimmerman contributed some of the photographs.

We thank Jim Henrickson for allowing us to consult a tentative treatment of Cactaceae that was prepared for the forthcoming *Chihuahuan Desert Flora*. The manuscript prepared by A. Zimmerman, C. Glass, R. Foster, and D. Pinkava (cited as Zimmerman et al., forthcoming) was often consulted, especially with regard to distributions in the Chihuahuan Desert in Mexico.

Many years of dedication earned Dave Ferguson special knowledge about *Opuntia*. He was generous in communicating to us much useful information about cacti of the northern Chihuahuan Desert Region. His contributions are liberally infused in the present treatment of *Opuntia*.

Both Jack Brady and Patty Manning maintained numerous cultivated cactus plants for ontogenetic observations, from seedlings to adults, and they potted many cacti in order to facilitate their long-term study. They helped to plant and maintain an experimental *Opuntia* plot. Jack also collected buds from numerous cultivated and field specimens to be used for obtaining chromosome counts. Patty assiduously planted, recorded, and mapped vouchered specimens in a cactus garden.

We appreciate the interest demonstrated and the study specimens provided by Gerald Raun, Richard Worthington, Kelly Bryan, Linda Hedges, Jackie Poole, Jim Talbot, Shane Lee, B. L. Turner, Teresa Weedin, and Don Campbell. Jerry Raun is a keen student of Trans-Pecos cacti. Over the years he has been kind enough to contribute cuttings of "strange" opuntias, specimens of other genera, and astute field observations regarding many cactus species. Richard contributed living specimens and photographs of cacti in El Paso County. Kelly had the foresight to salvage cactus plants from a construction site in the Franklin Mountains and made them available for our evaluations. Linda helped with critical literature and photographs, provided data concerning *Ferocactus wislizeni* and *Ancistrocactus,* and in general encouraged the project. Jackie thoughtfully delivered living specimens of certain South Texas cacti and offered a photograph of *Astrophytum.* Jim was an avid field observer, especially in Terrell County, and was generous in sharing his unique knowledge with us. Shane enthusiastically collected *Opuntia* cuttings from throughout the Trans-Pecos. Billie sent or brought pertinent literature, provided distributional information for some Texas cacti, smoothed out awkward phraseology (when asked), and was otherwise supportive of the project. Teresa secured beautiful specimens of *Coryphantha pottsiana.* Don Campbell, a cactus enthusiast from Colorado, heard that we needed specimens of *Opuntia basilaris,* and they soon arrived in the mail.

Numerous other individuals made important contributions to the current work. Bruce Parfitt kindly shared his knowledge about nomenclatural and other cactus matters. Scooter Cheatham and Lynn Marshall generously made available special photographs. James H. Everitt filled a critical need in allowing the use of some of his photographs of South Texas cacti. Jean Hardy made useful observations regarding *Thelocactus, Epithelantha,* and *Echinocereus viridiflorus* in the Solitario Dome. Ike and Sue Roberts shared information about Marathon Basin cacti. Kirsten Lund provided photo documentation of an *E. viridiflorus* var. *russanthus* morphotype in the south Chisos Mountains. Matthew Johnson and Rick and Nora Bowers sent photographic evidence of a range extension for *E. fendleri.* Dennie Miller nurtured important cactus specimens in the Chihuahuan Desert Research Institute (CDRI) greenhouse. Seth and Bonnie Warnock allowed access to field observation of certain cacti. Justin Allison took special care of *Pereskia aculeata.* Joe and Joyce Mussey arranged access for field investigation in Hudspeth County and have encouraged our work. Michael Chamberland offered a photo of *P. aculeata* in flower. Jim Scudday was on the alert for unusual cacti in unusual places when he found *O. davisii* in eastern Pecos County. Jim also reviewed part of the manuscript and offered helpful suggestions. Michael Lange kindly offered information about *Echinocereus.* Patrick Griffith contributed literature, observations, and specimens of *Opuntia.* Miller Talbot, always hunting Easter eggs, provided useful specimens of *Opuntia* from Terrell County and collected an unusual *Coryphantha* in south Brewster County. Jim Richerson made available needed clerical help. Bonnie McKinney generously shared her considerable knowledge about the distribution of certain cactus species. Gretchen Kliem pointed out critical literature, made editorial corrections, and shared information from her studies of the *Opuntia macrocentra* complex. She and her mother, Dora, also made available specimens of *Coryphantha missouriensis* from north Texas.

The Honorable Dudley Harrison, Terrell County judge, and Roby Golden contributed information about the Cactus Capital of Texas, Sanderson. Stanley

Jones provided nomenclatural information. The infectious enthusiasm shown by Chris Seabury made the collection of data more enjoyable. Teresa Davis, Joanne Calatayud, Shirley Powell, and Dorothy Angrist contributed word-processing skills and more. Dorothy was a persistent, skillful editor through each draft of the manuscript. Jennifer White, patient and efficient in the Biology Department office, cheerfully solved one computer-related problem after another and contributed many hours toward completing text and graphics. Ed McRae resolved word-processing problems. Computer programmer Karen A. Walker developed a process for electronic compilation of the index. Michael W. Nickell documented the existence of *Epithelantha micromeris* in Howard County, Texas, and brought the record to our attention. Sharon Yarborough, valued colleague and friend, avid supporter of floristic studies in Trans-Pecos Texas, offered advice, encouragement, and subsidy from the Edwill Fund. Martin Terry provided crucial impetus in raising subsidy funds and was generally supportive of the project. Martin also obtained a provisional radiocarbon date for an archeological cactus specimen. Steve Brack and Mesa Garden helped in obtaining certain photographs. We are grateful to Keith Sternes for his support of the project. Scott and JJ Lerich shared information about cacti at Elephant Mountain. JoAnn Klingemann and Kendall Craig used their graphics skills to produce Map 1. Pam Spooner, Mike Robinson, and Eleanor Wilson provided superb assistance in gathering biographical information and other literature. Sheri Tripp was helpful in many respects.

We thank the graduate students in the cactus and succulents classes at Sul Ross for intellectual and field participation in the study of Trans-Pecos cacti. These classes reminded us of how much we have learned from students and have been inspired by their interest. The 1997 class, including especially the most persistent "flower sniffers" (Bonnie Warnock, Marcia Roberts, Sally Roberts, Kirsten Lund), undertook olfactory tests for the presence or absence of a lemon scent in the flowers of *Echinocereus viridiflorus*. The 1999 class excelled in observations regarding floral phenology in most all the Trans-Pecos genera.

Shirley Powell contributed countless miles and hours in the mountains and desert, an expert field assistant, equally adept at sacking a bristly prickly pear, climbing out of a steep canyon, or locating the most cryptic specimen. The first author is deeply grateful for her companionship and for her expert help.

David Schmidly and Clyde Jones are champions of natural history studies in the Southwest. Their support of the current project led directly to its publication.

Support for fieldwork over the years was provided (to A. M. Powell) through scientific permits from Big Bend National Park, Big Bend Ranch State Park, and Guadalupe Mountains National Park and through the Faculty Research Enhancement Grant from Sul Ross State University. Jim Weedin wishes to acknowledge the following people for contributing to his studies of Cactaceae in the Trans-Pecos: Frank and Sue Bell; Lyman Benson; Mary Bridges; William Campbell; David Ferguson; Charles Glass; Bill and Dawn Grether; Donald Kolle; Ed and Beth Leuck; Dennie and Maggie Miller; Donald Pinkava; Gerald Raun; Robert Ross; Barton Warnock; Mike Lockhart; Marlene Hagerman; Jim Scudday; Rod Haenni; and Teresa Weedin. Weedin gratefully acknowledges grants/support from Chihuahuan Desert Research Institute, Colorado Cactus and Succulent Society, Community College of Aurora, Community Colleges of Colorado, Houston Cactus and Succulent Society, and Sul Ross State University.

Donald J. Pinkava and Stephan L. Hatch carefully reviewed the bulky manuscript, noted necessary changes and corrections, and offered copious suggestions for improving the work. We are very grateful to Don and Steve for their considerable efforts. We also appreciate the contributions of two anonymous reviewers. Gale Turner kindly provided the Latin diagnoses.

Cacti of the Trans-Pecos & Adjacent Areas

Introduction

TRANS-PECOS Texas is one of the most cactus-rich regions in the United States. Approximately 109 taxa, 76 species, and 33 varieties of cacti are found in the Trans-Pecos. Physiographically, the Trans-Pecos is a mountain and basin region, and it is occupied in part by the Chihuahuan Desert Biome. Two-thirds of the great Chihuahuan Desert Region (CDR) are in Mexico, while the northern one-third extends into the United States, through the Trans-Pecos and into New Mexico, along the Pecos and Rio Grande drainages, and barely into southeastern Arizona.

Cactaceae, the cactus family, includes 1,500–1,600 species in about 100–122 genera (Nobel, 1988; Hershkovitz and Zimmer, 1997), all of which are restricted to the New World with the exception of epiphytes in the genus *Rhipsalis* Gaertn. *Rhipsalis baccifera* (J. S. Muell.) Stearn is found in tropical forests of Kenya, Africa, and Sri Lanka and other islands of the Indian Ocean, including Madagascar (Nobel, 1994), where there are two additional endemic species of *Rhipsalis*. The Cactaceae make up the second-largest family that is essentially restricted to the New World, barely exceeded in diversity by the pineapple family Bromeliaceae (Gibson and Nobel, 1986). Cacti are widely distributed in the Americas, from Canada south to Patagonia in southern South America. Multiple lines of evidence suggest South America as the continent of origin for the family (Nobel, 1988).

Cacti in general are well-known components of North American deserts, particularly the Sonoran and Chihuahuan deserts. Trans-Pecos Texas (Map 1) at lower elevations is mostly desertic. Low basins and arid mountains and mesas extend from El Paso County east to the Pecos River and beyond (Plate 1). The desertic mountains are abundant in the southern Big Bend region. Mid-elevation basins of mesic mountains support extensive plains grasslands on the Diablo Plateau in Hudspeth County and in the Davis Mountains area in Jeff Davis, Presidio, and Brewster counties (Plates 2 and 3). In the central Trans-Pecos three mesic mountain ranges are dominated at middle and upper elevations by oak-

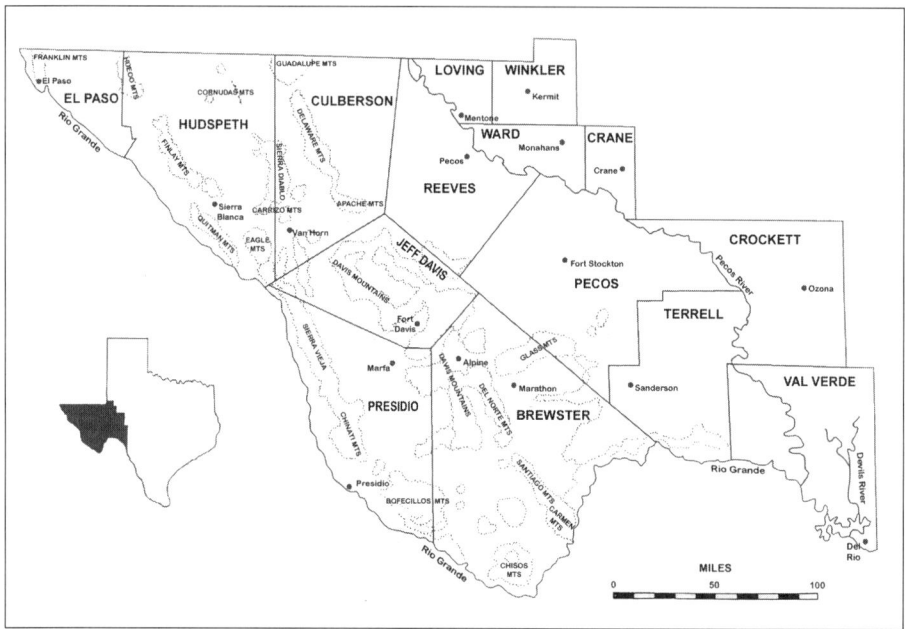

Map 1. Trans-Pecos Texas and adjacent counties, with some major geographic features shown.

juniper-pinyon woodlands (Plates 4 and 5). Relictual montane woodlands or conifer forests are found at the tops of these mountains: 8,749 feet at Guadalupe Peak; 8,382 feet at Mount Livermore in the Davis Mountains (Plate 6); 7,835 feet on Mount Emory in the Chisos Mountains (Plate 7). Another major mountain range in Presidio County is the moderately wooded Chinati Mountains, with its highest point, Chinati Peak, at 7,730 feet.

The relatively arid climate and variable desert and mountain physiognomy in the Trans-Pecos has favored the establishment of a diverse cactus flora. In the Trans-Pecos the average annual precipitation is about 12 inches, with annual precipitation usually increasing from about 8 inches in the west at El Paso to the east where an average 13–20 inches are recorded in the central mountains and nearby areas, depending upon elevation. Wide fluctuations in seasonal and annual precipitation are common, and severe periodic droughts have been recorded in the region since the late nineteenth century. In the twentieth century noted droughts have occurred in the 1930s, 1950s, and 1990s.

Miocene volcanic activity and other geologic dynamics have produced extensively varied substrates for the establishment of vegetation in the northern CDR. Igneous and sedimentary rock habitats are common, as are those featuring gravel, sand, and clay alluvium (Plates 8–11). Gypseous and saline deposits are exposed in many desert basins. More extensive discussion regarding the physical and vegetative characteristics of the Trans-Pecos can be found in Schmidly (1977), Diamond et al. (1988), and Powell (1998a, 2000).

In the United States the two major cactus provinces are Arizona and the Trans-Pecos region of Texas, each with similar numbers of cactus species. Benson

(1982) tabulated the following numbers of taxa for these areas: Texas, 126 taxa (73 species and 53 additional varieties); Arizona, 130 taxa (70 species and 60 varieties). We estimate that there are currently 135 taxa (94 species and 41 varieties) found in Texas. Approximately 109 taxa (76 species and 33 varieties) of the Texas cacti are distributed in the Trans-Pecos. Thus, about 79% of all Texas cacti are present in the Trans-Pecos. Other cactus-rich states listed by Benson are New Mexico with 88 taxa (53 species and 35 varieties) and California with 57 taxa (32 species and 25 varieties). The Chihuahuan Desert Region in Mexico includes substantially more cacti than either Texas or Arizona, approximately 256 taxa (194 species and 62 varieties; Zimmerman et al., forthcoming). In the country of Mexico there are more than twice as many taxa of cacti (563 species in 48 genera; Hernández and Bárcenas, 1995) than there are in the United States (Benson, 1982).

Actually, only about 20 taxa are strictly endemic to the Trans-Pecos. This means that most of the cactus species that occur in the Trans-Pecos also extend into other parts of Texas and/or into neighboring states and into northern Mexico. Depending upon taxonomic interpretations, only 4 genera, 15 species, and 11 varieties occur in Texas strictly east of the Pecos River. We have no doubt that future study will result in the description or identification of additional taxa for Texas, particularly in the genus *Opuntia*.

Large columnar cacti do not occur in the northern CDR, probably because it is too cold in the winter. Candelabriform species 2–6(–10) m tall do occur in the southern CDR (Henrickson and Johnston, 1986; Zimmerman et al., forthcoming): *Myrtillocactus geometrizans* (Mart.) Console, *Stenocereus griseus* (Haw.) F. Buxb., and *Pachycereus marginatus* (DC.) Britton & Rose. The existence of large columnar cacti in the Sonoran Desert of Arizona has been well publicized. In fact, the signature cactus species for Arizona is the saguaro, *Carnegiea gigantea* (Emory) Britton & Rose. Conversely, Trans-Pecos Texas is known for its diminutive cacti. Some of these plants, such as *Coryphantha minima* and *Echinocereus davisii* (Plate 12) are marble-size, with showy flowers larger than the stems.

The cacti of Trans-Pecos Texas have not been treated before in a single publication that was designed to feature the species of the area. Previous works of broader, statewide, or southwestern regional perspective have included some of the Trans-Pecos cacti or have attempted to include all of them. The most pertinent modern previous publications were the books by Benson (1982), *The Cacti of the United States and Canada,* and Weniger (1984), *Cacti of Texas and Neighboring States.* Other relevant previous cactus floristic treatments include Griffiths (1908–11); Britton and Rose (1919–23); Schulz and Runyon (1930); Earle (1963, 1980); Benson (1969b, 1970); Weniger (1970); Bravo-Hollis (1978); Fischer (1989); Bravo-Hollis and Sánchez-Mejorada (1991a, 1991b); and Zimmerman et al. (forthcoming). The Trans-Pecos cacti also have been treated in various floras and checklists, including Cory and Parks (1937); Gould (1962); Correll and Johnston (1970); Worthington (1989); Hatch et al. (1990); Kartesz (1994); Jones et al. (1997); and Diggs et al. (1999).

Biology of Cacti

REVIEW of the following subjects enhances appreciation for many aspects of cacti. The cacti are indeed remarkable (Nobel, 1994), particularly in some of their morphological and physiological adaptations to dry climates and specific habitats. In the following discussions, we emphasize features that contribute to identification and classification.

Morphology and Anatomy

In the Trans-Pecos region, and elsewhere, desert plants with spiny stems or pointed leaves collectively have been confused with "cacti" by some of the uninitiated. Actually, there are several families of desert plants that sometimes are misconstrued as cacti, particularly the yuccas (*Yucca* spp.; Plate 13), agaves or century plants (*Agave* spp.; Plate 14), sotols (*Dasylirion* spp.; Plate 15), and ocotillo (*Fouquieria splendens* Engelm.; Plate 16).

Cacti may be distinguished by fleshy (succulent), usually leafless stems, with spines localized in numerous specific areas along the stems called areoles. Often it is evident that the areoles are spirally arranged. Prickly pears and chollas may produce short-lived slender (cylindroid) leaves on young pads. Some primitive cacti produce broad, persistent leaves (*Pereskia* spp.; Plates 17 and 18), but these species are restricted to tropical and subtropical areas. Cacti also are distinguished by their flowers, which have numerous petal-like structures that intergrade with sepal-like structures to the outside. These petal- and sepal-like parts are referred to as tepals, petaloid and sepaloid tepals, or inner and outer tepals (Plate 19). In the flower center the stamens are numerous, and the pistil consists of an inferior ovary and four or more united carpels (Boke, 1980). The ovary has one seed chamber, usually with many ovules on several marginal (parietal) placentae. A pistil exhibits a single style with several stigma lobes at the apex. The ovules develop into seeds in fleshy ("berries") or dry fruits. In different cacti the color of the inner tepals may be white, yellow, purple, pink, magenta, red, or a combination of colors.

Habit

The largest cactus plants in the Trans-Pecos are certain prickly pears and chollas of the genus *Opuntia* (subfamily Opuntioideae). Trans-Pecos members of *Opuntia*, such as *O. engelmannii* and *O. imbricata* (Plate 20), may range to 1–3 m tall. Different species of *Opuntia* are branched, usually with one or a series of joints forming prostrate (Plate 21), declined, spreading, or upright shrubs. Prickly pears exhibit flattened joints, and chollas have cylindroid joints. Prickly pears and chollas may or may not have short trunks. In Mexico and some other areas, prickly pears may be arborescent, up to 10 m tall, with rather massive trunks.

In the Trans-Pecos the tallest members of the subfamily Cactoideae are the barrel cacti, specifically *Ferocactus wislizeni* (Plate 22) and *F. hamatacanthus*. The larger of the two is decidedly *F. wislizeni*, with the tallest known specimens, in El Paso County, approaching 2 m high. Some multistemmed and mounded specimens of *Echinocereus* are large in diameter (Plate 23). The biggest plants of *E. coccineus*, *E. stramineus*, and *E. enneacanthus* may form hundreds of stems and exceed 1 m in diameter. The large columnar cacti (e.g., cardon and saguaro in the Sonoran Desert) reach 15–23 m tall with trunks to nearly 1 m in diameter; equally large specimens grow in southern Mexico and South America.

Trans-Pecos members of the Cactoideae may be branched or unbranched. Branches may form near ground level or at almost any point up the stem. Many species, such as those in *Ferocactus* and *Echinomastus*, are unbranched. Single-stemmed plants of the Cactoideae may be columnar, cylindroid, obconic, globular, or hemispheric. Multistemmed plants may form caespitose mounds or mats. The plants of many Trans-Pecos species are cryptic, either because they are diminutive or because the stems retract almost completely underground (Plate 24). *Ariocarpus fissuratus* is commonly known as living rock cactus because the flattened stems resemble rocks, except for their pale gray-green color. When desiccated, the stems of *A. fissuratus* may shrink, pulling the stem apex below ground level and further obscuring the plant among surrounding limestone rubble.

In the ontogeny of cactus plants, from seedlings through sexually mature adults, different growth habits and spine features may be exhibited. Some of these growth stages or "age classes" are so different that they may appear to represent separate taxa. Cactus literature includes examples of growth stages having been formally described as species (see Benson, 1982).

Roots

The root systems of Trans-Pecos cacti have not been studied extensively. In general, many cactus taxa, such as prickly pears, chollas, and barrel cacti, are understood to have numerous shallow, horizontal roots that extend in all directions from the plant, out to 2–10 m or more (Benson, 1982; Gibson and Nobel,

1986). The usually branched shallow roots may be only 1.5–10(–15) cm below the surface, and they can take advantage of light precipitation. Many of the smallish Trans-Pecos cacti have more compact systems, with lateral roots that develop closer to the plant. This type of root system takes advantage of water runoff or collection from nearby objects, such as rocks, or from the plant itself. Some small Trans-Pecos cacti, such as *Coryphantha duncanii* and *Lophophora williamsii*, have fleshy taproots that store water and few lateral roots. Some prickly pears, such as *O. pottsii* and *O. macrorhiza*, produce branched, or "tuberous," fleshy roots. Very enlarged taproots, basketball-size or larger, are formed by old plants of *Peniocereus greggii*. Even the youngest permanent roots of cacti are sheathed by an orange or brown flaky bark; absorption happens mostly through ephemeral "spur root" tips, produced in response to moisture.

Adventitious roots are produced readily from cactus tissues, particularly after stem surfaces (including whole fruits of cholla cacti) have formed callus tissue in contact with the soil. Asexual reproduction after the dissemination of *Opuntia* stem joints is common.

Stems

"Shoot" is the anatomical term for the stem and its appendages collectively, as opposed to the root. Cacti produce three different kinds of shoots (Zimmerman, 1985): sterile long shoots (the stems); fertile long shoots (the flowers); and sterile short shoots (the areoles). The stems of most cacti appear to be leafless, but in many species rudimentary leaves about 0.1 mm long are produced at the lower edge of areoles that are developing near the growing tip of the shoot. The rudimentary leaves soon may wither and fall away. In Opuntioideae ephemeral, cylindrical (to flattened) leaves are produced. Prominent broad leaves are produced on the stems of *Pereskia* Miller and allied primitive cactus genera. The stems of most cacti exhibit indeterminate growth, except in *Opuntia* and several other genera, because an apical meristem remains active. Cacti have the largest known apical meristems among the angiosperms (Nobel, 1988). The flowers of cacti actually are reduced inflorescences, and their foliar appendages are the bracts and tepals. Flowers, the fertile long shoots, are determinate in growth, except for occasional bud production from pericarpel areoles, usually in fruit stage (proliferation). Flowers and stem branches are secondary long shoots that are produced from meristems in or near areoles. Rare exceptions include the adventitious buds formed on roots of *Opuntia polyacantha* var. *arenaria* (Boke, 1979) and dichotomous branching through bifurcation of the apical meristem in some *Mammillaria* (Zimmerman, 1985). Areoles, the sterile short shoots, are borne on long shoots in or near the axils of usually rudimentary leaves (in Cactoideae). Among Trans-Pecos cacti the axillary nature of areoles is most evident on young stems of prickly pears and chollas, where the slender succulent leaves persist for several weeks until the stem fully develops. In areoles the spines are derived from leaves. In most cacti areolar growth is determinate, but it is indeterminate in *Pereskia* and some other genera.

Stem Joints

Joints and jointed branches are produced by species of *Opuntia*. The joints of opuntias are cylindrical, clavate, globose, or flattened. Cylindrical joints are characteristic of species in the *Opuntia* subgenera *Corynopuntia* F. M. Knuth and *Cylindropuntia* (Engelm.) F. M. Knuth, commonly known as club chollas, chollas, or cylindropuntias. The cylindrical joints are not necessarily truly cylindrical or circular when viewed in cross section, because tuberculate epidermal projections are prominent in many species. The flattened stem joints of prickly pears (subgenus *Opuntia*) are formally known as cladodes but more commonly called "pads." The prickly pears also are known as the platyopuntias, where the Greek root word *platy,* "flat," is a reference to the pads. Cladodes, as flattened green stems, function in place of leaves. They are also termed "phylloclads" (leaflike stems).

Anatomically, a "joint," in our present terminology, is a shoot segment or stem that arises from an areole of an older stem. A joint usually is abruptly demarcated by a narrow base. Prickly pears and chollas are characterized by series of discrete segments that form branches. The segments or joints may develop in a relatively straight line or in a zigzag fashion. These are the branches or individual joints of prickly pears and chollas that most commonly are involved in rather prolific vegetative reproduction (e.g., Bobich and Nobel, 2001), after detachment and dispersal of the joints and subsequent establishment of adventitious roots. Joints are produced by some cacti other than opuntias, including *Rhipsalis* spp. and the Christmas cactus *Schlumbergera truncata* (Haw.) Moran.

Tubercles and Ribs

In many Trans-Pecos cacti, important identifying features include protuberances of the long-shoot surface commonly known as tubercles or ribs. For example, stems of *Coryphantha* and *Mammillaria* are tuberculate, and stems of *Echinocereus* are ribbed. Fewer than half of the cactus species of the world have ribs. Technically, "podaria" (singular, podarium) is the collective term for these surface protuberances. Individually protruding podaria are called "tubercles," and confluent podaria forming longitudinal files on the stem surface are "ribs" (Zimmerman, 1985). In some species the tubercles are confluent only near their bases, resulting in low ribs and tuberculate projections in longitudinal files. Ontogenetically, podaria originate and develop on stems in intricate helical patterns (Gibson and Nobel, 1986), but the ribs in some species may appear to be nearly vertical on the sides of a stem. Anatomically, podaria are similar to surrounding parts of the stem and are simply swellings of the stem upon which the actual leaves and their modified axillary buds (areoles) are elevated. Podaria, translated unsatisfactorily into English as "leaf bases," are bases in the sense of pedestals for the leaves, not basal portions of the actual leaves. Extensive cell division in the podarium contributes to tubercle formation (Gibson and Nobel, 1986).

Gibson and Nobel reported that the formation of ribs in cacti involves more than the fusion of tubercles into vertical series. A vertical strip of tissue develops under each row of tubercles so that the tubercles are elevated above the special strip of tissue.

Areoles

Areole morphology, including the spination, provides some of the most important characters in cactus taxonomy. Cactus areoles are axillary buds (or modified nonphotosynthetic short shoots) borne at the tip or adaxial (upper) end of each podarium (tubercle). In Trans-Pecos cacti the subtending leaves usually are not evident, because they are either rudimentary (for all practical purposes, absent) or ephemeral (subfamily Opuntioideae, where present at areoles on the new growth for a few weeks). The short-shoot (areole) leaves are highly modified as spines. Experimental studies (Mauseth and Halperin, 1975; Mauseth, 1977) have demonstrated that leaf primordia in areoles develop into leaves when cytokinin levels are high, and they form spines when gibberellic acid predominates.

Cactus areoles are bilaterally symmetrical with the spine cluster (usually dead at its maturity) developing toward the abaxial (lower) end and the long-lived meristematic region (if any) localized at the adaxial end. The spine cluster first differentiates into spine primordia; then each spine grows from a basal meristematic zone. Flowers or branches can develop only from the meristems near the adaxial margins of the areoles.

Traditionally, three types of areoles are recognized (Zimmerman, 1985) in the kinds of cacti that occupy the Trans-Pecos. These are (1) simple areoles, circular or elliptic in shape, with the adaxial floriferous meristem and spine cluster closely juxtaposed; (2) elongated areoles, with the abaxial spine cluster at the tubercle tip connected by a recessed linear isthmus, with the adaxial floriferous meristem toward the tubercle axil (as in *Coryphantha*); and (3) two-parted areoles, with the abaxial spine cluster at the tubercle tip, and with no recessed isthmus or other visible connection with the floriferous meristem that is located adaxially near the tubercle axil (as in *Mammillaria*). The simple areole type is said to be the most common type in the Cactaceae and is the primitive type. These simple areoles have been labeled "short monomorphic areoles" (see Boke, 1980). The elongated areole type has been labeled the "elongate monomorphic areoles." The elongate type obviously occurs mostly in cacti having tubercles instead of ribs. The linear, "connecting" part of the areole, the conspicuously sulcate or recessed adaxial surface of the podarium, is known in the cactus literature as the "tubercle groove," the "areolar groove," or the "meristematic groove." The term "meristematic groove" was deemed appropriate because the presence of trichomes, glands, or even branches suggests meristematic activity. The two-parted areole type has been called the "dimorphic areoles," where two distinct parts, the spine cluster and the flower meristem, are not visibly part of the same areole.

Zimmerman (1985) observed that the three traditional areole types described above are "artificial" and that in reality there are 10 or more natural types of areole morphology. Zimmerman described 11 types of areoles, based upon developmental sequences (whole plant ontogeny) as well as morphology. He pointed out that all cacti initially produce monomorphic areoles, that is, a single structure at the tip of each podarium, the primitive areole where both spine and meristematic potential are localized. Then in many cacti, later in life, ontogenetic development of areoles changes during whole plant ontogeny from seedling to oldest adult. Zimmerman proposed that the areole types he recognized be named after the taxa that are characterized by the particular type of areole development. Six of the eleven areole types outlined by Zimmerman are pertinent to the Trans-Pecos cacti:

1. Acharagma-type (the "primitive type"): unspecialized circular to elliptical areoles; flowers produced only at tips of podaria (Pereskioideae, Opuntioideae, and Cactoideae except for the tribe Cacteae)
2. Ferocactus-type: plants as they mature producing progressively longer and longer areoles; flowers produced from the adaxial margin of new areoles, but plants beginning to flower in short areoles (like type 1) before longer and longer but relatively wide ones are produced (*Ferocactus*, *Echinomastus*, *Ancistrocactus*, some *Ariocarpus*)
3. Macromeris-type: like type 2, but plants flowering only after narrower grooves attain maximum length, after a transitional sterile phase (*Coryphantha macromeris*)
4. Escobaria-type: like type 3, flowering only from adaxial end of full-length areolar grooves, in podaria axils, after a transitional sterile phase (most species of *Coryphantha*)
5. Neolloydia-type: flowering only at sexual maturity, when there is an abrupt and complete transition from short sterile areoles to flowering areoles, those having linear areolar grooves that extend to podaria axils (these cacti completely lack the transition phase involving successively longer grooves; *Neolloydia conoidea*)
6. Mammillaria-type (Zimmerman's type 9): flowering at isolated axillary meristems; areolar groove absent (*Mammillaria*, some *Ariocarpus*)

Zimmerman considered the *Epithelantha* areole described by Boke (1955) as essentially the same as the Acharagma-type.

Spines

Cactus spines are nonvascularized, modified leaves that develop individually or collectively from distinct spine meristems nearest the abaxial (lower) margin of the areole. Each spine is formed from several subepidermal cells (Gibson and Nobel, 1986) that undergo divisions. Very young spines may grow at the tip until they reach approximately 0.1 mm in length, after which spine elongation is the product of cell division from a basal meristematic zone and subsequent cell elongation. While they are developing, spines become hardened from the tip down, as the mature cells produce lignified cell walls. A cuticle is deposited on the outer surfaces of the appressed epidermis. Completely sclerified (hardened) and mature spine clusters are attached to the areolar tissue and to each other by cork cells that arise near the base of each spine (Gibson and Nobel, 1986). In many cactus species the spines abscise together in one group, giving rise to the term "spine cluster," which is applied to clusters that are still attached or separated. Spine clusters are extremely valuable in cactus identification, even detached from the plant or as subfossils preserved in pack rat middens or elsewhere. Determinate areoles produce species-specific numbers and arrangements of sclerified, durable spines (Zimmerman, 1985). Some genera have indeterminate areoles, which sprout supplementary new spines for many years, eventually obscuring the more tightly predictable pattern of the first-formed spines.

The shorter and thinner spines of *Opuntia*, the glochids, are formed in the same manner as larger spines, but they are not persistent because the base does not become sclerified. Glochids ultimately are deciduous, and they are easily dislodged from the areole. Sharp glochids readily penetrate human skin but are not easily removed. Microscopic observation of glochids reveals one reason why they are so obnoxious. Characteristically, glochids are retrorsely barbed (harpoon-like), the result of epidermal cells whose downward-pointing proximal tips extend at an angle from the main spine axis.

Spine variation in areoles of different ages on the same plant or in seedlings, juvenile sporophytes, and adults is understandable because each spine of the areole grows from a basal meristem, like the blade of lawn grass. The daily growth increments result in dozens of thickenings on large spines of *Ferocactus*, which reach full length only before the end of the annual growing season. Larger and smaller spines in the same areole may attain their size differences because some spines live longer or grow faster than others. In seedlings, spines usually are smaller than on adults because of shorter growth periods from more narrow primordia. The size of spines also may be related to environmental factors, such as nutrition, habit, and climate (best elaborated in full sunlight), and to interspecific differences in spine initiation and growth patterns. Such knowledge about potential spine development should be taken into account when evaluating the significance of intrapopulational and intraspecific spine variation. Spine variation is genetically controlled at the levels of physiological development of spines in the same areole, development in different areoles elsewhere on the same plant, and characteristics inherited by different individuals. The arrangement and sequence of spine initiation has been reviewed by Gibson and Nobel (1986).

There are other important aspects of spine variation (Robinson, 1974). In shape, spines of a single areole may be different or all alike. Spines may be stout, slender, or hairlike; cylindroid, acicular, angled, flattened and stout, or flattened and thin; smooth, scabrous, plumose, or corrugated with transverse ridges; straight, twisted, curved, or strongly hooked on the end (apically recurved), usually leading to the common name "fishhook." Spines may be slightly or much broader at the base, or basally swollen (or bulbous). Terminology for spine orientation (Zimmerman, 1985) varies. Appressed spines extend parallel to the stem contour nearly in contact with the stem surface, or at most their tips protrude at a tangent with respect to the rounded stem outline, leaving the plants easy to handle without gloves. Stems with all appressed spines appear smooth and "innocuous" to the touch. Erect or ascending spines angle upward, as in the tufts of connivent spines defending the stem apices of many species, more or less intermediate between appressed and porrect. Porrect spines are perpendicular to the surface. Descending spines angle downward. Deflexed spines are abruptly bent downward from near the base.

Spines may vary in color, often the result of water-soluble pigments called betalains (Gibson and Nobel, 1986), which, however, cannot be leached from the sclerified dead tissue of mature spines. Spines may be white, gray, black, yellow, golden, brown, orange, pink or rose, red, or multicolored. Colored spines may be opaque or somewhat translucent. With age spine colors usually fade to dull shades or change colors (e.g., from white to gray). The soft, living bases of actively growing (nascent) spines can be brilliantly colored.

The spines of a stem may be concolorous, as in the whitish spines of *Echinocereus pectinatus* var. *wenigeri*, or the central and radial spines may be of different colors. In some taxa different spine colors are manifested in broad or narrow bands or rings around the stem, reflecting the beginning and/or end of each growth season, as in the rainbow cacti, including *Echinocereus pectinatus* and many similar or related species of *Echinocereus*. The spines of North American cylindropuntias may have sheaths. The sheath forms from the epidermal layer and separates from the lignified spine core during development (Gibson and Nobel, 1986). The paper-thin sheaths may be yellowish, silver, or whitish and may fit tightly or loosely over the spine. Sheaths may cover the spines completely to the base, much of the spine, or only the distal portion. In the Trans-Pecos, sheaths are prominent in *Opuntia tunicata, O. imbricata,* and several other taxa.

RADIAL AND CENTRAL SPINES

Characters of spine clusters, including features and numbers of central and radial spines, are of major significance in cactus identification. In most cacti the distinction between radial and central spines is evident. Radial and central spines are not always discrete near the periphery of spine clusters. These two descriptive categories of spines genuinely intergrade with each other, and the actual homology of a given spine-tuft or individual spine is not always obvious unless both types are present together in areoles of the same plant (Zimmerman, 1985).

Radial spines may be defined as the series that radiates from the periphery of the spine cluster in a plane that is more or less parallel with the stem surface. Central spines protrude from the central region of the spine cluster, originating anywhere inside the radial spines. Central spines may protrude at almost any angle, including perpendicular to the stem surface. Central spines usually are more robust than radial spines. Even near-peripheral centrals that are similar to radials in size are slightly larger in caliber and have other features, such as enlarged basal structure, that allow rather positive differentiation when the two spine types are compared under magnification in the same spine cluster. Radial spines, like spokes of a wheel, often are similar in appearance, although perhaps of different lengths at poles or sides of the spine cluster. Central spines often are dissimilar when more than one per areole is present, and most of the variation in spine morphology is displayed by the centrals of different cactus species.

When identification of cacti is attempted through the evaluation of spine morphology, attention usually is drawn first to the central spines because the most useful comparative spine characters typically are those of the centrals. Also, distinction between central and radial spines, for the purpose of counting spine number or evaluating other characters, often is most reliably accomplished by first distinguishing the centrals, particularly any that may be similar in size and position to the radials. In the present treatment, this "centrals first" orientation is followed in descriptions and discussion.

Central and radial spines are distinguished in our descriptions of the Opuntioideae, as well as those of the Cactoideae. This is a departure from the traditional cactus literature, where centrals and radials usually are not differentiated in the chollas and prickly pears. We have observed that two or more discrete peripheral spines often are present in species of *Opuntia*. We have referred to these smaller spines as "radials" even though they usually occupy only the lower periphery of the areoles. The radial spines are not present in certain specimens or species of *Opuntia*. In *Opuntia* we have referred to the larger spines as centrals. The centrals may be positioned from the center of the areole to the near periphery. Glochids occur along with the spines in the areoles of most opuntias, and one or more bristles may be present as well. Zimmerman (1985) recognized at least four concentric series of spines in some members of the Cacteae. The outermost series consists of the radial spines. When there are three series of spines, the two series of centrals were treated as inner centrals and outer centrals, the latter being a middle series between the inner centrals and radials. When four series of spines are present, the layer between the outer centrals and the radials was interpreted as the subcentral spines. Subcentrals usually occur only in the adaxial parts of the spine cluster.

BRISTLES

Another spine type in cacti is known as the bristle (Gibson and Nobel, 1986). Bristles are slender, relatively long, pointed, and stiff but flexible. They are somewhat like the bristles of a brush, not rigid like spines. For example, golden-colored bristles are evident on stem tips and reproductive structures of certain Mexican columnar cacti. In Trans-Pecos cacti, bristles often accompany spines and glochids in the areoles of *Opuntia* species, where usually they are more slender than spines but longer and more strongly persistent than glochids. There seems to be a continuum from bristles through hairlike spines to multiseriate trichomes, one step away from the ordinary, soft, uniseriate trichomes that clothe the limited surfaces of the areoles in between the bases of the individual spines.

TRICHOMES

Cactus trichomes are hairlike, multicellular structures (or a protruding individual cell, if it protrudes far enough to be like an *Opuntia rufida* epidermal cell).

Each is formed from a single epidermal cell that divides and gives rise to a basal cell, which undergoes further division if a chain of elongate cells is to form the trichome (Gibson and Nobel, 1986).

Among cacti, trichomes in the broad sense potentially could be found on the surfaces or edges of most organs, including spines. In *O. rufida* the stem surfaces are densely short-pubescent, but these hairs are totally different from those in the areoles or those on the stem surface of *Astrophytum*. Trichomes are most conspicuous in new, actively growing areoles, where they may be densely matted together like wool, obscuring whole spines and flower buds at stem tips of certain species. In areoles the densely arranged trichomes may be straight, curved, or curled, giving rise to use of the popular term "wool," especially when the trichomes are whitish. Densely packed, straight trichomes may be referred to as "pile," as in a carpet, or as "felt" (velutinous). Areolar trichomes may be white, yellow, brown, gray, nearly black, or other colors, and the color may fade or change with age. In cacti, trichomes also may be conspicuous at shoot apexes, where the closely arranged young areoles collectively form a dense apical mat of hair (Plate 25).

Glands

Multicellular glands (highly specialized secretory spines) are produced in the areolar grooves of some Cacteae (e.g., *Coryphantha, Ancistrocactus, Thelocactus, Ferocactus*) and in the areoles of some Opuntioideae, such as the larger cholla cacti. The glands apparently arise from special primordia (at first similar to ordinary spine primordia) that develop among the trichomes in areolar grooves and/or adaxial edges of the spine clusters. The glands extend slightly from the areolar grooves, where they are visible to the naked eye as red, orange, yellow, or brownish structures, often surrounded by trichomes. Such glands usually are ephemeral, tending to be most obvious on younger tubercles on the upper half of the stem. In some *Coryphantha echinus*, for example, spheroidal red glands buried up to the equator in a peripheral ring of white trichomes are produced in the tubercle axils of juveniles, before any areolar groove reaches them. In *Ferocactus* the elongate or cylindroid areolar glands are relatively persistent. Dead and deflated glands may be visible as impressions among trichomes in the areolar grooves. In pressed specimens the detection of dead glands may depend on orientation of the pressed and dried areolar grooves. Under certain conditions glands can be seen to have slowly exuded a spherical viscous, colorless droplet. The droplets, containing common sugars, are sweet to the taste (Gibson and Nobel, 1986), qualifying these glands as extrafloral nectaries. We suspect that attracted ants in some cases may serve as agents of seed dispersal; in any case the ants inevitably encounter any other creatures on the cactus (presumably interfering with them to some extent). Extrafloral nectaries have been studied in *Opuntia* (Pickett and Clark, 1979) and *Ferocactus* (Ruffner and Clark, 1986). In the Trans-Pecos we have observed what appear to be extrafloral nectaries in *Opuntia imbricata* and *O. kleiniae* but not yet in *O. leptocaulis*.

Stem in Cross Section

The basic structure of a succulent cactus stem is easily observed after preparing a simple cross section (or longitudinal section). The stems of most members of Cactoideae have the same general anatomical organization as do taxa of the Opuntioideae, except that the stems of prickly pears are flat (distorted, as if run over by a steamroller), and the stems of some Trans-Pecos chollas have well-developed secondary xylem (wood).

In macroscopic view it is easy to see that there is a thin (but tangibly hard) outer rind or "skin" of the stem. Beneath the rind is a relatively thick cortex, which comprises several layers of dark green chlorenchyma cells near the rind, and a thick tissue of colorless, water-filled parenchyma cells. In many cactus species, very large, clear, mucilage-bearing cells can be seen in the cortex. Inside the cortex is a vascular cylinder, which in cross section is seen as a nearly solid or incomplete ring of tissue that appears more densely compacted than that of the cortex. To the inside of the vascular cylinder, in the center of the stem, is the parenchymatous pith, a tissue that resembles the cortex and that also is succulent. Secondary xylem, or wood, may be produced from the vascular cambium in an interrupted ring between the pith and to the inside of the vascular cylinder. The cambium is a single layer of cells that is not visible except through microscopic examination.

Microscopic observation of the stem tissues allows for a more accurate understanding of the tissue organization. This approach has revealed certain connections between structure and function in the stem.

RIND

The rind of the cactus stem more precisely consists of the cuticle, epidermis, and hypodermis. The epidermis usually is a single layer of cells; its cuticular (secreted, noncellular) coating is unusually thick compared to the surfaces of average plants. The substance of the cuticle, cutin, both impregnates and coats the outer walls of epidermal cells. Cuticular thickness can be estimated by scraping the outer surface of a stem. Stomates subtended by guard cells, much like those in other flowering plants, are distributed uniformly across the stem surface or localized on the sides of ribs. The stomata involve the epidermis and perhaps the underlying hypodermis. Stomata are functional in the stem, which is the primary photosynthetic organ of leafless cacti. The hypodermis is a layer of cells about 1 mm thick immediately inside the epidermis. It consists of up to 10 layers of collenchyma cells. Collenchyma cells are characterized by thick cell walls. The walls are unevenly thickened and contain high concentrations of the water-holding substance pectin and hemicellulose, but the walls are not lignified (Gibson and Nobel, 1986). Anatomically, the hypodermis is the outermost region of the cortex. The hypodermis extends the length of the stem, but it does not occur in roots. It functions mechanically in adding support to the outer stem. A thick hypodermis may be observed through a simple thin section of the outer stem. In all the rind is tough, flexible, and impervious to water when the stomates are closed. It contributes to protection against loss of water and deters herbivores, at least the insects.

CHLORENCHYMA

Chlorenchyma cells are the dark green parenchyma cells that form the highly visible outer part of the cortex immediately inside the hypodermis. The cells are green because chloroplasts are present in the cytoplasm. The chlorenchyma consists of 10–20 cell layers that are responsible for essentially all of the photosynthetic activity available to the leafless cacti. This compares with 3–5 layers for most other plants (Nobel, 1988). The raw material carbon dioxide enters the stem through stomata and diffuses to the chlorenchyma through substomatal canals and intercellular spaces in the collenchyma and chlorenchyma.

CORTEX AND PITH

In most cacti both the cortex and the pith consist of similar ground tissue (i.e., large, spheroidal parenchyma cells with thin walls and large central vacuoles that store water). Collectively, turgid cells of the cortex and pith provide considerable support for erect stems. Cactus stems that are wrinkled, flaccid, or of smaller diameter after prolonged drought have cortical tissue with reduced turgidity.

Many Trans-Pecos cactus species produce mucilage cells in the stem. Depending upon the species, such cells often are present in the cortex and perhaps in the pith as well. Mucilage is a complex carbohydrate substance that is very slippery and indigestible. The presence or absence of mucilage may be useful in distinguishing between morphologically similar species. For example, the tissues of *Coryphantha macromeris* are mucilaginous, and those of *C. ramillosa* are not. Mucilage can be detected after cutting into a fresh stem or often by simply removing a tubercle. Mucilage is slimy to the touch and visibly viscous (forming long strands, breaking when stretched to lengths of typically 1–10 mm), a sticky nuisance to the botanist when preparing specimens of many species. Reportedly, the functions of mucilage include the sealing of wounds and the deterrence of herbivory. Its absorption and tenacious retention of water is obvious to anyone who has, with difficulty, rinsed cactus mucilage from skin or tools.

The large mucilage cells are interestingly different from "normal" cortical cells from which they are formed. Dictyosomes develop in the protomucilage cell, migrate to the plasma membrane, and secrete mucilage that is released and stored between the plasma membrane and primary cell wall. The mucilage cell enlarges as mucilage deposition continues and as the hydrophilic mucilage continues to take up water. Eventually, the enlarged mucilage cell dies but remains loaded with the slippery substance. The production of mucilage in single cells is widespread in Cactaceae. Some species also produce the substance in mucilage reservoirs, where the collective products of many cells are localized. A different substance, a sticky, whitish latex, is produced in canals of certain species of *Mammillaria*, including *M. meiacantha* and *M. heyderi* of the Trans-Pecos. The canals are formed from rows of cells that disintegrate to form long tubes (see Gibson and Nobel, 1986). The "mucilage" of *Ariocarpus*, literally usable as glue, in specialized "mucilage canals" that extend through stem tissue, is yet another substance; *Ariocarpus* tissue feels dry to the touch (lacking mucilage cells) but contains a few ducts and cavities filled with translucent, slow-flowing or gelatinous latex that slowly hardens upon drying.

Probably most cactus species produce calcium oxalate crystals (Rivera and Smith, 1979; Gibson and Nobel, 1986). These crystals are formed in cell vacuoles of idioblastic cells from secreted, supposedly excess calcium oxalate ions. Depending upon the species these crystals may accumulate in any of the living tissues (except the vascular cambium), from the epidermis to the pith. Dense crystal layers may be formed in the rind, usually in the hypodermis, and the crystals may be scattered throughout the internal tissues. The gritty texture of the tissues in many Trans-Pecos cactus species is the result of abundant, aggregated calcium oxalate crystals, particularly in older internal tissues. Large aggregate crystals known as druses, usually whitish or transparent, are formed in many species. In the tribe Cacteae druses are spheroidal, oblong, or lens-shaped, and most are only 0.1–0.4 mm in diameter, but in some species, druses 0.5–1 mm in diameter can be seen with the naked eye. "Giant" lenticular druses to 1 mm in diameter are characteristic of *Coryphantha sneedii* and *C. vivipara*. Jones and Bryant (1992) discovered that calcium oxalate crystals as phytoliths have taxonomic utility.

VASCULAR SYSTEM

Structure of the vascular system in all three subfamilies of Cactaceae has been reviewed by Gibson and Nobel (1986). In most cacti the main vascular system is a hollow cylinder of soft, whitish wood sandwiched between the cortex and the internal core of pith. In some cacti vascular bundles also extend through the pith (medullary bundles); and in other cacti, through the cortex. Vascular "traces" extend outward from the vascular cylinder to ribs, tubercles, and areoles. Cactus spines usually are not vascularized. A separate "medullary vascular system," threads of wood in the pith, is present only in certain species (e.g., *Coryphantha vivipara*), whereas vascular strands in the cortex (at least the "areole traces" passing through) are ubiquitous. A complex three-dimensional network of threads of soft wood, the "cortical vascular system," innervates the vast succulent cortex in subfamily Cactoideae; however, the relatively thin layer of cortex in Pereskioideae and Opuntioideae may lack this supplementary vasculature. In simple, thin sections of the stem, vascular bundles can be seen without magnification as opaque lines in the transparent parenchyma tissues of the pith and cortex (Zimmerman, 1985).

Woody vascular tissues are produced by many cacti after activity of the vascular cambium and formation of lignified cell walls, particularly in secondary xylem to the inside of the cambium. It may seem unusual that the succulent stems of cacti have hard or soft woody parts inside, but it is the wood that provides much support for the stems. In many types of cacti, after the stem dies and dries, the central woody cylinder is exposed. The woody stems perhaps are most impressive in the cylindropuntias and in the tribe Pachycereeae, but woody cylinders are also present in many other cacti, such as in some *Mammillaria* and *Ferocactus* (Gibson and Nobel, 1986). Relatively thin, netlike vascular systems develop in prickly pear pads, and more extensive wood forms in the trunks of some platyopuntias. The wood skeletons differ between species, particularly in the shape and pattern of holes (which never contained lignified cells) in the wood, and they provide useful characters in cactus taxonomy. In the Trans-Pecos the most prominent woody skeletons are produced by *Opuntia imbricata* and its allies (Plate 26).

AGING

Regions of older cactus stems may form bark (periderm) on the outer surface. In cacti the cork cambium (phellogen) forms from the epidermis, perhaps from age, mechanical stress, freezing, sunscald, or injury, giving rise to several layers of cork cells to the outside. The cork cambium and cork cells may develop from cortical tissues following injury. Cork cells have thick walls impregnated with suberin, a waxy substance that effectively seals the stem. In cacti, bark characteristically develops in patches rather than envelops the whole stem, particularly near the base of the plant (Gibson and Nobel, 1986). Bark is usually brown at first and then turns gray with age. In the Trans-Pecos, bark is evident on the basal stems of larger plants of *Opuntia imbricata*.

FASCIATION

In cacti and plants of other families, fasciated or "crested" stems may develop as a result of injury or genetic anomaly in the apical meristem of the shoot. Orientation of growth from the apical meristem is altered so that a fanlike, eventually undulate-margined apex is formed at the tip of a normal cylindrical stem (Boke and Ross, 1978). Crested cacti are rare, and they are prized by collectors who appreciate anomalous plant forms. In the Trans-Pecos we have observed or had reports of occasional fasciated individuals in several species of *Echinocereus* and in *Thelocactus bicolor* (Plate 27), *Coryphantha tuberculosa*, *C. duncanii*, *Ancistrocactus tobuschii*, *Mammillaria meiacantha*, and *Opuntia imbricata*.

Flowers, Fruits, and Seeds

In general the areolar meristem has the potential to produce a flower. More specifically, in each cactus species, flowers are formed in areoles that are in characteristic position on the stem, perhaps very near the apex, one or more centimeters away from the apex, or on the sides of the stem. Flower position in some cases may be influenced by ecological conditions (Gibson and Nobel, 1986); for example, more flowers may tend to develop on the warmer sides of the stem.

Flowers may be formed from the areolar meristem just adaxial (above) to the spine cluster (many cacti), in an areolar groove toward the stem (e.g., *Coryphantha*), or in the tubercle-stem axis (*Mammillaria*), or the flower may erupt through the rind of the stem above the areole (*Echinocereus*). In prickly pears, flowers typically form in areoles on the apical margin of flattened stems, but they may develop in upper lateral areoles as well.

In all cacti the floral meristem is a determinate leafy shoot. The cactus flower actually is a single-flowered inflorescence that develops within the distal end of the determinate shoot (Zimmerman, 1985). The meristematic potential of the areole is expended with the formation of a single flower, unless as rarely occurs in cacti, a new bud is formed in the areole after the original flower develops (e.g., in *Myrtillocactus* Console there are two or more flowers per areole).

Cactus flowers have an inferior ovary, at least in subfamilies Opuntioideae and Cactoideae, and the ovaries are deeply sunken into the fertile shoots. In Cactaceae the inferior ovary is a derived condition. A few primitive cacti (*Pereskia* spp., Pereskioideae) have more or less superior ovaries.

The structure traditionally referred to in cactus literature as "flower" is not equivalent in all parts to the structure known as a flower in other angiosperms. Terminology preferred by Zimmerman (1985) is followed mostly in the current work. A cactus flower is a leafy shoot, with areoles, in which an ovary is embedded and other parts of the pistil are enclosed. The style extends through tissue, known as the "column" (Zimmerman, 1985), above the top of the ovary and is exserted into or from the receptacular tube. Stigma lobes or "style branches" are found at the distal end of the style. The ovary is multicarpellate with a single locule and parietal placentation. Consideration of the cactus "flower" in longitudinal section allows for the delineation of "flower parts" involving the stem. At the proximal end of the "flower," the tissue precisely surrounding the ovary is known as the pericarpel (Plate 28). The relatively thin expanse of tissue from the "ceiling" of the ovary to the floor of the nectar chamber is known as the "column." The nectar chamber is the recess around the base of the style and below the lowest stamen insertion, which receives any sugary exudate from the adjacent nectary tissue. Distal to the pericarpel and "column" is the receptacular tube. Precisely, the externally bracteate receptacular tube extends from the floor of the nectar chamber to the base of the outer tepals. The sepaloid outer tepals are transitional to the usually showy and petaloid inner tepals. The receptacular tube is fertile for the relative long distance from the nectar chamber (insertion of the lowermost stamens) distally to the insertion of the uppermost stamens. In cactus flowers the stamens usually are numerous, all with filaments of about equal length. The receptacular tube is sterile for the relatively short distance from the insertion of the uppermost stamens to the base of the outer tepals. The receptacular tube is equivalent to the floral tube or hypanthium used by some other authors. The pericarpel plus the receptacular tube is equivalent, respectively, to the inferior and superior floral cup or tube, the terminology favored by Benson (1982).

Even though the cactus "flower" as explained above is not homologous with the four-organ floral structure of most angiosperms, the term "flower" remains the most practical for use in Cactaceae at the present time. In the present work the word *flower* alludes to the specialized tip of the shoot, from the base of the ovary to the tepals, including the female and male reproductive parts. Sometimes we use the term "floral tube" loosely in reference to the generally funnelform section of the cactus flower.

FRUIT

Those acquainted with the structure of the cactus flower are aware that the cactus fruit involves the ovary and accessory tissues. A mature cactus fruit enlarges in size and changes color. The pericarpel at anthesis matures into a ripened pericarpel. Inside a mature fruit is an ovary that contains seeds. Outside the ovary is a rind that consists of the pericarpel tissues (cortex, surface tissues), including the superficial areoles and their subtending bracts. A fleshy cactus fruit is berrylike. The dry fruits in some cactus species are more like capsules than berries.

A mature fruit is one that contains fully formed seeds. A mature fleshy fruit usually, but not necessarily, has achieved a characteristic succulence, color, or color change and encloses fully formed seeds. A mature dry fruit has changed from a relatively fleshy developing pericarpel to a structure with dry internal tissues and fruit wall, and it contains fully formed seeds. A ripe fruit is one that has achieved its characteristic color and tissue succulence or dryness and is easily removed from the stem, without consideration of seed maturity.

FLORAL REMNANT

The shriveled remains at the apex of a cactus fruit are best referred to collectively as the floral remnant (Zimmerman, 1985). Following successful pollination and subsequent fertilization in the ovules of the ovary, flower closing involves a collapsing, drying, and sometimes longitudinal twisting of the tepals and receptacular tube. As the fruit matures, the apical remnants persist as brownish or blackish structures. The floral remnant is more or less persistent in different species of Trans-Pecos cacti (characteristically deciduous in Opuntioideae). Persistent floral remnants (e.g., in *Ancistrocactus* spp.) may be functional in providing leverage for the dissemination of fruits and seeds, as might the beaks (necks) in other taxa (e.g., *Peniocereus greggii*). A beak of a fruit is protruding necklike tissue between the main body of the fruit and the floral remnant, which is attached to the apex of the beak. Fruit morphology is very useful in the identification of cacti.

SEED

Seed morphology, particularly size, shape, color of the testa, and ornamentation of the testa, is useful in cactus identification. The cactus embryo is markedly curved or is completely bent around the central nutritive tissue. The nutritive tissue that is synthesized and stored in cacti is unusual. The reserve "food" in the seed is perisperm, derived from diploid nucellar tissue, rather than the triploid endosperm that is produced in most angiosperms. In Cactoideae a prominent, complex hilum-micropylar region is formed through fusion of the hilum and micropylar tissues. In some Cactoideae, outgrowths of the hilum-micropylar region are arillate or strophiolate.

In Opuntioideae, the seed typically is completely enclosed by the funicular envelope, the main vein forming the funicular girdle (Stuppy, 2002). The funicular girdle usually hardens, forming a tan or brownish bony aril-like envelope surrounding the seed (Anderson, 2001). In the taxonomic section for convenience we sometimes use the terms "aril," "arillate," "aril-rim," or "pseudoaril" in reference to the funicular girdle. In some taxa (*Nopalea* Salm-Dyck, *Consolea* Lemaire, *Pereskiopsis* Britton & Rose, etc.) the girdle is pubescent. Underdeveloped seeds sometimes show testa characters when the funicular envelope does not completely cover the seed.

Physiology (= Primary Compound Chemistry)

In North America the diversity of cacti is greatest in the desert regions. All desert plants have evolved morphological-anatomical and physiological adaptations that promote successful growth and reproduction under dry environmental conditions. Among the types of arid adaptations is the storage and maintenance of water in succulent tissues. Cacti are good examples of succulent plants in the North American deserts. In times of drought cacti preserve water, and when needed, water is drawn from the storage tissues. There are a number of morphological-anatomical and associated physiological arid adaptations of North American and Trans-Pecos cacti. Essentially all of the water-preserving features discussed by Gibson and Nobel (1986) and Nobel (1988, 1994) are summarized in the following discussion. The subjects are organized as morphological-anatomical (structure) or physiological (function), even though the correlation between structure and function usually is more evident than any distinction between the two. Generally, the anatomical feature is mentioned as a way of alluding to the associated function. The interrelationship of cactus physiology with ecology is discussed in a later section.

Morphological-Anatomical

The habit of cactus plants can be important in efficient physiological activity. In some cases the habit features influence rather subtle effects, and in other instances they are more apparent. Stem height, diameter, and branching are related to the exposure of cactus plants to sunlight, heat, dry air, and cold. Smaller stems are more likely to be shaded, at least during part of the day, by nearby physical features, such as other plants and rocks. Stem branches shade other branches. Stem branches also may spread the surface area available for interception of light required for photosynthetic productivity. Three-dimensional features of the stem surface, such as ribs and tubercles, increase the potential surface area available to receive sunlight. The ribs of cactus stems also are associated with surface-area exposure under conditions of ample water, when stems swell and ribs spread, and under conditions of limited water, when stems shrink and ribs close. Stem orientation, such as east-west or north-south alignment of cladodes in prickly pears and equatorial tilting of certain barrel cacti (Geller and Nobel, 1987; Nobel, 1994), also may enhance photosynthetic production if stem position allows optimal reception of light. Cladode positioning and stem tilting have not been much investigated in Trans-Pecos cacti. The shading effect of spines and trichomes has the amphibological effect of reducing photosynthesis but cooling the stem.

The shallow roots of many cacti spread from the base of the plant in all directions and function collectively in absorbing water even from minimal precipitation. As a general rule the shallow roots are deeper than about 3 cm, probably because summer soil temperatures may be hot enough to kill roots that are too close to the soil surface. In the Sonoran Desert, summer soil temperatures have been measured at 70°C (158°F) at the surface and as high as 55°C (131°F) at a depth of 3 cm (Nobel, 1994). Soil temperatures in some parts of the northern Chihuahuan Desert probably are about as high as those measured in the Sonoran Desert.

Many Trans-Pecos cacti, including some of the smaller species, produce a large taproot beneath the stem, perhaps in addition to lateral roots. Species with large taproots characteristically may not produce an extensive lateral root network, as reportedly is the case in peyote (*Lophophora williamsii*). In peyote and other species, such as *Coryphantha duncanii,* the storage roots typically are much larger than the stem that extends only a short distance aboveground when the stem is turgid. In times of water stress the stem may shrink to ground level or below. The fleshy taproots can store water and carbohydrates away from the hottest temperatures.

Another adaptive feature of cacti is the production of "rain roots" in relatively rapid response to moisture that enters the soil after a long dry period (Gibson and Nobel, 1986). When the soil begins to dry around cactus roots, some of the copious branches may abscise (Nobel, 1994), theoretically reducing the root surface area through which water could be lost from the roots to the soil. Cactus roots in drying soil also develop insulating layers of bark. Thus, in dry soil the cactus root system essentially is "sealed" from losing water. Rain roots develop from root primordia that form in the cortex of older main roots. When the soil becomes wet through precipitation, the parenchyma cells of the main roots quickly absorb water, and growth of the branching rain roots begins within hours or a few days. When the soil starts to dry again, the white rain roots shrink and eventually separate from the main roots, a process that may prevent water loss during drought.

The leafless succulent stems of most Trans-Pecos cacti are effective in storing and conserving water. All the parenchyma cells of the stem store water, particularly the cortex, the most abundant stem tissue, and the pith. Water conservation is effected through several aspects of stem structure and function. Obviously, there is no water loss through transpiration from leaves, and the gray-green stem surface of many species reflects some light and heat. The skin of the cactus stem with its thick cuticle is so impervious that little water is lost. Also, little water is lost through stomata that are closed during the heat of the day. In addition to the waterproof outer layer of a cactus, the living cells of the cactus skin, the epidermis and collenchyma, apparently have the ability to withstand high temperatures that denature cell proteins in most other plants. The surface temperature of cactus stems has been measured to exceed 50°C (122°F), up to a high of nearly 70°C

(158°F), without lethal consequences. In fact, mature cactus tissues in general, and some of the diminutive species of the northern Chihuahuan Desert in particular, are known to be among the most heat tolerant in all plants. The water-storage parenchyma cells of cacti can tolerate water losses of up to 70–95% by volume, whereas water loss of 30% usually is lethal in noncactus species. These physical characteristics of cacti, having to do with drought and heat tolerance and photosynthetic efficiency (Gibson, 1998), are discussed extensively elsewhere (Gibson and Nobel, 1986; Nobel, 1988, 1994), where additional features pertaining to "water relations" of cacti are emphasized.

The microclimate of individual cactus plants, particularly in the summer, is influenced by their spines and trichomes. There is little doubt that the shading effect of spines covering much of a stem results in lower temperatures at the stem surface. This in turn influences the water relations of the plant. Trichomes function in the same manner, most commonly at the stem apexes where dense coverings of trichomes are found. Lower temperatures at the plant apex logically would protect young cells at and near the meristem. Dense apical mats of trichomes also may provide insulation from cold. Spines and trichomes intercept light that could be used in photosynthesis.

Physiological

Approximately 98% of all cacti use a photosynthetic pathway known as Crassulacean acid metabolism (CAM). The leafy pereskias and species in a few other genera (e.g., *Maihuenia* Philippi ex K. Schum.) carry out C_3 photosynthesis (Nobel, 1994). In *Pereskia*, CAM occurs in the stems, and C_3 photosynthesis occurs in the leaves (Anderson, 2001). In the plant world three photosynthetic pathways are known. Most plant species on the earth, about 93% of the 300,000–350,000 species, utilize the C_3 pathway. A second pathway is referred to as C_4 photosynthesis, and it is operative in only about 1% of all plants. The third pathway, CAM metabolism, is found in about 6% of plant species, mostly succulents and tropical, nonparasitic epiphytes.

In all kinds of photosynthesis the light-reaction part of the processes is the same. The initial fixation of carbon dioxide (CO_2) is what differs and is the aspect after which the photosynthetic processes are designated: (1) In C_3 photosynthesis the first stable compound produced after CO_2 fixation is a 3-carbon substance (phosphoglycerate), which is the first step in the production of glucose and other molecules via the Calvin-Benson cycle. (2) In C_4 photosynthesis the first compound produced after CO_2 fixation is a 4-carbon substance (oxaloacetate), a process involving the Hatch-Slack pathway. In C_4 photosynthesis both the C_4 and C_3 pathways ultimately are employed. Carbon dioxide is fixed initially in the day in nonchlorenchymatous leaf mesophyll (in most plants) by the more efficient Hatch-Slack pathway, and then during the same day CO_2 is transferred to the Calvin-Benson cycle, which is isolated spatially in chlorenchymatous bundle sheath cells. (3) In CAM metabolism CO_2 is fixed initially at night by the C_4 pathway, and then during the next day CO_2 is processed via the C_3 pathway. Two of the photosynthetic pathways, C_4 and CAM, utilize both C_4 and C_3

carbon-fixation cycles. The two processes differ in the time of initial CO_2 fixation. In CAM plants, stomata are closed during the day and open during the night, thus facilitating CO_2 uptake from the atmosphere at a time when water is less likely to be lost through stomatal openings. In cacti the abundant CO_2 is fixed during the night, specifically in the cytosol of chlorenchyma cells. In cacti C_4 fixation results mainly in the 4-carbon organic acid malate, which is stored in the cell vacuoles (Nobel, 1994). Morning light of the following day stimulates C_3 photosynthesis in chloroplasts of the chlorenchyma cells. Stored malate and perhaps other organic acids are decarboxylated, releasing CO_2 that ultimately is fixed in the C_3 pathway. The C_3 process continues through the light of day. The CAM pathway in cacti and other succulents is regarded as a physiological process that contributes importantly to water conservation and photosynthetic efficiency in plants that occupy desert habitats.

Conditions are not always hot and dry in the northern Chihuahuan Desert. During periods of sufficient moisture, stomates may open during the day, allowing CO_2 uptake in addition to the uptake that occurs at night. When CO_2 is taken in during the day, usually it is fixed directly through the C_3 pathway, skipping the initial C_4 process. Daytime stomatal opening is most likely during cooler early morning hours or later in the day if it is cloudy (Nobel, 1994). At the other environmental extreme, during periods of prolonged drought and heat, CAM plants may undergo an emergency process known as "CAM idling," where stomates remain closed during the day and night and where CAM metabolism continues to operate mainly with CO_2 that is internally recycled (Pearcy et al., 1987). Normal CAM metabolism returns as soon as water is again available to the plants.

In most CAM plants the nightly buildup of organic acids can be detected by a simple taste test. Tissues crushed and tasted in the early morning will be more sour or acidic than those sampled later in the day after much of the stored malate or other organic acids have been decarboxylated.

One of the most important aspects of photosynthesis in cacti is the amount of photosynthetically active radiation (PAR) or photosynthetic photon flux (PPF) that reaches chloroplasts in the stem chlorenchyma (Gibson and Nobel, 1986; Nobel, 1994). Photosynthetically active radiation is a reference to the parts of light in the visible spectrum that are absorbed by chlorophyll and used in photosynthesis. Thus, PAR alludes to the red and blue spectra and the photons in those spectra. Photosynthetic photon flux essentially is the same thing as PAR, except that PPF refers to only those photons with wavelengths in the visible spectra that are absorbed by chlorophyll and used in the light reaction of photosynthesis (Nobel, 1994). There are many factors that influence PAR/PPF interception of cactus stems, including specific habitat, time of day, time of year, ecological associates (nurse plants or shade plants), plant spacing, stem architecture, stem orientation (e.g., cladodes facing the sun), and density of spines and trichomes on the stem. There is also a correlation of PAR/PPF with other aspects of photosynthetic efficiency (e.g., CO_2 metabolism). In CAM plants, most CO_2 uptake is at night, and CO_2 fixation is during the day. Thus, the enzyme phosphoenolpyruvate carboxylase, which picks up CO_2 at night, relates to the amount of carbohydrate (glucose) synthesis during the day and ultimately to the dry weight gain of the plant. Furthermore, CAM plants depend upon total daily PAR/PPF for the productive fixation of CO_2, whereas C_3 and C_4 plants utilize instantaneous PAR/PPF (Geller and Nobel, 1987). Photosynthetic efficiency in CAM plants is the result of interconnected physiological events. Nighttime conditions that facilitate the uptake of CO_2 through open stomates provide the circumstances for optimal CO_2 fixation and storage as part of organic acids in internal tissues. Timely and complete conversion of the chemically stored CO_2 to carbohydrates throughout the following day, which is facilitated by optimal PAR/PPF, allows maximum storage of additional CO_2 the next night. Optimal growth of the plant depends upon optimal photosynthetic production.

Water-use efficiency is another physiological process that is correlated with photosynthesis and plant vigor, particularly in arid regions. Water-use efficiency is a way of measuring the amount of carbon converted to photosynthetic products compared to the amount of water lost by transpiration (Nobel, 1994). Measurements from cultivated species have demonstrated that CAM plants have a significant advantage in water-use efficiency, about three times greater than C_4 plants, and six times greater than C_3 plants. If the same advantage exists in native species, as is expected, then cacti have another physiological edge in their typically arid habitats.

The growth rate of cacti in the Trans-Pecos and other regions is influenced by the daily thermoperiodicity that is most pronounced in arid environments (Benson, 1982). Daily thermoperiodicity refers to the warmer daytime temperatures and cooler nighttime temperatures. For CAM plants the highest CO_2 uptake and fixation are correlated with lowest nighttime temperatures during the growing season. Optimal growth in cacti reportedly is stimulated by hot day temperatures and cool nights, with the widest temperature extremes and coolest temperatures being of greatest benefit. Thermoperiodic conditions such as these are characteristic of the northern Chihuahuan Desert Region.

Apparently, seasonal thermoperiodicity plays a role in the production of flowers in Trans-Pecos cacti. Plants maintained through the winter in a temperature-controlled greenhouse, where there is little difference between day and night extremes, rarely produce flowers in the spring. However, we have observed that plants maintained through the winter in an unheated greenhouse, or those that are transferred from a heated greenhouse to the outside before winter is over, typically produce flowers to an extent that is indistinguishable from plants that are cultivated outside or those that exist in natural habitats. The duration of thermoperiodic conditions required before flower induction is not understood in most Trans-Pecos cactus species.

Some species of Trans-Pecos cacti are known to produce flowers in response to rainfall. This phenomenon has been observed over a period of several years in a population of *Coryphantha hesteri,* where at least some plants in the population always produced flowers 7–9 days after "significant" rainfall, at any time of the growing season, from spring through early November (A. M. Powell and S.

Powell, unpub.). Numerous other Trans-Pecos species are known or suspected to respond opportunistically to precipitation in flower production. Some of this information, where available, is included in the current work under the phenology section prepared for each taxon. Additional study is needed to understand the flowering and other phenological responses of most Trans-Pecos cacti, with respect to such factors as temperature, precipitation, and photoperiod (Bowers, 1996).

Most Trans-Pecos cacti are frost resistant and cold hardy. The northern Chihuahuan Desert is a "cold desert" with freezing temperatures to be expected at all elevations from November through March. At middle to higher elevations, common winter nighttime lows are −9° to −4°C (15° to 25°F). Temperatures frequently decline to −12°C (10°F), dip to −18°C (0°F) or below for several hours in some winters, and, rarely, drop to near −23°C (−10°F) for short periods. Winter temperatures, even at low elevations in the Trans-Pecos, commonly reach −7°C (20°F) or lower. Cactus species are like most plants in that they tolerate low temperatures best when they undergo low-temperature hardening (i.e., when the air temperatures gradually decrease over a period of time, perhaps weeks). One of the functions of mucilage in cactus stems might be in conferring some protection against freezing (Nobel, 1994). Hardening should be effected before transferring cultivated cacti from a warm greenhouse to outside during the winter. In the Trans-Pecos the warmest average winter temperatures, and the highest low temperatures, probably are south of the Chisos Mountains, and from there along and near the Rio Grande from the lower canyons toward the mouth of the Pecos River to northwest of Presidio below the Sierra Vieja rim. Some cacti from these warmer areas (e.g., *Echinocereus chisoensis*) are not as frost resistant as others in the region.

Many aspects of stem anatomy contribute to photosynthetic efficiency in cacti where leafless stems have evolved as the major photosynthetic organs. Slender leaves of young *Opuntia* cladodes may provide some photosynthetic productivity but only for a short period of time. The photosynthetic chlorenchyma tissue, perhaps 10–20 cell layers thick, is positioned just under the skin of the cactus stem. The skin, consisting of the cuticle, epidermis, and collenchyma, is translucent, so sufficient PAR/PPF easily reaches the chloroplasts. Photosynthetic efficiency is further enhanced by the anatomical substructure of the outer cortex, including the chlorenchyma (in Cactoideae). Cells of the outer cortex are arranged in columns, forming a palisade cortex, similar to the palisade mesophyll parenchyma of leaves in C_3 plants (Sajeva and Mauseth, 1991). The chlorenchyma also is interrupted by canals and intercellular spaces that facilitate gaseous diffusion between the chlorenchyma and the atmosphere. Another anatomical advantage of most cacti is the presence of vascular bundles in the cortex, extending outside to very near the chlorenchyma. Reportedly, the cortical bundles have evolved in Cactoideae, where in the leafless stem anatomy, the central vascular cylinder is distant from the chlorenchyma.

Germination of cactus seeds may be inhibited by substances in fleshy or juicy fruits (Benson, 1982). Washing and drying such seeds before sowing supposedly is effective in removing germination inhibitors. Deno (1994) maintained that in many cacti, gibberellins are an absolute requirement for germination. Treatment with gibberellic acid-3 is sufficient to effect germination. In our experience the seeds of most Trans-Pecos cacti germinate readily without pretreatment after washing, drying (if from fleshy fruits), and storing in packets for two weeks or more. Moist stratification or after-ripening may promote germination in some species, including *Echinocereus dasyacanthus* and *Opuntia* spp.

Secondary Compound Chemistry

This is a brief review of the secondary compounds produced in Cactaceae, especially those compounds that have been useful in classification of southwestern cacti. Primary compounds usually are interpreted as those involved in basic biochemical pathways that are common to most organisms, whereas secondary products are offshoots of basic metabolism that are suspected or known to have important functions, such as deterring herbivores, in particular organisms. The secondary compounds that appear to be the most significant in cacti are betalain pigments, flavonoids, and alkaloids (Meyer and McLaughlin, 1982).

Betalains

The betalains are a special class of water-soluble vacuolar pigments found only in Cactaceae, in nine other families of the Caryophyllales, and in some Basidiomycetes (*Amanita*). The betalains are special because they have a chemical structure, containing nitrogen, that is very different from that of the widespread anthocyanins (Mabry and Dreiding, 1968) and because they are found in relatively few families of flowering plants (Cronquist, 1988). Betalains are the same color as anthocyanins and anthoxanthins (colored flavonoids).

Betalains apparently do not co-occur with anthocyanins, but betalains and anthoxanthins sometimes coexist in the same plants. Betalains form two color groups, betacyanins (red, magenta, pink, etc.) and betaxanthins (yellows and orange-red). These colors are comparable to those of the anthocyanins and anthoxanthins except that no blue betacyanins are known, but blue anthocyanins are common.

In cacti, betacyanins are responsible for cyanic (blue end of the spectrum) colors of the stems, flowers, and fruits. Most yellow colors in cacti probably result from betaxanthins. The purple stems of the long-spined (purple) prickly pear (*Opuntia macrocentra*) are the result of betacyanins (Plate 29), and its yellow flowers are colored by betaxanthins. Other yellow pigments present in cacti, although not much investigated, are anthoxanthins, carotenoids, and certain alkaloids (Gibson and Nobel, 1986). Carotenoids are ubiquitous, membrane-bound pigments in plants and plantlike organisms. Because carotenoids occur with chlorophylls in chloroplasts, yellowish stem coloration in unhealthy cacti probably is the result of unmasked carotenoids after chlorophyll degeneration. Although some alkaloids are yellow, tissue coloration in cacti is probably rarely the result of abundant alkaloids.

Flavonoids

During the heyday of biochemical systematics, phenolic compounds such as flavonoids were considered to be among the most useful classes of organic compounds (Alston and Turner, 1963). Flavonoids are widespread in all kinds of plants. There are many different kinds of flavonoids, they are rather easily isolated and identified (Mabry et al., 1970), and they have been extensively studied (Geissman, 1962; Harborne, 1964). Flavonoids are the biosynthetic precursors of anthocyanins, which are present in most kinds of plants but absent in cacti and other betalain-producing families. It is believed that betalain-producing plants have lost the enzymatic capability to synthesize anthocyanins from flavonoids.

Flavonoids appear to be abundant in Cactaceae, but the distribution and taxonomic value of these compounds have not been widely investigated in the family (Wallace, 1986). Most of the previous studies of taxa native to the United States have involved *Opuntia* (e.g., Walkington, 1966; Rösler et al., 1966; Clark and Parfitt, 1980; Clark et al., 1980) and *Echinocereus* (Kolle, 1978; Breckenridge and Miller, 1982; Miller and Bohm, 1982; Leuck and Miller, 1982). Flavonoids also have been reported in the cactus genera *Mammillaria, Pereskia, Pereskiopsis, Quiabentia* Britton & Rose, *Echinopsis* Zucc., and *Epiphyllum* Haw. Flavonoid analyses have provided significant taxonomic information in the *Echinocereus viridiflorus* (Kolle, 1978; Leuck and Miller, 1982) and *E. enneacanthus* (Breckenridge and Miller, 1982) complexes.

Flavonoids appear to be most abundant in the flowers, particularly the tepals, of *Echinocereus* and *Opuntia*, although these compounds are present in stem tissue as well. Both Walkington (1966, in *Opuntia*) and Kolle (1978, in *Echinocereus*) extracted flavonoids from stems. The more sophisticated analyses by Miller and associates (1982) and Clark and colleagues (1980) utilized the abundant flavonoids extracted from flowers. Miller and Bohm (1982) determined that only trace flavonoids were present in the epidermis and spines of *Echinocereus triglochidiatus* var. *gurneyi* (= *E.* × *roetteri* var. *neomexicanus*), and no flavonoids were detected in cortical tissue. Because Walkington (1966) and Kolle (1978) extracted flavonoids from stem tissue, these compounds in cactus stems must be found from the chlorenchyma outward.

Although it appears evident that flavonoids are most abundant in tepals, which in most cacti are available for only a short period each year, it would be convenient in future studies to anticipate the possibility of using stem flavonoids as well. The inconvenience of having to deal with the abundant mucilage present in the cortex of many cacti could be avoided by attempting flavonoid extractions from superficial stem tissues, chlorenchyma outward, that have been sliced away from the cortex. It is anticipated that flavonoid studies might be useful in helping to resolve infrageneric taxonomic problems in Cactaceae.

Alkaloids

Certain cacti are noted for their alkaloid content. Perhaps the most famous alkaloid-producing cactus is peyote (*Lophophora williamsii*), a species that occurs at a few sites in the Trans-Pecos, southeast along the Rio Grande to South Texas, and in northern Mexico. Peyote contains about 57 alkaloid and alkaloid-like substances, the most notable of which is mescaline, a potent hallucinogenic compound (Anderson, 1996a). Mescaline is known to occur in at least eight other genera of North American cacti and three or more South American genera. Alkaloid chemistry has been useful in evaluating the intra- and intergeneric relationships of *Lophophora*. It is illegal to possess peyote, which is regarded as a Schedule I controlled substance, without a special permit.

In chemical structure alkaloids are usually heterocyclic rings with at least one nitrogen. In pure form they are generally a white powder that is bitter-tasting and basic. Some of the most effective pharmaceutical drugs are alkaloids, which accounts for the fact that cacti and other plant families have been rather thoroughly screened for the presence of alkaloids (Gibson and Nobel, 1986; Anderson, 1996a).

Other Chemical Considerations

Additional types of substances are relatively poorly known in the Cactaceae, but some of them are at least known to be characteristic of the family. These include an unusual phenolic alcohol that was shown to be responsible for a stem-darkening wound response in *Carnegiea gigantea*, di- and triterpenes, and sterols. The di- and triterpenes (Ponsinet et al., 1968) are known to be present in at least four genera of Mexican columnar cacti, in other cacti such as *Mammillaria*, and in many other kinds of plants. The presence of triterpenes in plant tissues can be suspected if suds are produced when the tissues are agitated in water, as was done by Native Americans and copied by European pioneers, who used such plants as soap (sapogenins) and, in some cases, as a way to stun fish in their aqueous habitats. The biologically active triterpenes may damage red blood cells. Mexican columnar cacti, including *Stenocereus thurberi* (Engelm.) Buxb. (organ pipe cactus), which extends into southwestern Arizona, apparently have abundant sterols. The sterols have received limited study but offer potential in taxonomic and pharmacological studies.

Other chemical substances produced by certain cacti are those referred to as mucilage and latex. Of the two substances, mucilage is much more common in the Cactaceae, probably being present in a majority of species, including most Trans-Pecos taxa. In substance the usually viscid mucilage is complex, consisting of different kinds of monosaccharides linked together in many-branched chains (Gibson and Nobel, 1986). Stem mucilage accounts for about 3% of the stem volume. Mucilage usually exudes from cuts or other deep wounds in the stem, coating the surface near the wound. Several functions are attributed to the presence of this usually sticky material, including wound sealant, deterrence of predators, protection of the stem from high temperatures, a secondary agent in water storage, and calcium storage. Both water molecules and calcium ions are bound to mucilage molecules. It is suspected that there is extensive interspecific variation in mucilage structure.

Latex is produced by relatively few species of cacti. It is best known in *Mammillaria*. Latexes are produced by many different families of angiosperms, where they vary extensively in color, viscosity, and chemical content. Practically any fluid that flows from the cut surfaces of plants may be referred to as latex. But in different taxa, latex may vary from white to watery or be yellow, orange, or red in color. In plant organs latex characteristically is produced in specialized tissues (i.e., in elongated cells or canals called laticifers). The most famous latex-producing plants include *Hevea* Aubl. (Euphorbiaceae) and *Papaver somniferum* L. (Papaveraceae), where rubber molecules or alkaloids dominate the complex makeup of the latex. In different species of *Mammillaria*, where latex is present at all, it may be white, translucent, or clear. Among Trans-Pecos taxa, the white, "milky" latex produced by *M. heyderi* is composed mostly of polysaccharides and chemically is similar to mucilage (Gibson and Nobel, 1986). In *M. heyderi* the latex is produced from specialized laticifers that ancestrally were derived from mucilage cells or tissues (Wittler and Mauseth, 1984).

Calcium oxalate crystals are characteristic in cacti, potentially occurring in vacuoles in cells of all tissues from the epidermis inward, except for the vascular cambium (Gibson and Nobel, 1986). The crystals are not known to occur in spines or trichomes. In cacti, calcium oxalate crystals may take on many different compositions and shapes, consisting of single or aggregate structures. The most common calcium oxalate formation in cacti is the druse, a spheroidal or flattened aggregate that may vary in size and shape even in different tissues of the same stem, and druse morphology may vary with physiological conditions of the plant. Druses already are of taxonomic significance in some groups (e.g., *Coryphantha sneedii*), and their taxonomic utility should increase after further study of calcium oxalate crystal variation.

Crystals of silicon dioxide, known as silica bodies, occur in species of *Stenocereus* (A. Berger) Riccob., a genus of Mexican columnar cacti (Gibson and Horak, 1978). Most Trans-Pecos cacti have not been examined for the silica bodies that in *Stenocereus* occur in epidermal and hypodermal cells of the cactus skin and may function to deter predation.

Cactaceae and other members of the Caryophyllales (Centrospermae) are characterized by a peculiar type of sieve-tube plastid (Cronquist, 1988) that is not known to occur in other angiosperms (Behnke, 1976). The plastids are distinguished by protein inclusions, which in most families of the order include a large, central crystalloid protein of spherical (globular) or polyhedral structure surrounded by a bundle of proteinaceous filaments. All families of the Caryophyllales have sieve-tube plastids with the peripheral fibrous protein, but 2 of the 12 families, Chenopodiaceae and Amaranthaceae, lack the central crystalloid protein.

Gibson and Nobel (1986) have reviewed the elemental contents of cacti, both macronutrients and micronutrients. It was determined that they are similar to those of most plants, except that cacti had a much higher amount of calcium, tend to have higher manganese content, and have much lower sodium levels than most other plants. Higher calcium levels might be expected in cacti because of the prevalent calcium oxalate crystals, which apparently increase with the age of the plants. The significance of these elemental differences in cacti, if any, is not understood. Experiments have demonstrated (Nobel et al., 1984; Gibson and Nobel, 1986) that cacti vary in their tolerance to sodium chloride, and their salt tolerance always seems to be low.

The elemental content of cacti, particularly prickly pears, is relevant to their use as food by humans and as fodder for cattle. In general, cacti are relatively low in nitrogen, as are the tissues of many other kinds of plants, and comparatively low in total nutrients/equal weight because of the high water content of the cortex and pith.

Among the more interesting future chemical studies of cacti would be those that investigate the role of secondary compounds in deterring or attracting other species of organisms in their natural habitats. The elemental tolerances or requirements of certain cactus species might help explain the ecological restrictions and the distributions of some species.

Ecology and Biogeography

In evaluating the distributions of cacti on a regional basis, it is of conceptual significance to understand that ecogeographic ranges have genetic constraints and are, therefore, important to consider in identification and classification of taxa. Topographic, edaphic, and vegetational diversity in the Trans-Pecos (Henrickson and Johnston, 1986; Powell, 1998a) provides many different habitats for cacti. The lowest elevation (ca. 1,800 feet), which is perhaps the hottest and driest area (Hernández and Bárcenas, 1995) in the CDR, is near Boquillas Canyon in southern Brewster County, Texas, and in adjacent Coahuila, Mexico. Even lower elevations (ca. 1,000 feet) are found near the mouth of the Pecos River near the eastern edge of the Chihuahuan Desert, but conditions there are more mesic. Desert vegetation, in general referred to as "desertscrub," occurs throughout the Trans-Pecos north of the Rio Grande at elevations up to about 3,500–4,000 feet, or higher in desert mountains and on the south slopes of mesic mountains. Some Chihuahuan desertscrub extends east of the Pecos River into southern New Mexico and into extreme southeastern Arizona. Desert grasslands or grasslands with desertscrub occur throughout much of the Trans-Pecos. Mid-elevations (4,000–5,500 feet) in the Trans-Pecos, in mountain basins and on plateaus, are dominated by more mesic grasslands that intergrade upslope with woodland in the Chisos, Davis, and Guadalupe mountains. Higher elevations (ca. 7,000–8,000 feet) support some montane woodland (Henrickson and Johnston, 1986).

A complex mosaic of habitat exposures and substrate types exists in the Trans-Pecos (Johnston, 1977; Powell, 1998b, plates 2–12). The mesic mountains are large enough to form ecologically different north-south exposures on peaks, ridges, and slopes and in canyons. Many smaller arid mountains and mesas also form significantly different exposures. Thin, rocky soils are found on the mountain slopes, and deeper alluvial substrates are formed in the basins and valleys. Alluvial deposits in different areas of the Trans-Pecos are variable in particle composition and derivation from igneous-rock and sedimentary parent materials. Different geologic strata are exposed in the various limestone and volcanic landforms. Contributing to the substrate mosaic in the Trans-Pecos are heavy clays, gypsum exposures (stable gypsum and dunes), saline flats, and silica sand (stable deep sand and dunes). One or more cactus species occupy essentially all natural dryland habitats and substrate types in the Trans-Pecos, including the harshest habitats for cacti—clay, gypsum, the edges of salt lakes, and the highest mountain peaks.

Ecology

The remarkable tolerance of cacti to harsh environmental conditions has been explored in considerable detail by Gibson and Nobel (1986) and Nobel (1988, 1994). For cacti in general, anatomical/physiological and ecological aspects are well understood. What is not well understood is the autecology of cacti, that is, the adaptations of individual taxa and populations to their particular environments (Kinraide, 1978). In the Trans-Pecos, for example, some taxa are relatively widespread in distribution, and others are limited to certain habitats and ranges. At present only general ecological observations are available for most Trans-Pecos cacti.

Although some cactus species can be found almost everywhere in the Trans-Pecos, more taxa and more individuals tend to be concentrated in certain habitats. Cacti tend to occur in thin, rocky soils or in soil-filled rock crevices on south and west exposures. These conditions tend to favor the establishment of cacti and tend to negate the growth of other, competing vegetation.

In mid-elevation grasslands of Trans-Pecos Texas, cacti tend to occupy rocky habitats with thin soils where grass cover is minimal. The reduction or elimination of grass cover by overgrazing allows for range expansion of cacti in grasslands (Anthony, 1954; Fraser and Pieper, 1972; Burger and Louda, 1995). Prickly pear and cholla species have the ability to spread rapidly because of vegetative reproduction, that is, the dissemination of stem joints and proliferating fruits (Anthony, 1954). In the Trans-Pecos there are numerous (usually small) areas of degraded grassland that are dominated by prickly pear (*Opuntia* spp.), cholla (*O. imbricata*, *O. leptocaulis*), or both.

Cacti that normally occur at lower to mid-elevations tend to extend up the south and west slopes of higher mountains. Again this is attributed to the ability of cacti to outcompete other forms of vegetation under hot and dry conditions. Certain cactus species, however, are well adapted to the more heavily vegetated north and east slopes and protected canyons of the higher mountains (e.g., *O. chisosensis* in the Chisos Mountains and *O. polyacantha* in the Davis Mountains, along with *Echinocereus viridiflorus* and *E. coccineus*).

Cacti that have encroached into grazed rangelands are sometimes at risk if grazing is suspended and a dense growth of grass is allowed to return. We have observed this circumstance over a period of about 15 years in an open grassland setting within oak-juniper-pinyon woodland at about 5,200 feet, approximately nine miles south of Alpine in the southern Davis Mountains. The site was part of a cattle ranch that had undergone continual grazing until the mid-1980s, at which time all grazing activity was suspended. Large plants of *Echinocereus coccineus*, some with more than 100 stems, had become established in the heavy soils of the continually cropped grassland site, apparently having migrated from adjacent rocky habitats. After grazing was suspended, a dense grass cover returned, closely enveloping the numerous plants of *E. coccineus*. Within 10 years most of the *E. coccineus* plants had died, apparently from rotting in heavy soils where periodic abundant moisture was maintained by the dense grass cover. Larger cacti, such as prickly pears and chollas, seem not to be similarly affected by heavy grass cover and seem to persist in grasslands once established.

Although desert cacti are remarkably equipped to cope with dry conditions, some cacti, even prickly pears, die during severe droughts. During excessively long dry periods we have observed desiccated cactus plants that looked as though they might have expired because water supplies were not replenished. This assumption usually is reached when nearby individuals of other desert-adapted species also appear to have died from desiccation, but we have not carefully examined the question about the possible role of internal stem insect damage or other obscure factors that might have contributed to the death of cactus plants during dry periods.

During droughts in the Trans-Pecos we have observed even more cacti that were harmed by secondary causes, not the dryness per se. In very dry habitats cacti are often eaten by herbivores, including rodents, rabbits, javelina, and deer. When animals are very hungry and thirsty, the noted physical and chemical herbivore deterrents of cacti seem to be less effective. We have observed tops, sides, or whole stems that have been removed, apparently by herbivores. After the spine defense is breached in cacti with larger stems (e.g., *Ferocactus hamatacanthus*), herbivores may revisit the plant until the entire plant is eaten.

When severe dry spells coincide with the flowering periods of cacti, relatively unprotected buds and flowers often are eaten by herbivores. In Pecos County we have seen hundreds of plants of *Echinocereus* × *roetteri* var. *neomexicanus*, virtually every plant in localized populations, with flowers eaten, leaving just a rim of chewed flower bases at the stem apex and tepal "crumbs" on the ground at the plant bases. South of the Chisos Mountains we have seen the same situation with *E. chisoensis*, where flowers and flower buds were removed by herbivores during a particularly dry spring. For cacti secondary effects of drought include the diminution or near loss of sexual reproductive potential for at least a year. Such circumstances emphasize the significance of asexual reproduction in *Opuntia* (Anthony, 1954; Bowers, 1997). Fortunately, at least for some cactus species, "seed hydration memory" (Dubrovsky, 1996) is an adaptive strategy after seeds are produced on the soil during drought episodes. Seed hydration memory is a phenomenon that allows seeds to remain viable under discontinuous hydration and dehydration and to retain during dehydration the physiological changes accrued by the seeds during hydration. Seeds benefiting from seed hydration memory would be expected to germinate earlier than seeds that did not have the experience, regardless of the length of an intervening dehydration period, thus leading to a higher seedling survival rate (Dubrovsky, 1996).

A pattern of edaphic endemism has been demonstrated for species of numerous families in the CDR (Johnston, 1977; Powell and Turner, 1977; Turner and Powell, 1979), including the Cactaceae (Hernández and Bárcenas, 1995). Although many Trans-Pecos cacti are relatively widespread in distribution and seem to have broad ecological tolerances, many other taxa are delimited in distribution, rather strictly associated with substrates or habitats and presumably with narrow ecological parameters. Some Trans-Pecos cacti are limited (essentially) to either limestone (e.g., *Coryphantha tuberculosa* var. *tuberculosa, C. duncanii, C. ramillosa, C. sneedii,* etc.) or igneous substrates. *Echinomastus warnockii* is a common species in the southern Big Bend region, occupying limestone and igneous soils of many types, especially in relatively unstable derived substrates. The related species *E. mariposensis* is known to be abundant in numerous disjunct populations, but it is always found in certain stable limestone strata. *Echinocereus enneacanthus* is common in alluvium near the Rio Grande, whereas its tetraploid relative *E. stramineus* is mostly saxicolous in nearby habitats at slightly higher elevations. *Coryphantha minima* and *E. davisii* are restricted to certain Caballos Novaculite (siliceous) exposures in the Marathon Basin (Plate 11). *Toumeya papyracantha,* normally found in grama grasslands in New Mexico and Arizona, has been reported to occur in saline conditions at the margin of a salt lake in northern Hudspeth County (Reeves, 1994). *Opuntia polyacantha* var. *arenaria* apparently occurs only in deep sand in El Paso County and other nearby areas. *Opuntia densispina* may be limited to certain clay soils in southern Brewster County. Many other examples of ecologically restricted Trans-Pecos cacti could be cited. Specific ecological information relating to restricted distributions is lacking for all of them (Hernández and Bárcenas, 1995), although some kinds of ecological studies are available for taxa that occur elsewhere in the CDR (e.g., Mandujano et al., 1996).

We speculate that desert cacti in the Trans-Pecos, if not those in the grasslands and mountains as well, become established mostly in the protection of nurse plants or similarly effective microhabitat modifications, such as boulders, ledges, or other shading exposures. The phenomenon of cactus plant establishment under nurse plants has been studied most extensively in connection with the saguaro (*Carnegiea gigantea*) and other cacti in the Sonoran Desert (see Gibson and Nobel, 1986; Hutto et al., 1986; Suzán et al., 1994). Nurse plants seem to protect and nourish (Nobel, 1994) cactus seedlings, which probably are more vulnerable to temperature extremes, desiccation, and nutrient deficiencies than are adults. Supposedly the essential nature of the nurse plant diminishes with advancing age of the juvenile cactus (Vargas-Mendoza and Gonzalez-Espinosa, 1992). Through many years of studying Trans-Pecos cacti of all ages in their natural habitats, we have observed that juveniles and young adults characteristically are associated ecologically with other plants (Plate 30). We presume that these "other plants" function as nurse plants. There are no specific ecological studies that go beyond statistical correlation of Trans-Pecos cacti with closely associated plants.

Seedlings of *Echinocereus chisoensis* are known to be associated with creosotebush [*Larrea tridentata* (Sessé & ex DC.) Coville], acacia (*Acacia* spp.), lechuguilla (*Agave lechuguilla* Torr.), dog cholla (*Opuntia aggeria*), and various grasses. In limestone hills west of the Dead Horse Mountains, we have observed seedlings and adults of *Glandulicactus uncinatus* characteristically associated with cespitose grasses, often actually growing in the grass clump. Anthony (1954, 1956) alluded to the general paucity of seedlings observed during her study of Big Bend opuntias but found them in the shade of aggregated taller plants. We have rarely observed cactus seedlings in habitats open to the sun, but their occurrence is predictable in wet periods during the warm season, always in shade at the base of plants, perhaps among organic litter, in shaded crevices with *Selaginella* spp., or in other shaded microhabitats. A notable exception to the "rule" of perennial nurse plants was seen on a bare limestone ridge in southern Brewster County. Juveniles and young adults of *Echinomastus mariposensis* were found to be so abundant at one site, poorly vegetated but covered with small limestone rocks 1–5 cm high, that it was nearly impossible to avoid stepping on the cacti when walking across the site. The small rocks are sufficient microhabitat for initial establishment of certain species.

Biogeography

Most modern authors agree that the Cactaceae originated in South America. Over geologic time, cacti have been dispersed to their present locations in Central America, North America, and islands of the New World (see Benson, 1982; Gibson and Nobel, 1986; Mauseth, 1990; Anderson, 1996b; Hicks and Mauchamp, 1996; Mauseth et al., 2002). Various members of the Cactaceae, particularly species of *Opuntia*, subsequently have been distributed by human activity over most of the temperate and warm regions of the world (see Benson, 1982; Forstner, 1996).

In present-day North America the largest number of cactus species (Hernández and Bárcenas, 1995) are found in Mexico, and most of these are in or around the Chihuahuan Desert. In fact, after multiple lineages of the Cactaceae reached North America, Mexico became the greatest center of diversity for the family. Cacti are well represented in the western United States as well (Benson, 1982), with the greatest concentrations of taxa in Trans-Pecos Texas, Arizona, New Mexico, and California. Although the geographic distribution of cacti in general is limited by freezing temperatures, some species are cold hardy enough to have occupied northern latitudes, including much of the northern tier of the United States and into Canada. The northernmost extension of cacti in North America is reported to be in the Peace River district of Alberta, Canada (Speirs, 1989). *Opuntia fragilis* (Nutt.) Haw. gets credit for the northernmost distribution record, but not far to the south in the southern Alberta prairie, it occurs together with *O. polyacantha* and *Coryphantha vivipara*.

The CDR is the largest and least-studied desert area in North America, and it is the center of cactus diversity on the continent. The CDR cactus flora and the CDR desert climate (Schmidt, 1986) extend into the United States, in Trans-Pecos Texas and adjacent areas, but most of its biotic diversity remains in Mexico, primarily in the states of Chihuahua, Coahuila, Nuevo León, Tamaulipas, and San Luis Potosí (Hernández and Bárcenas, 1995).

There is convincing evidence that climatic events and vegetation changes that led to the current CDR had their genesis in the Pleistocene glacial-interglacial fluctuations. Most impetus in shaping current vegetation began with the last glacial maximum at about 11,000 years ago, and regional dominance in xeric vegetation began during late Holocene, about 4,000 years ago. The documentations for these assumptions about the development of vegetation in the CDR have come from a series of ingenious paleoecological studies of plant macrofossils in pack rat middens, conducted by Wells (1966) and Van Devender (see Van Devender, 1986, 1990; Van Devender and Burgess, 1985) and his associates (see Betancourt et al., 1990). McCarten (1981) also has offered phytogeographic interpretations from the macrofossil data published by the authors cited above.

Data from pack rat middens suggest that the northern CDR, including the Trans-Pecos, was covered with oak-juniper-pinyon woodland at least until about 8,000 years ago. Macrofossil evidence from the central CDR, in the vicinity of Bolsón de Mapimí, indicates that the woodland extended well into Mexico but that woodland was increasingly less dominant to the south and was mixed with more xeric vegetation, including some cacti. In the northern Chihuahuan Desert the pack rat midden data are consistent in showing a gradual change (ca. 11,000–8,000 years ago) from mesic climatic conditions and woodlands in the late Wisconsin to maximum expansion of grasslands in the middle Holocene (8,000–4,500 years ago), to mostly xeric conditions with developing desertscrub in the last 4,000 years during late Holocene (Van Devender, 1986). Chihuahuan desertscrub, including cacti, expanded to its present geographic extent during modern climatic regimes, and the Chihuahuan Desert climate seems to be expanding generally northward at the present (Powell, 1998b). Van Devender and associates have speculated that the southern Big Bend area and the great *bolsones* of the CDR, particularly the Bolsón de Mapimí of the west-central CDR, served as refugia for desert vegetation, including cacti, at least during the late Wisconsin glacial maximum. The Bolsón de Cuatro Ciénegas in the east-central CDR contains relictual and/or endemic species indicative of yet another desert plant refugium. One of the best reviews of the CDR climate and biogeography is couched in discussion by Hernández and Bárcenas (1995) of distribution patterns for endangered cactus species in the Chihuahuan Desert.

The present-day Trans-Pecos cactus flora appears to have phytogeographic connections in all directions. The most obvious and most extensive relationships are with the central CDR cactus flora in Mexico, particularly true desert entities in the southern Big Bend region, such as *Echinocereus, Coryphantha, Echinomastus, Thelocactus, Ariocarpus, Lophophora,* and *Opuntia.* Some Trans-Pecos species of *Echinocereus, Coryphantha, Mammillaria, Ferocactus, Echinomastus, Opuntia,* and other genera clearly are related to taxa in northwestern Mexico and the southwestern United States, usually Arizona and New Mexico. Some individual species of *Opuntia, Echinocereus, Coryphantha,* and other genera have wide ranges, extending from the Trans-Pecos mountains into the Rocky Mountains and elsewhere in the western United States. Still other taxa of *Opuntia, Coryphantha,* and other genera that appear to have their southern limits in the Trans-Pecos extend across the Texas Panhandle and eastern New Mexico and into the Great Plains, sometimes all the way to Canada. Other prominent Trans-Pecos phytogeographic connections involving several genera such as *Echinocereus, Coryphantha, Mammillaria,* and *Opuntia* are with the Edwards Plateau region of Central Texas and with the Tamaulipan Scrub Province of South Texas. Relatives of species that occur in South Texas and adjacent Mexico east of the Sierra Madre Oriental (e.g., *Ancistrocactus, Echinocereus, Coryphantha,* and *Opuntia*) often reach at least the southeastern Trans-Pecos region.

Long-distance dispersal (relatively speaking) surely has been important in the dispersal of *Opuntia* species with wide ecological tolerances. The significance of dissemination of stem joints and proliferating fruits was noted by Anthony (1954) during ecological observations of Trans-Pecos *Opuntia*. Seeds of *Opuntia* spp. and other genera undoubtedly are disseminated by many kinds of animals that eat juicy sweet fruits (Mandujano et al., 1997).

Range discontinuities of many Trans-Pecos cacti suggest that current distributions in many genera are remnants of formerly widespread distributions followed by range contraction and subsequent isolation of populations. For example, it appears that this kind of geographic isolation is responsible for ongoing diversification of the *Echinocereus viridiflorus* species group. Ecological specializations, such as edaphic restriction to rare soils, sharply restricts the populations of some species, especially near the margins of their ranges. Earlier we have suggested that edaphic adaptations appear to have been a major factor in the isolation and genetic differentiation of Trans-Pecos cactus populations. Vicariant taxa, separated by circumstances and now visibly differentiated, appear to be commonplace in *Echinocereus, Coryphantha, Echinomastus, Opuntia,* and other Trans-Pecos genera.

Natural interspecific hybridization is suspected between several Trans-Pecos opuntias and in *Echinomastus,* and it is documented in *Echinocereus* (Powell et al., 1991; Powell, 1995, 1998c). In the *E.* × *roetteri* complex, hybridization has occurred where the desert-adapted *E. dasyacanthus* has expanded into the range of *E. coccineus* in concert with increasing desertification in the Trans-Pecos.

Evolution

Among the flowering plants, Cactaceae is positioned in the order Caryophyllales, along with 11 other families (Cronquist, 1988). Various interfamilial relationships have been proposed for Cactaceae, including sufficient distinctiveness to rate a monotypic order (Benson, 1982). Relatively close relationships to Cactaceae have been suggested for Basellaceae, Didiereaceae (Nowicke, 1996), and Portulacaceae. Modern evidence, including chloroplast DNA information, currently points to Portulacaceae as closest to Cactaceae (Hershkovitz and Zimmer, 1997).

Three subfamilies, Pereskioideae, Opuntioideae, and Cactoideae, have long been recognized in Cactaceae. The Pereskioideae are regarded as the basal or most primitive cacti (Hershkovitz and Zimmer, 1997). The Opuntioideae are believed to be an early offshoot from the basal cacti, and in the modern world they have achieved the widest distribution in the family. Nine tribes of the derived and leafless Cactoideae are recognized (Gibson and Nobel, 1986), with the tribes Notocacteae (mostly South America) and Cacteae (North America) representing perhaps the most specialized evolutionary lines in the family (Mauseth, 1990).

The fossil record for Cactaceae is meager (Speirs, 1978; McCarten, 1981), and it has not been useful in attempts to resolve the evolutionary origin of the family. Based mostly upon the evidence of continental drift and climate (Mauseth, 1990) since the breakup of Gondwana, speculation is that the earliest possible time for the origin of the Cactaceae was during the late Cretaceous, perhaps 100 million–65 million years ago (see Hershkovitz and Zimmer, 1997). Mauseth (1990) reasoned that cacti first evolved and began to diversify 70 million–40 million years ago (i.e., the Pereskioid and perhaps the Opuntioid types) and that most of the Cactoid lines did not develop until suitable habitats (the rain shadow of the Andes) became available about 20 million years ago. That there should be a long hiatus between the evolution of cacti from the basal Pereskioid line and the more recent types is supported by anatomical evidence. The structural and metabolic features of leafy pereskias are so different from those in leafless cacti that many anatomical and physiological modifications had to occur during the evolution of the cacti with stem photosynthesis (Sajeva and Mauseth, 1991). An even later, post-Cretaceous origin for cacti in the mid-Tertiary period about 30 million years ago was suggested by Hershkovitz and Zimmer (1997), based upon the evaluation of chloroplast DNA data. Whenever cacti originated, diversification and invasions of North America possibly did not begin until seasonally dry and rocky habitats were available (Boke, 1980), probably not more than 20 million years ago. North America has had desertlike habitats since at least 13 million years ago, in the Miocene epoch of the Tertiary period (Mauseth, 1990).

Although we have no petrified fossils of ancient extinct cactus genera, there are remarkable late Quaternary paleoecological data from pack rat middens (e.g., Betancourt et al., 1990). These records from the last 40,000 years show climate and vegetation changes in western North America that indicate contraction and expansion of arid regions during glacial and interglacial periods. Cacti and other arid-adapted plants evidently were restricted to small arid refugia during the Quaternary pluvial episodes. According to paleoecological data, regional expansion of the northern Chihuahuan Desert vegetation, with its diversity of cacti, began only about 4,000 years ago (Van Devender, 1986).

Chromosomal Evolution and Hybridization

A base chromosome number of $x = 11$ has been established for Cactaceae through reports from the primitive genus *Pereskia* (Leuenberger, 1986) and numerous species of Opuntioideae and Cactoideae from western North America. Pinkava and colleagues (1992, 1998) have published a series of papers dealing with chromosome numbers in western North America, concentrating on species in the Sonoran and Mojave deserts. Weedin and associates have concentrated on chromosome numbers of cactus species in the northern Chihuahuan Desert (e.g., Weedin and Powell, 1978). The chromosome number reports overwhelmingly support $x = 11$ for the family; absence or extreme rarity of aneuploidy (Pinkava et al., 1985); and the importance of polyploidy, especially in certain genera, such as *Opuntia, Echinocereus* (Cota and Philbrick, 1994), and *Mammillaria*.

In Cactaceae the percentage of polyploidy is greatest by far in Opuntioideae. Pinkava et al. (1985, 1998) estimated that 64.3% of the opuntioid taxa are polyploid, 60% in North America and 72.7% in South America. In all sections of the largest genus, *Opuntia,* polyploid percentages were found to be higher in the Southern Hemisphere. In Cactoideae only 12.9% of the taxa were found to be polyploid (Pinkava et al., 1985, 1998). Polyploidy estimated by these researchers for the whole Cactaceae was 28%. For Trans-Pecos Texas in the northern CDR, Weedin et al. (1989) found 55.1% polyploidy in the Opuntioideae, 15.2% in the Cactoideae, and 26.1% in the family. Polyploidy is not known for the Pereskioideae.

James F. Weedin (unpub.) has found that polyploid taxa in *Opuntia* and *Echinocereus* (Trans-Pecos Texas) generally have a greater biogeographic distribution in multiple plant communities (mountains, grasslands, desert) than do diploid taxa, which are most often found in single-plant communities. In *Opuntia,* approximately 70% of polyploids occur in multiple communities, whereas 100% of *Echinocereus* polyploids are found in multiple communities.

Ploidy levels in Cactaceae range from $2x$ to $8x$ in most taxa, but as high as $24x$ to about $30x$ in certain *Mammillaria* and Opuntioideae (Pinkava et al., 1985, 1998). In Trans-Pecos Cactaceae, most polyploidy in *Opuntia* is tetraploid ($4x$) or hexaploid ($6x$), so far as known, and polyploidy in *Echinocereus* and *Mammillaria* is $4x$. It is obvious that polyploidy has played a major role in the evolution of Cactaceae. Polyploidy works as a mechanism of speciation because taxa with different ploidy levels do not form fertile hybrids.

Interspecific hybridization is believed to be another important evolutionary process in Cactaceae (Pinkava et al., 1985, 1998; Cota and Philbrick, 1994), but its evolutionary role in the family is yet to be adequately investigated (Powell, 1998c). In the Trans-Pecos, interspecific hybridization is known or suspected to have occurred in *Opuntia, Echinocereus,* and *Echinomastus*. Hybridization in Cactaceae often is correlated with polyploidy and apomixis (Pinkava et al., 1985, 1998; Ross, 1981; Garcia-Aguilar and Pimienta-Barrios, 1996).

Pollination Biology

Although the pollination ecology of most cactus genera and species has not been formally investigated, there have been sufficient studies to provide a general framework of carefully documented information. Most of this has involved taxa that are native to or near the western deserts in the United States or arid regions in Mexico (e.g., Ganders, 1976; Grant and Grant 1979a, 1979b, 1979c, 1979d; Parfitt and Pickett, 1980; Rowley, 1980; Parfitt, 1985; McFarland et al., 1989; Johnson, 1992; Suzán et al., 1994; Scobell, 1999). Several of the large columnar cacti are pollinated by nectar-eating bats or moths (e.g., Tuttle, 1991; Fleming et al., 1994, 1996; Valiente-Banuet et al., 1997; Fleming and Holland, 1998), including species in Oaxaca, Mexico (Casas et al., 1999), and South America (Nassar et al., 1997). Aspects of pollination biology have been investigated in the epiphytic shrub of Brazil, *Schlumbergera* Lem. (Boyle et al., 1995); in some tropical prickly pears (Spears, 1987); and in the tropical American night-blooming climbing genera *Hylocereus* (A. Berger) Britton & Rose and *Selenicereus* (A. Berger) Britton & Rose (Lichtenzveig et al., 2000). Investigations of particular species of Chihuahuan Desert cacti have been few (Grant et al., 1979; Breckenridge and Miller, 1982; Leuck and Miller, 1982; Osborne et al., 1988; McFarland et al., 1989; Powell et al., 1991), but overviews of pollination biology in Cactaceae by Grant and Grant (1979d), Grant and Hurd (1979), Gibson and Nobel (1986), Cota (1993), and Nobel (1994) have touched upon some Chihuahuan Desert taxa.

General Information

The flowers of most cactus species are relatively large and showy. Their sizes, shapes, and colors and/or scents in general provide long-distance signals that attract animal life. Probably 80–90% of the cactus species are pollinated by bees (Nobel, 1994). No wind pollination has been documented in Cactaceae (Cota, 1993).

Most cacti have perfect, herkogamous flowers, conditions that promote outcrossing, and presumably most species are self-incompatible (Ganders, 1976), as they appear to be in *Echinocereus* (Taylor, 1985; Cota, 1993). In herkogamous flowers there is spatial separation between the stigma and stamens (i.e., the stigma extends above the anthers). Some cacti are autogamous or geitonogamous, but the extent of self-pollination in Cactaceae has not been studied (Gibson and Nobel, 1986). In the Trans-Pecos, *Epithelantha* spp. (Grant and Grant, 1979d) and *Lophophora* appear to be self-compatible. Dioecy is rare in Cactaceae (Parfitt, 1985), as are related gynodioecious and trioecious conditions. *Echinocereus coccineus,* a species that occurs in the Trans-Pecos, is morphologically gynodioecious and functionally dioecious (Ferguson, 1989; Powell et al., 1991; Hoffman, 1992; Powell, 1995), except in the western portion of its range (Hoffman, 1992; Scobell, 1999).

The stamens of some cacti are thigmotropically sensitive (i.e., they respond to contact by moving in toward the style). This phenomenon is best known in *Opuntia*, where most all the species have thigmotactic stamens (Toumey, 1899; Grant and Hurd, 1979). When touched, the stamens rapidly close tightly around the style, or according to Grant et al. (1979), they also may move outward toward the perianth if touched on the outside, or toward the point of contact (Grant and Hurd, 1979). In our experience, touching the anthers invariably causes fully extended stamens to bend toward the style and curl around it in all but one of the species of *Opuntia* we have tested (the stamens of *O. ellisiana* seem not to be particularly sensitive, as determined from one test) and in the flowers of some other cacti as well (e.g., *Coryphantha vivipara* and *Ancistrocactus* spp.). The function of the sensitive stamens in *Opuntia* and other cacti has not been satisfactorily explained (Grant and Hurd, 1979; Parfitt and Pickett, 1980), although logically it should be related to small herbivores, pollen predators, and/or the behavior of bees that pollinate the species.

Nectar is produced in the flowers of many cactus species, particularly those that are adapted for hummingbird and bat pollination. Many bee flowers also produce nectar, but some apparently do not (Grant and Hurd, 1979). In cacti the correlation between the morphological aspects of pollination syndromes and nectar has not been investigated extensively (McFarland et al., 1989), but Scogin (1985) has provided information about differential nectar composition in 51 species representing all three traditional subfamilies of Cactaceae, and Scobell (1999) has compiled nectar data for *Echinocereus coccineus*. There are correlations between characteristic types of pollinators (bees, hawkmoths, bats) and the sugar composition, sugar concentration, and caloric content of the nectar.

Visitors to cactus flowers are not necessarily pollinators. Small bees, for example, may visit medium-size or large cactus flowers and not contact the stigma when entering or leaving a flower. Beetles and ants are frequent visitors in cactus flowers, but they are not effective pollinators (McFarland et al., 1989). Flies, wasps, and small butterflies less frequently visit cactus flowers. Grant and Connell (1979) demonstrated that cactus flowers are the brood sites of *Carpophilus* beetles, which chew the flowers and only occasionally effect pollination.

As previously stated, most cactus species in the southwestern United States are pollinated by bees, and the bee pollination system is considered to be primitive (Grant and Grant, 1979d; Cota, 1993) except for secondary reduction of "night-blooming cereus" flowers to budlike bee flowers within each of numerous taxonomic groups, sometimes within a genus (e.g., *Echinopsis* in the broad sense). Reportedly, the remaining species (perhaps 10%) are pollinated by birds, hawkmoths, and bats, and all of these are derived conditions relative to the general condition in Pereskioideae, Opuntioideae, *Maihuenia*, and Portulacaceae. The various pollination syndromes are related to several aspects of floral morphology (Grant and Grant, 1979d; Gibson and Nobel, 1986; Cota, 1993).

Bee flowers are of various sizes and colors and may or may not produce nectar. Grant and Grant (1979d) determined that bee flowers are not necessarily distinct from generalized, promiscuous flowers. Promiscuous flowers do not exhibit in their syndrome (morphology, nectar, scent) any preference for a particular pollinator. The three classes of bee flowers identified by Grant and Grant (1979d) were (1) large bowl-shaped flowers, (2) medium-size to small bowl-shaped flowers, and (3) medium-size or small disc-shaped flowers. These flower types are common in cacti of the Southwest. The large bowl-shaped flowers are known to be pollinated by medium-size and large bees. The medium-size to small bowl-shaped flowers probably are bee flowers. The medium-size or small disc-shaped flowers probably are bee flowers as well, but in general this flower class has been the least studied. Pollinators of the disc-shaped flowers logically could consist of a wide range of small insects with short tongues, although bees may be attracted by pollen presentation and not nectar (McFarland et al., 1989). Ultraviolet floral patterns potentially are important in the future study of bee flowers, because bees can see in the ultraviolet spectrum. In North America about 90 bee species are known to visit taxa of *Opuntia* (Grant and Hurd, 1979). The nectar of bee flowers is dominated by glucose-fructose in some angiosperms (Spira, 1981).

The hummingbird floral syndrome usually involves (by plant standards) large and showy flowers or dense masses of smaller ones. Usually they are tubular, unscented, diurnal, and reddish flowers. The elongated receptacular tube provides a deeper nectar chamber in these flowers. Hummingbird flowers usually contain abundant nectar and pollen, and the nectar is rich in sugars, notably sucrose (Scogin, 1985; Spira, 1981). Cacti with red hummingbird flowers are fairly common in tropical regions of the Americas, but they are rare in the American Southwest, where only a few species are known to be pollinated by these birds (section *Triglochidiatus* of *Echinocereus*, Taylor, 1985; Scobell, 1999). *Echinocereus coccineus* of this section often is pollinated by bees as well as hummingbirds.

The hawkmoth floral syndrome is characterized by whitish nocturnal flowers with long, slender floral tubes, abundant nectar, and heavy sweet scent. In Texas and the United States in general, only *Peniocereus* and *Acanthocereus* display the hawkmoth syndrome, with nocturnal white flowers and floral tubes 10 cm or more long, although many other "night-blooming cereus" genera are in American horticulture. Other types of moths are known to be involved in pollination of certain cacti (Fleming and Holland, 1998).

The bat flower syndrome is typified by large funnel-shaped flowers that are whitish and nocturnal, with abundant nectar and pollen and foul or musky odors. About 160 cactus species in 37 genera with bat flowers (Grant and Grant, 1979d) are widespread among the columnar cacti in the tropical and subtropical Americas. A much smaller number of columnar cactus species in arid regions of Mexico and in the southwestern United States are pollinated by bats. The most famous bat-pollinated species in North America are the Sonoran Desert columnar cacti *Carnegiea gigantea* (saguaro), *Stenocereus thurberi* (organ pipe), and *Pachycereus pringlei* (S. Watson) Britton & Rose (cardon). Bats, moths, birds (white-winged dove and eight other species), and bees all visit the flowers: bats and moths at night, birds and bees during the day. Fleming et al. (1996) determined that birds and bees were the most effective pollinators for the saguaro, birds were most effective for the organ pipe, and bats accounted for most of the fruit set in the cardon.

For the Texas cactus flora Grant and Grant (1979d) estimated that 27 cactus species produced large bee flowers (diameter >5.5 cm), and 39 species produced medium-size to small bee flowers (diameter <5 cm). These authors tabulated only one Texas cactus species with hummingbird flowers and only two Texas cactus species with hawkmoth flowers. No Texas cacti are visited by bats, as far as we know.

In the flowers of *Opuntia engelmanii* var. *lindheimeri* Grant et al. (1979) observed two herkogamous arrangements during the blooming season. Flowers produced during the height of the blooming season exhibited the stigma well elevated above the anthers. Flowers produced at the end of the blooming season did not have elevated stigmas but instead had stigmas and anthers on the same level. The early and mid-season flowers were observed experimentally to require insect-vectored cross-pollination for seed set, and at least some of the late-season flowers were autogamous. The significance of this phenomenon in *Opuntia* requires further investigation. In the Trans-Pecos we have observed flowers of numerous *Opuntia* species, and it is apparent that flowers with elevated stigmas and those with near-equal stigmas and anthers both exist. It has not been determined if there is a seasonal pattern in floral herkogamy of the Trans-Pecos opuntias.

Numerous southwestern cactus species, including many in the Trans-Pecos, produce flowers that stay open, or open again, for several days. These flowers allow access to copious pollen and receptive stigmas over a period of days. These conditions inevitably expose flowers to an expanded number of pollinators (and herbivores), but they lose water rapidly for as long as they are open.

Trans-Pecos Cacti and Pollination Biology

Most Trans-Pecos cactus species have not received detailed study with respect to pollination ecology. Through casual observations of natural and cultivated specimens, we have noticed, as expected, that bees of many sizes, including European honeybees, appear to be the most common visitors to cactus flowers. Beetles also are relatively common visitors to the flowers of Trans-Pecos cacti, especially in *Opuntia*, where they seem ubiquitous.

The systematic studies of two Trans-Pecos species complexes of *Echinocereus* have included considerable information concerning pollination biology. In the *Echinocereus enneacanthus* complex, involving mostly *E. enneacanthus* and *E. stramineus*, Breckenridge and Miller (1982) found that 27 hymenopteran species visited the flowers. Pollination was carried out by solitary (or semisocial) halictid bees (Halictidae) and anthoporid bees (Anthoporidae), including *Diadasia rinconis, D. afflicta, Lasioglossum forbesii, Paralictus* sp., and *Dialictus* sp. In all, 20 of the 27 bee species collected in flowers of the *E. enneacanthus* complex belonged to the four genera listed above. The other seven bee species reported from *E. enneacanthus* flowers were leaf-cutting bees (Megachilidae: *Luthurge bruesi, Osmia subfasciata, Ashmeadiella meliloti*) and plasterer bees (Colletidae). Other pertinent data involved extensive flower phenological considerations, pollinating bee behavior, and breeding system information. The pollinating bees did not distinguish between flowers of sympatric *E. enneacanthus* and *E. stramineus*, usually alighted on the stigma when moving from flower to flower, exhibited foraging behavior among the anthers and deep in the flower, and took flight from the inner tepals after emerging from the androecium. The halictid bees were observed to visit several flowers in succession before leaving the immediate area.

The work of Leuck and Miller (1982) concerning the *Echinocereus viridiflorus* complex involved long-wave UV absorbing and reflecting floral patterns, as well as floral phenology and bee behavior studies. They discovered, among other things, that two consistent UV floral patterns existed among seven varieties of the *E. viridiflorus* complex and that pollination is performed by halictid bees. Pollination in the *E. viridiflorus* complex usually occurred within the first two hours after anthesis, involving bees that alighted on the stigma, crawled down the style and foraged in the pollen, or moved to the small amounts of nectar (less than 1 µl) produced near the base of the stamens. It was determined that nectar was available only on the first day that a flower was open but that flowers opened again for several days. Individual bees usually visited several flowers in close proximity before leaving the vicinity.

Zimmerman (1993, p. 271) stated that the mostly dioecious *Echinocereus coccineus* provided "the best example of ornithophilous flowers in the United States cactus flora" (see Grant and Grant, 1979d, 1967). This species and its closest relatives, the only other bird-worthy taxa of northern *Echinocereus*, range from Texas to Arizona and the Californias, and *E. coccineus* is common in the Trans-Pecos, where it has hybridized with *E. dasyacanthus* to produce the hybrid taxon *E.* × *roetteri*. Zimmerman reported repeated observations in New Mexico of the black-chinned hummingbird (*Archilochus alexandri*) feeding in nectar-rich *E. coccineus* flowers. The feeding behavior of the hummingbird brought the bird's forehead in contact with both the tufts of purplish anthers and the large green stigma. Zimmerman was convinced that hummingbirds are the primary pollinators of *E. coccineus* in those habitats supporting resident birds, but he also observed medium-size native bees acting as pollinators in this species. In the Trans-Pecos we have observed black-chinned hummingbirds repeatedly visiting flowers on plants of *E. coccineus* in cultivation and in the field. Medium-size bees and European honeybees also visit the flowers, and under cultivation they may fly from *E. coccineus* (red flowers) to *E. dasyacanthus* (yellow flowers), or vice versa. Also, we have observed black-chinned hummingbirds visiting small-flowered cultivated plants of *E.* × *roetteri* var. *neomexicanus*.

Peniocereus greggii is widely distributed throughout much of the Trans-Pecos, but plants usually are not in dense populations. The plants produce large (to 9.5 cm diameter), white, scented, nocturnal flowers with long floral tubes (ca. 15 cm) that supposedly attract hawkmoths as pollinators. In Texas we have observed numerous cultivated plants in flower at dusk and dawn but rarely at night, and we have yet to see a hawkmoth visit this species. We have, however, observed hawkmoths that were visiting columbine flowers (*Aquilegia chrysantha* A. Gray) in the vicinity of the cultivated *P. greggii*. Large "tomato hornworm" hawkmoths (*Manduca* sp.) visit the wild population of *Acleisanthes longiflora* on the Sul Ross campus, probing almost up to their eyes in the 10 cm long tubular flowers, but they are not often seen; they would visit rare *Peniocereus* flowers too, we assume, but encounters with *Acleisanthes* are much more predictable. Linda Coleman (pers. comm., 1999) also has observed cultivated flowers of *P. greggii* in Alpine. She was studying flower phenology and observed several flowers at dusk, on some occasions at night, and in the early morning. She reported seeing only a medium-size bee visit the open flowers of *P. greggii*. Like many others who think of breeding this species, we have been successful in producing fruits of *P. greggii* through artificial cross-pollination, but never through selfing; this is an obligate outcrosser. The pollination biology of *P. greggii* needs additional study with reference to the work by Suzán et al. (1994) on the Sonoran *P. striatus* (Brandegee) Buxb.

Uses and Other Ethnobotany

Archeological evidence documents the last 9,000 years of human association with cacti (Nobel, 1994). Prehistoric human coprolites containing residues of cactus stems, flowers, and fruits reflect extensive reliance on cacti as food by early humans in North America since at least 6,500 B.C. (Bryant, 1974; Sánchez-Mejorada, 1982), and presumably ever since the arrival of humans here, thousands or tens of thousands of years earlier. Archeological, anthropological, and ethnobotanical studies have revealed numerous other uses of cacti by pre-Hispanic native North Americans. Various uses of cacti have persisted into modern times, particularly in Mexico among the rural populations and Indians, and also by Native Americans and some agriculturalists in the southwestern United States. Prickly pears might have been cultivated in Mexico since about 5,000 B.C. (Meyer and McLaughlin, 1981). In the early 1900s great interest was generated through evaluation of the potential commercial value of prickly pears (*Opuntia* spp.) for livestock feed and many other uses. Involved were famous breeding experiments of Luther Burbank aimed at developing improved strains of prickly pears (Burbank, 1907), including spineless forms that could be used in arid regions of the world as supplemental forage for livestock (see Benson, 1982). Interest in these projects waned in 1917 for various reasons, including the widespread sexual reproduction of pernicious spiny prickly pears from cloned spineless forms that had been established on rangelands in many parts of the world. Interest in the economic development of prickly pears boomed again in the late 1900s largely because use of these plants was continued by rural people in Mexico and the southwestern United States. Also, various workers (especially Peter Felker) used new ingenuity, experimentation, and modern technological approaches in developing special agricultural practices for dealing with improved strains of prickly pears (Felker and Moss, 1995; Professional Association for Cactus Development, 1996). In connection with the new emphasis on prickly pears, the same type of scientific investigation and promotion has been applied to selling the products as was used in promoting previously accepted agricultural crops (Johnson, 1993).

International Uses of Cacti

Prehistorically and historically, humans in southwestern North America have consumed the "pads" or stems of prickly pears (*Opuntia* spp.) as a major source of vegetable nutrition. Historically, the young stems (*nopalitos*) were cleaned of their spines, sliced in strips or chunks, and blanched, marinated, or boiled in water (Mizrahi et al., 1997). Alternatively, the stems were fried with other vegetables, meat, or eggs and seasoned with chili peppers, onions, tomatoes, chocolate, pumpkin seeds, or other materials (Sánchez-Mejorada, 1982). The nopalitos also were used in stews with the meat of various animals. In modern times boiled or fried nopalitos are recommended for use in the traditional ways and as salad greens (Beesley, 1982; Edwards, 1994). A natural pickling process, using acids stored in CAM metabolism, has been developed for young stems. Fresh and pickled nopalitos are available in stores in Mexico and the southwestern United States (Pimienta-Barrios et al., 1993). The details for modern food preparation of nopalitos are discussed by Nobel (1994).

The ephemeral leaves of prickly pears and chollas (largest in some chollas; Benson, 1982) were boiled, fried, or pickled. The gum exuded from stems of some opuntias (e.g., *O. fulgida* Engelm.) was gathered by the Seri Indians of the Sonoran Desert in Mexico, then toasted, ground into a powder, and mixed with water to produce a beverage (Sánchez-Mejorada, 1982).

Pre-Hispanic people also prepared and ate the young stems of *O. bigelovii* Engelm., *Acanthocereus* (Engelm. ex A. Berger) Britton & Rose, *Echinocactus*, *Ferocactus*, and *Melocactus* Link & Otto. *Echinocactus* and *Ferocactus* stems were cut into pieces, cooked in water to produce a syrup, perhaps with sugar and flavoring added, and cooled to form a viznaga candy. Reportedly, harvest for the cactus candy industry (historically, and in Mexico) is one of the major threats to some species of *Ferocactus*, *Melocactus*, and possibly other genera.

Apparently few cacti produce edible roots. The tuberous roots of *Ancistrocactus scheeri*, a taxon of south Texas and adjacent Mexico, are reported to have a pleasant starchy taste when eaten raw (Cheatham and Johnston, 1995). The Seri Indians cooked and ate the tuberlike swellings of *Peniocereus striatus*, although they were thought to cause a skin rash (Sánchez-Mejorada, 1982).

Pre-Hispanic humans ate the flowers and buds of certain cacti, including those of *Opuntia*, *Ferocactus*, *Stenocereus*, *Myrtillocactus*, and *Pachycereus* (A. Berger) Britton & Rose (Sánchez-Mejorada, 1982). Preparation of flowers, sometimes just the tepals, involved cooking them as a vegetable component of stews, boiling them with syrup to make candied flowers, or adding honey to boiled flowers. Flowers also were used to color drinks and foods. Reportedly, still today the buds of *Opuntia versicolor* Engelm. ex J. M. Coult. and other chollas are cooked and eaten by the Papago, Hopi, and Pima (Meyer and McLaughlin, 1981), and the Seri eat fresh flowers (excluding the pericarpel) of *Pachycereus pringlei* and *Ferocactus* species.

Edible fruits are produced by a large number of different kinds of cacti (Mizrahi et al., 1997). Fruits were used extensively in pre-Hispanic times, and the variety of uses seemingly has been expanded through the twentieth century. The Seri are known to have harvested the fruits of about 20 species of cacti, and fruits may have been their most important food source (Nobel, 1994). Relatively few cacti produce dry fruits with no palatability. Most species form fleshy or juicy fruits that taste pleasant, and many have a high sugar content and sweet flavor. Historically, fruits were eaten fresh or sun-dried, pickled, cooked as a vegetable in stews, or cooked to derive syrups, jellies, or preserves (Griffiths and Hare, 1907; Sánchez-Mejorada, 1982; Benson, 1982). The Papago and Seri strained fruit juice and allowed it to ferment, forming a wine. Fruit juice or pulp, mostly from prickly pears, also was used to flavor the alcoholic beverage pulque, then resulting in *pulque curado de tuna* (or *nochote*), or the juice simply was mixed with water to produce a sweet drink. The Seri mashed fruit pulp into flat cakes. In southern Africa, certain people prepared a beer from the peeled fruit of prickly pear species.

Prickly pear species with large, sweet, juicy fruits have been most widely utilized in North America. These prickly pears include numerous species of the subgenus *Opuntia*, representing at least six taxonomic series of the subgenus (Sánchez-Mejorada, 1982). Apparently the fruits were a major food source for the pre-Hispanic peoples of many different regions, each tending to use the one or more species with the best fruits. It is evident, as inferred from today's diversity of cultivars, that the ancient people must have artificially selected for different strains of some of the species. Perhaps the best-known and most widely distributed cultivar today is O. *ficus-indica* (Pimienta-Barrios et al., 1993). Its fruits in Europe are known as Barberry figs or Indian figs. The fruits of O. *ficus-indica* and other prickly pears are called *tunas* in Mexico, a term of Spanish origin. In season the ripe tunas were eaten fresh, but large amounts of the fruits were preserved for later consumption by drying them in the sun or by various cooking processes to form a paste, syrups, or a marmalade-type substance. About 70–80% of the fruits by dry weight is sugar, mostly fructose, which is better tolerated by diabetics than glucose and sucrose (Nobel, 1994). In historic and modern Mexico the cheeselike *queso de tuna* (made mostly from O. *streptacantha* Lemaire) perhaps is best known, along with a thick syrup, *melcocha,* and a thin honeylike syrup, *meil de tuna* (Benson, 1982). In the United States, prickly pear jelly, preserves, a red margarita drink (one type is served on the San Antonio Riverwalk), cactus candy, and a salsa (from stems) appear to be gaining acceptance (Tate, 1972; Beesley, 1982; Edwards, 1994). Snapple Beverage Company is presently marketing an agave-cactus fruit drink that contains natural prickly pear cactus flavor, agave nectar, ginseng, astragalus, prickly pear cactus puree, and grape juice. In 1992 about 150,000 acres were planted internationally for the production of prickly pear fruits, most in Mexico, and over 10,000 tons of prickly pear fruits were imported into the United States from Mexico (Nobel, 1994). The increasing popularity of prickly pear products can be estimated from a 1999 report that 138,000 tons of fresh and preserved prickly pear materials, fruits, and stems, were imported into the United States through the border station at Nogales, Arizona, alone.

Few species of *Opuntia* subgenus *Cylindropuntia* produce edible fruits. Among them are *O. imbricata*, with fruits that were cooked by pre-Hispanic Indians (Chichimec tribes), and *O. leptocaulis,* with red fruits that were eaten fresh. Both of these species are widespread in the southwestern United States and northern Mexico.

Among the cacti that produce edible fruits, other than the opuntias, are certain species (Anderson, 2001) of *Echinocactus, Ferocactus, Pachycereus, Echinocereus, Stenocereus, Carnegiea* Britton & Rose, *Lophocereus, Mammillaria,* and *Coryphantha*. The small red fruits of *Mammillaria* and *Coryphantha*, known as *chilitos*, are sold in Mexican markets, as are the fruits of many other species.

Pre-Hispanic use of cactus seeds as a source of nutrition has continued into modern times among the Seri (Sánchez-Mejorada, 1982). Although seeds are small, they are produced in large numbers in many cactus species, and they can be stored for relatively long periods. Seeds were ground into a flour or meal and added to water to make a drink, cooked as a gruel, added to stews, or spread on tortillas. In the Sonoran Desert, Native Americans used the seeds of *Opuntia, Carnegiea, Pachycereus, Stenocereus, Ferocactus* (Felger and Moser, 1985), and *Echinocereus* for food. A cookbook devoted to the preparation of cactus dishes was compiled by Tate (1972).

Prickly pears and certain other cacti have been recognized since 1857 (Nobel, 1994) as a valuable potential source of feed for livestock and wildlife. Prickly pears are best known as an emergency forage and fodder source for cattle during droughts. Prickly pears were used as range forage throughout the southwestern United States and northern Mexico, but reportedly nowhere was this drylands range resource more utilized than in South Texas. In the early 1900s David Griffiths and coworkers (e.g., Griffiths, 1905; Griffiths and Hare, 1908) produced a series of reports dealing with the use of hardy succulent cacti on dry Texas ranges. During subsequent droughts ranchers learned to use propane burners to singe prickly pear plants and burn the spines (Benson, 1982). The activity was referred to as "burning pear." In South Texas during droughts, cattle eagerly consumed disarmed prickly pear. Cattle even became conditioned to the sound of propane burners and would rush from considerable distances to eat freshly singed plants. Once habituated to prickly pear, some cattle continued to eat spined plants after the drought had passed. Such "pear eaters" often suffered external and internal injuries from the spines, one of the several disadvantages of using prickly pears as emergency range forage (Hanselka and Paschal, 1991). Sheep are even more seriously affected by eating prickly pear, particularly if they consume abundant fruits, which they relish. Sheep may develop "pear mouth," swelling of the lips and tongue, resulting from glochids of the fruit areoles. Fruit-eating sheep may be harmed or even die after seeds from numerous fruits become compacted in a rumen compartment (Hanselka and Paschal, 1991). Texas ranchers may use the term "pear" in reference to the prickly pear fruit or to the whole plant, or they may differentiate the fruit as "pear apples."

The nutritional aspect of prickly pear and other cacti has been rather carefully evaluated (Gibson and Nobel, 1986; Everitt and Alaniz, 1981; Hanselka and Paschal, 1991; Nobel, 1994). In general, cacti are low in protein (ca. 5–9%) and phosphorous but high in vitamin A, fiber, and ash. The plants also are high in water and energy. For human use, the nutrition of prickly pear stems compares to that of other vegetables. A single cow may need as much as 100–200 pounds of prickly pear per day to obtain enough nutrition from this type of plant alone. On Texas ranges prickly pear is regarded today as an excellent natural resource that can be used as emergency forage in supplementing beef cattle (Hanselka and Paschal, 1991) and one that is about three times more efficient in converting water to organic matter than grasses and legumes (Pimienta-Barrios et al., 1993). Because of their ecological and physiological advantages in arid and semiarid habitats, prickly pears might well become part of worldwide strategies for maintaining productive ecosystems, especially in dry regions where atmospheric CO_2 levels are increasing. Worldwide over 1,482,000 acres already are devoted to growing prickly pears for cattle fodder (Nobel, 1994).

Benson (1982) alluded to the tendency of some range animals to eat chollas, and once having started, to become "addicted" to these bristly cacti. Benson referred to a bull eating *Opuntia bigelovii* Engelm. (teddy bear cholla) in southwestern Arizona, its face "covered with clinging joints," and to cattle and deer eating cholla in Sonora, Mexico (p. 223). Although the *O. imbricata* of the Chihuahuan Desert Region is much less spiny than the teddy bear cholla of the Sonoran Desert, we have been amazed to observe, on occasion, range cattle in Trans-Pecos Texas eating *O. imbricata*. In 1993 Stephanie Randell photographed a steer eating *O. imbricata* near the roadside. In October 1994 Jean Hardy reported watching an elk eating cholla 10 miles west of Marathon in Brewster County. At the same site in April 1995, Johnny Hamilton related seeing five elk cows browsing cholla.

Cacti are widely regarded as providing valuable food and cover for numerous wildlife animal species. Prickly pear stems are eaten by deer, javelina, rodents, and other animals. In South Texas, studies have demonstrated that prickly pear furnish up to 21% of the animal food supply of white-tailed deer (Hanselka and Paschal, 1991). In the Trans-Pecos, the 1990s drought brought desert mule deer and white-tailed deer into towns where they browsed on many types of plants, including cultivated *O. ellisiana*. Throughout the Trans-Pecos during this drought period, plants of many different prickly pear species were heavily browsed by wild animals. The fleshy fruits of prickly pears also are eaten by many animals, including game species. Scaled quail are particularly fond of the small red fruits of *O. leptocaulis*. During dry periods in the Trans-Pecos, small animals eat the stems and fruits of various cactus species, including those of *Echinomastus, Ferocactus, Echinocereus, Coryphantha, Mammillaria,* and *Ariocarpus*.

The many worldwide medicinal uses of *Opuntia*, both prickly pears and chollas, were reviewed by Meyer and McLaughlin (1981) and Nobel (1994). Medicinal applications have involved roots, stems, mucilage, leaves, spines, flowers, and fruits. We do not necessarily endorse any of the medical functions reported for

Opuntia. Examples include use of the roots of *O. bigelovii* by the Seri to prepare a diuretic tea through boiling the internal tissues. Stems were employed in South Africa to derive a poultice for treating open sores and boils and in making a drink, with sugar added, to treat whooping cough. In Mexico a water drink infused with pulverized, peeled prickly pear stems was given to women in difficult labor. There are many reports that extracts from prickly pear stems have the effect of lowering blood sugar and therefore are of value in treating diabetes. Saponins in some cactus stems have medical and/or dietary significance. Efficacious treatment of diabetes using prickly pear has been reported in South Africa, Australia, and Mexico. Data from Mexican studies of *O. streptacantha* regarding diabetes have been evaluated by a group in San Antonio, Texas (Felker and Moss, 1995). Some studies have shown that eating nopalitos before meals can control diabetes (Pimienta-Barrios et al., 1993). Extracts from prickly pears also have been reported to be effective against cancer (Meyer and McLaughlin, 1981). This report has not been supported by medical investigation.

The mucilage from prickly pears has been used in Colombia to produce a diuretic function. The Seri powdered dried sap of *O. fulgida* Engelm., mixed it with water, and used it to treat diarrhea. Curiously, in Hawaii the mucilage from *O. megacantha* Salm-Dyck purportedly had laxative effects. There is a report that ingested mucilage from prickly pear causes reduction in levels of LDL (the "bad cholesterol") in the blood (Pimienta-Barrios et al., 1993). Mucilage from various cactus species has been used topically to treat burns and sores and internally as a cure for stomach ulcers and kidney disease (Nobel, 1994). Local tour guides in Guatemala have suggested that headaches could be cured by slicing open prickly pear pads and placing the mucilage side on the forehead (J. F. Weedin, pers. comm.). Native Americans applied sliced pads to bruises and wounds.

Leaf extracts from prickly pears supposedly have been used to treat diabetes, although "leaves" in the following reports might have been confused with stems. In Australia and South Africa a strong decoction from *O. inermis* DC. was mixed with sodium bicarbonate and ingested, or minced leaves were covered with sodium bicarbonate overnight, and the resulting exudate was swallowed. In New Caledonia a poultice from leaves of *O. vulgaris* Miller was applied to treat burns. A quaint remedy for colds in the Bahamas involved the fixture of peeled prickly pear "leaves" to the inside of a foot.

The Blackfoot reportedly (Meyer and McLaughlin, 1981) removed warts and moles by slicing the protuberances in several directions, then rubbing in glochids and inserting spines of *O. polyacantha*. Setting fire to the glochids and spines subsequently initiated degeneration of the wart or mole. This incredible technique was tested by Jack Brady, biology graduate student at Sul Ross State University, who claimed that the process was successful.

Tea made from the flowers of *O. ficus-indica* has been used in Sicily to treat kidney problems, and a paste from the flowers was applied topically to combat measles. Italians drank a tea from *Opuntia* flowers for its diuretic effect.

Fruits were used more for food than for medicine, although the Seri reportedly treated persistent diarrhea in children with a fruit preparation from *O. fulgida*. The fleshy fruit rind was cooked, tartly flavored with a small amount of the pulp, and ingested.

The alkaloids produced in a number of cactus genera have been used for medicinal as well as religious purposes (Nobel, 1994). Perhaps most famous in this regard is peyote (*Lophophora*), which produces over 50 alkaloids, some of them with hallucinogenic effects in humans (see *Lophophora*). Some North American species of *Ariocarpus, Opuntia, Carnegiea, Pachycereus, Pelecyphora* Ehrenb., *Polaskia* Backeb., *Stenocereus, Echinocereus, Epithelantha*, and *Mammillaria* are known or suspected to produce alkaloids, as are South American species of *Opuntia, Trichocereus* (A. Berger) Riccob. (Nobel, 1994), *Echinopsis* (Anderson, 2001), *Gymnocalycium* Pfeiffer, and *Stetsonia* Britton & Rose. Some of the alkaloids of these taxa are hallucinogenic. The ethnobotanical use of many species for medicinal purposes suggests that at least some of the secondary compounds made by cacti have bona fide curative properties. *Ariocarpus fissuratus*, for example (see *Ariocarpus*) has been used to treat fever, tuberculosis, and superficial wounds. Most of the alkaloids produced by cacti have no known pharmacological properties.

Because of the potential economic value of prickly pears, an International Cactus Pear network has been established by researchers from Mexico, Italy, the United States, and Chile (Pimienta-Barrios et al., 1993). Although the human food, animal forage, medicinal, and ecological values perhaps are the principal reasons for propagating prickly pears, there are additional uses attributed to cacti of all kinds that are worthy of brief mention (e.g., Meyer and McLaughlin, 1981; Sánchez-Mejorada, 1982; Pimienta-Barrios et al., 1993; Nobel, 1994; Felger and Moser, 1985).

Use of a carmine red dye (cochineal) dates to pre-Columbian times. The dye is produced in the females of *Dactylopius*, a scale insect genus specializing on *Opuntia*. Individuals are only 2–3 mm long, but collectively they are conspicuous from the brilliant white masses of soft, waxy, cottonlike webs on the stems as they parasitize certain prickly pear species. Some chollas in Baja California also host *Dactylopius* (Vega-Villasante et al., 1994). Indians learned to cultivate the insect in prickly pear clones maintained for this purpose (Benson, 1982). The dye was used to paint skin and to color fabric, feathers, and other materials. Aztec rulers used the dye for their special robes. The Spanish noticed the colors and the insect source of the dye and transported the insects along with their host plants to Europe (Sánchez-Mejorada, 1982). From about 1520 to 1850 cochineal was behind only gold and silver as the most valuable export from New Spain. The most important cochineal-producing state was Oaxaca, but production extended north to Jalisco. In Turkey the red dye is used in oil paints (A. Kirk, pers. comm., 1999). The dye specifically stains chromosomes in cells and thus is used in cytogenetic research. The dye once was used to color the red jackets of Royal Canadian Mounted Police, and it was widely used in cosmetics, including lipstick. The demand for cochineal is likely to increase along with the worldwide trend toward natural products in foodstuffs. Today in Jalisco, Mexico, there are *Opuntia* plantations that use specialty sheds (*nopalotecas*) to protect cultured cochineal insects. In the sheds individual cladodes, infested with cochineal insects, are hung in rows away from rain and wind. The role of carminic acid in nature (Eisner and Nowicki, 1980) appears to involve deterrence of predators, such as ants that feed on *Dactylopius*. The carnivorous caterpillar of a pyralid moth is not deterred by the carminic acid in *Dactylopius*, however, and instead sequesters the substance for its own defensive functions; *Laetilia coccidivora*, in subfamily Phycitinae, is an enemy of various Coccidae in the United States (Mann, 1969), closely related to the specialized cactus-moths. Cochineal insects have been used for biological control of prickly pears in the United States (Nobel, 1994; Anderson, 2001).

Prickly pears have been exported from North America at least since the late nineteenth and early twentieth centuries. This activity was driven by the potential of economic benefits (fruits, nopalitos, ornamentals, cochineal dye) from plants that thrive in otherwise marginal agricultural areas. Prickly pear colonies were established in Australia, South Africa, India, Ceylon, Madagascar, Fiji, Hawaii, New Caledonia, Mauritius, and other countries, as well as in Mexico and Texas. Ultimately, prickly pears became pernicious pests in all or most of these areas of introduction, particularly as spineless forms gave rise to spined forms through sexual reproduction. Perhaps the worst infestation of prickly pears developed in Australia, where by 1925 approximately 60 million acres of rangeland (Ramaley, 1940) in Queensland and New South Wales were rendered essentially useless for both agriculture and the native biota. Many efforts to control the prickly pears were attempted, but by far the most successful one was the carefully researched introduction of a natural biological enemy, the *Cactoblastis cactorum* moth, a phycitine pyralid from South America. The larvae of *C. cactorum* develop inside prickly pear stems and eventually destroy the plant (Meyer and McLaughlin, 1981).

The use of barrel cacti (*Ferocactus, Echinocactus,* and allies) as an emergency water source in the desert perhaps has been overdramatized in western movies. These large succulent plants do store water that can be pounded or squeezed from the cortical tissue by using hard tools and considerable effort, but the resulting liquid is not dependably palatable (Benson, 1982) and may cause muscle pain and diarrhea (Nobel, 1994).

The mucilage from cactus stems has been used as a control for scale insects in water-handling systems, as a sizing agent for dyes in fabrics (Appalachian tribes and the Blackfoot), and in a mixture with stucco; and mucilage has been boiled together with tallow to make candles. The Blackfoot added chopped *Opuntia* stems to clarify turbid drinking water, a process that was accomplished by the stem pulp, the mucilage, or both. Stem mucilages of various cactus species are used as adhesives. Most notably the fine-quality mucilage in stems of *Ariocarpus kotschoubeyanus* (Lemaire) K. Schumann, a species of northeastern Mexico, is used by the native Indians as a glue for repairing pottery. Native Americans added boiled mucilage to mortar or whitewash to enhance adhesiveness. Mucilage also has been used to thicken food and pulque. A useful cholla "gum" exudes from the stems of *O. imbricata*.

The Seri fashioned containers from the hollowed despined stems of *Ferocac-*

tus species (Felger and Moser, 1985). The corklike "boot" of callus tissue that forms in stems of *Pachycereus* and *Carnegiea* in response to woodpecker burrowing also was used to make small containers.

Other potentially economic substances that can be extracted or produced from prickly pear stems include sugar, potash, oxalic acid, tannin, vinegar, alcohol, and triterpenes. Triterpenes have been used as soap and as an agent to stupefy fish in ponds and streams. A white, lignin-free cellulose can be extracted from prickly pear stems (Meyer and McLaughlin, 1981).

The woolly masses of hairs found at the stem apexes in certain cactus species (e.g., *Cephalocereus* spp., *Echinocactus* spp.) were used by Indians to stuff pillows. It has been reported but not documented that these trichomes also were mixed with other fibers in weaving cloth (Sánchez-Mejorada, 1982).

The heavy straight spines and hooked spines of various cacti, including those of *Ferocactus* spp. and *Echinocactus* spp., were used as fishhooks and as other simple tools, including toothpicks. The densely bristly pericarpels of *Pachycereus pecten-aboriginum* (Engelm.) Britton & Rose were used to make hairbrushes. Cactus spines were used as the "needles" on some early phonographs.

An extract from the stems of *O. vulgaris* Mill. has been used in the preparation of cosmetics (Pimienta-Barrios et al., 1993) and has been investigated for rubber production. In Algeria stems of *O. vulgaris* were used in mosquito control. Chopped stems were soaked in water, and the resulting extract was then spread over mosquito breeding sources, with the objective of inhibiting larval development. Other reported innovative uses of prickly pear stems include their employment as shade for garden vegetables, as emergency lunch plates, and as agents to reduce friction produced through moving heavy stones. The corklike tissue formed by some cacti in response to injury can be used in the manufacture of sound and heat insulation, as can the pulp from some other cacti, including prickly pears. A useful oil can be extracted from prickly pear seeds. Prickly pear fruits are used to color ice cream and drinks.

Some larger cacti produce enough wood to be used in light construction (Anderson, 2001); in the crafting of harpoons, spears and poles (columnar cacti), tool handles, and torches; and as firewood. The wood of large chollas may be used for some of the same purposes; in modern times it has been used in the production of novelties and in the construction of picture frames.

Many different cactus species have been used functionally in constructing living hedges and fences. These "fencing materials" include prickly pears, chollas, barrel cacti, columnar cacti, *Acanthocereus,* and *Pereskia* (Sánchez-Mejorada, 1982). This use is widespread in modern Mexico and probably dates back to pre-Hispanic times as well. Native Americans cultivated flowering plants, including cacti, as ornamentals. Currently, many kinds of cacti are used as components in xeriscapes.

Today prickly pear farms are managed in several countries, including Mexico and the United States (Texas), and in Sicily (Pimienta-Barrios et al., 1993). In Sicily prickly pear fruit farmers accidentally discovered that removing the first flowers of the season caused the cacti to flower again. Fruits developed from the second flowering were fewer in number, compared to those produced from a first flowering period, and invariably they were larger and sweeter than first fruits. This led to a harvesting practice called *scozzolatura* (to take the buds away) aimed at securing a more profitable fruit crop (Nobel, 1994).

According to a newspaper report in June 2000, a Texas company, Safe Solutions of Seymour, about 130 miles west of Fort Worth, has launched national and international sales of prickly pear products. Included among their specialties are Cactus Juice products that were sold originally in the Caribbean islands but now are also sold in supermarket chains and other retail stores across the United States, including Kmart and Sam's Club. Safe Solutions also was expecting to market an insect repellent and a skin-care moisturizing gel, both produced from prickly pear extracts. Safe Solutions obtains its prickly pear plants from ranchers who are contracted to harvest the native cactus.

Future cultivation of prickly pears and other cacti may be encouraged if economic uses in addition to human food and livestock forage become viable. The ecological sense of growing cacti becomes more cogent with the predictions of global warming and increasing aridity in many parts of the world.

One of the most bizarre uses of cacti is evident in Figure 1. The photograph is of Mary Thorbecke Ericson wearing a cactus swimsuit. The photo was published as a publicity stunt in *Life* magazine in 1941, according to A. J. Flick, staff writer for the *Tucson Citizen,* who researched the identity of the "mystery" model. (Another source has cited a different name for the model.) The article by Flick was published in 1997 in the *Tucson Citizen* with the 1941 "cactus model" photo opposite a 1997 photo of Mrs. Ericson. In 1998 the cactus model photo was used on a flyer designed to advertise the Prickly Pear Pachanga in Sanderson, Texas (Figure 2).

Sanderson was officially named the Cactus Capital of Texas after Governor George W. Bush signed a resolution communicating that designation on 14 May 1999. Earlier, the Terrell County Commissioners Court had proclaimed 10 October 1998, a Saturday, "as Prickly Pear Pachanga Day in Sanderson: Cactus Capital of Texas." The Terrell County proclamation was carried forward in the Texas legislature by Senator Frank Madla, who in 1999 introduced Senate Concurrent Resolution No. 1 (SCR–1), which, if passed, would make Sanderson and Terrell counties the official Cactus Capital of Texas. SCR–1 was passed in the Texas Senate on 2 March 1999. The Prickly Pear Pachanga (Spanish for "big party") is an annual celebration of cactus that entertains and informs visitors with cactus-themed food, folklore, costumes, arts and crafts, music, and more (Figure 2).

Uses of Trans-Pecos Cacti

Human coprolite evidence (Bryant, 1974) strongly suggests that early inhabitants of the Trans-Pecos region, at least those in southeastern portions, depended heavily on cactus stems, flowers, and fruits for food. Archeological materials from the southern Davis Mountains in the central Trans-Pecos also tend to substantiate the early use of cacti for food. These approximately 8,500-year-old collections, housed in the Museum of the Big Bend at Sul Ross State University, are

Fig. 1. A cactus swimsuit, modeled by Mary Thorbecke Ericson.

Fig. 2. The "cactus model" was used as the logo on this flyer advertising the Prickly Pear Pachanga in Sanderson, TX, the official Cactus Capital of Texas.

represented by about 70 stem fragments of what seem to be *Echinocereus viridiflorus* var. *cylindricus*. Most of the stems appear to have been hollowed out, although cortical tissue remains in some of them, suggesting that the stems might have been baked and eaten much as we do potatoes today (Jim and Teresa Weedin, pers. comm.). Prickly pear pads probably were used as bedding in Chihuahuan Desert archeological sites (Turpin, 1997).

It appears that the fleshy fruits of all Trans-Pecos cacti are edible (i.e., they are not known to be toxic). The fruits of some species, as determined from present-day observations, certainly are more palatable than others. One report evaluated the red fruits of *Glandulicactus uncinatus* as "bland to faintly sweet with an essence of watermelon or dewberry" (Cheatham and Johnston, 1995, p. 341). Fruits of the strawberry cactus (*Echinocereus stramineus*) have been described by locals as "absolutely delicious." The juicy fruits of many *Opuntia* spp. are sweet (e.g., *O. dulcis*, *O. engelmannii* var. *engelmannii*). The ripe fruits of other species (e.g., *Coryphantha scheeri*) have the smell, but not the taste, of delicious, or at least exotic, tropical fruits. The fruits of many Trans-Pecos cacti, both Opuntioideae and Cactoideae, are not individually "tasty" but might have been eaten by indigenous peoples after grinding, cooking, and flavoring. The fruits of Trans-Pecos cacti are eaten by a wide array of wildlife species, but very few animals are likely to covet the tiny, hidden, nearly dry bags of seeds produced by *Neolloydia* and similar genera.

Present-day uses of Trans-Pecos cacti in many cases parallel those discussed previously in the international context. One special value of Trans-Pecos cacti is that most species, at least potentially, are desirable ornamentals. For many years native cactus populations in the Big Bend region have been exploited and sold to collectors all over the world. But the new practicality of xeriscaping has given a boost to the cultivation of Trans-Pecos species, partly because most species of the region are attractive or otherwise functional as ornamentals, but mainly because plants of the northern Chihuahuan Desert Region, in general, are cold hardy. This means that most Trans-Pecos cacti tolerate subfreezing temperatures at more northerly latitudes and at higher elevations.

Optimal use of Trans-Pecos cacti depends upon knowing about their habitats and life histories as well as flower and fruit characters. For example, diminutive cacti that never grow tall (e.g., *Echinocereus davisii, Coryphantha minima, C. hesteri*) are best presented in elevated beds of specialty gardens with a soil surface so stable that many years of erosion and entropy will not see the tiny plants uprooted. It is desirable to keep cacti a safe distance from pedestrian traffic because of their dangerous spines, and thus it is useful to know the ultimate size that may be attained by chollas and prickly pears in cultivation. Popular ornamental species, such as *Opuntia imbricata, O. engelmannii* var. *engelmannii, O. engelmannii* var. *lindheimeri,* and *O. engelmannii* var. *linguiformis,* have a large spreading habit; whereas other species, such as *O. macrocentra* and *O. pottsii,* remain relatively small in cultivation. Species in all three subgenera of *Opuntia* have ornamental potential. The low mats or mounds of the dog chollas (subgenus *Grusonia,* e.g., *O. aggeria*) produce numerous yellow flowers and provide attractive ground cover in xeriscapes, but one of the species breaks into clonal dispersal units instead of staying put.

The Trans-Pecos chollas (subgenus *Cylindropuntia*), particularly *O. imbricata,* already are widely used in xeriscapes, except for *O. davisii*. All of the species are easily propagated from detached joints. Once established, *O. imbricata* requires little maintenance as it develops into a sturdy shrub with abundant magenta flowers in the late spring, and colorful yellow fruits persist for much of the year. *Opuntia imbricata* var. *argentea* is not as widely used in landscaping as is var. *arborescens,* but it has several advantages over var. *arborescens* in that the plants are more compact and the stems are covered with silvery spines. Most species of Trans-Pecos prickly pears (subgenus *Opuntia*) have some desirable ornamental features: for example, size; stem color (purple, blue-gray, or bright yellow-green in *O. azurea* and purplish to green in *O. macrocentra*); spine density, color, and length; flower color (red-centered yellow flowers in several species, and red tepals in *O. pottsii*); and fruit size and color. The tall, spineless *O. rufida* (a visually striking "accent plant," albeit with pernicious glochids) and the low, densely spined *O. polyacantha* are long-lived curiosities in xeriscapes.

Some specific examples of other useful features of Trans-Pecos opuntias include the woody skeletons of *O. imbricata,* which have been employed in decorating and in making various art items, such as the cross arms for agate wind chimes and standards for table lamps.

Some parts of O. leptocaulis are used in Mexican folk medicine. Earache is treated by squeezing juice from the fruit into the affected ear. In children, blockage of the small intestines or fever in the intestines, a condition called *empacho*, is treated by feeding them crushed roots (Sally Roberts, unpub.). The large prickly pear O. engelmannii has been used as emergency forage in the Trans-Pecos after the spines were burned away.

Practically all species of subfamily Cactoideae can be used in some fashion as ornamentals. The widely distributed *Mammillaria prolifera,* which barely occurs in the Trans-Pecos near Langtry, is one of the most cultivated mammillarias (Pilbeam, 1981).

Native Americans and Mexicans are known to have eaten the "tubers" (taproots) of *Peniocereus greggii* (Tate, 1972) after slicing and frying pieces in fat. The flavor is said to be turniplike. Slices of these roots also have been used to treat congestion by tying them to the chest of the patient. The roots of P. greggii are reported to contain triterpenes. Roots in excess of 40 pounds have been weighed (Benson, 1982).

The fruits of all claret-cup cacti probably were used as food by early Native Americans in the Trans-Pecos region. The juicy red fruits of *Echinocereus fendleri* are edible, as are those of the strawberry cacti. The sweet fruits of *E. stramineus* particularly, widely reported to have the smell and flavor of strawberries, must have been a favorite food source for early Native Americans. Modern humans in the Trans-Pecos region also have used the fruits for food, eaten fresh or in the form of jellies, and even for production of fermented beverages. In years when there are relatively few fruits in a given locale, humans who wish to eat the fruits (or photograph them or preserve a botanical specimen) often find themselves competing ineffectively with a legion of both nocturnal and diurnal ants, birds, and other animal life, which collectively seem to have the ecological advantage.

An unusual ornamental feature is displayed by young plants of *Echinocereus viridiflorus* var. *neocapillus* and *E. viridiflorus* var. *canus*. Here the seedlings and juveniles are covered with long, hairlike white spines. The adult plants of var. *canus,* with their shaggy white stems to 15 cm long and unusual green flowers, are exceptionally attractive.

The plants of *Echinocactus texensis* (horse-crippler) reportedly are despised by ranchers because the massive central spines are harmful to livestock. Britton and Rose (1919–23) reported that in northern Texas, fences were made from uprooted plants of *E. texensis*. When covered by a thin layer of snow or a dense growth of herbaceous plants, a desert population of *E. texensis* can stop humans, beasts, and small rubber-tired vehicles in their tracks. East of the Pecos, outside the desert, *E. texensis* is relatively spineless and merely suffers a bad reputation.

Ferocactus wislizeni is one of the large barrel cacti that has been used under emergency conditions as a source of water (Benson, 1982). The pulp of *F. wislizeni* was found to have a water content and flavor similar to that of watermelon rind. Large barrel cacti similar to *F. wislizeni* have been used as stock feed during drought periods in Baja California, and our species might have been used in this manner in the foothills of the Franklin Mountains where the species occurs in the Trans-Pecos. A popular type of cactus candy is made from the stems of larger barrel cacti, including *F. wislizeni,* down to the size of *Melocactus* and including *Echinocactus horizonthalonius* (Weniger, 1984). Pieces of the cortex and pith are boiled to remove the mucilage, the water being replaced several times during the process, and then candy is made with the stem pulp by adding desirable ingredients, such as sugar, flavoring, and coloring. A printed report (J. Casey, *Sheep and Goat Raiser,* March 1947, p. 26) showed a photo of "devil's head cactus," clearly *E. texensis,* with the explanation that "apples" (fruits) are eaten by "man, beast, and birds." The article further stated that a "palatable candy is made from the plant, and Mexicans make a brownish-colored cheese from the plant which they greatly esteem."

Uses and other ethnobotany associated with *Lophophora williamsii* (peyote) and *Ariocarpus fissuratus* are relatively extensive and are included in the text under these species. Biologically active alkaloids are produced by both species, with peyote by far being the best known in the Trans-Pecos region.

Yet another species, reportedly *Mammillaria heyderi* but probably a related species such as *M. gummifera,* was known to be used by Native Americans for its chemical effects, especially to enhance the performance of long-distance runners. Claret-cup stems may contain alkaloids that were consumed by Native Americans for the purpose of enhancing endurance and visual perception (Nobel, 1994).

In the past *Coryphantha minima,* a species with stems usually only 1–2 cm long, has been regarded by many cactus buffs as a collector's item. In the 1970s an enterprising person in Trans-Pecos Texas enclosed the plants in solid-plastic hemispheres and marketed them as paperweights. The fruits of *C. pottsiana, C. scheeri* (now = *C. robustispina*), and *C. echinus* are edible, although this aspect is not widely known for the species.

Horticulture

The Cactaceae collectively have been propagated and pampered all over the world. Cactus plants are fancied in part because they are succulents, and these kinds of plants basically are easy to transplant, grow, and maintain without any particular tender loving care. The cacti also are favored because their species exhibit many different sizes and shapes; most species have spines arranged in discrete clusters, these exhibiting characteristic color, number, and position, and collectively reminding us to keep a respectable distance; there is great variation in flower number, size, and color, and the flowers of most species are relatively showy; the fruits often are colorful and persistent. Once established in the yard or in containers, cacti are very forgiving of neglect. These low-maintenance plants can be abandoned without watering or other care during long absences from home, because they are preadapted to long dry spells. As Benson (1982, p.

216) put it, "Having a few plants brings about a creeping addiction to cultivating cacti."

Information about the propagation and cultivation of cacti has been widely disseminated for a long time (e.g., Schulz, 1932). Such information has increased over the years as hobbyists and even a few scientists have made contributions to knowledge about growing cacti (e.g., Potter et al., 1984; Hartmann et al., 1990; Deno, 1994; McIntosh, 1994; Tufenkian, 1999; Laras, 1999). Most comprehensive works emphasizing cacti include a section that deals with culture of the plants (Rowley, 1978; Earle, 1980; Cullmann et al., 1984; Graham, 1987; Haustein, 1988; Innes and Glass, 1991; Hewitt, 1993). The work by Johns (1993) is devoted entirely to growing cacti from seeds, cuttings, or grafts.

It is relatively easy to culture cacti. Most species from North America grow best in warm temperatures under full sun, but there are limits. Optimal growth conditions also include a growth substrate that allows drainage away from the plant base. One can expect most cacti to thrive outside in garden soil, or both outdoors and indoors in containers when rooted in artificial medium.

Most cacti are easily propagated from seed or asexually from stems. In many species of *Opuntia*, partially buried fruits also may strike adventitious roots and then sprout branches (normal, indeterminate, sterile long-shoots from the areoles). Most members of the subfamily Cactoideae produce seeds that germinate readily in high percentage after a few days or weeks of warm and consistently wet conditions. The critical role of gibberellins in the germination of cactus seeds has been emphasized by Deno (1994), although in general it is not necessary to pretreat seeds with gibberellins in order to obtain germination.

Stem cuttings in general are propagated under conditions similar to those that promote seed germination. Stems are placed in contact with or partially buried in moist substrate. Adventitious roots then may be produced within 1–3 weeks, except in cold weather.

The use of grafting procedures for culturing cacti is also popular among hobbyists and commercial growers. Grafting techniques are included in some of the references cited above. Cactus growers have learned that sterile conditions help to prevent the spreading of viruses when using asexual techniques. Sterilization of hands, tools, and bench surfaces may be accomplished effectively with 10% Clorox. Propagators should be mindful that viruses also can be transmitted by feeding insects (aphids, whiteflies, thrips, leafhoppers, and leaf beetles), soil fungi, nematodes, and contaminated water (Mayhew and Wiens, 1994) and directly from systemically infected grafting stock.

Trans-Pecos Horticulture

Essentially all of the Trans-Pecos cactus taxa have been propagated by seed, cuttings, or both. In the Trans-Pecos this has been accomplished through the native plant propagation program at Sul Ross State University and by Dennie Miller and the Chihuahuan Desert Research Institute (CDRI), where a collection of cacti encompassing the entire Chihuahuan Desert Region is maintained. Other growers in Texas, New Mexico, Arizona, and elsewhere have propagated many of the cacti of the northern CDR. At Sul Ross the Trans-Pecos cacti have functioned as ornamentals, most prominently in a cactus garden on campus, and in various types of scientific investigation. The cold hardiness of many Trans-Pecos cacti has been tested in Denver and Grand Junction, Colorado (J. F. Weedin, pers. comm.).

Specifically, seeds are removed from ripe fruits and washed to clean away any fruit juice or pulp that may contain germination inhibitors (Deno, 1994). The seeds are allowed to dry and then stored in paper envelopes.

Seed germination is facilitated by scattering the seeds on the level surface of a sterile peat medium in small plastic trays. The trays are filled about one-half full with the medium, which is kept moist by covering the trays with glass until the seedlings are well developed. The propagation efforts are carried out in a temperature-controlled greenhouse, with a specially constructed plastic tent to maintain warm and humid conditions for seed germination and seedling growth. After the tightly clustered seedlings reach more or less 1 cm tall, the glass is removed from the tray. The seedling trays are retained in the tent until the young cactus plants achieve vigorous growth and increase in size. Vigorous seedlings are transplanted singly or in clusters to larger trays or to three- to six-inch pots, where they are maintained in the tent for a short time and then moved to benches in the shaded greenhouse. After the juvenile cactus plants have grown in size, they are transplanted again, usually as single plants in three- to six-inch pots or in one-gallon containers. They are kept in the greenhouse until they are established, and then most plants are moved to another greenhouse that is not artificially cooled in the summer or heated in the winter.

Seed germination percentages in Trans-Pecos *Opuntia* species are low, typically 10–50% or less, compared to the usually high germination percentages in species of Cactoideae. A few seed pretreatment protocols, including aging, have been demonstrated to increase the rate of germination in certain species of *Opuntia* (Potter et al., 1984; Wang et al., 1996; Mandujano et al., 1997; Rojas-Aréchiga and Vázquez-Yanes, 2000). In preliminary nonscientific trials, using several species of Trans-Pecos opuntias, we have observed 0–34% results from untreated fresh and briefly stored seeds. We have also seen increased germination percentages after acid scarification and months of after-ripening. Inclusive seed germination experiments involving numerous Trans-Pecos species of *Opuntia* were initiated at Sul Ross in 1999 (Pendley, 2001).

Trans-Pecos cacti appear to grow best under greenhouse conditions when the temperature is allowed to fluctuate from hot days to cool nights, conditions that may be optimal for the CAM metabolism utilized by cacti (Gibson and Nobel, 1986). In the propagation program at Sul Ross, some cacti are maintained in the temperature-controlled greenhouses and some outside the greenhouses, but by far the most rapid growth is exhibited by cacti maintained in a greenhouse with no temperature controls other than partial shade and ventilation in the summer. Similarly, Chihuahuan Desert cacti maintained at the relatively low-tech CDRI greenhouse near Fort Davis, with profound nightly temperature extremes all year long, grow with exceptional vigor.

At Sul Ross a potting medium has been designed for growing cacti in containers. Many different formulations have been tested, but the best results have been achieved with a medium that consists of roughly one-half sharp sand and one-half commercial medium containing relatively large organic particles in peat. Commercial heat-expanded perlite, well known as a soil lightener for greenhouses, sometimes is added to achieve slightly increased drainage. A rock "mulch" or "topdressing" is added to the top of the container after cactus plants have been firmly planted upright in the growth medium; this layer keeps the lightweight particles of peat and perlite partially contained and moderates variations in the soil-surface conditions.

Cacti (Cactoideae) grown from seed in the manner described above usually reach the flowering stage in 1–5 years, depending on the taxon (saguaros, for example, cannot be forced to mature in a single-digit number of years). Both *Echinocereus davisii* and *Ancistrocactus tobuschii* have produced flowers toward the end of the first year of growth from seed. Most species of *Echinocereus* produce flowers after 4–5 years. In the wild, maturation usually requires two to several times longer than in optimal horticulture.

We have observed that flower initiation in most Trans-Pecos cacti requires daily thermoperiodicity and/or a "physiological winter" temperature threshold. In general, adult cacti maintained in a temperature-controlled greenhouse through the winter do not produce flowers. If these cacti are removed from the greenhouse and placed outside during late winter and early spring (for about three months), usually they produce flowers even if they are returned to the warm greenhouse before flower buds are evident. After bud formation outside, flower development continues even if plants are returned to a controlled greenhouse. Cactus plants of all Trans-Pecos species flower normally in spring in the Sul Ross greenhouse if left unheated during the winter, through temperatures likely to freeze a *Pereskia* or an *Epiphyllum*.

At Sul Ross, stem cuttings are rooted successfully in the same type of potting medium used for transplanted seedlings. In our experience, large and small cuttings from prickly pears and chollas take root equally well, but relatively large branches consisting of at least three cladodes or joints are more likely to produce flowers during the next season. The wounds of stem cuttings should be allowed to "heal" for 1–3 weeks before being exposed to moist potting medium.

Although the general techniques for cactus propagation are well established and successful, aspiring propagators should keep in mind that variations result from different physical settings. Propagation techniques followed at Sul Ross have been refined over the years, with certain aspects having been adapted to the specific greenhouses and other facilities available for growing plants in our particular climate. We have found that understanding the principles of cactus anatomy and physiology (Gibson and Nobel, 1986; Nobel, 1988, 1994) best provides the basis for successful innovations in techniques used to grow these plants.

Conservation

Many have discussed cactus conservation, some such as Carey (1980) choosing to organize the chief threats in two general categories, overcollecting and habitat loss. There are ancient, intrinsic biological reasons why many species of cacti are vulnerable to extinction (Benson, 1982). These include climate change and high frequency of "narrow endemism" in geographically small areas and/or in rare or "narrow" ecological niches (e.g., the novaculite endemics such as *Echinocereus davisii*). Other threats to whole species of cacti are modern or anthropogenic (Anderson, 1999a). Benson (1982) drew attention to commercial exploitation, overzealous collecting, housing developments, and habitat alteration by plowing, grazing, and fire.

Overcollecting

For most of the past century there have been tens of thousands of cactus collectors. This point is supported through estimates, and some documentation, that millions of cactus plants have been harvested from the wild and shipped to collectors in other countries. This point also is supported by the numerous societies centered around cacti and (often secondarily) other succulent plants. The popularity of such organizations is emphasized by Nobel's (1994) calculations of the approximate membership in several countries: 12,000 in the Czechoslovakian republics; 10,000 in Germany; 10,000 in Japan; 7,000 in Great Britain; and 7,000 in the United States. Some whole species of cacti consist of fewer individuals than the membership indicated here. Commercial harvesting is not limited to specialized exploitation of rare species (Westlund, 1991). At one time the nations dealing the most cacti were thought to be the United States, Japan, and West Germany (Kurlansky, 1980), and perhaps this is the case today, except that a broader Europe and Asia have become involved. Much of the world demand for cacti has been supplied through commercial harvesting of natural populations.

Overharvesting

In North America wholesale cactus harvesting has taken place mostly in the deserts of the southwestern United States and in northern Mexico, where cacti are most abundant and diverse. In Trans-Pecos Texas, especially the Big Bend area, uncountable tens of thousands of cactus plants have been harvested for wholesale trade (Plate 31). Many of the cacti gathered for sale in the Big Bend or for shipment out of the Big Bend were harvested in adjacent Chihuahua and Coahuila, Mexico. Individuals and teams of laborers have been digging large numbers of cacti in the desert and mountain habitats since before 1930, and probably for much longer.

In the past a common way of doing business among Trans-Pecos cactus dealers was that absentee buyers made arrangements with locals to procure the cacti. Some of the locals would gather plants themselves, on land where permission was obtained or not apparently needed, and other locals found it profitable as intermediaries to hire "diggers." Common remuneration for diggers was 5–10 cents each for cacti of prescribed size and type. Cacti were harvested and dumped in piles at strategic locations in anticipation of sales (Harrington, 1980) and in the process of filling orders. Some diggers took their burros into the hills and returned with load after load of cacti, day after day, until a desired number of uprooted plants was assembled. To fill certain commercial orders, a sufficient quantity of plants would be gathered, sorted into boxes according to directions of the buyer, and shipped away after prearranged transport arrived at the scene. To fill other commercial orders, workers would scoop piles of different kinds of cacti into waiting trucks to be hauled away (Harrington, 1980; Demarest, 1981).

Another kind of enterprise involving the removal of cacti and other saleable desert plants from their natural habitats is exemplified by a family business. The Desert Plant Company was established by its family patriarch in about 1930, in Marfa, Texas, and featured a catalog, "Cacti of the Southwest U.S.A." Advertised in the catalog for mail-order purchase were 29 taxa of cacti, entities found in all parts of the Trans-Pecos and beyond, as well as century plants, cholla wood, and pottery imported from Mexico. The business operated for over 50 years, with several family members as well as cowboys and additional hired hands being involved in search and collection activities. The founder was also a commercial pilot who had repeatedly observed regional topography with cactus distributions in mind. His airborne activity coupled with years of off-road experience in the field facilitated what was perhaps unexcelled knowledge of cactus populations in the Trans-Pecos. His written statements reflected awareness and concern about the problems of overcollecting cacti. In 1981 the patriarch stated that he personally had been gathering *Mammillaria lasiacantha* "off the same mountains for 50 years" and that after 50 years there were more plants than before. One of his convictions was that cactus populations were replenished through field collection of their numbers, because soil disturbed in the process of digging cacti provided favorable circumstances for natural propagation by seed.

Our field observations in the Trans-Pecos have revealed numerous sites, particularly in the southern Big Bend area and south of the Sierra Vieja rim, that are peculiarly devoid of cacti other than *Opuntia*. These locations seemingly are ideal for certain cacti, but at the times of observation, we have encountered only a few scattered adult plants and relatively few juveniles. The low density of these cactus populations, relative to those in nearby equivalent habitats, is surprising; usually there is a natural explanation, but in some cases a history of commercial harvesting is likely to be responsible.

Despite a great deal of attention to the matter in the United States, perhaps led by the Conservation Committee of the Cactus and Succulent Society of America, there is no unanimity about how to conserve cacti. Legislative attempts include CITES, the Lacey Act, the Endangered Species Act, state laws, and the regulations in various parks or preserves. The Convention of International Trade in Endangered Species of Wild Fauna and Flora (CITES), an international treaty, was negotiated in 1973 under United States leadership, and it has been ratified by at least 59 countries. In essence CITES applies only to international trade in species listed for protection under the treaty (Lyons, 1979; Hunt, 1992), which is a considerable number, but it does not apply to the voluminous local or interstate cactus trade within the United States or any other country.

The Lacey Act is a statute that can impose federal, civil, and criminal penalties for violations of state "wildlife" conservation laws, including plant laws. It involves, among other entities, the U.S. Fish and Wildlife Service and state governments in legal protection for listed cactus species. The Endangered Species Act (ESA) is at present the most visible federal enforcement vehicle relating to the protection of federally listed ("Threatened" or "Endangered") species of cacti. Many states, including Texas, also have laws protecting state-listed species. Arizona, for example, restricts the removal of all cacti and other specialized wild plants from natural populations.

National, state, and local parks theoretically protect natural populations of cacti along with other natural resources. Organizations such as the Nature Conservancy also control significant preserves. In Texas most populations of cacti are off-limits to outsiders behind the fences of private property, although there are some relatively large ranches and many smaller landholdings where the owners themselves have allowed wholesale exploitation of cacti.

All of the conventions, treaties, laws, and land preserves have obvious shortcomings with respect to cactus management. Lyons (1999) contended that CITES is dysfunctional in its present form. The Lacey Act has supported wildlife laws very well with respect to birds, for example, but its relatively recent extension to plant life has been awkward; it has little effect on worldwide prices for rare cacti. The ESA in the United States has drawn national attention with spectacular successes in restoration of previously endangered animal species, such as the bald eagle and peregrine falcon, but at the same time it has been criticized as contributing to the erosion of private property rights. In Texas the popular press has ridiculed the ESA and/or those responsible for enforcing it (McDonald, 1993), for the same general reason pertaining to private property rights. The ESA has been referred to as unworkable and even counterproductive in regions dominated by the perspective of the private property owner (Raun, 1998), such as is the case in the Trans-Pecos. Benson (1982, p. 245) cautioned that any laws bearing on the problem of cactus conservation "will not be enforced unless they have the support of the public."

National, state, and local parks, together with the National Wildlife Refuge system, are dedicated to the conservation and/or preservation of natural resources within their boundaries, but because they are open to public visitation, they cannot offer complete protection against unscrupulous individuals who want rare cacti. Some parks have freely admitted frustration that under their watch, populations of some cactus species have dwindled through "poaching," a term normally used with reference to game animals.

Any long-term success in cactus conservation must be international in scope, in addition to having the support of the local people where the cacti grow wild. Lyons (1999) suggested an international cactus and succulent society, presumably different from the existing IOS (International Organization for Succulent Plant Study; Smith et al., 1999). At present, the closest thing to a consensus seems to be that cacti should be propagated and made available in such large numbers and at such reasonable prices that the incentive for commercial harvesting will diminish (Carey, 1980; Benson, 1982; Hernández and Bárcenas, 1995; Bach, 1998; Breckenridge, 1999). Although propagation is the obvious method for relieving pressure on natural populations, it does not deal with overzealous collectors who focus on new and rare species (Lyons, 1999). The "collectors' syndrome" is one where propagated plants are seen as artificial, and there is a desire to find the "real thing" in its native habitat. Another factor, the "Easter egg syndrome" (S. Powell, pers. comm.), leads many people to collect small cacti or other kinds of cryptic objects. Others concerned with cactus conservation believe that the main focus should be on habitat protection (Benson, 1977; Lyons, 1979). In the CDR, Hernández and Bárcenas (1995) recommended that efficient habitat protection might be achieved through the organization of strategic quadrants that contain the most significant numbers and kinds of rare species.

Part of the eventual success in general cactus conservation also depends upon clear vision of precise objectives in a particular case. In some instances "conservation" is the objective, but in other circumstances "preservation" is desirable. These terms often seem to be used synonymously, and yet they have different meanings to most environmental scientists. Conservation refers to wise use of natural resources (Chamberland, 1996). Preservation means the protection of resources to the point that they are not used at all.

Cacti are renewable natural resources, but most species renew themselves only slowly or sporadically, unless they are artificially propagated. Weedy species of cacti, such as many prickly pears, chollas, and dog chollas, are in the minority.

Classification of Cacti

RECENT estimates place the number of legitimate species in the family at 1,500–1,600, in 100–22 genera (Nobel, 1988; Hershkovitz and Zimmer, 1997). An excessive number of invalid binomials have been published in the Cactaceae, approximately 11,000 in a total of about 350 genera (Gibson and Nobel, 1986).

Gibson and Nobel (1986) and Anderson (2001) reviewed the historical classification of Cactaceae. Twentieth-century authors consistently recognized three subfamilies in Cactaceae since the first comprehensive classification for the family was proposed (Schumann, 1898–1902). These subfamilies (used at the rank of "tribe" by some authors, but circumscribed almost identically) are Pereskioideae (Pereskioideae; Schumann, 1898–1902), Opuntioideae, and Cactoideae (Cereoideae; Schumann, 1898–1902). According to Gibson and Nobel (1986), the Pereskioideae include two genera, *Pereskia* (16 spp.; Leuenberger, 1986) and *Maihuenia* Phil. (2–3 spp.). *Pereskia* is a genus of leafy trees, shrubs, and vines indigenous from southern Mexico to Argentina and in the Greater Antilles. The species of *Maihuenia* are highly specialized terete-leaved, spiny, succulent "cushion plants" restricted to Patagonia. *Maihuenia*, previously placed in either Opuntioideae or Pereskioideae, has been segregated in its own subfamily, Maihuenioideae (Anderson, 2001).

At the time Gibson and Nobel (1986) examined the intrafamilial classification of Cactaceae, the subfamily Opuntioideae consisted of seven genera, namely *Opuntia* (ca. 165 spp.), *Pereskiopsis* (9–11 spp.), *Austrocylindropuntia* Backeb. (ca. 15 spp.), *Pterocactus* K. Schum. (9 spp.), *Quiabentia* (3–4 spp.), *Tacinga* Britton & Rose (102 spp.), and *Tephrocactus* Lemaire (ca. 50 spp.). Species of both *Pereskiopsis* and *Quiabentia* produce broad, flat, succulent leaves and grow in tropical deciduous forests. *Austrocylindropuntia* is Andean with terete, succulent, cylindroid leaves reaching 7–10 cm long in domesticated *A. exaltata* (A. Berger) Backeb. Species of *Opuntia* produce relatively short ephemeral, cylindroid, or conical leaves. Of the opuntioid genera, *Opuntia* is by far the most widespread in natural distribution, with species occurring from southern Canada to southern South America and on many American islands, including the West Indies and the Galápagos archipelago. When the generic name *Opuntia* is restricted in application to the smooth-seeded prickly pears (*Opuntia* subgenus *Opuntia*, sensu stricto), the distribution loses Patagonia but still qualifies as the most widespread genus in its subfamily. *Pereskiopsis* is distributed in Mexico and Guatemala. Many of the other genera in this subfamily are found in different countries and regions of South America. A more recent taxonomic consensus (Anderson, 2001) allowed 15 (or more) genera in Opuntioideae, including *Brasiliopuntia* (K. Schum.) A. Berger, *Consolea*, *Cumulopuntia* F. Ritter, *Cylindropuntia*, *Grusonia* F. Rchb. ex Britton & Rose, *Maihueniopsis* Speg., *Miqueliopuntia* Fri[c] ex F. Ritter, *Tunilla* D. R. Hunt & Lliff, and *Nopalea* (not in Anderson, 2001), often taxonomically segregated from the other North American smooth-seeded prickly pears because its flowers are specialized and the type species is ornithophilous. Members of the Opuntioideae are characterized by glochids in the areoles and a bony tan or whitish funicular envelope ("aril") encasing the seed coat.

About 80% of all the cactus species are classified in subfamily Cactoideae. Gibson and Nobel (1986) divided this subfamily among nine tribes: Browningeae (2 genera, 6–7 spp.); Cacteae (25 genera, ca. 385 spp.); Cereeae (9 genera, ca. 112 spp.); Echinocereeae (1 genus, ca. 50 spp.); Hylocereeae (22 genera, ca. 140 spp.); Leptocereeae (8 genera, ca. 20 spp.); Notocacteae (20 genera, ca. 340 spp.); Pachycereeae (14 genera, ca. 70 spp.); Trichocereeae (13 genera, ca. 175 spp.) Species of only three of these tribes are native in Texas. Most of the Texas cacti belong to the tribe Cacteae, with 14 genera represented in Texas: 12 genera represented in the Trans-Pecos, plus *Hamatocactus* and *Astrophytum*. Species of the single genus *Echinocereus*, tribe Echinocereeae, are widely distributed in Texas. Two genera frequently placed in the tribe Hylocereeae occur in Texas, *Acanthocereus* and *Peniocereus*, with only the latter occurring in the Trans-Pecos. Five of the other tribes of Cactoideae are predominantly South American in distribution. One of these, the tribe Cereeae, was thought to be represented in Mexico and the West Indies only by *Melocactus*, but today we know that *Pilosocereus* Byles & G. D. Rowley (formerly misclassified as part of *Cephalocereus* Pfeiff.) is another close ally of *Cereus*, sensu stricto. Four genera of tribe Leptocereeae are distributed in Mexico, the Galápagos Islands, West Indies, or Hispaniola. Tribe Pachycereeae is mostly Mexican and Central American in distribution, with a few species that extend into the United States in the Sonoran Desert (*Carnegiea gigantea*, *Stenocereus thurberi*, *Pachycereus pringlei*) and a few taxa that reach into northern South America and the West Indies. Tribe Notocacteae is exclusively South American unless circumscribed to include the epiphytic *Rhipsalis* (40–50 spp., the weediest of which, *R. baccifera*, is widespread from eastern Mexico to Florida and southern South America, and even in the Old World Tropics). The more recent taxonomic delineation of Cactoideae in Anderson (2001) also includes nine tribes, but with some differences (from Gibson and Nobel, 1986) in the tribes recognized (e.g., Calymmantheae and Rhipsalideae) and considerable rearrangement of genera within the tribes (e.g., *Echinocereus* is placed in the Pachycereeae, and not alone in Echinocereeae).

Currently we estimate that in Texas there are 135 taxa: 94 species and 41 varieties in addition to the nomenclaturally typical ones. In the Trans-Pecos 109 taxa are recognized: 76 species and 33 varieties. The present work includes keys, descriptions, taxonomic discussion, distribution maps, and color photographs for the Trans-Pecos taxa. Also included are appendixes, color photographs, and discussions of Texas cacti from north and east of the Pecos River, except for most of the eastern opuntias and the few taxa of Cactoideae that we have not studied. For example, taxa unaccounted for in the present manuscript include *Echinocereus reichenbachii* varieties *baileyi/albispinus, perbellus,* and *albertii;* the cylindroid *"setaceus"* race of *Hamatocactus bicolor;* the endemic race *roemeri* of claret-cup cacti in the granite of Central Texas; and the Coke County race of *E. viridiflorus.*

Names of Cacti

The Trans-Pecos cacti are listed by Latinized scientific names under their respective genera. The species names are binomials, consisting of the genus name, always capitalized (e.g., *Opuntia*) and specific epithet, always lowercase (e.g., *aggeria*). Taxa below the rank of species, which zoologists would call "subspecies," but which plant nomenclature has traditionally called "varieties," bear trinomials. The designation of scientific names follows the *International Code of Botanical Nomenclature.* The *Code* dictates the recognition of a single legitimate binomial name for each plant species or a single trinomial for each infraspecific variety. Any other scientific names that have been applied to a species, through nomenclatural superfluity or taxonomic misinterpretation, are regarded as synonyms. Some (rarely all) of the pertinent synonyms are listed under each species.

Common names (sometimes more accurately referred to as popular names, vernacular names, English names, or just the most frequently used names) are also listed for each species. The common name that we deem most appropriate, based upon historical, taxonomic, or other reasons, appears in capital letters at the right of each Latin name. Any other common names are listed under that heading at the end of the discussion of each species. There are no international rules, not even common sense, for designating common names. Any author can attempt to popularize a name that seems appropriate for a given species. In our case the common appellations are usually of English or Spanish derivation. Such names and listings are useful, according to some, because there are large numbers of amateurs and other cactus enthusiasts who tend to refer to the plants by common name. Almost certainly, some common names have been overlooked.

Key to the Genera

A TAXONOMIC key is an artificial device constructed to contribute to the identification of an unknown organism, herein an unknown cactus plant. Keys in the present work, as in most taxonomic literature, are known as dichotomous keys, whereby a choice is provided between two different "leads." The leads are arranged in pairs, here indicated by numbers (1,1; 2,2; 3,3; etc.). A pair of leads is known as a couplet. Some of the keys are limited to a single couplet, when identification of only two taxa is required, and other keys are composed of a more extensive series of couplets, depending upon the number of taxa that have been grouped for identification. Where possible, our cactus keys contrast macromorphological characters, but the leads often include microscopic and sometimes highly technical features that are required for the distinction of closely related or morphologically similar taxa.

To use the keys, evaluate characters of the unknown cactus plant that are required to make a choice between leads in the first couplet (number 1). If the second lead is chosen, a parenthetical number at the end of the lead directs the user to the next couplet. A number in parentheses at the beginning of a first lead reminds the user about which couplet was previously consulted.

1. Plants erect, sprawling shrubs or low mats (cholla and prickly pear); stems jointed, the segments abruptly separated, cylindroid, globular, or laterally compressed (pads); glochids present, usually in tufts, in areoles of stems, flowers, and fruits; leaves present on new growth, these usually terete, fleshy, and caducous; seeds often discoid, always enclosed by a bony tan or pale pseudoaril; cotyledons foliaceous, ca. 1 cm long, usually deciduous; subfamily Opuntioideae

   1. *Opuntia*, page 73

1. Plants cespitose, few-branched, or of solitary stems; stems cylindroid, globular, dorsiventrally flattened, or angular, with ribs or tubercles, not jointed (except seemingly in *Peniocereus*); glochids absent; leaves absent (or stems perhaps with microscopic scales or protrusions); seeds generally spherical or elongated in shape, not discoid, pseudoaril absent; cotyledons deltate to hemispheric and short, much less than 1 cm long to vestigial, persistent; subfamily Cactoideae (2).

2(1). Stems slender, 10–50 times longer than thick, erect or clambering in shrubs, prominently 3–6 angled; spines 1–5 mm long, appressed and arachnid-like on the stem angles; epidermis velvetlike with microscopic unicellular trichomes; flowers nocturnal, white, 11–17 cm long; root tuberlike, often deep underground; tribe Pachycereeae

   2. *Peniocereus*, page 193

2. Stems thicker, erect, globular, or dorsiventrally flattened, rarely elongate (inherently prostrate in some *Echinocereus*); stems ribbed or tuberculate, these features sometimes inconspicuous or small and hidden by spines; spines various or absent; epidermis glabrous or merely papillate; flowers diurnal (but perhaps remaining open at night), white or colored, shorter; roots various (3).

3(2). Receptacle surface bearing five or more spiny areoles, these comparable to those of the stem but smaller; stems 5–19 ribbed; stem tissue strongly mucilaginous; flower buds erupting through epidermis above areoles; seeds black, testa cells strongly convex, in some species forming irregular ridges; tribe Echinocereeae

   3. *Echinocereus*, page 197

3. Receptacle surface spineless, bearing spine-tipped bracts with woolly axils in *Echinocactus*; stems ribbed or tuberculate; stem tissue mucilaginous or not; flower buds forming in areoles with or without spines, in areolar grooves, or in tubercle-stem axils (the origin often obscured under wool at stem apexes); seeds black, brown, reddish, or yellowish, testa cells convex, flat, concave, or papillate but not forming ridges; tribe Cacteae (4).

4(3). Plants spineless, or appearing so (short or rudimentary spines perhaps sporadic and hidden by trichomes, or in seedlings) (5).

4. Plants with spines in the areoles (these perhaps short, hairlike, or flattened, but usually many in a characteristic pattern and readily visible) (6).

5(4). Plants soft to the touch; stems weakly tuberculate or weakly ribbed, the areoles with tufts of trichomes to 1 cm long

8. *Lophophora*, page 309

5. Plants hard to the touch (cuticle many times thicker than epidermis); stems strongly tuberculate, the tubercles pyramidal, conical, cylindroid-acute, or dorsiventrally flattened, fissured, forming a rosette

9. *Ariocarpus*, page 317

6(4). Spines, at least one or some of them, hooked (7).

6. Spines not (any of them) hooked (12).

7(6). Stems of adults usually greatly exceeding 15 cm in diameter, the larger barrel cacti in the Trans-Pecos (*F. wislizeni*, Franklin Mts, and *F. hamatacanthus* east of the Franklin Mts; juveniles of *F. hamatacanthus* most likely to be confused with *Glandulicactus*)

5. *Ferocactus*, page 288

7. Stems of adults usually less than 15 cm in diameter (except 10–30 cm wide in *Echinocactus*) (8).

8(7). Tubercles, if evident, without areolar grooves (9).

8. Tubercles with areolar grooves (10).

9(8). Stems tuberculate; flowers rose to magenta (*M. grahamii* and *M. wrightii*)

16. *Mammillaria*, page 371

9. Stems ribbed; flowers yellow with red centers

6. *Hamatocactus*, page 297

10(8). Lower 3(–5) radial spines hooked, in addition to the hooked central spine; flowers dark reddish

11. *Glandulicactus*, page 327

10. No radial spines hooked; flowers yellow, white, or greenish with brown or rose-pink (11).

11(10). Tubercles usually 1.5–3 cm long; central spine terete, straight, curved, or hooked; pith and cortex nonmucilaginous; flowers yellow (*C. scheeri* var. *uncinata*)

7. *Coryphantha*, page 392

11. Tubercles usually less than 1 cm long; central spines dorsiventrally flattened, the lower one hooked; pith and cortex strongly mucilaginous; flowers white or greenish with brown or rose-pink (yellow in *A. tobuschii*)

12. *Ancistrocactus*, page 332

12(6). Spines robust and strongly annulate or cross-ribbed (spines of some

Ferocactus also annulate); floral bracts spinose-tipped, their axils conspicuously woolly

                   4. *Echinocactus*, page 280

12. Spines relatively slender or less robust, not annulate or cross-ribbed; floral bracts not spinose-tipped or with woolly axils (13).
13(12). Central spines, at least some of them, strongly flattened and bladelike (14).
13. Central spines, terete, angled, or flattened usually on one side, not bladelike (15).
14(13). Central spines 1–4(–5) per areole, 1–3 upper ones (in near radial position) flat and straight or curved, 1.3–7.5(–10) cm long, 0.2–1.5 mm wide; radial spines 10–20, largest ones 1–1.5 cm long; flowers rose-pink to magenta

                   7. *Thelocactus*, page 301

14. Central spines 1–3(–4) per areole, usually one porrect, flat, and often curled (resembling a dried grass leaf blade), 1.2–2.7 cm long, 0.4–2 mm wide; radial spines 6–9, largest ones 2–5 mm long; flowers white

                   13. *Toumeya*, page 344

15(13). Plants small, adult stems usually 1.5–3.5(–5) cm in diameter near ground level, to ca. 3 cm long, rarely longer; tubercles less than 2 mm long, ungrooved; spines all equally thin; dense covering of whitish spines obscuring the stem, the spines 2–3(–6) mm long on the sides of the stem but appressed and collectively forming a smooth surface, spines slightly longer (to 7 mm) at the stem apex; flowers pink to white, rarely yellowish, produced immediately adjacent to spine clusters in a woolly apical region

                   15. *Epithelantha*, page 364

15. Plants usually larger, with adult stems exceeding 3.5 cm in diameter and 3 cm long, except for *Mammillaria lasiacantha*, which resembles *Epithelantha* in habit, and a few smaller species of *Echinomastus* and *Coryphantha* with stems covered by whitish spines but these different in habit; tubercles longer than 2 mm, if present (3–6 mm long in *M. lasiacantha*), grooved or ungrooved; spines usually tapering to a point; spines whitish or not, covering the stem or not, usually longer than 5–6 mm on sides of stem, forming a bristly surface except in *M. lasiacantha*; flowers of various colors, borne apically or laterally (16).
16(15). Tubercles without areolar grooves; flowers and fruits lateral (i.e., away from the stem apex, and perhaps conspicuously on the sides of the stems, as in *M. pottsii*); flowers emerging from tubercle axils, often subtended by trichomes, spines, or bracts, which also are present in sterile condition

                   16. *Mammillaria*, page 371

16. Tubercles with areolar grooves; flowers and fruits apical; flowers emerging from the areoles or areolar grooves, although sometimes at the base of the groove near the tubercle axil, spines or bracts not present in the tubercle axils (17).

17(16). A thin, yellowish, mucilaginous layer beneath older stem surface tissue (bark), usually evident in longitudinal sections of living plants; pith and cortex not mucilaginous; flowers pink to magenta; fruits dry, thin-walled, green to tan or whitish when mature; seeds like those of *Ariocarpus*

                              10. *Neolloydia*, page 322

17. A thin, yellowish, mucilaginous layer not present beneath old bark in living plants, pith and cortex mucilaginous or not; flowers white, yellow, pink, or magenta; fruits scarcely succulent to succulent, green to red when mature; seeds very different (18).

18(17). Stems ribbed, the ribs well or poorly defined, the podaria decurrent and confluent to some degree; areolar glands absent; fruits green, scarcely succulent at first, quickly drying after ripening

                              14. *Echinomastus*, page 348

18. Stems strongly tuberculate; areolar glands present except in subgenus *Escobaria*; fruits green or red, usually succulent (often juicy), in most species remaining succulent after ripening

                              17. *Coryphantha*, page 392

Descriptive Cactus Flora

1. OPUNTIA Mill.
Club Chollas, Chollas, Prickly Pears

Plants low with creeping stems or erect and shrublike, 6 cm to 2–4 m high, with or without a short trunk or caudex. Stems of cylindroid or flattened joints, 5–28 cm long (to ca. 1 m long in the cultivar *O. engelmannii* var. *linguiformis*), the cylindroid joints 0.5–5 cm in diameter, flattened stem joints (pads) 5–28 cm broad. Ribs none (or well-defined in a few Mexican species, especially *O. bradtiana*). Tubercles absent (or reduced to swellings, especially in subgenus *Opuntia*, the prickly pears) or present but low and strongly decurrent in several chollas (e.g., *O. imbricata*) and club chollas (e.g., *O. emoryi*). Leaves cylindroid or awl-shaped, 0.5–1.5 cm long, usually present only on new stems and fruits, falling after the stems are a month or more old. Areoles circular or elongated. Spines white, gray, yellow, brown, pink, reddish, purplish, to black, or more than one color on same spine, (0–)1–10 or more per areole, smooth, straight, or curved, usually needlelike and 0.25–1 mm in diameter or awl-shaped and flattened (elliptic or angled in cross section) and 1–2 mm or more wide at the base, sometimes with full-length or partial sheaths; glochids (only in *Opuntia*) in areoles with (or without) spines. Flowers 1–9(–12) cm in diameter; floral tube short, shallow, cuplike, so that stamens are borne a short distance above the ovary; inner tepals usually not white (our species yellow, red, magenta, pinkish, brown, or greenish); in some species the inner tepal bases, in the flower center, contrastingly colored; stamens of many species are sensitive to touch, quickly closing around the style. Fruits succulent or dry, indehiscent, or (rarely) rupturing irregularly, obovoid, turbinate, elliptical, or spheroidal, 1–8 cm long, 0.8–6 cm in diameter; areoles mostly naked or bearing spines, glochids, or hairs, round to elongate, 1.2–5.5 mm long, 1.2–4.5 mm in diameter. Seeds whitish, gray, tan, brown, or a mixture of colors, flattened and discoid, or longer than broad, 1.5–6(–7) mm in largest measurement; funicular envelope with girdle (midvein) developed over seeds (pseudoaril). $x = 11, 22, 33, 44$ (our species).

Opuntia is a genus of over 160 convenient taxonomic species (Gibson and Nobel, 1986) distributed over much of the Western Hemisphere from southern Canada to southern South America. Some species of *Opuntia* occurs in every state of the contiguous United States except Maine, Vermont, and New Hampshire (Benson, 1982). Many species are native to Mexico, Central and South America, and the West Indies; one species-group of prickly pears occupies the Galápagos Islands. *Opuntia* species have been introduced worldwide in warm climates, including the Hawaiian Islands. The genus name is from Opus, an ancient town in Greece, where the name was inspired either from a cactuslike plant or from a prickly pear of early introduction there (Meyer and McLaughlin, 1981).

In 1993 an experimental *Opuntia* garden was established at the Sul Ross State University in Alpine in an attempt to investigate different morphotypes of *Opuntia* under uniform conditions, as did Griffiths. Many of these individually numbered plants or clones have provided flower buds for chromosome counts, linked to the flowers and fruits that followed, and have served as relatively complete research specimens.

Classically, two great subgenera are recognized in *Opuntia* (Britton and Rose, 1919–23) along with a number of series and sections (Benson, 1982) that further organize the genus. One subgenus is *Cylindropuntia*, the species with cylindroid stem joints, and the other is *Opuntia* (= "platyopuntia") with flat stem joints. The cylindropuntias are known collectively as the chollas; the platyopuntias are known as prickly pears (U.S.) or *nopales* (Mexico).

Certain species of *Opuntia* are notable for their economic value. This subject is developed further in a separate section. The economic value of prickly pears has led to their deliberate introduction in most arid parts of the world. Surely the spiny plants were also introduced in many areas by accident, but once established, prickly pears may readily spread and become naturalized. In many areas they have escaped from agriculture and become naturalized, often becoming vigorous pests.

Opuntia is notorious for its taxonomic complexity. Natural interspecific hybridization (e.g., McLeod, 1975; Mayer et al., 2000; Bobich and Nobel, 2001) and vegetative reproduction are responsible for at least some of the complexity, and there are relatively few exomorphic characters to reflect the underlying phylogenetic relationships. Because prickly pears often are difficult to identify from herbarium specimens, most type specimens are difficult to compare with present-day collections or populations. To reliably interpret the type collections, often it is necessary to visit type localities to study the habit and other characters that are evident only from living plants and/or populational samples.

The taxonomy of *Opuntia* remains poorly understood today, although greatly clarified in recent years through the efforts of D. Pinkava and his students and D. Ferguson. Our treatment of *Opuntia* is tentative, pending much additional populational, chromosomal, genetic, and other biological investigation of the taxa. We suspect that we have overlooked as many as 10–20 potentially recognizable taxa of cholla and prickly pear in the Trans-Pecos. Here are two examples: (1) An unidentified population of short-spined prickly pears grows in and near

Boquillas Canyon. Their spines number 4–6 per areole: 2–3 short (to 1.7 cm long), stout, pale reddish-brown, deflexed spines and 2–3 even shorter spines. They have relatively small yellow flowers with red centers, and green stigmas. (2) Unidentified populations of low plants bristling with reddish and white spines and having red-centered yellow flowers occur from Reeves County west to El Paso County. Still other populations across the Trans-Pecos, with various spine colors and patterns, some resembling *O. engelmannii*, have not yet been identified.

Currently, it is impossible to count the number of *Opuntia* species (either prickly pears or chollas) in North America, because no one knows enough about them. Benson (1982) recognized 50 species of *Opuntia* in North America north of Mexico, including the only three taxa that extend north into Canada. For Mexico, Bravo-Hollis (1978) reported 114 species of *Opuntia* and close relatives, including most of the Texas species. We estimate that there are, by our current standards for "specific" rank, about 25 "normal" taxonomic species of native *Opuntia* in Trans-Pecos Texas, plus four introduced species and seven or more, potentially dozens, of taxonomic species of hexaploid prickly pears. At present we recognize about six additional taxa as varieties within certain polytypic species, but the hierarchical classification of opuntias into species and their component subspecies remains little more than a goal for the future.

Texas prickly pear species that are excluded from this book because of their easterly distributions in the state are (1) *Opuntia humifusa* (Raf.) Raf. (eastern prickly pear), [*O. compressa* J. F. Macbr. (low prickly pear)], widespread in the eastern half of the United States and present in the interior of southeast Texas, according to Weniger (1984) and Ferguson (1987); (2) *O. stricta* Haw. (pest pear), a coarse, yellow-spined (sometimes spineless) seashore prickly pear common in Florida and the Caribbean region, repeatedly but unreliably identified on every shore of the Gulf of Mexico, including the Texas coast, especially in the vicinity of Galveston Bay; (3) Texas coastal endemic taxa that are the basis for erroneous reports of *O. stricta*, mostly very coarse, yellow-spined plants with unusually tiny seeds for this genus (*O. alta* Griffiths, *O. gilvo-alba* Griffiths, etc.); (4) *O. pusilla* (Haw.) Nutt. ("little" or "crow-foot" prickly pear, = *O. drummondii* Graham), restricted to sand behind the beaches along the Gulf Coast, in Texas near the tip of Bolivar Peninsula in front of Galveston Bay (Weniger, 1984).

Three subgenera of *Opuntia* are recognized for the Trans-Pecos species. Subgenus *Grusonia* (F. Rchb. & K. Schum.) Bravo is distinguished by cylindroid to obovoid stem segments, prostrate or mat-forming habit (in our species), and rudimentary spine sheaths (only at the spine tips). Subgenus *Cylindropuntia* Engelm. has cylindroid stem segments, is shrubby or at least erect (except for *O. tunicata* and stunted *O. whipplei*), and has well-developed spine sheaths, forming by the epidermis separating completely or almost completely during the first year into spine sheaths. Subgenus *Opuntia* is delimited by flattened stem segments, absence of spine sheaths, and uniquely shaped seeds. Glochids are produced in all *Opuntia*, but they are most pernicious in subgenus *Opuntia*.

The Cactaceae Working Party of the International Organization for Succulent

Plant Study (IOS) has decided to subdivide the subfamily Opuntioideae into 14 genera, mostly restricted to South America (Anderson, 1999b). In the United States and northern Mexico, the prickly pears make up part of *Opuntia*, the chollas compose *Cylindropuntia* (Engelm.) F. M. Knuth, and the club/dog chollas (with miscellaneous others) are placed into a newly expanded concept of *Grusonia* F. Rchb. & K. Schum. The elevation of traditional subgenera, such as *Cylindropuntia*, to generic level was proposed earlier by Robinson (1973). Generic status for the prickly pears, *Opuntia*, sensu stricto, is being supported by current DNA studies (Wallace and Dickie, 2002) and the slowly growing morphological database. Generic status for *Cylindropuntia* and *Grusonia* was accepted by D. J. Pinkava for the *Flora of North America* (Pinkava, forthcoming), but these two groups continue to be regarded as closest relatives of one another, the North American "cylindropuntias," sensu lato (Wallace and Dickie, 2002). The Trans-Pecos club/dog chollas might best be placed in the genus *Corynopuntia* F. M. Knuth (P. Griffith, 2002).

Key to the Trans-Pecos Species

1. Stem segments (joints) cylindroid; unweathered young fresh spines enclosed by evident paperlike sheaths, or the sheaths only at the tips of some unweathered spines in the mat- or mound-forming dog chollas; seeds not encircled by a projecting rim (2).
1. Stem segments (cladodes or pads) strongly flattened (except in *O. polyacantha* var. *arenaria*, and in seedlings of all species); spines without sheaths; seeds encircled by a projecting (or visibly) thickened specialized rim, the funicular envelope, 0.1–1 mm or more high (subgenus *Opuntia*) (11).
2(1). Spine sheaths rudimentary; plants mat- or mound-forming; (subgenus *Grusonia*) (3).
2. Spine sheaths full-size and persistent for at least one year; plants erect or low, compact, intricately branched shrubs, not mat- or mound-forming, except for *O. tunicata* and stunted *O. davisii* (with especially conspicuous papery sheaths) (subgenus *Cylindropuntia*) (7).
3(2). Stem segments 7–17 cm long, 2.5–5 cm in diameter; tubercles 2.5–3.5 cm long; largest spines 1.5–3 mm in diameter; NW Presidio Co., near the Rio Grande

4. *O. emoryi*.

3. Stem segments 3–7 cm long, 1.5–3.5 cm in diameter; tubercles 0.8–2 cm long; largest spines 0.4–2 mm in diameter; widespread in Trans-Pecos (4).
4(3). Roots tuberous (including the large adventitious roots, at least those more than a year old); central spines usually whitish-gray, stramineous to pinkish, less often red-brown, bladelike to nearly terete (5).

4. Roots all diffuse (including even the primary roots of the oldest plants); central spines red-brown or tan (to pinkish or white), bladelike (6).

5(4). Radial spines (0–)2–4; central spines usually whitish-gray, 0.8–1.3 mm in diameter; stem segments all firmly attached, clavate or narrowly obovoid

                              1. *O. aggeria.*

5. Radial spines 6–8; central spines usually stramineous to pinkish or red-brown, 0.4–0.8 mm in diameter; stem segments (at least distal ones) weakly attached, ovoid or obovoid to cylindroid

                              2b. *O. schottii* var. *grahamii.*

6(4). Central spines red-brown; glochids of second-year and older areoles rarely more than 14 per areole, in tufts 5 mm or shorter; plants usually trailing or creeping, forming open mats or loose chains; stems readily disarticulating and rooting

                              2a. *O. schottii* var. *schottii.*

6. Central spines tan (or pinkish to white); glochids of second-year areoles typically 15–40, in tufts 6–8(–12) mm long; plants usually in dense mats or mounds with an obvious central root system; stems relatively woody, not disarticulating

                              3. *O. densispina.*

7(2). Plants densely branched, either low and compact or erect, the stems obscured by dense spines; spine sheaths loose; flowers yellowish-green or brownish, pinkish in some *O. davisii* (8).

7. Plants erect, openly branched, the stems not obscured by spines; spine sheaths usually tight; flowers magenta, purplish, or yellow-green (9).

8(7). Plants low, compact; spine sheaths silvery-white; restricted distribution mostly near S slopes of Glass Mts

                              5. *O. tunicata.*

8. Plants erect, usually at least one or more main axes bearing horizontal branches held aloft above the ground; spine sheaths tan, yellowish, or golden; wider distribution, mostly in grassland of mountain basins

                              6. *O. davisii.*

9(7). Plants short-trunked treelets or relatively large shrubs; ultimate stem segments 1.5–3 cm in diameter; flowers bright rose-pink to magenta; fruits yellow, 2–3.5 cm in diameter

                              7. *O. imbricata.*

9. Plants usually smaller shrubs; ultimate stem segments 0.3–1.5 cm in diameter; flowers cream-colored or yellowish-green to dull pink or magenta; fruits red or orange, 0.9–2 cm in diameter (10).

10(9). Ultimate stem segments 0.6–1.5 cm in diameter; flowers magenta to purple-red, violet, brownish, bronze, or greenish-red; fruits 1–2 cm in diameter, tuberculate

             9. *O. kleiniae.*

10. Ultimate stem segments 0.3–1 cm in diameter; flowers cream-colored or yellowish-green, never pinkish; fruits 0.9–1.5 cm in diameter, smooth

             8. *O. leptocaulis.*

11(1). Plants spineless (but glochids abundant); epidermis (including that of the fruit) microscopically hairy (velutinous)

             10. *O. rufida.*

11. Plants bearing at least a few persistent spines, not merely glochids; all surfaces glabrous (12).

12(11). Plants usually less than 30 cm high; spines usually whitish to gray, less often darker colored proximally, distally, or throughout (in O. mackensenii var. minor), rarely yellowish (13).

12. Plants usually more than 30 cm high; spines brown, reddish-brown, dark purple, blackish, yellow, golden, bi- or multicolored, or whitish (17).

13(12). Pads not wrinkling; plants slightly larger, taller (see text); main spines usually 2–5 per areole; tetraploid

         18. *O. mackensenii* (in part, when less than 30 cm tall).

13. Pads wrinkling transversely under stress; plants lower, usually less than 20 cm tall (14).

14(13). Branches usually upright from a caudex, not forming long chains; healthy pads glaucous blue-green; usually 1–3 whitish spines in distal areoles, absent proximally; flowers (in our region) red; roots tuberous-thickened

             17. *O. pottsii.*

14. Branches longer (chains of pads usually prostrate or sprawling); healthy pads usually brighter green; 1–17 whitish spines in distal areoles or in all but the lower areoles; flowers (in our region) yellow or yellow with red centers; roots diffuse or weakly/sporadically tuberous (15).

15(14). Spines 6–17 per areole; fruits dry, tan, very spiny; largest seeds of any Trans-Pecos *Opuntia*; all diploid in the Trans-Pecos

             25. *O. polyacantha.*

15. Spines usually 1–5 per areole; fruits fleshy, reddish, usually spineless or with few spines; seeds smaller; polyploids; relatively similar to O. phaeacantha, except for smaller and wrinkling habit (16).

16(15). Main spines usually 1–3 in distal areoles; NE periphery of Trans-Pecos, in Ward Co. and adjacent counties; tetraploid; most similar to O. pottsii

             16. *O. macrorhiza.*

16. Main spines usually 2–5 in all but the lower areoles; throughout much of the Trans-Pecos (central Trans-Pecos west to El Paso Co.); hexaploid
                                        15. *O. tortispina*.

17(12). Pads "officially" purplish (can be glaucous when fresh); flowers red-centered (18).

17. Pads typically greenish (purplish only during stress, if at all); flowers all yellow or red-centered (21).

18(17). Pads more purple, orbicular to broadly ovate or broader than long (19).

18. Pads less purple, basically blue-green or blue-gray, obovate to orbicular (20).

19(18). Distal pads usually orbicular or broader than long; distribution widespread in the Big Bend region, N at least to Jeff Davis Co.; diploid
                                14a. *O. azurea* var. *diplopurpurea*.

19. Distal pads usually broadly ovate, sometimes orbicular; Trans-Pecos distribution El Paso Co. E to at least Culberson Co.; diploid/tetraploid
                                      13. *O. macrocentra*.

20(18). Pads mostly obovate, especially on upper branches; largest spines 5–12 cm long; diploid
                                      14b. *O. azurea* var. *parva*.

20. Pads mostly orbicular to obovate; spines usually 3–5 cm long; tetraploid
                       18. *O. mackensenii* (in part, when both tall and purplish).

21(17). Spines to 10.5 cm long, whitish, yellow, golden to reddish, often curved and twisted
                                      14d. *O. azurea* var. *discolor*.

21. Spines to 7.5 cm long, usually shorter, yellow, reddish, reddish-brown, brown, to blackish or white, sometimes bi- or multicolored (22).

22(21). Spines to 4 cm long, reddish to blackish proximally, yellow distally; pericarpel relatively small and spheroidal; inner tepals uniformly yellow; flowers widely opening; diploids (23).

22. Spines usually fewer than 7–13 per areole, with one or more main central spines; pericarpel narrower or longer; inner tepals either uniformly yellow or red basally; flowers less widely opening; polyploids except for *O. chisosensis* and *O. azurea* var. *aureispina* (24).

23(22). Spines 7–13 per areole, with one central longer, typically straight main spine; Terrell Co. northward and westward
                                             11. *O. strigil*.

23. Spines usually 1–2 per areole, 1–2 longer, typically with one central down-curved or straight; central Terrell Co. eastward to SW Edwards Plateau
                                           12. *O. atrispina*.

24(22). Plants low, spreading, trailing, or weak-stemmed, 30–60 cm high, with 1–3 whitish spines

                21. *O. phaeacantha.*

24. Plants usually erect or spreading with stout stems, usually 30–60 cm or more high, with 1–11 spines of various colors, if mostly whitish then usually showing some brownish at least basally (25).

25(24). Spines typically yellow (red-spined color-phases or forms occur in some populations) (26).

25. Spines various colors, white to reddish, reddish-brown, to nearly black, not yellow; polyploids (28).

26(25). Plants usually over 1 m high; lower Pecos (Sheffield, Pandale) and Boquillas area downstream to the Gulf of Mexico; hexaploids

              24b. *O. engelmannii* var. *lindheimeri.*

26. Plants usually less than 1 m high; distribution mostly otherwise; diploids (27).

27(26). Spines 1–5 per areole, yellow in younger pads but darker with age; fresh flowers pale yellow, lacking red centers; fruits succulent, juicy, spineless; Chisos Mts, in the oak zone

                 19. *O. chisosensis.*

27. Spines 4–11 per areole, yellow to orange, brown, or nearly black; fresh flowers relatively bright yellow, with sharply defined red centers; fruits fast-drying, spiny; near Rio Grande, Mariscal Mt to Boquillas, in rocky desert

               14c. *O. azurea* var. *aureispina.*

28(25). Spines typically whitish (may have dark bases) or color-banded, but not all-brown; pads averaging relatively large (spines appearing relatively short; hexaploids) (29).

28. Spines (at least the largest ones) usually reddish-brown to nearly black, other colors; pads averaging smaller (spines appearing relatively long); tetraploids and hexaploids (30).

29(28). Spines usually 3–4 per areole arranged in a "bird's-foot" pattern; inner tepals all-yellow (may redden prior to wilting); hairy seedlings

             24a. *O. engelmannii* var. *engelmannii.*

29. Spines usually 2(–4) per areole, not in a "bird's-foot" pattern; inner tepals yellow with red bases; ordinary (bristly) seedlings

                  23. *O. dulcis.*

30(28). Fruits typically spiniferous distally, drying rapidly; erect shrubs; near Rio Grande, vicinity of Hot Springs, S Brewster Co.; tetraploid

                20. *O. spinosibacca.*

30. Fruits spineless or rarely few-spined near the apex, remaining succulent

long after ripening; low shrubs; widespread in the Trans-Pecos; hexaploid

22. *O. camanchica.*

1. **Opuntia aggeria** Ralston & Hilsenb., Madroño 36: 226–28, f. 2. 1989. CLUMPED DOG CHOLLA. Plates 32, 33. Loosely consolidated desert alluvium, gravel to silt, often gypseous, igneous, or limestone. Presidio Co., extreme SE portion; Brewster Co., southern half, most common within 10–20 mi of the Rio Grande. 1,800–3,500 ft. Flowering late Mar–Apr (to early May in cultivation). $2n$ = 22. Mexico, S into Coahuila (where replaced by O. *moelleri* A. Berger near Cuatro Ciénegas), presumably in NE Chihuahua. Map 2.

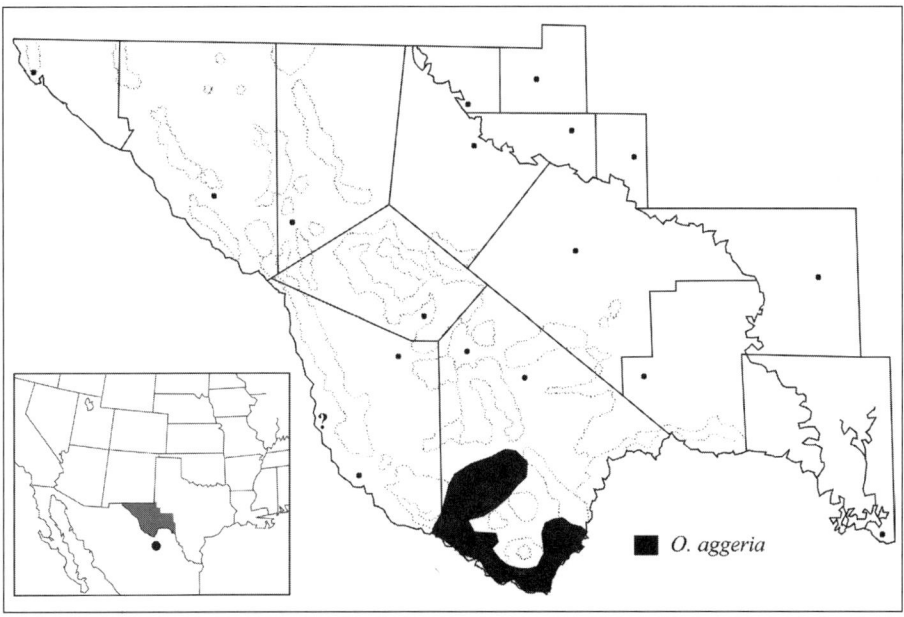

Map 2. Generalized distribution of *Opuntia aggeria* (clumped dog cholla).

The type locality is Tornillo Flats at 2,800 ft, Big Bend National Park, Brewster County, Texas. The specific epithet refers to the aggregated growth habit of O. *aggeria*, after the Latin *aggestus*, "mound."

Opuntia aggeria is a recently named member of a small group of closely related taxa, the "O. *schottii* complex" (Ralston and Hilsenbeck, 1989). In the Trans-Pecos these low, matted, mound-forming, or sprawling plants are known as dog chollas: "club cholla" is an international name for these and the larger "devil chollas" in the taxon *Corynopuntia*, currently included in the subgenus or segregate genus *Grusonia*. The dog chollas in general are disliked by some people because the readily dislodged stems of O. *schottii* and its closest relatives easily disperse as dangerously spiny nuisances embedded in shoes, tires, or hapless animals. However, *O. aggeria* has strongly persistent stems and is not at all a pest.

Identifying Characters. Morphologically, *O. aggeria* can be distinguished by its combination of (1) roots thickened and tuberous; (2) joints (stems) clavate, strongly persistent; (3) central spines (1–)3–4(–6), these usually grayish-white, less often reddish-brown, flattened or terete, divergent, 3–5(–9) cm long, the widest 0.8–1.3 mm in diameter near the base; (4) radial spines 2–4, these grayish-white, slender, deflexed, 0.6–2(–2.5) cm long; and (5) glochids numerous, 0.4–1 cm long.

In the northern part of its range and near Terlingua, *O. aggeria* is marginally sympatric with tetraploids ($2n$ = 44; *O. schottii*, sensu lato). The var. *grahamii* is tetraploid, both at the type locality (El Paso County) and in the Big Bend, its region of intergradation with var. *schottii* (northwest of Terlingua; Powell and Weedin, 2001), and *O. aggeria* is diploid ($2n$ = 22). Even without a chromosome count, the generally similar *O. schottii* var. *grahamii* is identifiable mostly by smaller features, different stem shapes, spine number and shape, readily disarticulating branches, and other subtleties discussed further under var. *grahamii*. In aspect the smaller stems of var. *grahamii* appear more densely spiny, a product of their slightly larger numbers of centrals, radials, and glochids. Both var. *grahamii* and *O. aggeria* have thickened, tuberous roots.

In the northeastern part of its range, where *O. schottii* var. *schottii* is the only other dog cholla present, identification should be easier: the root, spine, and other characters discussed under var. *schottii* are those of the weedy, most readily disarticulating extreme of that species. Although the central spines are grayish in some plants identified as var. *schottii* in Brewster County (e.g., *Raun 93–52*), they usually are reddish-brown in populations farther east (e.g., *Ralston 104–110*).

Opuntia aggeria is regionally sympatric with (but edaphically isolated from) *O. densispina* in extreme southern Brewster County east of Mariscal Mountain near the Rio Grande. *Opuntia aggeria* is relatively ubiquitous on gypsum and unconsolidated substrates within its range, whereas *O. densispina* may occur mostly if not entirely in clay soils (Ralston and Hilsenbeck, 1992). *Opuntia densispina* is distinguished by its (1) larger size, (2) roots that are fibrous, (3) ca. nine central spines that are off-white to pinkish-white to pale tan, and (4) longer glochids (up to 12 mm long on old stems). *Opuntia densispina* is tetraploid ($2n$ = 44) and produces flowers in May–June, later than *O. aggeria* (March–April).

The flowers of all club chollas are virtually identical, except that in our populations of *O. emoryi* and *O. schottii* var. *schottii*, the filaments are rarely (if ever) red. The flowers of *O. aggeria* are yellow, 5–7 cm long, and 4–5 cm wide when fully open, with the tepals in 3–4 whorls. The inner tepals are bright yellow, spatulate-apiculate, ca. 2.5 cm long, and ca. 2 cm wide. The outer tepals are yellow-green, perhaps pinkish-tinged in the midregion. The five or more stigma lobes of *O. aggeria* flowers are cream-yellow to pale green and ca. 5 mm long. The relatively thick style is very pale green to cream-colored, to 2.5 cm long. The anthers are yellow to cream; the filaments are up to 0.8 cm long. The innermost filaments are yellow-orange to reddish in all of the Trans-Pecos club chollas, so far as known, except in *O. emoryi* and most flowers of *O. schottii,* where all of the filaments are greenish. In at least some plants of *O. aggeria,* even the outer filaments are reddish. The outermost filaments can be green (dimorphic with respect to color) in *O. aggeria* but usually are red. In *O. schottii* var. *schottii,* most plants appear to have only greenish filaments, but relatively few flowers from the northern portion of the range have been examined. The pericarpel is narrowly obconic, to 5.5 cm long, 2 cm in diameter, with glochids and woolly hairs in the areoles.

In *O. aggeria* the fruits are light yellow, to 5.5 cm long, becoming gray and dry with age, the areoles white-woolly with tufts of glochids and bristles 3–7 mm long. The floral remnant is persistent as in all club chollas. The subdiscoid seeds are brown to cream-colored, smooth, 5–6 mm in diameter, and weakly pointed on one margin.

Phenology. The principal flowering period for *O. aggeria* is in March and April. In cultivation in Alpine, blooming may continue into mid-May. The flowers open about midday, close in the late afternoon, and do not open the next day. In fact, the flowers open for a single day in all dog chollas. Fruit maturation appears to require one to several months, with yellow (soft) or gray (dried) fruits persisting into late summer or early winter.

Sterile and Immature Specimens. The plants of *O. aggeria* occur in mats or low mounds reaching 15 cm or more high and 0.8–1 m wide. New stem growth is from lateral or lower areoles. The clavate stems are prominently tuberculate, with the tubercles 1–2 cm long, 0.8–1 cm wide, and 0.5–0.7 cm high. The areoles are circular, 3–6 mm in diameter, usually densely covered with white wool.

The spines in areoles of *O. aggeria* and other dog chollas are interpreted here as either central or radial. Usually the distinction between central and radial spines is evident to the naked eye, and the spines are clearly delimited under magnification. The central spines, borne inside the periphery of the areole, are prominent and divergent; they vary from flattened or angular to terete. The central spines potentially are of several sizes, shapes, and colors, but they are always larger and usually differ in color from the radial spines. The radial spines originate at the areole periphery, they are slender (terete or compressed) and whitish, and all or some of them (toward the base or front of the areole) are severely deflexed. The glochids are restricted to the upper part or back of the areole, toward the tubercle surface. In *O. aggeria*, spines are produced in areoles on the upper two-thirds or more of the joints.

In *O. aggeria* the typical number of central spines is 3–4 per areole, but there can be 1–6. Typically, there is one deflexed lower central spine that is conspicuously flattened on the upper surface and rounded or angled on the lower surface. The lower central usually is the largest spine of the areole, in length and width. Typically, in each areole there are two upper or lateral centrals that are slightly (subterete) to conspicuously flattened and one upper central that usually is the most flattened central spine. Characteristically, the younger central spines of *O. aggeria* are grayish-white with pinkish bases. In age the central spines become gray or purplish-brown with whitish margins. In areoles of some plants the lower central and perhaps other centrals may be reddish-brown, a color that is characteristic in other dog chollas of the Trans-Pecos except O. *densispina* and usually O. *schottii* var. *grahamii*. The central spines have translucent yellow tips. In O. *aggeria* there are typically 2–4 deflexed, whitish radial spines. The radials are slender and terete or somewhat flattened. The spine bases in O. *aggeria*, particularly the central spines, are enlarged-bulbous. The radials have small-bulbous bases. The spine surfaces usually are roughened with closely set, irregular transverse ridges that give the surfaces a cross-wrinkled appearance characteristic of the dog chollas, but not as evident in O. *schottii* var. *grahamii*. In O. *aggeria* the glochids have been described as erect and numerous (Ralston and Hilsenbeck, 1989). We have counted 15–40 glochids in the areoles of O. *aggeria*. Glochid numbers may vary in the different species of dog chollas. The glochids are whitish, stramineous, to rusty-brown in all of the Trans-Pecos dog chollas.

Juvenile and immature plants of O. *aggeria* have not been evaluated carefully for the current work, but they are seen frequently in outdoor horticulture at Tucson, where this species reproduces spontaneously from seed (unlike O. *schottii*). On immature plants the spines are shorter than those of adults.

Biosystematics. *Opuntia aggeria,* as described by Ralston and Hilsenbeck (1989), mostly is restricted in distribution to low altitudes in southern Brewster County, extending into adjacent southeastern Presidio County along and near the Rio Grande and across the Rio Grande in adjacent Mexico. Specimens cytologically revealed as O. *aggeria* have been collected near Ruidosa, Presidio County ($n = 11$, *Ralston 130*). Other interesting populations (no chromosome counts) that resemble O. *aggeria* are found southwest of the Guadalupe Mountains near the Hudspeth-Culberson County line (Heil and Brack, 1986, photograph, p. 167) and at Hueco Tanks, El Paso County (*Warnock 7774*). In the southern half of Brewster County (e.g., at Terlingua and south of Santiago Peak), O. *aggeria* sporadically is sympatric with O. *schottii* var. *grahamii* and/or O. *schottii* var. *schottii*. Typical O. *schottii* var. *grahamii* is poorly documented outside of El Paso County and one or two sites closely adjacent in southern New Mexico, and typical O. *schottii* var. *schottii* occurs mostly from Terrell and Val Verde counties southeast into the Rio Grande valley.

Occasional dog cholla plants (O. aff. *aggeria*) are essentially spineless. Taxonomically, these are considered by us to be mere forms (e.g., within the populations near Study Butte), although some exist as small populations, for example, at a site 3.3 miles east of the crest of the Big Hill in southern Presidio County.

Anthony (1956) observed that the dog cholla populations in southern Brewster County are intermediate geographically and in some morphological features between O. *grahamii* and O. *schottii*, as she called them at the time, aspects which led to description of the morphologically intermediate forms (some, not necessarily all of them) as a hybrid between O. *grahamii* and O. *schottii*. A different interpretation of some specimens resulted from the biosystematic work by Ralston (1987) and Ralston and Hilsenbeck (1989). Only diploids ($2n = 22$) were observed among the putative hybrids of the sort described and collected by Anthony, and both of the supposed parents were already reported as tetraploid

($2n = 44$). Ralston and Hilsenbeck concluded that the diploid southern Brewster County dog chollas were readily distinguished from both *O. grahamii* and *O. schottii* by a number of morphological characters and that they were best treated as a distinct species, which they formally named as *O. aggeria*. Possibly this is a rediscovery of *O. moelleri*, a poorly known Mexican species, but that is another project. We agree with Ralston and Hilsenbeck that the diploid dog cholla population in southern Brewster County represents a taxon (= *O. aggeria*) that is distinct from both *O. grahamii* and *O. schottii*, irrespective of the controversial relationship between the latter two.

We emphasize, however, that intermediate morphology in Brewster County dog chollas is commonplace. During study of the *O. schottii* complex in preparation for the current work, we encountered dog cholla specimens from southern Brewster County that were not easily identifiable. The chromosome numbers of morphologically ambiguous dog chollas in southern Brewster County have so far been determined as either diploids, now assigned to *O. aggeria* (sensu Ralston and Hilsenbeck), or "possibly hybrid" tetraploids (6.5 miles north-northwest of Terlingua, *G. G. Raun 93-52;* Powell and Weedin, 2001) assigned to either *O. schottii* var. *grahamii* or *O. schottii* var. *schottii*. No triploids have been reported. The morphologically ambiguous tetraploids approach the appearance of *O. aggeria*, but the roots are not always tuberous-thickened, and the branches disarticulate more readily.

Opuntia aggeria is one of five closely related species recognized for Texas by Ralston and Hilsenbeck (1989, 1992). The dog chollas (or club chollas) belong with the subgenus *Grusonia* in series *Clavatae* (sensu Benson, 1982), a group of about 17 taxa in North America. Benson (1982) recognized *O. schottii* var. *schottii* and *O. schottii* var. *grahamii* as varieties that intergraded in Brewster County; specimens of *O. aggeria*, then unnamed, were arbitrarily cited among either *schottii* or *grahamii* in *Flora of Texas* (Benson, 1969b). In Benson's 1982 work, the photograph (fig. 367, p. 369) labeled "*Opuntia schottii* var. *schottii*" is of *O. aggeria*. Weniger (1984) treated *O. schottii* and *O. grahamii* as distinct species, and the southern Brewster County populations, including all of *O. aggeria*, as part of *O. schottii*.

Synonym. *Grusonia aggeria* (Ralston & Hilsenb.) E. F. Anderson.

Common Names. Mound-forming opuntia. Clumped dog cholla is an appropriate name, in that we prefer the use of dog cholla instead of club cholla for the smallest Chihuahuan Desert opuntias in the series *Clavatae*.

2. **Opuntia schottii** Engelm., Proc. Amer. Acad. Arts 3: 304. 1856. COMMON DOG CHOLLA. Plates 34–37. Igneous and limestone-derived alluvium, desertscrub to Tamaulipan scrub. El Paso Co. E to Val Verde Co. 1,000–4,000 ft. Flowering May–Jul. $2n = 44$. South-central NM, E of the Rio Grande; in TX SE to Cameron Co., Rio Grande valley, E to a few localities in central TX. Mexico, NE states, Chihuahua to Tamaulipas. Map 3.

The type locality is "on the arid hills near the Rio Grande, between the San Pedro and Pecos Rivers (Wright, Schott)" presumably somewhere in Val Verde

County. The specific epithet is after Arthur Schott who collected the type of this species and many others during the United States and Mexican Boundary Survey of 1851–53.

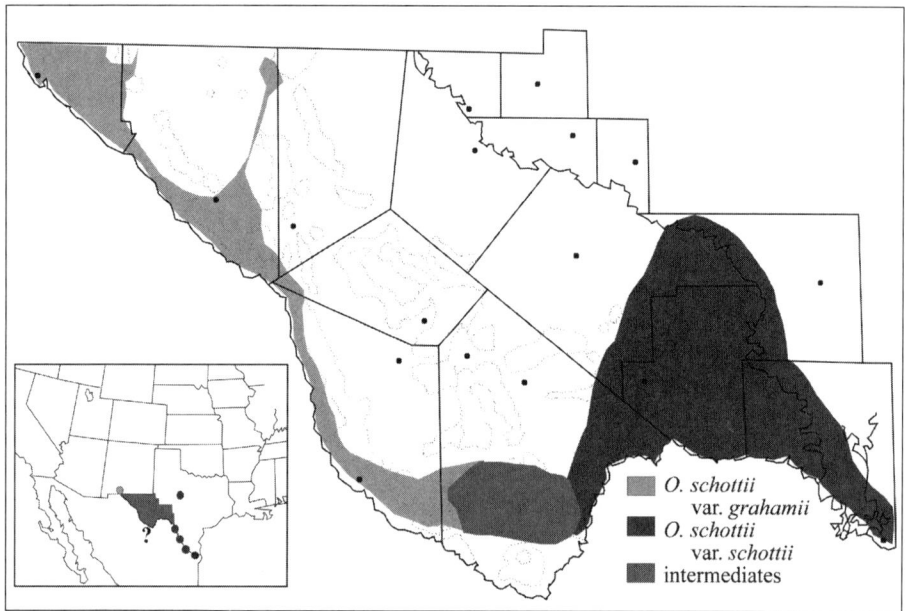

Map 3. Generalized distribution of *Opuntia schottii* var. *schottii* (Schott's dog cholla) and *O. schottii* var. *grahamii* (Graham's dog cholla).

Key to the Varieties

1. Easternmost in distribution: Roots all adventitious (never taprooted); always-disarticulating chains of joints; major spines dark- or reddish-brown, flat and bladelike with translucent edges
    2a. *O. schottii* var. *schottii*.

1. Westernmost in distribution: Usually taprooted, roots becoming tuberous after the first year; easily disarticulating joints three in a file; spines gray or brown, terete or moderately flattened, not bladelike
    2b. *O. schottii* var. *grahamii*.

    2a. Opuntia schottii Engelm. var. **schottii**. SCHOTT'S DOG CHOLLA. Plates 34, 35. Alluvial substrates, desertscrub to Tamaulipan scrub. Southeastern Brewster Co., Terrell Co., Pecos Co., Crockett Co., and Val Verde Co. 1,000–4,000 ft. Flowering Jun–Jul. $2n = 44$. Southeast to Cameron Co., Rio Grande valley; eastern "outposts" in Schleicher and Brown counties (Benson, 1982). Mexico, adjacent in Coahuila; presumably on the Mexican side of the Rio Grande in Nuevo León and Tamaulipas. Map 3.

Dense populations of var. *schottii* in some parts of the southeastern Trans-Pecos, particularly in Val Verde County, hinder cross-country foot traffic. It appears that increasing desertification enhances the spread of these plants, as readily separated and dispersed spiny joints quickly strike root in bare soil.

Identifying Characters. Morphologically, var. *schottii* is distinguished primarily by its diffuse roots; secondarily by clavate joints, central spines (4–)6–8(–11), these reddish-brown, flattened, divergent, 3.8–7 cm long, the widest 1.5–2 mm in diameter near the base, radial spines 4(–6), these grayish-white, slender, deflexed, 1–1.9(–3) cm long, and 10–14 glochids, ca. 5 mm long. Variety *schottii* is tetraploid ($2n = 44$), a generalization at first seemingly well documented (Ralston and Hilsenbeck, 1989, 1992) but upon investigation seems to derive from only one count (Weedin et al., 1989).

In southern Brewster County var. *schottii* is sympatric with *O. aggeria*: var. *schottii* has more and flatter central spines that are wider at the base and typically reddish-brown in color, and fewer glochids. Variety *schottii* also co-occurs with var. *grahamii* in southern Brewster County.

The flowers of var. *schottii* vary considerably in size. Like those of *O. aggeria*, they may be slightly larger than those of var. *grahamii*, at least in the southern Big Bend region. The staminal filaments are greenish in flowers of most of the plants examined and, so far, in all of the plants from as far east as Langtry.

The fruits of var. *schottii* are narrowly obconic or clavate, fleshy, light yellow, and up to 4.5 cm long. The fruit surface displays numerous narrow tubercles and areoles bearing white wool, some bearing about four short central spines and perhaps numerous bristlelike spines to ca. 7 mm long, and all bearing glochids to ca. 3 mm long. The subdiscoid seeds are cream to brown, 5–6 mm in diameter, smooth, and weakly pointed on one margin.

Phenology. The principal flowering period for var. *schottii* is from mid-June to early July (Ralston and Hilsenbeck, 1989). Under cultivation in Alpine, plants have bloomed as early as 5 May. The flowers open about midday, close in the late afternoon, and do not open the next day. Fruit maturation seems to require about one month to several months, but this aspect has not been carefully evaluated. Yellow fruits seem able to persist on plants for several months.

Sterile and Immature Specimens. The plants of var. *schottii* form medium-size to extensive mats to 5 m in diameter but barely ankle high, usually surrounded by their clonal progeny. The branches sprawl, forming small or long chains, with new stem joints emerging from lateral areoles. The elongated, clavate stem joints are 3–6.5 cm long, ca. 3 cm in diameter, straight, or curved upward. The stem surfaces are covered with prominent tubercles 1.5–2 cm long, 6–8 mm wide, and 6–8 mm high. The circular areoles are 5–7 mm in diameter, usually with a dense covering of short white or dirty-white wool. Spines are produced in areoles on the upper two-thirds of the joint.

In var. *schottii* each areole is dominated by about 3–4 longer and wider reddish-brown, "burnt-red," or brown "main" centrals. The lower central spine is the largest one, 3.8–7 cm long, 1.5–2 mm in diameter near the base. Flattened central spines are convex, rounded, or triangular (in cross section) on the lower surface. In addition to the 3–4 flattened main centrals in each areole, usually there are three or more additional centrals that are more slender and only dorsiventrally compressed or subterete. In each areole, after the deflexed lowermost central, the next 3–4 largest centrals usually are divergent and borne in the lower to upper portion of the areole. Some of the more slender (1–1.5 mm wide) centrals are lower and lateral in position, and in some plants they may be gray in color. Young spines may have caducous membranous sheaths to 5 mm long. In var. *schottii* the central spines have expanded bases but not bulbous bases as in *O. aggeria* and var. *grahamii*. The expanded bases of centrals in var. *schottii* are evident, especially in younger areoles, but often not in older areoles with mature centrals. In var. *schottii* usually there are four front (or lower) and lateral radials that are strongly deflexed, slender, and gray-white, and 1–1.9(–3) cm long, and perhaps two radials toward the back of the areole that are pale brown, terete, and divergent, with the aspect of small centrals. The spine surfaces are distinctly cross-striated (Weniger, 1984), a character that results from numerous closely set, microscopic, irregular transverse ridges. The glochids in areoles of var. *schottii* are relatively few, as in *O. emoryi*.

Juvenile and immature plants of var. *schottii* have not been studied for the present work. Observations of immature stems have revealed that spines are shorter than on mature stems, and the dirty-white woolly hairs are longer than in older stems.

Biosystematics. In the Trans-Pecos var. *schottii* occurs as pure populations in Terrell, eastern Pecos, and Val Verde counties. In Brewster County it occurs sporadically as isolated or taxonomically ambiguous specimens all the way to near the Presidio County line south of Agua Fria Mountain. Ralston and Hilsenbeck (1989) maintained that var. *schottii* was ecologically segregated from var. *grahamii* and *O. aggeria* in southern Brewster County, with var. *schottii* favoring more mesic habitats primarily east of Brewster County. Preliminary meiotic and pollen observations of var. *schottii* plants from Val Verde (near Langtry) and Terrell (near Sanderson) counties, and var. *schottii*-like plants from Heath Canyon in Brewster County, indicated that the Val Verde and Terrell county plants were completely sterile and that the Heath Canyon plants were highly sterile (A. M. Powell, unpub.).

Variety *schottii* is distinguished from both var. *grahamii* and *O. aggeria* by its thin, unspecialized roots and wider, flattened, often bright reddish-brown central spines. When growing in typical mesic habitats, the largest stems of var. *schottii* resemble the smallest stems of *O. emoryi*, another tetraploid.

Synonyms. *Corynopuntia schottii* (Engelm.) F. M. Knuth; *Grusonia schottii* (Engelm.) H. Rob.

Common Names. Schott's Dwarf Cholla; clavellina; dog-turd cactus; dog cactus; dog cholla; devil cactus; devil cholla.

2b. **Opuntia schottii** Engelm. var. **grahamii** (Engelm.) L. D. Benson. GRAHAM'S DOG CHOLLA. Plates 36, 37. [*Opuntia grahamii* Engelm., Proc. Amer. Acad. Arts 3: 304. 1856]. Loosely consolidated alluvium, igneous or limestone, in desertscrub. El Paso Co. SE to (depending upon classification of intermediates)

Brewster Co. 1,800–4,500 ft. Flowering May–Jun. $2n = 44$. South-central NM, E of the Rio Grande. Mexico, from eastern Chihuahua S possibly to Durango, Zacatecas, and San Luis Potosí, depending upon taxonomic circumscription of related endemic taxa. Map 3.

Opuntia schottii var. *grahamii* is one of the three dog chollas that were described from collections made during the United States and Mexican Boundary Survey of 1851–53 (Engelmann, 1856). The type locality is from "near El Paso," specifically "bottoms of the Rio Grande, and downriver about 100 miles in sandy soil." The specific epithet honors Colonel James D. Graham, head of the scientific corps of the United States and Mexican Boundary Commission that provided many plant collections for study by Engelmann (Crook and Mottram, 1998).

Identifying Characters. Morphologically, var. *grahamii* can be distinguished by its thickened and tuberous roots (like those of *O. aggeria*): joints obovoid, ovoid, or cylindroid, central spines (3–)4–7, these stramineous to brown, often tinged pinkish or reddish, to dull reddish, gray with age, terete or only slightly flattened, divergent, 2–3.5(–5) cm long, the widest 0.4–0.7(–0.8) mm in diameter near the base, radial spines 6–8(–9), these grayish-white, slender, deflexed, 1–1.2 cm long, and glochids numerous, 5–7 mm long. Variety *grahamii* is tetraploid ($2n = 44$; Pinkava et al., 1985).

Some plants of *O. schottii* var. *grahamii* superficially are similar to some plants of *O. aggeria*, but the species always are distinguishable by the suite of characters listed above. In aspect the stems of var. *grahamii* appear to be more densely spined, a product of the slightly smaller stems with a few more spines per areole than in *O. aggeria*.

The yellow flowers of var. *grahamii* are virtually identical to those of the other dog chollas, as discussed under *O. aggeria*. In some plants we have observed, the flowers of *O. schottii* var. *grahamii* were slightly smaller than those of *O. aggeria*, but this size difference is not yet substantiated by populational measurements.

The fruits of *O. schottii* var. *grahamii* are similar to those of the other dog chollas. Fruits we have measured were ovoid or oblong (2–3.5 cm long), less elongate than fully developed yellow fruits of var. *schottii*. The fruits of *O. schottii* and *O. aggeria* are light yellow when ripe, prominently tuberculate and with the areoles dominated by glochids or bristlelike spines, or both. In *O. schottii* var. *grahamii* the areoles exhibit white wool, several slender, white, bristlelike spines to ca. 7 mm long and tufts of whitish glochids to ca. 3 mm long. The subdiscoid seeds are 5–6 mm in diameter, brown, tan, or cream-colored, smooth, and weakly pointed on one margin.

Phenology. *Opuntia schottii* var. *grahamii* flowers from May through early June (Ralston and Hilsenbeck, 1989, 1992). Under cultivation in Alpine, var. *grahamii* may start to bloom in late April. There is potential overlap in the flowering periods of all Trans-Pecos dog chollas. The flowers open about midday, close in the late afternoon, and do not reopen the next day. Fruit maturation appears to require one to several months.

Sterile and Immature Specimens. The plants of var. *grahamii* occur in low mats or mounds 8–20 cm high and to 30 cm across. Usually the branches are creeping, three joints in a single file, with new growth originating from apical or upper lateral areoles. The obovate to cylindroid joints are prominently tuberculate. The tubercles are 0.8–1.2 cm long, to 0.6 cm wide, and 0.4–0.6 mm high. The circular areoles are 3–4 mm in diameter, usually with a dense covering of white wool. Spines are produced in areoles on the upper one-half or two-thirds of the joint.

In *O. schottii* var. *grahamii* the lower central spine is deflexed and somewhat compressed dorsiventrally, and usually the longest spine of the areole is 2–3.5(–5) cm. The divergent lateral and upper centrals may be almost as long, and they are terete or only slightly compressed. The centrals have reddish-brown or golden bulbous bases. Young spines occasionally have caducous membranous sheaths to 3 mm long, but spine sheaths are obscure in all the Trans-Pecos dog chollas. The slender central spines, usually 0.4–0.7 mm in diameter, provide one of the best characters for distinguishing var. *grahamii* from *O. aggeria* and var. *schottii*. In var. *grahamii* the abruptly deflexed radial spines have small-bulbous bases. The spine surfaces, particularly those of the central spines, are microscopically roughened by numerous, irregular, short transverse ridges. Compared to those of *O. aggeria* and var. *schottii*, the spine surfaces of var. *grahamii* are less conspicuously cross-striated. According to Weniger (1984) the spines of var. *grahamii* are not cross-striated. The number of glochids in areoles of var. *grahamii* increases in areoles toward the bases of the joints.

Juvenile and immature plants of var. *grahamii* have not been studied for the current work. On immature joints, the spines are shorter than on mature stems, and they tend to be lighter in color.

Biosystematics. Typical tuberous-rooted *O. schottii* var. *grahamii* occurs only sporadically in the Big Bend; the easternmost similar plants are those near upper Reagan Canyon about 10 miles from the Rio Grande. In southern Brewster County, typical tuberous-rooted var. *grahamii* is regionally or "marginally" sympatric with *O. aggeria,* which is distributed mainly in deep south Brewster County (Ralston and Hilsenbeck, 1989) and adjacent Mexico. The principal distribution of *O. schottii,* sensu lato (the small tetraploid dog chollas), is from the "lower canyons" of the Rio Grande and the Dead Horse Mountains across central Brewster County south of Elephant Mountain and Santiago Peak to southeastern Presidio County, in and near the Solitario Dome. Farther northwest, dog chollas grow disjunctly along the Rio Grande valley to El Paso County (where rare) and a few miles into adjacent New Mexico.

Benson (1982) treated *O. grahamii* as a variety of *O. schottii,* apparently being convinced that the two taxa intergraded in southern Brewster County, presumably because of Anthony's (1956) formal description of supposed hybrids *O. grahamii* × *O. schottii.* Weniger (1984) was not persuaded that *O. grahamii* occurred at all in the Big Bend region of Texas, and he incorporated all the southern Brewster County dog chollas, including the diploid species, in his concept of *O. schottii.* For several years we followed Weniger (1984) and Ralston and Hilsenbeck (1989, 1992) in allowing specific rank for var. *grahamii,* after the putative intermediates described by Anthony were demonstrated by Ralston and Hilsenbeck to represent the distinct diploid entity O. *aggeria*. However, the real interface between the eastern and western tetraploids (after successfully "weeding out" the diploids) remains insufficiently documented. Ralston and Hilsenbeck (1989) somehow concluded that var. *grahamii* and var. *schottii* were marginally sympatric in southern Brewster County and that they did not intergrade. However, plants just south of Elephant Mountain, such as those near Terlingua Creek (*A. Zimmerman 2752*) and on Chalk Bluff (*A. Zimmerman 2546*), seemed to produce only diffuse roots but otherwise exhibited the characters of var. *grahamii*.

Synonyms. *Corynopuntia grahamii* (Engelm.) F. M. Knuth; *Grusonia grahamii* (Engelm.) H. Rob.

Common Names. Mounded dwarf cholla; Graham prickly pear; Graham dog cactus; Graham dog-cactus.

3. **Opuntia densispina** Ralston & Hilsenb, Madroño 39: 281–284, f. 1. 1992. BIG BEND DEVIL CHOLLA. Plates 38, 39. Bare clay deposits and clay overlaid by gravel or sand, endemic, along and near the Rio Grande, between Mariscal Mt and the Dead Horse Mts, Big Bend National Park, Brewster Co. 1,975–2,200 ft. Flowering May–Jun. $2n = 44$. Not yet reported from adjacent Mexico. Map 4.

Opuntia densispina, the most recently described of the five *Grusonia* taxa in the Trans-Pecos, has a limited distribution. The type locality is on the River Road near Solis Ranch, Big Bend National Park, Brewster County, Texas. The specific epithet refers to the dense appearance of the spine clusters, after the Latin *densus*, "dense," and *spina*, "spine."

Identifying Characters. The fibrous roots of *O. densispina* immediately distinguish it from the sympatric *O. aggeria* and allopatric *O. schottii* var. *grahamii*, both of which appear visibly smaller. The joints of *O. densispina* are strongly persistent, obovate to clavate, with central spines ca. nine, these usually white to pinkish-white or tan, flattened, divergent, 4–8 cm long, the widest 1–2 mm in diameter near the base, radial spines 2–4, these whitish, to 2.5 cm long or longer, and glochids relatively numerous, up to 11 mm long.

In its natural habitat *O. densispina* is not likely to be confused with any other cactus species except for the regionally sympatric *O. aggeria*, which is conspicuously smaller, fewer- and shorter-spined, and tuberous-rooted; and the chromosome number is different. *Opuntia densispina* resembles allopatric *O. schottii* var. *schottii* more closely; so closely that we are not certain which species is represented by one of our most important dog cholla specimens, the tetraploid at a site near the Dead Horse Mountains that has for many years been the only chromosome count attributed to *O. schottii* var. *schottii*. The central spines of *O. densispina* on average are longer, more numerous, more often twisted or curved, and of pale pink to tan color, different from the ashy-white aspect of *O. aggeria* and the bright chestnut-brown of typical (eastern) *O. schottii* var. *schottii*.

The flowers of *O. densispina* are like those of other Trans-Pecos club chollas.

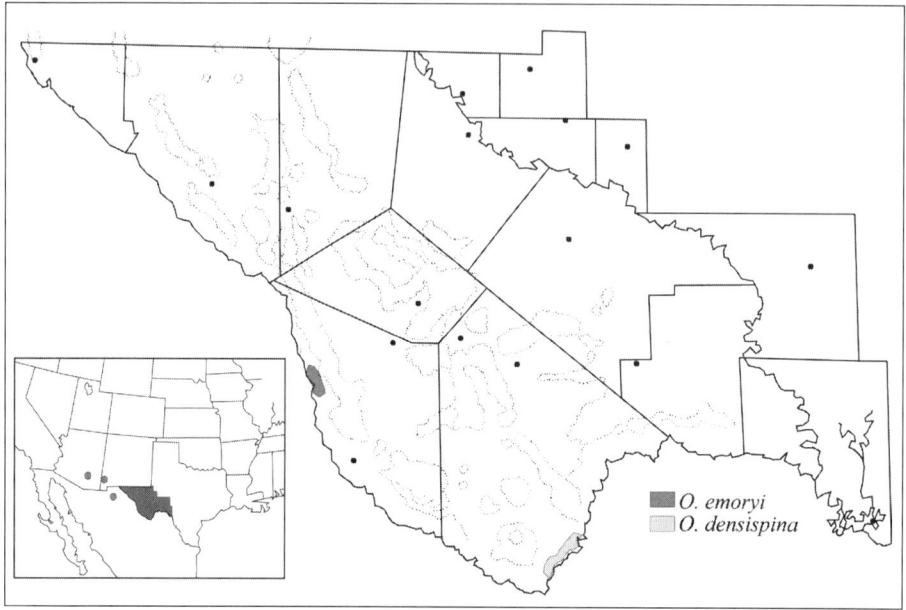

Map 4. Generalized distribution of *Opuntia densispina* (Big Bend devil cholla) and *O. emoryi* (common devil cholla).

The filaments of O. *densispina*, particularly the inner ones, are reddish-orange or reddish, as they are in most plants of O. *aggeria*.

Normal ripe fruits of O. *densispina* are clavate, uniformly pale lemon-yellow inside and out but quickly drying and browning, with white wool, and with accrescent tufts of bristlelike pale glochids 3–9 mm long in the bulging areoles. The subdiscoid seeds are cream-colored to tan, smooth, 5–6.5 mm in diameter, 2.4–2.9 mm thick, and weakly pointed at one margin.

Phenology. *Opuntia densispina* blooms from mid-May to early June (Ralston and Hilsenbeck, 1992). Under cultivation in Alpine, flowers on one plant began opening on 19 May 1998 and on 5 May in 1999. The flowers open about midday and close permanently in late afternoon. Fruit maturation (in late July in horticulture at 3,000 feet in Arizona) seems to require about one month to several months, but requires careful study because some fruits persist without normal ripening (also true in O. *emoryi* and related species).

Sterile and Immature Specimens. The low mats or mounds of O. *densispina* grow to 12 cm or more high and to 3 m in diameter. When stems are compactly arranged, the whole mound glistens because of the dense complement of whitish spines. The branches mostly sprawl, forming usually short chains of stems. New stems emerge from lateral areoles. The joints are 4.5–7 cm long and 3.5 cm in diameter. The tubercles are prominent, 1.5–2 cm long, 0.5–0.7 cm wide, and ca. 0.5 cm high. The ovate areoles are 3–4 mm in diameter and covered with short white wool. Spines are produced in areoles on the upper two-thirds of the joint.

In O. *densispina* the usually nine central spines per areole may be described as pink, white, off-white, tan, to soft deerskin tan. From a distance the central spine often appears silvery. In the nine–central spine pattern usually there is a central one that is flattened, 6–7 cm long, and 1–2 mm in diameter. Typically, there are three upper centrals that are flattened, or flat on the upper surface and rounded or angled (in cross section) on the lower surface. The upper lateral centrals usually are the longest spines in the areole at 7–8 cm; the upper middle central is shorter, to 5–5.5 cm long. The central spines positioned low in the areole, the lower centrals, are flattened, to 5 cm long. Usually there are three lower centrals, one in the middle and two in lateral position. The ninth central spine generally is in lateral to upper lateral position and is 4–5 cm long. One or more of the upper or lower central spines sometimes are twisted longitudinally, a trait possibly rare or absent in the other dog chollas. The radial spines are slender and strongly deflexed, as they are in other dog and devil chollas. The central spine surfaces are tangibly scabrous with microscopic, irregular transverse ridges. The spine bases are expanded, glistening, and yellowish, but these somewhat bulbous bases are essentially hidden by the areolar wool.

Juvenile plants of O. densispina have not been studied. Immature joints have shorter spines and no glochids.

Biosystematics. The known range of O. densispina is limited, involving discontinuous populations in a narrow region mostly in clay along the Rio Grande for about 18 miles. Few specimens other than the type collection have been available for study; Ralston and Hilsenbeck (1992) described the species only from its type locality, ca. 5.3 road miles from Mariscal Mine (not Solis Ranch as reported in the original description). *Opuntia aggeria* occurs in the same area as *O. densispina,* but the two taxa apparently are separated ecologically, with *O. densispina* occurring in clay substrates and *O. aggeria* occupying loosely consolidated alluvium (Ralston and Hilsenbeck, 1992). *Opuntia densispina* and *O. aggeria* differ from each other in root morphology, size of all vegetative parts, number of central spines per areole, spine color, fruit size, shape, color, and glochid-vestiture, ploidy level, and (so far as known) flowering phenology in the wild.

Opuntia densispina has some features in common with *O. schottii* var. *schottii* and *O. emoryi*, especially their fibrous roots and ploidy level (Ralston and Hilsenbeck, 1992). Geographically, *O. densispina* is closer to *O. schottii* var. *schottii,* but the two taxa differ significantly in known ecological preference, habit, woodiness, areole diameter, spine sizes (especially color, but also average length), glochid number and length, and perhaps in flowering phenology. Tentatively we accept *O. densispina,* in spite of our dismay at the increasing number of cryptic species in the literature; we await documentation of its ecogeographic interface with *O. schottii* and systematic study of variation in *O. emoryi,* which replaces *O. densispina* directly upstream as the rare woody-stemmed large club cholla of the valley floor.

Synonyms. None.

Common Name. Densely spined opuntia.

4. **Opuntia emoryi** Engelm., Proc. Amer. Acad. Arts 3: 304. 1856. COMMON DEVIL CHOLLA. Plates 40, 41. Sand or gravel flats, washes, and low hills, Presidio Co., below the Sierra Vieja rim, near the Rio Grande between Can-

delaria and Porvenir. 2,300–3,300 ft. Flowering (Apr–)May–Jun(–Jul). $2n = 44$. Main population in SE AZ and SW NM. The type locality is in northern Chihuahua, Mexico. Map 4.

Opuntia emoryi is easily distinguished from the other Trans-Pecos dog chollas by the larger size of all vegetative parts, except that *O. densispina* has longer spines. Benson (1982) and Weniger (1984) treated this taxon as *O.* "*stanlyi* Engelm.," an unaccepted provisional epithet (Pinkava and Parfitt, 1988; Crook and Mottram, 1996). The type locality is between "the sandhills" (i.e., the Samalayuca dunes) and Lake Santa Maria, south and west of El Paso, in Chihuahua, Mexico. The specific epithet honors Colonel W. H. Emory, U.S. Army Corps of Topographical Engineers, a major explorer of the southwestern United States, in charge of the United States and Mexican Boundary Survey, and collector of many species of cacti during the earliest expeditions through the Southwest.

Identifying Characters. Morphologically, *O. emoryi* can be distinguished from *O. schottii*, sensu lato, by its fibrous roots, woody (not disarticulating) stems, large clavate joints 7–17(–19) cm long and 2.5–5 cm in diameter, central spines 6–7, these tan to yellowish or reddish-brown, flattened, divergent, 4.5–7 cm long, the widest 1.5–3 mm in diameter near the base, radial spines 5–6(–10), these reddish-brown or pale yellow, slender, deflexed or divergent, 1.4–2.5 cm long, and glochids 3–30(–37) in number and to 5(–8) mm long.

The yellow flowers of *O. emoryi* closely resemble those of the smaller Trans-Pecos dog chollas, except that the pericarpel may be slightly more elongate (4–6 cm long, 1–1.3 cm in diameter). The flowers are 5.5–8 cm long and 4–7 cm wide. The bracts are long-attenuate, yellow-green basally and pinkish or pale reddish distally. The outer tepals are pinkish-tinged in their centers. The inner tepals are yellow, ca. 2.5 cm long, ca. 1.5 cm wide, broadly spatulate and apiculate. The 6–7 stigma lobes are cream-colored in the flowers of *O. emoryi* so far observed, while the stigmas in other Trans-Pecos dog chollas appear consistently to be light green. The cream-colored style is 1.8–2.5 cm long. The anthers are cream-colored and the filaments, ca. 1 cm. long, are greenish proximally and cream-colored distally. The related species all contain forms having red staminal filaments (at least the inner filaments in flowers of a few plants somewhere in each taxon), but thus far we have not seen any such form in *O. emoryi*.

The fruits of *O. emoryi* are light yellow, clavate or cylindroid-turbinate, 5–6(–8) cm long and 1.5–2(–3) cm in diameter. The fruit surfaces are tuberculate with spherical areoles exhibiting mounded white wool, numerous yellowish glochids 5–8 mm long, and rarely a few short spines. The fruits of *O. emoryi* may be somewhat larger but are otherwise similar to those of *O. densispina* and the common dog chollas, except that the areoles of the dog chollas sometimes support persistent bristles or spines in addition to glochids. The discoid, tan, smooth seeds of *O. emoryi*, weakly pointed on one side, are slightly smaller, 4.5 mm in diameter, 1.5–2 mm thick, than those of the related species.

Phenology. The main flowering period for *O. emoryi* appears to be in May to early June (Ralston and Hilsenbeck, 1989), with blooming in some years occurring as early as April and extending to as late as July. The flowers open about midday, close in late afternoon, and do not open again the next day. Fruit maturation appears to require a month to perhaps several months. In the field, fruits have been observed on plants in August. In cultivation, and perhaps in natural populations, fruits may persist throughout the winter and into the spring at least until the next flowering period, and perhaps longer.

Sterile and Immature Specimens. The plants of *O. emoryi* form low, sprawling mats to 15 cm high and to 4–6 m in diameter. New stems emerge from lateral areoles. The branches are many joints long; individual joints are prostrate, curving upward, or erect. The clavate joints are much larger (to 17 cm or more long) than those of other Trans-Pecos dog chollas, and they are not as easily detached (unlike those of *O. schottii*). The stem tubercles are 2.5–3.5 cm long, 1–1.5 cm in diameter, and 1–1.2 cm high. The circular areoles are 5–7 mm in diameter, with short white wool.

The central spines of *O. emoryi* are broader than those of other Trans-Pecos *Grusonia* taxa. The central spine pattern in each areole seems to be one where the largest central is centrally positioned and divergent or slightly deflexed. The swordlike "main" central is flat on the upper surface and convex underneath. There is one lower central, deflexed and not as wide as the main central, two lower lateral ones, these also flattened on the upper surface (perhaps convex underneath) and 0.7–1.5 cm in diameter. Usually three additional centrals are positioned in the upper (back) portion of the areole, these 4–6 cm long, stout, triangular in cross section or flat on the upper surface and broadly rounded underneath, or subterete. The central spine surfaces are cross-striated, the visual effect of microscopic, irregular transverse ridges. The bases of the central spines are expanded, in some cases with evident golden bulbous bases. The radial spines are of two types. The lower or front three radials are deflexed, and the others are positioned lateral and near the upper part or back of the areole, divergent to erect. The radials are terete, or nearly so, essentially the same color as the central spines, and they have small-bulbous bases. The glochids are rather sparse, compared to most of the related species.

Juvenile and immature specimens of *O. emoryi* are rarely encountered and have not been evaluated for the present work. Immature individual stems reach nearly full size before their spines attain full length.

Biosystematics. The distribution of *O. emoryi* in Texas is localized in northwestern Presidio County, with the plants occurring in alluvial substrates in a narrow strip about 20 miles long near the Rio Grande. *Opuntia emoryi* is much more widely distributed in southeastern Arizona, barely entering southwestern New Mexico, and has been collected at least once in Chihuahua. The total range of *O. emoryi* in Texas is roughly equivalent in size to that of *O. densispina*.

When compared to the other Texas *Grusonia* taxa, *O. emoryi* most closely resembles *O. schottii*, which Weniger (1984) described as having the characters of *O. emoryi* in miniature, and *O. densispina*. *Opuntia emoryi* is tetraploid, like the other Trans-Pecos dog chollas except for the diploid *O. aggeria*. Benson (1982) treated *O. emoryi* as one of four southwestern varieties of the one species, which he treated as *O. stanlyi*; however, one of them is a diploid species (*O. parishii* Orcutt), and var. *kunzei* (Rose) L. D. Benson is currently treated at specific rank, and var. *peeblesiana* L. D. Benson is a synonym.

Synonyms. *Cactus emoryi* Lem.; *Opuntia stanlyi* Engelm. var. *stanlyi*; *Corynopuntia stanlyi* (Engelm.) F. M. Knuth; *Grusonia stanlyi* (Engelm.) H. Rob.; *G. emoryi* (Engelm.) Pinkava.

Common Names. Devil's cholla; devil cholla; creeping cholla; Stanly's cholla; Emory's opuntia; cursed cholla; Stanly club cholla; Stanly dog-cactus.

5. **Opuntia tunicata** (Lehm.) Link & Otto. ICICLE CHOLLA. Plates 42, 43. [*Cactus tunicatus* Lehm., Nov. Minus Cog. Stirp. Pug. 16: 319. 1832]. Localized populations, upper margin of basin grassland, rocky slopes, associated with grasses, shrubs, and *Pinus remota* (Little) D. K. Bailey & Hawksw., eastern Glass Mts, S slope, Pecos Co.; smaller colonies in Brewster Co., outlier of the S Glass Mts, and limestone mesa W of Sanderson near the Brewster-Pecos county line. 4,500–5,000 ft. Flowering Jun–Jul. $2n = 22$. Mexico, S margins of the Chihuahuan Desert Region at 6,000–6,900 ft. Also Cuba and South America (Ecuador, Peru, and Chile). Map 5.

Populations of *O. tunicata* in Mexico are much more extensive than the one most conspicuous population in the United States, visible from the highway in probably less than one square mile of the southeastern Glass Mountains, Pecos County. The largest plants form low spreading clumps to ca. 1 m in diameter and ca. 35 cm high. On bright days, or when backlit through the glistening mass of spine sheaths, the plants glow as if coated with ice, as reflected in the common name. The type locality is "In Mexico." The specific epithet is after the Latin *tunica*, "garment," a reference to the sheathed (clothed) spines.

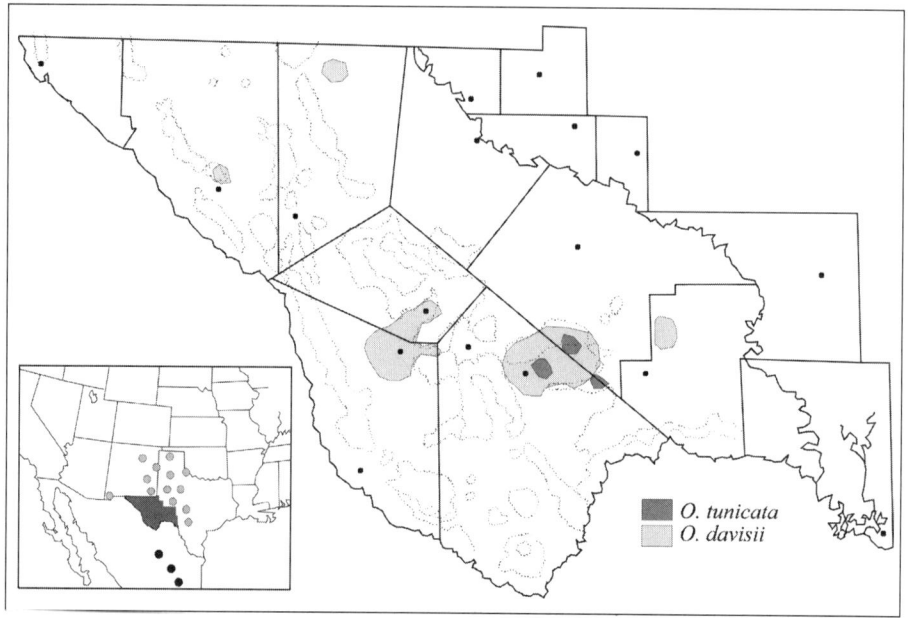

Map 5. Distribution of *Opuntia tunicata* (icicle cholla) and *O. davisii* (Davis' cholla).

Identifying Characters. The plants of O. *tunicata* typically are compact masses of 50 or more branching main stems covered with sheathed spines that almost obscure the stems. The papery-white or silvery-white spine sheaths are broader than the spines themselves, collectively exhibiting a sheen, or glistening mass, as if every spine were individually wrapped in a translucent toothpick-sleeve. In the vicinity of the Glass Mountains no other cactus species is likely to be confused with O. *tunicata*, except perhaps O. *davisii*, which is clothed with equally baggy but plain tan, yellowish, or golden spine sheaths. Plants of O. *davisii* occur in the alluvial basin grasslands adjacent to and south of the Glass Mountains and elsewhere in the Trans-Pecos. Typically, some plants in each population or colony of O. *davisii* are erect, with numerous low branches. There are many other vegetative characters by which O. *tunicata* and O. *davisii* can be distinguished (see following discussion).

The greenish-yellow flowers of O. *tunicata* are 3.5–5 cm in diameter. Both outer and inner tepals are greenish-yellow. The inner tepals are cuneate, 1.2–2 cm long, 0.9–1.2 cm in diameter, and mucronate apically. The 3–6 stigma lobes are greenish to greenish-yellow, thick, and ca. 3 mm long. The style is greenish, perhaps with a rose tinge. The anthers are yellow, and the greenish filaments are ca. 6 mm long. The pericarpel is ovate to obconic and tuberculate, the areoles with felty hairs and glochids only. Our information regarding flowers of O. *tunicata* is derived from observations of relatively few plants in the Pecos County population.

The fruits of O. *tunicata* are somewhat fleshy and yellow at maturity. The tuberculate, obconic fruits are ca. 3 cm long, 1.2–1.3 cm across, and concave at the apex. The felty areoles support brownish glochids, but usually no spines. The seeds are tan, somewhat obovate, 3–3.5 mm long, ca. 2 mm wide, and ca. 1 mm thick, with a thin aril. Sometimes the fruits develop without seeds (Anthony, 1956).

Phenology. The flowering time for O. *tunicata* is not well documented. Anthony (1956) observed that blooming in 1948 extended from about 24 June to early July. Our observations of plants in the field suggest that the plants may not bloom every year. Plants cultivated in Alpine have bloomed in June. In mid-May 2003, Marcos Rodriguez and Amy Causey photographed the plants in abundant bloom. Fruit maturation requires about two months (Anthony, 1956) for the green developing fruits to turn yellowish. The yellowish fruits may persist through the winter at least into March and April, and perhaps much longer.

Sterile and Immature Specimens. Seemingly mature mats of O. *tunicata* have produced a thick, woody, creeping stem, perhaps 30 cm long, that developed numerous erect main branches. The cylindroid main branches extend to 20 cm long and 5 cm in diameter and give rise to cylindroid or clavate ultimate joints to 9 cm long and 2.5 cm in diameter. The stems are green, with prominent tubercles to 2–3 cm long, 0.2–0.5 cm wide, and 0.3–1 cm high. The terminal joints are readily detached and tend to produce adventitious roots under favorable conditions, as do most stems of opuntias.

The areoles of terminal stems are oval-elliptic to oblong, 3–7 mm long, and ca. 2 mm in diameter. The areoles are covered with felty hairs that are dirty-white to yellowish. In more mature stem segments the areoles are ovate or spatulate, with the broader portion at the upper part of the areole toward the stem. Approximately the upper halves of elongated areoles are without spines, except for a tuft of tan to brown glochids, 1–2 mm long, these erect, densely arranged, and easily dislodged. The spines, when present, are located near the base of the areole. Mostly spines are produced in areoles on the upper one-half or one-third of the joint. Usually there are 3–5 sheathed central spines per areole. In areoles of young terminal joints, usually there are three centrals. In the youngest areoles there is 0–1 central spine. In young areoles with ca. three central spines, typically there are two deflexed radial spines at the base of the areole, these slender, terete, gray to tannish, and 0.5–1.1 cm long. In some areoles there may be 3–4 of these radial spines, two lower ones and 1–2 lower lateral. Radial spines are not sheathed. In longer, presumably older stem segments, in elongated areoles, the typical spine pattern for O. *tunicata* includes 4–5 sheathed central spines and no radial spines. Usually there is one lowermost, deflexed central, two lower lateral centrals, these deflexed or divergent, and 1–2 upper, erect or divergent centrals, these positioned one in front of the other near the front-middle portion of the areole. The 4–5 spines are crowded in the front half of the areole. The sheathed central spines are 2.5–5.5 cm long, (0.4–)0.7–1 cm in diameter, pale yellowish-tan to tan-brown (or whitish-yellow), often drying light reddish-brown. The centrals are flattened on both sides, flattened on the upper surface and rounded (in cross section) underneath, or the smaller ones may be subterete. The centrals are not enlarged or bulbous at the base, and at the distal end they are very sharp-pointed. The terete spine tips are microscopically retrorsely barbellate or scabrous. The outline of the central spines can be seen through the translucent spine sheaths. The usually silvery-white sheaths are flattened and 2–3 mm wide. The sheaths fit loosely around the spines, tend to tear near the base, and are easily slipped off the spines. Spine sheaths in some plants, particularly cultivated ones, may be very pale tan or cream in color.

Another areole character in O. *tunicata* involves the presence of 2–3 short peglike structures, one on each side at the upper periphery, these in younger areoles, barely erupting above the felty hair. By analogy these microscopic structures resemble emerging antlers of a deer. The "nubs" may be vestigial, stunted, or injured spines, or spine bases, or they may be glands (as well documented in some related species) whose secreted substances have been observed attracting copious numbers of small ants to the cacti.

Anthony (1956) described a dwarf form of O. *tunicata* that was found among typical plants in the Glass Mountains population. Similar dwarf forms are well known in stunted, windswept high-altitude populations of O. *whipplei* Engelm. & Bigelow in the northern deserts and O. *fulgida* Engelm. of the Sonoran Desert: the plants rise only 7 cm or less high; the spines are relatively slender; and as expected from the parallel situation in O. *fulgida*, Anthony did not find flowers or fruits on the dwarf plants. According to Anthony, the creeping stems of dwarf plants are only 10 cm long and 1.5 cm wide, with a short trunklet at the base. In dwarfs the joints are short-obovate and clavate, ca. 3.5 cm long, and 1 cm in diameter. The areoles exhibit 1–2 spines to 2.4 cm long and 2–4 bristles that are deflexed and to 5 mm long. Anthony speculated that the dwarf plants are both (1) seedlings, slow to reach maturity, and (2) the product of vegetative reproduction, developing from detached terminal joints and fruits of typical plants. She further speculated that the larger typical plants of *O. tunicata* seemed to be vegetatively reproduced from the larger detached main branches.

In juvenile stems of *O. tunicata* the areoles are circular, to 2 mm in diameter, and usually with 0–2 central (sheathed) spines, these located in the central to lower central portion of the areole. At the lower and lower lateral margins of the areoles, usually there are two deflexed radial spines, perhaps also with 1–2 rudimentary radials, the longest 3–4 mm. In juvenile areoles, in addition to the regular centrals and the radials, 1–3 bristlelike spines ca. 1 mm long may be present in a position near the front of the areole. The seedlings, seldom seen in this largely clonal species, remain beyond our investigative reach.

Biosystematics. The only known locality for *O. tunicata* in the United States was the one on a lower south slope of the Glass Mountains in Pecos County, until much smaller populations of the species were discovered in 1976 about 15–20 miles to the west, and in 1996 about 15–20 miles to the southeast, both sites in Brewster County. The existence of the recently discovered populations suggests that additional localities for *O. tunicata* await discovery.

Benson (1982), followed by Pinkava in Zimmerman et al. (forthcoming), extended the taxonomic circumscription of *O. tunicata* to include *O. davisii*. Bravo-Hollis (1978) placed the two taxa in separate series of *Opuntia*, but direct comparison shows they are closely related. We believe that these taxa are sufficiently distinct reproductively, ecologically, and morphologically that they should be recognized at specific rank, pending taxonomic revision of all the low-growing, baggy-sheathed, yellowish-flowered chollas together. *Opuntia tunicata* occurs in rocky limestone habitats, but *O. davisii* favors alluvial grasslands. Each species contains its own variation in ploidy, habit, spine, and other features, as does the apparently related northern *O. whipplei*. No intermediates have been reported.

Synonyms. *Opuntia stapeliae* DC.; *O. furiosa* J. C. Wendl.; *O. hystrix* Griseb.; *O. tunicata* (Lehm.) Link & Otto var. *tunicata*; *Cylindropuntia tunicata* (Lehm.) F. M. Knuth.

Common Names. Abrojo; clavellina; clavelina; tencholote; coyonoxtle; sheathed cholla; icicle cholla; thistle cholla.

6. **Opuntia davisii** Engelm. & Bigelow, Proc. Amer. Acad. Arts 3: 305. 1856. DAVIS' CHOLLA. Plates 44, 45. Widespread but scattered and uncommon, or several plants in local colonies, alluvial mountain basins in plains grasslands, other sites in alluvial soils, especially sand or loam. Hudspeth, Culberson, Presidio, Jeff Davis, Brewster, and Terrell counties. 3,500–5,000 ft. Flowering Jun–Jul. $2n = 22, 44$? Texas E to eastern Edwards Plateau, N through the Panhandle. Extreme W OK; NM. $2n = 22$? 44. Expected in NE Chihuahua (Bravo-Hollis, 1978). Map 5.

In the Trans-Pecos O. *davisii* most likely is to be observed from paved roads through the plains grasslands in the mountain basins near Marfa, Fort Davis, and Marathon. The erect shrublets or mounds, often filled with tall grass, are so densely covered with sheathed spines as to hide the branching pattern of slender stems. When backlit, the plants are conspicuous from great distances owing to their glistening golden spines. The type locality is the "Upper Canadian, about Tucumcari Hills, near the Llano Estacado," probably in northeastern New Mexico. The specific epithet honors Jefferson Davis, the U.S. Secretary of War at the time the Pacific railway surveys were conducted and also the namesake of Jeff Davis County.

Identifying Characters. Although O. *davisii* is described as a densely branched erect shrub 40–85 cm tall, some plants or stems of some plants may collapse into dense, untidy mounds 30–40 cm high. The slender, short trunk is woody and many-branched, potentially with several branches at each node. The larger terminal joints are 6–12 cm long and 1–1.5 cm in diameter. The terminal stem joints are readily detached, as they are in O. *tunicata,* but the older joints are firmly attached. The stems are obscured by the spines diverging from closely spaced areoles. The spine surfaces themselves are not readily visible, being covered by loose, flattened sheaths that are pale yellowish-tan or pale golden in color. Within its natural range, O. *davisii* is not likely to be confused with any other species.

The flowers of O. *davisii* have a firm, waxy appearance, similar in this respect to those of *Echinocereus coccineus*. The flowers are about 5 cm long and ca. 4 cm in diameter when fully open. The tepals are described as green to greenish-yellow or pale green. The 4–7 stigma lobes are cream-colored. The style is reddish. The anthers are yellow, and the filaments are pale purplish distally and greenish below. The flowers of one plant from near Fort Davis were different, with rose-pink stigma lobes and brown tepals. The pericarpel is obconic and tuberculate, with a few whitish, deciduous spines 2–2.5 cm long.

The fruits of O. *davisii* are slightly fleshy and yellow at maturity. The narrowly turbinate or obconic fruits are 2.5–3.5 cm long, ca. 1.5 cm wide, and concave at the apex. The fruit surface is tuberculate with spherical or oval areoles. The whitish areoles have light brown glochids around the periphery, along with 1–2 short bristles, these 0.7–1.5 mm long and deflexed like radial spines. The seeds are 3.5 mm in diameter with a thin beaked aril-margin (Anthony, 1956). There are only a few (1–3) seeds per fruit, or in some fruits there are no viable seeds.

Phenology. The flowering period for O. *davisii* is not well understood at present, although Anthony (1956) described it as short, only a few weeks in late June and early July. Two cultivated plants from southeastern New Mexico initiated flowering on 7 June one year and 24 May the next year. The flowers open by late morning or midday, close at night, and do not open again the next day. The fruits usually are still green in August, turning yellow in September, and persisting potentially through October or longer.

Sterile and Immature Specimens. At least some plants of O. *davisii* may produce tuberous roots just below the surface in sandy soil (Hester, 1939). The short trunk and slender stems of O. *davisii* are woody. In old stems with the wood exposed, large vascular gaps that are narrow and acute can be seen. The densely spined stem segments have laterally compressed tubercles, these 1–2 cm long, ca. 4.5 mm wide, and ca. 3 mm high. Stem segments are cylindroid, or less often slender-clavate. The areoles are spherical to oval, 4–5 mm in diameter, and with mounded felt that is yellowish or dirty-white to gray.

The spine sheaths are flattened and loose around the spines, adherent only near the bases, and tight-fitting near the apex. The sheaths are 1–2.5 mm across, glistening yellow-tan or pale golden, sometimes with a hint of very pale red. In some plants, including some cultivated specimens, the sheaths are papery-white or grayish-white with a pale reddish tinge. The typical spine pattern includes two groups of centrals; 4–5 wider and longer, flattened spines, and usually five additional more slender and shorter centrals; and 4–5 radial spines. Among the larger flattened centrals, located toward the front of the areole, there is one lowermost, deflexed spine (the most slender one), 2–3 lower lateral, deflexed spines, and one larger, divergent spine that is located in the central lower part of the areole. All of these are flattened on both sides or flattened on the upper surface and rounded or angled (in cross section) underneath. These larger "front" centrals are pale reddish-brown, 0.8–1.5 mm in diameter, and 3–5.5 cm long. In the lateral and back portion of the areole, usually there are five additional centrals that are more slender than the front or lower centrals, 0.4–0.6 mm in diameter, and compressed or angled. This group of slender centrals mostly are peripheral in position, except for the uppermost one (or two) that is positioned away from the edge of the areole. Typically, in this second group there are two lateral, two upper lateral, and one upper central, these all divergent and 1.3–2 cm long. This second group of centrals is reddish-brown, as are the lower (larger) centrals, except that the upper centrals may be slightly lighter in color. The spine tips are very sharp and scabrous, usually with microscopic barbs. In each areole there may be four radial spines. Two radials are at the front (lower) margin of the areole and two are lower lateral in position. The lowermost radials are deflexed, and the lateral ones are divergent or deflexed. The radials are acicular, gray and black-speckled, 1–1.3 cm long, and not sheathed. A tuft of numerous glochids is positioned at the back of each areole, with a short space of open felt between the central spines and the glochids. The glochids are ca. 2 mm long, erect, and light yellow-brown.

The areoles of immature stem joints are characterized by few spines, and the spines are all slender and shorter. We have not investigated the seedlings of O. *davisii*.

Biosystematics. In the Trans-Pecos O. *davisii* mostly is a species of plains grasslands in the mountain basins of northern Brewster and Presidio counties and in adjacent Jeff Davis County. The plants seem to be uncommon to rare at other sites, including in southern Hudspeth, northern Culberson, and Terrell counties. In the plains grasslands, where some plants can be observed in roadside pastures, the plants occur singly, scattered over many miles of roadside habitat, or in small colonies of about 10 plants, more or less. The occurrence of plants in colonies suggests that at those sites vegetative reproduction from dislodged joints has occurred. Scattered colonies appear to be most frequent in roadside pastures south of Marfa and north of Marathon, although we suspect that the prevalence of vegetative progeny increases with heavy grazing or similar disturbances.

Opuntia davisii is easily distinguished from its very close relative, *O. tunicata*, by its taller growth habit and yellow or tan spine sheaths, golden instead of silver when backlit. Benson (1982) "lumped" *O. davisii* into *O. tunicata*, at varietal rank, unlike other authors (Anthony, 1956; Bravo-Hollis, 1978; Weniger, 1984), who treated them as distinct. *Opuntia davisii* may be most closely related either to *O. tunicata* or to *O. whipplei* Engelm. & Bigelow of the southwestern United States, or perhaps it has hybrid origins.

Two ploidy levels have been found in *O. davisii*, thus far known from only three published chromosome counts. The diploid count ($2n = 22$), from root tips, came from northern Presidio County (Weedin et al., 1989). Several recent meiotic observations from plants of *O. davisii* in Presidio and Jeff Davis counties (A. Causey and A. M. Powell, unpub.) have suggested triploid or tetraploid numbers in highly irregular meiosis, apparently with numerous multivalent configurations. The tetraploid counts ($2n = $ ca. 44) were (1) from a very isolated colony near Animas, New Mexico, and (2) from plants collected near Roswell in southeastern New Mexico (Powell and Weedin, 2001; A. Causey and A. M. Powell, unpub.). Meiosis appeared to be regular in the Roswell plants examined. The New Mexico plants had yellow-green flowers. A plant of *O. davisii* with brown tepals and rosy stigma was observed in southern Jeff Davis County not far from the known diploid collection. We suspect that the ploidy levels in *O. davisii*, or at least the Trans-Pecos and New Mexico populations, are correlated with consistent morphological differences and that two taxa should be recognized. According to David Ferguson (pers. comm.) the northern (i.e., tetraploids) *O. davisii* plants produce fruits but never seeds. We found three seeds in one fruit of the confirmed tetraploid *O. davisii* population from southeastern New Mexico, but four other fruits from the same plant were devoid of seeds. Anthony (1956) reported that *O. davisii* plants of the central Trans-Pecos region, presumably the diploids, produce a few seeds in each fruit.

Synonyms. *Cylindropuntia davisii* (Engelm. & Bigelow) F. M. Knuth; *Opuntia tunicata* (Lehm.) Link & Otto var. *davisii* (Engelm. & Bigelow) L. D. Benson.

Common Names. Abrojo; Davis cholla; Davis's opuntia; Jeffdavis cholla; thistle cholla.

7. **Opuntia imbricata** (Haw.) DC. TREE CHOLLA. Plates 46, 47. [*Cereus imbricatus* Haw., Suppl. Pl. Succ. 70. 1821]. Mountains and desert, throughout most of the Trans-Pecos, from El Paso Co. to Val Verde Co., most common from Brewster Co. to the west. 2,000–7,300 ft. Flowering (Apr–)May–Jun(–Jul). $2n = 22$. South-central TX, NW through the Panhandle. OK, KS, W to CO, through NM to SE AZ. Mexico, throughout the Chihuahuan Desert Region, S to near Mexico City. Map 6.

After Anthony (1956) described *O. imbricata* var. *argentea*, most workers

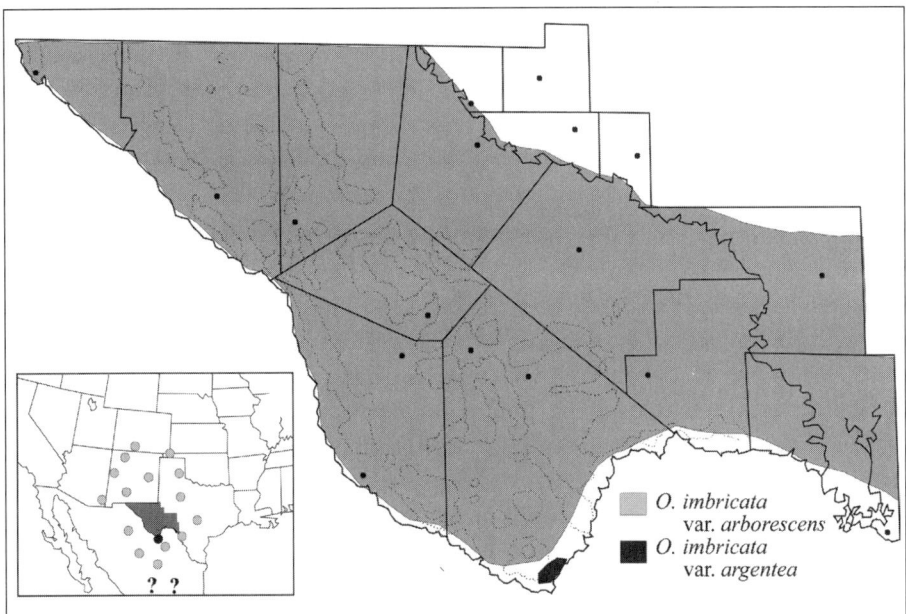

Map 6. Generalized distribution of *Opuntia imbricata* var. *arborescens* (tree cholla) and *O. imbricata* var. *argentea* (Big Bend cholla).

have recognized two varieties of *O. imbricata* to occur in the Trans-Pecos, and in the United States. Additional varieties occur in Mexico (Bravo-Hollis, 1978), one of which is grown in horticulture in the United States (Weniger, 1984). *Opuntia imbricata* is one of the most easily identifiable cacti in the Trans-Pecos, with its relatively large, many-branched shrublike or small treelike habit. The specific epithet alludes to the prominent overlapping appearance of the stem tubercles.

Key to the Varieties

1. Plants 1–2(–3) m tall; spine sheaths yellowish, tan, to silver; distribution widespread

    7a. *O. imbricata* var. *arborescens*.

1. Plants usually less than 1–1.2 m tall; spine sheaths silvery; distribution restricted to vicinity of Mariscal Mt

    7b. *O. imbricata* var. *argentea*.

7a. Opuntia imbricata (Haw.) DC. var. **arborescens** (Engelm.) A. D. Zimmerman, comb. nov. (forthcoming). TREE CHOLLA. Plates 46, 47. BASIONYM: (as cited by Benson, 1982, so requiring verification) *Opuntia arborescens* Engelm., Mem. Tour. N. Mex. 90. 1848. Mountains and desert, igneous and sedimentary substrates, in every county of the Trans-Pecos, most common from Brewster, Jeff Davis, and Presidio counties, W to El Paso Co. 2,400–7,300 ft. Flowering (Apr–)May–Jun(–Jul). $2n = 22$. South-central Texas N through the

Panhandle and W through the Trans-Pecos. Panhandle of OK, extreme SW KS, S CO, most of NM, rare and poorly documented in E and SE AZ. Mexico, common in the Chihuahuan Desert floristic region. Map 6.

The plants of O. *imbricata* var. *arborescens* are the tallest of any cactus species in the Trans-Pecos. Some rangeland sites in the Davis Mountains and elsewhere have become infested with thickets of var. *arborescens*, probably in most cases where human activities have facilitated the dissemination and growth of dislodged stem segments. The type locality is, for Engelmann's O. *arborescens* in effect, at Santa Fe, where the first specimens with definite locality data were secured (lectotype designation by Benson, 1982, p. 914), clearly the same variety common from Texas to Colorado. The varietal epithet, after the Latin *arbor,* "tree," and *-escens,* "becoming," is descriptive of the habit.

Identifying Characters. The arborescent cylindroid-stemmed var. *arborescens* is not likely to be confused with any other Trans-Pecos cactus species. The plants of var. *argentea* are smaller, densely silver-spined, and restricted in distribution. The stems with larger terminal joints 2–3 cm or more in diameter are much thicker than those of O. *kleiniae* and profoundly larger than those of O. *leptocaulis*. In southwestern New Mexico and adjacent Arizona, O. *imbricata* is geographically replaced by the more densely spined O. *spinosior* (Engelm.) Toumey, which it closely resembles, except for a few isolated reports. Apparently O. *spinosior* does not extend into Texas.

In the Trans-Pecos different spine forms of var. *arborescens* are recognizable. These include a darker-spined form in the central Trans-Pecos mountains and basins and in many surrounding desert habitats, and a silvery- or yellowish-spined form in populations of var. *arborescens* along and near the Rio Grande, from near Presidio in Presidio County west to El Paso County. In var. *arborescens* there are 6–17 central spines and 3–10 radial spines in the areoles of mature stems.

The flowers of var. *arborescens* are 5–6 cm long and 5–7.5 cm in diameter. The tepal color has been described as magenta, purplish, reddish-purple, rose-pink, and even lavender. Pale yellow or colorless flower-forms have been observed near Fort Davis. Here the outer tepals are yellowish, but the inner tepals are basically without pigments. The tepals in all flowers are ca. 3 cm long, 1.2–1.5 cm in diameter, and spatulate. The filaments are 0.8–1 cm long, greenish to greenish-red proximally, and reddish distally. The anthers are yellow to cream-yellow and ca. 2 mm long. The style is 2.5–2.9 cm long, 2–3 mm thick, and cream to reddish. The stigma is cream in color, with 5–8 lobes. The pericarpel is tuberculate with 2–3 whitish bristles in the areoles. Subulate leaves are borne in the upper areoles of young stems.

The fruits are yellow, somewhat fleshy at maturity, spineless, turbinate, short-cylindric, or hemispheric, 2.5–4.5 cm long, 2.2–3.5 cm in diameter, strongly tuberculate, with a deep umbilicus, and they readily proliferate, especially after being detached while still green. The tan seeds are discoid, 3–3.5 mm in diameter, 1.5 mm thick, with a narrow, slightly beaked aril-rim.

Phenology. Plants of var. *arborescens* bloom from late April through May and into June, and occasionally as late as July. Near Alpine the typical peak blooming period appears to be in late May. After a late freeze in 1997, however, the peak bloom was about 18 June, and after severe drought in 1998 and 2000, there was a second major bloom beginning in mid-July, three weeks after rain. Near Fort Davis in 1996, plants of var. *arborescens* bloomed in May and then were back in bud by 30 June following rains (Linda Hedges, pers. comm.). There are 2–6 flower buds borne at the apexes of ultimate joints, and one or more of these open during the same day. Flowers usually open in late morning and last one day. Some plants, at least in the Davis Mountains, may have fully open flowers before 9:00 A.M. It is not known if these flowers opened late the previous day and remained open or only partially closed all night, or if some flowers open very early. The pollination biology of *O. imbricata* has been studied in a population in southern Colorado (McFarland et al., 1989). Fruits mature in 2–3 months, as indicated by green turning to yellow, and they are persistent on the plants through the fall and winter so that the previous fruit crop may be present with the current crop (Anthony, 1956; Fraser and Pieper, 1972).

Sterile and Immature Specimens. Typical mature plants of var. *arborescens* have a short trunk ca. 10 cm or more long, 7–10 cm in diameter, and exhibit rough bark. In presumably older plants the trunk may be 25 cm across. At least three primary branches from the trunk give rise to verticillate secondary branches at the nodes. Usually the plants are 1–2(–3) m tall with thick branching crowns. In the Trans-Pecos, plants exceeding 3 m high occasionally are encountered. Woody skeletons of branches are evident in most populations. The skeletons are distinguished by a pattern of relatively long and wide vascular gaps (Anthony, 1956). Larger terminal joints, usually 10–25 cm long and 2–3 cm in diameter, exhibit prominent tubercles. The tubercles are 2–4 cm long, 4–9 mm wide, and 6–12 mm high. Usually 3–4 rows of tubercles are visible in side view of the stem. The areoles are 1.2–2 cm apart, elliptic, ovoid, obovate, to obspatulate, apparently enlarging with age, 5–6.5 mm long, ca. 4 mm across, felted, mounded, dirty-white or gray to tan or yellowish, or gray-felted in front and yellow-mounded in the upper portion. Usually the spines are located in the lower one-half or more of the areole, with a spineless felted area between the spines and a tuft of short (1–2 mm long) brown glochids rimming the top of the areole. Circular yellow patches or white, gray, or dark "nubs," 10–20 or more, often are evident at the upper margins of areoles. These are extrafloral nectaries, now well known to students of large cholla cacti in the Sonoran Desert (Pickett and Clark, 1979); viscid spherical droplets often are evident in the areolar region, and actively patrolling ants have been observed on some stems.

Usually central spines are easily distinguished from the radials. The central spines are positioned in the lower half of the areole, when they are 7–8 in number, or may occupy nearly the whole areole when 13–17 in number. The centrals are straight, diffusely spreading, with the lowermost 1–3 usually deflexed, the largest of the lowermost at 2–3 cm long and 0.5–0.8 mm in diameter. The centrals are acicular, or perhaps somewhat flattened, particularly the lowermost one, with retrorsely scabrous tips. The number and color of central spines varies from population to population in the Trans-Pecos. Central spines may be silvery proximally and stramineous, tan, pale reddish-tan to dark red-brown distally, or red-brown throughout. The sheaths are loose-fitting, ca. 1.5 mm in diameter, and persist for one year, more or less. In different populations the sheaths vary in color from silver to pale yellowish to tan proximally, and yellowish distally. The central spine pattern varies from population to population, but typically the spines are evenly distributed, positioned at the margins and in the center. Usually there are fewer central spines in apical or younger areoles and more spines with age. Younger areoles on terminal joints in some mountain populations may have 2–3 centrals and 2–3 radials, or perhaps just one spine in whole areoles.

Radial spines are positioned at the lower periphery of each areole. The radials usually are deflexed, subterete or angular, slender, ca. 0.3 mm wide, whitish to gray, and perhaps black-streaked or -speckled. The lower central radial usually is the longest at 1.2–2 cm. In most populations there are five radials, but in others there are 3–10, perhaps with five larger (to 1 cm or more long) and five smaller (ca. 2 mm long), or seven radials of gradually diminishing size from the base toward the top of the areole.

Anthony (1956) noted that seedlings of this variety were abundant by comparison with those of other cactus species. Seedlings are slender and erect, with slender conical-cylindroid leaves and 8–15 monomorphic, whitish, bristlelike spines in felty circular areoles ca. 1.5 mm in diameter. Vegetative reproduction follows the formation of adventitious roots from fallen stems, detached joints, or even fruits.

Biosystematics. Plants in the Davis Mountains south to the Chisos Mountains usually have darker spines and less conspicuous sheaths. In aspect the stems are green, not at all dominated by the spines. The green-stemmed aspect appears to be the result of darker, often shorter and fewer spines in relatively widely spaced areoles on large tubercles. Plants from southern Brewster County along and near the Rio Grande northwest to Culberson, Hudspeth, and El Paso counties have a bristly-stemmed habit, this seemingly the result of more closely spaced areoles on smaller tubercles and lighter-colored spines with conspicuous pale yellow to dirty-white or nearly silver sheaths. The stems are dominated by the pale spines. Although the bristly-stemmed form tends to have more spines per areole, some of the green-stemmed plants have as many spines, at least in some areoles, and otherwise the spine characters of the two forms are much the same.

The bristly-stemmed form of var. *arborescens*, particularly the plants with nearly silver spines, closely resembles O. *imbricata* var. *argentea* in spination. The occurrence of the bristly-spined form and var. *argentea* is correlated with hot, dry environmental conditions. The bristly-spined form extends from the desertic western Trans-Pecos southeast below the Sierra Vieja rim near the Rio Grande to southern Presidio County, while var. *argentea* is restricted to one of the harshest desert habitats in the northern Chihuahuan Desert Region.

The var. *arborescens* is reported to hybridize with O. *spinosior* ([as var. *imbricata*, in part] Pinkava in Zimmerman et al., forthcoming; Benson, 1982) where they are sympatric in southwestern New Mexico, adjacent Arizona, and northern

Chihuahua, Mexico, but this has not been explicitly documented; O. *imbricata* is extremely rare within the range of O *spinosior*, typically seen as isolated weeds invading from the east along highways and near corrals. The stems of O. *spinosior* are densely spiny, partly the effect of closely positioned areoles on relatively small tubercles. We have seen no evidence that either var. *argentea* or the bristly-stemmed form of var. *arborescens* in Trans-Pecos Texas is a product of ancient introgression from O. *spinosior*, but this might explain their spininess.

David Ferguson (pers. comm., 1998) plausibly asserts that O. *imbricata* in the Trans-Pecos and in most of the CDR should be recognized as var. *arborescens* Engelm. and not var. *imbricata*. Ferguson's concept of var. *imbricata* is restricted to taller Mexican form(s) with relatively smooth, juicy, sweet fruits that vary from the usual yellow to orange, or even reddish, distributed from near Laredo, Texas (but Weniger found it only in horticulture there), to east of Cuatro Ciénegas in Coahuila, Mexico. These may have smaller flowers than var. *arborescens*, but the vegetative parts are more robust than either of our northern varieties. Surprisingly, no one except for Weniger seems to have published this combination, and he failed to cite basionyms, so his recombinations do not officially "count."

Synonyms. *Cereus imbricatus* Haw.; *Cactus cylindricus* Lam.; *C. bleo* Kunth; *Opuntia imbricata* (Haw.) DC.; *O. decipiens* DC.; *O. exuviata* DC.; *O. cylindrica* DC.; *O. rosea* DC.; *Cylindropuntia imbricata* (Haw.) F. M. Knuth; *Opuntia imbricata* (Haw.) DC. var. *arborescens* Weniger, *nom. nud.*; *O. lloydii* Rose; *O. imbricata* (Haw.) DC. var. *lloydii* (Rose) Bravo; *Cactus imbricatus* Lem.; *Opuntia vexans* Griffiths; *O. imbricata* var. *vexans* (Griffiths) Weniger; *O. magna* Griffiths; *O. spinotecta* Griffiths.

Common Names. Tree cactus; walkingstick cholla; cane cactus; cholla; candelabrum cactus; coyonostle; coyonostli; coyonoxtle; coyonostole; chain-link cactus; tasajo; vela de coyote; velas de coyote; coyote candles; entrena; cardon; abrojo; cardenche; xoconostle; joconostli. Cane cholla is used in the Trans-Pecos as the common name of var. *arborescens*, but the name "cane cholla" also is well established in reference to O. *spinosior*.

7b. **Opuntia imbricata** (Haw.) DC. var. **argentea** M. S. Anthony, Amer. Midl. Naturalist 55: 225. 1956. BIG BEND CHOLLA. Plates 48, 49. Mariscal Mt, mostly on N- and W-facing limestone slopes, fewer on alluvial Rio Grande plain in *Prosopis* thickets, W of Solis Ranch, and NE to Rooney's Place, Big Bend National Park, extreme S Brewster Co. 2,000–2,400 ft. Flowering early Apr. $2n = 22$. Mexico in adjacent Coahuila and Chihuahua. Map 6.

In the Trans-Pecos, var. *argentea* is almost restricted to its type locality, Mariscal Mountain, Big Bend National Park, Brewster County. The varietal epithet alludes to the silvery aspect of the spine sheaths and the plant in general, after the Latin *argenteus*, "of silver."

Identifying Characters. In its natural habitat the most distinctive aspect of var. *argentea* is its dense clothing of silvery spine sheaths, contributing to a generally gray aspect to the entire plant. The stems are relatively densely and uniformly covered with spines because the areoles are comparatively close together. In habit the plants of var. *argentea* resemble the "bristly-stemmed" form of var. *arborescens,* particularly the bristly-stemmed plants with more silvery spines. The var. *arborescens* does not occur sympatrically with var. *argentea,* but a rare downhill waif from a Chisos Mountains population of var. *arborescens* was seen along River Road West in the 1980s, 0.6 miles east of the turnoff to Loop Camp (A. Zimmerman, field notes). The plants of var. *argentea* are the smallest in the species, tending to be "chubby" (Anthony, 1956), usually less than 1.2 m tall, to ca. 1 m wide.

The dark magenta flowers are 4.5–5 cm long and 5 cm across, with the tepals in four whorls or fewer. The outer tepals are pinkish with olive midregions, oblong, and apiculate. The inner tepals are deep or dark magenta, or reddish-purple, usually darker than in var. *arborescens.* The inner tepals are broadly spatulate, to 2.5 cm long, 1.5 cm in diameter, and apiculate at the apex. The filaments are deep reddish-purple and 7–9 mm long. The anthers are light to cream-yellow and 1.5 mm long. The style is reddish-purple, to 1.8 cm long, bulbous above the base, and supports 7–9 cream-colored stigma lobes to 5 mm long. The pericarpel is short-conic, truncate, ca. 1.5 cm long, 1.5 cm in diameter, tuberculate, with 1–2(–5) white, caducous bristles or slender spines to 1.5 cm long in each areole. The uppermost areoles bear subulate leaves on young stems.

The ripe fruits are yellow, turbinate, 2.5–3(–4) cm long and 1.8–2.3(–3.2) cm wide, with a deep umbilicus, and strongly tuberculate. The areoles are circular, ca. 3 mm in diameter, densely covered with gray-white felt, with a compact row of glochids along the upper margin, these ca. 1 mm long and pale yellow. The seeds are tan, discoid, ca. 3 mm across, 1–1.5 mm thick, with a narrow, slightly beaked aril.

Phenology. The principal flowering period for var. *argentea* appears to be in early April (Anthony, 1956). Cultivated plants in Alpine have bloomed in April, May, and mid-July. The July flowers were produced on plants that were in full bloom in early to mid-May, following a dry spring and moderate rain in late June. The flowers open about midday, close at night, and do not open again the next day. Green immature fruits turn yellow in 2–3 months, and they persist colorfully on the plants for several months after ripening.

Sterile and Immature Specimens. The plants of var. *argentea* are erect shrubs with spreading branches and fibrous roots. The short trunk, usually less than 8 cm thick, bears a few or many branches. The ultimate stem segments reach 20 cm long and 1.5–4 cm in diameter. The tubercles are ca. 2 cm long, 0.5–1.2 cm wide, protrude ca. 0.5 cm, and usually are silvery-green or gray. The relatively closely set areoles (2 cm or less apart) are ovate-elliptic, 5–7 mm long, 3–4 mm wide, densely felted, with the felt pale yellow and turning gray with age.

In var. *argentea* there are 11–23 spines per areole in all but the lowermost areoles. The spine sheaths are ca. 0.8 mm in diameter and silvery-white, turning gray with age. In each areole there are 6–14(–16) central spines, these acicular or slightly flattened (and elliptic in cross section), 2–3 cm long, and 0.5 mm in diameter. Older centrals are silvery to ivory with pinkish bases, sometimes with pale yellow or pale reddish tips. Young centrals are white or pinkish with a greenish base. The lowermost centrals are deflexed, often the lower 1–4 spines, and others are diffuse. The 5–7 radial spines are even more slender than the centrals, are 1.8–2 cm long, white, and are deflexed or spreading. The radials are positioned at the lower margin of the areole. Glochids are compactly arranged at the upper margin of the areole, these whitish with green bases when young, pale yellow and ca. 1 mm long in older areoles.

Immature specimens of var. *argentea* have not been available for evaluation. We suspect that in its harsh habitat the most successful mode of reproduction for var. *argentea* is asexual from detached joints.

Biosystematics. The var. *argentea* differs from var. *arborescens* by its shorter habit, silvery spines densely clothing the stems, darker tepals, and restriction to a different habitat. The var. *argentea* is obviously different from the "green-stemmed" mountain form of var. *arborescens,* but it closely resembles the "bristly-stemmed" forms that replace var. *argentea* a short distance upstream from southern Presidio County along and near the Rio Grande northwest to El Paso County. The spine sheaths and spines of var. *argentea* consistently are silver and white, while the sheaths usually have a yellowish hue in the bristly-stemmed form of var. *arborescens.* The spine sheaths are silver in some populations of var. *arborescens,* particularly in the southern portion of its range. We suspect that var. *argentea* originated from an isolated population of the bristly-stemmed form of the common variety.

Plants of var. *argentea* have been cultivated at several sites in Alpine. They vigorously survived several winters at 4,400–4,600 feet above sea level. Cultivated plants of var. *argentea* maintain their distinctive features, and they are easily delimited from all forms of var. *arborescens* except for the silver-spined, bristly-stemmed form.

Synonyms. None.

Common Names. Big Bend cane cholla; silver-spine cane cholla; silver tree cholla. The "silver cholla" of the Sonoran Desert is *O. echinocarpa* Engelm. & Bigelow.

8. **Opuntia leptocaulis** DC., Mem. Mus. Hist. Nat. 17: 118. 1828. CHRISTMAS CHOLLA. Plates 50–53. Widely distributed in desert and semidesert habitats, every county of the Trans-Pecos. 1,900–5,000 ft. Flowering (May–)Jun–Sep. $2n = 22, 44$. Throughout most of TX, especially western two-thirds. Southern OK, NM, and AZ. Mexico S to Puebla. Map 7.

The habit of *O. leptocaulis,* growing among the other low shrubs, has been discovered inadvertently by many hikers in the Trans-Pecos. The type locality is "In Mexico." The common name is taken from the tendency of the copious red fruits to persist on plants through the winter. The specific epithet appropriately refers to the slender stems of this species, after the Greek *leptos,* "slender," and *caulis,* "stem."

Identifying Characters. The plants of *O. leptocaulis* are compact low shrubs with many primary stems, or erect, subarborescent shrubs with one or two pri-

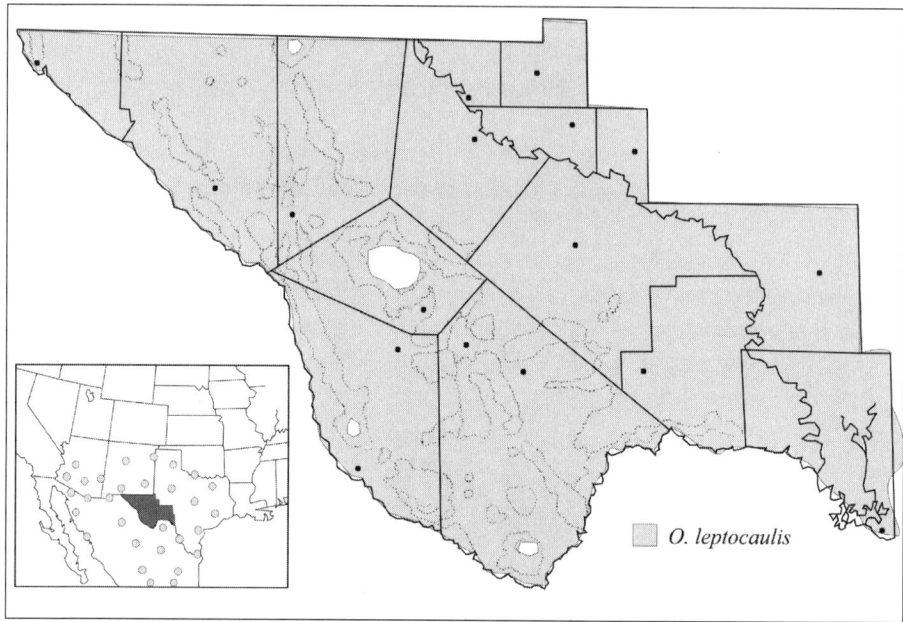

Map 7. Generalized distribution of *Opuntia leptocaulis* (Christmas cholla); short-spine and long-spine forms.

mary stems. Pencil-thin secondary branches arise ca. 8 cm above the ground, often contributing to an intricately branched habit. Typically, there is a single spine in each areole along the stems. The ultimate joints usually are spineless. The plants of O. *leptocaulis* are not likely to be confused with any other cactus species of the Trans-Pecos, except perhaps dehydrated plants of O. *kleiniae* or hybrids between O. *leptocaulis* and O. *kleiniae*.

The greenish-yellow flowers of O. *leptocaulis* are borne on the distal halves of longer joints. The flowers are 1.5–2 cm long and 1–2.2 cm in diameter. The tepals open widely, and the flowers are subrotate. The inner tepals are greenish-yellow, oblong, and pointed. The outer tepals are greenish-yellow with spinose tips. The outer bracts are linear-subulate and dark red. The anthers are ca. 1 mm long, pale yellow to cream-yellow, and the filaments ca. 6 mm long and pale green to yellow-green. The style is ca. 9 mm long, and the 3–6 short, thick, stigma lobes are cream to cream-white. The pericarpel is ca. 1.5 cm long and ovate to conical, not or only slightly tuberculate, with brown glochids in the areoles.

The fleshy fruits are bright red, less often pale to medium red, or rarely yellow, obovoid, pyriform, clavate, or obconic, 1.4–2.4 cm long, 0.8–1.3 cm in diameter, with a usually shallow umbilicus. Plants with yellow fruits occur rarely in south Texas (Benson, 1982), but yellow fruits have not been reported in the Trans-Pecos. The fruit surface is smooth between scattered circular to oval areoles, these 1–1.5 mm in diameter, felted, and often with a few or a tuft of brown glochids at the upper margin, occasionally also with 0–4 glochidlike bristles. Only the fruit wall is red, and the cavity is filled with seeds. The seeds are tan, discoid, angular, 3–4 mm in diameter, and ca. 1.5 mm thick, with a slender aril. Fruits may proliferate, especially while still green.

Phenology. The blooming period for O. *leptocaulis* may extend from May through September, but usually flowering occurs from June to August. It appears that blooming is influenced by precipitation. The timing of anthesis in O. *leptocaulis* requires additional study. Anthony (1956) reported that flowers do not open until late afternoon and do not close until late at night. This seems to be the usual pattern in the Trans-Pecos, although in 1995 Clark Champie (pers. comm.) observed different flowering times among short-spined plants in the Franklin Mountains (see following discussion). The flowers of one short-spined type opened at noon and closed about 7:00 P.M., and in the other short-spined type, the flowering time was 5:00–9:00 P.M. Other than the flowering pattern, according to Champie, the short-spined types were indistinguishable. Under cultivation in Alpine, the flowers of several long-spined and short-spined plants do not open until well after midday, probably about 2:00–3:00 P.M. on hot, sunny days. After closing, the flowers of O. *leptocaulis* do not open again the next day. The fruits ripen in about five months and are persistent, remaining on the plants through the winter, and potentially through the next summer, a year or more after formation.

Sterile and Immature Specimens. In the Trans-Pecos the plants of O. *leptocaulis* are many-branched, compact or straggling shrubs. The primary stems usually are 10–40 cm long, sometimes 8–9 mm thick, and give rise to ultimate stem segments that typically are 2–8 cm long and 5–6 mm or 3–5 mm in diameter. Each ultimate segment arises from the upper half of an areole, usually extending at right angles but may curve up sharply. Characteristically, the stems of O. *leptocaulis* are yellow-green, but plants with darker green stems are not uncommon. Often purple pigmentation is evident on the stems, especially in winter, particularly below the areoles. Woody stem skeletons exhibit small vascular gaps, these oval and surrounded by relatively thick wood (Anthony, 1956). The areoles are circular or oval, 2–3 mm across, often with shorter gray felt in the front portion of the areole and longer yellow felt in the back. The glochids are 2–3 mm long, erect, reddish-brown, positioned in large tufts at the back or upper margin of the areoles or in small tufts or single at the areole margin, or sometimes glochids emerge from the felt.

In the Trans-Pecos O. *leptocaulis* is characterized by a single central spine in each areole. There are no radial spines. The ultimate segments, and sometimes whole plants, often are spineless. Rarely there may be two spines (possibly three; Weniger, 1984) in some areoles. The single spine is borne in the front center of the areole, and in angle it is either deflexed or perpendicular (porrect). The spines are acicular or slightly flattened mostly on the upper surface. Two spine forms of O. *leptocaulis* are recognized in the present tentative treatment (see Biosystematics), one of them referred to here as the "long-spined" form and the other as the "short-spined" form. In the long-spined form the spines are at first perpendicular to the stem, later sometimes slightly deflexed, to 4.8 cm long, and to 0.8–1 mm in diameter. The spines are gray proximally, yellow at the tip, perhaps with a zone of reddish-brown below the yellow tip. The spines are conspicuously sheathed. The sheath is relatively loose, 1.4–1.9 mm in diameter, golden-yellow the full length, or silver proximally and golden distally. Typically, the long-spined form has slightly thicker stems than the short-spined form, with ultimate segments usually 5–6 mm or 2–5 mm in diameter. In the short-spined form the spines tend to be deflexed or perpendicular, (0.35–)0.6–3.2 cm long, and 0.3–0.4 mm in diameter. The spines are gray proximally, reddish-brown distally, then yellowish for 2–4 mm at the retrorsely scabrous tips. The spines are not as conspicuously sheathed in the short-spined form as in the long-spined form. Here the sheath may be absent on fully formed spines, and rather closely fitting and fugacious on other spines. When present, the sheaths are pale yellowish or silver proximally and golden-yellow distally, or golden throughout. The spine forms in Trans-Pecos populations are readily distinguished by spine length and diameter and by the sheaths.

Because of its slender stems *O. leptocaulis* is easily distinguished from other Trans-Pecos chollas, except for *O. kleiniae*. The largest turgid stems of *O. leptocaulis* approach the diameter (ca. 6 mm) of small stems of *O. kleiniae*. Vegetatively, the species are distinguished by habit, stem size, stem color, and spine number in populations away from the Rio Grande. *Opuntia kleiniae* along the Rio Grande typically has one spine per areole, as does *O. leptocaulis,* whereas *O. kleiniae* elsewhere in the Trans-Pecos exhibits 1–4 spines per areole, plus bristle-like radial spines.

Juvenile plants of *O. leptocaulis* typically have shorter spines, 2–7 mm long, than some adults. These spines have tight-fitting sheaths. In addition to the single central spine in younger areoles, there are 1–7 reddish-brown bristles in the position of radial spines, at the lower and lower peripheral margins of the areole. These bristlelike radials are 1–3 mm long, erect, and resemble the glochids in size and color, but they are not retrorsely scabrous like the glochids. The glochids are positioned at the upper margin of the areoles.

Biosystematics. The "strikingly different" (Benson, 1982, p. 345) long- and short-spined forms of *O. leptocaulis* have been given different names, including "longispina" and "brevispina," since the middle 1800s. In her treatment of Texas Opuntiae, Anthony (1956) regarded the two spine types as varieties of *O. leptocaulis*. Most recent workers have recognized a single taxon of *O. leptocaulis* in the United States (Benson, 1982; Weniger, 1984) and in the CDR (Zimmerman et al., forthcoming) presumably because the spine forms are sympatric, often occurring in mixed stands, and because the existence of intermediate forms had been reported (Benson, 1982). So far our observations of *O. leptocaulis* in the Trans-Pecos have not revealed intermediates between the long- and short-spined forms. Instead, our preliminary investigations (A. M. Powell, unpub.) suggest that the spine forms may be isolated cytologically, where the long-spined form is tetraploid ($2n = 44$) and the short-spined form is diploid ($2n = 22$). Jack Brady (M.S. thesis research) also found tetraploid and diploid cytotypes in Trans-Pecos *O. leptocaulis*. He concluded that ploidy level differences were somewhat correlated with slight differences in stem diameter (larger in tetraploids) but not with spine length. Previous studies (Fischer, 1962; Pinkava et al., 1973, 1977, 1985,

1992; Weedin and Powell, 1978) collectively suggested that *O. leptocaulis* was tetraploid in the Chihuahuan Desert Region and diploid in the Sonoran Desert region, but the report of $2n = 33$ (Pinkava et al., 1992) for *O. leptocaulis* in Arizona suggests that tetraploids might also occur in the Sonoran Desert region. Diploid counts now have been obtained for numerous populations of *O. leptocaulis* in the Trans-Pecos, as well as additional tetraploid counts, and so far the chromosome number seems to be correlated with the spine form (Powell and Weedin, 2001). If future studies substantiate the ploidy level difference between long- and short-spined morphotypes of *O. leptocaulis,* and the general absence of fertile intermediates, there would be ample justification for interpreting them as distinct species, but we do not know which one of them (if either) is the "real" *O. leptocaulis* that De Candolle named in 1828.

Opuntia leptocaulis hybridizes with *O. kleiniae* (or backcrosses with it depending upon semantics, if *O. kleiniae* itself is considered a hybrid) in the Davis Mountains (Anthony, 1956; Fischer, 1962; J. Brady, unpub.) and in Mexico (Pinkava and Parfitt, 1982). *Opuntia leptocaulis* also hybridizes with *O. spinosior* (Pinkava et al., 1985) and practically every other species of tall cholla cactus in Arizona (Marc Baker, pers. comm. to A. Zimmerman).

Synonyms. *Cylindropuntia leptocaulis* (DC.) F. M. Knuth; *O. vaginata* Engelm.; *O. frutescens* Engelm. var. *brevispina* Engelm.

Common Names. Desert Christmas cactus; Christmas cactus; pencil cactus; pencil cholla; tasajillo; tasajilla; tesajillo; tesajo; catalinaria; alfilerillo; slender stem cactus; aguijilla; garambullo. The most appropriate English names for *O. leptocaulis* are Christmas cholla and desert Christmas cactus. We prefer Christmas cholla because related species of *Cylindropuntia* in general are called "chollas," and the species is not restricted to the deserts. Other common names for *O. leptocaulis* are less desirable for the following reasons: "Christmas cactus" is the well-established name for *Schlumbergera truncata* (Haw.) Moran; "tesajo" is applied to other small chollas, including *O. tesajo* Engelm. ex J. M. Coult. of Baja California; "pencil cholla" is the well-established English name for *O. arbuscula* Engelm. of the Sonoran Desert; "garambullo" more commonly is used for *Myrtillocactus geometrizans,* a large columnar cactus in Mexico. In Mexico the name most commonly applied to *O. leptocaulis* (and *O. kleiniae*) is "tasajillo," and this name also is widely used in the Trans-Pecos.

9. **Opuntia kleiniae** DC., Mem. Mus. Hist. Nat. 17: 118. 1828. CANDLE CHOLLA. Plates 54, 55. Rocky slopes or alluvial flats, typically with *Prosopis* and other shrubs, Davis Mts, and rocky slopes, alluvial flats, or sandy loam, Rio Grande floodplain. Davis Mts in Jeff Davis and NW Brewster counties, SE Brewster Co. along and near the Rio Grande to El Paso Co.; also N Culberson Co. 1,800–5,000 ft. Flowering (Apr–)May–Aug. $2n = 33, 44$. Central TX (Eastland and Lampasas counties; Diggs et al., 1999). Scattered introductions in central and western OK (Benson, 1982). Central and SE NM, in the region of overlap between *O. leptocaulis* and *O. imbricata.* Mexico, S to Hidalgo (probable source of DC's original material, *O. kleiniae,* sensu stricto). Map 8.

A taxon similar to *O. kleiniae,* sometimes called *O. kleiniae* var. *tetracantha*

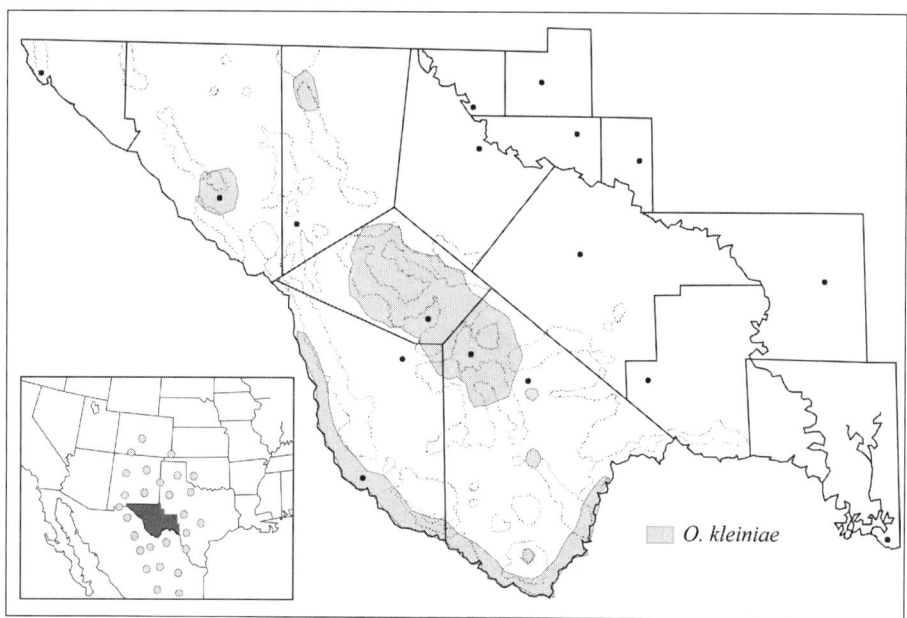

Map 8. Generalized distribution of *Opuntia kleiniae* (candle cholla), including the "Rio Grande *kleiniae*."

(Toumey) W. T. Marshall, is based upon Arizona plants, presumably from hybridization of *O. leptocaulis* with a western species instead of *O. imbricata*. The type locality of *O. kleiniae* is "In Mexico." According to Anthony (1956), the specific epithet was inspired by the resemblance of *O. kleiniae* to *Cacalia kleinia* L., a succulent in the family Asteraceae (Crook and Mottram, 1999).

Identifying Characters. The plants of *O. kleiniae* are larger than those of *O. leptocaulis* and smaller than those of *O. imbricata* (and, in Arizona, diverse other products of hybridization between *O. leptocaulis* and the larger chollas). In fact, *O. kleiniae* exhibits complete morphological intermediacy between *O. leptocaulis* and *O. imbricata*. The ultimate branches of *O. kleiniae* are more or less the size of a fountain pen (Benson, 1982) as opposed to pencil-size in *O. leptocaulis*, and broom handle–size (or larger) in *O. imbricata*. In *O. kleiniae* typically there are 1–4 central spines per areole, compared to one central spine per areole in *O. leptocaulis* and 6–17 centrals in *O. imbricata*.

The pinkish, purplish, to magenta or purple-tinged flowers of *O. kleiniae* contrast sharply with the smaller yellow-green flowers of *O. leptocaulis*, and they are smaller and usually paler than the larger magenta flowers of *O. imbricata*. In *O. kleiniae* the flowers are 3–3.5 cm long and 2.5–4(–5) cm in diameter. There are floral differences between the two forms of *O. kleiniae* in the Trans-Pecos. The Davis Mountains populations have larger flowers with the inner tepals pinkish, purplish, to purple-red or magenta. Populations along and near the Rio Grande typically have smaller flowers that are not brightly colored, but instead the inner tepals often are whitish-pink, or greenish-cream and maroon-tinged, more intensely so distally, or exhibit some other combination of pale greenish and cyanic colors. In flowers of *O. kleiniae* there are about five inner tepals and about five outer tepals with another series of outer bracts. The inner tepals are cuneate, 1.2–2 cm long, 0.5–0.9 cm in diameter, and truncate, rounded, or acute at the apex. The filaments are greenish to pinkish-red, and anthers are yellow and ca. 1.5 mm long. The style is pale green to pinkish or pale maroon, 1–2 cm long, and ca. 1.5 mm in diameter. The 5–7 stigma lobes are whitish or cream to pinkish, ca. 3 mm long and thick, the lobes persistently folded together in a capitate cluster.

The fully ripe fruits at maturity are red to red-orange, fleshy, obovoid, ovate, to subclavate, tuberculate or smooth, 2–3.5 cm long, and 1–2.2 cm in diameter. The fruit areoles have glochids and sometimes short, whitish bristles, but they are spineless. The fruits are persistent through the winter and proliferous when in contact with a favorable substrate. The seeds are tan, irregularly discoid, 3.5–5 mm in diameter, and 1–2 mm thick. We have observed different fruit morphologies in the two forms of *O. kleiniae*. Plants in the Davis Mountains typically have tuberculate fruits (although not comparable to those of *O. imbricata*), and the Rio Grande plants have smooth or inconspicuously tuberculate fruits. So far we have carefully observed the fruits of the Rio Grande form only, near Santa Elena Canyon and Heath Canyon.

Phenology. The principal flowering period for *O. kleiniae* in Texas is from late May through August. Probably the main flowering episodes are earliest in the season, May and June, in years when there is ample spring precipitation. This blooming pattern is particularly true for the Davis Mountains populations of *O. kleiniae*. In the Rio Grande populations the main flowering period may be earlier. We have observed blooming in Rio Grande *O. kleiniae* on 20 April near Castolon in Big Bend National Park and on 23 May near Heath Canyon. Under cultivation in Alpine, both the mountain and the Rio Grande *O. kleiniae* have bloomed in early to mid-May. The flowers of *O. kleiniae* open about midday to mid-afternoon, close at night, and do not open again the next day. Fruit maturation probably requires 1–3 months, and the fruits tend to persist through the winter and late into the next spring.

Sterile and Immature Specimens. Plants of *O. kleiniae* vary somewhat in habit and are 1–2.5 m tall. Some plants are sparingly branched, while others are profusely branched from a woody trunk to 4 cm in diameter. Plants of *O. kleiniae* in the Davis Mountains tend to reach maximum height when partially supported by other vegetation or by fences. The plants occur singly or form thickets of several plants together. Plants of *O. kleiniae* along the Rio Grande also may occur singly or in clusters. On the river delta near Santa Elena Canyon we have observed one thicket of *O. kleiniae* that was perhaps 50 m long and 30 m or more wide, comprising few-branched to many-branched plants 1–2 m tall. The ultimate segments are 4–25 cm long, 0.6–1.3 cm in diameter, and rather easily detached. The segments are verticillate, with as many as four arising at each node and then gradually curving upward. The branches are gray-green, often somewhat darker green in the Davis Mountains populations, with purplish pigment around the areoles and perhaps widely distributed on the stem surface. In

O. kleiniae the vascular gaps are oval, relatively short and wide, in stem skeletons (Anthony, 1956). The stem tubercles are low, 1.2–2.7 cm long, and 3–4.6 mm broad. The areoles are obovate, oval, or spherical, to 5 mm long and wide or 4.5 mm long and 3–4 mm in diameter. The areoles are densely felty, gray in front and yellowish toward the back, the yellowish posterior perhaps mounded, or the felt is yellowish throughout.

In Trans-Pecos populations of *O. kleiniae* we have examined, there are 1–4 central spines per areole. Other authors have reported 1–9 spines per areole (Anthony, 1956; Zimmerman et al., forthcoming). The central spines are loosely sheathed, with the sheaths ca. 1.5 mm wide, golden or silver proximally and golden distally, and early deciduous. The centrals are acicular, or flattened on the top or upper and lower surfaces, 1–3.3 cm long, 0.5–1.2 mm in diameter, gray, or gray proximally and red-brown distally, often with a translucent yellow tip. In the Davis Mountains populations, commonly there are four centrals per areole, these positioned in the front of the areole, with one lowermost and deflexed, two lower lateral, and the fourth in a middle position. All of the centrals usually are stout, of different sizes, and the lowermost and middle ones typically are the largest. Populations along the Rio Grande characteristically have one central spine per areole, 1–2 cm long and 0.5 mm in diameter, smaller than the lowermost deflexed central in the Davis Mountains populations. In this spine character the Rio Grande *O. kleiniae* resemble *O. leptocaulis*. In *O. kleiniae*, radial spines are absent, or in the Davis Mountains populations there are 1–3 radial bristles, these 0.5–3 mm long, slender, and borne in the lower periphery of the areoles. In the Rio Grande populations there are 3–6 "radial" spines to 4 mm long, these slender, pale yellow or gray, and borne in the front and front-lateral periphery of the areole. In the back portion of areoles in *O. kleiniae*, often there are 1–6 or more low, truncate "nubs," resembling those found in *O. imbricata*. At the back periphery of the areoles in *O. kleiniae*, usually there is a small tuft of glochids, these yellowish or brown and 1–2 mm long.

Dense stands of *O. kleiniae* frequently reproduce from seed underneath mature plants (Anthony, 1956), but seedlings were not studied during the current work. The leaves of young stems are cylindroid and 1–2 cm long.

Biosystematics. Fischer (1962) concluded that there were two distinct populations of *O. kleiniae*, the polyploid ($2n$ = 33, 44) *O. kleiniae* in the Chihuahuan Desert Region and the diploid ($2n$ = 22) var. *tetracantha* of the Sonoran Desert region. Subsequent chromosome counts (Pinkava et al., 1977, 1985; Pinkava and Parfitt, 1982; Weedin and Powell, 1978; Weedin et al., 1989; Powell and Weedin, 2001) have supported the ploidy level distinction between *O. kleiniae* populations in the Chihuahuan and Sonoran deserts.

Opuntia kleiniae is morphologically intermediate between *O. imbricata* and *O. leptocaulis,* and it is suspected of having arisen in multiple places and times through hybridization between these species. There is considerable evidence of interfertility between the subgenus *Cylindropuntia* species of the Chihuahuan and Sonoran desert regions. Hybridization was reported between the diploid *O. imbricata* var. *imbricata* and *O. spinosior* in Doña Ana County, New Mexico

(Pinkava et al., 1992), and between the diploid *O. leptocaulis* and *O. spinosior* in Pinal County, Arizona (Pinkava et al., 1985). In addition, Anthony (1956) described a hybrid between *O. kleiniae* var. *kleiniae* and *O. leptocaulis*. This hybrid combination was reported as triploid ($2n = 33$) by Fischer (1962; presumably in Texas) and Pinkava and Parfitt (1982; in San Luis Potosí, Mexico). Much additional unpublished information regarding putative hybrids and backcrosses involving *O. kleiniae*, *O. imbricata*, and *O. leptocaulis* has been obtained in separate but related studies in the Trans-Pecos by Carolyn Allred and Jack Brady. Allred's study, in the late 1980s, was mostly in the Davis Mountains of Jeff Davis and northern Brewster counties. Allred investigated a hybrid population resembling *O. kleiniae* that appeared to involve segregating later-generation hybrids and backcrosses to one or both putative parents. Allred obtained copious morphological and photographic evidence (Plates 56 and 57) of apparent complex hybridization and flower-color variation (pink to peach), but abandoned the project before completing her master's degree thesis. Brady continued and expanded the project in the mid-1990s to include populational data from throughout the Trans-Pecos and elsewhere in Texas, chromosomal investigations, and artificial hybridizations. Brady's data are as yet unpublished, but he did obtain significant evidence that *O. kleiniae* is of hybrid origin, located numerous hybrid plants of populations involving the Trans-Pecos *Cylindropuntia*, documented extensive diploidy and tetraploidy in Trans-Pecos *O. leptocaulis* and in putative backcrosses of *O. kleiniae* × *O. leptocaulis* and *O. imbricata* × *O. kleiniae*, and propagated artificial hybrids.

The population of *O. kleiniae* in the Rio Grande floodplain, from Brewster County northwest perhaps to El Paso County, is morphologically distinctive and appears to warrant formal taxonomic status (Weedin et al., 1989) as a separate variety. Tentatively, we refer to this apparently distinctive entity as the "Rio Grande *kleiniae*" (Plates 58 and 59). It differs from the "Davis Mountains *kleiniae*" by its slightly smaller plants and often smaller stems, more gray-green stems, single, smaller central spine in each areole, pale greenish-cream and maroon tepals, and nontuberculate fruits. The Rio Grande *kleiniae* appears to be tetraploid (Weedin et al., 1989; Weedin and Powell, 2001). We suspect that the Rio Grande *kleiniae* had the same kind of origin as did Davis Mountains *kleiniae* (i.e., hybridization between *O. imbricata* and *O. leptocaulis*) but that it is derived from a different hybridization event or a backcross. The single-spined Rio Grande *kleiniae* is similar in many characters to the *O. kleiniae* × *O. leptocaulis* described by Anthony (1956) and may have had similar origin. Further study of the *O. kleiniae* complex should involve more populations, especially in Mexico.

A small population of the Rio Grande *kleiniae* occurs away from the Rio Grande in the Chisos Mountains Basin campground, near campsite 19. This was the location of a Civilian Conservation Corps (CCC) camp that was active in the early years of Big Bend National Park. We suspect that the basin plants of Rio Grande *kleiniae* resulted from relatively recent introduction.

Synonyms. *Opuntia wrightii* Engelm.; *O. caerulescens* Griffiths; *O. perrita*

Griffiths; *O. recondita* Griffiths; *Cylindropuntia kleiniae* (DC.) F. M. Knuth; *C. caerulescens* (Griffiths) F. M. Knuth; *C. recondita* (Griffiths) F. M. Knuth; *C. recondita* (Griffiths) F. M. Knuth var. *perrita* (Griffiths) Backeb.

Common Names. Klein's pencil cholla; Klein pencil cholla; Klein cholla; candle cactus; tasajillo; tasajilla; tasajo; cardoncillo. Typically, we would recommend the use of "Klein's cholla" for any cylindropuntia having a proper-noun plus *-iae* ending, as if named for some Miss, Mrs., or Ms. Klein. However, because this particular "*kleiniae*" epithet has a different meaning, none of the "Klein"-based English names is valid.

10. **Opuntia rufida** Engelm., Proc. Amer. Acad. Arts 3: 298. 1856. BLIND PRICKLY PEAR. Plates 60, 61. Desert habitats, limestone and igneous, alluvium or rocks, flats, hills, canyons, mountain slopes. Hudspeth Co., S Quitman Mts; Culberson Co., Van Horn Mts; Presidio Co., below the rim, near the Rio Grande. Brewster Co., mostly S of Chisos Mts but N to near Agua Fria Mt, Chalk Bluff, and Black Gap. 1,800–4,000 ft. Flowering Apr–May. $2n = 22$. Mexico, Chihuahua, Coahuila, NE Durango. Map 9.

In the Trans-Pecos *O. rufida* is most common along south-facing cliffs in the hottest desert habitats near the Rio Grande in southern Brewster County, downstream at least to the Reagan Canyon drainage. The type locality is "About Presidio del Norte, on the Rio Grande." The specific epithet is after the Latin *rufulus*, "reddish," in reference to the tufts of reddish-brown glochids that dominate the spineless pads. The English name is associated with the blinding of cattle by dislodged glochids (Benson, 1982), corresponding to the Mexican name, *nopal cegador* (also applied to *O. microdasys*).

Identifying Characters. *Opuntia rufida* is the only truly spineless native

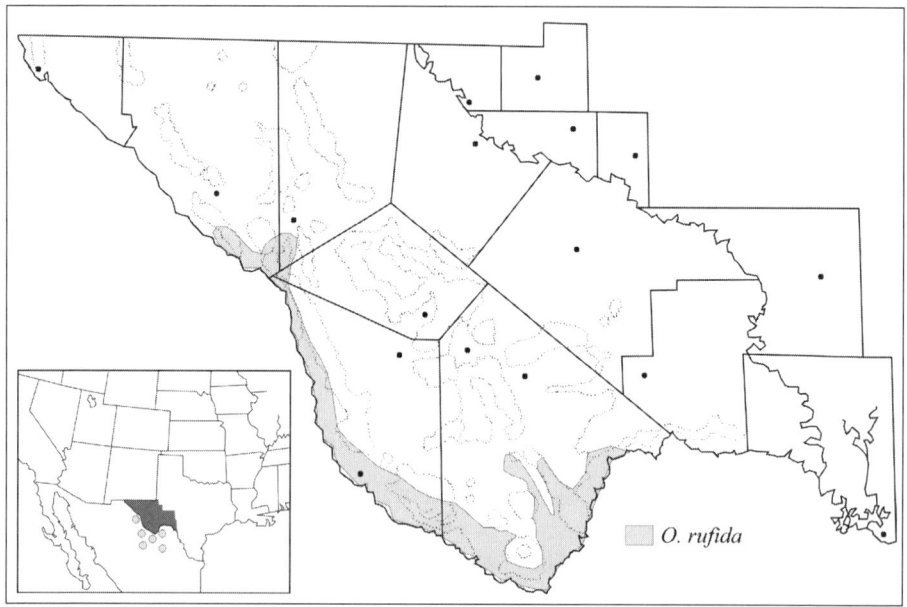

Map 9. Generalized distribution of *Opuntia rufida* (blind prickly pear).

Opuntia species in the Trans-Pecos. Others are basically spiny except for spineless freak individual plants, and there are clonal colonies of (100% spineless) *O. leptocaulis*. *Opuntia rufida* forms erect shrubs 0.5–1.5 m tall, with or without short trunks. The gray-green pads are circular, obovate, or elliptic, (7.5–)10–20(–25) cm long, and dotted with circular to oval, spineless areoles. The areoles are crowded with whitish or gray hairs and a bulging tuft of reddish-brown glochids. The areoles are 1.5–2.5 cm apart on the distal portion of the pad. The pad surfaces are densely short-pubescent (velutinous, i.e., velvety with straight short hairs) unlike any other native Texas *Opuntia*, a character that is detectable either (1) by touch (carefully between areoles) or (2) visually, using a powerful hand lens.

The yellow flowers of *O. rufida* are 5–7 cm long and 4.5–7 cm in diameter. The inner tepals are bright yellow to golden-yellow, or perhaps even orange-yellow (Weniger, 1984), without any red pigment at the bases. Orangish flowers probably are always the older ones. The inner tepals are obovate and erose, notched or apiculate at the apex. The outer tepals are linear to lanceolate. The anthers are yellow to cream, and the filaments are whitish. The style is 2–2.4 cm long, colorless, with 6–10 dark green stigma lobes. The pericarpel is closely low-tuberculate with prominent small, spherical areoles filled with white hairs and reddish-brown glochids.

The fruits are fleshy, subspherical to obovate, greenish-red to red, sometimes fading yellowish before turning red, 2–2.5 cm long, and 2–2.5 cm in diameter. The fruit surface has circular areoles 2–3 mm in diameter, these with white hairs and at least a few red-brown glochids. The fruit pulp is greenish. The seeds are discoid, irregular in outline, 2–3 mm in diameter, with a narrow arillate rim.

Phenology. The main blooming period for *O. rufida* appears to be in April, at least in populations in southern Brewster County. Flowering often extends into May. Plants cultivated in Alpine have formed large flower buds in early May with anthesis of the first flowers in mid-May. The flowers usually open about midday, close at night, and usually do not open again the next day, although in nature it is possible that late-opening flowers might open again the next day. The fruits begin turning reddish in June but may not be fully pigmented and ripe until late July or August.

Sterile and Immature Specimens. In desert habitats along and near the Rio Grande, *O. rufida* is one of the most easily recognizable prickly pears because of its spineless pads. In addition, the hemispheric areoles are closely spaced over the entire pad and bulging with reddish-brown glochids. Often the areoles are progressively larger toward the distal portion of the pad and largest on the apical margin. The glochids are readily dislodged, a trait that is evident in the field and on herbarium sheets. The areoles have a dense hemispheric tuft of red-brown glochids (with age they may fade to gray) emerging from the hairs and ultimately crowding most of the areole. The glochids are 1–2.5 mm long. The leaves are conical and 3–4.7 mm long.

Immature specimens of *O. rufida* closely resemble the adults. The seedlings are not hairy (D. Ferguson, pers. comm.).

Biosystematics. *Opuntia rufida* is a Mexican species that barely enters the

United States along and near the Rio Grande in Trans-Pecos Texas. Its closest relative is *O. microdasys* (Lehm.) Pfeiff., another spineless prickly pear, endemic to Mexico from southeast Coahuila south to Hidalgo. *Opuntia microdasys* reportedly hybridizes with *O. rufida* at their contact zone in southeastern Coahuila and in adjacent northeastern Zacatecas (Zimmerman et al., forthcoming); *O. macrocalyx* Griffiths apparently pertains to one of these intermediates.

Anthony (1956) described *O. rufida* var. *tortiflora* from southern Big Bend National Park, with the type locality at Hot Springs. According to Anthony the var. *tortiflora* is distinguished from the typical variety by its short-obovate to elliptic pads, relatively distant areoles, and outer tepals "swirled in imbrication and twisted sideways in anthesis" (p. 240). We follow Benson (1982) and Zimmerman et al. (forthcoming), who treated var. *tortiflora* as a mere form of *O. rufida*.

Synonyms. *Opuntia rufida* Engelm. var. *tortiflora* M. S. Anthony; ? *O. microdasys* (Lehm.) Pfeiff. var. *rufida* K. Schum.; *O. herrfeldtii* Kupper.

Common Names. Blind pricklypear; blind prickly-pear; blind pear; cinnamon pear; nopal; nopal cegador; prickly pear.

11. **Opuntia strigil** Engelm., Proc. Amer. Acad. Arts 3: 290. 1856. MARBLE-FRUIT PRICKLY PEAR. Plates 62, 63. Shallow soils, limestone hills, mesas, and canyons, W Stockton Plateau. Western Reeves, Pecos, Terrell and Val Verde counties. 2,400–4,100 ft. Flowering Apr–Jun. $2n = 22$ (many observations), 44 (only one tetraploid plant known). Also Upton Co., Crockett Co., and Nolan Co. (Benson, 1982). Probably adjacent Coahuila, Mexico. Map 10.

Opuntia strigil has a relatively limited distribution. It extends southeast in the Trans-Pecos to ca. 25 miles north of Langtry (Val Verde County) and northeast across the Pecos River. A specimen of *O. strigil* attributed to El Paso County (Benson, 1982) lacks good data (merely "Below El Paso" in 1852, Charles Wright). With its characteristic spination *O. strigil* is one of the most easily identified prickly pears in its range. The type locality is "western Texas, west of the Pecos, in crevices of flat limestone rocks." The word *strigil* (short *i*, soft *g*) is based upon the Latin *strigilis*, "scraper" or "skin-brush," used at baths in the days before soap, probably in reference to the appearance of the pads with their copious deflexed spines; thus, their vestiture is *strigose* (long *i*, hard *g*).

Identifying Characters. The plants of *O. strigil* are upright and compact, 0.5–1 m tall, or less commonly sprawling. The usually obovate pads have closely set areoles with spines in areoles over the entire pad, except sometimes at the base. The spines are strongly deflexed, almost flattened against the pad, except for the main central that is porrect or deflexed. There is a single main central spine that is reddish proximally and yellow distally, and several smaller spines borne in the lower portion of the areole.

The yellow flowers of *O. strigil* are 5–5.5 cm long and 5–6 cm in diameter. The inner tepals are lemon-yellow to pale yellow, fading to salmon. The tepal midregions may be pale reddish proximally, but the overall appearance of the flower usually is one without a target center. The tepals are ca. 3 cm long and

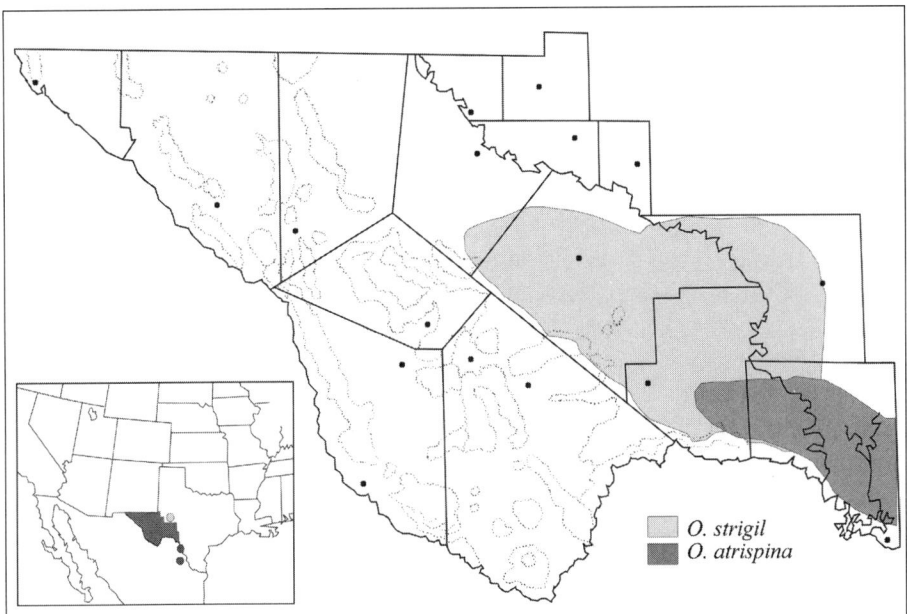

Map 10. Generalized distribution of *Opuntia strigil* (marble-fruit prickly pear) and *O. atrispina* (black-and-yellow-spined prickly pear).

broadly cuneate-obovate with subtruncate apexes. The filaments are ca. 1.5 cm long, pale green proximally and cream-colored distally. The anthers are ca. 1.5 mm long and yellow to cream-colored. The pale style is 1.5–2.4 cm long and slender-urceolate. The 6–8 stigma lobes are 2–4 mm long, and pale green to cream-colored. The pericarpel is 1.7–2 cm long, broadly obconic, with closely spaced, small, circular areoles with glochids and bristlelike spines.

The fruits of *O. strigil* are among the smallest of any *Opuntia* in the southwestern United States (Weniger, 1984). They are fleshy, not very juicy, subspherical to turbinate, 2–2.7(–3.6) cm long, and 1.3–2.2 cm in diameter. The outer layer is red at maturity, but the fruit wall and pulp are light green. The limited juice is clear and not sweet. The umbilicus usually is shallow, but relatively deep in the fruits of some plants. The areoles are armed with glochids and a few bristles to 4–8 mm long, especially at the fruit apex. The tannish-white seeds are irregularly reniform, 2.5–4 mm in diameter, ca. 1.5 mm thick, narrow-margined, not beaked at the hilum, with a slender aril.

Phenology. The blooming period of *O. strigil* extends from early April into June. The major blooming time seems to be April and May. Flowers have been noted as late as 20 June and may be produced even later in some years. Flowers open after midday, close at night, and usually do not open again. Fruit maturation in general seems to require about two months. Characteristically, ripe fruits are abundant in the field during July. Ripe fruits vary from caducous, as observed under cultivation, to sometimes persistent into August or even through most of the winter (Benson, 1982).

Sterile and Immature Specimens. The plants of *O. strigil* produce many branches from a single short trunk or basal area. Usually the green to yellow-green pads are obovate, but they also may be subelliptic to broadly obovate or suborbicular. The pads are 10–13(–19) cm long and 8–10(–16) cm in diameter. The areoles are 0.9–2 cm apart and have spines over most of the pad, with the spines reduced in length from apex to base, the more main central spines in the upper one-half or three-fourths of the pad. The areoles are oblong to oval, 4–5 mm long, 2.5–3.5 mm wide, or spherical and 5–7 mm in diameter. Oblong or oval areoles usually have gray felt in the front portion and one or more mounded yellowish-brown tufts of glochids in the back half. Spherical areoles have a hemispheric or subconical tuft of crinkled hairs in the center, surrounded by glochids at the lower portion where spines are borne. The glochids are erect, or the peripheral ones are spreading.

There are 7–13 spines, especially in upper areoles of mature stems. The most prominent spine is a porrect or deflexed main central that is borne in the middle or lower middle of each areole. On herbarium specimens the main central usually is pressed against the pad face and descending. The main central is 2–2.4 cm long, 0.6–0.9 mm in diameter, acicular to slightly flattened near the base, and usually reddish-brown, reddish, orange, or reddish-black proximally and yellow on the distal half. Below the main central in the lower portion of the areole are 6–12 deflexed subcentral spines. The lowermost, peripheral 2–4 of these also may be regarded as radial spines by position in the areole. All of the subcentrals or radials are similar in appearance except that they become slightly smaller from upper to lower ranks, with the largest to ca. 2 cm, but usually shorter. The subcentrals are slender, acicular, yellowish or red-brown proximally, and yellowish distally. Additional small bristlelike spines (long glochids) may be present in the lower lateral and peripheral portions of the areoles, and perhaps in the upper periphery as well. The glochids are 1–2(–6) mm long, yellowish to reddish tan, erect, and retrorsely barbed. In the areoles of some plants the glochids are bristlelike, 6–9 mm long, and mixed with shorter glochids at the upper portion of the areole.

Immature pads have fewer and smaller spines, including fewer areoles that exhibit the main central spine. The seedlings are not hairy (D. Ferguson, pers. comm.).

Biosystematics. *Opuntia strigil* is a distinct taxon without obviously close relatives (Weniger, 1984) except for *O. atrispina*. David Ferguson (pers. comm.) contends that *O. strigil* blends completely with *O. atrispina* from about Dryden east to near the Pecos River, and that the two taxa should be regarded as varieties. In hills just west of the lower Pecos River, we have observed what appears to be pure *O. atrispina* and very atypical plants of *O. atrispina*. It might be that *O. atrispina* hybridizes with species other than or in addition to *O. strigil*. The lectotype locality for *O. strigil* is "six miles west of the Pecos" River, probably in present-day Val Verde County.

Benson (1982), without explanation, extended the taxonomic circumscription of *O. strigil* with a second variety, *O. strigil* var. *flexospina* (Griffiths) L. D. Benson. Benson's var. *flexospina* is disjunct in South Texas in Webb and Zapata counties at ca. 450 feet elevation. The South Texas entity has larger pads, different spines, larger fruits, and smaller seeds, characters that led Weniger (1984), followed by Parfitt and Pinkava (1988), to treat it as a variety of a very different species, *O. engelmannii*. Ferguson (pers. comm.) asserts that it is the normal or wild counterpart of domesticated *O. aciculata* Griffiths, a tetraploid not closely related to either *O. strigil* or *O. engelmannii*.

Almost 10 chromosome counts have been made for *O. strigil* from throughout much of its range (Weedin and Powell, 1978; Powell and Weedin, 2001), and all of them have been diploid ($2n = 22$) except for one tetraploid meiotic count (Weedin et al., 1989) from an otherwise typical plant of *O. strigil* (possibly still extant) east of Fort Stockton in Pecos County.

Synonyms. None.

Common Names. Bearded prickly pear; marblefruit prickly pear.

12. **Opuntia atrispina** Griffiths, Annual Rep. Missouri Bot. Gard. 21: 172–73, t. 26. lower fig. 1910. BLACK-AND-YELLOW-SPINED PRICKLY PEAR. Plates 64, 65. Limestone hills, mesas, and canyons, east of the desert. Val Verde Co., just W of the Pecos River, SE across the county across the Devils River to the Anacacho "Mountains." Reported in Terrell Co. W as far as Dryden (Weniger, 1984). 1,000–2,100 ft. Flowering probably April. $2n = 22$. Mexico in eastern Coahuila (Bravo-Hollis, 1978), reported S to near Monclova (D. Ferguson, pers. comm.). Map 10.

Opuntia atrispina is a taxon of limited distribution in Texas, apparently always on limestone, in the extreme southeastern Trans-Pecos southeast to Uvalde County, according to Weniger (1984) in a strip only about 20 miles wide. The type locality is "Near Devil's River, Texas," in Val Verde County, about 35 miles to the east of the lower Pecos River. The specific epithet is after the Latin *atra* or *atrum*, "black," and *spina*, "spine," in reference to the basal color of the central spines.

Identifying Characters. The plants of *O. atrispina* are compact or spreading shrubs 0.5–1 m tall. The pads are obovate to subcircular and green to yellow-green. The most distinguishing vegetative feature involves the central spines. The 1–2 central spines per areole are black proximally and yellow distally, but sometimes just the tip is yellow, and often the bases are more brown than black. In aspect the areoles appear to be profusely glochidiate, particularly those crowded on the apical margin of the pads.

The flowers are pure yellow or with greenish centers, fading to apricot, and relatively small, to 2.5–6.5 cm in diameter. The inner tepals are yellow to chrome-yellow, 2.5–3 cm long, and ca. 1.5 cm wide. The outer tepals may be rose in the proximal midregions. The filaments and anthers are cream-colored, colorless, or greenish. The style is whitish, fading rose, and the 7–8 stigma lobes are cream-white to very pale green. The pericarpel is ca. 2.5 cm long, with the areoles on low tubercles, with glochids.

The fruits are fleshy, bright red to reddish-yellow, subspherical to obovate, of variable sizes but mostly small, 1.2–2.5(–4) cm long, and 1–1.5 cm in diameter.

The fruit pericarpel "rind" is red with some clear juice, and the pulp is relatively dry and greenish. In putative hybrids the fruit wall is green or partially red, with or without clear juice. The umbilicus is shallow to deep. The seeds are 3–4 mm in diameter, flattened but relatively thick, with a narrow margin.

Phenology. The major blooming period for O. *atrispina* is not known, but we suspect that it is in April, perhaps extending into May. Cultivated plants in Alpine have bloomed in early to late April. Flowers of these cultivated plants, from a suspected hybrid population near the lower Pecos River, opened for 1–2 days, with the tepals turning pale rosy or peach the second day. Bona fide O. *atrispina* in cultivation produces yellow flowers that open again for 2–3 days, with tepals turning peachy after the first day. The time required for fruit maturation is not known.

Sterile and Immature Specimens. The pads of O. *atrispina* are 10–17 cm long, 9–15 cm in diameter, with spines on the upper half of the pad. Reddish pigment may develop around the areoles when plants are stressed. The areoles are oblong, ovate to oval, and 7–8 mm long, 5–6 mm across, or nearly spherical and ca. 8 mm in diameter. The areoles are mounded with a dense pile of gray hairs, along with glochids. Areoles at the upper margin of the pads typically are larger, to 1.2 cm across, are close together, almost touching, and bulging with glochids.

In O. *atrispina* the areoles are few-spined, usually no more than 5–6 per areole, and usually with only 1–2 main spines. The main central spines are porrect or slightly deflexed, usually curving downward, or perhaps straight. In populations near the Pecos River, usually there are two centrals, with the first positioned in the lower center of the areole and the second borne immediately below. Either the upper or lower central may be the longest in an areole, with the centrals usually 2.5–3.7 cm long, 1–1.3 mm in diameter, and acicular or weakly flattened near the base. One of the centrals may be missing, or, according to Weniger (1984), a third central to 2.7 cm long may be present. The main centrals typically are black proximally, or perhaps dark red-brown proximally, with a zone of brown before the yellow tip. Central spines may be yellow at the tip or for the distal half. The spine tips may fade to whitish, and older spines may fade to gray. The lowest, peripheral 1–3 spines are interpreted by us as radials. In populations near the Pecos River, typically there is one strongly deflexed radial spine directly below the centrals. The radials are much smaller than the centrals, to ca. 1 cm long. When there are three radials, there is one in the lower front of the areole on either side of the slightly larger and centrally positioned radial. The radials are pale to medium reddish-brown proximally, and black-speckled or pinkish-gray, with the tips light brown to yellowish. One or more projecting, slender bristles may be found at the lower front periphery of the areole, these often mixed with the glochids and of similar caliber, but a little stouter and longer, and not retrorsely scabrous like the glochids.

The glochid configuration is variable in areoles of O. *atrispina*. There may be a subapical tuft of erect, yellow to light brown glochids to 8 mm long, or primarily an apical crescent of shorter glochids 2–3 mm long. In most areoles light brown glochids also are erect at the entire periphery of the areole, mixed with the hairs. In some areoles, a few glochids may project from the lower front periphery of the areole, in a position near the single deflexed radial spine. In other areoles the glochids occur in untidy clusters of different lengths, or the glochids are scattered in the areoles.

Immature pads have fewer and shorter spines. The seedlings are not hairy (D. Ferguson, pers. comm.).

Biosystematics. The distribution and description given for O. atrispina (see previous discussion) both coincide closely with those given by Weniger (1984). Benson (1969b, 1982) had a unique concept of O. atrispina that was considerably wider. Benson listed the distribution from southern Presidio County southeast in Texas, to Uvalde and Bee counties, and north to Taylor County. The description and certain descriptive characters used by Benson, and especially the specimens that he cited (Benson, 1969b), prove that he included black-spined plants of two or more species in his concept of O. atrispina. For example, he also considered O. macrocentra var. minor from Presidio County (Anthony, 1956) as synonymous with O. atrispina.

Opuntia atrispina is a distinctive taxon, but ultimately it may be recognized as only a distinct variety of O. strigil. Opuntia atrispina consistently is recognizable east of the Pecos River, and O. strigil is distinct from east of Dryden west to Pecos County. David Ferguson (pers. comm.) and A. Zimmerman (pers. comm.) have independently observed a seemingly complete morphological continuum between the two taxa east of Dryden (e.g., north of Langtry on the Pandale road). The apparent hybrids are also evident north of Dryden (*Powell and Powell 6319,* SRSC). On the west bank of the lower Pecos River we have observed rather large populations of lower, spreading plants, ca. 40 cm high, exhibiting long slender spines that are black with yellow tips (*Powell and Powell 6243,* SRSC). These relatively low plants resemble typical or "pure" O. atrispina (present at the same site) in spine characters, but in addition to lower habit they have smaller, less glochidiate areoles. Flowers of some of the low, spreading plants have red midregions of the inner and outer tepals, among other plants with pure yellow inner tepals. At the "west bank" site, it appears that O. atrispina is hybridizing with some taxon other than O. strigil, which was not seen in the immediate vicinity, rather than blending with O. strigil, as seen elsewhere in the same region.

The hexaploid chromosome counts published as O. atrispina by Weedin and Powell (1978) were the result of misidentifications. Opuntia macrocentra var. minor does not belong as a synonym of O. atrispina (Benson, 1982), and so all specimens mapped/cited as O. atrispina by Benson require reexamination and verification.

Synonyms. None.
Common Name. Dark-spined opuntia.

13. **Opuntia macrocentra** Engelm., Proc. Amer. Acad. Arts 3: 292. 1856. LONG-SPINED PURPLISH PRICKLY PEAR. Plates 66, 67. Various substrates and habitats, desert to mountain grasslands and intermediate slopes. El Paso Co.

E through Hudspeth, Culberson, and Reeves counties, NE Brewster Co. E of Alpine and adjacent Pecos Co. 3,000–5,000 ft. 2n = 22, 44. Flowering Apr–May. Southern NM and SE AZ. Mexico, NE Chihuahua S to near Ciudad Chihuahua (Ferguson, sight records), and barely into NE Sonora from AZ. Map 11.

Opuntia macrocentra is one of relatively few prickly pears noted for purple stems, the product of a betalain pigment (presumably one particular compound, betacyanin) most evident when the plants are stressed by cold or drought. Benson (1982) treated *O. macrocentra* as one of five varieties within *O. violacea*. According to Ferguson (1988) and Pinkava and Parfitt (1988), *O. violacea* was

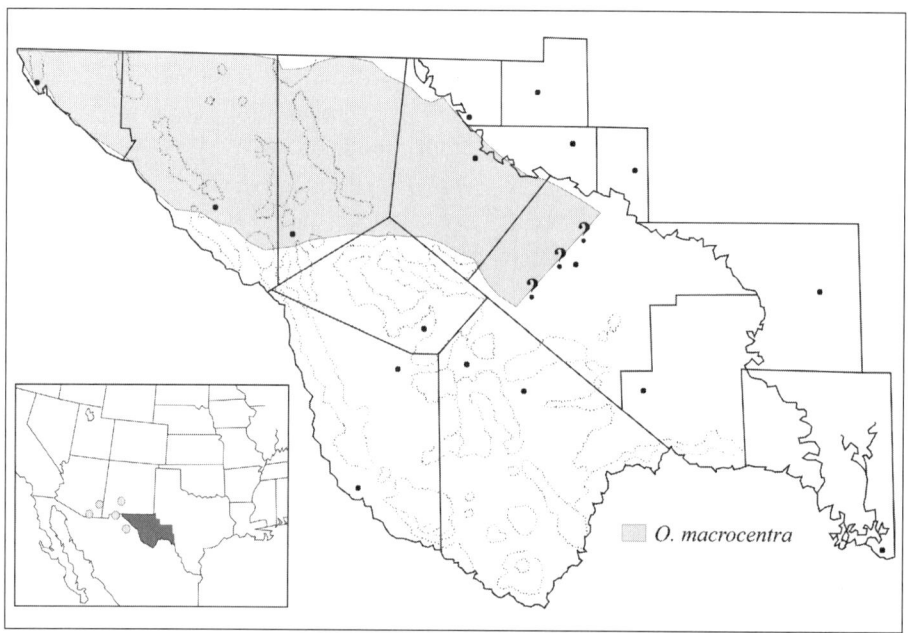

Map 11. Generalized distribution of *Opuntia macrocentra* (long-spined purplish prickly pear).

not validated until 1895, and then only accidentally (attributed to Engelmann by B. D. Jackson, Index Kewensis 2: 358); by then, in 1856, Engelmann had formally named *O. macrocentra*. Of the varieties recognized by Benson, Pinkava and Parfitt considered *violacea* and *castetteri* as synonymous with *O. macrocentra*; two others were removed as distinct species [*O. santa-rita* (Griffiths & Hare) Rose and *O. gosseliniana* F. A. C. Weber]. Thus, Pinkava and Parfitt recognized no varieties of *O. macrocentra* in the southwestern United States and northern Mexico. The variety of *O. macrocentra* described by Anthony (1956), *O. macrocentra* var. *minor*, was treated erroneously by Benson (1969b, 1982) as synonymous with *O. atrispina* and not reviewed by Pinkava and Parfitt in 1988. Ferguson erroneously equated *minor* with the small-padded endemic diploids of the lower Big Bend. Weniger (1984) presented *O. macrocentra* as a species without varieties; he did not comment on Benson's varieties of this taxon except to exclude one of them, as *O. violacea* var. *santa-rita*. Ferguson reviewed this taxonomic complex and divided Benson's five varieties of *O. violacea* between two species, *O. macrocentra* and *O. chlorotica* (see Biosystematics under *O. azurea* var. *diplopurpurea*).

Our review of the *O. macrocentra* complex in Trans-Pecos Texas and adjacent areas suggested that there are at least eight taxonomic entities mentioned in the literature on this complex, including Mexican and Sonoran Desert fringe taxa. One taxon, previously known as *O. macrocentra* var. *minor*, was paired by us with *O. mackensenii* var. *mackensenii*. In the present treatment we recognize *Opuntia macrocentra*, sensu stricto, and five related taxa now placed in *O. azurea*; among these, three are newly named and described herein.

Some populations of *O. macrocentra*, sensu lato, conspicuously fit the popular names of "long-spined prickly pear" and "purple prickly pear." The syntype locality for *O. macrocentra* first appeared in print as "Sand-hills on the Rio Grande near El Paso." Specifically (quote from *Cactaceae of the Boundary*), the syntypes were from "Sandy ridges in the bottom of the Rio Grande near El Paso, also on the Limpia (*Wright*)," later restricted (by lectotypification) to the El Paso locality. Engelmann's diagnosis might be based on a mixture of both El Paso and Limpia Creek populations. The specific epithet alludes to the long central spines, after the Greek *makros*, "long," and *kentron*, "spine."

Identifying Characters. The plants are spreading to nearly upright shrubs usually 30–60 cm tall, but to 1 m high, and only rarely with a short trunk. The "smooth" blue-gray, blue-green, or purplish pads are obovate to orbicular, 10–20 cm long, 10–20 cm in diameter, or the pads are slightly wider than long. The areoles are 1.5–3 cm apart, and spines are produced in areoles on the upper one-fourth to one-half of the pad, or only on the upper margin. The largest spines are notably directed upward. In most upper areoles there are 1–2(–4) central spines, with the upper projecting ones usually 5–10(–12) cm long, and black, dark brown, or reddish-brown, at least basally, sometimes tipped more or less extensively with yellowish or white. There are abundant reddish-brown to yellowish glochids, especially in the upper areoles.

The yellow flowers of *O. macrocentra* have sharply defined bright red centers. The flowers are 6–8 cm long, 5.5–8 cm wide, and in general not opening as widely as most other opuntias. The inner tepals are 3–5 cm long, 2–3.5 cm wide, obovate-spatulate, obtuse apically, sometimes emarginate and/or apiculate. The red bases of the yellow tepals extend upward to near midtepal, often tapering distally so that the red center of the flower is "star-shaped," and in some plants the red pigment extends along the midvein almost to the tepal apex, like a "midstripe" (rare in *Opuntia*). The filaments are ca. 1.5 cm long, pale green proximally and cream-colored distally. The anthers are 1.8–2 mm long, and yellow. The cream-colored style is 1.7–2 cm long. The ca. 6 stigma lobes are ca. 5 mm long, and cream-colored or pale green. The pericarpel is 2.5–3 cm long, slender, 1.2–1.5 cm in diameter, with scattered areoles (12–16), mostly on the upper half, and small, with brown glochids.

The fruit at ripening turns reddish-purple or purple and remains succulent. It is obovoid, ovoid, or ellipsoid, 3–4.3 cm long, 1.5–3 cm in diameter, not much, if at all, constricted below the apical rim. The umbilicus is deeply concave. The fruits usually have 12–16 areoles. The fruit rind is purple, and the juice and pulp are pale purple to clear. The seeds are flattened, tan, 3.5–4.5 mm in diameter, 1.5–1.9 mm thick, irregular in outline, with a broad notch on one side and a prominent raphe to ca. 1 mm wide.

Phenology. The major blooming period for O. *macrocentra* appears to be in April and May. Flowers may open as early as 9:00–10:00 A.M., close at night, and probably do not open again the next day, unless the "open time" is interrupted by cloudy or cool conditions. Fruits mature in July, at least in some years.

Sterile and Immature Specimens. Normal areoles are elliptic, obovate, or oval, or perhaps subcircular, and usually 1.7–3 cm apart, 3.5–5 mm long, and 2–3 mm across. Marginal areoles may be much larger, as in most other prickly pears. Each areole is filled with a tuft of short glochids and a pile of gray to tan hairs; some large marginal areoles are densely mounded with brown woolly hairs.

Spineless or near-spineless forms of O. *macrocentra* are at least occasional (*Griffith 103*). The spines of O. *macrocentra* typically include 1–3(–4) long centrals in the center and lower center portion of each distal areole, and one strongly deflexed central in the lower front center of the areole. The strongly deflexed lower central usually is to 1.2–3.5 cm long, acicular or slightly flattened, and milky- or ivory-white with a yellow-brown tip. This lowermost central is dark reddish-brown in some plants, or reddish-brown at the base and white distally. The longer centrals usually include a lower central, one that perhaps is deflexed and flattened or even grooved, and perhaps 1–2 upper ones that are projecting and acicular, flattened basally, or flattened for most of the length. The longest spines on each plant are 5–10(–12) cm, and 1–1.5 mm thick. The spines (all centrals) are black, dark brown, or reddish-brown, for either the full length or only proximally, and sometimes white to gray distally or only at the tip. Spine tips of some plants are reddish or honey-colored. In some plants the spines may be white for most of their length (see "var. *castetteri*"), but usually with a black or reddish base. Ferguson (1988) described the white color on such spines as epidermal, not evident when it is wet, and noted that the white coloration disappears with age.

Glochids are abundant in mid-pad and apical areoles. The glochids are most evident in an apical crescent, tuft, or brush, but some glochids 1–3 mm long usually are found also at the sides of the areoles mixed with hairs. Glochids in the apical tuft are 1–6 mm long, reddish-brown, brown, to yellowish and retrorsely scabrous. In mid-pad areoles the glochids are longest at the areole apex and shortest toward the middle of the areole.

On immature stems there are fewer and shorter spines. Seedlings of O. *macrocentra* are not hairy (D. Ferguson, pers. comm.)

Biosystematics. Three of the five varieties of O. *violacea* recognized by Benson (1982) were submerged in O. *macrocentra* by Pinkava and Parfitt (1988). These include var. *macrocentra* (Engelm.) L. Benson, var. *castetteri* L. Benson, and var. *violacea*, sensu stricto. The var. *violacea*, sensu Benson, is a morphotype with relatively short (4–6 cm long), dark reddish-brown spines that occurs mostly in Arizona and New Mexico and rarely in the western and northern Trans-Pecos. The var. *castetteri*, supposedly a white-spined taxon, is based on a type specimen from the Hueco Mountains of El Paso County at an elevation of 4,300 feet. Purple-skinned prickly pears with white spines occur sporadically as single plants or in small groups, not as obvious populations. Plants with half-white or partly white spines also are found, suggesting that the white spine color is merely an extreme of the spine-color variation segregating within population systems of one (or more) common dark-spined species.

Opuntia macrocentra of previous authors was a mixture of diploid and tetraploid taxa, extending in distribution from southeastern Arizona east to the Pecos River in Pecos and perhaps Val Verde counties, and south for an unknown distance into Mexico (where *O. azurea* complicates identification). Published chromosome counts for *O. macrocentra* in the western Trans-Pecos (Pinkava et al., 1985; Powell and Weedin, 2001), like those from New Mexico and Arizona (Pinkava et al., 1973, 1985, 1992, 1998), some of them reported as *O. violacea* var. *violacea*, have all been tetraploid.

One might assume, based upon previously published chromosome counts, that the name *O. macrocentra*, sensu stricto, pertains to the tetraploid component of this taxonomic species. However, eight recent chromosome counts were diploid (A. M. Powell, unpub.) from plants in Hudspeth and Culberson counties tentatively identified as *O. macrocentra*. Documented diploid and tetraploid plants of *O. macrocentra* are so similar that it has not been possible to accurately predict ploidy levels from morphology alone, even through observation of living specimens in the field. The lectotype specimen of *O. macrocentra*, which we have examined, might have come from a population of diploid or tetraploid plants. No chromosome number has been obtained from long-spined purple prickly pears at a site perfectly matching the lectotype locality of *O. macrocentra*. The closest population studied cytologically (tetraploid; Pinkava et al., 1985) is at 4,000 feet on the west *bajada* (outwash fan) of the Franklin Mountains, El Paso County, Texas (*Worthington 8093*).

One hypothesis is that tetraploid plants of *O. macrocentra* have predominantly oval, ovate, or obovate pads that are blue-gray and turn only slightly purplish, whereas diploids tend to have orbicular blue-gray pads that turn thoroughly purple under stress. In addition, the tetraploids tend to have almost-black spines (varying to the whitish "*castetteri*" form), but those of the documented diploids are reddish-brown (likewise varying to white-spined forms).

Synonyms. *Opuntia violacea* Engelm. in Emory (invalid provisional name); *O. violacea* Engelm. ex B. D. Jacks.; *O. violacea* Engelm. ex B. D. Jackson var. *macrocentra* (Engelm.) L. D. Benson; *O. violacea* Engelm. ex B. D. Jackson var. *castetteri* L. D. Benson; *O. violacea* Engelm. ex B. D. Jackson var. *violacea*; ? *O. phaeacantha* Engelm. var. *nigricans* Engelm. There is still some nomenclatural question about whether *O. macrocentra* or some other name is correctly applied to the species.

Common Names. Redeye prickly pear; longspine prickly-pear; longspine pricklypear; long-spined prickly pear; purple prickly pear; purple-tinged prickly pear; black-spine prickly pear; black-spined prickly pear; blackspine purple prickly pear; nopal; prickly pear; including the white-spined form, "var. *castetteri*," Castetter pricklypear cactus; Castetter purple prickly pear.

14. **Opuntia azurea** Rose, Contr. U.S. Natl. Herb. 12: 291, t. 24, f. 33. 1909. Various substrates and habitats, desert to mountain grasslands and intermediate slopes. Big Bend region of Trans-Pecos TX. 1,900–5,500 ft. $2n = 22$. Flowering Mar–May. Mexico in Coahuila, Chihuahua, Durango, and Zacatecas. Map 12.

Opuntia azurea was described as a Mexican species, and this name has not been used for any United States opuntias except by Anthony (1956). Our taxonomic concept of *O. azurea* is derived partly from information provided by D. Ferguson (pers. comm.). Herein we interpret *O. azurea* as a species with multiple forms and geographic races; four of them occur in the Big Bend region of the Trans-Pecos, and the typical variety is a Mexican taxon that closely approaches the international border south of Big Bend National Park. The type locality for *O. azurea* is in northeastern Zacatecas, Mexico. The specific epithet is after the Latin, *azureus*, "blue," in reference to one of the stem colors of this polymorphic, geographically variable, and seasonally variable species.

Key to the Varieties

1. Purple only at the areoles, if at all; spines usually yellow, golden, or reddish (2).
1. Uniformly purple in drought or winter, or greenish-purple between the areoles after moisture in warm seasons; spines usually blackish with white tips, reddish-black, or dark-reddish, often with pale tips (3).

2(1). Spines slender (0.7–1.3 mm wide at the base), numerous (4–12); pads small; fruit spiny

14c. *O. azurea* var. *aureispina*.

2. Spines robust (1–1.5 mm wide at the base), few (1–4); pads large; fruit spineless

14d. *O. azurea* var. *discolor*.

3(1). Plants taller (usually 1–2 m); Mexican

14. *O. azurea* var. *azurea*.

3. Plants shorter (usually 0.3–0.9 m); Trans-Pecos (4).

4(3). Distal pads larger (14–20 cm long, 14–20 cm wide), orbicular

14a. *O. azurea* var. *diplopurpurea*.

4. Distal pads smaller (6.5–19 cm long, 6–14 cm wide), obovate

14B. *O. azurea* VAR. *parva*.

14a. **Opuntia azurea** Rose var. **diplopurpurea** A. M. Powell & J. F. Weedin, **var. nov.** DIPLOID PURPLE PRICKLY PEAR. Plates 68, 69. TYPE: UNITED STATES. TEXAS: Brewster Co., Sul Ross Hill, Alpine, 21 Apr 1997, *J. F. Weedin 481*. $2n = 11$II. (HOLOTYPE: SRSC!).

Plantae humiles, patentes, 35–60(–70) cm altae. Caules rubri-purpurei aut purpurei, 10–20 cm longilatique, orbiculares, latiores quam longiores aut late obovati. Spinae rubrae-bruneae aut atrae, saepe cum apicibus albis, 5–11 mm longae, plerumque (0–)1–4 in quoque areola, 4–8 in aliquis areolis superis marginalibus, areolae spiniferae plerumque in one-third parte distali caulium. Flores lutei cum centris rubris. Fructus rubri, ovales aut ovati, 2–3 cm longi, aut carnosi aut desiccati, contracti, et rosei. Semina complanata, fulva, 4–5 mm diametro. $2n = 22$.

Common purple prickly pear of Big Bend region of the Trans-Pecos, from mountain grasslands down to the desert. Hudspeth, Jeff Davis, Presidio, Brewster, Pecos, and Crane counties. 2,500–5,500 ft. $2n = 22$. Mexico, adjacent to the Rio Grande, N Coahuila, and NE Chihuahua. Map 12.

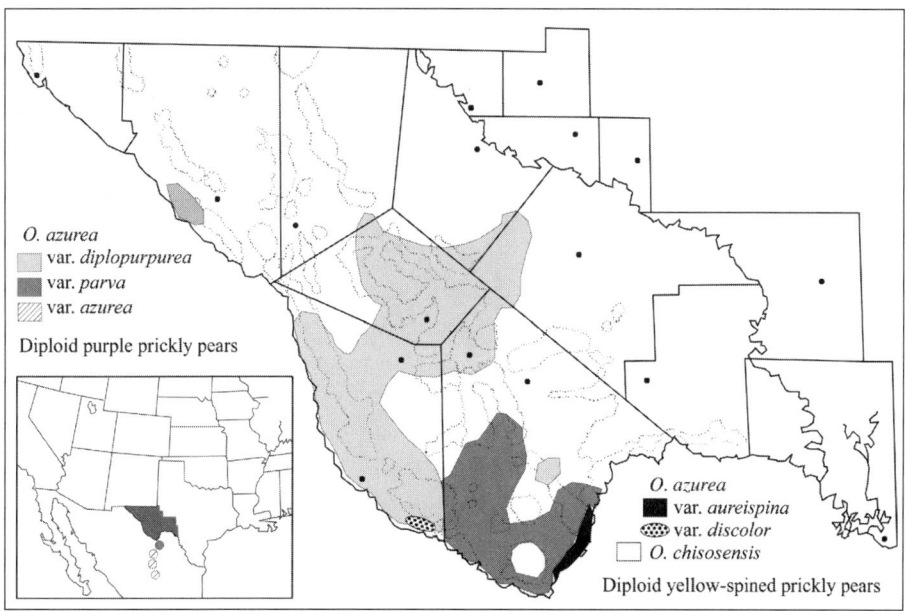

Map 12. Generalized distribution of diploid purple and purplish prickly pear and diploid yellow-spined prickly pears; *Opuntia azurea* var. *azurea*, *O. azurea* var. *diplopurpurea* (diploid purple prickly pear), *O. azurea* var. *parva* (Big Bend purplish prickly pear), *O. azurea* var. *aureispina* (golden-spined prickly pear), *O. azurea* var. *discolor* (Big Hill prickly pear), and *O. chisosensis* (Chisos prickly pear).

The pads of this taxon turn completely reddish-purple or purple under stress. The varietal epithet is descriptive of a diploid taxon with purple stems, after the Latin *purpureus,* "purple."

Identifying Characters. The plants of *O. azurea* var. *diplopurpurea* are relatively low, open, sprawling shrubs 30–60 cm high. The pads are reddish-purple (for most of the year), orbicular to broadly obovate, 14–20 cm long, 14–20 cm in diameter, or broader than long. Usually there are (1–)2–6(–9) dark spines per areole, more in the distal and marginal areoles. Spiniferous areoles occur on the distal half to one-fourth of the pads, often restricted to the distal margin, sometimes absent. The longest spines are 5–11 cm, nearly black, reddish-black, or reddish-brown, and white or at least lighter-colored distally. Occasional "white-spined" forms occur, in which only the bases are dark.

The yellow flowers with red centers are virtually identical to those of O. *macrocentra* and the other varieties of O. *azurea*. The flowers are 6–7.5(–8) cm long and wide; their inner tepals are 3–5 cm long, 2.5–3.5 cm wide, with broadly rounded apices that are minutely apiculate or emarginate. The filaments are 1.1–1.5 cm long, supporting yellow anthers 1.5–2 mm long. The style is 1–2 cm long with cream-colored to pale green stigma lobes ca. 5 mm long. The pericarpel is ca. 2.5 cm long and ca. 1.5 cm in diameter.

The fruit upon ripening turns red to reddish-purple; it either remains fleshy and juicy or dries and shrinks, fading to tan or pale red. The fruit is ovate to obovate, 1.5–3 cm long, 1.5–2.8 cm in diameter, somewhat constricted below the apical rim. The umbilicus is rather deep. The fruits (in at least some populations) have a green pulp and flesh, with very little clear, moderately sweet juice. The seeds are flattened, tan, 3.8–4.5 cm in diameter, ca. 1.5 mm thick, and irregular in outline.

Phenology. Variety *diplopurpurea* blooms in April and May, extending into early June at higher elevations in Jeff Davis and Brewster counties. Flowers open as early as 9:00–10:00 A.M., close at night, and usually do not open again the next day, unless the "open time" is delayed or interrupted by cloudy or cool conditions. Second-day flowers are not uncommon during cool periods. The time required for fruit maturation is not well understood. On cultivated plants in Alpine, fruits are ripe in July and August, but at lower elevations ripening apparently occurs in June and July.

Sterile and Immature Specimens. Variety *diplopurpurea* is not always easy to distinguish from var. *parva*. In the field var. *diplopurpurea* usually can be recognized by its low, sprawling habit that appears "open." The plants are branched at the base but they produce relatively fewer, larger pads, resulting in a more open plant. Typically, the pads are orbicular, but they may be broadly obovate, obovate-truncate, or broader than long. Plants of var. *parva* are dense, whether sprawling or upright. The relatively more numerous, smaller, obovate pads collectively contribute to a dense habit. Some dense, upright plants of var. *parva* have a short trunk. Herbarium specimens of var. *diplopurpurea* are difficult to distinguish from those of var. *parva* unless typical stems were selected for preservation. The stems of var. *diplopurpurea* are reddish-purple for much of the year, especially during winter and drought periods. Physiologically, vigorous stems are green, usually with the betacyanin pigment most abundant around areoles and at pad margins. The seasonal appearance of betacyanin throughout the skin, hiding the chlorenchyma inside, usually is correlated with cold or drought stress.

The elliptic to circular areoles are 1.5–3 cm apart and 3.5–5 mm long, 2–3(–3.5) mm in diameter, larger near the stem apex. The central spine pattern usually involves one strongly deflexed central in the lower front center of the areole, and (0–)1–3(–4) or more longer centrals in the lower and central portion of the areole. The lowermost central spine is 1.5–6.5 cm long, acicular or flattened, dark in color, or dark proximally and white distally. In typical areoles a second, longer spine is positioned just above the lowermost deflexed central. If additional centrals are present, as they usually are in more distal areoles, they are angular or terete and positioned in the lower half or central portion of the areole. Radial spines usually are absent. Glochids are abundant in mid-pad and apical areoles, where they form reddish-brown tufts in the whole areole, or in the upper half and at the margins. The glochids are 2–3 mm long in most distal areoles, 5–8 mm long in upper marginal areoles.

On immature stems there are fewer and shorter spines. Seedlings of var. *diplopurpurea* are not hairy.

Biosystematics. A few miles south of the Rio Grande, var. *diplopurpurea* is replaced by *O. azurea* Rose var. *azurea* (D. Ferguson, pers. comm.), a Mexican taxon that occurs mostly in desert flats in Coahuila, eastern Chihuahua, Durango, and Zacatecas (Map 12). Variety *azurea* apparently is diploid (reported as "*O.* aff. *chihuahuensis*" and "*O.* aff. *lindheimeri*" by Pinkava et al., 1985). A hexaploid count reported for *O. azurea* (Yuasa et al., 1973) is unsubstantiated, presumably from a misidentified specimen of some completely unrelated species. Plants of var. *azurea* are compact and upright, to 1 m or more high, with a short trunk, or branching from the base and somewhat spreading. Branches have orbicular to obovate blue-green pads 10–15 cm in diameter and 1–3 blackish spines 2–3 cm long in areoles on the upper half of the pad. Stems of var. *azurea* turn brilliant purple when sufficiently stressed, except that some populations contain "blond" individuals that lack betacyanin pigments in all parts of the stem, including spines, which consequently are bright yellow, approaching var. *aureispina*.

Variety *diplopurpurea* is a diploid taxon with seasonally brilliant reddish-purple stems almost endemic to the Big Bend region of the Trans-Pecos. It is most common at mid-elevations in the Davis Mountains, in the western foothills of the Chisos Mountains, and in southwestern Presidio County; probably these are the predominant purple prickly pears along the Rio Grande valley northwest to Hudspeth County. Benson (1982) included var. *diplopurpurea* in *O. violacea* var. *macrocentra*, which he treated as extending from the Big Bend region northwest through southern New Mexico and into southeastern Arizona. Benson's concept of var. *macrocentra* included both diploid and tetraploid taxa and misidentified specimens of other species (probably hexaploids).

The few-spined or spineless form of var. *diplopurpurea* is more prevalent in the Davis Mountains and on the western slopes of the Chisos Mountains, but plants with essentially spineless pads arc also found sporadically in the desert. Likewise, spiny plants like those in the desert are found sporadically in the mountains. Stems of the "spineless" form may be completely spineless or have 1–2 spines in a few areoles at the distal margin of the pads. The spineless form is diploid (Weedin and Powell, 1978; Weedin et al., 1989; Powell and Weedin, 2001). Both the spiny and spineless morphotypes of var. *diplopurpurea* have rather dry, shrunken mature fruits, much like those of var. *parva*, although the fruits in some populations remain fleshy and juicy.

The spineless form of var. *diplopurpurea* was misidentified as "*O. violacea* var. *santa-rita*," a relatively unrelated taxon, by Benson (1982), Weedin and Powell (1978), and other workers, based solely on the erroneous premise that *santa-rita* is the only spineless purple prickly pear, when in fact all are equally variable. *Opuntia santa-rita* (Griffiths & Hare) Rose is treated as a distinct species (Pinkava and Parfitt, 1988) or as *O. chlorotica* var. *santa-rita* Griffiths & Hare (Ferguson, 1988). *Opuntia santa-rita*, with its all-yellow flowers and densely areolate pericarpels, occurs in southeastern Arizona, in adjacent New Mexico (Hidalgo County), and in the Sierra Madre Occidental west of the desert in Mexico. All published reports of *O. santa-rita* in Texas and New Mexico (except in extreme southwest Hidalgo County), and many of those in Arizona, were based upon misidentified specimens of spineless plants of other purple-stemmed taxa (Ferguson, 1988). It does not occur anywhere near the Trans-Pecos unless the report by Weniger (1984) from between Presidio and Big Bend National Park is validated.

White-spined individuals or clones occur sporadically in the population of var. *diplopurpurea*. These white-spined plants are the Big Bend equivalents of "var. *castetteri*" L. D. Benson, herein considered to be a synonym of *O. macrocentra*. A chromosome number of $n = 11$ has been reported for *O. violacea* var. *castetteri* in Durango, Mexico (Pinkava and Parfitt, 1982). In Ferguson (1988, p. 156), the photograph labeled "*O. macrocentra* var. *minor*" appears to be var. *diplopurpurea*.

Synonyms. None?

Common Names. Virtually the same popular names that have been applied to *O. macrocentra* were also attributed to var. *diplopurpurea* when the two taxa were treated as conspecific.

14b. **Opuntia azurea** Rose var. **parva** A. M. Powell & J. F. Weedin, **var. nov.** BIG BEND PURPLISH PRICKLY PEAR. Plates 70, 71. TYPE: UNITED STATES. TEXAS: Brewster Co., clay soil, 2 mi N of Study Butte, 12 Aug 1993, *A. M. Powell and S. A. Powell 6007*. $2n = 11\text{II}$. (HOLOTYPE, SRSC!).

Plantae plerumque erectae, 0.4–1 m altae, fortasse cum truncis brevibus. Caules caerulei-grisei, glauci, suffusi colore purpureo, in ramibus superis obovati, (5.5–)8–19 cm longi, (5–)8–14 cm diametro. Spinae atrae aut porphyreae, plerumque (6–)10–12 cm longae, apicibus albis, 1–3 in quoque areola, areolae spiniferae in supera 1/2 parte caulium. Fructus rubri, ovati aut ovales, 2–3 cm longi, saepe desiccati, contracti, et rosei. Semina complanata, fulva, 5–5.5 mm diametro. $2n = 22$.

In the Trans-Pecos, restricted to desert habitats surrounding the Chisos Mountains, Brewster Co. 1,900–3,700 ft. $2n = 22$. Mexico, inferred to be in adjacent Coahuila, but replaced by var. *azurea*. Map 12. The varietal epithet is after the Latin *parvus*, "small," denoting the usually small distal pads of the taxon, relative to all of the other purple prickly pears.

Identifying Characters. The plants of *O. azurea* var. *parva* usually are many-branched, sprawling or upright shrubs 30–90 cm high, sometimes with a short trunk. The whole upright plants or individual branches of sprawling plants have a compact appearance. The pads are blue-green or blue-gray, glaucous, with an underlying purplish tinge most obvious around the areoles. In apparently stressed plants the reddish-purple betacyanin pigments increase to dominate whole pads. The pads, at least on terminal branches, typically are obovate, cuneate basally, (6.5–)8–19 cm long, and (6–)8–14 cm in diameter. The lower pads may be suborbicular and somewhat larger. The pads typically support long black to dark red-brown spines to 10–12 cm or more long (as seen in populational samples), with white tips in areoles on the upper one-half of the pad. Usually there are 1–3 spines per areole, potentially with a few more spines in marginal areoles.

The yellow flowers with red centers are essentially like those in *O. macrocentra* and in other members of the *O. azurea* complex. The flowers are 6–7.5 cm long and to 8 cm wide. The obovate inner tepals are (3.5–)4–4.5 cm long, 2–3 cm wide, and apiculate at the apex. The stamens have filaments 1.3–1.8 cm long and yellow anthers 1.5–2 mm long. The style is ca. 1 cm long, with cream-colored to pale green stigma lobes 3–5 mm long. The pericarpel is 2–3 cm long and ca. 1.5 cm in diameter.

Mature fruits of var. *parva* are pale red to purple and tend to dry and shrink with age. The fruit is ovate to obovate, 1.5–3 cm long, 1.5–2.7 cm in diameter, and constricted below the apical rim. The umbilicus is deep. The seeds are discoid, irregular in outline, large, 5–5.5 mm in diameter, and 1–1.4 mm thick. The seeds of var. *parva* and var. *aureispina* are similar in size and larger than those of var. *diplopurpurea* and var. *discolor*.

Phenology. Plants of var. *parva* are among the earliest-blooming opuntias in the Trans-Pecos. The major flowering period is in April extending into May, but flowers open in some years as early as 8 March. Flowers open in mid-morning, close at night, and may not open again the next day unless the "open time" is abbreviated by cloudy or cool conditions. Flowers after anthesis fade from yellow to orange-red by late afternoon, and by this time the flowers are almost closed. It is common to see both the young yellow flowers and older orangish flowers open or partially open on the same plant. Fruit maturation may begin in June and extend into July.

Sterile and Immature Specimens. Compact upright plants of var. *parva* with obovate-cuneate blue-green pads are easily distinguished, in the field and in good herbarium specimens, from var. *diplopurpurea* with spreading, less compact branches and orbicular, reddish-purple pads. Mature pads of var. *parva* have areoles that are 1.5–2.5 cm apart. The spine pattern in var. *parva* is the same as for other varieties of the complex, although typically there are fewer spines in lateral areoles, 1–2 or 2–3 per areole in var. *parva*, and maximum spine length is greater in var. *parva*, 11–12 cm or more. Benson (1982) reported that central spines of *O. macrocentra*, sensu lato, reach 17.5 cm long, possibly a reference to extraordinary specimens of var. *parva* or desert plants of var. *diplopurpurea*. Figure 474 in Benson (1982, p. 468) is an excellent illustration of var. *parva*, except that a maximum of two spines per areole is more typical of the taxon.

Radial spines occur in some specimens of var. *parva*, especially those near

Mariscal Mountain along the Rio Grande. When radials are present, there usually are two, slender, deflexed, ca. 6 mm long, and in the lower parts of areoles. Glochids of var. *parva* are similar to those of related taxa.

On immature stems there are fewer and shorter spines. Seedlings of var. *parva* are not hairy.

Biosystematics. Variety *parva* is regionally sympatric with var. *diplopurpurea* in the southern Big Bend region, but the "pure" forms of each apparently are not intermixed. Plants throughout the northern portion of the range of var. *parva* display some habit, stem, and spine features more typical of var. *diplopurpurea*. Experimental crossability between var. *parva* and var. *diplopurpurea* has been demonstrated (Griffith, 2000).

Variety *parva* is one of the most common prickly pears in desert habitats around the Chisos Mountains in Big Bend National Park. It extends north in the park to near Persimmon Gap, and north of the park in Brewster County to near Nine Point Mesa. Variety *diplopurpurea* occurs in the Davis Mountains sporadically south to Persimmon Gap in Brewster County, south of Alpine to the Christmas Mountains, and sporadically south of Marfa in Presidio County to the Rio Grande and up the Rio Grande valley past Candelaria. Variety *diplopurpurea* also occurs in the western foothills of the Chisos Mountains above the ecogeographic range of var. *parva*, near Sotol Vista and perhaps elsewhere. Plants of var. *parva* throughout the northern part of their range tend to be profusely branched with blue-green to purplish obovate pads. Those in the northeast (i.e., from Panther Junction northward) are relatively low and sprawling. In the northwestern part of its range, mostly south of Study Butte, var. *parva* tends to be upright and compact. These populations, in at least the northeastern and northwestern part of the range, appear to be introgressed from var. *diplopurpurea*. Introgression of these taxa might extend into Presidio County at the lowest altitudes. The ecogeographic range of var. *parva*, near the Rio Grande, stands between var. *azurea* in nearby Mexico and var. *diplopurpurea* in most of the Big Bend region.

Synonyms. Previous workers have not given independent taxonomic recognition to var. *parva*. Benson (1982), Weniger (1984), and Ferguson (1988) included var. *parva* in a geographically widespread and morphologically diverse concept of *O. macrocentra*.

Common Names. Essentially the same popular names that have been used for *O. macrocentra* have been applied to var. *parva* when the taxa were treated as conspecific.

14c. **Opuntia azurea** Rose var. **aureispina** (S. Brack & K. D. Heil) A. M. Powell & J. F. Weedin, **comb. nov.** GOLDEN-SPINED PRICKLY PEAR. Plates 72, 73. BASIONYM: *Opuntia macrocentra* Engelm. var. *aureispina* S. Brack & K. D. Heil, in Heil and Brack, Cact. Succ. J. (U.S.) 60: 17–34. 1988. TYPE: UNITED STATES. TEXAS: Brewster Co., near the Rio Grande River, Big Bend National Park (no collector or date cited; HOLOTYPE: SJNM 3777). Fractured limestone (Boquillas), desert hills and canyons along and near the Rio Grande. Brewster

Co. near Mariscal Mt and in Boquillas Canyon. 1,600–2,800 ft. Flowering Mar–May. $2n = 22$. Mexico, to be expected in Coahuila and Chihuahua; to ca. 30 mi S of Big Bend National Park (D. Ferguson, pers. comm.), in the same type of hot and exposed limestone habitats as in Texas. Map 12.

The specific epithet is descriptive of the yellow spines, after the Latin *aureus,* "golden," and *spina,* "spine."

Identifying Characters. The plants of var. *aureispina* are upright shrubs 0.3–1(–1.5) m high. The orbicular to broadly obovate pads are 8–16 cm long, 8–15 cm wide, light blue-green to yellow-green, glaucous, and with spines in areoles from the apex to the base of the pad. The spines are yellow in many plants, but in others they may be light brown, orange-brown, to nearly black with only yellow tips. There are 4–12 spines in the upper areoles on mature pads, only 1–3 spines in younger areoles and areoles on the sides of the stems.

The yellow flowers with red or orange centers are like those of other members of the *O. azurea* complex. The flowers are 5.5–7 cm long and 5–7 cm in diameter. In all there are 19–23 tepals, the inner ones yellow, red, or orange on the basal half, broadly obovate and apiculate, 2–4 cm long, and ca. 2–3 cm wide. The filaments are pale green or yellow with pale green bases, to 1.5 cm long, and the light yellow anthers are ca. 1.5 mm long. The style is ca. 2 cm long, narrowly urceolate, usually 5–7 mm wide, yellowish or pink at the base, supporting 6–8 pale green or pale yellow stigma lobes 3–8 mm long. The pericarpel areoles are small and bristly yellow-spinose.

The fruits at first are fleshy and pale green to slightly reddish-tinged, but at maturity they become dry without pulp or juice, shrivel and turn tan, yellowish-tan, to greenish-tan. Some shriveled fruits are purplish, but these do not appear to be typical of the taxon. The fruits are at first short-cylindroid or ovoid, 2.5–4 cm long, 1.7–2.5 cm in diameter, but later constricted below the hard apical rim, which does not shrivel much during drying. The 13–24 small spherical to oval areoles 1.5–2 mm in diameter are filled with white or gray felt and 0–4 yellowish spines to 1–2 cm or more long. Usually there are spines in areoles toward the fruit apex and at the upper rim. In addition to spines, areoles may support a few yellow bristles 3–4 mm long and yellowish glochids. The rapidly drying spiny fruits of var. *aureispina* help distinguish the taxon from its closest allies with fleshy, spineless fruits (Pinkava and Parfitt, 1988). However, fruits of var. *parva* and var. *diplopurpurea* may shrink. The tan to light brown seeds are flattened, large, 4–6 mm in diameter, ca. 1.5 mm thick, irregular in shape, and with a wide rim/raphe, this accounting for 2 mm or more of the diameter of the seed.

Phenology. Variety *aureispina* blooms in late March and April, extending into May (to 7 June in 1987). The flowers open about midday, close at night, and usually do not open again the next day. During cool periods second-day flowers are not uncommon. The fruits appear to develop slowly, requiring about four months, often in mid-August, to turn tan and shrivel. Under cultivation in Alpine, fruits in 1998 matured to dryness during 7–25 August.

Sterile and Immature Specimens. The compact plants of var. *aureispina* produce several ascending stems from a short trunk. Although spines are present in areoles over the entire pads, the upper portions of the pads appear densely spined because the spines are more numerous and longer in upper areoles. The areoles are oval-elliptic, 4–6 mm long, 2.5–4 mm in diameter, and 1.5–3 cm apart on the pad.

Both central and radial spines can be distinguished in var. *aureispina*. The spine pattern in upper areoles usually includes 1–3 deflexed spines in the lower front portion of the areole and 3–8 mostly divergent spines in the central, lateral, and back portion of the areole. The longest spines are up to (2–)3.5–6 cm, and 0.7–1.3 mm wide, often with the longest and stoutest spines in the back of the areole. In some areoles a lower deflexed central is the largest spine. The largest spines usually are flattened at least proximally, while short spines are flattened or acicular and as little as 0.05 mm thick. The basic spine color in var. *aureispina* seems to be yellow, although reddish bases are not uncommon, and plants with darker spines also are not uncommon. In fact, D. Ferguson (pers. comm.) has stated that in his recent experience the dark-spined plants are common enough to dominate in some populations of the taxon. The radial spines are positioned at the front periphery of the mature areoles. Usually there are two strongly deflexed slender radials, occasionally with a third radial in a lower lateral position, these typically 1–1.5 cm long. Often one of the flattened central spines, as interpreted here, is borne in the lower center of the areole barely above the level of the radials. In addition to spines, a few to numerous slender, erect, yellow bristles (transition forms between spines and glochids), these 0.4–1.4 cm long, may be present in the areoles. In upper areoles, normal glochids form a semicircular tuft in the upper rim or form a tuft in the back half of the areole. These normal glochids are yellowish and 1–3 mm long. In areoles at midpad the glochid arrangement is relatively dispersed, not forming tufts, a pattern that is useful in distinguishing other varieties of *O. azurea* and *O. macrocentra*, which has glochids densely tufted in areoles at midpad (Pinkava and Parfitt, 1988).

Immature pads have fewer and shorter spines per areole. The seedlings of var. *aureispina* are not hairy.

Biosystematics. In Mexico there are some intermediates between var. *aureispina* and *O. azurea* var. *azurea* Rose (D. Ferguson, pers. comm.). According to Ferguson, var. *azurea* occurs in flats and var. *aureispina* in hills. They are mostly separated ecologically. Possibly var. *azurea* extends across the Rio Grande into the Trans-Pecos, but to date pure var. *azurea* has not been documented north of the river. The report of *O. azurea* by Anthony (1956) was based on the type population of var. *aureispina*, which had not been named yet, as demonstrated by a specimen probably collected as a voucher for *O. azurea* at this locality (*Anthony A–6*, SRSC).

At the site north of Solis some plants designated as var. *aureispina* have yellow spines, and some have light brown to orange-brown spines, or even darker spines. We suspect that some spine-color variation is intrinsic within var. *aureispina*, but var. *aureispina* may hybridize with vars. *azurea*, *diplopurpurea*, and/or *parva*; all of these varieties occur in the general area, and they are mostly dark-spined. The hybrids between var. *aureispina* and var. *diplopurpurea/parva* were described as *O.* × *rooneyi* (Griffith, 2001).

There has been some confusion about the fruits produced by var. *aureispina*. Mature fruits originally were described as tan and dry. We have cultivated yellow-spined plants from the type locality, and they have produced tan and dry fruits, as has a yellow-spined plant from Boquillas Canyon. Ferguson (pers. comm.) asserts, however, that the spiny ripe fruits of var. *aureispina* usually are bright reddish and juicy, although they may be greenish, especially on yellow-spined plants. Presumably the darker-spined plants with juicy red fruits are hybrids with a taxon that produces juicy reddish fruits.

Synonym. *Opuntia aureispina* (S. Brack & K. D. Heil) Pinkava and B. D. Parfitt (Pinkava and Parfitt, 1988; Pinkava et al., 2001).

Common Name. Golden-spined opuntia.

14d. **Opuntia azurea** Rose var. **discolor** J. F. Weedin, **var. nov.** BIG HILL PRICKLY PEAR. Plates 74, 75. TYPE: UNITED STATES. TEXAS: Presidio Co., Big Hill, ca. 12.5 mi NW of Lajitas, 12 Aug 1993, *A. M. Powell and S. A. Powell 6004*, 2n = 11II. (HOLOTYPE, SRSC!).

Plantae plerumque erectae et patentes 0.5–1 m altae. Caules viridules aut glauci, plerumque sine pigmentis betacyaneis, late obovati, maxima 20 cm longi. Spinae luteolae, aureae, aut rufae, fortasse curvatae et torquatae, maxima 10.5 cm longae, plerumque 3–4 in quoque areola, areolae spiniferae in 3/4 caulium. Flores lutei cum centris rubris. Fructus rubri, ovales aut ovati, 2–3 cm longi, plerumque desiccati, contracti, et rosei aut fulvi. Semina complanata, fulva, 4–5 mm diametro. 2n = 22.

Known distribution restricted to a small area on the slopes of "Big Hill," a desertic high ridge approximately 12–13 mi NW of Lajitas that is traversed by the paved road between Lajitas and Redford (the Camino Real, or Hwy 170), and a small population 3.3 mi. E of Big Hill, S of Colorado Canyon along the Rio Grande, Presidio Co. 2,500–3,000 ft. 2n = 22. Mexico, visible on desertic hills directly across the Rio Grande in Chihuahua. Map 12. The varietal Latin epithet, *discolor,* means "of different colors" and refers to the different spine colors and lack of purple pigment in the stems and spines.

Identifying Characters. The plants are upright and spreading, with broadly obovate to suborbicular pads to 20 cm long, which are pale green to pale blue-green and not much, if at all, colored with betacyanins. Spines are borne in areoles over the upper three-fourths or more of the pad. The spines are strongly projecting, some of them irregularly curved or twisted, usually (1–)3–4 per areole, to 10.5 cm long, and yellow, white, or golden, to reddish.

The yellow flowers with red centers appear identical to those of *O. azurea* var. *diplopurpurea* and other members of the *O. macrocentra* complex. The stamens have pale green filaments with yellow anthers, the style is cream-colored, and the stigma lobes, on different plants, are pale green to cream-colored. The slender pericarpels have 18–26 areoles.

The fruits are pale red to pinkish and tend to dry and shrink at maturity. Fruit shape and size, and the seeds, are like those of var. *diplopurpurea.*

Phenology. Variety *discolor,* like the plants of var. *diplopurpurea* alongside it at the type locality, blooms in April, perhaps extending into May. In some years at least a few plants may flower in late March, as inferred from observation of large buds on 26–28 March 1997 on several plants at Big Hill. Flowers open about midday, close at night, and do not open again the next day. Fruit maturation occurs in June and July.

Sterile and Immature Specimens. Without fruits and seeds, these plants superficially resemble O. *camanchica* as much as O. *azurea* or O. *macrocentra.* Vegetative features alone distinguish var. *discolor* from the related purple prickly pear, var. *diplopurpurea.* In general the plants of var. *discolor* are slightly more upright, with pale green or blue-green (never purple) pads, stouter spines (1–1.5 mm across at the base) that are yellow, white, or golden in most individuals of the only known population. Some of the plants there have darker spines that are orange, red-orange, or reddish, but so far as known, they do not produce the blackish or blackish-and-white spines that characterize var. *diplopurpurea.* In var. *discolor* the areoles are farther apart on the stems, 4–6.5 cm, than in var. *diplopurpurea.*

Biosystematics. Variety *discolor* and var. *aureispina* are green-stemmed, yellow-spined members of the otherwise dark-spined (or blackish-and-white-spined), purple- or purplish-skinned O. *macrocentra/azurea* species group. Variety *discolor* is readily distinguished from var. *aureispina* by its fewer and much longer spines, which are thicker and not as straight.

The Big Hill plants of var. *discolor* occur within a few meters of plants of var. *diplopurpurea,* and they bloom together. At present we have no evidence that these taxa have ever directly exchanged genes.

Synonyms. None.

Common Names. None other than Big Hill prickly pear.

15. **Opuntia tortispina** Engelm. & Bigelow, Proc. Amer. Acad. Arts 3: 293. 1856. TWISTED SPINE PLAINS PRICKLY PEAR. Plates 76, 77. Igneous or limestone alluvium of mountain basins, in mountain woodlands, or less often in sand or sandy loam, rarely in desert mountains. Present in much of the Trans-Pecos, but mostly in grasslands of mountain basins, Davis Mts W to El Paso Co., not known in Terrell Co. 3,400–6,500 ft. Flowering (Apr–)May–Jun. $2n = 66$. From Trans-Pecos TX, the Panhandle, and NM, to CO; apparently also in N AZ, UT, NV, and S CA, and rare in SE AZ (D. Ferguson, pers. comm.). Mexico (according to D. Ferguson, pers. comm.), in NE Sonora, N Chihuahua, and NW Coahuila. Map 13.

Opuntia tortispina is a low, sprawling, often slightly wrinkled prickly pear that is common in the grassy plains from the Davis Mountains northward and westward. *Opuntia tortispina* is very similar to O. *cymochila* Engelm., and slightly less similar to O. *phaeacantha* or O. *macrorhiza.* *Opuntia tortispina* has been confused with all of these species in some previous publications (e.g., Ben-

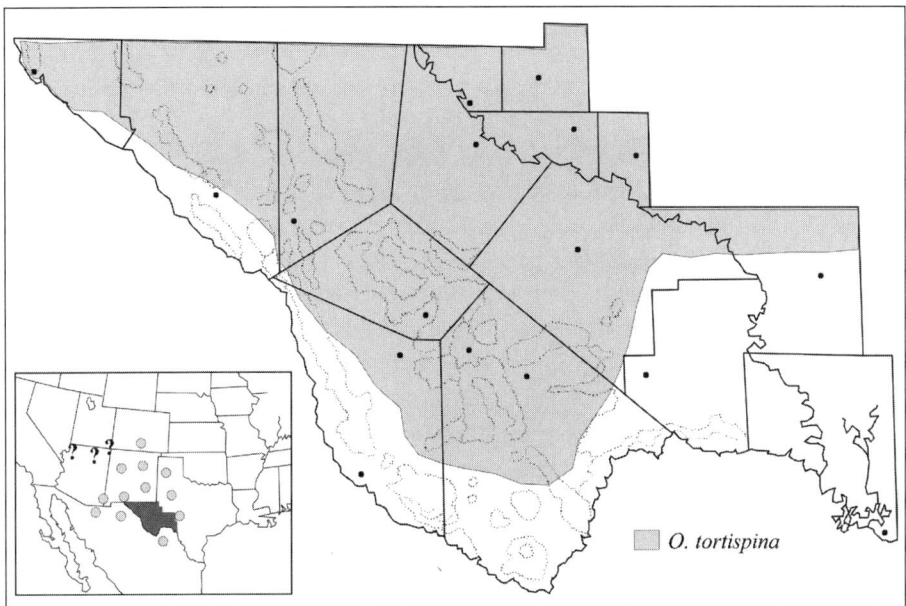

Map 13. Generalized distribution of *Opuntia tortispina* (twisted spine plains prickly pear).

son, 1982; Weniger, 1984; D. Ferguson, 1987). The type locality is "On the Camanchica Plains near the Canadian River." The specific epithet refers to twisted spines, after the Latin *tortus*, "twisting," and *spina*, "spine."

Identifying Characters. The habit of *O. tortispina* is low, creeping, prostrate, and sprawling, with some branches 2–3 pads high; whole plants usually are less than 30–40 cm high. At some sites the plants may be sympatric with *O. macrocentra* (the tetraploids), *O. azurea*. var. *diplopurpurea*, *O. phaeacantha*, or *O. polyacantha*, but in the central mountain basin grasslands, relatively pure stands of *O. tortispina* are common. The pads of *O. tortispina* are obovate to long-obovate, 9–19(–24) cm long, 7–15 cm in diameter; spinose areoles cover all or the upper three-fourths of the pad, usually with 4–5 white central spines per areole. The green stems may exhibit some purplish pigmentation, especially around the areoles, as in *O. phaeacantha*, much less purplish than *O. macrocentra*. *Opuntia tortispina* has more closely set areoles than does *O. phaeacantha*, and its pads tend to be slightly wrinkled after their first winter, as in *O. macrorhiza*, *O. pottsii*, and *O. polyacantha* var. *trichophora* (all of which are smaller plants).

The yellow flowers of *O. tortispina* are 4.5–7 cm long and 4–6 cm in diameter. The flowers in most populations have pale to prominent red or rusty-red centers, but in some plants the inner tepals are yellow to the base. This flower-color variation does not appear to have geographic significance. Plants with "peach color" flowers occur sporadically among the yellow-flowered plants in northwest Brewster County (*P. Zelazny, 129, 131*; SRSC). The tepals are up to ca. 3 cm long and 1.5–1.7 cm in diameter, oblong-obovate, and obtuse or apiculate at the apex. The filaments are 1–1.5 cm long, reddish distally and greenish proximally, or reddish from distal end to base. The pale yellow anthers are 2–2.5 mm long.

The cream-colored or whitish style is slender-urceolate and 1.5–2.3 cm long. The 5–7 stigma lobes are ca. 3 mm long and green. The pericarpel is 2–3 cm long, with nearly spherical areoles, these 1.5–2 mm in diameter, and with glochids and perhaps a few bristles.

The fruits are obovate to elliptic, to 3.5–4.5 cm long, 2–2.8 mm in diameter, and dull reddish, purple-red, rose-purple, or pale red on tan. The fruits are somewhat juicy and sweet with a light red to pale green pulp and fleshy rind thickened toward the top of the fruit. The seeds are 4–6 mm in diameter, flattened but irregular in shape, ca. 2 mm thick, with a beakless aril-rim ca. 1 mm wide.

Phenology. The main flowering period is from about mid-May to mid-June, but in some years peak flowering occurs by early May, and flowering may continue into mid-July (Anthony, 1956). At Limpia Crossing north of Fort Davis, plants of O. tortispina were back in bud on 20 June 1996 after a drought and then rain in June (L. Hedges, pers. comm.). In the spring of 1998, during a severe drought, populations in northern Brewster County, and perhaps elsewhere, produced very few flowers. Flowers open about midday to 2:00 P.M. and usually remain open for only one day. The fruits mature in about one month (Anthony, 1956) or more. In the vicinity of Alpine, ripe fruits are common in mid- to late August or early September.

Sterile and Immature Specimens. Commonly, plants of this species have smallish pads, 9–15 cm long and 7–11 cm in diameter. The areoles are 1.5–2.5(–4) cm apart, oblong, elliptic, to nearly spherical, especially on the apical margin. The mid-pad and upper areoles are 4–6 mm long, 2–5 mm in diameter, with a dense pile of brownish or gray hairs especially evident in the front half of the areole.

The spines of O. tortispina are interpreted here as either centrals or radials. Usually there are 4–5 centrals and two radials. The central spine pattern typically involves one lowermost deflexed central, a second deflexed central in the lower center of the areole just above the lowermost central, and 2–3 projecting centrals in the center to upper portion of the areole. The lowermost central is the shortest one at 1.4–2.5 cm long. The lowermost central usually is chalky-white, often gray- or black-speckled, flattened proximally or acicular, and strongly deflexed. The second lowest central also is chalky-white, flattened, twisted (in most plants examined), 3–5 cm long, and 0.7–1.1 mm wide. The 2–3 projecting centrals are light brown to reddish-brown proximally, fading to nearly white distally, or all white and often black-speckled. These longest centrals are flattened to subterete, twisted, 4.5–7.5 cm, and ca. 1 mm in diameter. Usually all of the centrals have yellowish retrorsely scabrous tips. Populations of O. tortispina often deviate from the central spine pattern described above, particularly in the Trans-Pecos central mountain region. Here the spines often are white, with no centrals in the upper part of many areoles, but with one lowermost central, two lower lateral centrals, and 1–2 upper centrals in some areoles. In other areoles there may be one lowermost central, two lower lateral, and 1–2 lower central spines positioned above the lowermost spine. Usually there are two radial spines in the areoles, these strongly deflexed and positioned at the lower periphery of the areoles on either side of the lowermost central. The radials are white with a yellowish tip, acicular, 0.2–1.2 cm long, and distinct in appearance from the lowermost centrals.

The ephemeral subulate leaves, borne in the lower front portion of the areole, are ca. 8 mm long. On young pads, typically there is only one spine per areole in distal areoles. In mid-pad areoles of slightly older stems, usually only the 1–2 lower deflexed spines are present, with 1–3 projecting upper spines appearing in distal areoles. In some distal areoles of young stems, 1–2 erect, bristlelike, yellowish projecting spines occur. These spines are ca. 1 cm long, stout proximally, and probably they represent the juvenile form of the longer projecting centrals. The seedlings are not hairy (D. Ferguson, pers. comm.)

Biosystematics. Benson (1982) synonymized *O. tortispina* (among diverse other taxa) under *O. macrorhiza* var. *macrorhiza*. Resultantly, Benson's uniquely broad concept of *O. macrorhiza* var. *macrorhiza* was a taxon distributed from Mexico north to Michigan and west to California.

Benson (1969b, 1982), working mostly from herbarium specimens, frequently misidentified *O. tortispina* specimens (and *O. pottsii*) as *O. macrorhiza* var. *macrorhiza*.

Opuntia macrorhiza (sensu stricto, excluding *pottsii*) is from Central Texas; it extends west only to about western Ward County, Texas, so far as known (D. Ferguson, pers. comm.), and north into the Great Plains region. *Opuntia macrorhiza*, like *O. pottsii*, is tetraploid ($2n = 44$), as based on chromosome counts from several states (Pinkava and McLeod, 1971; Pinkava et al., 1973, 1977, 1992, 1998). Weniger (1984) recognized *O. tortispina*, at least in part, as *O. cymochila*, and possibly also *O. compressa* (Salisbury) Macbride, at least in part. Anthony (1956) treated this wide-ranging low prickly pear as *O. tortispina* and also *O. tenuispina*. Like D. Ferguson (pers. comm.) and Weniger (1984, in part) we have interpreted the wide-ranging, low, slightly wrinkled prickly pears like these as composing at least two species: (1) *O. tortispina* is the common species in the grasslands of Trans-Pecos, and it extends through New Mexico into parts of Colorado, Utah, Nevada, and California, and south through Arizona into northern Mexico; (2) *O. cymochila* extends from Central Texas west to central New Mexico, and northward where it is one of the most common prickly pears on the Great Plains north at least to South Dakota. According to our interpretation of the species, *O. cymochila* is not known to occur in the Trans-Pecos or in Mexico.

Compared to *O. tortispina*, *O. cymochila* creases more deeply from wrinkling in the winter, has "softer" pads with more areoles, usually has fewer main central spines that often are terete and more slender, has more bristlelike lower central spines, and has distinct flowers and fruits (D. Ferguson, pers. comm.). *Opuntia tortispina* is consistently hexaploid ($2n = 66$) in the Trans-Pecos and Hutchinson County, Texas (Powell and Weedin, 2001), as are similar plants near Las Cruces, New Mexico (voucher at ASU).

The type locality for *O. tortispina* is near the Canadian River, east of the plateau of the Llano Estacado (Texas Panhandle near Borger), later lectotypified as "Comanche Plains," the Llano Estacado. The type locality for *O. cymochila* was "Comanche Plains east of the Llano Estacado near 100° longitude, and from there to Tucumcari hill 80 miles east of the Pecos," now lectotypified "Plaza Largo south of Tucumcari, New Mexico" (Benson, 1982). See Ferguson (1987) for a more detailed explanation of the localities.

In the Trans-Pecos *O. tortispina* stands visually intermediate between (1) its fellow hexaploid *O. phaeacantha* (equal-size or often much larger) and (2) *O. pottsii*. It is identified by its low habit with relatively small pads (which wrinkle transversely under stress, as in *O. pottsii* and *O. phaeacantha*), usually white spines, yellow flowers with weakly defined reddish centers (not a bold "target" pattern), and smallish red fruits. In some populations a few spines on the plants may be brown, reddish-brown, or even yellowish, particularly on proximal portions of the projecting centrals in distal areoles. In some western populations identified as *O. tortispina* by D. Ferguson (pers. comm.), for example, in Nevada, Utah, and Colorado, the flowers may be orange, magenta, or pink as well as yellow.

Synonyms. *Opuntia tenuispina* Engelm. & Bigelow; ? *O. charlestonensis* Clokey.

Common Names. *Opuntia tortispina* is one of several common prickly pears of the plains country in the central United States, some of which have been referred to as "plains prickly pear." Its specific epithet implies twisted spines, hence its additional name, "twisted spine plains prickly pear."

16. **Opuntia macrorhiza** Engelm., Boston J. Nat. Hist. 6: 206. 1850. PLAINS PRICKLY PEAR. Plates 78, 79. Sand and limestone, Ward, Winkler, Crane, and Ector counties, eastern periphery of the Trans-Pecos. 2,500–3,000 ft. Flowering Apr–Jun. $2n = 44$. In TX, almost throughout the state except for the Trans-Pecos, W nearly to the Pecos River. Widespread E of the Rocky Mts, but its true range obscured by misidentifications; rare in arid Southwest. Map 14.

Although *O. macrorhiza* technically is not known to occur in the Trans-Pecos, its near approach to the western boundary of Ward County and adjacent counties suggests that it might also exist west of the Pecos River in localized, probably sandy habitats. The type locality for *O. macrorhiza* is on the upper Guadalupe River in south-central Texas, with a lectotype designation from between the Pedernales and Guadalupe rivers. The specific epithet refers to the fleshy roots, after the Greek *makros*, "large," and *rhiza*, "root."

Identifying Characters. The low, sprawling plants of *O. macrorhiza*, usually less than 30–40 cm high, closely resemble those of a northern species, *O. cymochila* (Ferguson, 1987). They are in the size range of Davis Mountains *O. polyacantha*, slightly smaller than *O. cymochila* and larger than the tiny bluish-gray clumps (or single erect pads) of *O. pottsii*. The pads of *O. macrorhiza* (and others except for *O. pottsii*) tend to root where their edges touch the ground. Individual pads of *O. macrorhiza* are only 7.5–13 cm long and 8–12 cm wide. Usually there are only 1–3 main spines in areoles on the upper part of the pads in *O. macrorhiza*. The lowermost spine usually is flattened and deflexed, and the

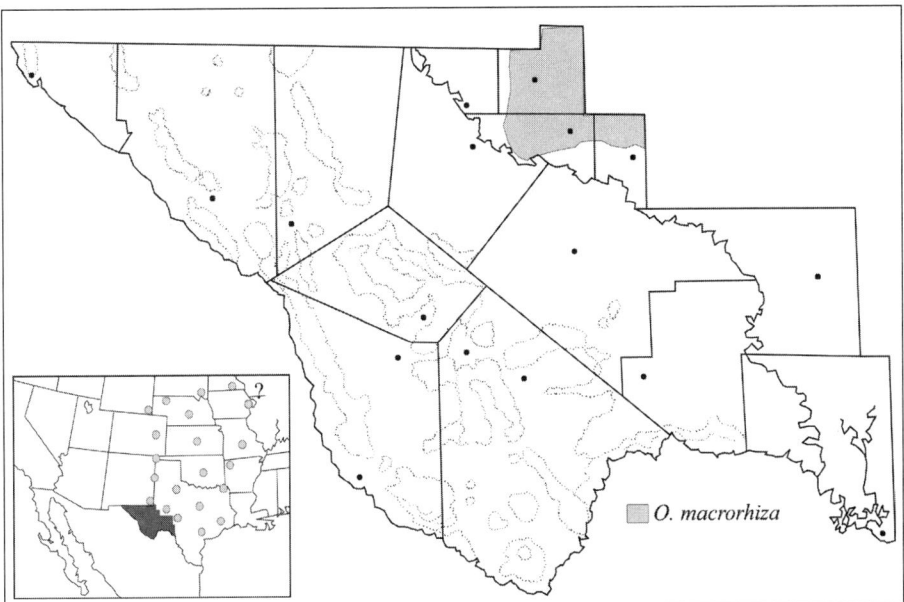

Map 14. Generalized distribution of *Opuntia macrorhiza* (plains prickly pear).

other one or two usually are subterete and project outward or upward. Typically the spines are chalky-white to whitish (perhaps rarely yellowish), and often their bases are yellow, orange, reddish, or brown.

The flowers of *O. macrorhiza* have fewer tepals than those of *O. cymochila*. Typically, the flowers of *O. macrorhiza* are yellow with sharply defined orange or red centers, whereas the flowers of *O. cymochila* are all yellow, or rarely with red centers. An additional prominent floral difference between the two species is the cream-colored or pale green stigma lobes in *O. macrorhiza* and the deep green stigma lobes in *O. cymochila* (Ferguson, 1987).

Flowers of *O. macrorhiza* vary in size; measurements include 6 cm long and 5–8 cm in diameter; the tepals often are 3–3.5 cm long, ca. 2 cm wide, narrowly obovate, and obtuse. The filaments are cream distally and greenish basally. The anthers are yellow. The cream-colored style closely matches the stigma. The stigma is barely elevated above the anthers, at least in some flowers. The pericarpel is slender, ca. 3 cm long; its few areoles are small and circular with brown hairs and relatively few glochids.

Ferguson (1987) reported that the fruits of *O. macrorhiza* are variable throughout the range of the species, with some of them being greenish, orange, to purplish-red and sour to the taste. The fruits of *O. macrorhiza* are dull reddish-purple in Ward County; there they are obovoid, 2.5–3 cm long, 1.5–2 cm in diameter, with a few small areoles. Not many glochids and rarely any spines are present in the areoles. The fruits have a clear juice with a taste like unsweetened apple, according to one of our tests. Fruits of the confusingly similar *O. cymochila* are dull purplish-red to brownish, ovoid, more slender below the apex; the areoles usually are prominent, with abundant glochids and often with slender spines. The fruit pulp of O. *cymochila* is juicy and purple, and the fruits are said to be among the sweetest of any prickly pear fruits in the United States (Ferguson, 1987). The seeds of O. *macrorhiza* are light brownish to tan, irregularly shaped, 3–3.5 mm in diameter (the seeds of O. *cymochila* are 5–6 mm in diameter), 1.5–2 mm thick, notched at the hilum, with a narrow margin to 0.5 mm wide (O. *cymochila* seeds have a broad margin).

Phenology. The main blooming period for O. *macrorhiza* appears to be April–May. After closing at night, the flowers usually do not open again. The time required for fruit maturation probably is 2–3 months, but seemingly ripe fruits have been observed in the field to persist through August in some years.

Sterile and Immature Specimens. Roots of the low, creeping plants of O. *macrorhiza* often form tuberlike swellings up to 2.5 cm thick, particularly when rooted in sandy substrates (Ferguson, 1987). The orbicular, obovate, or elongate-obovate pads are green, yellowish-green, or glaucous bluish, typically wrinkled after the first winter. The mid-pad areoles are 1.5–3 cm apart, narrowly elliptic to nearly round, 3–5 mm long, 2.5–4 mm wide, with the marginal areoles larger. The areoles support dense brownish to gray felt in the lower part and center.

The spines of O. *macrorhiza* areoles in a Ward County specimen (*Weedin 2065*, SRSC) are 1–2 in number and located one above the other in the lower central portion of the areole. The largest central spines are 3–5.5 cm long, 0.6–0.9 mm thick, usually flattened proximally or for one-half or more of the length, twisted, chalky-white, perhaps rust-colored proximally, and with yellowish tips. Both centrals are subequal, or the lower central is slightly shorter. The lower central typically is deflexed and more flattened; the upper one may be slightly deflexed in lateral areoles, or projecting in distal areoles.

The typical spine configuration in O. *macrorhiza* in its wider range includes 1–3 central spines, with one usually flattened and deflexed, and 1–2 usually terete and projecting. The centrals are whitish to yellowish with yellow, orange, or reddish-brown bases (Ferguson, 1987). Radial spines are present or absent. When present, the 1–2 radials are slender, deflexed, and located at the lower periphery of the areole. A dense apical tuft of reddish brown glochids, 2–4 mm long, occupies the mid-pad areoles, or the glochids may fill the upper one-half or more of the areoles.

Immature pads have 1–2 spines to ca. 1.5 cm long in the lower central portion of the areole, these arranged one above the other, with both of them deflexed or with the upper one projecting. Spines are produced in areoles located on the distal one-fourth of juvenile pads. The ephemeral leaves of young pads are ca. 7 mm long or more, these often glaucous and bluish-green. The seedlings of O. *macrorhiza* are not hairy (D. Ferguson, pers. comm.).

Biosystematics. Our concept of O. *macrorhiza* is a small subset of the O. *macrorhiza* as identified by Benson (1982). Benson divided his concept of O. *macrorhiza* between O. *macrorhiza* var. *macrorhiza* and O. *macrorhiza* var. *pottsii*. Unfortunately, Benson's perception of var. *macrorhiza* included many misidentified specimens of other species, including practically all of O. *tortispina* and O. *cymochila*, both recognized as distinct in the present treatment. Unlike Benson, we treat var. *pottsii* as a distinct species.

Our present concept of *O. macrorhiza* follows that of Ferguson (pers. comm., 1987). *Opuntia macrorhiza* is a tetraploid ($2n = 44$) species that is very closely related to *O. humifusa*, the common eastern prickly pear that occurs from the Atlantic coast of the United States west to extreme eastern Texas.

Opuntia macrorhiza has small, wrinkled pads, like *O. cymochila* and *O. tortispina*, but both of those taxa appear to be hexaploid ($2n = 66$; D. Ferguson, pers. comm.; Powell and Weedin, 2001). *Opuntia macrorhiza* occurs sympatrically with *O. cymochila* over a wide area (not in the Trans-Pecos), and occasional interspecific hybridization reportedly occurs between them (Ferguson, 1987). *Opuntia cymochila* occurs from Central Texas north to northern Colorado, Nebraska, and perhaps southern South Dakota, and west through much of New Mexico, but not west of the Continental Divide. As compared to *O. cymochila*, *O. macrorhiza* typically occurs in deeper soils and more mesic habitats, including tall-grass prairie (Ferguson, 1987). Neither *O. macrorhiza* nor *O. cymochila* is known to occur in the Trans-Pecos, but both taxa eventually might be found there.

Synonyms. *Opuntia mesacantha* Raf. var. *macrorhiza* (Engelm.) J. M. Coult.; *O. compressa* J. F. Macbr. var. *macrorhiza* (Engelm.) L. D. Benson; *O. fusiformis* Engelm. & Bigelow; *O. mesacantha* var. *greenei* J. M. Coult.; ? *O. grandiflora* Engelm.; *O. xanthoglochia* Griffiths; ? *O. fuscoatra* Engelm.

Common Names. Chain prickly pear; grassland pricklypear; grassland prickly pear; plains prickly-pear; tuberous-rooted prickly pear; nopal; prickly pear.

17. **Opuntia pottsii** Salm-Dyck, Cact. Hort. Dyck. [ed. 1849] 236. 1849 [1850]. POTTS' PRICKLY PEAR. Plates 80, 81. Alluvium in basin grasslands, sand, gypsum, or limestone away from the mountains. Northern Presidio and Brewster counties, reported in the Chisos Mts of S Brewster Co. (Wauer, 1980), Jeff Davis Co., basins of the Davis Mts; Culberson Co.; Reeves Co.; probably Pecos and Terrell counties; Winkler Co. and other counties adjacent to the Pecos River NE of the Trans-Pecos. 2,800–5,500 ft. Flowering (Apr–)May. $2n = 44$. Texas, reported in Maverick Co. (Benson, 1982); southern Panhandle S to Ector Co., and sand hills. Across southern NM and into SE AZ; central and NW AZ into SW Utah (D. Ferguson, pers. comm.). Mexico, NE Sonora and Chihuahua. Map 15.

Opuntia pottsii is one of the smallest prickly pears in the Trans-Pecos, along with *O. polyacantha*. *Opuntia pottsii* often goes unnoticed among grasses in the basins of the Davis Mountains until the plants produce red flowers in May. The type locality is near Chihuahua City, Mexico, which is at an elevation of about 4,000 feet. The species was named after John Potts, who managed the mint in Chihuahua, Mexico, collected the original material of the species, and between the years 1842 and 1850 sent many cactus collections to F. Scheer at the Royal Botanic Gardens, Kew, England (Britton and Rose, 1919–23).

Identifying Characters. The typically small plants of *O. pottsii* are upright and usually not more than 30 cm high. The pads are relatively small and glaucous blue-green when healthy. The usually 1–3 whitish to gray spines per areole are produced on the upper part of the pads. *Opuntia pottsii* is the only Trans-

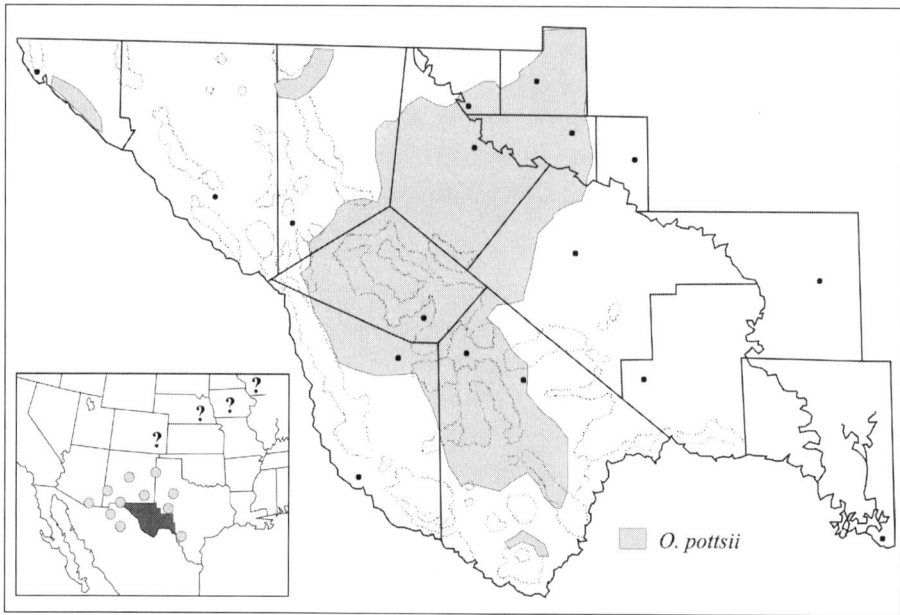

Map 15. Generalized distribution of *Opuntia pottsii* (Potts' prickly pear).

Pecos prickly pear with totally red flowers. In other species the flowers are yellow or yellow with red or orange centers, or the tepals may change to orange or light reddish if the flowers last a day longer than usual.

The flowers are 5–9 cm long and 4–6 cm in diameter. In some flowers the slender pericarpel alone often is ca. 5 cm long. The tepals are red to rose-red, orange-red, or pink, the inner ones broadly spatulate-apiculate, 3.5–4.5 cm long, and ca. 2 cm in diameter. The outer tepals are obovate-acuminate, and the outer bracts are ovate-lanceolate and attenuate. The filaments are 1.2–1.5 cm long, pale yellow, greenish-yellow, to purplish. The anthers are bright yellow and ca. 2 mm long. The style is pinkish, 1.7–2.3 cm long, and 2–3 mm wide near the base. The 5–8 stigma lobes are cream-colored, 3–4 mm long, and stout. The pericarpel is attenuate at the base, expanded distally, smooth, with scattered small areoles, these with white hairs and a few yellow or brown glochids.

The fruits are narrowly obovoid to clavate, 2.5–5 cm long, ca. 2 cm in diameter distally, with a slender base, and deep umbilicus. Typically, the fruits are glaucous and pinkish-purple (pigments confined to the skin). In some populations in Reeves County, and perhaps elsewhere, the ripened fruits are dull to light red or reddish-brown. The fruits are juicy with a pale greenish pulp and not very sweet. The seeds are 3.5–5.5 mm in diameter, thick-discoid, irregular in outline, wide-margined, and beaked at the hilum.

Phenology. The peak flowering time for *O. pottsii* is mid- to late May, at least in the Davis Mountains. In some years a few plants may bloom as early as mid-April, and in some years flowering probably extends into June. In localized populations most plants tend to produce flowers at much the same time, and the flowers may be finished in about one week. Individual flowers open about midday, close by 2:30–4:30 P.M., and do not open again. After limited observations it appears that fruits mature in 2.5–3 months, in some cases perhaps requiring four months or even longer. In the Davis Mountains region, mature fruits appear on plants from mid-August to October.

Sterile and Immature Specimens. In habit the plants of *O. pottsii* are low but with usually upright stems and often a short trunklet. Commonly, plants produce only 6–10 pads. The roots are fleshy and exude a milky sap when injured. The tuberous-enlarged, or at least strongly succulent, taproot is single or branched, smooth or constricted (Weniger, 1984). The literature always describes the roots as tuberous (Anthony, 1956; Ferguson, 1987). The pads are orbicular, obovate, or even subrhombic or stipitate, the smallest mature ones 4–5 cm long and 3–5 cm in diameter, the largest ones 10–13 cm long and 8–10 cm in diameter. The pads are glaucous blue-green when healthy and yellow-green or purplish when stressed. Healthy pads may be purplish around the areoles or on larger portions of the stem. The areoles are 1–2.6 cm apart, narrowly obovate to nearly circular, 3–3.5 mm long, 2–3 mm across, and covered with wrinkled whitish hairs and a tuft of glochids.

Spines are produced in areoles near the stem apex, usually the upper one-third or one-fourth of the pad, including the upper margin. There are 1–4 central spines per areole, these commonly twisted, whitish or gray and black-speckled, the longest 1.4–6.4 cm, ca. 0.2 mm thick, and acicular or somewhat flattened. In each areole usually one spine is longer, while the others are shorter and of different lengths. In each areole one or two spines are deflexed, often the lower 1–2 spines, and the middle and upper 1–2 spines may be projecting, particularly in areoles on the pad margin. Often the central spines are purplish-black proximally and whitish distally, or purplish-black throughout or nearly so. Radial spines are absent, or in some areoles there may be two or more brownish, bristle-like radials in a lower peripheral position. Rarely plants of *O. pottsii* are spineless, or some pads are spineless. Glochids are present in most mature areoles, especially those on the upper half of the pads. The glochids are dirty-yellow, 3–5 mm long, and numerous in a tight tuft.

On young pads the ephemeral leaves are small, only 3–4 mm long. The areoles of young pads have kinked white hairs and a tuft of glochids ca. 1 mm long. In addition to having glochids, young areoles may support 3–9 projecting bristles, these 2–3 mm long and brownish like the glochids but not retrorsely barbellate like the glochids. The areoles of immature stems have 1–2 short spines, or none. In juveniles the areoles are smaller and circular, with very short hairs and few glochids. Seedlings of *O. pottsii* are not hairy (D. Ferguson, pers. comm.).

Biosystematics. Benson (1982) treated *O. pottsii* as a variety of *O. macrorhiza*. Our concept of *O. macrorhiza* is narrower than that published by Benson (see *O. macrorhiza* and *O. tortispina*). In our opinion *O. pottsii* clearly is distinct as recognized by Weniger (1984) and other workers.

In the present treatment we view *O. ballii* Rose as a synonym of *O. pottsii* (Benson, 1982), as opposed to Weniger (1984) who maintained *O. ballii* as distinct. Future study of *O. ballii,* named for the original collector, C. R. Ball, might show that it should be treated as a distinct taxon.

Although we have observed only red flowers in Trans-Pecos *O. pottsii,* D.

Ferguson (pers. comm.) states that flower color is not of taxonomic significance in *O. pottsii* throughout its range, where red, magenta, pink, orange, yellow, white, and combinations of these colors are to be found, sometimes together in the same populations. When yellow, the flowers often have red centers. According to Ferguson, some of these different flower colors occur in the Trans-Pecos, and thus future populational studies of the taxon should be timed so that flower color can be observed. Ferguson also believes that a second variety of *O. pottsii*, yellow-flowered and as yet not formally recognized, occurs in the mountains from New Mexico and eastern Arizona northward into Colorado and Utah, and possibly in the Guadalupe and higher Davis mountains of the Trans-Pecos. Weniger (1984) too recognized such a taxon (he equated it with *O. stenochila* Engelm.) but viewed it as a variety of *O. compressa* (along with *O. macrorhiza*, sensu stricto) instead of *O. pottsii*.

Synonyms. *Opuntia filipendula* Engelm.; *O. ballii* Rose; *O. delicata* Rose; *O. plumbea* Rose; *O. loomsii* Peebles; *O. macrorhiza* Engelm. var. *pottsii* (Salm-Dyck) L. D. Benson; ? *O. stenochila* Engelm.

Common Names. Potts prickly pear; Potts pricklypear cactus.

18. **Opuntia mackensenii** Rose, Contr. U.S. Natl. Herb. 13: 310, t. 67. 1911. MACKENSEN'S PRICKLY PEAR, SHORT-SPINED PURPLISH PRICKLY PEAR. Plates 82–85. Rocky or alluvial substrates, limestone or igneous, desertscrub, grassland or woodland. Presidio and Brewster counties E to Terrell and Val Verde counties. 1,000–4,600 ft. Texas, Edwards Plateau N to the Panhandle. Presumably Mexico in Coahuila. Map 16.

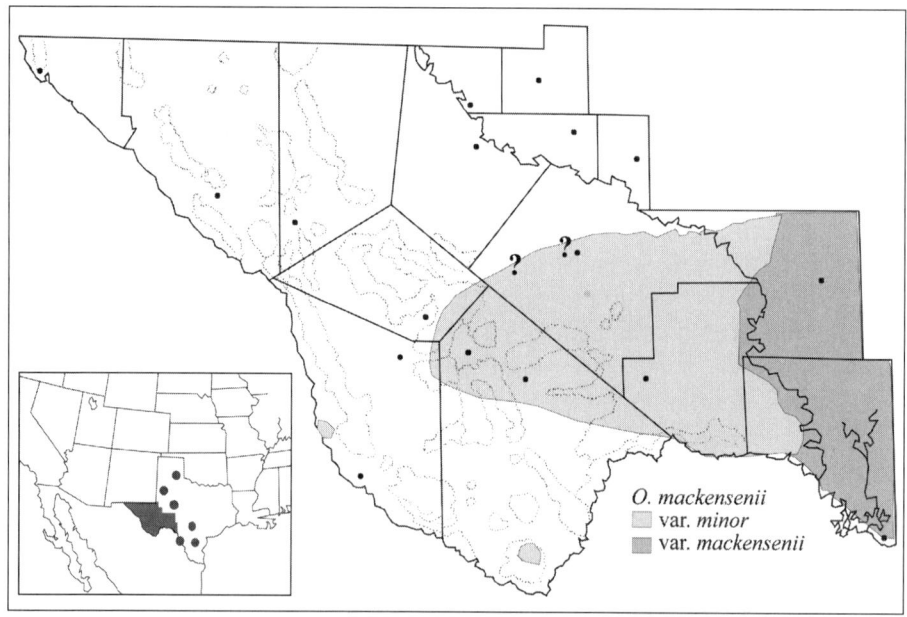

Map 16. Generalized distribution of *Opuntia mackensenii* var. *mackensenii* (Mackensen's prickly pear) and *O. mackensenii* var. *minor* (short-spined purplish prickly pear).

Recent workers have not recognized *O. mackensenii*, except possibly D. Ferguson (unpub.) and A. D. Zimmerman (pers. comm.). Benson (1982) included *O. mackensenii* (among diverse other species) in his concept of *O. macrorhiza* var. *macrorhiza*, with the statement that plants described as *O. mackensenii* were intermediate between *O. macrorhiza* and *O. phaeacantha* var. *major*. Weniger (1984) did not mention *O. mackensenii*. The type locality of *O. mackensenii* is "near Kerrville," Kerr County, Texas, where it was originally collected in 1909 by Bernard Mackensen, a collector and author of *Opuntia* species in the vicinity of San Antonio, Texas. The species was named in honor of Mackensen in 1911 by J. N. Rose (see Britton and Rose [1919–23] for a photo of the type).

Tentatively, we recognize *O. mackensenii* as a relatively widespread and common species with two varieties in Texas. We have conducted only superficial studies concerning the populations we interpret here as var. *mackensenii*, and virtually no research east of the Pecos River. The inclusion of Anthony's (1956) *O. macrocentra* var. *minor* and the similar "grassland tetraploids" (discussed later) with *O. mackensenii* is merely one way to organize these poorly understood populations for the present time.

Key to the Varieties

1. Spines usually whitish; eastern distribution, relatively mesomorphilic
   18a. *O. mackensenii* var. *mackensenii*.
1. Spines usually dark purple or reddish-brown to nearly black, often with whitish tips; western distribution, relatively xeromorphic
   18b. *O. mackensenii* var. *minor*.

18a. **Opuntia mackensenii** Rose var. **mackensenii** MACKENSEN'S PRICKLY PEAR. Plates 82, 83. Alluvial or rocky substrates, mostly limestone, among grasses in shrubland, woodland, or grassland habitats. Terrell Co., from W of Sanderson E through Val Verde Co. 1,000–3,000 ft. Flowering spring. $2n = 44$. Crockett Co. E throughout the Edwards Plateau, N through rolling plains and high plains to Randall Co. in TX Panhandle (Grant and Grant, 1979b). Presumably in adjacent Coahuila, Mexico. Map 16.

The type locality for the recent synonym *O. edwardsii* (Grant and Grant, 1979b) is Pedernales Falls State Park, Blanco County, Texas. Despite ending in -*ii* instead of in -*ensis*, the epithet *edwardsii* was taken after the Edwards Plateau, an area of major distribution for *O. mackensenii* var. *mackensenii*.

Identifying Characters. The plants of var. *mackensenii*, like most of its allies, are low, sprawling subshrubs 20–45 cm high with slightly ascending branches. Sometimes the roots include tuberous thickenings. In habit and other morphology var. *mackensenii* closely resembles *O. phaeacantha* var. *phaeacantha*, which has a more westerly distribution and slightly thinner pads. In var. *mackensenii* spines are produced over three-fourths to nearly the entire pad, with 2–5(–6) usually white spines in the distal areoles. In the lower part of the areole usually there is one larger deflexed spine and one or more shorter deflexed spines. In the upper part of the areole spines may be absent, or there may be 1–2 projecting spines.

The flowers of var. *mackensenii* are 7–8 cm in diameter, and basically yellow or (in our experience) yellow with red, pale red, or reddish-brown centers. The filaments are colorless, and the anthers are light yellow. The style is whitish to cream with pale green to cream stigma lobes. The flowers on some plants have longer styles, exceeding the length of stamens by 5–7 mm, but on other plants the stigma barely extends above the anthers.

The fruits of var. *mackensenii* are purple, rose, to pale red, obovate, fleshy, and 3–6 cm long. The fruit pulp, as determined from one specimen in Terrell County, is dull to bright green with the seeds in a small amount of clear juice, matching the original description. In profile the fruits are truncate apically, and the areoles are spineless. The seeds are discoid, suborbicular in outline, and 4–6 mm in diameter ("5 to 6 mm" in the original diagnosis).

Phenology. Under cultivation in Alpine, plants of var. *mackensenii* have bloomed from early to mid-May. In the Edwards Plateau region, they bloom 2–3 weeks earlier than a sympatric species (presumably *O. tardospina* Griffiths) reported as "the hill-country race of *O. lindheimeri*" (Grant and Grant, 1979b). Flowers open about midday and usually do not open again after closing at night.

Sterile and Immature Specimens. In var. *mackensenii* the sprawling branches rest with their lower edges on the ground. Healthy mature pads reportedly are darker or more bluish-green than those of *O. engelmannii*, while young pads may be pale green and glaucous. The pads are narrowly to broadly obovate to orbiculate or obovate-rhombic, 10–20 cm long and to 20 cm in width. Mid-pad areoles are 2–3.5 cm apart, oblong, 4–6 mm long, and 2.5–4 mm wide. The apical portion of each areole is densely packed with glochids, these perhaps extending to mid-areole and even farther along the margins. The lower to mid-areole space is filled with wavy hairs, these dark or gray.

We have examined relatively few specimens of var. *mackensenii*, but the spine pattern may involve both central and radial spines, according to our interpretation, and the spines typically are white. Usually there are two deflexed lower central spines, these positioned one above the other. The lowermost central is the shortest, 1–2.5 cm long, and the upper one of the lower pair is the largest, 4–5.5 cm long and 0.7–1 mm in diameter. The lower deflexed centrals are flattened proximally and terete distally. In addition to the two lower deflexed centrals, which may be the only centrals in some areoles, there may be a projecting central in the upper portion of the areole. Or there may be 1–2 lowermost deflexed centrals plus two lower lateral and deflexed spines. Another pattern exhibits two lower deflexed, two lower lateral deflexed, and 1–2 upper projecting centrals. Other spine patterns include two lower lateral deflexed plus two upper projecting spines, or two deflexed lowermost lower lateral, one lower central deflexed, and two upper projecting centrals. The central spines usually are ashy-white or gray, these perhaps black-speckled, perhaps brownish to brownish-black at the base, with reddish-brown tips. The projecting upper centrals, when present, may be slightly flattened proximally but terete for most of their lengths. Upper centrals are most commonly found in distal areoles. The longer centrals may be twisted lengthwise. Spines in hybridized plants may not be all white but may vary from pinkish to brownish (Grant and Grant, 1979b), at least at the base. One or two short (3–6 mm) radial spines are present in some areoles, particularly in distal areoles, in a lower peripheral position. They are slender, terete, and white with yellow tips. The glochids are 2–3 mm long, reddish to reddish-brown, brown, or yellowish, or with yellowish bases.

Immature plants of var. *mackensenii* have not been examined for the present study, but young pads usually have areoles with only the 1–2 lower deflexed central spines. Seedlings of var. *mackensenii* are not hairy (D. Ferguson, pers. comm.).

Biosystematics. Grant and Grant (1979b) regarded var. *mackensenii* as a member of the *O. phaeacantha* complex. These are juicy-fruited prickly pears that do not wrinkle transversely under stress. They are larger than any of the hexaploid wrinkling species, including *O. tortispina*, and much larger than *O. macrorhiza* and *O. pottsii*, with which it is sympatric over a large region.

Variety *mackensenii* is mostly tetraploid ($2n = 44$), although Grant and Grant (1979b, 1982) also reported a hexaploid ($2n = 66$) count for the taxon. The morphologically similar *O. phaeacantha* var. *phaeacantha* apparently is consistently hexaploid (Pinkava and McLeod, 1971; Pinkava et al., 1973, 1985; Pinkava and Parfitt; 1982; Weedin et al., 1989). Variety *mackensenii* is not known to be sympatric with *O. phaeacantha* var. *phaeacantha*, but east of the Pecos River var. *mackensenii* is sympatric with *O. camanchica* ($6x$; = part of *O. phaeacantha* "var. *major*" of authors) and *O. engelmannii* var. *lindheimeri* ($6x$), and var. *mackensenii* is reported to occasionally hybridize with these taxa, forming pentaploids ($2n = 55$; Grant and Grant, 1982; Powell and Weedin, 2001; Pinkava et al., 2001). In the western portion of its range, var. *mackensenii* is sympatric with *O. engelmannii* ($6x$), but it is not known to hybridize with this taxon. Additional study of var. *mackensenii* is needed, particularly in its western territory.

Synonym. *Opuntia edwardsii* V. E. Grant & K. A. Grant.

Common Name. Edwards Plateau opuntia.

18b. **Opuntia mackensenii** Rose var. **minor** (M. S. Anthony) A. M. Powell & J. F. Weedin, comb. nov. (under consideration). SHORT-SPINED PURPLISH PRICKLY PEAR. Plates 84, 85. [*Opuntia macrocentra* Engelm. var. *minor* M. S. Anthony, Amer. Midl. Naturalist 55: 244, f. 21. 1956]. Sand and sandy loam, *Larrea* and *Prosopis* flats, Presidio and Brewster counties; grasslands and slopes of igneous hills, S Davis Mts, N Brewster Co.; limestone hills and mesas, Terrell and Val Verde counties. 2,500–4,600 ft. Flowering Apr–May. $2n = 44$. Presumably in adjacent Chihuahua and Coahuila, Mexico. Map 16.

Opuntia mackensenii var. *minor* was described as *O. macrocentra* var. *minor* by Anthony (1956) as a product of her study of the Opuntiae of the Big Bend region of Texas. The type locality is 1.4 miles southeast of Ruidosa in southern Presidio County, Texas. The varietal epithet is after the Latin *minor*, "smaller," supposedly a reference to the smaller characters of var. *minor* as compared by Anthony to *O. macrocentra* var. *macrocentra*.

Identifying Characters. The plants of var. *minor* are sprawling and many-branched, often with a low, creeping habit, ca. 25–60(–70) cm high, with some of the branches ascending. The stems are glaucous and blue-green, with reddish-purple pigmentation increasing in plants (at least those near the type locality) under environmental stress. The pads are obovate, 12–16 cm long, and 8–12.5 cm wide. The spines are 3–4.5(–6.2) cm long, dark purple to dark reddish-brown, some with whitish tips. Usually there are 4–9 spines in the areoles, and spiny areoles are present on the upper one-half to three-fourths of the pad.

The yellow flowers with red centers, or at least a reddish blush, are similar to those of *O. macrocentra* and *O. azurea*. The flowers are up to 7.5 cm long and 7.5 cm wide when fully open. The yellow inner tepals have reddish midstripes (extending to the tepal tips), not merely the red bases. The filaments and anthers are cream-yellow. The style is cream-colored, as are the stigma lobes.

The fruits are red-brown to dull red and fleshy at maturity. The base of the fruit often remains green long after the distal portion turns reddish. The fruits are short ovate-truncate to oblong-elliptic, 2.5–3 cm long, 1.5–1.7 cm in diameter, with a shallow umbilicus and without a constriction below the upper rim. The rind and pulp are green. The juice is moderate in amount, sweet, and clear. The seeds are tan, flattened, 4–5 mm in diameter, including a prominent margin that is 0.5–0.6 mm wide.

Phenology. Variety *minor* blooms in April and May at the type locality near the Rio Grande. There, we once observed a few plants in flower on 19 June, following a very dry winter and spring in 1999. Under cultivation in Alpine, plants were in full bloom in early to late May, and fruits matured in July, usually earlier than those of most prickly pears except for *O. azurea* and other species native to the southern Big Bend. Flowers usually remain open for one day.

Sterile and Immature Specimens. The plants of var. *minor* are without trunks. The roots are fibrous, but sometimes fleshy (Anthony, 1956). The branches are 3–5 pads long. Areoles are 2–3.5 cm apart on the pads, and they are oval to spherical, ca. 4–5 mm long and wide, larger and crowded on the distal margin of the pads.

The spines, as interpreted here, are distinguishable as centrals and radials. Usually there are 4–5 centrals. The lowermost central spine in the lower central portion of a typical areole is deflexed, whitish, and 1–2.2 cm long. Just above it usually there is a single, deflexed, dark, flattened second central that typically is 3–3.5 cm long. Usually there are 2–3 strongly projecting dark upper centrals (up to seven upper centrals are present in some distal areoles), which are 3.5–4.5(–6.2) cm long. The upper centrals are flattened or terete and dark purple to reddish-brown, nearly black in populations away from the Rio Grande, and often with whitish tips. Usually there are two slender, bristlelike radial spines at the lower front periphery of the areole; they are yellowish, to 6–8 mm long. The glochids are brown to light brown, tufted in the apex of the areole, and extend down the areole periphery. The glochids are up to ca. 3 mm long in mid-pad areoles and up to 1.5 cm long in distal, marginal areoles.

Immature pads of var. *minor* have fewer and shorter spines. The lower two central spines are present in proximal areoles and up to three centrals appear in progressively more distal areoles. The seedlings of var. *minor* are not hairy.

Biosystematics. Anthony described O. *mackensenii* var. *minor* as a variety of O. *macrocentra*, the purple prickly pear. Variety *minor* resembles O. *macrocentra* and O. *azurea* in its habit, pad shape, purplish stem pigments, spine configuration (but the spines are shorter), and the flower morphology.

Plants most closely matching the original description of var. *minor* occur from near Candelaria to east of Ruidosa in southern Presidio County, and near Santa Elena Canyon in southern Brewster County (Anthony, 1956). The tetraploid chromosome number ($2n = 44$) for var. *minor*, at or near the type locality, was first reported by Weedin et al. (1989) and later substantiated by counts from three individual plants (Powell and Weedin, 2001). Tetraploid plants (Powell and Weedin, 2001; the "grassland tetraploids") in Pecos County, Presidio County, northern Brewster County, near Dryden in Terrell County, and near Langtry in Val Verde County closely resemble var. *minor* in spine and flower morphology, although in general the nearly black to dark reddish-purple spines are darker proximally than are those of var. *minor*, and they are more typically white-tipped. These tetraploid midsize prickly pears only infrequently form dense populations near Alpine, where commonly they are sporadic and sympatric with various hexaploid species, including O. *tortispina*, which is locally abundant in the grasslands. The tetraploids east of Fort Stockton and near Dryden form more extensive populations where they are sympatric with O. *engelmannii*, O. *camanchica* (both hexaploid), and O. *strigil* (diploid). The tetraploid Langtry plants (A. M. Powell, unpub.), here tentatively assigned to var. *minor*, appear to have more densely spined pads and more prominent, white lower central spines than the tetraploids farther west, and at least some plants do not have flowers with red centers (*Powell and Powell 6249*). The tetraploids in question closely resemble each other under cultivation in the same *Opuntia* garden at Alpine, except that the Ruidosa plants tend to be more purplish. The Ruidosa tetraploids may prove to be more closely related to the O. *macrocentra* tetraploids or to the O. *azurea* diploids than to the O. *mackensenii* tetraploids, or even be distinct at the species level. And currently we suspect that the "grassland tetraploids" may represent one or more taxa that are either undescribed or ones for which we have not yet associated a previous name(s).

Ferguson (pers. comm.) questioned whether the type collection (*Anthony 1081*, in 1948, at MICH) represents the tetraploid population discussed here or a diploid purple prickly pear. The type collection has smaller pads and slightly larger spines than average in the tetraploid population near Ruidosa in Presidio County, but otherwise the type specimen and the recent collections are similar. Our repeated visits to the type locality of var. *minor* convince us that Anthony's type collection came from the tetraploids that were the dominant prickly pear taxon present there in 1999. We did not find any plants of the diploid purple prickly pear along the road or across the fences in the area 1–3 miles southeast of

Ruidosa. The diploid *O. azurea* var. *diplopurpurea*, however, is conspicuous in the surrounding region. Plants resembling *O. engelmannii, O. dulcis, O. camanchica,* and possible hybrids between these hexaploids also are present at and near the type locality of var. *minor*. The pad shape, pad color, and shorter spine length of var. *minor* are different from the diploid purple prickly pears in the region. The photograph of a plant at Presidio, published as var. *minor* by Ferguson (1988, fig. 1), exhibits the long, sparse spines typical of diploid var. *diplopurpurea*, not the var. *minor* described by Anthony (1956).

Synonym. See basionym under *O. mackensenii* var. *minor*.

Common Names. There was no popular name for this taxon, which even specialists have barely noticed among all of the other prickly pears.

19. **Opuntia chisosensis** (Anthony) Ferguson. CHISOS PRICKLY PEAR. Plates 86, 87. [*Opuntia lindheimeri* Engelm. var. *chisosensis* M. S. Anthony, Amer. Mid. Naturalist 55: 252, f. 26. 1956]. Oak-juniper-pinyon woodland and grassy meadows, middle to upper mountain slopes and canyons. Brewster Co., Chisos Mts. 5,200–7,200 ft. Flowering May. $2n = 22$. Mexico: Coahuila, Maderas del Carmen (igneous, wooded portion of the Sierra del Carmen); to be expected elsewhere in Mexico (Ferguson, 1986). Map 12.

Opuntia chisosensis is a distinct species with an interesting taxonomic history (see Biosystematics). The type locality is the Basin, Chisos Mountains, Big Bend National Park, Brewster County, Texas. The species is named after the Chisos Mountains, using the Latin suffix *-ensis*, meaning "belonging to." In the United States *O. chisosensis* appears to be restricted to the Chisos Mountains.

Identifying Characters. *Opuntia chisosensis* is readily identified in the Chisos Mountains as the medium-size prickly pear with bright yellow spines. A few of them have reddish spines instead, especially on lower pads, but are otherwise identical. The species is common in the wooded areas. The plants are upright shrubs to ca. 1 m tall, with branches spreading and ascending from a thick base, but not a trunk. The orbicular to short-obovate pads are pale gray-green or blue-green and glaucous to medium- or yellow-green, with spines in the upper two-thirds of the areoles. The spines are yellow on whole plants or on upper stems, but spines on lower stems might be reddish to nearly black, apparently a function of age. There are 1–5 spines per areole, these mostly deflexed on the face of the pads, to upright and spreading at all angles on the upper margin of the pads. The largest spines are about 3–6 cm long.

The flowers of *O. chisosensis* are about 5–6.5 cm long and wide. The tepals are an unusual yellow color, described by Ferguson (1986, p. 124) as "pale yellowish-buff." The pale yellow color fades to pale salmon at the end of the day. The filaments are pale green, and the anthers are yellow. The style is yellow, and the stigma lobes are green. The pericarpel areoles are small and distant.

The fruits are relatively small, compared to those of most other prickly pears in the Trans-Pecos, and conspicuous in smaller size compared to those of other prickly pears in the upper Chisos Mountains. The small fruits are variable in size and shape. In a transect from the Basin to Boot Spring we observed fruits that were 2–3.5(–4.5) cm long, 2–3.5 cm in diameter, globose, depressed globose and truncate at both ends except for the apical umbilicus, short-campanulate, cup-shaped, short oblong, and obovate. One plant had obconic fruits to 4.5 cm long. The umbilicus is saucerlike to 4–7(–9) mm deep and 0.7–1 cm wide. The areoles are distant, bearing only a few short glochids, and one or several areoles at the upper rim bear one to several slender yellow spines or bristles ca. 0.5–1.2 cm long. The fruits are juicy, red to beet red, or dark red to reddish-purple, and often glaucous. The rind and pulp are beet-red, and the juice is sweet and clear, stained with red pigment from the rest of the fruit. The seeds are yellowish, compressed or flattened and irregular in shape, 3–4.5(–5) mm long, 3–4 mm wide, and 1.3–1.5(–2) mm thick. The aril-margin approaches 1 mm wide, and appears to be beaked on some seeds and not beaked on others. The seeds are notched on one side at the hilum.

Phenology. The main blooming period for O. *chisosensis* is May. Flowers open first on plants at elevations near that of the Basin and progress to higher elevations at the top of the mountains. At higher elevations blooming extends into June. Under cultivation in Alpine, plants of O. *chisosensis* produced their first open flowers on 9 May 1998 and continued to produce flowers for 2–3 weeks until all the buds were spent. Flowers open during late morning, on one occasion by 10:20 A.M., or by midday, close by late evening, and may or may not open again the next day. In May 1999, Patrick Griffith observed a cultivated plant in Alpine to produce flowers that were open for one or one and one-half days. We have seen flowers of O. *chisosensis* in cultivation open for 1–2(–3) days. Fruits on cultivated plants were observed to ripen in ca. 1.5 months, beginning in late June. In the Chisos Mountains fruits may persist on plants at high elevations at least into mid-September, whereas fruits at Basin elevations by this time have fallen from the plants.

Sterile and Immature Specimens. The plants of O. *chisosensis* have fibrous roots and up to six pads per branch (Anthony, 1956). The largest orbicular to ovate pads are 16–29 cm long, 13–22 cm wide, and 0.6–1.4 cm thick. Most plants also support numerous smaller pads. The areoles are relatively distant, 3–3.5 cm apart, and they are narrowly elliptic, oblong, oval, to circular in shape. The areoles are smaller in size on the surfaces of the pads and larger on the apical margin of the stem. Surface areoles are ca. 4 mm long and 2–3 mm across, and areoles on the stem crest are 7–8 mm long and 5–6 mm wide. Areoles usually support abundant tan wool.

The younger spines of O. *chisosensis* are bright yellow with pale yellow tips. Spines on older stem joints may be yellow, reddish-orange perhaps with yellow bases and tips, or nearly black. Apparently the spines turn dark after several years (Ferguson, 1986). The 1–5 central spines are borne at the lower margin and in the center of the areole. The spines are subulate, flattened, curved, and mostly deflexed in areoles on the sides of the stems. Usually there is one larger spine, to 6.7 cm long and 1.5 cm wide at the base, positioned at the lower center margin of the areole. The 1–4 shorter spines usually are 1–5.5 cm long, with 0–2 of them positioned at or near the center of the areole. In herbarium specimens these center spines usually are broken near the base. One severely deflexed lower central spine, usually 1–2 cm long and acicular, is in the position of a lone radial spine. This "radial" spine is most noticeable when there are 3–4 other, obvious central spines in the areole. Radial spines are either absent, or there may be the one radial mentioned above. In some areoles, particularly those on the upper stem margin, there are one or a few, or numerous, slender, erect, bristles to 1 cm long in radial position at the margin of the areole. These bristles, probably not to be regarded as radial spines, appear to replace glochids in some areoles. Yellowish glochids are present in most areoles, seemingly fewer in surface areoles, and in dense tufts throughout or in a semicircle at the upper areole margin. The glochids are retrorsely scabrous and 2–3 mm long.

Immature stems of *O. chisosensis* usually have 1–2 spines per areole, these to 2.5 cm long. When there are two spines, often the lower one is deflexed, and the middle one is porrect. Juvenile pads often are obovate. Seedlings are not hairy (D. Ferguson, pers. comm.).

Biosystematics. Opuntia chisosensis was described originally (Anthony, 1956) as a variety of *O. lindheimeri,* which also has yellow spines. Benson (1982) incorporated *O. chisosensis* into *O. lindheimeri* var. *lindheimeri,* believing it to be part of a hybrid swarm involving *O. lindheimeri* and his *O. phaeacantha* var. *major* (= *O. camanchica*) and *O. phaeacantha* var. *discata* (= *O. engelmannii* var. *engelmannii*). The interpretation of *O. chisosensis* by Weniger (1984) is not clear, but according to Ferguson (1986), Weniger may have equated the Chisos Mountains plants with both *O. engelmannii* var. *cyclodes* Engelm. & Bigelow and *O. engelmannii* var. *cacanapa* (Griffiths & Hare), *nom. nud.* Ferguson elevated *O. chisosensis* to species status and further discussed the taxonomy of var. *cyclodes,* var. *cacanapa,* and other species with historical or biological connection to *O. chisosensis.* We agree with Ferguson that *O. chisosensis* is a distinct species that has its closest relationship to the other diploids, such as *O. aureispina* and *O. azurea,* and related taxa, and not the hexaploid *O. engelmannii* species group (including var. *lindheimeri*), which has larger pads, flowers, and purple fruits, along with smaller seeds and hairy seedlings.

The uncommon presence of certain yellow-spined prickly pears in the Davis, Guadalupe, and Franklin mountains suggests that *O. chisosensis* occurs outside the Chisos Mountains. Although the extra-Chisos yellow-spined plants are not well understood at present, D. Ferguson (pers. comm.) speculates that they are not *O. chisosensis* but possibly *O. cyclodes* Rose, a hexaploid taxon mostly restricted to the upper Pecos and Canadian River area in northeastern New Mexico. Ferguson's concept of *O. cyclodes* extends into Texas in the Panhandle west of Amarillo, and into the Trans-Pecos through the moist New Mexico mountains east of the Rio Grande (D. Ferguson, pers. comm.).

Synonym. See basionym under *O. chisosensis.*

Common Name. Chisos prickly pear is the only common name that has been used by those who recognized the taxon as distinct.

20. **Opuntia spinosibacca** M. S. Anthony, Amer. Mid. Naturalist 55: 246, f. 22–23. 1956. SPINY-FRUITED PRICKLY PEAR. Plates 88, 89. Limestone hills, slopes, and canyons. Brewster Co., W of Hot Springs E along and near Rio Grande to near Reagan Canyon. 1,600–2,300 ft. Flowering (Mar–)Apr(–May). $2n = 44$. Not documented from Mexico, but probably in limestone near the Rio Grande, vicinity of Boquillas. Reports from elsewhere require special documentation because of uncertainties concerning the limits of variation in O. *phaeacantha*, sensu lato, O. *azurea*, and the partially explored *Opuntia* diversity in Mexico. Map 17.

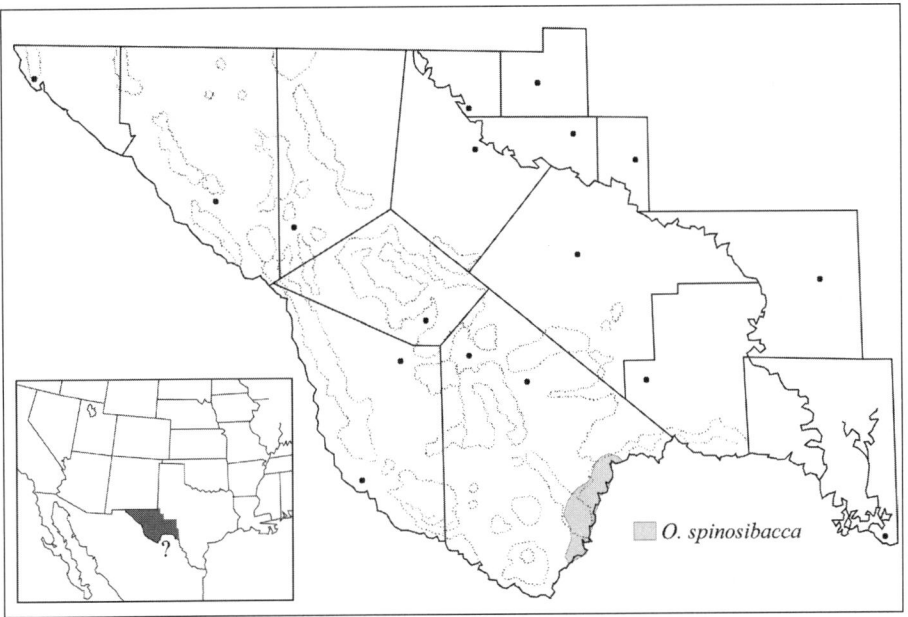

Map 17. Distribution of *Opuntia spinosibacca* (spiny-fruited prickly pear).

Opuntia spinosibacca is one of several cactus species with its distribution limited to the hot, dry, limestone substrates in the southern Big Bend region. Good places to observe the species are at Hot Springs and near the Boquillas Tunnel in Big Bend National Park. The type locality is on the slopes of a limestone hill just west of the rangers' quarters, Boquillas, Big Bend National Park, Brewster County, Texas. The specific epithet is descriptive of the spiny fruits, after the Latin *spina*, "spine," and *bacca*, "a small round fruit."

Identifying Characters. The plants of O. *spinosibacca* basically are upright compact shrubs to 1–1.5 m tall, but some spreading plants are found near Heath Canyon and at other sites. These "spreading growth forms" of O. *spinosibacca* closely resemble O. *camanchica*, unless the characteristic spiny flowers and fruits of O. *spinosibacca* are present. The pads are light green to yellowish-green with spines in areoles across most of the pad except near the base. Expect a purple blotch near each areole. The areoles are elevated on low but conspicuous tubercles, an unusual feature among upright prickly pears. The areoles number relatively few per pad; the pads are smaller than might be expected from the size of the shrub. The spines usually are reddish-brown and number 4–8 in upper areoles, where they are up to 2–7 cm long and divergent.

The flowers of *O. spinosibacca* are bright to golden-yellow or orange-yellow, with red centers, 5–7.5 cm long and 5–7 cm in diameter. The inner tepals are spatulate or obovate, often broadly so, 3–5 cm long, ca. 2.5 cm broad, often with an erose, apiculate apex. The yellow inner tepals are red proximally, with the red pigmentation often extending distally up the midregion to the apex, as in *O. macrocentra* and *O. mackensenii* var. *minor*. With age the red centers may fade to a washed-out red-pink. The filaments are pale green to cream-colored and 0.6–1.4 cm long. The anthers are pale yellow and 1.5–2 mm long. The whitish to pinkish style is slender-urceolate, 1.5–2.2 cm long, supporting 7–9 pale green to cream-yellow stigma lobes to ca. 3 mm long. The pericarpel is ca. 2.5 cm long, broadly urceolate-campanulate, slightly tuberculate, the areoles with bristles and glochids, those at and near the apex with spines to 1.3 cm long.

The spiny fruits of *O. spinosibacca* are fleshy and greenish-yellow at first but at maturity are dry, shrunken, and tan to yellowish or reddish. On some plants the fruits remain fleshy, somewhat juicy, and mottled red, but it is possible that some of the plants with fleshy fruits are the misidentified lowland (tall) red-spined form of *O. camanchica*, a taxon that sometimes is difficult to distinguish without a chromosome count from *O. spinosibacca*. In putative hybrids (3x) *O. spinosibacca* × *O. azurea* the ripe fruits fade to mottled reddish and yellow, and they become shrunken. Other plants (e.g., the color plate in Benson, 1982) have succulent (but quickly drying) brilliantly red, very spiny fruits, of a characteristic red tint seen in *O. azurea*, sensu lato, but not among any of the confused hexaploid "phaeacantha" of the Big Bend. The fruits in *O. spinosibacca* are slightly tuberculate, 2.5–4.5 cm long, 1.3–2.5 cm in diameter, basically ovoid with a flared apical rim, with a prominent umbilicus. Spines appear in scattered areoles, but they are particularly evident at and near the rim, where usually reddish-brown spines to 2.5 cm long project mostly horizontally around the entire fruit apex, or nearly so. The yellowish seeds are flattened, irregularly shaped, 4–6.5 mm in diameter, 1.5–2 mm thick, notched at the hilum, with a prominent aril-margin 1–1.5 mm wide.

Phenology. The principal blooming period for *O. spinosibacca* is in April. In some years anthesis may occur in late March. Reportedly, flowering extends into the middle of May with mature fruits developing about one month after the flowers (Anthony, 1956). Mature fruits apparently are not long persistent, but they have been seen to last through the end of June. In Alpine cultivated plants of *O. spinosibacca* have bloomed from late April through mid-May.

Sterile and Immature Specimens. The plants of *O. spinosibacca* have fibrous roots and usually many branches ascending from the base or a short trunk. The pads are ovate to orbicular, glaucous, but quickly weathering to yellow-green, 10–20(–25) cm long, and 7.5–15 cm in diameter. Often the pads are conspicuously checkered by sharply defined purplish blotches near the areoles. The margins of the pads may be drawn downward between the marginal areoles. Standard areoles are 2–4 cm apart, oval to elliptic or circular, 6–7 mm long, ca. 2 mm across on older pads, mounded with a dense pile of gray hairs.

There are 1–8 central spines per areole. In fully-spined areoles near the stem apex, typically there are 4–8 centrals, 2–7 cm long, to 1.5–2 mm in diameter, these stout, flattened on upper and lower surfaces or only on the upper surface, sometimes angular, often twisted and curved. The central spines are reddish, reddish-brown, dark brown, or reddish-black proximally or nearly throughout. Distally the spines often become lighter red-brown, white, or gray with yellow translucent tips. These central spine colors usually apply to the upper divergent spines, not the 1–4 downward-angled lower spines that are reddish to stramineous or brown-mottled. The typical central spine pattern in each upper mature areole involves 1–3 spines at the lower central margin, these curved and deflexed; one larger lower spine just above the lowermost spine, this one usually flattened and down-curved; and 2–4 mid-lateral to backlateral spines, these stout, dark, erect, and straight to divergent and curved. The largest and darkest spines typically are toward the upper part of the areoles, which are rather noticeably elevated above the stem surface on short tubercles. In most mature areoles there are (1–)2(–4) radial spines strongly deflexed from the lower periphery. These lower radials are slender, acicular, yellowish to gray, and 0.5–1.1 cm long. In addition to the lower radials, there is one or several erect or laterally projecting bristles or radials borne in the upper periphery of the areoles. These bristles are yellowish and clearly not retrorsely barbellate as are the shorter glochids, but they emerge from among the glochids. The glochids in mid-pad areoles are in dense subapical tufts or narrow apical crescents. The glochids are light brown, ca. 2 mm long, with yellow tips.

The ephemeral leaves on young pads are subulate, mucronate, and either green or pinkish with bronze tips (Anthony, 1956). The seedlings are not hairy.

Biosystematics. Opuntia spinosibacca was described by Anthony (1956) in connection with her comprehensive study of the opuntias of the Big Bend region of Texas. Benson (1969a) portrayed a close relationship of O. *spinosibacca* to O. *phaeacantha* by treating it as a variety of O. *phaeacantha* (Benson, 1982). Subsequent study led Pinkava and Parfitt (1988) to the formal recognition of O. *spinosibacca* as a hybrid species, after a tetraploid ($2n = 44$) chromosome number was reported for the taxon (Weedin and Powell, 1978). The putative parents listed by Pinkava and Parfitt were the diploid O. *aureispina* ($2n = 22$) and the hexaploid O. *phaeacantha* var. *major* ($2n = 66$). Geographic, morphological, and chromosomal evidence all support such a hybrid origin for O. *spinosibacca*, but herein we recognize O. *phaeacantha* var. *major*, sensu Benson (1982), as O. *camanchica* and O. *dulcis*. A hexaploid red-spined form of O. *dulcis* is widespread and common in the southern Big Bend region of the Trans-Pecos and is an ideal candidate for one of the putative parents of O. *spinosibacca*. This red-spined form is almost identical to O. *spinosibacca* in vegetative and flower morphology, even the tuberculate pads, but it produces fleshy red fruits. The existence of fleshy reddish fruits on some plants identified as O. *spinosibacca* could be explained if the red-spined O. *dulcis* were a parent of O. *spinosibacca*. It is possible, however, that plants with fleshy reddish fruits are misidentified as *O. spinosibacca* and are instead red-spined hexaploids. Additional chromosomal investigations may help resolve the question.

Synonyms. *Opuntia* × *spinosibacca* M. S. Anthony; *O. phaeacantha* Engelm. var. *spinosibacca* (M. S. Anthony) L. D. Benson.

Common Names. Big Bend prickly pear; spinyfruit prickly pear.

21. **Opuntia phaeacantha** Engelm. var. **phaeacantha**. BROWN-SPINED PRICKLY PEAR. Plates 90, 91. [*Opuntia phaeacantha* Engelm., Mem. Amer. Acad. Arts 4: 52. 1849]. Infrequent in grasslands and woodlands of the central mountains, mid- to higher-elevations, less frequent to rare in rocky and woodland areas of more desertic mountains to the W, S, and E, usually above 4,000 ft, very rare in the desert at elevations below 4,000 ft, El Paso, Hudspeth, Culberson, Jeff Davis, Presidio, and Brewster counties. (2,400–)4,000–7,500 ft. Flowering (May–)Jun(–Jul). $2n = 66$. In NM N to N CO and N UT, W to AZ and CA. Mexico, N Sonora, Chihuahua, and Coahuila. Map 18.

Opuntia phaeacantha and its allies, collectively called "the hexaploid juicy-fruited northern prickly pear cacti," comprise one of the most poorly understood groups of prickly pears in the Trans-Pecos. Benson (1982) recognized 10 varieties of *O. phaeacantha* in the United States, with four of these taxa represented in the Trans-Pecos. Bravo-Hollis (1978) listed six varieties of *O. phaeacantha* for Mexico, four of them the same taxa as treated by Benson. Weniger (1984) also recognized six varieties of *O. phaeacantha,* all of them occurring in Texas, and at least two of them different by name from those treated by the other authors. Benson

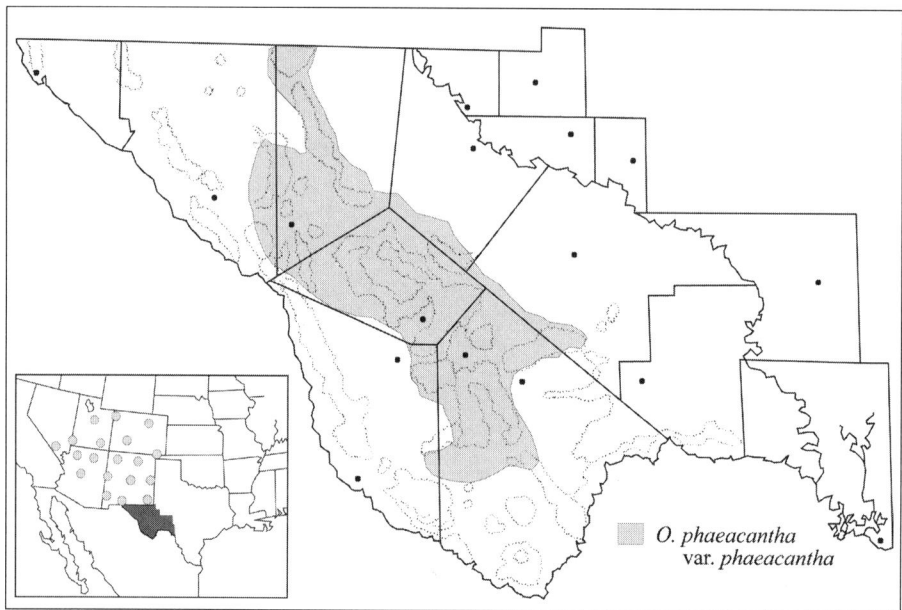

Map 18. Generalized distribution of *Opuntia phaeacantha* var. *phaeacantha* (brown-spined prickly pear).

and Weniger evidently held very different views of the *O. phaeacantha* complex. Our concept of the *O. phaeacantha* group in the Trans-Pecos incorporates generally unpublished information from D. Ferguson (pers. comm.) as well as our own observations. The type and lectotype localities of var. *phaeacantha* are near Santa Fe, New Mexico, near the Rio Grande. The epithet *phaeacantha* is after the Greek *phae*, "dark," and *acanth*, "spine," a misleading descriptor because the spines of *O. phaeacantha* var. *phaeacantha* usually are light-colored, usually whitish, at least in our region.

Identifying Characters. One of the best distinguishing features of *O. phaeacantha* var. *phaeacantha* is its winter habit of weak stems and pads that sag or even lie flat on the ground, as in *O. macrorhiza* and *O. tortispina*, probably because the wood is relatively poorly developed (D. Ferguson, pers. comm.). A sometimes-subtle difference from *O. tortispina* (and *O. macrorhiza*, *O. pottsii*) is that there are no sharply defined transverse wrinkles in winter conditions and no corresponding visible markings on pads that have passed through the winter. The turgid, smooth-skinned summer habit is more nearly erect but still sprawling, to ca. 30–60 cm high (taller than *O. tortispina*). The pads are obovate to broadly elliptic or suborbicular, 10–22 cm long, and 9–18 cm wide, averaging larger than those of *O. tortispina*. Spines (fewer than in *O. tortispina*) are produced over the upper one-half to three-fourths of the pad.

Plants of var. *phaeacantha* often are found in the same vicinity, if not side by side, with *O. tortispina*, *O. engelmannii*, *O. macrocentra* and/or *O. azurea*, sensu lato, *O. polyacantha*, and *O. pottsii*. The var. *phaeacantha* is more easily distinguished from the other associated prickly pears than it is from *O. tortispina*. In the central Trans-Pecos mountains, var. *phaeacantha* is more likely to be found on rocky hillsides and in woodlands than in alluvial grasslands that are the typical habit of *O. tortispina* and *O. pottsii*.

The flowers of var. *phaeacantha* are yellow with pale to prominent reddish centers. The flowers are 5–7 cm long and wide. The inner tepals are 2.5–3 cm long, obovate-obtuse, and perhaps toothed at the apex. The filaments and anthers are pale yellow. The style is cream-colored to pinkish, and the short stigma lobes are yellowish to pale green, or perhaps deep green. The pericarpel is 2–4 cm long with reddish-brown glochids and white trichomes in scattered areoles.

The reddish fruits of var. *phaeacantha* are 2.4–5.5 cm long, 1.9–2.8 cm wide, turbinate or broadly clavate, with a shallow or cup-like umbilicus. The fruit pulp is bright red to pinkish with a sweet taste. The seeds are 4–5 mm in diameter, tan, discoid-reniform or irregularly orbicular, with a hilar notch, and a margin ca. 0.5 mm wide. The fruit morphology of var. *phaeacantha* from throughout its range requires further study.

Phenology. The main flowering period for var. *phaeacantha* has not been clearly established, but it appears to be in May and June. Flowers open at midday or in early afternoon, close at night, and seldom reopen the next day. Osborne et al. (1988) determined that individual flowers remained open for 7–11 hours; flowers may open in the morning and close in the afternoon, while those opening in the afternoon are more likely to reopen the next morning. These authors also demonstrated that the breeding system of O. *phaeacantha* potentially involved entomophilous allogamy and self-compatibility, but is not automatically self-fertilizing. The taxon was found not to be agamospermous. Fruits ripen in 1–2 months, with the red fruits appearing on plants from July to August. Fruits may persist into the early winter.

Sterile and Immature Specimens. In var. *phaeacantha* the subapical areoles are ca. 2–5 cm apart, oblong to orbicular, 5–6 mm long, 3–5 mm wide, and filled with brownish kinky hairs. The usual spine pattern in subapical areoles includes two spines. The lowermost spine, positioned in the lower center of the areole, is deflexed, whitish, somewhat flattened proximally, terete distally, and 2–3 cm long. The second spine is positioned just above the lowermost spine, or less often in the middle or upper part of the areole, and it is also deflexed, but less severely so than the lowermost spine. The second spine is flattened proximally, terete distally, and larger than the lowermost spine at 3.5–5(–6.3) cm long and 1–1.3 mm in diameter. The larger spine usually is twisted, off-white to tan throughout, perhaps brownish or reddish at the base or on the lower half, or in some plants reddish-black proximally, pinkish-tan distally, and with red-brown tips. In distal areoles the second spine usually is more projecting, and a projecting third spine often is present. The third spine, if present, is positioned near the center of the areole, is like the second spine in appearance, or is shorter and brown to red-brown for most or all of its length. In some plants, especially in distal areoles, there may be three lower, deflexed spines and two upper spines in the areole, or two lower and two upper spines. Two radial spines are characteristic in areoles of var. *phaeacantha* we have examined. The radials are positioned at the lower periphery of the areole, and these slender, acicular spines are 4.5–8 mm long, whitish to yellowish, especially distally, and are distinct from the lower central and other central spines. Glochids occur in dense tufts in the upper part or upper half of the areole. In distal areoles on a pad the yellowish or yellow-brown glochids are 2–3 mm long. In apical areoles the glochids may be ca. 1 cm long and bristlelike.

Immature specimens are characterized by pads with a single deflexed spine in the lower portion of distal areoles. Additional spines are added with age. The seedlings are not hairy (D. Ferguson, pers. comm.).

Biosystematics. In the Trans-Pecos O. *phaeacantha* var. *phaeacantha* typically is found in medium to higher altitudes in the central mountain region and in the more xeric mountains to the west and south of the central mountains. The var. *phaeacantha* is rare in desert habitats, and it is relatively uncommon among other prickly pear species everywhere in the Trans-Pecos (Benson, 1982; D. Ferguson, pers. comm.). The var. *phaeacantha* is more common in the southwestern mountains north and west of the Trans-Pecos, and in Mexico.

Benson (1982, p. 470) described the spines of var. *phaeacantha* as "brown, dark or rarely light" in color. In the Trans-Pecos, plants we have interpreted as var. *phaeacantha* mostly have whitish spines or spines that are off-white, light tan, or other light colors, often with brown, red-brown, or reddish-black bases.

The var. *phaeacantha* has relatively fewer spines than *O. camanchica*, but unfortunately the spinier extreme within var. *phaeacantha* cannot be so conveniently identified. A distinguishing feature of var. *phaeacantha*, compared to *O. camanchica*, is its floppy-stemmed habit, but this information seldom is evident from herbarium specimens and not always apparent in summer after rainfall and vigorous growth.

Our concept of the "*phaeacantha* group" in the Trans-Pecos, excluding most of the Benson era "varieties" of *O. phaeacantha* and tetraploid species such as *O. mackensenii* and *O. spinosibacca*, comprises three closely related hexaploid (6x) taxa (*O. phaeacantha* var. *phaeacantha*, *O. camanchica*, and *O. dulcis*). Throughout the Trans-Pecos the *O. phaeacantha*-type plants are noted for their variable vegetative characters, including habit and spines. Nonetheless, these three taxa, our region's last taxonomic remnants of the *O. phaeacantha* "complex," are given taxonomic recognition mostly on the basis of these minor habit and spine characters. We suspect that some of the morphological variation in the Trans-Pecos is the result of interspecific hybridization (Anthony, 1956; Benson, 1982), perhaps involving taxa of other species groups. We also suspect that there are additional taxa in the Trans-Pecos, with affinities to the *O. phaeacantha* alliance, that are yet to be delimited.

Now we have interpreted *O. phaeacantha* var. *major* as a synonym of var. *phaeacantha*. For nearly 30 years we used the name "*phaeacantha* var. *major*" as our designated polyphyletic receptacle for otherwise unidentified medium-size prickly pears everywhere in the southwestern United States. This was Benson's (1969b; 1982) concept of "var. *major*," and by extension, of (polyphyletic) *O. phaeacantha* in general. Our opinion is based directly on D. Ferguson's (pers. comm.) observations of the type population (four miles east of Santa Fe, New Mexico; although recently, June 2003, Ferguson has had reason to reassess his interpretations). It may not be the widespread taxon perceived by Benson (1982) or Weniger (1984). Populations in New Mexico identified as var. *major* are barely distinguishable from var. *phaeacantha*, and they intergrade (D. Ferguson, pers. comm.); it may or may not deserve taxonomic recognition apart from var. *phaeacantha*, but in either case it appears to be endemic to north-central and central New Mexico, not part of our Trans-Pecos (or CDR) flora.

Almost nothing captioned as *O. phaeacantha* in the literature is reliably identified even to species, let alone to variety or subspecies. Approach the literature with caution.

Synonyms. ? *Opuntia phaeacantha* Engelm. var. *nigricans* Engelm.; *O. phaeacantha* Engelm. var. *major* Engelm., sensu Benson (in part); ? *O. phaeacantha* Engelm. var. *piercei* Fosberg; ? *O. rubiflora* Davidson.

Common Names. New Mexico prickly pear; tulip prickly pear; brownspine prickly pear; brown-spine prickly-pear; purple-fruited prickly pear; nopal; prickly pear. Although the epithet *phaeacantha* is misleading because the spines usually are white (at least here in the Trans-Pecos), English translation into "brown-spined prickly pear" remains in use despite its inaccuracy.

22. **Opuntia camanchica** Engelm. & Bigelow, Proc. Amer. Acad. Arts 3: 293. 1856. COMANCHE PRICKLY PEAR. Plates 92, 93. Mostly desert or semidesert habitats, throughout most of the Trans-Pecos. 2,000–4,500 ft. Flowering Apr–Jun. $2n = 66$. Texas, throughout most of western half, infrequent in deep S TX; OK (W portion including Panhandle), CO (SE corner and N to Aurora), NM, AZ, SW UT, S NV, and SE CA. Mexico, Coahuila S to Durango. Map 19.

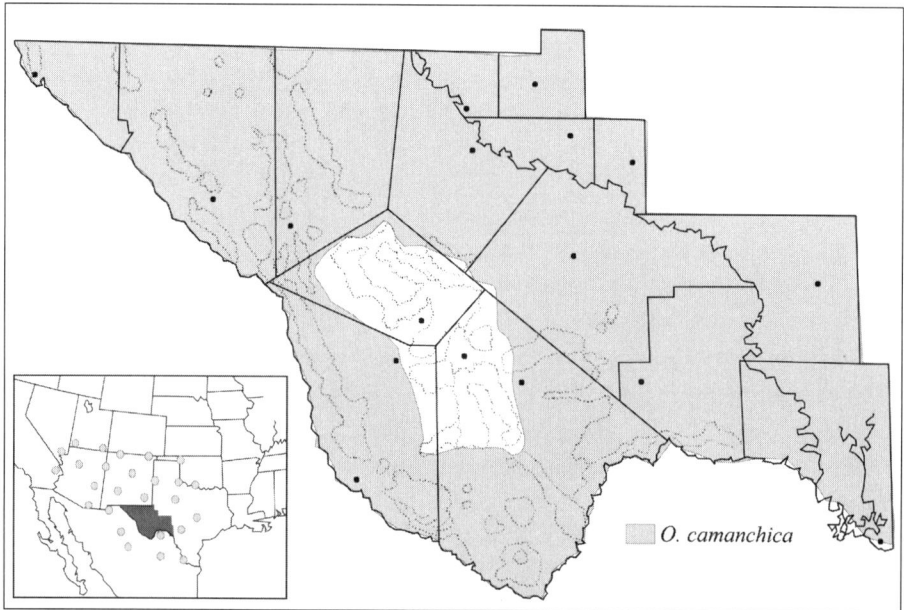

Map 19. Generalized distribution of *Opuntia camanchica* (Comanche prickly pear).

Opuntia camanchica is one of the most common prickly pear species in the Trans-Pecos (D. Ferguson, pers. comm.). Both Benson (1982) and Weniger (1984) interpreted O. *camanchica* as a variety of O. *phaeacantha,* but no two authors thus far have agreed upon the morphological and geographic criteria considered diagnostic for var. *camanchica*. Neither author considered *camanchica* to be a Trans-Pecos entity. Our concept of O. *camanchica* in the Trans-Pecos includes mostly specimens/sites cited by Benson as O. *phaeacantha* var. *major*. Fortunately, the epithet *camanchica* seems adequately typified: the type locality for O. *camanchica* is on the Llano Estacado near the Canadian River, the Comanche Plains country. Presumably the specific epithet is a geographic reference to the Comanche Plains.

Identifying Characters. Opuntia camanchica is similar in habit to O. *phaeacantha* var. *phaeacantha,* but it has more substantial supportive tissues in the stems, that is, it is woodier (D. Ferguson, pers. comm.), and so under good conditions it grows taller and remains relatively erect in winter. The plants are low and spreading to mostly erect, hemispheric or lower shrubs usually less than 70 cm high. In O. *camanchica* there is much variation in spine number and color, including red and black, whereas *O. phaeacantha* var. *phaeacantha* is basically brown and white. The plants may be nearly spineless or with four or more main spines in each areole. The distal areoles usually exhibit the maximum number of spines. When present (as they usually are, in abundance in most Trans-Pecos populations), the spines typically are stout, divergent in several directions, and often curved.

The flowers of *O. camanchica* are yellow with red to pale reddish centers. The flowers are 5–7 cm long and wide, usually slightly smaller and opening less widely then those of *O. phaeacantha* var. *phaeacantha* and *O. dulcis* (D. Ferguson, pers. comm.). The inner tepals of *O. camanchica* are 2.5–3 cm long, obovate-spatulate, obtuse-truncate at the apex, or even emarginate. The filaments are 1.5–1.7 cm long, cream-colored, and the yellow anthers are ca. 2 mm long. The style is cream-colored, ca. 2 cm long, and the stigma lobes are cream-colored to pale green or deep green. The pericarpel is 3–5 cm long with scattered areoles and no spines except perhaps in distal areoles near the rim.

The reddish to cherry-red fruits are 3.7–5(–6.7) cm long, 2.3–3.3 cm wide, usually with a deep umbilicus. The fruits are spineless or with 1–2 spines in areoles near the rim. The relatively few areoles are cottony-white with brownish glochids ca. 1 mm long. The fruit pulp is green, with clear juice, but the juice is not abundant and not very sweet or flavorful, but with a light, sweet scent. Only the outer layer of the pericarpel is red, with most of the fruit rind being colorless. Although red fruits with greenish pulp seem to be typical for *O. camanchica*, many plants in the Trans-Pecos with spine characters matching *O. camanchica* produce red fruits with reddish pulp and juice. The seeds are 4–5 mm in diameter, tan to light brown, irregularly discoid with a broad hilar notch, and a prominent aril 0.5 mm or more wide.

Phenology. The flowering period for *O. camanchica* in the Trans-Pecos seems to be from late April to mid-June. Flowers open by 10:00–11:00 A.M. We suspect that under natural conditions flowers that open early the first day usually do not open the next day after closing or partially closing at night. Under greenhouse cultivation, however, flowers may last one or two days. Mature fruits are present from June through August on plants under cultivation in Alpine. In the field, fruits seem to be gone by August.

Sterile and Immature Specimens. The pads of *O. camanchica* usually are broadly obovate to suborbicular, 15–25 cm long, and 15–20 cm wide. Areoles on the upper two-thirds to three-fourths of the pad have spines; rarely, spines are restricted to the upper one-third or less of the stem. The mid-pad areoles are elliptic, 5–8 mm long, 3–5 mm wide, and filled with woolly or felty gray hairs.

The usual spine pattern in *O. camanchica* involves 3–7 large spines. There are 1–2 lowermost central spines, positioned in the lower center of the areole, that are deflexed, 1–2.7 cm long, acicular or flattened proximally, whitish to gray with a yellowish to red-brown tip, or dark proximally and gray on the distal portion. There is one central spine positioned immediately above the lowermost central or upward toward the middle of the areole. This central is deflexed or projecting, 3–5.5 cm long, 1–1.5 mm in diameter at the base, flattened, subulate, sometimes curved and twisted, blackish proximally and gray distally with a red-brown tip, or blackish, red-brown, reddish throughout or brown at the base and light brown to whitish distally. In the upper portion of the areole, 1–4 central spines project in different directions, these (3–)4–5.5(–6.5) cm long, subterete, usually curved, and black to red-brown or reddish. Spines of other colors, including orange, yellow, white, or mixes of colors, are found in some populations of O. camanchica (D. Ferguson, pers. comm.). Characteristically, there are more curved projecting spines in distal areoles. Another spine pattern in some plants interpreted by us as O. camanchica includes one lowermost, two lower lateral, and 1–2 upper central spines.

Radial spines are produced in the areoles of many specimens interpreted by us as O. camanchica. Usually there are two radials positioned at the lower periphery of areoles, especially distal areoles, when they are present at all. The radials are 3–6(–9) mm long, acicular, whitish to gray or reddish-brown, and distinct even from the smallest lower central spine.

Immature specimens of O. camanchica usually have 1–2 deflexed lower spines in the areoles, without upper centrals, or sometimes with one short upper central or one upper central just erupting through the felty areole. The seedlings are not hairy (D. Ferguson, pers. comm.).

Biosystematics. *Opuntia camanchica* is accepted here as the correct name for the morphotype in the Trans-Pecos that Benson (1969b, 1982) usually identified as *O. phaeacantha* var. *major*. The reason we use the epithet *camanchica* is that *O. phaeacantha* var. *major* (type from northern New Mexico, near Santa Fe) is now considered a synonym of, or at most weakly varietally distinct from, *O. phaeacantha* var. *phaeacantha*, also from near Santa Fe (D. Ferguson, pers. comm.). The entity typified as var. *major* is barely distinguishable, and it occurs only in or near population systems of *O. phaeacantha* in north-central and central New Mexico. The Trans-Pecos populations that Benson included in var. *major*, and those widespread elsewhere in the southwestern United States and part of northern Mexico, are accurately represented by the type of *O. camanchica*, a name that also has nomenclatural priority at the species level. We recognize *O. camanchica* as a widespread and morphologically variable taxon. In the Trans-Pecos it overlaps ecogeographically with *O. phaeacantha* var. *phaeacantha* in the Davis Mountains, but most of the populations are segregated: *O. camanchica* is often abundant in hot desert and semidesert habitats in addition to prairie and woodland; it occurs infrequently in the mesic mountains at lower to middle elevations, whereas var. *phaeacantha* is mostly restricted to more mesic habitats in the central mountains and some semidesert habitats in the western Trans-Pecos.

Future studies might reveal taxa comparable in distinctiveness to *O. phaeacantha*, sensu stricto, within the variable entity recognized here as *O. camanchica*. For example, already we have noticed that spine morphs interpreted here as *O. camanchica* tend to form consistent populations in different parts of the Trans-Pecos.

Opuntia camanchica is one of the most common prickly pears in the Trans-

Pecos and one of the most heavily represented in herbaria (mostly as "*O. phaeacantha* var. *major*"). The species is consistently hexaploid ($2n = 66$) as revealed through numerous chromosome counts (Powell and Weedin, 2001). We cannot yet refute the popular hypothesis that *O. phaeacantha*, sensu lato (including *O. camanchica*), hybridizes with other Trans-Pecos taxa such as *O. engelmannii*. Plants vegetatively resembling *O. camanchica* but with circular areoles and red-pulped fruits may be interspecific hybrids. Also, we have observed some specimens resembling *O. camanchica* that have been annotated by cactus specialists as hybrids between *O. phaeacantha* and *O. engelmannii*. Such specimens exhibit in distal areoles deflexed spines that are whitish or at least gray on the distal half, an *O. engelmannii* trait, and 3–4 projecting upper centrals, an *O. camanchica* character. In our opinion such specimens do exhibit spine features that would be expected in hybrids between *O. camanchica* and *O. engelmannii*. A thorough field-based study of the entire *O. phaeacantha* group remains warranted.

Several photographs and illustrations labeled "*O. phaeacantha* var. *major*" in Benson (1982) appear to be of *O. camanchica*.

Synonyms. *Opuntia phaeacantha* Engelm. var. *brunnea* Engelm.; *O. phaeacantha* Engelm. subsp. *camanchica* (Engelm.) Borg; *O. phaeacantha* Engelm. var. *camanchica* (Engelm. & Bigelow) L. D. Benson; *O. phaeacantha* Engelm. var. *camanchica* Weniger, *comb. nud.*; *O. phaeacantha* Engelm. var. *major* Engelm., in part, sensu Benson; *O. camanchica* Engelm. & Bigelow vars. *albispina, longispina, major, gigantea, minor, pallida, rubra,* and *salmonea* Hort ex Borg, *nom. nud.*; the above varieties and *forma orbicularis* published earlier as forms by Schelle (see Benson, 1982); ? *O. chihuahuensis* Rose.

Common Name. Camanchican opuntia.

23. **Opuntia dulcis** Engelm., Proc. Amer. Acad. Arts 3: 291. 1856. SWEET PRICKLY PEAR. Plates 94–96. Common in desert habitats, often along and near the Rio Grande. El Paso Co., SE at least to Terrell Co. Restricted mostly to the Trans-Pecos. 2,200–3,500 ft. Flowering Apr–May $2n = 66$. To be expected in adjacent NM and Mexico. Map 20.

Although *O. dulcis* appears to be relatively common in the Trans-Pecos, it has not been interpreted consistently as a distinct taxon. The syntype locality was "near the middle course of the Rio Grande, near Presidio del Norte, etc.;" and "frequently observed near Presidio del Norte and Eagle Pass." The lectotype is a Wright collection with the ambiguous data, "El Paso? West Texas? Probably Presidio del Norte," *Wright in 1852*, MO. The specific epithet is after the Latin, *dulcis*, "sweet," a reference to the very sweet fruits as noted on the label of the type specimen.

Identifying Characters. *Opuntia dulcis* overlaps the geographic range of *O. camanchica*, but their contact zone is undocumented, at least in the Trans-Pecos. Like most prickly pears, *O. dulcis* is most easily identified using the growth habit. It is larger and more upright in habit than *O. camanchica* (i.e., almost as large as *O. engelmannii*). In *O. dulcis* the spines usually are fewer in each areole, and typically they are more slender than in *O. camanchica*. Usually there are

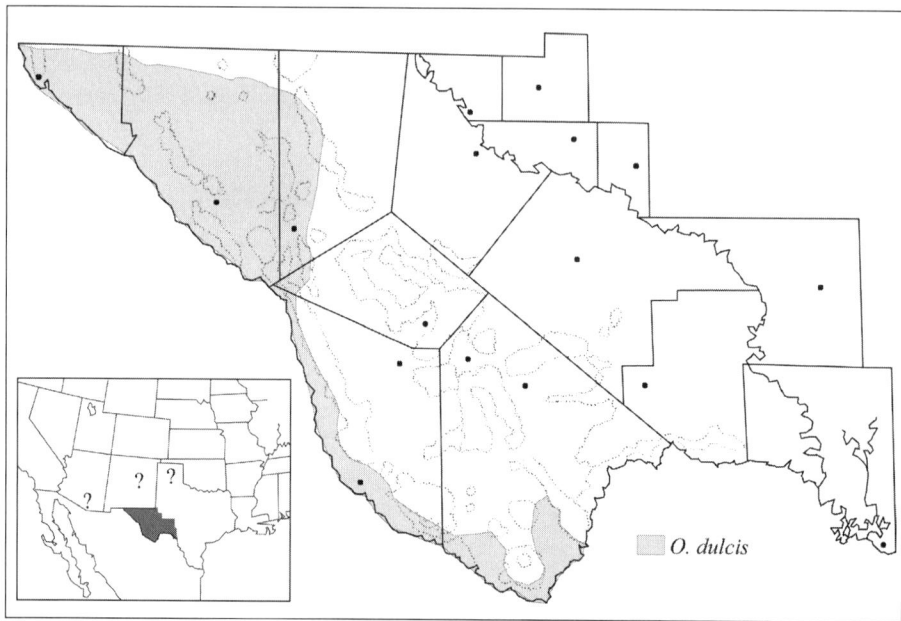

Map 20. Generalized distribution of *Opuntia dulcis* (sweet prickly pear).

2(–4) spines per areole in O. *dulcis*. The spines in many distinct populations of O. *dulcis* are light brown proximally and nearly white on the distal portion, but the spines may be almost any color in some populations of O. *dulcis* (D. Ferguson, pers. comm.).

We have observed the flowers of relatively few plants identified as O. *dulcis*. Their flowers are yellow with conspicuous or inconspicuous red centers, and on average they seem to be slightly larger (ca. 5–7 cm), open wider, and have more tepals than those of O. *camanchica* (D. Ferguson, pers. comm.). The inner tepals are ca. 3 cm long, ca. 2 cm wide, broadly spatulate, and truncate-apiculate or obtuse at the apex. The yellow tepals are suffused, not sharply marked with red basally, or with merely an inconspicuous reddish flush in the midregion. Usually the outer tepals are red basally or in the midregion. The filaments are pale green to cream-colored or colorless, ca. 1.5 cm long, and the yellow anthers are ca. 2 mm long. The colorless or rosy style is 2–2.5 cm long, with the stigma lobes usually light green.

The ripe fruits of O. *dulcis* are red to purplish, obovate to obconic, (4–)5–5.5 cm long, 2.5–3 cm in diameter, with a shallow or deep umbilicus. The fruit surface is smooth with few areoles bearing a small number of glochids and no spines. The fruit rind is purple, and the pulp is either pink, purple, to red, or greenish. Apparently, mature fruits are always juicy; the deep purple, pink, to clear juice is sweet. The seeds are tan, irregularly discoid, 3.5–4.5 mm in diameter, with a narrow hilar notch and a prominent aril-margin 0.7–1 mm wide.

Phenology. Preliminary indications are that flowers usually open in early to mid-April, near the Rio Grande, and that flowering may continue through May.

Flowers open at midday and at least partially close at night for 1–3 days in cultivated specimens. Fruits mature in June and July. A cultivated specimen in Alpine produced fruits as late as early August. At Tucson, the possible counterpart of *O. dulcis* peaked after 19 August 2000, later than any other prickly pear except *O. ellisiana*.

Sterile and Immature Specimens. In aspect *O. dulcis* could be confused with *O. phaeacantha*. The spines in *O. dulcis* usually are not as stout as they are in other Trans-Pecos *phaeacantha* types. The woody pads of *O. dulcis* are not distinctive, being obovate to suborbicular, 15–18 cm long, and 10–15 cm in diameter. Areoles on the distal half of each pad have spines. Distal areoles are elliptic to hemispheric, 5–7 mm long, 4–7 mm wide, with dense, crinkled, light brown hairs.

In most areoles there are two spines. Larger third and fourth spines may be present in some distal or apical areoles. The two-spined pattern includes (1) one lowermost spine that is deflexed, chalky-white with a yellowish tip, 1.2–2.5 cm long, to ca. 0.7 mm wide at the base, flattened on the proximal half, and distally terete; (2) a second spine that is usually positioned immediately above the lowermost spine, in the lower to middle portion of the areole. The second spine is porrect or deflexed, typically off-white to tan-white or tan distally except for the yellowish tip, and brown, red-brown, or reddish at the base, 4–7 cm long, and 1–1.2 mm wide at the base. The second spine is flattened for most of its length. A third spine, if present, projects from the middle portion of the areole. The third spine (when present) is 2–3 cm long, the color of the second spine, and terete or flattened. A fourth spine, resembling the third spine, is present in the apical areoles of some specimens. Although the spine colors described above are typical in some populations of *O. dulcis*, the spines in other populations may be almost any color (D. Ferguson, pers. comm.). Typical living specimens of *O. dulcis* are easily distinguished from those of *O. camanchica* by their habit, but some herbarium specimens tentatively identified as *O. dulcis* approach *O. camanchica* in spine morphology, and these are difficult to distinguish.

Radial spines are absent in some specimens of *O. dulcis* and present in others. Usually there are two radials; when present, at the lower periphery of the areole. The radials are deflexed, whitish to yellowish, acicular, 2–3 mm long, and readily distinguished from the central spines. Glochidlike bristles are present along the lower areole margin in some specimens, these either projecting or deflexed and with the appearance of slightly longer glochids. The typical glochids are in dense apical crescents or tufts; additional scattered glochids occur along the lateral margins of some areoles.

Immature specimens have one or two spines in each distal areole, and the spines are shorter than those on mature pads. The seedlings are not hairy (D. Ferguson, pers. comm.).

Biosystematics. Benson (1982) listed *O. dulcis* as a synonym of *O. phaeacantha* var. *phaeacantha*. Without explanation Weniger (1984) employed the name *O. engelmannii* var. *dulcis* (Engelm.) J. M. Coult. ex K. Schum. for a perceived taxon restricted to South Texas and adjacent Nuevo León, Mexico. Possibly the entity Weniger perceived as *O. engelmannii* var. *dulcis* is different from the relatively common Trans-Pecos *O. dulcis* of interpretation. The photograph labeled "*Opuntia engelmannii* var. *dulcis*" in Weniger (1984, p. 256) does not appear to be the *O. dulcis* we have recognized in the Trans-Pecos. *Opuntia phaeacantha* and *O. engelmannii* belong to separate but related species groups in *Opuntia*. *Opuntia dulcis* in the Trans-Pecos seems to be closely related to *O. phaeacantha* var. *phaeacantha*, which differs mainly in habit and ecological preference. The Trans-Pecos *O. dulcis* also is similar to and probably closely related to *O. camanchica*, which differs in habit, flower, and fruit character (D. Ferguson, pers. comm.).

Synonyms. *Opuntia lindheimeri* Engelm. var. *dulcis* (Engelm.) J. M. Coult.; *O. engelmannii* Salm-Dyck var. *dulcis* (Engelm.) J. M. Coulter ex K. Schum.; ? *O. expansa* Griffiths; ? *O. eocarpa* Griffiths.

Common Name. Sweet opuntia.

24. **Opuntia engelmannii** Salm-Dyck ex Engelm., Boston J. Nat. Hist. 6: 208. 1850. ENGELMANN'S PRICKLY PEAR, TEXAS PRICKLY PEAR. Plates 97–100. Widespread and common in the southwestern U.S., from S CA E through TX. Sea level (in CA) to 5,000(–7,500) ft. Flowering spring. $2n =$ "44" (reported by Grant and Grant, 1979b) and 66 (all other reports). Northern Mexico. Maps 21, 22.

After two decades of nomenclatural and taxonomic confusion (see Benson, 1982; Weniger, 1984) concerning *O. engelmannii* (*O. phaeacantha* var. *discata*) and *O. lindheimeri*, Parfitt and Pinkava (1988) reinstated the correct application of the name *O. engelmannii*. Secondarily, they revised some of the several closely related taxa within *O. engelmannii*, sensu lato. Parfitt and Pinkava recognized six taxa as varieties of *O. engelmannii*, three of which, var. *engelmannii*, var. *lindheimeri*, and var. *linguiformis*, are known to occur in the Trans-Pecos. Two of these taxa are fully treated in the following discussions as natural populations, and var. *linguiformis*, a cultivar with no known natural populations, is given only abbreviated discussion at the end of the *Opuntia* section. The specific epithet honors George Engelmann, one of the early authorities in Cactaceae, and a prolific worker who described numerous southwestern cacti, including many of those in the Trans-Pecos region.

Key to the Native Trans-Pecos Varieties

1. Spines mostly white; mostly western
   24a. *O. engelmannii* var. *engelmannii*.
1. Spines all yellow; mostly eastern
   24b. *O. engelmannii* var. *lindheimeri*.

24a. **Opuntia engelmannii** var. **engelmannii** ENGELMANN'S PRICKLY PEAR. Plates 97, 98. Probably the most common prickly pear species in the Trans-Pecos, and certainly the most conspicuous; desert habitats to mountain

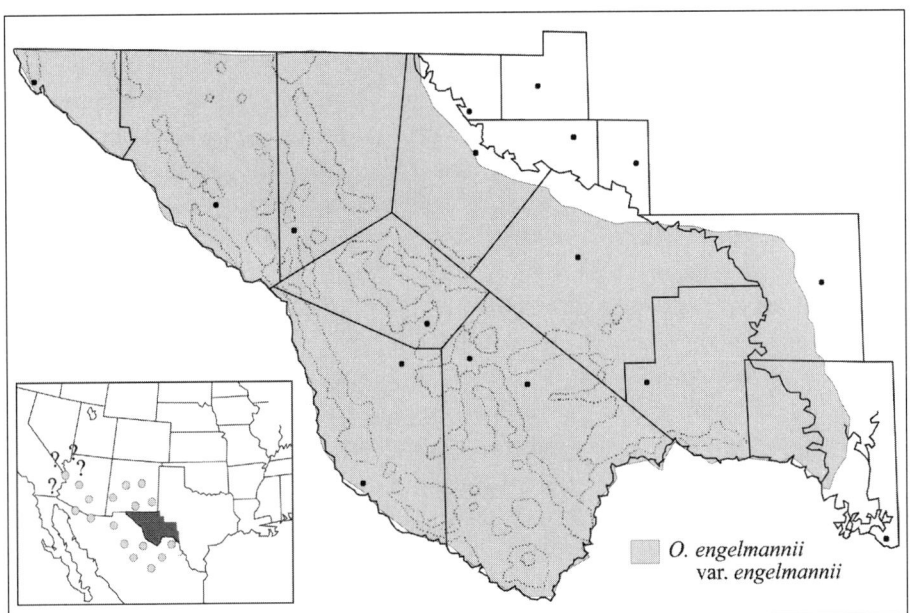

Map 21. Generalized distribution of *Opuntia engelmannii* var. *engelmannii* (Engelmann's prickly pear).

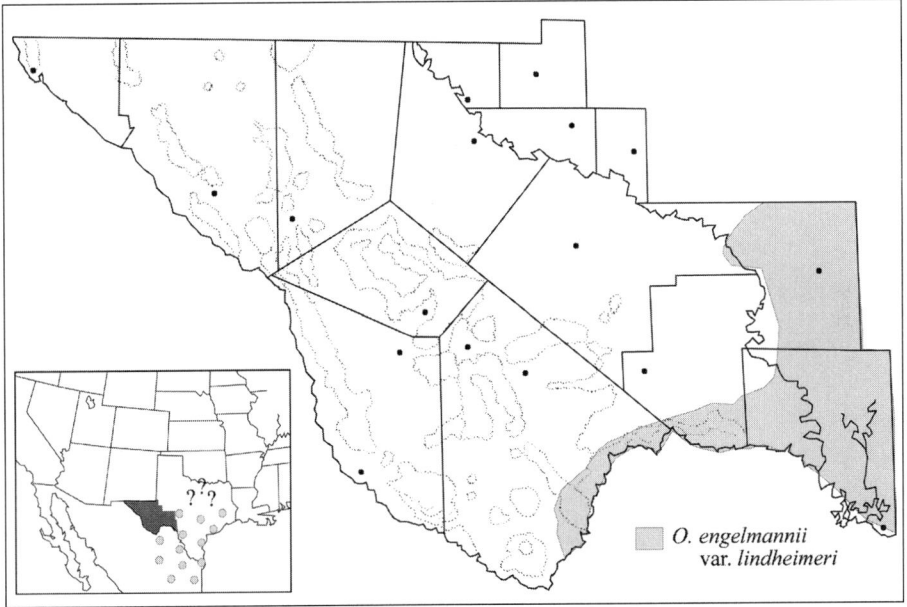

Map 22. Generalized distribution of *Opuntia engelmannii* var. *lindheimeri* (Texas prickly pear).

grasslands and woodlands. 1,000–5,000(–7,500) ft. Flowering Apr–Jul. $2n = 66$, with one report of a tetraploid (Grant and Grant, 1979b). In TX rare and poorly documented E of the Pecos River, SE reportedly to Bexar and Travis counties, scattered NE to the S plains and probably into OK. West through central and S

NM to NW AZ, and S CA. Mexico in northernmost Sonora, Chihuahua, and Coahuila. Map 21.

This is the taxon called *O. phaeacantha* var. *discata* by Benson (1982) and *O. engelmannii* var. *engelmannii* by Weniger (1984). A recent taxonomic evaluation of the complex (Parfitt and Pinkava, 1988) resolved some questions about the group. The type locality of var. *engelmannii* is north of Chihuahua City in Chihuahua, Mexico.

Identifying Characters. The plants of var. *engelmannii* usually are 90–140 cm high, with upright and spreading branches. Plants at some localities, particularly in deep soil along and near the Rio Grande, may be taller than 1.4 m. The plants are variable in shape and size of the pads, but usually they are obovate, broadly elliptic, or orbicular, 20–30 cm long, and 15–25 cm in diameter. Under optimal growth conditions the pads of some plants get large, thick, and heavy. The larger plants of var. *engelmannii* are the largest wild prickly pears in the Trans-Pecos. In var. *engelmannii*, characteristically the spines are distributed over most of the pad, usually all but the lower few areoles. There is a great deal of spine variation in var. *engelmannii*, but the typical spine pattern is distinctive. This distinctive spine pattern includes more or less three flattened bone-white or chalky-white spines deflexed in a "bird's foot." Spineless and weakly spined plants are also found in the Trans-Pecos. The glochids of var. *engelmannii* and its immediate allies also are distinctive (Parfitt and Pinkava, 1988). The glochids of mid-stem areoles are distributed around much of the areole periphery or throughout the areole, they are spaced apart (not touching each other and basically not densely crowded), conspicuous, stout, 3–5(–9) mm long, and of unequal lengths. In other taxa, including the commonly sympatric *O. camanchica*, the glochids tend to be more organized in dense apical tufts or crescents where the glochids mostly are of equal or subequal lengths.

The flowers of var. *engelmannii* are uniformly clear yellow, without red centers. Orange venation may be evident upon close examination of the inner tepals. At the end of a day the flowers may change to pale orange. The normal yellow color of the tepals is bright and clear but not as rich and brilliant as in *O. engelmannii* var. *lindheimeri*. The relatively large flowers, 7–8 cm in length and diameter, are widely funnelform in afternoon sun. The inner tepals are 4–5 cm long, 2–2.5 cm in diameter, obovate-spatulate, and obtuse-apiculate at the apex. The filaments are cream-colored to pale green and 1–2 cm long. The yellow anthers are 2–3 mm long. The style is cream-colored, and the stigma lobes are green. The pericarpel typically is ca. 5.5 cm long, ca. 3.2 cm in diameter, with scattered, circular areoles filled with tan hairs and bristlelike glochids projecting from all parts of the areole.

The fruits are deep purple, reddish-purple, or dark beet-red; older fruits appear almost black from a distance. The relatively large fruits typically are barrel-shaped, or they may be oval or obovate, 5.5–8 cm long, 3.3–5 cm in diameter, with a shallow umbilicus. The fruit pulp is deep red, very juicy, and very sweet to the taste and smell. The juice is beet-red and tends to stain any surface. Usually the fruits are spineless, but slender spines 5–8 mm long are found in some upper areoles on the fruits of some plants. The areoles are dominated by white wool and glochids. The seeds of var. *engelmannii* and its allies are smaller and more numerous than those of most other Trans-Pecos prickly pears. The seeds are tan, irregularly discoid, (2–)3–4 mm in diameter, with a narrow hilar notch, and narrow beakless aril-margin.

Phenology. The flowering period of the widespread var. *engelmannii* in the Trans-Pecos is correlated with elevation. In the lower desert of the Big Bend, flowering may commence in late April, and in the mountain grasslands, anthesis may not occur until late June or July. Flowering may persist into late July. At lower elevations in the central Trans-Pecos mountains, the principal flowering period is in June. Most flowers open in late morning or at midday, close at night, and do not open again the next day. Flowers that open late on day one may open again on day two. The yellow tepals change to orangish late in the day. In the lower desert the first ripe fruits may appear in May. In the mountains it is not unusual to find plants of var. *engelmannii* in peak fruit in August or September. For example, on 14 September 1997, we observed peak fruit at high elevations in the Chisos Mountains and in the Davis Mountains south of Alpine. In 1999, near Alpine, the peak fruiting period was in early August. Fruit maturation requires about two months. Ripe fruits are eagerly consumed by many wild animal species, and they may not persist long on the plants after full maturity.

Sterile and Immature Specimens. The pads of var. *engelmannii* are pale green and stay that way even in drought and winter. The areoles mostly are 3–6.5 cm apart on the pads. Mid-pad areoles are elliptic to circular, 6–8 mm long, 4–6 mm in diameter, with a dense covering of felty or curly brown, tan, or gray to black hairs. As in most prickly pears, the areoles may be larger at the stem margins.

The characteristic spine pattern of var. *engelmannii* in mature areoles includes about 3–4 deflexed, bone-white or chalky-white, central spines in a "bird's-foot" arrangement. Fewer than three spines or as many as six spines are common in the areoles of many plants in the Trans-Pecos. In the typical pattern there is one lowermost, deflexed central spine that is white, 1–2.5 cm long, 0.4–0.8 mm in diameter at the base, and usually acicular except for a flattened base. In some plants the lowermost spine may be yellow-white or have a yellowish tip. A second central spine is positioned just above the lowermost central. It is deflexed, 1.5–4.5 cm long, 1–1.6 mm in diameter, flattened, sometimes twisted, opaque white, and sometimes brown, red, or nearly black at the base. Usually 1–4 additional central spines are positioned in mid-lower, lateral, and/or upper lateral portions of areoles. These centrals are deflexed, angled downward, or projecting. In apical areoles one or more of the upper centrals are often projecting. The upper centrals usually are 1–4 cm long, 1–1.3 mm in diameter, flattened proximally and terete distally, and perhaps twisted. Younger, shorter upper centrals usually are terete. The upper centrals are opaque, white, pale yellow or honey color, with a reddish or red-brown base, or often they are gray to reddish throughout. Usually there are 1–2 radial spines located at the lower periphery of most mid-stem and distal areoles. The radials are acicular, 0.35–1 cm long, yellow or white, and readily distinguished from even the smallest central spine. The glochids and glochid pattern in var. *engelmannii* and its allies are distinctive (Parfitt and Pinkava, 1988). The glochids are distributed around the periphery of the areole, and perhaps throughout the areole, and they are of different lengths, as opposed to a dense apical tuft or crescent of equal or subequal glochids in many other prickly pears. The yellowish glochids in var. *engelmannii* are 1.5–9 mm long, longest in distal areoles. Some of the "glochids" are bristlelike (i.e., without prominent scabrosity, but instead smooth-surfaced, slightly longer, slightly larger in caliber, and a deeper yellow color). The bristles in some plants may be distributed throughout the areole, whereas in areoles of other plants bristles and glochids may be peripheral as well as localized in a subapical circular area.

Immature pads of var. *engelmannii* usually have one deflexed central spine in mid-pad areoles and 2–3 spines in apical areoles. Slightly older stems are characterized by mid-pad areoles with two deflexed centrals. The seedlings are hairy (D. Ferguson, pers. comm.).

Biosystematics. Opuntia engelmannii var. engelmannii is a hexaploid taxon with wide distribution from California east to Texas and in northern Mexico. The var. *engelmannii* is sympatric throughout much of its range with many other prickly pear taxa of hexaploid or other ploidy levels, including the hexaploids *O. camanchica* and *O. phaeacantha* var. *phaeacantha*. The var. *engelmannii* exhibits considerable morphological variation throughout its range. Plants of var. *engelmannii* with typical "bird's-foot" patterns of white, flat spines are common in the Trans-Pecos, and plants with the appearance of var. *engelmannii* but with atypical spine patterns have been observed at numerous sites in this region. At present it is not always known (1) if such plants are part of the intraspecific variation in var. *engelmannii*, (2) if they are products of hybridization with other taxa, or (3) if they represent taxa that have not been recognized in the current study. The spine character recombinations in certain plants strongly suggest hybridization between var. *engelmannii* and *O. camanchica* (Anthony, 1956; Benson, 1982). *Opuntia camanchica* (reported as *O. phaeacantha* var. *major*) is also known or suspected to hybridize with *O. engelmannii* var. *lindheimeri* (Grant and Grant, 1979b, 1982) in the eastern Trans-Pecos and in south-central Texas. Experimental studies are needed to help clarify the suspected hybridization wherever it is reported (Pinkava et al., 2001). We also expect that further investigations will demonstrate the existence of additional taxa of the *O. engelmannii* complex in the Trans-Pecos and elsewhere in its range. According to D. Ferguson (pers. comm.) there are 15 or more distinct taxa in Arizona, New Mexico, and Texas that are routinely confused with *O. engelmannii* (e.g., *O. valida* Griffiths). Many of the prickly pear taxa described by early workers, such as Griffiths (1908–11), were treated as synonyms of *O. engelmannii* by Benson (1969b, 1982) and Benson's followers.

The nomenclatural and taxonomic history of *O. engelmannii* was reviewed by Parfitt and Pinkava (1988). Benson (1982) recognized var. *engelmannii* (*O. phaeacantha* var. *discata*) as 1 of 10 varieties of *O. phaeacantha* and treated *O. lindheimeri* as a distinct species. Parfitt and Pinkava did not accept any varieties of *O. phaeacantha*. Weniger (1984) considered *O. lindheimeri* to be synonymous with *O. engelmannii* and recognized 10 varieties of *O. engelmannii*, most of them traditionally associated by other authors with *O. lindheimeri*. Parfitt and Pinkava abundantly refuted Benson's claim that *O. engelmannii* was a redescription (junior synonym) of *O. ficus-indica*. Thus, *O. engelmannii* was the legitimate name after all for the common wild plants that Benson called *O. phaeacantha* var. *discata*, described by Salm-Dyck and Engelmann in 1850, with the type locality north of Chihuahua City, Chihuahua, Mexico. Parfitt and Pinkava extended their taxonomic circumscription of *O. engelmannii* to include *O. lindheimeri* as one of six varieties. Parfitt and Pinkava explained that *O. engelmannii* and *O. lindheimeri* shared a similar habit, hairy seedlings, sparse glochids, purple fruit color, and all-yellow inner tepals (although all-red or all-orange tepals occur in some eastern *O. lindheimeri*) and that they had only subtle differences between them. The glochid characters described and illustrated by Parfitt and Pinkava for *O. engelmannii* and *O. lindheimeri* are not shared by *O. phaeacantha* or *O. ficus-indica*. For the present work we have elected to follow the taxonomy of Parfitt and Pinkava in treating *O. engelmannii* and *O. lindheimeri* as varieties, although we suspect that the eventual consensus might be that they are best treated as distinct species, each with several allied taxa (D. Ferguson, pers. comm.). Three of the six varieties of *O. engelmannii* recognized by Parfitt and Pinkava occur in the Trans-Pecos. The other three varieties, var. *flavispina* (L. D. Benson) Parfitt and Pinkava, var. *cuija* Griffiths & Hare, and var. *flexospina* (Griffiths) Parfitt & Pinkava, occur respectively in Arizona, north-central Mexico, and South Texas.

Anthony (1956) tentatively identified plants from two populations in southern Brewster County as *O. engelmannii* var. *wootonii* (Griffiths) Fosberg, = *O. phaeacantha* var. *wootonii* (Griffiths) L. D. Benson. The type locality for *O. wootonii* Griffiths is the Organ Mountains in Doña Ana County, New Mexico, just north of the Franklin Mountains, which are in the western corner of Trans-Pecos Texas. The populations alluded to by Anthony do not match the *O. wootonii* in the Organ Mountains, but they may represent another taxon that is distinct from *O. engelmannii* (D. Ferguson, pers. comm.). *Opuntia wootonii* may extend from central and south-central New Mexico into the Trans-Pecos in the Franklin and Guadalupe mountains. It is not yet clear if *O. wootonii* is best treated as a species, a nothotaxon, or a variety of *O. engelmannii*. Its distribution in the Trans-Pecos, if it occurs here, is not understood.

Synonyms. *Opuntia engelmannii* Salm-Dyck; *O. discata* Griffiths; *O. engelmannii* Salm-Dyck var. *discata* (Griffiths) A. Nelson; *O. phaeacantha* Engelm. var. *discata* (Griffiths) L. D. Benson & Walk.; *O. dillei* Griffiths; *O. arizonica* Griffiths; *O. gregoriana* Griffiths; *O. procumbens* Engelm. & Bigelow.

Common Names. Engelmann pricklypear; Engelmann prickly-pear; Engelmann's opuntia; purple-fruited prickly pear; tulip prickly pear; discus cactus; discoid opuntia; desert prickly pear; nopal; prickly pear.

24b. **Opuntia engelmannii** var. **lindheimeri** (Engelm.) B. D. Parfitt & Pinkava. TEXAS PRICKLY PEAR. Plates 99, 100. [*Opuntia lindheimeri* Engelm., Boston

J. Nat. Hist. 6: 207. 1850]. Rare in relatively deep soil mostly along and near the Rio Grande, Brewster Co. in Big Bend National Park; also one documented site and two sight records by A. Zimmerman in northern Brewster Co.; more common in the southeastern Trans-Pecos, near the Pecos River. 1,000–1,800(–3,500) ft. Flowering May–Jun. $2n = 66$. South TX, NE to central TX, E to SW LA, N to S OK. $2n = (22, 44)$ 66. Mexico, N Tamaulipas E to Durango and Chihuahua (Bravo-Hollis, 1978). Map 22.

Traditionally, var. *lindheimeri* has been treated either as a distinct species with several varieties (e.g., Benson, 1982) or as a variety of *O. engelmannii* (Weniger, 1984; Parfitt and Pinkava, 1988). The cultivar *O. engelmannii* var. *linguiformis* (Griffiths) B. D. Parfitt & Pinkava is not known from any native populations, and it is not included here with the formal treatment of *O. engelmannii*. A brief discussion of var. *linguiformis* is included below under cultivars and escapees. The type locality of *O. lindheimeri* is near New Braunfels in Comal County, south-central Texas. The epithet honors Ferdinand Lindheimer, who obtained significant plant collections from southern Texas from 1843 to 1852.

Identifying Characters. The plants of var. *lindheimeri* typically are relatively large, sprawling shrubs 0.5–1.2(–3) m high, with obovate to orbiculate pads 15–27(–30) cm long, 12–22(–30) cm in width, and ca. 2 cm thick. Usually the (0–)1–5(–8) spines per areole are clear yellow, sometimes whitish-yellow, or with reddish-brown or blackish bases. Typically there are one or two, perhaps more, prominent deflexed or porrect spines not arranged in the "bird's-foot" pattern that is characteristic of var. *engelmannii*. In Trans-Pecos populations of var. *lindheimeri*, 1–4 spines per areole are usual. And, in var. *lindheimeri*, some of the spines on the lower half of the pad are as long, or nearly as long, as spines on the upper half of the pad. The spines of var. *lindheimeri* usually are more slender than those of var. *engelmannii*. The glochid pattern is the same in var. *lindheimeri* and var. *engelmannii* (i.e., unequal, often bristlelike glochids positioned at the periphery and perhaps throughout the areole).

The var. *lindheimeri*, although abundant eastward, is rare in the Trans-Pecos except for the southeastern Big Bend along the Rio Grande and associated drainages in deep soil. Probably the westernmost robust population is near Rio Grande Village in Big Bend National Park. Reportedly, plants are scattered along and near the Lower Scenic River to Val Verde County, where the plants become more common east across the Edwards Plateau and into South Texas. In Texas, plants of var. *lindheimeri* are most common southeast of the Balcones Escarpment, in the deep soils of the coastal plain. In the southeastern Trans-Pecos and on the Edwards Plateau, var. *lindheimeri* is sympatric with var. *engelmannii* and *O. camanchica*. When typical spine configurations are exhibited, all three of these taxa are easily distinguished.

Several records attributed to var. *lindheimeri* (e.g., Benson, 1982) in New Mexico, and possibly in the Trans-Pecos, are instead *O. cyclodes* Rose, *O. subarmata* Griffiths, or perhaps another taxon (D. Ferguson, pers. comm.). Certain yellow-spined plants, or sometimes spineless plants, reported in the Davis, Guadalupe, and Franklin mountains might belong with one or more of the above taxa, but not with var. *lindheimeri*. Yellow-spined diploids in the Chisos Mountains are now recognized as *O. chisosensis*.

The flowers of var. *lindheimeri* are brilliant yellow, without red centers. In the Trans-Pecos populations, we have observed only the rich yellow flower color, except that the tepals of late first-day or second-day flowers turn orange or reddish. In South Texas the flowers of var. *lindheimeri* may be yellow, orange-red, orange-yellow, various shades of yellow and orange-red, and rarely red. It might be that most orange or reddish flowers have opened for more than one day. The flowers are relatively large at 5–8 cm long and 5–7.5(–10) cm in diameter. The inner tepals are obovate-spatulate, obtuse-apiculate, 3–4(–5) cm long, and 1.2–2.5(–4) cm in diameter. The filaments are ca. 1.5 cm long and cream-colored, as are the anthers, which are ca. 2 mm long. The stamens are sensitive, perhaps more so than in most other prickly pears. The style is greenish-yellow to whitish, 1.7–2 cm long, with a bulbous base. The 6–8 heavy stigma lobes are ca. 5 mm long and usually dark green. The pericarpel is 4–6 cm long and 2–2.3 cm wide.

The fruits of var. *lindheimeri* are purple to reddish-purple, often pyriform, 3–7 cm long, 2.5–3(–4) cm in diameter, spineless, with a shallow umbilicus. The fruit pulp and juice are beet-red, very juicy, sweet, and edible. The seeds are tan, irregularly discoid, 3–4 mm in diameter, 1–2 mm thick, with a narrow aril. The seeds from some populations in the Trans-Pecos and in South Texas were measured at ca. 3 mm in diameter, slightly but consistently smaller than in most populations of var. *engelmannii*.

Phenology. The principal flowering period of var. *lindheimeri* near Boquillas appears to be from mid-April to mid-May. Flowers that open at mid- to late morning may turn orange by late in the day, while flowers that open later in the day may remain partially open at night and open again the next day with orange or red tepals. Fruits ripen in about six weeks and may not persist long because they are eaten by various animals. In the Trans-Pecos usually there are no fruits remaining on plants by September. More detailed phenological studies of var. *lindheimeri* are needed.

Sterile and Immature Characters. The shrubs of var. *lindheimeri* may or may not produce a short trunk. The largest plants, particularly those in south Texas, may reach 5 m or more in diameter and 3 m high. The erect or ascending branches are dominated by relatively large, heavy, green to blue-green pads. Spines are produced in all but the lowest areoles of the pads. The areoles are 2.5–5 cm apart, elliptic to orbicular, 4.5–6 mm long, 3–6 mm wide, and filled with short gray to tan wool.

In Trans-Pecos populations of var. *lindheimeri* some plants are spineless, or nearly so, or produce 1–2 spines per areole, but in most plants the spine pattern consists of 3–4 spines, with 1–3 longer spines deflexed, angled downward, or porrect, but usually not in a "bird's-foot" pattern. In most Trans-Pecos specimens of bona fide var. *lindheimeri*, there are 3–4 central spines. There are 1–2 lowermost centrals that are yellow, deflexed, acicular except flattened at the base, 1.3–2.5 cm long, and 0.5–0.7 mm wide. One central is positioned immediately above or slightly to one side of the lowermost one. This is a larger spine that is yellow, deflexed or porrect, flattened, twisted or straight, sometimes curved and twisted, (3–)4–6(–7.5) cm long, and 0.7–1 mm in diameter. Usually a third or fourth mostly terete yellow spine occurs in mid-areole, particularly in distal and marginal areoles. This spine is 1.8–5 cm long. In some Trans-Pecos specimens there is another "lowermost" central below the type of partially flattened lowermost spine described above (i.e., the spine closest to the lower periphery of the areole). In these specimens the lowest spine is deflexed and acicular and is below the basally flattened spine that is lowermost in other specimens, and the longest, flattened spine is third-ranked from the bottom of the areole. In these areoles usually there are at least four spines. In specimens east of the Trans-Pecos, the central spine number varies from one to eight, but commonly is 3–5. One such spine pattern we have observed includes one acicular lowermost central and 1–4 deflexed or porrect lower and lower lateral flattened spines, these all yellow or with reddish bases. Plants resembling var. *lindheimeri* but with reddish central spines may be hybrids. In var. *lindheimeri* there may be 1–2 radial spines at the lower periphery of the areole, or the radials may be absent. Radial spines are acicular, yellow, deflexed, and readily distinguished from the smallest centrals in shorter length and smaller caliber. In var. *lindheimeri* the dirty-yellow glochids occur around the periphery of the areole, or they may appear almost anywhere in the areole, including clustered in the upper half of the areole. The glochids may be unequal in length, 3–6 mm long, sometimes interspersed with slightly larger bristles that have smooth surfaces. Marginal areoles especially may be dominated by bristles to ca. 1 cm long. The glochid pattern in var. *lindheimeri* is like that in var. *engelmannii* (Parfitt and Pinkava, 1988).

Immature specimens are characterized by fewer spines. One or two spines are common in mid-pad areoles of young stems. The lower centrals appear first in areoles. The seedlings are hairy.

Biosystematics. The authors of several recent works concerning *Opuntia* have presented different concepts of var. *lindheimeri* and its allies. Bravo-Hollis (1978) followed Benson: she treated *O. lindheimeri* as a distinct species with three varieties and otherwise essentially followed Weniger (1970, 1984), who incorporated *O. lindheimeri* and a number of taxa traditionally associated with *O. lindheimeri* and *O. engelmannii* under the species *O. engelmannii*. Weniger (1984) did not recognize a var. *lindheimeri* under his *O. engelmannii*. Benson (1982) treated *O. lindheimeri* as a distinct species with five varieties, and he anomalously lumped the closely related *O. discata* (*O. engelmannii*) with *O. phaeacantha* instead. All except one, var. *cuija* (Griffiths & Hare) L. D. Benson, occur in Texas. The so-called var. *cuija*, apparently a diploid species, is in northeastern Mexico.

Different ploidy levels have been reported for certain populations of var. *lindheimeri* and its allies (Grant and Grant, 1979b; Pinkava et al., 1982; Weedin and Powell, 1978). The most reliably identified var. *lindheimeri* populations, including form *linguiformis*, all apparently are hexaploid, after the exclusion of misclassified less-than-closest peripheral taxa such as *O. ellisiana*. Certain populations along the Gulf Coast are diploid (Grant and Grant, 1979b), possibly those referable to *O. cacanapa*, *O. alta* Griffiths, or *O. lindheimeri* var. *lehmannii* L.

D. Benson. Those in part of the Edwards Plateau region are tetraploid (referable to *O. subarmata* Griffiths or *O. tardospina* Griffiths). Some collections from the Davis Mountains and reports from the Guadalupe Mountains in Texas and New Mexico suggest that *O. subarmata* may extend as occasional populations into the Trans-Pecos (D. Ferguson, pers. comm.). A chromosome number is known for one Trans-Pecos collection that A. Zimmerman suspects is *O. subarmata* ("tetraploid *discata*," from near Dryden; Grant and Grant, 1979b).

Populations of var. *lindheimeri* and var. *engelmannii* overlap in the eastern Trans-Pecos, the adjacent Texas Hill Country, and northeastern Mexico. At some sites in these areas these closely related taxa, both presumably hexaploid, occur together without apparent hybridization. Anthropogenic dispersal of the "cow tongue" mutant form of var. *lindheimeri,* far into the ranges of many other species, including *O. engelmannii*, sensu stricto, has not resulted in even one putative hybrid of *linguiformis* being reported.

Synonyms (misidentifications; mistakes in lumping). *Opuntia lindheimeri* Engelm. var. *lindheimeri; O. texana* Griffiths; *O. ferruginispina* Griffiths; ? *O. sinclairii* Griffiths; ? *O. griffithsiana* Mackensen; ? *O. reflexa* Mackensen; ? *O. winteriana* A. Berger; *O.* ? *haematocarpa* A. Berger; ? *O. longiclada* Griffiths.

Common Names. Nopal; tuna; flaming prickly pear.

25. **Opuntia polyacantha** Haw., Syn. Pl. Succ. 82. 1819. PLAINS PRICKLY PEAR. Plates 101–4. In the Trans-Pecos this species has an unusual distribution, spanning a wide altitudinal range yet disjunct and unpredictable. The central mountains, western plateau grasslands, and sandy desert in far western portions all support various forms of this species, from 3,600 to 8,382 ft. Flowering May–Jun. $2n = 22$. Widespread in the western U.S. except for the W coast and southern AZ, but mostly northern or at moderate to high elevations (to ca. 10,000 ft). Canada in Alberta and Saskatchewan. $2n = 22, 44, 66, 88$. Mexico in N Chihuahua and (one collection) N Coahuila. Map 23.

This is a diploid-polyploid complex with chromosome numbers ranging from diploid (as in the Trans-Pecos) to octaploid and higher in some hybrids, and it is distinguished by dry, often spiny fruits, large flat seeds with a wide aril-margin, and green stigmas. Only three named members of the *O. polyacantha* complex occur in Texas, two of them in the Trans-Pecos, and one of them only in the Panhandle (where *O. fragilis* also grows).

Parfitt (1991) recognized five species (counting a hybrid species or nothospecies) and five varieties in a biosystematic study of the *O. polyacantha* complex. The species are *O. polyacantha, O. fragilis* (Nutt.) Haw., *O. aurea* E. M. Baxter, *O. pinkavae* B. D. Parfitt, and *O.* × *columbiana* Griffiths, pro. sp. (Parfitt, 1998). Five varieties of *O. polyacantha* recognized by Parfitt are var. *polyacantha,* var. *arenaria* (Engelm.) B. D. Parfitt, var. *erinacea* (Engelm. & Bigelow) B. D. Parfitt, var. *hystricina* (Engelm. & Bigelow) B. D. Parfitt, and var. *nicholii* (L. D. Benson) B. D. Parfitt. The type locality of *O. polyacantha* is officially restricted to near Fort Vanderburgh on the Missouri River in Mercer County, North Dakota. Consequently, *O. polyacantha* var. *polyacantha,* the nomenclaturally

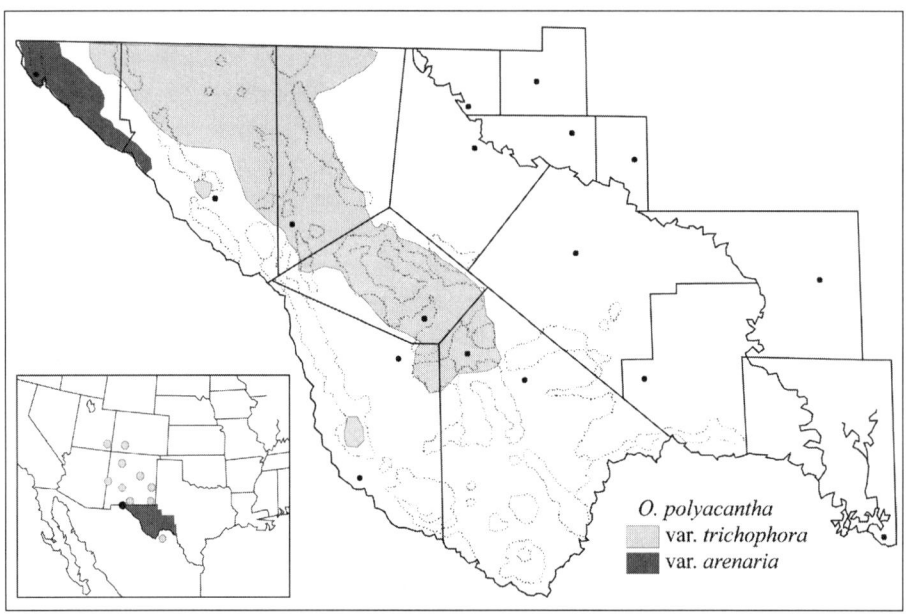

Map 23. Generalized distribution of *Opuntia polyacantha* var. *trichophora* (southern plains prickly pear) and *O. polyacantha* var. *arenaria* (sand prickly pear).

typical part of the species, refers to the northeastern (presumably tetraploid) plants. The specific epithet is after the Greek *poly*, "many," and *acantha*, "spine," in reference to the many-spined stem segments.

Key to the Varieties

1. Plants 10–25 cm high; pads all flat, usually obovate or orbiculate, 7–13 cm long, 5.5–11 cm wide, ca. 1 cm thick; pericarpel elongated but not "slender," roughly 1.5 times longer than thick; habitat mountain woodlands, grasslands, desert sand, or gypsum

   25a. *O. polyacantha* var. *trichophora*.

1. Plants lower shrubs to 5–15 cm high; pads, or some of them, weakly compressed or terete, superficially resembling joints of a cholla cactus, 4–7(–10) cm long, 2–3(–5) cm wide, often to 2 cm thick; pericarpel slender, sometimes 2–3 times as long as thick; habitat deep sand

   25b. *O. polyacantha* var. *arenaria*.

25a. **Opuntia polyacantha** Haw. var. **trichophora** (Engelm. and Bigelow) J. M. Coult. SOUTHERN PLAINS PRICKLY PEAR. Plates 101, 102. [*Opuntia missouriensis* DC. var. *trichophora* Engelm. & Bigelow, Proc. Amer. Acad. Arts 3: 300. 1856]. Clay, gravel, or sandy soils, oak-juniper-pinyon woodlands, grasslands, less often desert sands or gypsum. Expected in E extreme of El Paso Co., Hudspeth Co., mostly Sierra Diablo Plateau in sandy grassland and desert, Culberson Co., Apache Mts N to Guadalupe Mts. Presidio Co., Chinati Peak. Jeff

Davis Co., Davis Mts. Brewster Co., NW portion in Davis Mts. 3,600–8,382 ft. Flowering May–Jun. $2n = 22$. South-central NM, from the Guadalupe Mts. Coahuila, Sierra del Carmen; to be expected in Chihuahua adjacent to the Trans-Pecos. Map 23.

Opuntia polyacantha var. *polyacantha*, whether in the broad sense of Parfitt (1991) or in the narrowest possible sense of only the tetraploid Great Plains and Rocky Mountains part of *polyacantha*, is documented in herbaria as the second most widespread taxon in the *O. polyacantha* complex, after *O. fragilis*. They share basically the same range except that *O. fragilis* extends farther west, northwest, northeast, and farther into Canada and higher in the mountains than does *O. polyacantha*. *Opuntia fragilis* is unknown in the Trans-Pecos but does occur rarely in the Texas Panhandle. *Opuntia polyacantha* var. *polyacantha* (tetraploid?) is known for Briscoe, Armstrong, and Potter counties of the Texas Panhandle (Parfitt, 1991).

Benson (1982) accepted four varieties of *O. polyacantha*, but his concept of the species was relatively arbitrary. Parfitt (1991) recognized five varieties, including var. *arenaria* (which Benson had left at its original specific rank). According to Benson, the Trans-Pecos varieties of *O. polyacantha* were var. *rufispina* and var. *trichophora*. Parfitt (1991) treated both var. *rufispina* and var. *trichophora* as synonyms of var. *polyacantha*. Benson's concept of var. *polyacantha*, in the somewhat narrower sense than was Parfitt's, held that it was widespread from central New Mexico northward (corresponding, we now know, with the range of the tetraploids, east of the Rockies and in the far north), but that it did not occur in the Trans-Pecos. Weniger (1984) recognized only one variety of this species in the Trans-Pecos, and he called it var. *polyacantha*; Weniger treated *O. arenaria* as a species.

Identifying Characters. The apparently nonrhizomatous plants of *O. polyacantha* var. *trichophora* in mountain woodlands and grasslands are too low to be called "shrubs": 10–25 cm high with relatively small pads. They have closely spaced areoles and usually 6–17 rather slender spines per areole. The longest spines, usually 1–3 per areole, may be 4–7 cm, while the other spines often are appressed and less than 2 cm long. The longest spines are whitish to gray, or less often reddish to brown or nearly black. The prostrate stems of var. *trichophora* often produce roots where pads touch the ground, especially in sandy substrates, but they rarely "root-sprout" from rhizomes as does var. *arenaria*. Usually the plants of var. *trichophora* are readily distinguishable from other prickly pears in the Trans-Pecos, except perhaps var. *arenaria*; the small pads of var. *trichophora* are partially obscured by thin spines produced in all but the lowermost areoles.

The flowers are yellow in all Trans-Pecos populations of var. *trichophora*, so far as known, although they may darken slightly before wilting (late in the first day after anthesis). In New Mexico and farther north, *O. polyacantha* may have yellow flowers with red centers or may be pale yellow to magenta, rarely white, or combinations of these colors. The flowers usually are 4–7 cm long and 4–5 cm wide, although smaller flowers are produced by some plants. The inner tepals are less brilliant yellow than in fresh flowers of *O. phaeacantha*, but not as pale as

O. chisosensis, 3–4 cm long, ovate-spatulate, and 1–2 cm wide. The filaments are white or barely green-tinted to cream-colored or pale yellow (commonly yellow or magenta north of the Trans-Pecos, but some or all of these might be the northern [tetraploid] variety), ca. 1 cm long, and the anthers are cream-yellow. The style is cream-colored to pinkish, ca. 2 cm long, and only slightly expanded at the base (less bulbous than that of *O. azurea* var. *aureispina*, for example). The ca. eight stigma lobes are ca. 3 mm long and dark green. The slender (1.47 times longer than thick at anthesis) pericarpel is notable for its spiny areoles.

The fruits of var. *trichophora* are dry, relatively small, and very spiny at maturity. As fruit maturation proceeds, the green surface turns dull red, and then the whole fruit rapidly dries, even the previously green cortex, and turns tan or cream-colored. The dried mature fruits are somewhat obconic, 1.9–2.5 cm long, 0.8–1 cm (to 1.7 cm immediately before drying) in diameter, ultimately with a shriveled or wrinkled tan surface. The fruit surface supports numerous (ca. 12–42) uniformly distributed, whitish, cottony areoles, with (4–)6–15 spines per areole, these 0.4–1.2 cm long. The papery, dry fruit surface is partially hidden by the spines. The umbilicus reaches ca. 8 mm deep (Parfitt, 1991). After abscission from the plant, the fruit base from some individuals in the Davis Mountains is covered by dried, hard pericarpel tissue. Fruits that have been lying on the ground for some time have a circular basal pore through which the lowermost seed can be seen. In such cases, the seeds are too large to fit through the basal pore. Evidently, the basal pore forms over time through disintegration of the covering tissue, or as a wide, perforate abscission scar, and detachment of a circular basal plate. In fruits of other plants the basal pore is large enough for seeds to fit through and open (nonseptate) at the time of abscission, so that one or a few seeds are left on the stem where the fruit base was located. The flattened seeds of *O. polyacantha* are the largest of any Trans-Pecos prickly pear species and few per fruit (only four ovules in flowers of *A. M. Powell 5442*). The seeds are irregularly discoid, tan or cream-colored, 6–6.8 mm in maximum diameter, 1.5–2 mm thick, with a beaked aril-margin (raphe) 1–2 mm wide. In some fruits examined, the seeds positioned near the top of the fruit were noticeably larger than seeds near the bottom. The horizontally oriented seeds are either numerous or as few as 8–10 per fruit.

Phenology. The principal flowering period for var. *trichophora* is from early May through June, later at higher elevations of mesic mountains. Flowers open after mid-morning for one day, close at night, and seldom reopen the next day. Osborne et al. (1988) noted that late-opening individual flowers close at night and open again the next morning until individual flowers had exhibited an open time of 7–11 hours. These authors also demonstrated that *O. polyacantha* (presumably the tetraploid var. *polyacantha*) was allogamous and obligately entomophilous in southeastern Colorado. Fruits mature in July and August and are easily detached from stems.

Sterile and Immature Specimens. The pads of var. *trichophora* are broadly ovate, orbiculate, obovate, to narrowly obovate, flat, (4–)7–13 cm long, (4–)5.5–11 cm in diameter, and ca. 1 cm thick. The areoles are 0.6–1.3 cm apart, oblong to circular, 2–3.5 mm long, and 1.5–3 mm wide. Spines are present in all but the lower areoles.

An evaluation of the spine pattern in Trans-Pecos populations of var. *trichophora* revealed that usually there are 4–7(–8) central (or "major") spines and 4–6 radial (or "minor") spines per areole on mature stems, particularly in distal areoles. The terms "major" and "minor" spines were employed by both Anthony (1956) and Parfitt (1991). We prefer to use the terms "central" and "radial" spines; "centrals" and "radials," as nouns, are already well established in the cactus literature. In Trans-Pecos specimens usually there are one lower and two lower lateral centrals that are descending, 0.9–1.6 cm long, acicular, slightly flattened, or angular, and whitish with pale reddish tips. Above the lowermost centrals, often there is one longest central, positioned near the center of the areole. The longest spine is 4–7 cm, 0.5–1 mm wide at the base, filiform, deflexed or porrect, gray with black speckles to pale reddish, sometimes twisted, and somewhat flattened, at least near the base. The color of the longest central may vary from white to gray, tan, or translucent yellowish, sometimes with reddish bases, or reddish to dark reddish-brown throughout, especially in apical areoles. A fifth shorter "wild" spine may be present at the periphery in the upper part of the areole. Or there may be a second longer spine positioned immediately above the longest central, along with the "wild," obliquely ascending upper peripheral spine. Areoles with 6–8 centrals may have three lowermost and two mid-lateral centrals in addition to the longest and other upper centrals. Usually there are 4–6 radial spines originating at the lower periphery (lower one-fourth or one-third) of the areole, these obviously smaller than the smallest central. The radials are appressed, 3–6 mm long, acicular, and whitish with pale yellow or reddish tips. Glochids are tightly packed in an apical tuft or crescent in the areole apex. The glochids protrude 1.5–5 mm, and they are light yellow. One or more longer bristlelike glochids may stand at the edges of the apical tuft.

Immature specimens usually have fewer and shorter spines per areole. The spines include 2–4 radials and ca. 3 lower and lower lateral centrals, these usually only 2–8 mm long and white or yellowish with reddish-yellow tips. In other juvenile areoles there may be three or more short centrals in various positions in the areole, these appressed or porrect. The seedlings often are hairy (D. Ferguson, pers. comm.).

Biosystematics. The var. *polyacantha*, in the sense of Parfitt (1991), is a morphologically variable taxon including both diploid and polyploid populations. The tetraploids, var. *polyacantha*, sensu stricto, are by far the most wide-ranging taxon or subtaxon of the species (Parfitt, 1991), extending from New Mexico north to Canada. Populations of diploid (not nomenclaturally typical) *O. polyacantha* in the Trans-Pecos, which we identify as var. *trichophora*, vary in habitat and in morphology. Plants in Culberson County occur in desert or semidesert habitats, including sand and gypsum, and have mostly white spines. Plants in the Davis and Chinati mountains occur in mesic habitats at mid- to high elevations and often have more slender, translucent, pale yellow-brown or white spines. Plants in Hudspeth County occur in grasslands, often in sandy soil, and usually have longer, darker, slightly thicker spines, as well as more spines, particularly in the upper parts of the distal areoles. All chromosome number reports from Trans-Pecos populations of *O. polyacantha* are diploid (see Parfitt, 1991), including the Hudspeth County plants (Powell and Weedin, 2001) that otherwise resemble *O. polyacantha* var. *hystricina*. According to Parfitt, var. *hystricina* occurs from northwestern New Mexico west to eastern California, where it is entirely polyploid.

Benson (1969, 1982) attempted to distinguish *O. polyacantha* var. *trichophora* primarily by the presence of remarkably long, hairlike, white, flexible, curly, or down-curving spines, which almost obscure the pads in some populations. Benson (1982) described the spines as resembling the hair of an Angora goat. The pads often exhibit a shaggy appearance (Anthony, 1956), particularly on their lower portions. This unusual spine morphology is sporadic almost throughout the range of *O. polyacantha*, even in some tetraploid plants of Wyoming, and should not be recognized taxonomically (Parfitt, 1991).

Parfitt (1991) believed that the widespread *O. polyacantha* species-group originated in relatively xeric areas of the southwestern United States and northern Mexico, where the diploid taxa still grow. Farther north and west, the polyploid taxa extend into colder and wetter habitats, including sagebrush, boreal forest, and the Great Plains. Diploid populations assigned by Parfitt to var. *polyacantha*, but here "split" as var. *trichophora*, occur in Trans-Pecos Texas, New Mexico, and southeastern Utah, in neither the driest nor the wettest of habitats occupied by the dry-fruited prickly pears (the *O. polyacantha* species-group in general). Except for the diploid vars. *arenaria* and *trichophora* all other named taxa of the *O. polyacantha* complex are polyploid (tetraploids through octoploids), and higher ploidy levels occur in some hybrids (Parfitt, 1991). The *hystricina*-like populations in Hudspeth County with dark spines and off-yellow tepals are receiving more deliberate study now than in the past; at least we have a vouchered diploid chromosome count. Variety *arenaria* might deserve specific status, because no synapomorphies unite the diploids as a group. However, the peculiar "root sprouts" characteristic of var. *arenaria* were found sporadically in tetraploid *O. polyacantha* var. *polyacantha* (suggesting a closer relationship than had been evident from the general appearance of the plants).

Synonyms. ? *O. missouriensis* DC. var. *rufispina* Engelm. & Bigelow; *O. polyacantha* Haw. var. *rufispina* (Engelm. & Bigelow) L. D. Benson, in part; *O. trichopohora* (Engelm.) Britton & Rose.

Common Names. Plains prickly pear; plains prickly-pear; plains pricklypear; hunger cactus; starvation cactus; starvation prickly pear; nopal; prickly pear; the brown-spined morphs, erroneously identified as *rufispina*, have been called red-spined prickly pear; the "hairy" morphs, easily and accurately identified as *trichophora*, have been referred to as hair-spined prickly pear and bristlehair prickly pear cactus. The ordinary white, "nonhairy" morph, is most likely to be misidentified as plains prickly pear, the appropriate name for tetraploid, northern var. *polyacantha*, sensu stricto.

25b. **Opuntia polyacantha** Haw. var. **arenaria** (Engelm.) B. D. Parfitt. SAND PRICKLY PEAR. Plates 103, 104. [*Opuntia arenaria* Engelm., Proc. Amer. Acad. Arts 3: 301. 1856]. Relatively uncommon in deep sand and silt, usually in dune areas. El Paso Co., in the Rio Grande valley from NM boundary at Anthony downstream through Fabens and the S Hueco Mts to SW Hudspeth Co. near McNary. 3,600–4,500 ft. Flowering May–Jun? $2n = 22$. Doña Ana Co., NM, from E of Columbus E to the Rio Grande valley. Mexico, Chihuahua, near Samalayuca S of Juarez, presumably W to near Palomas. Map 23.

Prior to the research by Parfitt (1991, 1998) this taxon was treated as the distinct species *O. arenaria*. It is a sand-loving prickly pear that is almost entirely restricted to the deep sands in the Rio Grande valley near El Paso and Juarez. The type locality is "Sandy bottoms of the Rio Grande near El Paso." The varietal epithet is after the Latin *arenarius*, pertaining to sand, in reference to this habitat specialization.

Identifying Characters. These are the prickly pears that superficially resemble dog chollas. The plants of var. *arenaria* form low creeping mats that are relatively small or loosely and irregularly spread to 1–3 m in diameter. Branches growing in shifting sands often are partially buried in the substrate and can be misinterpreted as rhizomes. Glochid-filled areoles, pads, and ultimately whole clumps are produced ("root-sprouting") as adventitious buds develop directly from horizontal, undersand, rhizomelike roots (Boke, 1979), extending up to 2 m. The small pads are narrower and less flattened than those of any other prickly pear in the Trans-Pecos, at 4–7(–10) cm long, 2–3(–5) cm wide, and 1–2 cm thick. The largest pads are oval, ovate, obovate-elongate, subcylindroid, or clavate, or narrowly obovoid, and almost terete in cross section. The pad faces have the areoles raised on short tubercles. Spines are produced in all but the lowermost areoles, but the rather full spine cover does not obscure the glaucous or light green stem surface. The spines are white, or mostly so, usually with one spine much longer (to 4 cm) than the 5–10 others in each areole. The largest, flattest individual pads of var. *arenaria* closely resemble those in the other varieties of *Opuntia polyacantha* (which do not grow wild at El Paso).

The yellow flowers are 4–6.5 cm long and 4–6 cm wide. The tepals are obovate, 2–3 cm long, 1–1.5 cm broad, and apiculate or entire. The filaments are white to pink-tinged, 0.6–1 cm long, and support yellow anthers that are ca. 1.5 mm long. The style is white to pale green, 1.2–2 cm long, and 2–2.5 mm in diameter at its bulbous base. The 5–8 stigma lobes are ca. 2 mm long and green. The pericarpel is uniquely slender in the *O. polyacantha* complex, perhaps 2–3 times as long as thick (Parfitt, 1991).

The fruit turns reddish and then tan and dry at maturity. The fruit surface is somewhat obscured by 3–6 white spines 6–9 mm long in each of the 10–14 areoles. The fruit is narrowly obovate-obconic, constricted below the apex, 2.5–3 cm long, and 0.9–1.2 cm in diameter. The umbilicus is relatively deep. The seeds are tan, shiny, irregularly discoid, and large, like those of var. *trichophora*. They are 6–6.5 mm in largest diameter (7–8 mm according to Benson, 1982), 1–2 mm thick, and with a wide margin.

Phenology. So far as known, the major blooming period occurs during late May and perhaps most of June. The fruits turn tan and dry after 2–2.5 months and may persist through the summer (Benson, 1982).

Sterile and Immature Specimens. The plants of var. *arenaria* form low mats in the sand. Although the individual pads and clumps are small, the whole plant may form conspicuous clonal colonies 1–3 m in diameter, with individual creeping branches to 2 m long, partially buried in the sand. The most slender and/or terete pads of var. *arenaria* resemble those of *O. fragilis* (Nutt.) Haw., but *O. fragilis* is not documented to occur south of the Texas Panhandle near Amarillo, northern New Mexico near Santa Fe, and northern Arizona near Flagstaff; it is not likely to grow within hundreds of miles of the Trans-Pecos. The rhizomelike roots of var. *arenaria*, to 1 cm or more in diameter and bearing branch roots, aboveground pads, and sporadic areoles with glochids 3 mm or more long, are not unique to this taxon. "Root-sprouting," or the production of adventitious stems from underground roots, is discussed by Parfitt (1991) as a general phenomenon in two or more taxa of the *O. polyacantha* complex, found sporadically in var. *polyacantha* far from El Paso; it also happens frequently in *O. pottsii*.

Areoles on the slender pads are 0.5–1.1 cm apart. The small areoles are ca. 2 mm in diameter and raised on tubercles 1–2 mm high.

The 6–11 spines per areole in var. *arenaria* have been differentiated by some authors as major and minor spines. The major spines are the 1–4 longer ones, often positioned in the middle to upper portion of the areole. Or there may be one toward the lower middle, two lateral, and one in the upper areole. The major spines usually are present only in the distal areoles, and they are acicular, straight, white, gray, to tan or reddish-brown with yellowish tips, 2–4(–6) cm long, 0.5 mm in diameter, and more or less porrect or angled downward. Among the major spines, usually one is much longer than the others. There are 5–7 minor spines that are depressed-deflexed at the lower edge of the areole. The minor spines are 0.4–2 cm long and white with yellow tips. The middle minor spine usually is much longer than the others. We prefer to recognize central and radial spines. By our interpretation there are 1–5 centrals (i.e., the "major" spines), plus the one or more larger "minor" spines near the lower areole margin. There are 2–6 radial spines in var. *arenaria*. These are the smaller minor spines, 3–4(–7) mm long, and chalky-white with reddish-yellow tips. The glochids are relatively inconspicuous or present in dense apical tufts or crescents. In mid-pad areoles the glochids protrude 1–2 mm and are tan to yellowish.

Immature stems are light green, rather glossy, and more or less cylindroidal or subclavate. The whitish spines are fewer and shorter in areoles on young pads. The seedlings are not hairy (D. Ferguson, pers. comm.).

Biosystematics. The var. *arenaria* clearly is a diploid basal lineage of the *O. polyacantha* species-group, and we accept the treatment of this taxon as a variety of *O. polyacantha* (Parfitt, 1991). All workers prior to Parfitt maintained var. *arenaria* as a distinct species. Ecologically and in superficial appearance, var. *arenaria* is strikingly different from most (but not all) populations of *O. polyacantha*.

The var. *arenaria* is locally common at some deep sand sites and silted basins in El Paso County, but otherwise seems to be uncommon; agriculture and urban development have claimed much of the habitat. Populational studies from throughout its range might clarify both its ecological requirements and its relationship with *O. polyacantha*.

Synonym. See basionym under var. *arenaria*.

Common Names. El Paso prickly pear; sand-loving opuntia; sand-loving prickly pear.

Appendix 1

The four prickly pear taxa briefly treated below are cultivars or escapees in the Trans-Pecos but have not become naturalized in significant wild populations.

Opuntia basilaris Engelm. & Bigelow var. **basilaris**, Proc. Amer. Acad. Arts 3: 298. 1856. BEAVERTAIL CACTUS. Plates 105, 106. Plants low-growing, clumps 15–30 cm high (but usually only one or two pad heights), 0.3–1.5 m across. Branches short, from a vaguely defined trunk, not forming long chains. Stems green to blue-green or irregularly purplish, minutely canescent or velvety as in *O. rufida* and *O. microdasys*, usually obovate, 5–18(–25) cm long, 4–9(–14) cm wide; areoles usually 1–1.5 cm apart, circular, 1.5–3 mm across. Spines absent; glochids brown to tannish, ca. 3 mm long. Flowers purplish-pink to reddish (cerise), 5–7.5 cm long and wide; inner tepals 2.5–4 cm long, 1.2–2.5 cm across; filaments reddish, anthers yellow; style pale pink to reddish, stigma white (unlike the dark green stigmas of *O. rufida*, *O. microdasys*, the *O. phaeacantha* species-group and many other opuntias). Fruits green at first, turning tan or gray, dry at maturity, spineless, narrowly urceolate, 2.5–3 cm long, 1.5–2.3 cm in diameter; seeds irregularly discoid, ca. 6 mm in largest diameter, 2–3 mm thick, bone-white or grayish. 3,600–3,800 ft. Flowering in spring. Fruits persist for several months. Occasionally seen as an escape in El Paso (D. Ferguson, pers. comm.). The var. *basilaris* is native to southern California and adjacent Arizona, Nevada, and southern Utah. The specific epithet, after the Latin *basilis*, "basal," and *aris*, "belonging to," in reference to the new growth that develops only from the plant base (Crook and Mottram, 1995). The plants are recognizable by spineless obovate pads that are microscopically pubescent (like our native *O. rufida* and domesticated *O. microdasys*) and purplish-red flowers, white stigma lobes, and dry fruits with unique large seeds. The spineless pads are not harmless because, as in *O. rufida* and *O. microdasys*, the areoles are loaded with pesky glochids.

Opuntia ellisiana Griffiths, Annual Rep. Missouri Bot. Gard. 21: 170–71, t. 25. 1910. ELLIS' PRICKLY PEAR. Plates 107, 108. [*Opuntia lindheimeri* Engelm. var. *ellisiana* (Griffiths) K. Hammer]. Plants 1–2 m high, usually twice as wide as tall (contrasting with *O. ficus-indica,* which is erect). Stems (pads) typically obovate, 15–23 cm long, 10–15 cm across; areoles 2.5–3.5 cm apart; areoles small, circular, 2–3 mm across, with very short white or gray wool; areoles slightly elevated on low tubercles; glochids few, in center of wool, ca. 1 mm long. Spines of cultivar *ellisiana* usually absent. Flowers plain yellow (no red centers);

filaments and anthers cream-yellow, style cream-colored, stigma lobes green. Fruits pyriform or turbinate, turning pink to rose (partially to mostly), then reddish-purple, ca. 3 cm long, 2.5 cm in diameter, with a shallow umbilicus, fleshy, pulp reddish-purple, moderately juicy, juice purple; areoles few, mostly distal, these with a few glochids and perhaps bristles; seeds discoid, dark gray to tan, 2.5–3.6 mm in diameter, aril-margin 0.3–0.6 mm wide, tan. In Trans-Pecos horticulture, 1,800–5,000 ft. Flowering June; in the Sul Ross Cactus Garden usually the last prickly pear species to bloom. Apparently the only prickly pear in the Trans-Pecos with stamens that are nonsensitive or barely so. Fruits maturing in late September through November in Alpine. $2n = 22$. *Opuntia ellisiana* is commonly planted as an ornamental in Alpine and some other towns of the Trans-Pecos. Its distribution in the Trans-Pecos is not fully known. The type specimen was from a garden in Corpus Christi, Texas (Crook and Mottram, 1996). It is widely grown in the south half of Texas. Benson (1969b, 1982), followed by Bravo-Hollis (1978), erroneously dismissed *O. ellisiana* as a synonym of unrelated *O. subarmata* Griffiths (type from "near Devils River, Texas" in 1908), which Benson erroneously dismissed as a "hybrid population" involving *O. lindheimeri* and *O. ficus-indica*. David Ferguson (unpub. list, 1999) considered *O. ellisiana* and *O. tricolor* Griffiths [= *O. lindheimeri* var. *tricolor* (Griffiths) L. D. Benson, type from "near Laredo, Texas" in 1907] to be synonyms of *O. cacanapa* Griffiths & Hare (type from Encinal, LaSalle Co., Texas, in 1906). *Opuntia ellisiana* plants have absolutely spineless stems, but wild *O. tricolor* is very spiny. Both types are unusual for having strongly recurved leaves, including the bracteoles on the pericarpel. In Alpine, plants of *O. ellisiana* were heavily browsed by mule deer that came into the town during the drought of the 1990s. The species was named after Prof. J. Coswell Ellis, who collected the plants for Griffiths.

Opuntia engelmannii Salm-Dyck cv. **linguiformis** (Griffiths) A. D. Zimmerman, comb. et stat. nov. (forthcoming). COW-TONGUE PRICKLY PEAR. Plates 109, 110. [*Opuntia linguiformis* Griffiths, Annual Rep. Missouri Bot. Gard. 19: 270–71, t. 27, lower fig. 1908; *O. lindheimeri* Engelm. var. *linguiformis* (Griffiths) L. D. Benson; *O. engelmannii* Salm-Dyck var. *linguiformis* (Griffiths) Weniger, *nom. nud.*; *O. engelmannii* Salm-Dyck ex Engelm. var. *linguiformis* (Griffiths) B. D. Parfitt and Pinkava]. Plants erect, 1–2 m high, often with spreading branches. Stems typically lanceolate to ca. 1 m long, usually shorter, varying from narrowly ovate to linear, and perhaps subcrescentiform, 10–15 cm across at the widest point near the base of the pad. All vegetative and floral characters are those of *O. engelmannii* var. *lindheimeri*, sensu stricto, except for the elongated stems that result from continued growth of the apical meristem. Spines usually 1–3 per areole, 1–2 cm long, yellow to reddish-brown. Flowers in Trans-Pecos plants are yellow, changing in late afternoon (or if reopening a second day) to yellow-orange, orange, orange-red; flowers reportedly orange or red in Bexar County (Weniger, 1984); filaments and anthers yellow; stigma green. Fruits purplish, fleshy, juicy; seeds irregularly discoid, 3–4 mm in diameter. 1,800–5,000(–6,000) ft. (cultivated for more than 43 years at Silver City, New

Mexico). Flowering May (near the Rio Grande) to June. In Alpine, fruits usually mature in mid- to late August but may not ripen until mid- to late September. $2n = 66$.

Opuntia engelmannii var. *linguiformis* supposedly is a "mutant form" of *O. engelmannii* var. *lindheimeri* that originated by a single-gene mutation. This freak lineage of the ordinary Texas prickly pear was reported growing wild only in Bexar County near San Antonio (Schulz and Runyon, 1930). Weniger (1984) contended that there were plants still growing "wild" near Sayers in Bexar County east of San Antonio, although there are no known native populations of var. *linguiformis* (Schulz and Runyon, 1930) in the sense of an endemic wild taxon. In the Trans-Pecos, plants have been widely propagated vegetatively and dispersed as ornamentals. Such plants are particularly common in Alpine, where they easily survive the winters. In Big Bend National Park scattered colonies near the Rio Grande are all presumed to be historical/archeological remnants persisting from otherwise inconspicuous ruins of human habitation. Anthony (1956) predicted that var. *linguiformis* might eventually become naturalized along the Rio Grande, but to date none of the colonies that we know of in the lower Big Bend has produced seedlings. In contrast, the colony of "cow tongue" in Alpine has been reproducing "true" from seed since the 1980s.

The varietal epithet is after the Latin *lingua*, "tongue," and *forma*, "shape," a reference to the tongue-shaped stems. The characteristic stem shape also inspired the English name and the Mexican name, *lengua de vaca*.

According to D. Ferguson (pers. comm.) any clone of var. *linguiformis* may revert to a growth form of typical var. *lindheimeri*, or the "*linguiformis*" stem shape may be produced from seed. The var. *linguiformis* has been treated as a species (Anthony, 1956) or as a variety (Benson, 1982; Weniger, 1984; Parfitt and Pinkava, 1988), and we retain use of the name, although we consider the entity to be a mere cultivar. It is not a part of the natural prickly pear flora of the Trans-Pecos.

Opuntia ficus-indica (L.) Miller. INDIAN FIG. Plates 111, 112. [*Cactus ficus-indica* L., Sp. Pl. 1: 468. 1753]. Plants large shrubs or trees, potentially 3–5 m or more high; trunk large, to 60–120 cm long; stems (pads) green, typically obovate to oblong, the largest pads (22–)28–60 cm long, 15–40 cm in diameter, 2–2.5 cm thick (smaller, immature pads frequently have been selected by collectors as herbarium specimens), areoles 2–6 cm apart, elliptic-oblong, 2–5(–7) mm long, 3–4 mm broad (areoles perhaps oval to subspherical on young pads). Spines absent, present in a few areoles, or 1–6 present in most of the areoles on each pad; spines usually white, or tan to pale brown, deflexed and spreading, straight, flattened, subulate, to 1.2–2.5(–4) cm long, 0.6–0.9 mm wide at the base; glochids yellow, protruding 1–2 mm, numerous, early deciduous. Flowers 5–7 cm long and wide, yellow to orange-yellow with reddish or greenish centers, the inner tepals sometimes pink-tinged externally, 2.5–3 cm long, 1.5–2 cm in diameter; filaments and anthers yellow; style and stigma lobes greenish. Fruits potentially of several different colors, including red to purplish and yellow to orange, fleshy, juicy, edible, typically oval, 5–10 cm long, 4–9 cm in diameter, with a shallow umbilicus, spineless, or sometimes with spines; seeds irregularly discoid, gray or tan, 3–4(–5) mm in largest diameter. Cultivated from sea level to (in the Trans-Pecos) 1,800–3,600(–4,500) ft. Flowering in spring. Fruit persistent for several months. $2n = 88$. Reports of $2n = 22$ and 66 were misidentifications; $2n = 77$ is an F1 hybrid with a native hexaploid in California (McLeod, 1975). *Opuntia ficus-indica* is a species that has been widely introduced and naturalized throughout warm areas of the world. Spineless cultivars have been the most widely dispersed, even in prehistoric times, because the pads are edible as well as the fruit. Feral populations of this (and similar) species have been prolific in some areas; in Australia and parts of Africa, they are pests worthy of extermination. *Opuntia ficus-indica* has been cultivated in Mexico for about 8,000 years (Crook and Mottram, 1997) and probably is native in Mexico. Historically, many cultivars were developed there and disseminated elsewhere. In Mexico the spiny forms ("reversions") of *O. ficus-indica*, widely but imprecisely known as *O. megacantha*, are difficult to distinguish from the wild type. *Opuntia ficus-indica* can hybridize with other species. In southern California, where extensive populations exist, a heptaploid F1 (McLeod, 1975) documents local compatibility with a native hexaploid prickly pear. In Texas and northern Mexico, these are the tallest prickly pear cacti seen as occasional single plants or in colonies, as relicts or escapees in towns or near isolated dwellings, and mostly as casually maintained hedges or dooryard resources in the suburbs. In the Trans-Pecos, *O. ficus-indica* is practically restricted to rural and suburban gardens along the Rio Grande. *Opuntia ficus-indica* was reported from northwestern Brewster County a few miles north of Alpine (Benson, 1969b, 1982), but this was a misidentification of some native species, *O. engelmannii*, sensu lato (fortunately, Benson [1969b] cited an herbarium specimen). *Opuntia ficus-indica* is to be expected at any old or Spanish-speaking family residence from El Paso to Brownsville, but above 2,000 feet altitude, mostly they are frozen-down and inconspicuous below rooftop level against south-facing walls.

Benson (1969b, 1982) misinterpreted many native prickly pears in Texas and New Mexico as hybrid populations involving *O. ficus-indica*. These included *O. dillei* (= *O. engelmannii*, sensu lato) and *O. subarmata* (= *O. tardospina*, probably tetraploid). Benson even misidentified the type specimen of *O. engelmannii* as *O. ficus-indica* (see Parfitt and Pinkava, 1988), causing the popular epithet *engelmannii* to be displaced by *discata* in general taxonomic usage during the 1970s–90s. The specific epithet is after the Latin *ficus*, "fig tree," and the Greek *indikos*, "Indian," a reference to the figlike fruit that was used by North American Indians.

2. PENIOCEREUS (A. Berger) Britton & Rose
Desert Night-Blooming Cereus

A genus of about 15 species distributed mostly in Mexico and Central America. Benson (1982) retained *Peniocereus greggii* in *Cereus*, where it was placed originally by Engelmann. Benson recognized, however, that the large genus *Cereus* Mill. (with about 1,000 species) probably should be split into smaller, more natural groups such as those recognized by all other cactus specialists since Engelmann's time (1850s). *Cereus* is strictly a South American taxon (in a different tribe, the Cereeae) of perhaps 30 species. The genus name *Cereus* is from Latin *cereus* (adj., "waxen," "a torch"; see *ceraceus*, "like candle wax"; and/or, as noun, applied to the earliest known group of these cacti because they resembled either wax candles or candelabras), here combined with the Latin *penis*, "tail," in reference to the slender stems of *P. greggii*. $x = 11$.

1. **Peniocereus greggii** (Engelm.) Britton & Rose var. **greggii**. DESERT NIGHT-BLOOMING CEREUS. Plates 113, 114. [*Cereus greggii* Engelm., Mem. Tour N. Mex. 102. 1848]. Plants slender with angular (almost "winged") stems typically growing hidden among the branches of shrubs. Roots solitary, thicker than stems, carrotlike or turniplike, in old plants 25–60 cm in diameter. Stems grayish-green to dull purple, microscopically velutinous (having a velvety sheen when fresh), unbranched or branched, erect or sprawling, 15–30(–60) cm long (rarely 2–3 m long, if sheltered inside large shrubs), 1–2(–2.5) cm in diameter (lower sterile parts of stems slender, ca. 6 mm in diameter), strongly ribbed (angled) above; epidermis densely and finely velutinous with microscopic unicellular trichomes, somewhat like skin of *Opuntia rufida*. Ribs (3–)4–6. Areoles elongate-elliptic, 1.5–4.4 mm long, usually 4.5–7 mm apart, on tiny tubercle-like projections along the rib crest, white wool in those on the new growth. Spines small, (8–)10–13(–15) per areole; central spines 1–2 porrect or deflexed, ca. 1 mm long; radial spines 6–9(–13), the spines blackish, becoming grayish with age, most of them 0.8–1 mm long, the three lower radials to 3.2 mm long, appressed and straight, or slightly curved, basally swollen and stout, acicular, pubescent. Flowers nocturnal, fragrant, salverform, to 5–8(–10) cm in diameter, to 17(–21.5) cm long, borne laterally from areoles at least one year old; floral tube 10–15(–17) cm long, bearing scalelike bracteoles that subtend bristlelike spines; inner tepals white, largest narrowly oblanceolate, to 2.5 cm long, 1.2 cm wide, mucronate, entire; outer tepals with whitish margins and red, greenish, or brownish midribs, largest lanceolate-linear, to 5 cm long and 6 mm broad, recurved at full opening, attenuate or acuminate, entire or minutely ciliate; filaments cream-colored, ca. 2.5 cm long, anthers pale yellowish, tan, or creamy-yellow, different from many of the diurnal cacti, 1.5–2.5 mm long; stigma lobes whitish, ca. 10, 10–15 mm long, slender. Fruit abruptly turning shiny bright red or red-orange at maturity, ultimately turning darker red with age, ellipsoid, 4.5–7.5 cm long, 2.5–3.9 cm in diameter, attenuated apically to form a sterile "beak," weakly tuberculate, each tubercle bearing an areole with short spines; floral remnant persistent. Seeds

black, obovoid, ca. 3 mm long, 2.2–2.3 mm broad, 1.5 mm thick, tuberculate-roughened (testa cells convex), hilum "basal." $2n = 22$.

Desert flats, gravel benches, *bajadas,* and in degraded grassland and desertscrub in alluvial basins between mountains, documented or expected in every county of the Trans-Pecos. 3,500–5,000 ft. Flowering early to mid-May, sometimes as late as early to mid-Jun; fruits ripening in midsummer to early fall. $2n = 22$. West across southern NM to the E edge of Cochise Co., AZ. Adjacent Mexico S to Zacatecas. Map 24.

Peniocereus greggii has long been considered by some to be rare in Texas, having been listed as a Category 2 species in the Federal Register. Our observa-

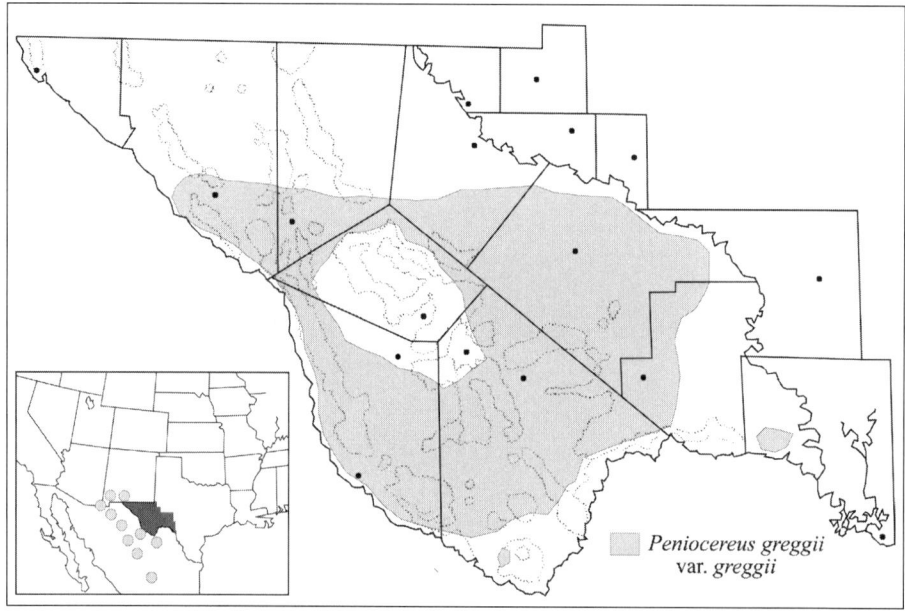

Map 24. Generalized distribution of *Peniocereus greggii* var. *greggii* (desert night-blooming cereus).

tions suggest that *P. greggii* is widespread in the Trans-Pecos and common in certain areas, although very rare in Big Bend National Park. In Presidio County, for example, the species is widespread in desertscrub in the southern part of the county. At one site south of the Chinati Mountains over 50 plants were observed, all growing in shrubs, in an area judged to be less than 200 m across. Also, numerous plants have been observed in an area near Hovey in northeast Brewster County near the Pecos County line. Most plants of *P. greggii* apparently begin under *Larrea, Acacia, Prosopis,* or other shrubs, which serve as nurse plants during their establishment. The slender stems of *P. greggii* resemble the stems of shrubs in which they grow and are thus difficult to locate by casual observation.

Identifying Characters. The slender, gray-green stems of *P. greggii* are unlike those of any other cactus species (or other plant species) in the Trans-Pecos. The stems, usually 1–2 cm in diameter, are sparsely branched or unbranched, qualifying weakly as "shrubs" in the botanical sense, but not bushy. Older stems and plant bases tend to shrink in size, become terete, and turn brownish or grayish, after losing ribs and their areoles, which are characteristic of the gray-green younger stems.

The plants produce fleshy taproots that range from turnip- or carrot-size in younger plants to much larger in older plants. According to Weniger (1984) the taproots may reach two feet in diameter and weigh up to 125 pounds. After growing numerous plants from seed, we have observed that plants in containers may produce turnip-size taproots in a few years. In the field in northeast Brewster County we have seen a partially excavated taproot of *P. greggii* that was in excess of 30 cm in diameter. In this case the aerial stem was only about 20 cm high.

The flowers, borne well below the stem apex, are quite large, about 15 cm long and 8 cm wide when fully open, with a long floral tube that supports scales and bristlelike spines. The flowers appear to attract hawkmoths, which serve as primary pollinators, although the pollinators of *P. greggii* have not been thoroughly studied. The throat of an open flower is filled with numerous spreading, exserted stamens. The style is slender with white stigma lobes.

The circular areoles of the pericarpel bear short, rigid spines. Mature fruits are the approximate size and shape of a small hen egg, but with a prominent apical beak.

Phenology. The flowers of *P. greggii* are strictly nocturnal. Flowers begin opening at about 7:00 P.M. (DST) and are fully open by dusk. The flowers remain open all night for one night and often close just before, or soon after, sunrise. If temperatures are cool, shaded flowers may remain open until about 9:00 A.M or rarely, under these apparently optimal conditions, until after noon. Plants grown in containers in Alpine become sexually mature after five or six years. Flower buds are produced in late March or early April. Flowers open in late April to early mid-May. Scott and JJ Lerich recorded by photograph flowers opening 19–24 May 2003 on one plant at Elephant Mountain in Brewster County. Benson (1982) reported that the flowering time for *P. greggii* extends into late June. On a single plant most of the buds mature and flowers open simultaneously. When several plants are involved, some buds mature and flowers open in preceding or succeeding nights, so flowering times in natural populations may be scattered out over several days. Regionally, flowering episodes presumably are distributed over a month. Fruit maturation requires 2–4 months.

Sterile and Immature Specimens. The unique stem morphology described above allows for easy identification of *P. greggii* in sterile condition. Areoles with distinctive "arachnoid" spine clusters, best seen under magnification, appear "plastered" to the narrow rib crests. Younger areoles are copiously white-woolly, almost hiding some of the very short (ca. 1 mm), acicular, bulbous-based radial spines that are tightly appressed against the stem surface. Farther down a stem in older areoles, hairs of the elliptic-elongate areoles may be darker and less evident in subtending, usually 8–12, black radial spines and 1–2 black central spines. All of the spines are only ca. 1 mm long, except for the lower three radials that are longer (2–3.2 mm). The middle one of the three lower radials is usually stout, and the lateral ones typically are slender and lighter in color. In older areoles, spines may become jet-black.

Immature specimens of *P. greggii* are seldom located in the field or preserved as herbarium specimens, but they are not likely to be confused with any other cactus species in the Trans-Pecos. The first stems of juvenile plants grown from seed quickly take on the unique morphology previously described. Juvenile stems, however, may have only three ribs, instead of the usual four; the areoles are closer together, relatively white-woolly, and their spine clusters are smaller, seemingly with fewer and shorter radial spines, and 0–1 central spine that is slightly smaller than the always short central spines of fertile areoles on adult plants.

Biosystematics. *Peniocereus greggii* var. *greggii* occurs from Terrell County, in the southeastern Trans-Pecos, northwest across southern New Mexico to extreme southeastern Arizona, where it is replaced by a weakly defined geographic race, var. *transmontanus* (Engelm.) Backeb. The var. *transmontanus* (Benson, 1982; Bravo-Hollis, 1978) ranges across southern Arizona, below the Mogollon Rim, west to Yuma County, and south into adjacent Sonora, Mexico. According to Bravo-Hollis, var. *greggii* occurs in Coahuila, Zacatecas, Chihuahua, and Sonora, Mexico; Zimmerman and others have seen it in northeast Durango. The var. *transmontanus* is distinguished from var. *greggii* by its smaller, circular areoles, ca. 1.5 mm in diameter, thinner stems, narrower floral tube, larger flowers to 7.5 cm in diameter and (rarely) up to 21 cm long, and other weak or unconfirmed floral characters.

The most closely related species of *Peniocereus* are *P. johnstonii* Britton & Rose, of Baja California, and *P. marianus* (Gentry) Sánchez-Mejorada, of Sonora and Sinaloa, Mexico. The other, much less similar, species of this genus all occupy the Pacific lowlands (and Tehuacan area) of tropical and subtropical Mexico, with one species [*P. striatus* (Brandegee) Buxb.] reaching extreme southern Arizona.

Synonyms. *Cereus pottsii* Salm-Dyck; *Cereus greggii* Engelm. var. *cismontanus* Engelm.

Common Names. Night-blooming cereus; night blooming cereus; desert nightblooming cereus; queen of the night; queen-of-the-night; Arizona queen of the night; Texas night-blooming cereus; chaparral cactus; sweet-potato cactus; deer-horn cactus; huevo de venado. Desert night-blooming cereus is preferred because "night-blooming cereus" is used for cultivated species of *Epiphyllum*, *Cereus*, *Selenicereus*, *Hylocereus*, *Harrisia* Britton, and other genera.

3. ECHINOCEREUS Engelm.
Hedgehog Cactus

Plants usually 8–50 cm high, rarely diminutive and 1–2 cm high, or with longer (mostly reclining) stems to ca. 1 m long. Roots diffuse to tuberous. Stems 1–500, single or branching from sides and base, globose to cylindroid, 1–10 cm in diameter, most erect, some weakly clambering (*E. poselgeri*) or prostrate. Ribs 4–25, prominent to inconspicuous. Tubercles usually prominent, coalescent with and projecting from ribs. Areoles circular to linear. Spines smooth (in Trans-Pecos spp.), acicular or subulate, terete or elliptic in cross section, black, white, gray, yellow, tan, brown, pinkish, or reddish; central spines 0.1–10 cm long, straight or curved but never hooked; radial spines 0.3–3 cm long, straight or curved, rarely hairlike. Flowers typically developing on sides of stem from areoles already a year old, in some species lower (rarely down to soil level) on the stems, rarely terminal (*E. poselgeri*); buds of most species erumpent (rupturing the stem epidermis above the areole), otherwise superficial from the outset (developing at upper edge of the areole like normal cactus buds); flowers 1.5–13 cm in diameter; inner tepals often brightly colored: red, magenta, orange, pink, or yellow, or greenish, brown, or reddish-brown, rarely white, the throat in many species contrastingly colored and perhaps aiding in attraction of insect pollinators; floral tube short (bee-pollinated species) to long (hummingbird-adapted species), narrowly to broadly funnelform, areolate up to the bract-tepal transition level with conspicuous spine clusters (these often woolly) subtended by mostly triangular small scalelike bracts ("scales" of subsequent text); stamens usually included; stigma lobes green, rarely white or red. Fruit in most species succulent at maturity, globose to ovoid, usually 2–3 cm long, green to red, in some species weakly dehiscent by one or a few longitudinal slits, the pulp white, pink, or magenta, rarely nearly dry; spiny areoles either somewhat persistent, deciduous, or easily brushed away; the dried floral remnant persistent at fruit maturity. Seeds usually black, obovoid to nearly globose, 0.8–2 mm long, strongly tuberculate or rugose; hilum "basal." [*Wilcoxia* Britton & Rose (in part, as to type)]. $x = 11$.

A genus of 44–49 (Taylor, 1985) or up to 70 species (Blum et al., 1998), throughout west Texas, and ranging from south-central Texas to Oklahoma, north to the Black Hills of South Dakota, west to the Pacific Ocean in Baja California, and south to northern Oaxaca, Mexico. Numerous species occur in the Sonoran and Chihuahuan desert regions, in true desert habitats and in mesic mountains. Most of the species are Mexican endemics; ca. 26 species occur in the United States, ca. 16 species in Texas, and ca. 12 of these in the Trans-Pecos. The genus name is derived from *echino-*, a combining form denoting spiny or bearing spines, from the Greek *echinos,* a sea urchin or hedgehog, and the previously named cactus genus *Cereus* (from the Latin *cereus,* "waxen," presumably in reference to the upright, wax candle–like habit of many species).

Key to the Species

1. Stems of mature plants 1–2 cm tall (aboveground); habitat restricted to Caballos Novaculite substrate in the Marathon Basin, Brewster Co.
   11. *E. davisii.*

1. Stems of mature plants 4–30(–40) cm tall; habitats in the Trans-Pecos various (2).

2(1). Flowers 1.5–2.5 cm in width and length; inner tepals green, dull yellow, bronze, or reddish-brown, rarely dull red
   2. *E. viridiflorus.*

2. Flowers (2.5–)4–13 cm in width and length; inner tepals brilliantly colored: red, magenta, pink, yellow, or multicolored (if dull red or brownish, in some *E.* × *neomexicanus*, then always sterile) (3).

3(2). Areoles elliptic to linear (elongate vertically) and close-set (2–6 mm apart); spines all short but often dense, obscuring the stem; stems often solitary (multistemmed in older plants) (4).

3. Areoles basically circular, the spacing either close or distant but often far apart; spines relatively long but widely divergent, not obscuring the stem (except as seen from a distance in *E. stramineus*); stems characteristically clumped (likely to form large mounds of numerous stems) except for *E. fendleri,* which is smaller (7).

4(3). Fruit and flower areoles with long, conspicuous white wool and relatively slender spines; bases of inner tepals may be contrastingly marked with dark red-brown; relatively small plants (5).

4. Fruit and flower areoles usually with much shorter, inconspicuous white wool and relatively rigid spines; tepal bases green or white (6).

5(4). Stems soft, relatively conical; central spines usually present and similar to the radials; spines not tightly appressed (plants appear bristly); Big Bend National Park
   9. *E. chisoensis* var. *chisoensis.*

5. Stems thick, firm, compact, rounded or short-cylindroid; central spines absent or very short; radial spines tightly appressed (plants appear smooth); not in Big Bend National Park
   10. *E. reichenbachii.*

6(4). Spines all whitish, not forming seasonal dark bands down the stem; inner tepals three-colored, pinkish distally, whitish in the middle, greenish proximally; distribution near Sanderson E to Del Rio; diploid
   5. *E. pectinatus* var. *wenigeri.*

6. Spines reddish-brown, gray, or whitish, often color-zoned down the stem (a "rainbow" cactus); inner tepals variously colored (yellow, red, pink, magenta, etc.), but never white-centered; distribution widespread in the Trans-Pecos; tetraploid
   4. *E. dasyacanthus.*

7(3). Flowers (inner tepals) deep pure red (rarely orange), 2.5–5 cm long and wide, the inner tepals stiff, usually rounded (or mucronulate) apically; flowers remaining constantly open for two or more days and nights

                                                  1. *E. coccineus*.

7. Flowers (inner tepals) variously colored: rarely red or orange (mostly in hybrids with *E. coccineus*, rarely in *E. viridiflorus* var. *cylindricus* and var. *russanthus*), rarely white (in *E. stramineus*), 5–13 cm long and wide, rounded, acute, or acuminate apically; flowers either ephemeral or at least closing at night (except in some *E.* × *roetteri*), shorter-lived than those of *E. coccineus* (8).

8(7). Radial spines (often whitish) usually 5–6(–9); central spines usually one, usually light gray to black, or with light gray tips; plants usually with 1–5 stems, to a maximum of ca. 18 stems, these typically 7.5–18 cm long

                                                  6. *E. fendleri* var. *fendleri*.

8. Radial spines usually 7–14; central spines (1–)2–4; plants usually many-stemmed, to a maximum of ca. 350, these typically 30 cm or more long (9).

9(8). Central spine(s) 1.2–8.7 cm long, usually a main one or two, these somewhat flattened in older areoles (10).

9. Central spines usually less than 1.2 cm long, the 2–5 of them all much alike, needlelike, dark gray, reddish to whitish; polyphyletic nothospecies (11).

10(9). Plants in compact (ultimately hemispheric) mounds, these cloaked in straw-colored or silvery spines obscuring the green surfaces of stems as seen from a distance in the field (usually on hot, steep, rocky slopes); average 12 ribs per stem; tetraploid

                                                  8. *E. stramineus* var. *stramineus*.

10. Plants loosely clumped, these with gray, dark gray to reddish, straw-colored, ashy-white, or gray-brown spines, widely spaced, either obscuring the stem surfaces or not (usually in deep soils of *bajadas* and valley floors); average 8–9 ribs per stem; diploid

                                                  7. *E. enneacanthus*.

11(9). Central spines 4–6; radial spines 13–16; spine tips bright purplish-red, giving the plant a red-and-white aspect (similar to some *E. viridiflorus*); flowers produced at variable heights on the stem (as in *E. viridiflorus*), relatively small (ca. 5 cm long, 2.5–3 cm in diameter); inner tepals red at least distally, or greenish-yellow proximally, or entirely brownish-pink

                                                  3. *E.* × *neomexicanus*.

11. Central spine, radial spine, and plant aspect very diverse, exceeding the limits of the rare F1 hybrid in the opposing lead; flowers produced on

distal parts of the stems (not like *E. viridiflorus*), relatively large and showy

                                                                            2. *E.* × *roetteri*.

1. **Echinocereus coccineus** Engelm., Mem. Tour N. Mex. 93. 1848. CLARET-CUP CACTUS, TEXAS CLARET-CUP CACTUS. Plates 115–18. Relatively mesic habitats in igneous and limestone substrates throughout much of the Trans-Pecos, occasionally in alluvium. 1,000–7,500 ft. Flowering in early spring, Mar–Apr. $2n = 44$. East across the Edwards Plateau to Burnet, Blanco, Hays, and Comal counties, S to Uvalde Co. Northwest across much of NM, into S-central CO, W across AZ below the Mogollon Rim and in the Grand Canyon region; extreme SW UT. Mexico, mostly adjacent N Coahuila and Chihuahua, S to Durango (*E.* "*polyacanthus*"), W to Sonora (*E.* "*santaritensis*") and Baja California (*E.* "*pacificus*"). Map 25.

Two varieties of this wide-ranging and morphologically variable species are recognized here for the Trans-Pecos. The specific epithet is after the Latin *coccineus*, "scarlet," the flower color.

Benson (1982) included *E. coccineus* within *E. triglochidiatus* Engelm., as did Taylor (1985) and Bravo-Hollis and Sánchez-Mejorada (1991a). The tentative taxonomic reassessment of this complex by Ferguson (1989) resulted in the recognition of *E. triglochidiatus* (presumably diploid, $2n = 22$) as a species distinct from *E. coccineus* (provisionally circumscribed so as to include only tetraploid populations, $2n = 44$). Documentation for identification of the diploid populations, *E. triglochidiatus* and *E. arizonicus* Rose ex Orcutt, is mostly

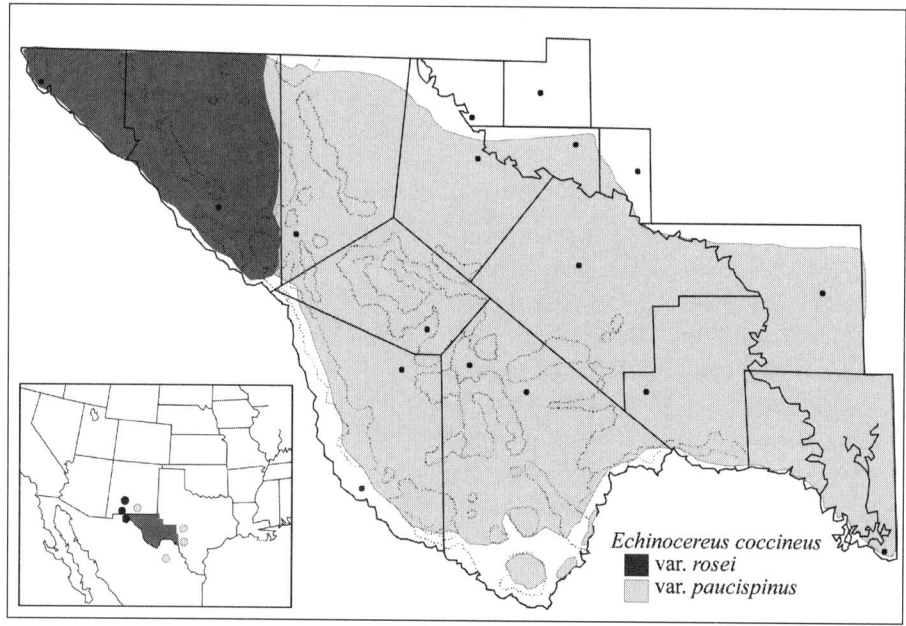

Map 25. Generalized distribution of *Echinocereus coccineus* var. *rosei* (claret-cup cactus) and *E. coccineus* var. *paucispinus* (Texas claret-cup cactus).

unpublished (A. Zimmerman, M. Powell, B. Parfitt, M. Baker, J. Weedin), although the consistently tetraploid status of *E. coccineus* is easily inferred from the literature (Pinkava et al., 1973; Weedin and Powell, 1978, 1980; Weedin et al., 1989; Pinkava et al., 1985). Weniger (1984) and Blum et al. (1998) also recognized both *E. triglochidiatus* and *E. coccineus* at specific rank, but Blum et al. used specific rank for many related taxa in addition. The current treatment assumes that all of the dioecious tetraploids (together with perfect-flowered tetraploids common in southeastern Arizona mountains) are one species, *E. coccineus* (including *E. polyacanthus* Engelm.).

Ferguson (1989) recognized two varieties of *E. triglochidiatus*; neither occurs in the Trans-Pecos region. *Echinocereus triglochidiatus* var. *triglochidiatus* occurs from south-central to northern New Mexico (south to White Sands National Monument, Otero County) and into south-central Colorado. The more widespread *E. triglochidiatus* var. *mojavensis* (Engelm. & J. M. Bigelow) L. D. Benson intergrades with var. *triglochidiatus* and extends west across Utah, southern Nevada, northern Arizona, and southern California.

Echinocereus coccineus hybridizes occasionally with several different species, including *E. dasyacanthus* (Powell et al., 1991; Powell, 1995), *E. viridiflorus* (see *E.* × *neomexicanus*), and reportedly *E. fendleri* (Ferguson, 1989). As many as two species or four varieties of the *E. triglochidiatus*–*E. coccineus* complex have been recognized for the Trans-Pecos (Benson, 1982; Weniger, 1984).

Benson (1982) recognized eight taxa (all treated as varieties of *E. triglochidiatus*) in the United States. Bravo-Hollis and Sánchez-Mejorada (1991a) listed six varieties in Mexico. Weniger (1984), despite excluding Arizona from the geographic scope of his book, recognized eight taxa in the United States taxonomically disposed between *E. triglochidiatus* and two segregate species, *E. coccineus* and *E. polyacanthus*. Taylor (1985) accepted eight taxa of the *E. triglochidiatus* complex for the United States, five of them also occurring in Mexico. Blum et al. (1998) elevated the *E. triglochidiatus* species-group to the rank of a subgenus; correspondingly, their taxonomic treatment of the components was finely divided. Their concept of the *E. triglochidiatus* species-group was divided into six species (nine taxa). Despite this extensive taxonomic "splitting," the Trans-Pecos populations all fall within a single one of their taxonomic species: *E. coccineus* in a relatively restricted sense (divided into subspecies). Like both Ferguson and Blum et al., irrespective of taxonomic controversies outside the Trans-Pecos, we perceive a single variable species of this complex in the Trans-Pecos. Herein we treat the two morphological (and geographic) extremes as "centers" of two arbitrarily delimited geographic races ("varieties"), pending further study.

Benson (1982) recognized four varieties of *E. coccineus* (under *E. triglochidiatus*) to occur in the Trans-Pecos: var. *neomexicanus*, var. *melanacanthus*, var. *gurneyi*, and var. *paucispinus*. Benson's distribution maps (pp. 604–5) revealed his interpretation that these four varieties were regionally sympatric across much of the Trans-Pecos. The varieties of *E. coccineus* (part of his *E. triglochidiatus*, sensu lato) were distinguished by Benson mostly on the basis of stem number and size, rib number, and spination.

Weniger (1984) recognized three species and several varieties in this complex, using names different from those in Benson (1982), Taylor (1985), and Ferguson (1989). According to Weniger, *E. triglochidiatus* var. *octacanthus* was the most widely distributed member of the *E. coccineus* complex in the Big Bend, closely corresponding with our concept of *E. coccineus* var. *paucispinus*, except that Weniger segregated the "extreme" form of *paucispinus*.

Still another taxonomic interpretation was offered by Blum et al. (1998). These authors recognized five subspecies of *E. coccineus*, two of them in the Trans-Pecos [subsp. *aggregatus* and subsp. *rosei* (Wooton & Standl.) W. Blum & Rutow]. The principal Trans-Pecos taxon was labeled "*E. coccineus* subsp. *aggregatus* (Engelm. ex S. Watson) W. Blum, Mich. Lange, & Rutow." Two other species were mapped from the eastern periphery of the Trans-Pecos into Central Texas [subsp. *roemeri* (Muehlenpf.) W. Blum, Mich. Lange, & Rutow north of Interstate Highway 10, and subsp. *paucispinus* (Engelm.) W. Blum, Mich. Lange, & Rutow restricted to south-central Texas and adjacent Mexico]. Their concept of subsp. *coccineus* is that it is distributed from central New Mexico to south-central Colorado.

In the Trans-Pecos mountains, plants of *E. coccineus* exhibit considerable morphological variation, often in individuals that occur within a few meters of each other. Many-stemmed plants, often with 100 or more smaller stems, may occur in rocky habitats, whereas plants with larger and usually fewer stems tend to grow in deeper soil. Plants with longer or shorter spines and with different spine numbers often grow in proximity to each other. We agree with Ferguson (1989) that the Trans-Pecos varieties of *E. coccineus* are highly variable, clinally intergrading races. We differ in that (1) we divide the varieties into two, Ferguson into three; and (2) the epithet *gurneyi* L. D. Benson pertains to part of the nothospecies *E.* × *roetteri*, but Ferguson used it as the valid epithet for the whole regional variety of *E. coccineus*. Part of the variation seen in Trans-Pecos populations of *E. coccineus* may have resulted from introgression (through the "*gurneyi*" phenotype of detectable hybrids) after interspecific hybridization between *E. coccineus* and *E. dasyacanthus* (Powell et al., 1991; Zimmerman, 1993; Powell, 1995).

Dioecy is rare in the Cactaceae (Parfitt, 1985), but the gynodioecious condition (functional dioecy) is now well documented for most of the geographic races of *E. coccineus* (not for *E. triglochidiatus*). Exceptions are (1) the southern Arizona populations (Scobell [1999] found *E. coccineus* in the Chiracahua Mountains of southeastern Arizona to be bisexual and self-compatible) newly named *E. santaritensis* W. Blum & Rutow (Blum et al., 1998; where misclassified as a member of the ostensibly all-diploid *E. triglochidiatus* complex), and (2) the southwestern New Mexico populations, including near topotypes of "subsp." *aggregatus* (Blum et al., 1998). Hoffman (1992) studied the breeding system of *E. coccineus* (presumably *E. coccineus* var. *rosei*) at two sites in southern New Mexico, and verification of dioecy for *E. coccineus* var. *paucispinus* in Trans-Pecos Texas was published by Powell et al. (1991) and Powell (1995). Male and female plants of *E. coccineus* apparently are not distinguishable on the basis of vegetative morphology. The flowers of male plants are morphologically bisexual, but only superficially so: the "brush" of stamens with pollen-laden anthers closely surrounds relatively small, infolded, functionless stigma lobes. The flowers of female plants average slightly smaller than male flowers and exhibit sterile stamens with abortive, pollenless anthers convergent around the style several millimeters below relatively large and (often the second day) widely spreading stigma lobes. The trained eye can distinguish male and female flowers of *E. coccineus* at a glance, or from photographs, unless insect-damaged or stripped of pollen by foraging bees.

In both sexes, the lowermost (inner) stamens are twice as long as the uppermost so that all of the anthers are displayed at one level, unlike any other Trans-Pecos species of *Echinocereus*. Large nectar chambers are typical of claret-cup flowers and thus complete a syndrome of characters that appear adapted to pollination by hummingbirds. Flowers do not close at night but remain open for 3–4 days or longer. After pollination, flowers may close before the normal open period of 3–4 days.

Key to the Varieties

1. Central spines (1–)3–4(–6); radial spines (7–)8–11(–13); ribs (9–)10–12; El Paso and Hudspeth counties
   1a. *E. coccineus* var. *rosei*.

1. Central spines 0–1(–4); radial spines 4–7(–8); ribs (5–)7–8; Hudspeth Co. E to Terrell Co., and Val Verde Co. E across the Edwards Plateau
   1b. *E. coccineus* var. *paucispinus*.

1a. **Echinocereus coccineus var. rosei** (Wooton & Standl.) A. D. Zimmerman (forthcoming). CLARET-CUP CACTUS. Plates 115, 116. [*Echinocereus rosei* Wooton & Standl., Contrib. U.S. Nat. Herb. 19: 457. 1915]. Infrequent to common in arid mountains, rocky outcrops, and in alluvial deposits. El Paso and Hudspeth counties. 3,700–5,700 ft. Flowering Mar–Apr. $2n = 44$. Arid lowlands of S-central NM. Northern Chihuahua, Mexico, from Juarez and Porvenir to SW of Samalayuca. Map 25.

This variety occurs in the Trans-Pecos only in the far western portion, from El Paso east to near Sierra Blanca and the grasslands north of Sierra Blanca. It is replaced by other geographic races of *E. coccineus* north into Colorado, west to Arizona, and into northwestern Mexico.

Identifying Characters. The typical habit of this variety is loosely cespitose, with 5–20 stems. The oldest plants form hemispheric mounds of potentially 100 or more stems. Stem size is 4–22(–27) cm high and 4–10 cm in diameter, typically about 8 cm in diameter. Stem size is somewhat smaller in cespitose plants with numerous stems. Rib number is 8–12 (typically 9–11), and areoles are closely spaced at (10–)14–19(–21) mm apart, the spination partly obscuring the stem. The spines are light (ashy-gray) to dark brown or black in color, darkest at the tips. Radial spine number is 7–12(–13) in mature areoles, and usually there are 3–4 central spines.

The dimorphic flowers of var. *rosei* are narrowly to broadly funnelform, 4–7 cm long, and 2.5–5 cm in diameter. The crimson tepals typically exhibit a "waxy," stiff appearance, with rounded, entire, perhaps mucronulate apexes. The anthers are pink or purplish (rarely yellow) and produce distinctive gray-purple pollen in the functional anthers of male flowers. The 5–10 stigma lobes are green. Nectar chambers, including the tube formed by connate stamen bases, are 4.5–10.5 mm long and 2.8–4(–6.5) mm wide.

Fruits occur only on female plants. They are green and have spine clusters on the pericarpel until they ripen and turn pale orange, dull red, brick-red, to pinkish. The fruits are fleshy and juicy with white pulp. The seeds are black, globular, strongly papillate, and 1.5–2 mm long.

Phenology. In Trans-Pecos Texas, flowers of var. *rosei* open in March and April. Usually the earliest flowers appear first in plants at lower elevations. Ordinarily, flowering is over by mid-April. In El Paso County, Texas, Worthington (1986) observed that the usual flowering time for *E. coccineus* is from early to late April. In one year when the first crop of buds was killed by a late freeze, another crop of buds led to blooming in mid-May. We, too, have witnessed delayed flowering of *E. coccineus* (var. *paucispinus*) after late freezes in Brewster and Pecos counties. Worthington reported that the flowers of var. *rosei* remain open at night and last for 4–5 days.

Sterile and Immature Specimens. In the field in the Trans-Pecos, *E. coccineus* is identifiable from a distance, without flowers, from consistent subtleties of habit, stem, and spine characters. Relatively small plants of *E. coccineus* with relatively few stems most resemble larger plants of *E. fendleri* or medium-size plants of *E. enneacanthus*. The 1–6 (usually 3–4) central spines of *E. coccineus* var. *rosei* are 2.5–3.5 cm long, or to 6 cm long in many plants. When there is more than one central spine, the main (larger) central is the lowest one in the circular areoles. The main central spine often is nearly black in color and somewhat flattened or elliptic in cross section. Both the central spine(s) and the 7–12(–13 or more) radial spines, these 1.5–2.8 cm long, have enlarged, bulbous bases. Young areoles have short, whitish, matted hairs that are not present in older areoles.

Juvenile specimens of *E. coccineus*, depending upon the age, have a reduced number of ribs and fewer spines in spherical, woolly areoles. The areoles of younger plants have seven or fewer radial spines and no central spine. In the areoles of slightly older plants there may be a single central spine that is terete.

Biosystematics. The var. *rosei* (type locality: Las Cruces, New Mexico) occurs in extreme western Texas, in El Paso County east to about Hudspeth County where we arbitrarily delimit our concept of var. *paucispinus* (type locality: lower Devils River region). The var. *paucispinus*, as thus defined, is the only variety of *E. coccineus* across most of the Trans-Pecos, from the arbitrary eastern limit of var. *rosei* east across the Edwards Plateau nearly to Austin.

Synonyms. Echinocereus triglochidiatus var. rosei W. T. Marshall; Cereus coccineus (Engelm.) Engelm. var. melanacanthus Engelm., in part; Echinocereus Engelm. coccineus var. conoideus Engelm. ex Weniger; E. polyacanthus Engelm. var. rosei (Wooton & Standl.) Weniger nom. nud.; E. polyacanthus Engelm. var. neomexicanus (Standl.) Weniger, in part; E. triglochidiatus Engelm. var. rosei

(Wooton & Standl.) W. T. Marshall, in part; *E. triglochidiatus* Engelm. var. *coccineus* (Engelm.) W. T. Marshall; *E. triglochidiatus* Engelm. var. *melanacanthus* (Engelm.) L. D. Benson; *E. triglochidiatus* Engelm. var. *neomexicanus* (Standl.) Standl. ex W. T. Marshall, in part, excluding type.

Common Names. Red-goblet cactus; pitahaya; southwest claret-cup; Mexican claret-cup hedgehog; claret-cup hedgehog cactus.

1b. **Echinocereus coccineus** var. **paucispinus** (Engelm.) D. J. Ferguson. TEXAS CLARET-CUP CACTUS. Plates 117, 118. [*Cereus paucispinus* Engelm., Proc. Amer. Acad. Arts 3: 285. 1856]. Mountains, hills, and mesas, igneous and limestone, oak-juniper-pinyon woodland or juniper woodland on limestone mesas, mostly rocky habitats but also in alluvial basins, grasslands, or among mesquite or other shrubs. In every county of the Trans-Pecos from Culberson Co. SE to Pecos and Val Verde counties. 500–7,500 ft. Flowering Mar–Apr. $2n = 44$. Southeast NM; E across the Edwards Plateau to San Saba, Burnet, Williamson, Blanco, Hays, Comal, and Uvalde counties. Mexico, NE Coahuila. Map 25.

The populations of *E. coccineus* in southeastern New Mexico, south into the central Trans-Pecos are intermediate between the densely spine-covered western *E. coccineus* var. *rosei* and the green-looking, few-spined eastern plants. The geographic cline between these two extremes is sporadically interrupted by populations having (on average) more ribs and shorter spines than any other claret-cup cacti. One of the variable populations in the central region was described as var. *gurneyi* by Benson (1969a). Because var. *gurneyi* has been interpreted as part of the nothospecies *E.* × *roetteri* and has been brought into synonymy with *E.* × *roetteri*, the name *gurneyi* is no longer available for use in reference to any claret-cup populations, only to products of hybridization. In the current treatment tentatively we place the central Trans-Pecos claret-cups with var. *paucispinus*, and we interpret var. *paucispinus* as the variable entity that extends west to Culberson and Hudspeth counties, where it is replaced by var. *rosei*.

Weniger (1984) recognized the same taxon that we are calling var. *paucispinus*, but he identified this variety as *octacanthus*. Benson (1982), Taylor (1985), Ferguson (1989), and Bravo-Hollis and Sánchez-Mejorada (1991a) all recognized var. *paucispinus* but restricted its application to the specimens with fewest spines. Plants with no central spines, occurring in populations with plants that have one central spine, occur as far west as the Davis Mountains. The type specimen of var. *paucispinus*, with no central spines, was collected at a site between the Pecos and Devils rivers. The varietal epithet refers to the characteristically fewer spines compared to other varieties of *E. coccineus*.

Identifying Characters. The var. *paucispinus* closely resembles var. *rosei* except that there are usually fewer spines and ribs in var. *paucispinus*. Typically (mostly east of the Pecos River), var. *paucispinus* has 0–1 central spine and (1–)4–8(–9) radial spines. Plants of more westerly distribution in the Trans-Pecos occasionally have two central spines. In eastern Pecos and Terrell counties it is common to find claret-cup plants in close proximity with no central spines and with one central spine.

The flowers of var. *paucispinus* are crimson in color with a stiff waxy appearance and rounded tepals. Orange-flowered plants in the central Trans-Pecos, populationally predominant in the Marathon Basin, sporadic elsewhere, possibly are introgressed from *E.* × *roetteri*. The var. *paucispinus* is gynodioecious apparently throughout its range (Powell et al., 1991; Powell, 1995), with dimorphic flowers, those of the male plants (Plate 119) being slightly larger than those of female flowers (Plate 120).

The fruits of var. *paucispinus* are dull to bright red at maturity. The pulp is white, or reddish near the fruit wall. Spine clusters on ripe fruits are easily dislodged. Fruits are produced only on female plants that have received cross-pollination from male plants. The seeds are black, irregularly globose, 1.3–2 mm in diameter, and strongly papillose.

Phenology. Typically, flowers of var. *paucispinus* open in March and April. In some years, apparently when there is ample winter moisture, flower buds erupt and may open in late February. Usually the flowering period is over by mid-April, but if the regular blooming time is delayed by late freezes, plants in some populations may not produce their last flowers until early or mid-May. Flowers remain open at night and last usually for 2–3 days, or perhaps a day or two longer. Fruit maturation requires about one month, but may take longer. See Powell (1995) and Scobell (1999) for additional phenological data concerning *E. coccineus* and some related taxa.

Sterile and Immature Specimens. Mature plants usually are multistemmed, forming loosely aggregated or tightly packed mounds 1–10 dm across and 1–4 dm tall. Most mature plants have fewer than 20 stems, but apparently older plants with more than 100 stems, or as many as 500 stems, are not uncommon. Stem size varies considerably, apparently in response to substrate conditions. In shallow soils or in rock crevices, the stems of mounded plants may be smaller, perhaps ca. 5 cm long and wide. Plants with 10–50 stems are common in all habitats. In the southern part of the range of var. *paucispinus*, the plants are characterized by relatively few, robust stems (to 20 cm long, 10 cm wide) whose green surfaces are slightly more evident at a glance because of the relative paucity of spines.

The stems have (5–)7–9 prominent ribs. The spines in circular areoles are relatively heavy, 0.9–1.3 mm thick, and they vary in color from pale gray to black, rarely dark brown or tan. In southeastern populations of var. *paucispinus*, most commonly there is 0–1 central spine per areole. In the central Trans-Pecos, one central spine is typical. Plants with no centrals or 1–2 centrals also occur in the mountains of the central region. Central spines are either somewhat flattened or terete. Radial spines in each areole are (1–)4–7(–8) in the eastern regions, but 7–9 or even 8–10 radials are not uncommon in the central mountains. In eastern Pecos County, near the type locality of *E.* × *roetteri* var. *neomexicanus*, plants of var. *paucispinus* usually exhibit one central spine and 6–8 radial spines per areole. In claret-cups there is notable variation in spine length throughout the Trans-Pecos. Central spines of var. *paucispinus*, when present, measure (1.2–)2–3.2(–5.5) cm long, and radial spines are (1.2–)1.7–2.8(–3.2) cm long. Both central and radial spines have enlarged or bulbous bases. Marathon Basin claret-cup populations, formerly known as var. *gurneyi*, now interpreted as part of the nothospecies *E.* × *roetteri*, typically have both central and radial spines 1–1.2 cm long, considerably shorter than spine lengths exhibited by claret-cups elsewhere in the central mountains.

Juvenile plants produce areoles with no central spines and fewer radial spines. Slightly older areoles of juveniles produce a single central spine, in populations of the central region, even if two centrals ultimately are produced in older areoles. Younger areoles produce a whitish or gray felty cover of hairs. Under greenhouse cultivation var. *paucispinus* juveniles may produce branches from the stem base, erupting just above and through an areole, as early as four years and nine months from seed. In grassland areas of the central Trans-Pecos, immature plants of var. *paucispinus* are most likely to be confused with *E. fendleri,* which is most common in grassland habitats of the west-central Trans-Pecos. Mature stems of *E. fendleri* are smaller than mature stems of var. *paucispinus* and have 5–7(–9) usually ashy-gray or nearly white radial spines and one black central spine.

Biosystematics. The var. *paucispinus* varies extensively in plant size and considerably in spine variation. Stem numbers from a few to 500 have been reported (Benson, 1982). Length of the main central spine varies from about 1 cm to 5.5 cm or more. The number of central and radial spines varies modestly within populations in the central Trans-Pecos; spine numbers average fewer in the eastern and more in the western portions of the range. In var. *paucispinus* we interpret most of the morphological variation to be the result of genetic diversity expected in any widespread population system, combined with individual phenotypic response to microhabitat. Both types of variation in *E. coccineus* are observable in any mountain system of the Trans-Pecos.

Some of the variation in var. *paucispinus* may be the result of repeated hybridization with *E. dasyacanthus* (Powell et al., 1991; Zimmerman et al., forthcoming), having resulted in the formation of scattered hybrid populations collectively known as *E.* × *roetteri*. There is considerable evidence of introgression of genes from each species into the other through *E.* × *roetteri* (Powell, 1995, 1998c). In the Marathon Basin, for example, it is not unusual to find plants of Benson's "var. *gurneyi*" (= *E.* × *roetteri*) with orange flowers and/or shorter spines (1–1.2 cm long), characters that suggest backcrossing of *E.* × *roetteri* into var. *paucispinus*. If the short spines of "var. *gurneyi*" in the Marathon Basin were some kind of adaptation to the Caballos Novaculite substrate, where other groups of cacti have evolved endemic taxa, then we would not expect them elsewhere. The type of "var. *gurneyi*" is an unusually short-spined specimen from south of Marathon in novaculite habitats where claret-cup plants are predominantly orange-flowered. A few claret-cup plants with yellow or cream-colored flowers have been observed south of Marathon.

Weniger (1970) and Ferguson (1989) both pointed out that var. *paucispinus* is only superficially similar to *E. triglochidiatus* var. *triglochidiatus,* a diploid taxon that occurs in New Mexico and Colorado. Both taxa share unusually low rib and spine counts. The tetraploid var. *paucispinus* is subtly distinct in vegetative and floral characters. The spines usually are strongly angular in cross section and sometimes sinuous or contorted in *E. triglochidiatus* var. *triglochidiatus*, whereas the radial spines in *E. coccineus*, and sometimes the centrals as well, are terete and straight. The flowers of *E. triglochidiatus* supposedly do not open as widely as those of *E. coccineus* (Ferguson, 1989), but *E. coccineus*, sensu lato (tetraploids), includes the complete gamut of variation from salverform to nearly tubular flowers.

The claret-cup cacti on granite outcrops and elsewhere in the northern Edwards Plateau region (e.g., in San Saba County) deserve further study; some of these are the basis for the taxon *roemeri*, of unknown ploidy level. The distinction is not strictly environmental; the endemic *roemeri* remains distinctive in horticulture and may deserve varietal designation (Ferguson, 1989), subspecific rank (Blum et al., 1998), or other rank.

Synonyms. ? *Cereus roemeri* Muehlenpf.; ? *C. roemeri* Engelm.; ? *Echinocereus roemeri* Rümpler; ? *E. roemeri* (Muehlenpf.) Rydb.; *E. triglochidiatus* var. *paucispinus* (Engelm.) W. T. Marshall; *E. coccineus* Engelm. subsp. *aggregatus* (Engelm. ex S. Watson) W. Blum, Mich. Lange, & Rutow (in part, excluding type locality and other western populations); *E. octacanthus* (Muehlenpf.) Britton & Rose; *E. triglochidiatus* Engelm. var. *octacanthus* (Muehlenpf. ex W. T. Marshall) Britton & Rose; *E. coccineus* Engelm. var. *octacanthus* (Muehlenpf.) Boissev.

Common Names. Strawberry cactus; Langtry claret-cup; Langtry claret-cup cactus; claret-cup; claretcup echinocereus; claret-cup hedgehog; red-flowered hedgehog cactus; aggregate cactus; bunch-ball cactus; Turk's head cactus; heart twister; little claret-cup; black-spine claret cup hedgehog cactus. Langtry claret-cup cactus was usefully applied to the "extreme" form of *E. coccineus* var. *paucispinus*, which is concentrated in the lower Pecos region near Langtry, Texas. Texas claret-cup cactus has been narrowly applied to *E. coccineus* var. *gurneyi*, but this entity is part of the nothospecies *E.* × *roetteri*.

2. **Echinocereus × roetteri** (Engelm.) Engelm. ex Rümpler. ROETTER'S HYBRID HEDGEHOG CACTUS. Plate 121. [*Cereus roetteri* Engelm., Cactaceae Boundary 33. 1856]. Mesquite and desertscrub of degraded grasslands, to be expected throughout much of the Trans-Pecos where *E. coccineus* and *E. dasyacanthus* overlap in distribution, best known in eastern Pecos Co., TX, and Eddy Co., NM, and near El Paso to NM in Doña Ana and Otero counties. 2,500–4,500 ft. Flowering Apr–May. $2n = 44$. Mexico, N Chihuahua. Map 26.

The nothospecies (a hybrid "species") *E.* × *roetteri* exists as a series of scattered individual plants, mostly small populations of relatively recent hybrid origin. The total known distribution of *E.* × *roetteri* lies within the sympatric ranges of the two parental species, *E. coccineus* and *E. dasyacanthus*. One of the parents, *E. coccineus*, is gynodioecious, a trait that also occurs in *E.* × *roetteri*. A growing accumulation of morphological and experimental evidence (Powell et al., 1991; Zimmerman, 1993; Powell, 1995; Zimmerman et al., forthcoming) assures us that introgression from fertile hybrids has affected both parental species. According to article H.4.1. of the *International Code of Botanical Nomenclature* (ICBN), a hybrid taxon consists of all individuals derived from the

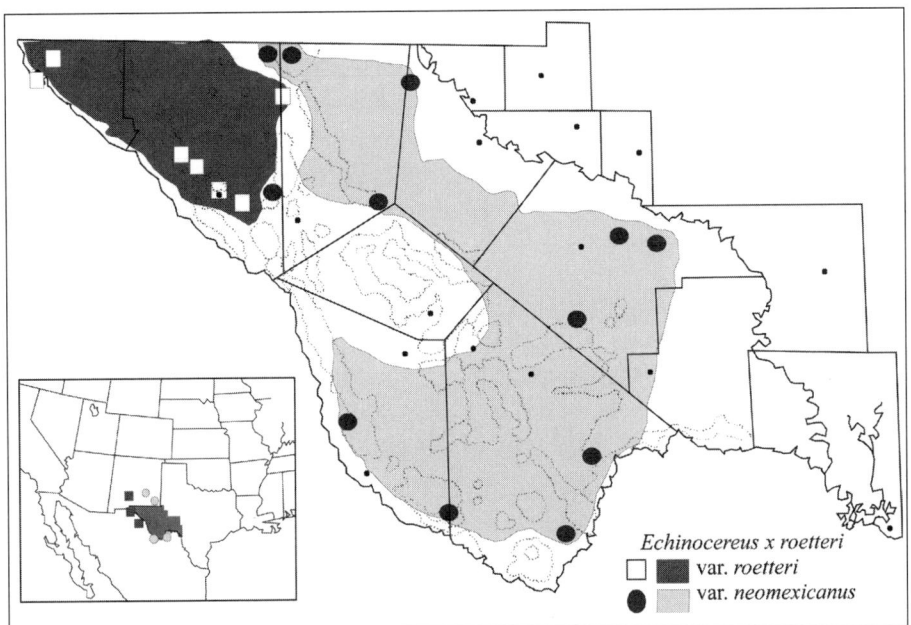

Map 26. Generalized distribution and some documented sites for *Echinocereus* × *roetteri* var. *roetteri* (Roetter's hybrid hedgehog cactus) and *E.* × *roetteri* var. *neomexicanus* (Lloyd's hedgehog cactus).

crossing of the parental taxa, including F1 and later-generation hybrids and backcrosses. For example, the type of *E. triglochidiatus* (= *coccineus*) var. *gurneyi* is regarded (Zimmerman et al., forthcoming) as a product of introgression. Under the ICBN, therefore, the epithet *gurneyi* is treated here as a synonym of *E.* × *roetteri* var. *neomexicanus*, not a variety of *E. coccineus*. The epithet *roetteri* was chosen by Engelmann, who described the taxon after Paulus Roetter, the artist who illustrated cacti for the Mexican Boundary Survey. Zimmerman (1993) determined that two varieties of *E.* × *roetteri* should be recognized, western var. *roetteri* and eastern var. *neomexicanus*, each having slightly different parentage.

Key to the Varieties

1. Western (El Paso Co. and adjacent areas); associated with *E. coccineus* var. *rosei* (all parts smaller and more numerous); stems typically 11–19 cm long (ranging 8.2–27 cm long), 6–7.5 cm in diameter (ranging 5.3–8.3 cm thick); ribs (10–)12–14(–15); areoles 7–14 mm apart; fruit 2–3 cm long

   2a. *E.* × *roetteri* var. *roetteri*.

1. Eastern (see text); associated with *E. coccineus* var. *paucispinus* (all parts larger and fewer); stems typically 19–33 cm long (ranging to 42 cm long), (6–)8.5–11 cm in diameter; ribs (9–)12(–14); areoles typically 13–17 mm apart (ranging 9–19 mm apart); fruit 2–5 cm long

   2b. *E.* × *roetteri* var. *neomexicanus*.

2a. **Echinocereus × roetteri** var. **roetteri**. ROETTER'S HYBRID HEDGEHOG CACTUS. Plates 122, 123. [*Cereus roetteri* Engelm., Cactaceae Boundary 33. 1856]. Sporadic in desert hills, grassland, and desertscrub. El Paso Co., near El Paso, E at least to Culberson Co. (depending upon the taxonomy of *E. coccineus* var. *rosei*). 3,200–4,500 ft. Flowering Apr–May. $2n = 44$. Adjacent NM, Otero and Doña Ana counties. Adjacent Chihuahua, Mexico. Map 26.

The best-known population of *E.* × *roetteri* var. *roetteri* is in the Jarilla Mountains near Orogrande in Otero County, New Mexico. As a nothotaxon, var. *roetteri* includes all hybrids between *E. coccineus* var. *rosei* and *E. dasyacanthus*.

Identifying Characters. Familiarity with the parental species facilitates recognition of the hybrid taxon var. *roetteri*. The plants of var. *roetteri* basically are intermediate in vegetative and floral characters, or in backcross plants they may exhibit greater similarity to one parent or the other.

The flowers of var. *roetteri* are 5–9 cm long and 4.5–8.5 cm wide. Flower color varies from bright yellow to creamy-white or bright magenta to pink or yellow-green, with most of each inner tepal with one of these colors, and with or without green, pink, or brown at the bases, pink or orange at the tips or centers, or perhaps with a green, pink, or brown midregion. The inner tepals are 2.8–4.4 cm long and 0.7–2.1 cm wide, with acute or rounded apexes. The receptacular tube is 1.6–2.6 cm long with numerous areoles, spines, and hairs as in *E. dasyacanthus*. The nectar chamber is 4–5 mm long.

The fruits are indehiscent, 2–3 cm long, dull reddish or with a brighter pinkish color at the base. The juicy pulp is white or pinkish near the fruit wall. The seeds are black, globular, 1.3–2 mm in diameter, and papillate.

Phenology. The peak flowering time for var. *roetteri* in the vicinity of El Paso, Texas, and Orogrande, New Mexico, is mid- to late April. The flowering time for the claret-cup parent, var. *rosei*, usually is in March, whereas the other parent reaches peak flowering in late April or May. *Echinocereus* × *roetteri* is expected to hybridize with *E. fendleri*, which usually blooms in May, and with *E. viridiflorus*, which usually has its peak bloom in April.

Sterile and Immature Specimens. Plants of var. *roetteri* are single- or multi-stemmed, with the stems ranging from 8–27 cm long and 5.5–8.5 cm in diameter, and with usually 12–14 ribs. In each areole there are (2–)4–5(–8) central spines and (8–)10–16(–18) radial spines. The longest central spines are 1.1–1.8(–2.6) cm, and the longest radial spines are (1–)1.2–1.5(–1.8) cm. The spines are all straight and (0.25–)0.4–0.6(–0.7) mm thick. Spine color varies from white to pale pink, or gray to black, brown, or stramineous. The areoles are 7–14 mm apart.

Juvenile specimens of var. *roetteri* have fewer ribs and fewer spines per areole than in sexually mature plants. Juvenile plants may be difficult to distinguish from those of the parental species.

Biosystematics. Experimental documentation regarding the origin of var. *roetteri* has not been attempted, although Zimmerman (1993) found no evidence to refute the widespread consensus that this is a nothotaxon produced by hybridization between *E. coccineus* and *E. dasyacanthus*. The distinction between var. *roetteri* and var. *neomexicanus* is that each is produced by a different geographic race of *E. coccineus* hybridizing with *E. dasyacanthus* (Zimmerman et al., forthcoming).

Synonyms. *Echinocereus dasyacanthus* var. *minor* Engelm.; *E. pectinatus* var. *minor* (Engelm.) L. D. Benson, in part, as to type.

Common Name. We propose using the name Roetter's hybrid hedgehog cactus, as opposed to Lloyd's hybrid hedgehog cactus, as a means of reflecting the parallel relationship between var. *roetteri* and var. *neomexicanus*.

2b. **Echinocereus × roetteri** (Engelm.) Engelm. ex Rümpler var. **neomexicanus** (J. M. Coult.) A. D. Zimmerman, comb. nov. (forthcoming). LLOYD'S HEDGEHOG CACTUS. Plates 124, 125. [*Cereus dasyacanthus* Engelm. var. *neomexicanus* J. M. Coult., Contr. U.S. Natl. Herb. 3: 384. 1896, non *E. × neomexicanus* Standl., *pro sp.*]. Rocky hillsides or brushy alluvial habitats, isolated individual plants or localized small populations in Pecos, Brewster, Presidio, Culberson, and Hudspeth counties; a relatively large population in E Pecos Co. 2,500–4,500 ft. Flowering Apr–early May. $2n = 44$. Eddy Co., NM. Map 26.

In the Trans-Pecos *E. × roetteri* var. *neomexicanus* has long been known as *E. lloydii* (Benson, 1982; Weniger, 1984; Taylor, 1985) or *E. × lloydii* after its hybrid status was documented (Powell et al., 1991). The nothotaxon *E. × roetteri* var. *neomexicanus* includes all hybrids between *E. coccineus* var. *paucispinus* and *E. dasyacanthus*. The epithet *lloydii* is after F. E. Lloyd, who collected the type specimen. Lloyd's hedgehog cactus was accorded federal status in 1979 as an Endangered species and also listed as Endangered by the state of Texas in 1983. In 1996 the U.S. Fish and Wildlife Service proposed to delist Lloyd's hedgehog cactus because it was determined to represent an assortment of dynamic hybrid populations rather than an established evolutionary lineage. In June 1999 Lloyd's hedgehog cactus was officially delisted through publication in the Federal Register.

Identifying Characters. In overall aspect *E. × roetteri* var. *neomexicanus* looks more or less intermediate between the parental species *E. coccineus* and *E. dasyacanthus*. The details of individual plants of var. *neomexicanus* depend upon the morphology of the parental species at the particular site where the hybridization occurred and upon whether the hybrids are of F_1, F_2 or later, or backcross generations. In habit they are at first single-stemmed, potentially forming clumps of 20 or more stems. The stems are about 8–11 cm thick and reliably show 12 ribs unless the plants are products of backcrossing with *E. coccineus* var. *paucispinus* (which has fewer ribs). Spine clusters of 4–6 centrals and 12–16 radials are typical.

The flowers of F_1 hybrids are orange (Plate 124). Later-generation and backcross hybrids may produce flowers that are orange, red-orange, red, to pinkish, or yellow (Plates 126–28). Flower-color variations, as well as variation in vegetative morphology, are hallmarks of the extensive population of var. *neomexicanus* in eastern Pecos County. The flowers of var. *neomexicanus* are about 4.6–7 cm in diameter, usually smaller in backcrosses to *E. coccineus*. The flowers of var. *neomexicanus* are either bisexual or male sterile (with nonfunctional anthers). The anthers usually are yellow in bisexual flowers and purplish but abortive in male sterile flowers.

Mature fruits are purplish-maroon to brick-red, or even greenish-orange to pinkish-green, with the juicy pulp white to pink in color. The black seeds are strongly papillate and about 1–1.5 mm in largest diameter.

Phenology. The peak flowering time for the var. *neomexicanus* population in eastern Pecos County is usually in mid-April. This falls in between the typical blooming times of *E. coccineus* (March) and *E. dasyacanthus* (May). Over the years, probably in response to variable weather conditions, there has been some overlap in flowering time between the two parental species, explaining how hybridization can occur. The distribution of var. *neomexicanus* overlaps with both of its parental species, providing annual opportunity for backcrossing and subsequent introgression. Individual flowers of var. *neomexicanus* remain open during the night, or only partially close, and stay open for 3–5 days. The fruits of var. *neomexicanus* are mature in July.

Sterile and Immature Specimens. In the field *E.* × *roetteri* var. *neomexicanus* is likely to be confused with either of its parental species, *E. coccineus* or *E. dasyacanthus*, particularly if the plants in question are later-generation segregates or backcrosses that resemble the parental species in overall aspects or in one or a few critical characters. Other species that resemble var. *neomexicanus* are *E. fendleri, E. viridiflorus,* and *E. enneacanthus,* but all of these are easily distinguished by spine characters. These taxa usually are not commonly topographically sympatric with var. *neomexicanus*.

The characters discussed here for the purpose of distinguishing var. *neomexicanus* are those of the obvious intermediates such as the F_1 generation. Typical var. *neomexicanus* has one to several stems with 12 ribs or, exceptionally, 11–13 ribs. The spines are reddish-gray, stramineous, tan, or whitish, and collectively they are rather dense but only partly obscure the stem. There are (2–)4–6 central spines, the longest 1.2–1.9 cm, and (11–)12–16 radial spines, with a maximum spine diameter of (0.4–)0.5–0.6 mm. *Echinocereus coccineus* typically has more stems (to 100 or more), which are thicker, with fewer ribs (5–12), typically fewer, thicker spines (0–4 centrals, 4–13 radials), and longer central spines. Plants of *E. dasyacanthus* typically exhibit solitary stems or 2–3 basal branches (occasionally more), more slender stems, more ribs (15–19), and more spines (8–12 centrals, 17–25 radials), which usually are only 0.25–0.45 mm thick.

Juvenile specimens of var. *neomexicanus* have fewer ribs and fewer spines in each areole than in sexually mature plants. Tests involving the propagation of var. *neomexicanus* and artificial hybrids from seed have demonstrated that these plants develop more ribs and spines over the years, up to the typical number (Powell et al., 1991; Powell, 1995).

Biosystematics. Experimental studies involving artificial hybridization have documented the belief by several workers (see Zimmerman, 1993) including Taylor (1985), that the mostly orange-flowered *E.* × *roetteri* var. *neomexicanus* is of hybrid origin, and the parental species are the mostly red-flowered *E. coccineus* and the mostly yellow-flowered *E. dasyacanthus* (Powell et al., 1991; Powell, 1995, 1998c). The results of these studies conducted over about 14 years suggested that var. *neomexicanus* in eastern Pecos County (and by inference all *E.* × *roetteri*) is a complex, dynamic population comprising first- and later-generation hybrids as well as backcross hybrids. The inference that introgression into both parents has occurred is supported by the discovery that all of the hybrids are highly fertile outcrossers. Thus, later-generation hybrids are expected, and backcrosses with the parental populations are practically inevitable.

It seems clear that the dioecious condition characteristic of *E. coccineus* (very rare among cacti) was inherited by var. *neomexicanus* from *E. coccineus*. Natural var. *neomexicanus*, like *E. coccineus*, has separate female and perfect-flowered plants. Unlike *E. coccineus*, however (in which the perfect flowers are functionally male), the morphologically bisexual flower-form in var. *neomexicanus* is truly hermaphroditic (functionally bisexual), hence gynodioecious instead of truly (but cryptically) dioecious.

No two populations of *E.* × *roetteri* are exactly alike for two reasons. First, the parents of this hybrid species differ somewhat at each site where interspecific hybridization has occurred and has become established (and detected). Both parental species are widespread and morphologically variable, even clinally variable in the case of *E. coccineus*. The well-documented population of *E.* × *roetteri* var. *neomexicanus* in eastern Pecos County, for example, was derived from a cross involving *E. coccineus* var. *paucispinus* as the claret-cup parent. The western nothotaxon *E.* × *roetteri* var. *roetteri* in El Paso County and adjacent south-central New Mexico, and Chihuahua, Mexico, originated through hybridization between different, western varieties or races of the same parental species that gave rise farther east to *E.* × *roetteri* var. *neomexicanus*. Consequently, the western hybrids exhibit certain distinctions that resulted in separate taxonomic histories. Additional varieties of the nothospecies *E.* × *roetteri* eventually might be described as the infraspecific variation in the parental species is gradually better documented. Second, the differences between eastern and western hybrids are much more subtle than the differences between individual plants within populations, owing to unique histories of backcrossing into the parental species (one or both at each site).

The introgression discussed earlier (Zimmerman, 1993; Powell, 1995, 1998c) is remarkably apparent in flowering plants of *E. dasyacanthus* in Pecos County. *Echinocereus dasyacanthus* normally has yellow flowers throughout its range. Diverse additional colors of *E. dasyacanthus* flowers are commonly observed in eastern Pecos County from the Pecos River west through the *E.* × *roetteri* var. *neomexicanus* population (ca. 20–25 miles E of Fort Stockton) to at least 20–30 miles west of Fort Stockton. Reddish-flowered *E. dasyacanthus* plants, sometimes abundant and sometimes rare, are found at various other sites in the Trans-Pecos, including sites in southern Brewster County and in Reeves, Culberson, and Hudspeth counties; in some areas, the anomalous flower color is associated with features such as long, sparse spines, presumably inherited through introgression from *E. coccineus*.

Examples of introgression of *E. dasyacanthus* into *E. coccineus* probably outnumber first-generation hybrids, but they are detected mostly by relatively subtle features, such as orange instead of red flowers, higher rib number, and shorter spines. Orange flowers occur sporadically in some Trans-Pecos populations of the typically red-flowered *E. coccineus,* especially in the Marathon Basin, where orange-flowered plants predominate. Because the type specimen of *E. "triglochidiatus var. gurneyi"* is regarded as a product of introgression, although not a first-generation hybrid (Zimmerman et al., forthcoming), var. *gurneyi* is treated here under ICBN rules as a synonym of the nothotaxon *E.* × *roetteri* var. *neomexicanus* (Zimmerman, 1993). The rare occurrence of orange flowers in otherwise normal *E. coccineus* might be dismissed as simple flower-color variation, but the Marathon population displays additional evidence of introgression, such has high rib counts, closely spaced areoles, short central spines, yellow anthers, and/or short nectar chamber.

Synonyms. *Echinocereus* × *roetteri* (Engelm.) Engelm. ex Rümpler; ? *E. pleiogonus* (Labour.) Croucher; ? *E. multicostatus* Cels; *E. hildmannii* Arendt (non "*E. dasyacanthus* var. *hildmannii*" Weniger, *nom. nud.*); *E. rubescens* Dams; *E. triglochidiatus* var. *gurneyi* L. D. Benson (in part, as to type); *E. coccineus* Engelm. var. *gurneyi* (L. D. Benson) K. D. Heil & S. Brack (in part, as to type); *E. lloydii* Britton & Rose; *E. roetteri* var. *lloydii* Backeb.

Common Names. Lloyd Hedgehog Cactus; Lloyd's hedgehog. Since about 1980 the U.S. Fish and Wildlife Service has used Lloyd's hedgehog cactus.

3. **Echinocereus** × **neomexicanus** Standl., *pro sp.* TRIPLOID HYBRID HEDGEHOG CACTUS. Plate 129. [*Echinocereus neomexicanus* Standl., Bull Torr. Bot. Club. 35: 87. 1908, non *E.* × *roetteri* var. *neomexicanus* (Engelm.) Engelm. ex Rümpler]. El Paso Co., Franklin Mts along Trans-Mountain Hwy. Culberson Co., Guadalupe Mts, Pine Spring Canyon. 4,000–5,600 ft. Flowering spring. $2n = 33$? New Mexico, Doña Ana, and (sight records) in Otero, Sierra, and Eddy counties. Map 27.

Echinocereus × *neomexicanus* is a nothospecies of sporadic occurrence, mostly in south-central New Mexico, but there are two sight records by D. Ferguson from near the New Mexico state line in the northwestern Trans-Pecos region (Zimmerman, 1993). The specific epithet is a geographic reference to its discovery site in New Mexico.

Identifying Characters. Familiarity with the parental species, *E. coccineus* and *E. viridiflorus,* is a prerequisite for the ability to recognize the very rare and usually isolated individuals of the F1 hybrid *E.* × *neomexicanus*. As F1 hybrids they combine the characters of both parents in infertile, sometimes visibly "ill-looking" individuals.

Flowers are produced at different heights along the stem, as in *E. viridiflorus*. The flowers are ca. 5 cm long, 2.5–3 cm wide, tubular-funnelform, with inner tepals red distally and greenish-yellow toward the base, or they are entirely red or entirely brownish-pink. The filaments are crowded and support yellow to pinkish anthers that may be abortive and apparently do not produce pollen. Fruits and seeds have not been reported.

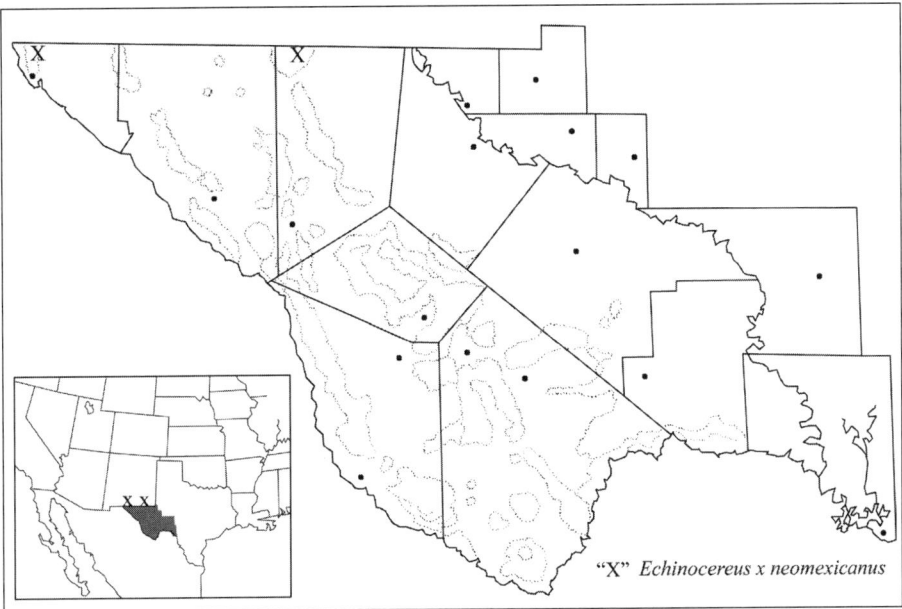

Map 27. Generalized distribution of *Echinocereus* × *neomexicanus* (triploid hybrid hedgehog cactus).

Phenology. It is not clear if all plants of *E.* × *neomexicanus* produce flowers. In some plants at least, flowers apparently form in the spring at about the same time flowering occurs in nearby parental species.

Sterile and Immature Specimens. Plants are single-stemmed or with 2–8 branches, and the stems are 17–25 cm long, 5.5–7 cm in diameter, with 11–12 ribs. The spines partially obscure the stem and lend an overall red-and-white aspect to the plant, much as is characteristic of *E. viridiflorus* var. *cylindricus*. Areoles are 10–15 mm apart on the ribs, each bearing about 17–22 spines. There are 4–6 central spines and 13–16 radial spines in each areole. One central spine, the lowermost one, is noticeably longer (ca. 3 cm) than the others (shortest ca. 0.6 cm long), angled downward in position, and is pale in color. The other central spines are usually reddish or pinkish-brown. Radial spines, the longest 1.1–1.5 cm, are more slender than the centrals.

Juvenile plants of *E.* × *neomexicanus* have not been available for study. In other species of *Echinocereus*, immature plants characteristically have fewer ribs and spines, both centrals and radials, on their smaller stems.

Biosystematics. Hybridization between *E. coccineus* var. *rosei* and *E. viridiflorus* vars. *cylindricus* and/or *chloranthus* appears to occur sporadically where these two parental species are closely sympatric, but mostly in regions where *E. coccineus* is introgressed from *E. dasyacanthus*. The Davis Mountains, despite an abundance of both *E. coccineus* and *E. viridiflorus,* exemplify a well-explored area in which no *E.* × *neomexicanus* have ever been reported. So far as is known, *E. coccineus* is always tetraploid ($4n$), and *E. viridiflorus* is always diploid ($2n$). Hybrids between these taxa would be expected to be triploid ($3n$)

and predictably sterile. Sight records of about 17 plants of *E.* × *neomexicanus* have been reported by several cactus experts after many years of fieldwork within the range of the species (Zimmerman, 1993). It is not surprising that relatively few of these supposedly triploid plants have been reported. Zimmerman observed three plants that all lacked pollen but noted that complete sterility was not experimentally demonstrated. His conclusion regarding *E.* × *neomexicanus* is that the nothospecies "is merely a collection of 'dead-end' first generation hybrids" (p. 282). Benson (1982) misapplied Standley's epithet *neomexicanus* to the common Las Cruces race of claret-cup cactus, as *E. triglochidiatus* var. *neomexicanus* (Standl.) Standl. ex W. T. Marshall.

Synonym. *Echinocereus triglochidiatus* var. *neomexicanus* (Standl.) W. T. Marshall, in part, as to type.

Common Name. The common name triploid hybrid hedgehog cactus reflects the suspected origin of the nothotaxon.

4. **Echinocereus dasyacanthus** Engelm., Mem. Tour N. Mex. 100. 1848. TEXAS RAINBOW CACTUS. Plates 130, 131. Rocky slopes of arid mountains or desert floor, in desertscrub or desert grasslands, opportunistically in more mesic habitats, occurring in every county of the Trans-Pecos except Val Verde, not extending much E of Sanderson in Terrell Co. 1,700–4,700 ft. Flowering Mar–May. 2*n* = 44. In TX extending E to about Mitchell Co., disjunct in Maverick Co. near Eagle Pass. Southeast and S-central NM; reports from Cochise Co., AZ, are based on specimens of *E. bristolii* W. T. Marshall. Mexico, NE Chihuahua and N Coahuila. Map 28.

Echinocereus dasyacanthus is commonly known in the Trans-Pecos region as

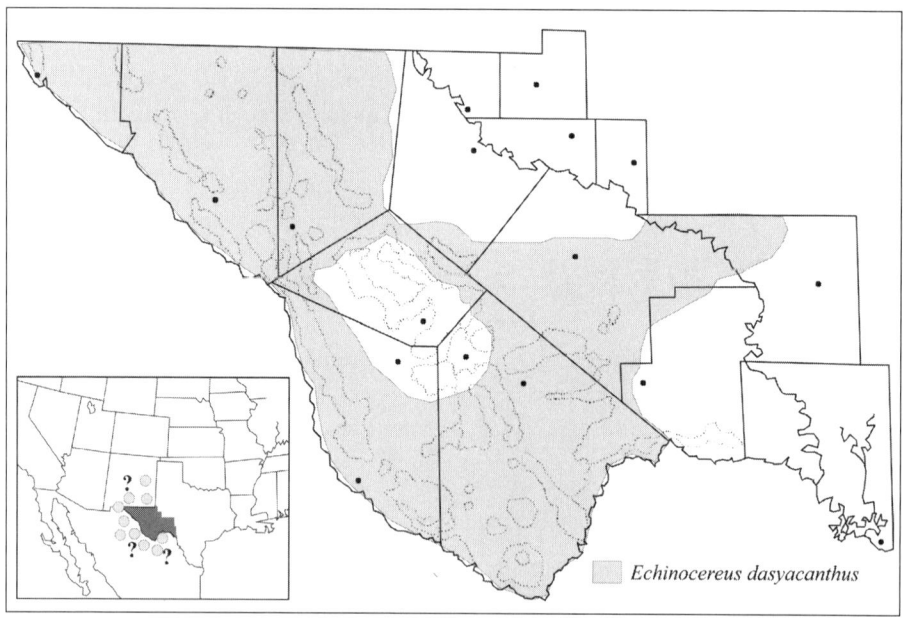

Map 28. Generalized distribution of *Echinocereus dasyacanthus* (Texas rainbow cactus).

the rainbow cactus because of rather subtle rings caused by bands of contrasting spine coloration along the length of the stems. Not all plants and, indeed, not all populations regarded here as belonging with *E. dasycanthus* exhibit the "rainbow" coloration of the stems. *Echinocereus dasyacanthus* exhibits considerable variation in spine color, number, length, and configuration. In addition, some previous workers (e.g., Benson, 1982) have been confused by flower-color variation associated with *E. dasyacanthus* and related taxa. As interpreted here, *E. dasyacanthus* is a tetraploid, predominantly yellow-flowered species with betacyanic-flowered individuals and populations scattered throughout most of its range, at least in the Trans-Pecos and in southern New Mexico. Recent experimental evidence (Powell et al., 1991; Powell, 1995, 1998c) suggested that the cyanic flower color and possibly some of the spine variation are the result of past or present introgressive hybridization of *E. dasyacanthus* with *E. coccineus*. The specific epithet is after the Greek *dasys*, "hairy" or "shaggy," and *acantha*, "thorn," or "spines" in this case, which is descriptive in reference to the predominant forms of this species with rather long and divergent, overlapping spines and a "shaggy" aspect.

Identifying Characters. Plants of *E. dasyacanthus* typically have solitary stems, or 2–3 basal branches. Plants with more than three stems (even 8–10 or more) are not uncommon, however, in populations containing particularly old and vigorous plants. The stems are 11–24(–40) cm long, (4.5–)5.5–7(–10) cm in diameter, with 15–18(–19) ribs. The areoles are 5–8(–11) mm apart. The spines in most plants are overlapping and obscure the stem surface. From the stem apex downward, the spine color shows annual growth increments, but these are vague or absent in some plants, marked by rings of lighter and darker colors. Spine colors, numbers, lengths, and position (given in more detail in following discussion) vary considerably, but in general 4–12 central spines 0.5–1.2 cm long, and 14–25 radial spines to 0.7–2 cm long, are characteristic of the species in the Trans-Pecos.

The large, showy flowers of *E. dasyacanthus* erupt from above areoles at the sides of the stem but usually nearer the stem apex, in contrast with the generally lateral flowers of *E. viridiflorus*. The sweet-scented, bee-pollinated flowers are 8–12 cm long and 7–11 cm wide, with larger dimensions reported (e.g., Taylor, 1985). In most Trans-Pecos populations the flowers are yellow with a green throat, or more specifically, the inner tepals have narrow greenish bases and broad blades that are polymorphic with respect to color: dark to pale lemon-yellow, golden-yellow, or canary-yellow, or rose-pink to deep red or magenta, rarely with an orange tinge, but often salmon-pink, rose-pink, or orange-red with age. Populations containing flowers of the various betacyanic colors and orange are the minority, in a mostly yellow-flowered taxon. Usually the cyanic-flowered plants are mixed among yellow-flowered plants, instead of forming pure populations. The sporadic occurrence of plants with cyanic flowers is seen from end to end of the Trans-Pecos. Extreme flower-size variation seems to be characteristic in some populations that have mostly cyanic-flowered plants, with flowers one to several centimeters smaller or larger than the measurements given above. The tepals are rounded to attenuate-acute, often erose or mucronate, relatively thin and soft in comparison with *E. coccineus,* but relatively thick and durable compared to *E. reichenbachii* tepals. In flowers of *E. dasyacanthus* the nectar chamber is 2–4 mm long, much smaller than in *E. coccineus,* a trait that appears to be correlated with the pollination syndromes of these species. The spreading stamens in *E. dasyacanthus* flowers have filaments of approximately equal length, resulting in a floral throat that is filled with a funnel of yellow anthers. The style is 1.5–3 mm thick, whitish, with 12–22 deep green stigma lobes.

The fruit of *E. dasyacanthus* is at first green or greenish-purple but typically matures dark, dull purplish, with a juicy pulp that may be white to purplish-pink. The globose to ellipsoid fruit is relatively large, to 6 cm long and 4.5 cm in diameter. Spiny areoles on the fruits are ultimately deciduous. Mature fruits usually are indehiscent but may split lengthwise. The black seeds are globular, papillose-rugose, and 1–1.4(–1.5) mm in largest diameter.

Phenology. In populations near the Rio Grande, where it is warmer earlier in the year, flowering begins in March. In slightly higher elevations (e.g., in eastern Pecos County), flowering is delayed until late April or early May, except in extraordinary years when the blooming period may overlap with that of *E. coccineus.* In the Franklin and Hueco mountains, El Paso County, Worthington (1986) reported the typical flowering time for *E. dasyacanthus* to be between 24 April and 16 May, with most plants flowering in late April. In Alpine, cultivated plants have flowered as early as 6 April, following an exceptionally warm winter and spring. The flowers of *E. dasyacanthus* almost completely close at night, reopen during the day, and last for 3–8 days.

Sterile and Immature Specimens. In vegetative condition *E. dasyacanthus* superficially resembles certain varieties or spine forms of *E. viridiflorus,* although close examination of areoles and spination almost always allows positive identification. In the central mountain region of the Trans-Pecos, *E. dasyacanthus* might be confused with *E. viridiflorus* var. *cylindricus* at lower mountain elevations and on southern exposures where the typically desert-dwelling *E. dasyacanthus* may be brought into sympatry with var. *cylindricus.* The var. *cylindricus* often exhibits a "rainbow" banding of contrasting spine colors up and down the stem, albeit usually with snowier white and brighter red colors than those of *E. dasyacanthus.* Stems of *E. dasyacanthus* might average larger than those of the *E. viridiflorus* varieties with which they grow sympatrically. The short-spined extreme of *E. viridiflorus* var. *russanthus* found near the Chisos Mountains deceptively resembles *E. dasyacanthus;* the real *E. dasyacanthus* in that area is the short-spined extreme of its own species, looking deceptively like *E. pectinatus.* The large shaggy-looking, magenta-flowered *E.* × *roetteri* from eastern Brewster County also resemble *E. dasyacanthus.* In Terrell County east of Sanderson, *E. dasyacanthus* might be confused temporarily with *E. pectinatus* var. *wenigeri,* but the ashy-white spines and spine pattern of var. *wenigeri* are distinctive.

The characteristic but variable spine pattern of *E. dasyacanthus* includes the occurrence of (3–6–)8–12(–15) central spines and (14–)17–25(–28) radial spines.

The shortest central spines are 2–5 mm long, and the longest ones are 0.4–1.1(–1.4) cm. If there are relatively few central spines (e.g., 3–4), they may appear to be in one vertical row, but usually the 8–12 central spines are not in discernible rows, and they spread in all directions, giving the stem a bristly aspect. In short-spined forms of *E. dasyacanthus* the stem aspect may approach the relatively smooth appearance of *E. pectinatus*. In *E. dasyacanthus* the largest radial spines, 0.9.–1.8(–2.3) cm long, are the lateral and lower ones in the areoles. The radial spines are mostly appressed and conspicuously overlapping (or intermingled) with those of adjacent areoles. The thickness of the mostly acicular central and radial spines is about equal, at (0.15–)0.25–0.45 mm in diameter, with that of the central spines, in some forms only, having rather prominent bulbous bases. The central spines usually are darker than the radials. The spine color in *E. dasyacanthus* varies considerably but basically is tan or pale yellow to pink, or less often ashy-white to reddish-brown. The spine tips often are darker, reddish to reddish-black.

Immature plants of *E. dasyacanthus* are difficult to distinguish from some certain juveniles of *E. viridiflorus*, and in western Terrell County, from young plants of *E. pectinatus* var. *wenigeri*. In *E. dasyacanthus* the areoles of juvenile plants, and younger areoles in general, have fewer spines than do areoles on older stems. Even areoles with fewer central or radial spines retain their character, as seen under a microscope, and can be used to distinguish *E. dasyacanthus*.

Biosystematics. Benson (1982) treated *E. dasyacanthus* as a variety of *E. pectinatus*. In fact, Benson submerged many taxa (not all closely related) into his broad concept of *E. pectinatus*: var. *pectinatus*, var. *wenigeri*, var. *minor*, and var. *neomexicanus*. His var. *neomexicanus* (mostly yellow-flowered specimens) is equivalent to our *E. dasyacanthus*, except that Benson misidentified some of the cyanic-flowered specimens of *E. dasyacanthus* as *E. pectinatus* var. *pectinatus* (which does not grow north of Mexico).

For the Trans-Pecos region Weniger (1984) recognized two yellow-flowered varieties of *E. dasyacanthus* [var. *dasyacanthus* and var. *hildmannii* (Arendt) Weniger], and two varieties of *E. pectinatus* [cyanic-flowered var. *wenigeri* and yellow-flowered var. *ctenoides* (Engelm.) Weniger, *nom. nud.*]. Bravo-Hollis and Sánchez-Mejorada (1991a) apparently followed Benson in presenting the taxonomy of this complex, recognizing five varieties of *E. pectinatus*, including the placement of *E. ctenoides* and *E. hildmannii* in synonymy of *E. pectinatus* var. *pectinatus* and *E. pectinatus* var. *neomexicanus* respectively. Taylor (1985) also recognized one species for the complex, *E. pectinatus*, with three varieties in the Trans-Pecos, var. *pectinatus*, var. *wenigeri*, and var. *dasyacanthus* (Engelm.) N. P. Taylor. Taylor (1985) was not the first to speculate that the var. *minor* (= *E.* × *roetteri*) could be a hybrid between *E. dasyacanthus* and *E. coccineus*, a disposition that is supported by recent experimental evidence (Powell et al., 1991; Zimmerman, 1993; Powell, 1995, 1998c).

A major problem that confronted previous workers has been to deal with the occurrence of both yellow- and cyanic-flowered forms of *E. dasyacanthus*. *Echinocereus pectinatus* is represented in Texas only by var. *wenigeri*.

Recent chromosomal and experimental data have helped to resolve some of the taxonomic problems concerning the *E. pectinatus* species-group. *Echinocereus pectinatus*, sensu stricto, appears to be a diploid species (Weedin and Powell, 1978; Ross, 1981; Pinkava and Parfitt, 1982; Pinkava et al., 1985; Weedin et al., 1989) with primary distribution in northeastern Mexico. *Echinocereus dasyacanthus* appears to be a tetraploid species with primary distribution in Trans-Pecos Texas and New Mexico, but extending into adjacent Mexico. *Echinocereus pectinatus* is relatively uniform in its spination (1–5 very short central spines in 1–2 vertical series, 12–16 pectinate radial spines, the spines ash-white to pink) and pattern of cyanic flower coloration (inner tepals pink to magenta distally, with a white band in the middle, and green proximally, or pink to magenta distally and a maroon, green, or white throat). *Echinocereus dasyacanthus* is relatively variable, particularly in stem size, spine numbers and length, and spine color, including forms with shorter spines and fewer central spines; it has mostly yellow flowers, but there are exceptions (see following discussion). In *E. dasyacanthus* cyanic flowers seem never to exhibit the *E. pectinatus* pattern of complex, concentric color-banding in the flowers.

Most of the *E. dasyacanthus* populations are yellow-flowered. There is some variation in flower size and yellow coloration. Patterns of morphological variation often are localized geographically, forming recognizable races of *E. dasyacanthus*. Some of the geographic races have been formally described. The var. *hildmannii*, plants identified by Weniger (1984) with fewer spines and spines of intermediate lengths, make up a potentially mappable morphotype from the Apache Mountains in southern Culberson County southeast through southern Reeves County and western Jeff Davis County (Barrilla Mountains), and into western Pecos County. This entity is an infrequent tetraploid (Weedin et al., 1989) considered by several authors to be a synonym of *E. dasyacanthus*. In the western foothills of the Chisos Mountains another spine form (shorter spines, especially the centrals, lending a smooth appearance similar to that in *E. pectinatus*) seems geographically isolated. This morphotype vegetatively resembles the Mexican taxon *ctenoides*, as illustrated by Weniger (1984, p. 45). Benson (1982) placed *E. dasyacanthus* var. *ctenoides* Backeb. in synonymy with *E. pectinatus* var. *pectinatus*, whereas Taylor (1985) related the entity, through synonymy, to *E. dasyacanthus*. The morphotype west of the Chisos Mountains that some have identified as *ctenoides* has smallish, yellow flowers (not the showy, orange-tinted flowers shown in Weniger's photograph of true *ctenoides* from Mexico). Two plants have been determined to be tetraploid (A. M. Powell and J. F. Weedin, unpub.), and we believe that it is just another short-spined geographic race within *E. dasyacanthus*. Blum et al. (1998) treated *E. ctenoides* (Engelm.) Rümpler as a distinct species. If the diploid count reported by Blum et al. for *E. ctenoides* is substantiated, then it would seem likely to be a truly yellow-flowered geographic race of *E. pectinatus*, but endemic to a small area in Coahuila, Mexico. Other geographic races or morphotypes of *E. dasyacanthus* occur south of the Chinati Mountains in Presidio County (with dark spines), west of Mariscal Mountain in Big Bend National Park (with slender, branching stems), and elsewhere in the Trans-Pecos. Further populational studies are needed to evaluate the taxonomic status of geographic races in *E. dasyacanthus.*

Cyanic-flowered populations of *E. dasyacanthus* are known to occur in several areas of the Trans-Pecos. In different plants the cyanic flower colors include red, orange-red, magenta, pink, and orange. A large, somewhat discontinuous population of these cyanic-flowered plants occurs sympatrically, or in partial sympatry, with *E. coccineus,* and in certain sites with *E.* × *roetteri,* from eastern Pecos County (Plate 132) west across southern Reeves County to near Kent in southern Culberson County, where only a few plants with red flowers have been observed. Other cyanic-flowered populations are known from eastern Big Bend National Park to near Reagan Canyon along the Rio Grande in southeastern Brewster County, and they have been reported from near Ruidosa in Presidio County and south and west of Sanderson in Terrell, Pecos, and Brewster counties (M. Rodriguez, F. Duran, and B. Warnock, pers. comm., photographic records, and flowering specimens). Another population occurs on Mesa de Anguila in western Big Bend National Park.

In all morphological characters except flower color, cyanic-flowered forms, so far as known, fit well within the range of variability seen in yellow-flowered forms. Powell (1995, 1998) suggested that the cyanic flower colors have entered the yellow-flowered populations of *E. dasyacanthus* through introgression from *E. coccineus.* Plants with cyanic flowers would in that case all belong taxonomically with the nothospecies *E.* × *roetteri,* but morphologically they belong with *E. dasyacanthus* in all visible respects. Rose-pink or magenta flowers are the common ancestral condition for the "rainbow cactus" group in general, and *E. dasyacanthus* might not be entirely "fixed" for its derived yellow character state.

Synonyms. *Cereus dasyacanthus* Engelm.; *Echinocereus pectinatus* (Scheidw.) Engelm. var. *neomexicanus* (J. M. Coult.) L. D. Benson in part, excluding type; *E. hildmannii* Arendt; *E. dasyacanthus* Engelm. var. *hildmannii* (Arendt) Weniger; *E. degandii* Rebut ex Schum.; *E. spinosissimus* Walton; *E. rubescens* Dams; *E. pectinatus* var. *dasyacanthus* (Engelm.) W. Earle ex N. P. Taylor; ? *E. ctenoides* (Engelm.) Rümpler; *E. pectinatus* var. *ctenoides* (Engelm.) Weniger ex G. Frank; ? *E. dasyacanthus* var. *ctenoides* Backeb.; *E. steereae* Clover; *E. dasyacanthus* var. *steereae* W. T. Marshall; ? *E. pectinatus* var. *texensis* Schelle.

Common Names. Rainbow cactus; golden rainbow hedgehog; Texas rainbow hedgehog; yellow-flowered pitaya; New Mexico rainbow hedgehog; Texas rainbow pitaya; hedgehog-cactus; slender-spined pitaya.

5. **Echinocereus pectinatus** (Scheidw.) Engelm. var. **wenigeri** L. D. Benson, Cact. Succ. J. (U.S.) 40: 124–25, f. 3. 1968. LANGTRY RAINBOW CACTUS. Plates 133, 134. Limestone soils, outcrops, slopes, mesas, degraded grasslands, desertscrub. East Pecos Co. Terrell Co. E of Sanderson. Val Verde Co. Reported but needing confirmation in E Brewster Co. 900–3,600 ft. Flowering Mar–May. $2n = 22$. Crockett and Sutton counties. Adjacent Coahuila, Mexico. Map 29.

Echinocereus pectinatus is a species of north-central Mexico, mostly in the Chihuahuan Desert Region, extending into the United States in and near the southeastern Trans-Pecos region. Two varieties of the species are recognized by

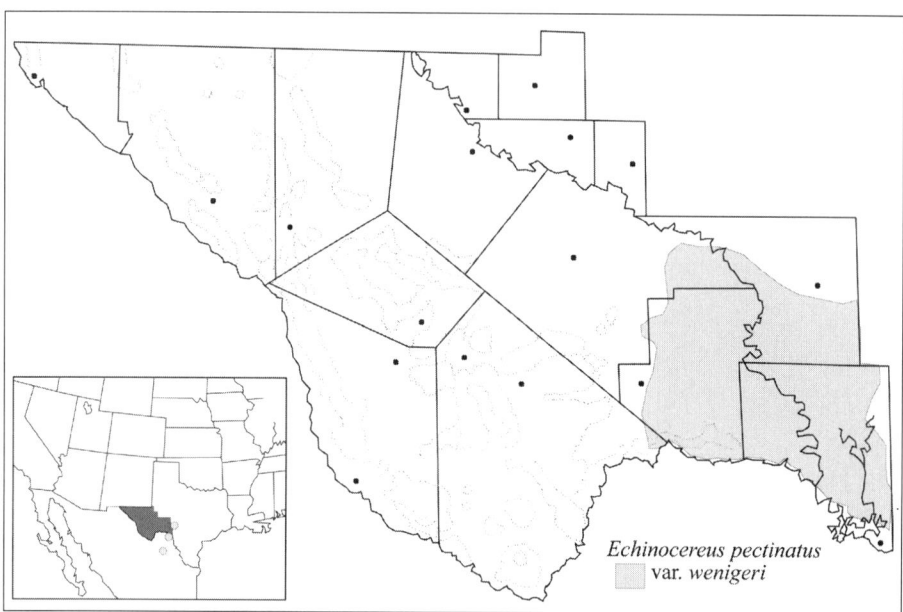

Map 29. Generalized distribution of *Echinocereus pectinatus* var. *wenigeri* (Langtry rainbow cactus).

most workers (Taylor, 1985): the more widespread var. *pectinatus* and var. *wenigeri*, which is more restricted in occurrence not far from the Rio Grande in Texas and adjacent Mexico. *Echinocereus pectinatus* is a morphologically distinct, strictly cyanic-flowered species, unless golden-orange-flowered *E. ctenoides* of northeast Coahuila proves conspecific, but this is pure speculation at present. Some previous workers included the similar *E. dasyacanthus* as a variety of *E. pectinatus*, partly because some members of the predominantly yellow-flowered *E. dasyacanthus* produce cyanic flowers, and partly because the short-spined extreme of *E. dasyacanthus* (e.g., near Burro Mesa) is vegetatively indistinguishable from Mexican *E. pectinatus* (although the flowers and chromosomes are typical of *E. dasyacanthus*). The specific epithet is after the Latin *pectinis*, "comb," in reference to the pectinate radial spines. The varietal name honors the late Del Weniger, author of two books on the cacti of Texas and the southwestern United States.

Identifying Characters. The plants of var. *wenigeri* consist of solitary stems, or in older plants, 1–2 branches at the base. The stems of sexually mature plants are cylindrical, 8–21(–25) cm long, 5.5–8(–10) cm in diameter, with an even covering of ashy-white spines that mostly obscure the stem surface. The radial spines are appressed to the stem, and along with relatively short (to 4–6 mm) central spines, give the plant a smooth appearance. In var. *wenigeri* the spines are ashy-white all along the stem without the rings of contrasting spine colors that mark stem coloration in certain related taxa. There are (13–)14–18 stem ribs with areoles 4–12 mm apart on the ribs.

The flowers of var. *wenigeri* are multicolored, giving a banded appearance inside. The inner tepals are pink, lavender-pink, to magenta on the distal one-third to one-half, then blending or abruptly whitish in the middle, and green at the base (Plate 135). Authors other than Weniger (1984) have not emphasized the pink-white-green rings inside the flowers of var. *wenigeri*. There may be significant variation from this color pattern somewhere in the range, but the banded pattern described above is the familiar flower color for this taxon in the Trans-Pecos. The tepals are somewhat rounded or obtuse. The whole flowers are 7–9(–11) cm long and 5.5–7.5(–10) cm wide. Cream-yellow anthers on whitish filaments are spread across most of the throat. The style is whitish with 9–12 dark green stigma lobes. The nectar chamber is 2–4 mm long. Trichomes (along with spines) in the receptacular areoles are 2–3(–5) mm long.

The fruits of var. *wenigeri* are broadly elliptical, 2–3 cm long and nearly as wide, dark dull purplish or reddish, juicy, and with whitish pulp. Older fruits may turn bronze or brown, sometimes drying, and the pericarpel may split. The pericarpel areoles are deciduous after the fruit ripens. The seeds are black, globular, 1–1.3 mm in diameter, and papillate.

Phenology. The flowering time of var. *wenigeri* in the Trans-Pecos overlaps with that of *E. dasyacanthus* from April to early May. The two taxa are partially sympatric in the eastern Trans-Pecos, but they are of different ploidy levels and apparently do not produce successful hybrids. Individual flowers of var. *wenigeri* open about 11:00 A.M., close at night, and open again for 2–3 days. The inner tepals of var. *wenigeri* are rather delicate and tend to be moved easily by wind but are more substantial than those of *E. reichenbachii* or *E. chisoensis*. Flower opening appears to be influenced by strong wind, cool temperatures, and cloudy days. Fruits mature in about 1.5–2 months or more.

Sterile and Immature Specimens. The var. *wenigeri* is most likely to be confused with the related *E. dasyacanthus* and the unrelated *E. reichenbachii*. In the Trans-Pecos, *E. reichenbachii* is not known to be commonly sympatric with var. *wenigeri*, but the two may occur together in eastern Pecos and Terrell counties. The ashy-white spines of var. *wenigeri* are consistent along the length of the stem and never pink or in contrasting bands of darker (pink) and lighter (whitish) spines that develop as annual growth increments (from the apex), as is characteristic of var. *pectinatus* in Mexico. Contrasting "rainbow" bands also characterize some individuals and populations of *E. dasyacanthus*. *Echinocereus reichenbachii* in the Trans-Pecos is distinguished by its 0–1(–2) central spines that are minute, ca. 1 mm long, if present.

The white spines of var. *wenigeri*, at least the largest ones, usually are tipped with pink or purplish-brown. The areoles are broadly oval, those at the apex with woolly hairs, becoming elongated down the stem and without hairs. The central spines are (1–)2–6(–7) per areole, the shortest ones 1.9–2.5(–3) mm long, and the longest ones (2.2–)4–6 mm. There are (13–)15–24(–26) radial spines per areole, the largest ones (lateral and lower) 7–10(–13) mm long. The classic spine configuration of var. *wenigeri* is 2–3 short central spines in one vertical row in the center of the areole, with about 14–20 radial spines that are pectinate and somewhat recurved between the ribs. These spine characters along with the consistent ashy-white spine color provide the best means for distinguishing nonflowering var. *wenigeri* from any other cactus species of sympatric occurrence. Plants of *E. dasyacanthus*, even plants with whitish spines, typically have 8–12 longer central spines spreading at all angles and not arranged in vertical rows. In *E. dasyacanthus* there are few-spined and shorter-spined forms, but these are not known to occur with var. *wenigeri* in the southeastern Trans-Pecos. A short-spined race of *E. dasyacanthus* overlaps the geographic range var. *wenigeri* in adjacent Mexico (Zimmerman et al., forthcoming).

Sterile specimens of var. *wenigeri* possibly could be confused with *E. reichenbachii*, which is very rarely sympatric with var. *wenigeri*. Adult plants of *E. reichenbachii* typically are smaller in size, in diameter (2.5–5 cm) if not in length, than are adults of var. *wenigeri*. In Trans-Pecos *E. reichenbachii*, areoles usually lack central spines, or there may be 1–2 minute central spines, and 12–16(–21) radial spines 4.5–6 mm long. The smaller, immature stems of var. *wenigeri* have fewer spines per areole, including 0–1 central spine, and thus resemble *E. reichenbachii* even more closely than the adults.

Biosystematics. The degree of taxonomic distinction between *E. pectinatus* and *E. dasyacanthus* apparently was not realized (Benson, 1982; Taylor, 1985) until recent years (Zimmerman, et al., forthcoming). It now seems clear that *E. pectinatus* is a diploid, cyanic-flowered entity of primarily Mexican distribution, whereas *E. dasyacanthus* is a tetraploid, basically yellowed-flowered species with its main distribution in Texas and New Mexico.

Two varieties of *E. pectinatus* (or three, if the yellow-flowered *E. ctenoides* of Coahuila proves diploid) occur in Mexico. Variety *pectinatus* seems widespread throughout the Chihuahuan Desert Region, geographically replaced by var. *wenigeri* in northeastern Coahuila and the lower Pecos River region of Texas. The vegetative morphology of var. *pectinatus* includes stems that are globular to short-cylindric, 8–15 cm long, near 6 cm in diameter, with (16–)18–20(–23) ribs, 3–5 central spines, these ca. 3 mm long and arranged in 1–2 vertical rows, and 19–30 radial spines appressed to the stem and not (or barely) interlaced. In aspect the stems of var. *pectinatus* have a smooth appearance with whitish and rather bright pink spines occurring in "rainbow" bands. The var. *wenigeri* is easily distinguished by its uniform ashy-white spine coloration and other characters described above. The flowers of both var. *pectinatus* and var. *wenigeri* are strikingly similar, with imperfect bands of three colors (pink-magenta distally, then white, green basally) displayed collectively by the inner tepals. The populational consistency of this flower coloration remains to be verified, particularly in var. *pectinatus*.

Echinocereus reichenbachii occurs sympatrically with *E. pectinatus* var. *wenigeri* in Pecos County and nearby areas of the Trans-Pecos, and presumably overlaps with var. *wenigeri* in northeast Mexico or on the Edwards Plateau in Texas. These taxa are not considered to be closely related (Taylor, 1985), but both species are diploid, and the possibility of hybridization between them should be considered in future studies of the taxa.

Synonym. *Echinocereus ctenoides* (Engelm.) Rümpler, in part, sensu Weniger, excluding type.

Common Names. Comb hedgehog; Weniger hedgehog cactus; ashy-white pitaya.

6. **Echinocereus fendleri** (Engelm.) Rümpler var. **fendleri**, Handb. Cacteenk. (ed. 2). 801. 1885. FENDLER'S HEDGEHOG CACTUS. Plates 136, 137. Upper desert grasslands and plains grasslands in the alluvial mountain basins, rare near Panther Junction in the Chisos Mts of southern Brewster Co., NW Brewster Co. W across N Presidio Co., Jeff Davis Co. and adjacent Reeves Co., W across SE and central Culberson and Hudspeth counties, and slopes of the Franklin Mts in El Paso Co. 3,500–5,500 ft. Flowering Apr–May (usually early May in higher mountain basins). $2n = 22$. Western two-thirds of NM, S CO, NE and SE AZ. Possibly NE Sonora, Mexico. Map 30.

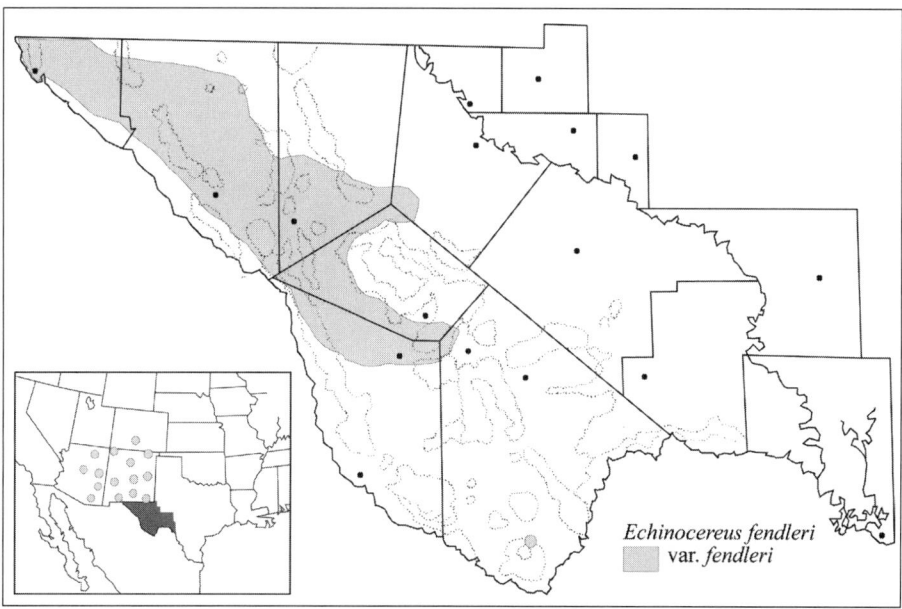

Map 30. Generalized distribution of *Echinocereus fendleri* var. *fendleri* (Fendler's hedgehog cactus).

Several authors (e.g., Benson, 1982; Taylor, 1985; Mellen, 1991) have recognized two varieties of *E. fendleri* in Trans-Pecos Texas, var. *fendleri* and var. *rectispinus*. Weniger (1984) seemingly recognized only var. *fendleri* for Texas and var. *rectispinus* for southwestern New Mexico and southeastern Arizona. We accept only var. *fendleri* in the Trans-Pecos region. The specific epithet honors the early collector in New Mexico, Augustus Fendler.

Relatively few herbarium specimens of *E. fendleri* from the Trans-Pecos have been preserved, making it difficult to understand populational variation of this taxon without further field investigations. Current field observations suggest that *E. fendleri* var. *fendleri* is the only variety of this species represented in the Trans-Pecos and that the records of var. *rectispinus* attributed to the area (Benson, 1982) are spine forms of the variable var. *fendleri*. *Echinocereus fendleri* var.

kuenzleri (Castetter, P. Pierce, and K. H. Schwerin) L. D. Benson occurs near the northern border of the Trans-Pecos in the Sacramento Mountains of New Mexico (Otero County) and recently has been applied to the disjunct populations long known from northern Chihuahua, Mexico (Taylor, 1985), based on *E. hempelii* Fobe, but it is not expected in the Trans-Pecos (see Appendix 2).

Identifying Characters. The grassland-dwelling *E. fendleri* var. *fendleri* is characterized by solitary or few stems but often forms loose clumps of 5–10 or more erect and perhaps flaccid stems. The stems are dark green, somewhat wrinkled, 7.5–17(–25) cm long, 3.8–7.5 cm in diameter, with 8–10 ribs. The circular areoles are about 15–17 mm apart with conspicuous swellings around the areoles. Collectively, the spines only partially hide the stem surface. The spines themselves are opaquely white or ashy-gray, often with contrasting black or brown spines in the same areoles, these weathering to gray. Or the central spine and largest radial spines may be nearly black proximally and whitish distally except for reddish-brown tips. The whitish spines often have a longitudinal dark stripe on the underside, or the dark stripe is brought to the upper surface in twisted spines, or the spines are sometimes variegated. The single central spine, often nearly black, 5 cm long, and curving upward, at least in some areoles on some stems, is perhaps the best identifying feature of var. *fendleri*, for example, when the superficially similar but larger-stemmed *E. coccineus* might be in the area. Usually there are 5–9(–10) radial spines in the areoles of var. *fendleri*.

The flowers of var. *fendleri* are relatively large and showy, 5–11 cm long, and 5–11 cm wide. The inner tepals are magenta, often a darker purplish-red at the base, 3–4.6 cm long, and 1.2–1.4 cm wide. Pink- and white-flowered forms of *E. fendleri* are less common or rare. Flower buds are conspicuously erumpent above areoles on the sides of stems near the apex. The receptacular tube of mature flowers is about 1.6–2 cm long. The nectar chamber is (2–)3.5–4.5(–8) mm long. The stamens have greenish filaments and light yellow anthers. The style is whitish, only slightly longer than the anthers, and supports 9–16 dark green stigma lobes.

The bright red to brick-red fruits are ellipsoidal, sometimes almost spherical, and 2–3 cm long. The fruits are juicy with a magenta pulp. At maturity the fruits may rupture. The areoles, with whitish spines and white wool, are readily dehiscent on ripe fruits. The seeds are black, irregularly globular or obovoid, reticulate-punctate, and 1–1.5 mm in largest diameter.

Phenology. The typical flowering time for *E. fendleri* appears to be from about mid-April to mid-May, possibly depending on temperature and the timing of spring moisture. Specimens in the Sul Ross Cactus Garden have bloomed as early as 14 April. Worthington (1986) noted that *E. fendleri* was in flower between 18 April and 12 May in the arid Franklin Mountains of El Paso County, and the species was observed in flower on 18 April in the nearby Florida Mountains of Luna County, New Mexico. One collection from the Chisos Mountains was in full flower on 24 April. Plants at this more southerly location, where average daily temperatures usually are higher earlier in the season, probably bloom earlier than plants in the central mountain region of the Trans-Pecos. Observations of var. *fendleri* over several years near Marfa, in the plains grasslands at ca. 4,700 feet, revealed that flowering usually occurs in early to mid-May. The flowers of *E. fendleri* partially close each night and reopen during the day for up to seven days (Worthington, 1986). Fruit maturation required about two months on plants cultivated in Alpine.

Sterile and Immature Specimens. In the Trans-Pecos *E. fendleri* is most likely to be confused with *E. coccineus*, which is similar in habit, stem shape, and spine pattern, but the two species seldom occur together. The stems of *E. fendleri* are darker green, smaller and more wrinkled-flaccid than are those of *E. coccineus*, and the spine patterns and colors are similar but distinctive. Other taxa of *Echinocereus* that have some resemblance to *E. fendleri* and might be found in the same general distributional range are *E. viridiflorus*, *E. enneacanthus*, *E. stramineus*, *E.* × *neomexicanus*, and *E.* × *roetteri*.

Both central and radial spines of var. *fendleri* are prominently bulbous at the base. The central spine is terete or somewhat flattened at the base, or for most of its length. Rarely the central spine is absent (the absence of a central spine is typical in var. *kuenzleri*, which is not known to occur in the Trans-Pecos), and there are up to three central spines reported in some plants. The second and third central spines, if present, are smaller than the single main central spine, difficult to distinguish from the upper radial spines.

The radial spines of *E. fendleri* var. *fendleri* may be whitish, ashy-gray, or multicolored. Spine-color variation, even in the same areole, is typical. The most characteristic color pattern is of whitish to brown radials and a dark central spine. The longest and sometimes most prominently pigmented radial spines are the two lower lateral ones and the lowermost one. The next-longest radial spines are those disposed laterally, those in upper lateral position, and the upper one or two. Radial spines are terete or usually flattened on the upper side or near the base. The largest radial spines and central spines are of equal thickness, (0.5–)0.9–1.5 mm.

Immature plants of *E. fendleri* have fewer spines. Usually the single central spine is present even in young areoles, and 5–7 radials are present, but smaller in diameter than those in mature areoles. The slender spines of the juvenile areoles are shorter than those in areoles of sexually mature plants. Their circular areoles are matted with woolly white hairs.

Biosystematics. *Echinocereus fendleri* and similar or closely related species are distributed from Trans-Pecos Texas north and west through New Mexico, Colorado, Utah, Nevada, California, and northern Mexico. Benson (1982) and Taylor (1985) recognized two varieties of *E. fendleri* to occur in the Trans-Pecos: var. *fendleri* and var. *rectispinus*.

Benson (1982) mapped both var. *fendleri* and var. *rectispinus* to occur in the Franklin Mountains of El Paso County, both varieties in Hudspeth County, and var. *rectispinus* in southeastern Culberson County. The distributions given for var. *fendleri* ("near El Paso") and var. *rectispinus* ("W Texas") by Taylor (1985) perpetuated Benson's concept of straight-spined and curved-spined taxa in regional sympatry. According to Weniger's (1984) interpretation, and ours, only one taxon (probably var. *fendleri*, sensu stricto) extends into the Trans-Pecos, whereas var. *rectispinus* is restricted to extreme southwestern New Mexico and adjacent southeastern Arizona. We have observed numerous specimens of *E. fendleri* from northern Presidio County and adjacent Jeff Davis County in the central mountain region of the Trans-Pecos. The "highland grassland" region around the towns of Marfa and Valentine appears to be a major distributional center for *E. fendleri* that was not realized by previous workers. Some plants in the populations exhibit the shorter porrect central spine supposedly characteristic of var. *rectispinus*, and other plants in the same populations, all of the populations we have examined, have the up-curved, longer central spine characteristic of var. *fendleri*. We infer that only one taxon, with variable spine morphology, occurs in the Trans-Pecos, including the population in the Chisos Mountains of southern Brewster County.

Ferguson (1989) mentioned sterile hybrids between *E. fendleri* ($2n = 22$) and *E. coccineus* ($2n = 44$) and other (presumably fertile) hybrids between *E. fendleri* and *E. triglochidiatus*. Allan D. Zimmerman (pers. comm.) has documented one occurrence near Marfa of a probable first-generation hybrid between *E. fendleri* and *E. viridiflorus* var. *cylindricus*.

Synonyms. *Cereus fendleri* Engelm.; *Cereus fendleri* var. *pauperculus* Engelm.; *Echinocereus albiflorus* Weing.

Common Names. Fendler's hedgehog; Fendler hedgehog; Fendler echinocereus; Fendler's pitaya; purple hedgehog; strawberry cactus; torch cactus; sitting cactus; pink-flowered echinocereus. Fendler's hedgehog cactus is in consistent use in New Mexico.

7. **Echinocereus enneacanthus** Engelm., Mem. Tour N. Mex. 111. 1848. STRAWBERRY CACTUS. Plates 138, 139. Diverse habitats in drainages and the Rio Grande delta; mostly within 40–50 mi of the Rio Grande from El Paso Co. SE to Val Verde Co.; also in Upton, Crockett, and Val Verde counties adjacent to the Pecos River. Sea level to 5,100 ft. Flowering Apr–mid-May. $2n = 22$. Also Edwards Plateau and S along and near the Rio Grande to Cameron Co., more localized in numerous other counties in Central and S TX. Southern NM in Ash Canyon, according to Benson (1982). Mexico, NE and N-central parts. Map 31.

This treatment follows Taylor (1985) in recognizing *E. enneacanthus* as the correct name for a taxon that includes as synonyms *E. dubius* and *E. enneacanthus* var. *dubius*, familiar names in the Trans-Pecos, and *E. enneacanthus* var. *brevispinus*, a familiar name in southern Texas. *Echinocereus enneacanthus* is one of the two (or more) species that are commonly known as strawberry cactus. In the Trans-Pecos, the best-known strawberry cactus is *E. stramineus*. The *E. enneacanthus* complex, traditionally involving four or more taxa, is one of the most thoroughly investigated cactus groups in the Trans-Pecos, considering not only the major cactus books by Benson (1982), Weniger (1970, 1984), Taylor (1985), and Bravo-Hollis and Sánchez-Mejorada (1991a) but also three systematic studies specifically dedicated to the complex (Moore, 1967; Breckenridge, 1981; Breckenridge and Miller, 1982). The specific epithet is after the Greek

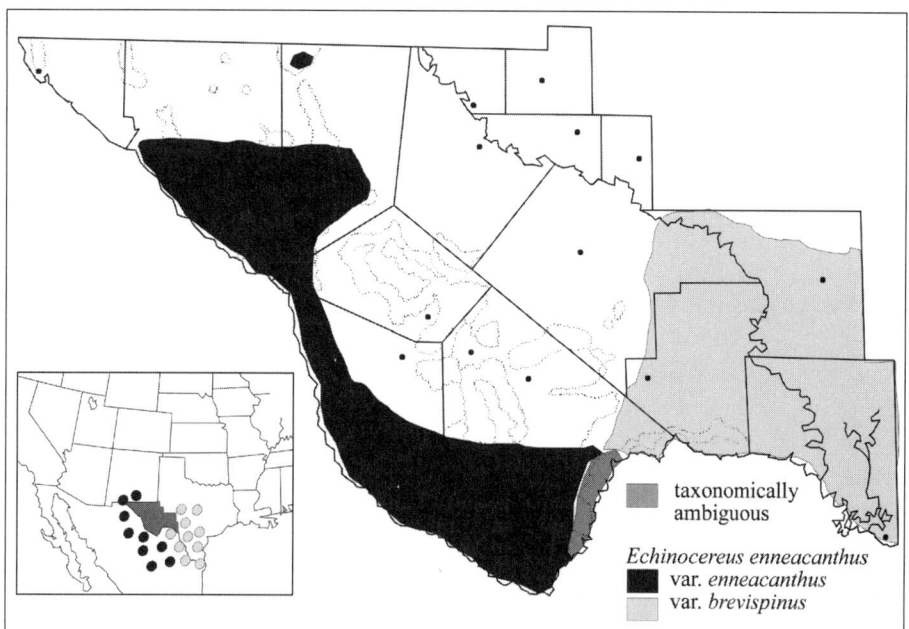

Map 31. Generalized distribution of *Echinocereus enneacanthus* var. *enneacanthus* and *E. enneacanthus* var. *brevispinus* (strawberry cactus).

ennea, "nine," and *akantha*, "thorn," presumably in reference to the ca. nine radial spines per areole.

Key to the Varieties

1. Stems 5–10(–15) cm in diameter; areoles 2.5–4.5 cm apart on ribs; central spines 1–4, to 9.2 cm long, divergent, often curved, stout, (1.2–)1.5–2(–2.5) mm at the base; radial spines to 4.2 cm long; distribution Big Bend region of the Trans-Pecos, and W to El Paso Co.

   7a. *E. enneacanthus* var. *enneacanthus.*

1. Stems to 5 cm in diameter; areoles 1–1.5(–2.4) cm apart on ribs; central spines 1–3, to 4.5(–5) cm long, porrect, straight, relatively slender, 1.0–1.5 mm at the base; radial spines 1–2 cm long; distribution mostly in S TX and Edwards Plateau, extending west along the Rio Grande to the Big Bend region

   7b. *E. enneacanthus* var. *brevispinus.*

7a. Echinocereus enneacanthus var. enneacanthus. STRAWBERRY CACTUS. Plates 138, 139. Gravel hills and benches, alluvial outwash fans and *bolsones,* also desert mountains, exposed to the sun or under shrubs, El Paso Co. along and near the Rio Grande, SE to Brewster Co. where it extends to ca. 50 mi N of Rio Grande (also reported in the Guadalupe Mts where disjunct). 1,800–4,000 ft. Flowering Apr–May. $2n = 22$. Mexico, E Chihuahua, central and S Coahuila,

NE Durango, N Zacatecas, N San Luis Potosí, W Nuevo León, and SW Tamaulipas. Map 31.

This is the entity that in Trans-Pecos Texas has long been known as *E. enneacanthus* var. *dubius* or *E. dubius*. Taylor (1985), however, has noted that the name *E. enneacanthus* originally was based on a specimen (from between Chihuahua City and Camargo in Chihuahua, Mexico) of a taxon that modern authors referred to as *E. dubius*. The type specimen of *E. enneacanthus* was collected by Wislizenus in 1847, and the species was described by George Englemann in 1848. Eleven years later Engelmann studied the cactus specimens sent to him from collections secured during the survey of the U.S.-Mexican boundary (Emory, 1859) and described two allied species, *E. dubius* and *E. stramineus*. The publication of misidentified plates in Engelmann's *Cactaceae of the Boundary* in 1859 (Emory, 1859) led to the use by modern authors of the name *E. enneacanthus* for the mostly South Texas entity with relatively slender stems. After it was realized that the correct name for the larger-stemmed Mexican and Trans-Pecos entity, long known as *dubius*, was instead *enneacanthus* (Taylor, 1985), a new name had to be selected for the South Texas taxon with more slender stems. The name selected by Taylor was var. *brevispinus*, and he deemed this epithet most appropriate because it was clearly typified, was (in his opinion) essentially identical and therefore synonymous with var. *enneacanthus*, sensu Benson (1982), and was characterized by shorter spines than var. *enneacanthus* (= *dubius*).

Identifying Characters. In habit sexually mature plants of var. *enneacanthus* have several to numerous stems in loose or rather tight clumps. Larger plants may have 50–100 or even several hundred stems. The stems are about 15 cm long, longer in shade forms, and 6–10 cm in diameter. The stems are medium to light green or somewhat yellow-green in color. There are 7–10 ribs, but on average there are 8–9 ribs supporting areoles that are (1.9–)2.5–4.5 cm apart. The 1–3(–4) central spines are flattened or angular, to 8–9.2 cm long, rather stout, often curved, and divergent from the stem. The 6–9(–13) radial spines are up to 4.2 cm long.

The magenta flowers of var. *enneacanthus* are funnelform, to 8 cm long, and 7–9(–10) cm in diameter. The tepals are in 2–3 series, oblanceolate, to 5.5 cm long, 1–2 cm wide, and acuminate-serrate at the apex. The throat of the magenta flowers deepens in color to reddish. The nectaries are 5–6 mm long. The filaments are greenish to pink, and the anthers are yellow. The style is about 3 cm long, is 1.5–2 mm in diameter, and supports 6–10 green stigma lobes.

The fruit is globular to ovoid, to 3.8 cm long and 2.5 cm in diameter. Moore (1967) reported that the fruit of var. *enneacanthus* is green or pinkish at maturity and has white pulp that is sweet but of no particular flavor. Our observations reveal that the fruit is green at first but matures red or brick-red, is pink inside, and has a strawberry smell and flavor. The seeds are black, ovoid, tuberculate, and 1–1.4 mm long.

Phenology. The typical flowering time for var. *enneacanthus* is April and May. Occasionally, plants produce single or a few flowers later in the year, possibly in response to late freezes or abundant rainfall following an exceedingly dry winter and spring. On 19 June 1999, following a very dry winter and spring, but after recent rains, several plants of E. *enneacanthus* were seen in full bloom between Presidio and Ruidosa in Presidio County. On 28 June 1999 in Alpine, a large cultivated plant was in full flower for the second time, following heavy rains 2–3 weeks earlier. Individual flowers stay open for 2–4 days. The flowers of var. *enneacanthus* are protandrous, and the stigma lobes are "closed" (in vertical position) on the first day (Breckenridge and Miller, 1982). On the second day the stigma lobes open to a horizontal position and are receptive to pollen. The flowers were reported to require about 30 minutes to open at mid-morning. Flowers close each night and may close during the day under cool, cloudy conditions. The fruits ripen in late May or June, rarely later in the summer when flowers open beyond the normal flowering time.

Sterile and Immature Specimens. In the field var. *enneacanthus* is most likely to be confused with the two other members of the E. *enneacanthus* complex, especially in areas where the taxa are known to be sympatric. Also, var. *enneacanthus* has been rarely confused with the superficially similar E. *coccineus* in Pecos and Terrell counties and east of the Pecos River. A careful analysis of spine patterns should allow distinction between all of these taxa.

In habit var. *enneacanthus* is most like var. *brevispinus*. The wider stems and longer spines of var. *enneacanthus* provide the best means of identification. Except for differences in stem size, the habits of both var. *enneacanthus* and var. *brevispinus* are similar, unless the typical form is modified seemingly as a result of ecological conditions (e.g., in shaded habitats). Plants of both taxa may have several to numerous loosely arranged stems so that the stems are clustered into flat-topped or slightly mounded plants. In both taxa the effect of a large number of stems is smaller stem size and a mounded habit.

A mounded habit with numerous compactly arranged small stems in rocky habitats is characteristic of E. *stramineus*. These traits along with rib and spine characters usually allow easy distinction between E. *stramineus* and E. *enneacanthus*. Moore (1967) and Breckenridge (1981) suggested that partial division of the ribs into low tubercles reliably distinguishes var. *enneacanthus* from other members of the complex, which lack raised tubercles. In E. *enneacanthus* the stem surface usually is only partially obscured by the spines. In E. *stramineus* the stem surface is mostly obscured by overlapping spines.

The circular areoles of var. *enneacanthus* have short matted or woolly hairs that gradually disappear with age. The stout central spines have prominent bulbous bases, are commonly 6.5–9.2 cm long, (1.2–)1.5–2(–2.5) mm wide at the base, and angular or flattened, at least on one side. The main central in some specimens is slightly sinuate or serpentine. Older centrals may split on the upper surface. The central spines are light to dark gray, reddish-brown, to opaque-glassy or stramineous, and usually are black-speckled. The radial spines are terete or somewhat flattened. The flattened radial spines usually are the lowermost 1–3 in the areole, and the lower radials are also the longest ones, to 4.2 cm. The other radial spines, those lateral or upper in the areole, are 1.7–3.3 cm long, or shorter than 1.7 cm in the uppermost positions and in younger areoles. The radial spines usually are paler and more slender than the centrals. Radials have bulbous bases but rarely are the swollen bases as prominent as those of the central spines.

Juvenile plants of var. *enneacanthus* are single- or few-stemmed. Some juvenile stems show circular areoles, these on low tubercles and with yellow or white hairs, and fewer central (ca. 1–2) and radial spines (ca. 6–7). The central spines may be black to dark reddish-brown or stramineous, and the radials are stramineous. In stems of juveniles, typical spine numbers (four centrals, nine radials) may occur, but the juvenile spines are of smaller caliber than in mature plants.

In younger areoles of *E. enneacanthus*, usually there is one clearly centrally oriented spine and 1–2 other centrals in peripheral position with the radial spines, especially at the upper areole margin. Such radially arranged central spines are recognizable by their longer length and by their angular or flattened shape, as seen in cross section.

Biosystematics. The *E. enneacanthus* complex is a relatively wide-ranging group of taxa extending from southern New Mexico near El Paso, Texas, southeast along and near the Rio Grande to the Edwards Plateau region, and on farther south to Cameron County at the southern tip of Texas. The complex also occurs in north-central Mexico. Plants of the *E. enneacanthus* complex exist in environmental conditions that range from rocky desert mountains and their alluvial outwash *bajadas*, to the wooded Edwards Plateau, and to the humid Rio Grande plains and "brush country" of South Texas. Considerable morphological variation is expressed in plants of the *E. enneacanthus* complex that occupy distributional and ecological extremes. In the southeastern Trans-Pecos region, variation (sometimes subtle) in populations of the *E. enneacanthus* complex suggests the occurrence of north-south clinal intergradation between morphological extremes or hybridization between taxa of the complex.

After Engelmann described *E. enneacanthus*, *E. dubius*, and *E. stramineus* (see Taylor, 1985), three additional species (*E. merkeri* Hildmann, *E. sarissophorus* Britton & Rose, and *E. conglomeratus* C. F. Först.) were aligned with the complex (Britton and Rose [1937] 1963). Subsequent workers reduced to synonymy some of these species while recognizing either three species and two varieties or forms (Moore, 1967; Weniger, 1984), two species plus varieties (Breckenridge and Miller, 1982; Taylor, 1985), or four varieties of a single species (Benson, 1982).

The study by Moore (1967) provided considerable new distributional, ecological, and morphological information for Texas members of the *E. enneacanthus* complex. Breckenridge (1981) and Breckenridge and Miller (1982) expanded upon that theme, included Mexican populations in their evaluations, and employed the additional systematic tools of experimental hybridizations, flavonoid chemistry, and pollination biology. They concluded that *E. stramineus* was a species distinct from the others but that there was no support for maintaining *E. enneacanthus* var. *enneacanthus* (= *dubius*) and *E. enneacanthus* var. *brevispinus* (= *enneacanthus* of some authors) as separate species. This taxonomic disposition was the same as that favored by Taylor (1985).

The specific distinctiveness of *E. enneacanthus* and *E. stramineus* was supported by a chromosome number report (Weedin et al., 1989) suggesting that *E. enneacanthus* was diploid and *E. stramineus* was tetraploid. Although Moore (1967) reported that *E. stramineus* was diploid like the other members of the *E. enneacanthus* complex examined, all other chromosome number counts obtained for *E. stramineus* (Powell and Weedin, 2001) have been tetraploid. In addition to the chromosomal data, our field observations and morphological evaluations tend to support the taxonomic conclusions of Breckenridge and Miller (1982) and Taylor (1985).

In the southern Big Bend region of the Trans-Pecos, *E. enneacanthus* commonly occurs in a variety of substrates, including alluvial gravels and soils in exposed areas, and in the shade of shrubs and small trees. *Echinocereus enneacanthus* var. *enneacanthus* (= *dubius*) occurs in habitats along and near the Rio Grande and in desert areas to about 50 miles north of the river. *Echinocereus enneacanthus* var. *brevispinus* occurs in southeastern Brewster County near the Rio Grande, usually in the shade of riparian vegetation. Farther east var. *brevispinus* occurs as far north as Upton County. The var. *brevispinus* also grows to the east on the wooded Edwards Plateau and south to near Brownsville.

Echinocereus enneacanthus var. *enneacanthus* and *E. stramineus* are regionally sympatric in the southern Trans-Pecos, where they occur in close proximity in many locations. The two taxa usually are ecologically separated, with var. *enneacanthus* characteristically occupying alluvial habitats and *E. stramineus* usually found in various exposed rocky habitats. It has been suggested that *E. enneacanthus* var. *enneacanthus* and *E. stramineus* might hybridize (Moore, 1967; Breckenridge and Miller, 1982) and that this might account for some of the morphological variation seen in these taxa. Breckenridge and Miller demonstrated crossability (at least as to seed-set) between the species, but no progeny were grown to maturity. Hybrids between *E. enneacanthus* ($2n = 22$) and *E. stramineus* ($2n = 44$) theoretically should be triploid and possibly sterile. We have observed numerous plants in the field that resemble the expected F$_1$ morphotype, combining the mounded habit typical of *E. stramineus* with the spine morphology, flower size, and fruit size of var. *enneacanthus*. Chromosomal analyses of some of the mounded forms of var. *enneacanthus* might help resolve whether or not their origin is through interspecific hybridization. One seemingly intermediate plant from Hudspeth County, counted as $2n = $ ca. 44 (Weedin and Powell, 1978) and originally identified as *E. enneacanthus* var. *dubius*, possibly was misidentified and is instead *E. stramineus*. We are inclined to speculate, however, contrary to statement by Taylor (1985) that var. *enneacanthus* is not mound-forming, that many plants of var. *enneacanthus* in the Trans-Pecos do indeed have the mounded habit of *E. stramineus*. So far, with the exception of the account in Weedin and Powell (1978), all chromosome number reports for *E. enneacanthus* suggest that it is diploid.

If only the extreme forms of *E. enneacanthus* var. *enneacanthus* (= *dubius*) and *E. enneacanthus* var. *brevispinus* were taken into account, then recognition of these entities as species might be justified. In general the western populations (i.e., var. *enneacanthus*) have stems that are larger in diameter, with longer and larger spines, and areoles farther apart on the ribs, than the eastern populations (i.e., var. *brevispinus*). The morphological differences between the western and eastern populations of *E. enneacanthus* do appear to be clinal, however, as recognized by Benson (1982), Breckenridge and Miller (1982), and Taylor (1985). Breckenridge and Miller reported that *E. enneacanthus* and *E. brevispinus* intergrade, at least in Maravillas Canyon and in the lower canyons of the Rio Grande between Maravillas and near Sanderson. It seems best at present to regard *enneacanthus* and *brevispinus* as conspecific.

Specimens of *E. enneacanthus* from Coahuila, Mexico, appear consistently to have a more "bristly" habit, approaching *E. stramineus* in their denser spination. The Mexican plants also appear to have areoles set slightly more closely together than in var. *enneacanthus*, more stramineous central spines, and an average of about nine radial spines. No chromosome number reports are available for the Mexican populations.

In Engelmann's (1859) *Cactaceae of the Boundary*, plate 46 is mislabeled (Moore, 1967) as "*Cereus* (= *Echinocereus*) *stramineus*," although it clearly is *E. enneacanthus* (= *E. dubius*). In plate 47, figures 2 and 4, both drawings of spine clusters appear to be those of *E. enneacanthus* (= *E. dubius*) and not "*E. stramineus*" as labeled.

Synonyms. *Cereus dubius* Engelm.; *Echinocereus dubius* (Engelm.) Rümpler; *E. enneacanthus* var. *dubius* (Engelm.) L. D. Benson; *E. merkeri* Hildmann; *E. sarissophorus* Britton & Rose.

Common Names. Strawberry hedgehog; pitaya; warty hedgehog; pitahaya; purple echinocereus; alicoche; purple pitaya; pitaya hedgehog cactus.

7b. **Echinocereus enneacanthus** var. **brevispinus** (W. O. Moore) L. D. Benson. STRAWBERRY CACTUS. Plates 140, 141. [*Echinocereus enneacanthus* f. *brevispinus* W. O. Moore, Brittonia 18: 93. 1967]. Mostly alluvial substrates, among riparian vegetation along and near the Rio Grande in SE Brewster Co., occasionally in limestone rocks, in various substrates including open rocky habitats. 1,000–3,000 ft. Flowering Apr–May. $2n = 22$. "Edwards Plateau," Tamaulipan scrublands, and on the lowland plains along and near the Rio Grande to Cameron Co., at near sea level. Mexico, E Coahuila, N Nuevo León, Tamaulipas. Map 31.

Echinocereus enneacanthus var. *carnosus* (Rümpler) Quehl, recognized by Weniger (1984), found near Laredo and Eagle Pass, is treated here as an unusually large and flabby-stemmed form of var. *brevispinus*. The name *brevispinus* refers to the short spines of *E. enneacanthus* var. *brevispinus*, as opposed to the long-spined var. *enneacanthus*. The root word *brevi* is after the Latin *brevis*, which means "short."

Identifying Characters. Mature plants form flat-topped or rounded clumps with 15–100 or more stems. In most populations the stems are slender and upright, to ca. 5 cm in diameter, but in some populations, particularly in the form recognized as var. *carnosus* (Weniger, 1984), the prostrate, decumbent stems may be 8–10 cm in diameter. The stems typically produce (6–)8–9(–10) ribs, with circular areoles, these 1–2(–2.4) cm apart. The green stem surface is readily visible through the relatively short spines. In each areole there are 1–2(–3) central spines to 4(–5) cm long, and (6–)8–9(–13) radial spines 1–2 cm long. The central spines in var. *brevispinus* are porrect and straight, compared to the typically curving central spines of var. *enneacanthus*.

The magenta flowers of var. *brevispinus* may be similar in size to slightly larger, to 8 cm long and 10 cm or more in diameter, compared to those of var. *enneacanthus*. The tepals are in 2–3 series, oblanceolate, acuminate, to 5.5 cm long, 1–2 cm wide, with the apexes serrate. The throat is a darker, reddish color. The filaments are greenish to pink, ca. 1 cm long, and the anthers are spread across the throat and are yellow. The style is ca. 3 cm long, 1.5–2 cm thick, and supports 6–12 green stigma lobes that are 5–6 mm long. Weniger (1984) maintained that in the flowers of var. *brevispinus* (reported as var. *enneacanthus*), the tepals are all in one whorl, whereas in the var. *carnosus* that he recognized, the tepals are in three whorls.

The fruit is globular to ovoid, 2.5–3.8 cm long, and ca. 2.5 cm in diameter. Mature fruits are greenish to reddish-brown, or ultimately turn red. The areoles with spines are easily detached from mature fruits. The pulp is pink and has a strawberry smell and flavor. The seeds are black, ovoid, 1–1.4 mm long, and prominently tuberculate.

Phenology. The flowering time of var. *brevispinus* in the Trans-Pecos is from early April through most of May. Individual flowers open for 2–4 days in succession after partially closing at night. In the Trans-Pecos, the flowering time overlaps with that of var. *enneacanthus* and *E. stramineus*. For populations of var. *brevispinus* in South Texas and on the Edwards Plateau, the peak flowering time probably is earlier in the spring.

Sterile and Immature Specimens. The var. *brevispinus* is distinguished from *E. stramineus*, with which it is sympatric in part of its range in the eastern Trans-Pecos, by its less mounded habit, fewer ribs, and shorter spines. In var. *brevispinus* the central spines are angular or somewhat flattened in cross section. The larger, main centrals may be grooved or split lengthwise, apparently a character of age, as in var. *enneacanthus*. The main central spines are yellowish, brownish, or even gray-blue, 1–1.5 mm in diameter at the base, to 4.5 cm or more long (noticeably smaller than the spines in var. *enneacanthus* or in *E. stramineus*). The radial spines are straight, whitish, to tan or brownish, and often with dark tips. The radials are 0.2–0.5 mm in diameter at the base. Especially the centrals but also the radial spines have bulbous bases.

Biosystematics. Previous workers, including Moore (1967), Benson (1982), and Weniger (1984), have attempted to recognize two or more taxonomic entities in the populations treated here as var. *brevispinus* (Taylor 1985). Moore formally segregated three forms: f. *enneacanthus,* f. *intermedius,* and f. *brevispinus,* based on spine characters, mostly the length of central spines. In Moore's taxonomic concept, populations containing form *brevispinus* were restricted to the lower Rio Grande valley, extending north to San Antonio. Benson (1982) rejected Moore's form *intermedius* but raised her form *brevispinus* to varietal rank (still restricted to South Texas, mostly near the Rio Grande). Benson and Weniger distinguished *E. enneacanthus* "var. *enneacanthus*" as the most widely distributed smaller-spined entity, extending from South Texas into the Trans-Pecos; the plants with the smallest spines, designated by other authors as var. *brevispinus* in the narrowest taxonomic sense, seemed to be a form displayed by unhealthy individuals. Weniger recognized *E. enneacanthus* var. *carnosus* as a decumbent- and flabby-stemmed entity that produced large flowers with tepals in three whorls, which was restricted in distribution to "near Laredo and Eagle Pass, TX." Taylor brought all these short-spined eastern forms and varieties except for var. *carnosus* into synonymy under the name *E. enneacanthus* var. *brevispinus*, the epithet that had been used by Benson in reference to the shortest-spined of the short-spined varieties.

The shortest-spined forms of var. *brevispinus* are so conspicuously different in spine morphology as to suggest a separate taxon. Previous collection records and field observations cited by various authors also suggest, however, that the small-spined forms are not geographically isolated from "regular" var. *brevispinus* and that the small-spined forms might be the result of environmental rather than genetic factors. Our microscopic examination of the small-spined forms revealed that at least in some specimens, the spines are minutely scabrous, a trait not observed by us in "regular" var. *brevispinus*. Additional field observations from Val Verde County southward, and appropriate experimental tests, would be useful in elaborating the ecological and possible taxonomic status of the small-spined forms of var. *brevispinus*.

In the Trans-Pecos var. *brevispinus* makes geographic contact with both var. *enneacanthus* and *E. stramineus* (Breckenridge, 1981; Breckenridge and Miller, 1982). Breckenridge (1981) reported the discovery of three locations where "var. *enneacanthus*" (misapplied to var. *brevispinus*) and *E. stramineus* were sympatric: (1) in Upton County near Rankin and just east of the Pecos River, (2) in Pecos County near Longfellow, and (3) in Brewster County in Maravillas Canyon near the Rio Grande. Breckenridge reported suspected natural hybrids with *E. stramineus* at the Upton County location. Such hybrids, if they exist, presumably would be triploid and possibly sterile.

Echinocereus enneacanthus var. *brevispinus* is more or less sympatric with var. *enneacanthus* in Maravillas Canyon near the Rio Grande (Breckenridge and Miller, 1982), but they appear to intergrade morphologically. It is the apparent intergradation in Maravillas Canyon, the floral flavonoid similarities between var. *brevispinus* and var. *enneacanthus,* and the experimental crossability (Breckenridge and Miller, 1982) of these taxa that have so far provided some of the best evidence for interpreting var. *brevispinus* and var. *enneacanthus* as varieties of the same species. Populations of small plants identified as var. *brevispinus* extend past Maravillas Canyon farther up the Rio Grande to west of Boquillas Canyon in Big Bend National Park.

The epithet *enneacanthus*, long used erroneously for var. *brevispinus* in south-central and South Texas, technically applies only to the western taxon that was previously known by the name *dubius*. The type locality of the entity Engelmann described first as *E. enneacanthus* is in the west, where only the familiar "*E. dubius*" grows. The epithet *brevispinus* is the oldest available name at varietal rank for the small eastern plants, as pointed out by Taylor (1985), except that the poorly typified epithet *carnosus* would have priority by many years if it could be reliably identified as the same taxon.

Synonyms. *Cereus enneacanthus* Engelm.; *Echinocereus enneacanthus* Engelm.; ? *E. carnosus* Rümpler; ? *E. enneacanthus* var. *carnosus* (Rümpler) Quehl; *E. enneacanthus* var. *enneacanthus*, sensu L. D. Benson.

Common Names. Pitaya; pitalla; Mexican strawberry; mound pitaya.

8. **Echinocereus stramineus** (Engelm.) Rümpler var. **stramineus**. STRAWBERRY CACTUS. Plates 142, 143. [*Cereus stramineus* Engelm., Proc. Amer. Acad. Arts 3: 282. 1856]. Mostly on exposed limestone rock outcrops, in the desert mountains and arid S slopes of higher mountains, El Paso Co. SE through the southern Trans-Pecos and Big Bend region, SE to Terrell Co. and the lower Pecos River, possibly in Val Verde Co. (1,800–)2,500–5,000 ft. Flowering late Mar–May, occasionally into Jun. $2n = 44$. East to Upton Co. across the Pecos River; reported from Tom Green Co. (Benson, 1982). South-central NM, Doña Ana, Otero, and Eddy counties. Mexico, E Chihuahua, Coahuila, W edge of Nuevo León, NE Durango, N Zacatecas and N San Luis Potosí. Map 32.

In Trans-Pecos Texas *E. stramineus* is the "real" strawberry cactus, as

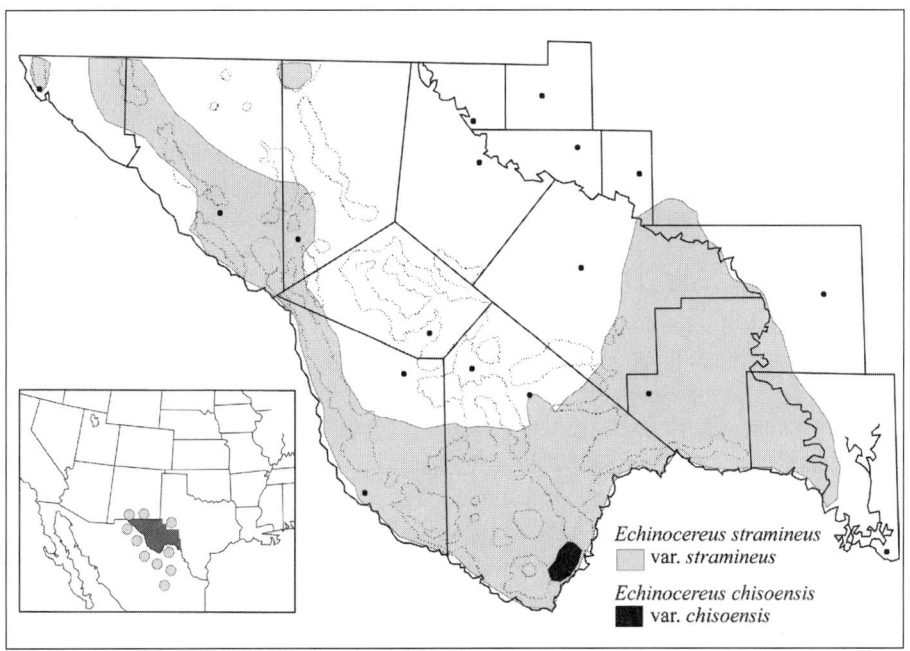

Map 32. Generalized distribution of *Echinocereus stramineus* var. *stramineus* (strawberry cactus) and *E. chisoensis* var. *chisoensis* (Chisos hedgehog cactus).

opposed to *E. enneacanthus*, because of its larger, juicier fruits that have the color, smell, and flavor of strawberries. Both *E. stramineus* and *E. enneacanthus* var. *enneacanthus* (= *dubius*) are of about equal abundance in deserts of the lower and eastern Big Bend area. The specific epithet is after the Latin *stramineus*, "made of straw," a reference to the dense covering of divergent, straw-colored spines that characterize the mounded plants.

Identifying Characters. The plants of *E. stramineus* are cespitose with numerous stems forming hemispheric mounds. In sexually mature plants, the number of stems may be 10–50, or in the largest plants, 100–350 stems. Plants ca. 1 m in diameter with up to 500 stems have been reported. The number of ribs per stem is one of the best distinguishing features of *E. stramineus*, especially when comparing this species with *E. enneacanthus*. *Echinocereus stramineus* has (10–)11–17 ribs per stem, but a populational average of ca. 12 ribs per stem (Breckenridge, 1981), compared to (6–)8–9 ribs per stem for *E. enneacanthus*. In *E. stramineus* the circular areoles are 7–15(–18) mm apart on the ribs. This relatively close spacing of areoles is one the factors that accounts for the stem surfaces being mostly obscured by spines. The areoles of *E. stramineus* display (1–)2–4 central spines that are (4–)5–9 cm long, terete to somewhat flattened, straight or slightly curved, and strongly projecting. There are 7–10(–15) radial spines that are (1.2–)2–3(–4.6) cm long and mostly acicular. The longest radial spines, to 4.6 cm, were observed in plants from lower Maravillas Canyon.

The magenta flowers of *E. stramineus* are noticeably larger than those of *E. enneacanthus*, at least in Trans-Pecos Texas. The flowers are 8.5–12.5 cm in length and diameter. Inner tepals are bright magenta with this color extending to the base of the throat, or in some plants the throat base may be darker, approaching deep red, or sometimes even lighter in color. Sporadic individual plants with pure white flowers have been reported from Big Bend National Park. Flowers develop relatively near the tips of the stems. Cespitose plants of *E. stramineus* are spectacular when in full sun and covered with completely open large flowers.

The tepals occur in about two whorls with showy inner tepals being broadly oblanceolate, to ca. 6 cm long, ca. 2.5 cm wide near the apex, and with the apical margins irregularly toothed. Each tepal is apiculate or notched. The filaments are 0.8–1 cm long and reddish. The anthers are yellow. The style is ca. 2.7 cm long and ca. 2.5 mm thick, reddish, and supports 10–13 green stigma lobes, these ca. 8 mm long. The nectar chamber is ca. 5 mm long.

The fruits are globular to broadly oval, (4.5–)5–6 cm long and nearly as broad, and reddish-brown when ripe. Bristly areoles on the fruit surface are easily dislodged. The copious pink to white flesh is a refreshingly cool treat on hot summer days of the fruiting season and usually has the strong smell and taste of strawberries. The seeds are black, ovoid, 1.2–1.5 mm long, and tuberculate. Normally the seeds are numerous in each fruit.

Phenology. The usual flowering time for *E. stramineus* is April to May, but occasional late flowering may occur in June or even July (Worthington, 1986). Individual flowers last for one to four days and partially close at night. In the Big

Bend area, plants closest to the Rio Grande (e.g., in lower Maravillas Canyon) usually initiate flowering earlier (e.g., early April) than plants at higher elevations away from the river. Flowering may be delayed by late freezes (observed but not thoroughly documented) or enhanced by timely precipitation after an unusually dry winter and spring. For example, plants were observed in full bloom at Big Bend Ranch State Park on about 22 June 1998, apparently prompted by earlier showers in the area (L. Hedges, pers. comm.). Fruits ripen one month or more after flowering. In 1993 mature fruits were abundant on 12 June in a population in the Christmas Mountains, southern Brewster County.

Sterile and Immature Specimens. *Echinocereus stramineus* is most frequently confused with densely clumped and mounded specimens of *E. enneacanthus* var. *enneacanthus* (= *dubius*). Where *E. stramineus* overlaps with *E. enneacanthus* var. *brevispinus* in the southeastern and eastern Trans-Pecos, identification of taxa in the field usually is easy because of distinct spine patterns that obscure the stem surface in *E. stramineus* and reveal the stem surface in var. *brevispinus*. *Echinocereus stramineus* is only superficially similar to *E. coccineus*, which has distinct spination, but forms equally large mounds of stems; these two species occasionally may be found together, but the differences are obvious in the field.

The typical mounded habit of *E. stramineus* is a good but not absolute field character for distinguishing the species from *E. enneacanthus,* which typically has fewer stems somewhat loosely arranged in sprawling, flat-topped, or low mounded clumps. We have observed numerous instances where individual plants of *E. enneacanthus* var. *enneacanthus* (= *dubius*) form dense, mounded clumps of smaller-than-normal stems, which brings their spines closer together, partially obscuring the stem surfaces. Such plants closely resemble *E. stramineus* until one counts the number of ribs or closely examines the spine pattern in individual areoles. *Echinocereus stramineus* features straight, mostly terete central spines, whereas var. *enneacanthus* usually has flattened or angular curved centrals.

The practice of examining several plants at any given site generally facilitates identification because the populations usually are segregated. The well-documented (Breckenridge and Miller, 1982; and others) ecological preferences of *E. stramineus* and *E. enneacanthus* var. *enneacanthus* provide another useful, but not absolute, aid to field identification. *Echinocereus stramineus* has been aptly described as saxicolous because of its typical association with stable, rocky substrates on desertic mountains, hills, slopes, mesas, canyons, or outcrops. Most commonly these are limestone, but in the most favorable climates its populations also thrive on basalt and other substrates of igneous origin. *Echinocereus enneacanthus* var. *enneacanthus* typically occurs in alluvial substrates originating from the *bajadas* or outwash fans of desertic mountains, to gravel hills and benches, and in silty or clayey *bolsones* or drainage deltas. The var. *enneacanthus* frequently occurs under the shade of shrubs or small trees, while *E. stramineus* is most commonly found in habitats open to full sun. Generally var. *enneacanthus* occurs at lower elevations than *E. stramineus,* but obviously the two taxa come in contact at numerous sites in the desert mosaic of the southern Trans-Pecos. One well-documented point of contact is in Maravillas Canyon near the Rio

Grande, where all three Texan taxa of the *E. enneacanthus* complex occur in close proximity. *Echinocereus stramineus* is the only member of the complex with its characteristic distribution above about 3,800 feet in the Trans-Pecos and in Mexico (Breckenridge and Miller, 1982).

In *E. stramineus* the spine bases are bulbous but not as prominently so as in var. *enneacanthus*, particularly the radial spines. The central spines of *E. stramineus* are smaller, ca. 1–1.8 mm in diameter above the bulbous base, and terete to somewhat flattened, but not angled or prominently flattened as in var. *enneacanthus*. Classically, the central spine color is "stramineous," as reflected by the specific epithet, but in age they bleach to glassy white, giving an overall stramineous or whitish aspect to the mounded plants. The central spines usually are microscopically black-speckled, and may even be brownish or rosy-pink with slightly darker tips, or even dark gray or brownish throughout on some older stems. Seemingly older central spines may crack or split lengthwise, a feature that is seen usually on the upper surface.

The radial spines of *E. stramineus* are obviously more slender, 0.3–0.4 mm in diameter at the base, than are the central spines. In Trans-Pecos populations usually there are 7–10 radials per areole, with the maximum number increasing to 7–15 radials in some Mexican populations. The radials are acicular or perhaps flattened, opaque-glassy, and minutely black-speckled. The longest radial spines are the lower 3–4 in each areole, with those in lateral and upper positions somewhat shorter.

Juvenile plants with one or a few stems are commonly found in populations of *E. stramineus* but may not be noticed among the larger, many-stemmed adult plants. The stems of these small plants look very much like the stems of adult plants. In younger areoles whitish matted or woolly hairs are more noticeable. In areoles of juveniles there are fewer central spines, or perhaps with one obvious central and 1–2 or more seemingly in the position of radial spines. In younger areoles there may be fewer radial spines, with these being shorter, especially the uppermost 1–3 in each areole.

Biosystematics. *Echinocereus stramineus* has been recognized by most workers as a distinct species, except by Benson (1982), who included the taxon as one of four varieties of *E. enneacanthus*. The most recent systematic study of the *E. enneacanthus* complex, including some of the Mexican populations, utilized morphological, ecological, and flavonoid chemistry data in supporting the specific status of *E. stramineus* (Breckenridge and Miller, 1982). This distinctiveness was further supported by the report of a tetraploid ($2n = 44$) chromosome number for the larger-flowered *E. stramineus* (Weedin et al., 1989) in a complex that otherwise appears to be diploid ($2n = 22$). Additional chromosome counts for *E. stramineus* in the Trans-Pecos (Powell and Weedin, 2001) have been consistently tetraploid.

In Mexico *E. stramineus* var. *occidentalis* N. P. Taylor replaces typical var. *stramineus* in a small region in Durango. Smaller-flowered plants from eastern Mexico (Coahuila and Nuevo León, between Saltillo and Monterrey) sometimes taxonomically segregated as *E. conglomeratus*, were allowed varietal rank by

Bravo-Hollis and Sánchez-Mejorada (1991a) as a variety of *E. stramineus*. Taylor (1985), Blum et al. (1998), and Zimmerman et al. (forthcoming) have dismissed *E. conglomeratus* as perhaps not worthy of even varietal status.

Plate 46 in *Cactaceae of the Boundary* (Engelmann, 1859) is labeled "*Cereus* (= *Echinocereus*) *stramineus*," but the taxon portrayed is *E. enneacanthus* (= *dubius*). Plate 47 of this publication, labeled "*stramineus*," includes four figures, two of which, figures 2 and 4, appear to be spine clusters of *E. enneacanthus* (= *dubius*).

Synonyms. *Echinocereus enneacanthus* Engelm. var. *stramineus* (Engelm.) L. D. Benson.

Common Names. Strawpile hedgehog; spinemound echinocereus; pitaya; organo; pitahaya; alicoche; straw-colored hedgehog. Strawberry cactus is widely used throughout the region. Pitaya is used for this species and for many unrelated genera, especially in Mexico.

9. **Echinocereus chisoensis** W. T. Marshall var. **chisoensis**, Cact. Succ. J. (U.S.) 12: 15, illus. cover, p. 1. 1940. CHISOS HEDGEHOG CACTUS. Plates 144, 145. Typically associated with nurse plants, these usually shrubs, but also dog cholla and tasajillo, gravel to sandy alluvium, flats, benches, small and larger drainages, near the margins or in the drainages protected by desertscrub. Known only from southern Brewster Co., Big Bend National Park, SE of the Chisos Mts. 1,900–2,600 ft. Flowering mid-Mar to mid-Apr. $2n = 22$. Although seemingly suitable localities abound in adjacent Coahuila, Mexico, *E. chisoensis* var. *chisoensis* has yet to be reported there. Map 32.

Taylor (1985) and Zimmerman et al. (forthcoming) recognized two varieties of *E. chisoensis*. Ours is var. *chisoensis*; the other is var. *fobeanus* (Oehme) N. P. Taylor, endemic to Mexico, documented only in the vicinity of San Pedro de las Colonias in southwestern Coahuila and sometimes speculatively mapped north of Gomez Palacio in nearby northeastern Durango. Benson (1982), followed by Bravo-Hollis and Sánchez-Mejorada (1991a), dismissed var. *fobeanus* as a synonym of var. *chisoensis*. Blum et al. (1998) recognized *E. fobeanus* as a distinct species, *E. fobeanus* Oehme, with one recently discovered additional subspecies, in Coahuila, Mexico.

The specific epithet is a reference to the Chisos Mountains whose impressive main peaks loom approximately 10 miles to the northwest of the only known range of var. *chisoensis*. *Echinocereus chisoensis* var. *chisoensis* occurs on the lower *bajadas* of the Chisos Mountains (i.e., alluvial outwash well below the base of the mountains near the lower Dead Horse Mountains, Lower Tornillo Creek, and the Rio Grande). *Echinocereus chisoensis* is listed as Threatened by both federal and state agencies.

Identifying Characters. Plants of *E. chisoensis* var. *chisoensis* are almost always associated with nurse plants, either under shrubs among grasses or in the mats of dog cholla, often among decaying stems toward the center of mats. The stems of older plants of var. *chisoensis* typically are branched, while smaller, presumably younger plants may have solitary stems. The stems are 5–20(–24) cm long, 3–5 cm in diameter, and are cylindroid or slightly conical. Healthy stems are light green to gray-green or bluish-green. Zones of darker and lighter spines lend a banded aspect to the stems of some plants. The stems have 11–16 ribs with slender tubercles 6–8 mm apart, these supporting small areoles with mostly divergent spines that only partially obscure the stem surface. Areoles at the stem apex usually are densely white-woolly, but most areoles down the stem are naked, or with some persistent wool. Scattered areoles downstem may be densely woolly, especially in enlarged flower-bearing areoles. The central spines are (1–)2–4 in number with the main 1–2 centrals dark reddish-brown, especially on the distal portions. Radial spines usually number 10–16, and are tan, ashy-white, or pinkish-gray, some of them with red-brown tips.

The magenta funnelform flowers are ca. 6 cm long, ca. 5 cm in diameter, and in the field sometimes do not open widely. We suspect that flowers may not open fully when they are shaded by nurse plants. Under cultivation flowers open widely in full sun. The receptacular tube has evenly spaced areoles with much white wool, and 8–14 bristlelike to hairlike spines, these white or red-brown at the tips. The woolly receptacles with hairlike spines resemble those of *E. reichenbachii*, which is closely related. The nectar chamber is 1–2 mm long and is mostly filled by the style base. The delicate inner tepals usually are in two whorls, oblong-oblanceolate, ca. 5 cm long and ca. 1.5 cm wide. The tepal color may be deep magenta or a lighter rose or pinkish, particularly toward the apex, perhaps with a darker midstripe. The throat of the flower usually has a broad or narrow, whitish (or at least pale) band above a dark red base. The stamens have white to pinkish filaments that are ca. 1.3 cm long, and pale yellow anthers. The style is ca. 2.5 cm long, whitish, exserted 4–5 mm above the stamens, with ca. 10 green stigma lobes, these 3–6 mm long.

The fruit is oblong to narrowly obovoid, 2.5–3.7 cm long, ca. 1.4 cm in diameter and more or less covered with wool and bristlelike spines. Maturing fruits remain covered by the woolly areoles, but ultimately the areoles are deciduous. The fruit is green to dull red and fleshy when ripe, but drying dull red or brownish-green, and sometimes eventually splitting open on one side. Fruit pulp is whitish and somewhat viscid or nearly dry. The seeds are ovoid, ca. 1.2 cm long, black, and strongly tuberculate.

Phenology. The typical flowering time for *E. chisoensis* var. *chisoensis* is from about mid-March to mid-April. Individual flowers open about mid-morning, partially close at night, and sometimes open again for 1–3 days. Plants in cultivation at Alpine produced flowers that opened for two days and then closed (April 1999). One survey conducted on 19 March 1996 (Denise A. Louie, National Park Service, Big Bend National Park) found that 64% of the *E. chisoensis* plants observed during a walking transect were in bud and 30% of the plants had mature flowers. In 1992 one of our field studies of *E. chisoensis* revealed that on 18 April all plants observed at one site had withered flowers and no flower buds. Fruits mature in about one month. During drought periods we have observed heavy herbivore damage of flower buds, flowers, and fruits.

Sterile and Immature Specimens. Without flowers or fruits *E. chisoensis* var.

chisoensis might be confused with *E. viridiflorus* var. *russanthus* and perhaps *E. dasyacanthus*. Two of the best field characters for *E. chisoensis* are the prominent white-woolly areoles, particularly those at and near the stem apex, and the spine features. Both *E. viridiflorus* var. *russanthus* and *E. dasyacanthus* may have some white wool in areoles near the apex, but it is not as prominent as the diffuse "puffy," white-woolly areoles in *E. chisoensis*. In *E. chisoensis* the (1–)2–4 central spines are of different sizes, with 1–2 main centrals decidedly more prominent than the other 1–2, and often only the main centrals are present. The centrals are 0.3 mm thick near the base, tapering evenly to a point. The 1–2 main central spines are deflexed or porrect, or both. The 10–16 radial spines are somewhat appressed in near-parallel position or somewhat spreading from the stem. The radials are slender, usually with only slightly enlarged bases, and they are of different lengths in even progression, with 1–3 of the longest spines at the bottom and the shortest spines at the top of the areole. The areoles are small, oval or elliptic, on the upper stem ca. 3 mm long and 2–2.5 mm in diameter, and usually bulging with white wool. Downstem the areoles are circular, ca. 2 mm in diameter, and naked, or some wool might be persistent or even abundant in a few areoles. In juvenile plants the slender, cylindroid stems are mostly solitary, and with conspicuous white-woolly areoles. In each areole there are 1–2 central spines, and there are fewer radial spines than in adults, although some of the upper radials are very short and barely visible through the wool. On younger stems the spines, particularly radials, are more slender and flexible. We have seen young stems 1–3 cm long established in dog cholla mats. Stems this size are more difficult to see under shrubs, in grasses, or in lechuguilla, all of which are abundant in the range of *E. chisoensis*.

Biosystematics. When W. Taylor Marshall described *E. chisoensis*, he characterized it as "very distinct" from any other species of the genus but speculated that it was related to *E. fendleri*. Benson (1969b, 1982) reduced *E. chisoensis* to a variety of *E. reichenbachii* without explanation, apparently being influenced by the common trait of woolly areoles at stem apexes and wool on the receptacular tube. Bravo-Hollis and Sánchez-Mejorada (1991a) followed Benson's classification without comment, but Weniger (1984), Taylor (1985), Blum et al. (1998), and Zimmerman et al. (forthcoming) considered *E. chisoensis* to be a distinct species.

Several unpublished surveys have been directed toward evaluating the populational status of *E. chisoensis*, with much of the information known about the taxon being discussed by Kennedy and Poole (1993), who prepared a Recovery Plan for *E. chisoensis* subsequent to its federal designation in 1988 as Threatened. The common name used in the federal listing was Chisos Mountain Hedgehog Cactus. The general consensus of cactus experts in 1993 was that *E. chisoensis* may be "losing ground" in its population size. Some experts have noted that there are fewer plants of *E. chisoensis* at sites where they were known to be more numerous before. Our own field observations in the mid-1990s concur with this consensus, in the sense that fewer plants of *E. chisoensis* are present at certain sites where we have informally monitored the species for years. Disappearance of the plants has coincided with a severe drought in the early and mid-1990s, but factors that are directly responsible for reduction in number of plants are not known. The population dynamics of E. chisoensis are in need of further study.

The original spelling of the epithet *chisoensis* by W. T. Marshall (1940) was "corrected" by L. D. Benson to reflect accurately the geographic basis of the name after the Chisos Mountains (*chisosensis*), but because Marshall's name was not considered to be an "orthographic error," the officially correct epithet remains *chisoensis* for now.

Synonym. E. reichenbachii var. chisosensis (W. T. Marshall) L. D. Benson.

Common Names. Chisos hedgehog; Chisos pitaya; Chisos Mountain hedgehog cactus.

10. **Echinocereus reichenbachii** (Terscheck) F. Haage. LACE CACTUS. Plates 146, 147. [*Echinocactus reichenbachii* Terscheck, Suppl. Cact. Verz. 2; in Walpers, Repert. Bot. Syst. 2: 320. 1843]. Eastern periphery of the Trans-Pecos in rocky limestone outcrops, or in shallow, rocky to sandy soils among shrubs, junipers, or grasses. Pecos Co., at extreme S point of the county and from Fort Stockton E to the Pecos River. Terrell Co., S of Sheffield, limestone mesas. Val Verde Co. near Pandale. 1,800–4,000 ft. Flowering Apr–May. $2n = 22$. In TX from Val Verde, Crockett, and Sutton counties NE to Taylor Co., N across the plains country and the Panhandle, and other varieties to sea level. Southeastern NM, SE CO, SW OK in the Great Plains grassland. Adjacent Mexico. Map 33.

Echinocereus reichenbachii was not reported by Benson (1982) and other workers to occur in the Trans-Pecos, although Weniger (1984) stated that var.

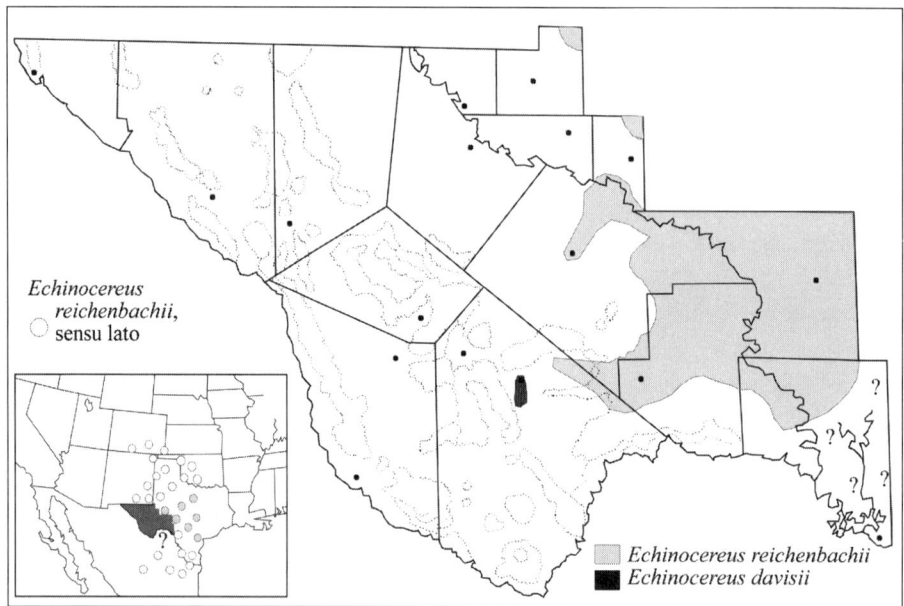

Map 33. Generalized distribution of *Echinocereus reichenbachii* (lace cactus) and *E. davisii* (dwarf hedgehog cactus).

perbellus (Britton & Rose) L. D. Benson occurred in Texas west to the Pecos River. The specific epithet honors the German naturalist H. G. L. Reichenbach.

Identifying Characters. The plants of *E. reichenbachii* are relatively small by comparison with most species in *Echinocereus,* with solitary or branching stems usually to 7–10(–15) cm long, occasionally longer, and 3–6.5(–7.5) cm in diameter. Mature plants often have 3–12 or more stems. The stems are subglobose to cylindroid and sometimes tapered toward the apex. The usually dark green surfaces are partially obscured by the spines, especially in dehydrated plants where the radial spine tips are drawn into an interlocking position. In *E. reichenbachii* as a whole there are (10–)13–15(–19) narrow ribs with low tubercles. The areoles are elliptic to oval, 2–3 mm long, and (2–)3–5 mm apart. Areoles at and near the stem apex are conspicuously white-woolly. Areoles downstem are naked, or mostly so. Perhaps the most distinctive vegetative feature is the pectinately arranged radial spines in a single row on each side of the elongated areole and tightly appressed against the stem. There are 15–21 radial spines and 0(–2) central spines. In most Trans-Pecos specimens examined so far, the central spine is absent, but in extreme south Pecos County 1–2 centrals are present in some plants. In the south Pecos County population, when two central spines are present, the lower one is porrect, and an upper one is ascending and often slightly longer at 1–2.2 mm long. The longest radial spines are 6–8 mm. Typically, the spines are ashy-white with pinkish, reddish, or black tips. According to Zimmerman et al. (forthcoming), white-spined populations occur on limestone, while other substrates support brown-spined plants.

The funnelform flowers of *E. reichenbachii* are 5–7.5 cm long and 5–6 cm wide. The receptacular tube is obscured by closely spaced areoles bearing much diffuse wool and dark brown, bristlelike spines. The white wool is also present on flower buds and persists through floral development to obscure the fruit as well. The nectar chamber is ca. 4 mm long and mostly filled by the style base. The inner tepals, in 2–4 whorls, are magenta or deep pinkish with dark red bases, are oblanceolate, to 4.5 cm long, 0.7–1.3 cm wide, and apiculate and entire or lacerate at the apex. The anthers are cream-yellow, on filaments to 2 cm long. The style is pinkish, to 3.5 cm long and ca. 3 mm in diameter, with 8–20 stigma lobes, these 5–6 mm long.

The fruits are globose to ovoid, 1.5–2.5(–3) cm long, and covered with wool and slender spines. The ripened pericarpel remains permanently green, ultimately splitting longitudinally to reveal a fleshy white pulp that quickly dries. Areoles on the fruit wall ultimately are deciduous. The seeds are ovoid, ca. 1.5 mm long, and strongly tuberculate.

Phenology. Although flowers usually open for one day only, anthesis is often staggered one or more days apart so that whole plants are in bloom over a period of a week or more. Individual flowers open about midday and close before sundown. In the Trans-Pecos the usual spring flowering time for *E. reichenbachii* appears to be from early April to mid-May. In cultivation, at least, plants may also flower periodically in the summer as if responding opportunistically to periodic rainfall.

Sterile and Immature Specimens. In the southeastern Trans-Pecos *E. reichenbachii* might be confused with *E. pectinatus* var. *wenigeri,* which also has pectinate white spines and woolly areoles at the stem apex. Without its woolly, finely bristly, thin-walled, ephemeral flowers, our geographic race of *E. reichenbachii* usually can be distinguished from *E. pectinatus* by its complete lack of central spines (or 1–2 very short ones present), thinner and shorter radial spines, and narrower, elliptic-oblong areoles. In our region the radial spines of *E. reichenbachii* are appressed (but elsewhere they sometimes project strongly, as in the Oklahoma populations), straight, slender, acicular, and only slightly bulbous basally. The longest radial spines (6–8 mm) are the lateral or lateral and lower ones. The lower 2–4 radial spines often are shorter than the lateral ones but longer than the uppermost radial spines, where one or more may be almost rudimentary. The largest plants we have seen in the Trans-Pecos were from extreme south Pecos County, where one stem measured 15 cm long and 7.5 cm in diameter.

Immature specimens, including seedlings only a few months old, already exhibit 8–12 ribs, the distinctive woolly apical areoles, and pectinate arrangements of glabrous radial spines. The spines in such young plants are shorter and sometimes fewer in number than in older plants.

Biosystematics. Recent workers list up to 13 (Blum et al., 1998) taxa for the *E. reichenbachii* complex, which extends geographically from Colorado and Oklahoma south through eastern New Mexico and much of Texas into northeastern Mexico (mostly Coahuila, Nuevo León, and Tamaulipas). Taxa of the *E. reichenbachii* complex grow from sea level (in southern Texas) to elevations above 4,000 feet.

Taylor (1985) explained the use by Weniger (1984) of the name *E. caespitosus* for *E. reichenbachii,* and he either recognized or listed in synonymy all of the names used by Benson (1982) and Weniger in reference to the complex. In Di Martino (1996) most of the varieties of *E. reichenbachii* are treated as species. The photographs of *E. reichenbachii* (Plates 146–49) represent flower, fruit, and vegetative morphology typical of the complex, as recognized by some authors (see previous discussion): Plate 146 resembles var. *reichenbachii* (of south-central Texas north to the Panhandle, adjacent Oklahoma, northeastern Mexico); Plate 147 is similar to var. *perbellus* (central and northern Texas, eastern New Mexico, and southern Colorado); Plate 148 is var. *fitchii* (South Texas); and Plate 149 is var. *baileyi* (eastern Texas Panhandle and adjacent southern Oklahoma).

The taxonomy used by Blum et al. (1998) draws an undefined taxonomic line at subspecific rank between seemingly identical populations a short distance east of the lower Pecos River. The Blum et al. concept of *E. reichenbachii,* sensu stricto, extends far beyond the restricted taxonomic circumscription of var. *reichenbachii* in *Chihuahuan Desert Flora* (Zimmerman et al., forthcoming), but not as far as that of Benson (1982).

According to Blum et al. (1998) and *Flora of North America* (Zimmerman and Parfitt, forthcoming), *E. reichenbachii* subsp. (or var.) *caespitosus* is the only taxon of the complex that occurs in Central Texas, and our Trans-Pecos plants appear to represent the common Central Texas taxon, which grows all across the Edwards Plateau and its surrounds, with only localized variation. The Blum et al. map-concept of subsp. *reichenbachii* extends far from its type locality near Saltillo, Mexico, north into Texas south of Interstate 10, but no evidence has been seen for the brownish-spined, weakly tuberous-rooted Saltillo taxon anywhere in the United States. In the Trans-Pecos *E. reichenbachii* appears mostly restricted to limestone habitats along and near the Pecos River in Pecos, Terrell, and Val Verde counties, although it has also been reported from "vacant lots in Ft. Stockton," about 30 miles west of the Pecos River. It was documented by Jim Talbot in 1997 and more recently by three field investigators who have shown us specimens near the Terrell and Brewster county lines in extreme south Pecos County, where the plants resemble var. *perbellus*, sensu Benson (1982).

Synonyms. *Echinocereus caespitosus* Engelm.; *E. perbellus* Britton & Rose; *E. reichenbachii* (Terscheck) Britton & Rose var. *perbellus* (Britton & Rose) L. D. Benson; *E. caespitosus* (Engelm.) Engelm. var. *perbellus* (Britton & Rose) Weniger; *E. reichenbachii* subsp. *perbellus* (Britton & Rose) N. P. Taylor; see Blum et al. (1998) for additional synonymy.

Common Names. Beautiful lace cactus; purple candle; Classen's cactus; lace hedgehog.

11. **Echinocereus davisii** Houghton, Cact. Succ. J. (U.S.) 2: 466. 1931. DWARF HEDGEHOG CACTUS. Plates 150–52. Trans-Pecos endemic, restricted to the Caballos Novaculite (and adjacent exposures) in the Marathon Basin in northern Brewster Co., TX; the small plants are often found growing in and partially hidden beneath mats of prostrate *Selaginella* P. Beauv. (inaccurately called "moss" in some of the literature). 3,900–4,400 ft. Flowering (Feb–)Mar–Apr. $2n = 22$. Map 33.

The substrate in which *E. davisii* grows is derived from (and filled with chips of) novaculite, a pale, very dense, fine-grained rock material that is almost pure silica and is essentially chert. Novaculite is late Devonian (about 370 million years) in age and is thus much older than the limestone and igneous formations that overlie the novaculite. In geologic history the novaculite stratum was tilted upward during the formation of a large dome and subsequently exposed as the dome eroded. The Marathon Basin is the name of the area approximately 30–40 miles in diameter inside the eroded dome. Caballos Novaculite is the geologic name for the whitish, tan, or rusty-colored formation that is exposed at various sites in the Marathon Basin, often as ridges, steep "hogback" ridgetops, or strata in hillsides.

The specific epithet chosen by A. D. Houghton honors A. R. Davis, who discovered the taxon. *Echinocereus davisii* is listed as an Endangered species by the state of Texas and the federal government.

Identifying Characters. *Echinocereus davisii* is one of the smallest cacti in the world. The globose to ovoid stems are usually solitary, 1–2(–3.5) cm long, and 0.8–1.2(–2.5) cm in diameter. Branched plants are found occasionally, although in the field it is not always apparent if adjacent stems are of separate plants or truly connected. The spines partially or mostly obscure a dark green stem surface. Usually all of the spines are radial, although Leuck (1980) found that sporadic solitary central spines (to 1–1.2 cm long) are present in areoles of some older plants.

The greenish-yellow flowers of *E. davisii*, which are 1.8–2.5 cm long and 1.5–2(–2.7) cm wide, frequently are larger than the stems. The flowers may not open wider than 1.5–2 cm, or the tepals may recurve and become revolute, apparently only in full sun. Leuck (1980) described the flowers as having a weak, lemony scent. Irrespective of its strength, the scent is always detectable (at least on warm, calm days) and always described as lemony. The inner tepals are glossy, 1–1.5 cm long, 2.6–3.8 mm wide, linear to oblanceolate, entire and rounded, and emarginate or acute at the apex. The outer tepals usually are reddish in the midline. Taylor (1985) contended that the flowers of *E. davisii* (and *E. viridiflorus* var. *viridiflorus*, and by inference var. *correllii*) are distinctive in that (1) they all share the lemony scent, (2) the inner tepals are reflexed, widely opened, and (3) the inner tepal tips are bluntly rounded or even emarginate. The filaments are pale green and 5–9 mm long. The anthers are yellow and 0.5–1.2 mm long. The style is pale green, ca. 1.4 cm long, and exserted above the anthers. The 5–7 short (1.5–2 mm long) stigma lobes are green and erect or spreading. A small nectar chamber, ca. 1 mm long, is mostly filled by the style base. The pericarpel has about 10–14 areoles each with ca. 12 slender, short, minutely plumose white spines, and no wool.

Mature fruits are green or reddish-brown, 6–9(–11) cm long, and 4–5.5 mm in diameter. At maturity the fruits split longitudinally and immediately dry out, and the areoles are deciduous. The black seeds are 0.9–1 mm long, with finely tuberculate surfaces. Each small fruit of *E. davisii* contains about 40 or fewer seeds.

Phenology. This is the earliest-blooming *Echinocereus* in our flora. The classic flowering time for *E. davisii* is mid-March, although in some years flowering may begin in February and extend to mid-April or a little longer. We have observed cultivated specimens, placed in a warm greenhouse following several days of cold temperatures, that initiated flower buds in mid-December, and plants in the Sul Ross Cactus Garden in first flower on 25 January 1999, during an exceptionally warm winter. Leuck (1980) studied flowering phenology for the *E. viridiflorus* complex, including *E. davisii*, and reported that the flowering period typically lasts 10–20 days but that this period may be extended under favorable climatic conditions, particularly moist periods. Plants in the Sul Ross Cactus Garden were in full bloom for a second time on 13 April 1999, 17 days after the first and only rainfall of the year to date (about one inch). Flowers open first on the south sides of the plants. Flowers open rapidly, within 30 minutes, in late morning, and remain open all day, but maximum expansion of flowers occurs only on warm days in full sun and may be delayed or retarded by cool or cloudy conditions. Flowers close at night and usually open again the next morning for 3–4 days.

Leuck (1980) reviewed pollination biology and experimental hybridization results for *E. davisii*. The flowers are self-incompatible. Stigma receptivity closely follows anthesis. Pollination is by solitary halictid bees that begin visits within two hours after anthesis.

Dennie Miller of the Chihuahuan Desert Research Institute reported that when he propagated *E. davisii* from seed in 1991, several very small plants produced flowers during the first year of growth (on 26 February 1992). Each of the flowers opened for 2–3 days (i.e., slightly shorter-lived than the flowers of older, larger plants).

Sterile and Immature Specimens. Sterile specimens of *E. davisii,* and even those in full fruit, are difficult to locate in the field because the stems aboveground are only 1–2 cm tall and because characteristically the plants are partially, at times completely, covered by mats of *Selaginella*. Most of the stem is underground anyway, with as little as 10% projecting aboveground (Leuck, 1980). Plants of *E. davisii* also are found among rocks and under shrubs. During dry periods the stems shrink, pulling plants even farther into the ground. Obviously, the best way to locate *E. davisii* in the field is to look for them at flowering time.

Only the healthy adults bearing traces of flowers or fruits can be recognized as the truly dwarf plants that they are. Otherwise, they are just plain small cacti, like seedlings of some ordinary larger species. In the novaculite substrates where it grows there are two other diminutive endemic species, *Coryphantha minima* and *C. hesteri,* which are easily distinguished by spine characters. Otherwise, identification requires comparison with the seedlings and immatures of all the other, larger species. In cultivation the stems of *E. davisii* can become much larger, to 5–8 cm high and 4–6 cm in diameter, but eventually they branch into compact clumps (a cespitose growth habit), never forming solitary large cylinders. The stems have 6–9(–10) low ribs that are almost completely divided into prominently raised tubercles 2–2.5 mm high. In fact, the ribs are so deeply divided into tubercles that the plant is not immediately recognizable as an *Echinocereus*. When sterile plants are located, they are identifiable only by very careful attention to detail. The areoles are elliptical, 1.5–2.5 mm long, and naked or with a little wool. The 8–15 radial spines are subpectinately arranged, with the upper ones being the shortest, and the lateral ones the longest at (0.8–)1.2–2.5(–1.9) cm. The spines are stout for such small stems, ca. 0.5 mm in basal diameter, terete or slightly flattened, gray to ashy-white with brown or reddish tips, or they are whitish throughout. Some spines in the apical areoles may be reddish throughout or nearly black. The radial spines are straight on seedlings and many of the adults, but those of the larger and/or older plants may curve in any direction, unlike most other small cacti in the region (but see *Ancisrocactus*). The spines are minutely plumose (barbellate), a trait that is most visible on younger spines at high magnification. Weathered spines, or those of relatively old individuals, may not show the minute barbs. Benson (1982) pointed out that the spine clusters in areoles low on the stem (i.e., those persistent from early in life) differ markedly from those on the upper stem (i.e., those produced after reaching full maturity). The first areoles formed, the lower ones, produce smaller spines, whereas the later formed areoles (the only ones visible aboveground, in the wild) produce the larger spines described above. Of course, to some extent this is true of almost any cactus, and the transition to adult-type spines is subtle and gradual by comparison with the dramatically heteromorphic growth of *E. viridiflorus* vars. *neocapillus* and *canus*.

Juvenile specimens have spine clusters all similar to those persisting in lower areoles of the youngest adults. Seedlings several months old have small, minutely plumose, pectinately arranged spines that are white and appressed to the stem.

Biosystematics. Clearly, *E. davisii* is part of the *E. viridiflorus* complex, either a closely related but distinct species or a variety of *E. viridiflorus*. Leuck (1980), who completed a biosystematic study of the *E. viridiflorus* complex, Weniger (1984), and Zimmerman et al. (forthcoming) treated *E. davisii* as distinct. Benson (1969, 1982), followed by Taylor (1985), reduced the rank of *E. davisii* to a variety of *E. viridiflorus*, presumably closest to *E. viridiflorus* var. *viridiflorus*, which has the most similar flowers. Leuck and Miller (1982) showed that all members of the *E. viridiflorus* complex share an identical profile of tepal flavonoids (i.e., that the floral flavonoid pigments are of zero utility for taxonomic resolution within this group).

In most years the early flowering time, March, for *E. davisii* does not overlap with later flowering times, April–May, for the other members of the *E. viridiflorus* complex or any other regionally sympatric species of *Echinocereus*. Taxa growing within six miles of *E. davisii* include *E. viridiflorus* var. *neocapillus*, *E. viridiflorus* var. *cylindricus*, *E. viridiflorus* var. *correllii*, *E. dasyacanthus*, *E. coccineus*, *E. stramineus*, and possibly *E. reichenbachii*. The earlier blooming season of *E. davisii* (resulting in reproductive isolation), diminutive stems, large underground portion of the stem, 6–9 ribs (rib number is 10–18 in other members of the complex), distinctive raised tubercles, and spine characters all separate *E. davisii* from other members of the complex. No natural hybrids of *E. davisii* (or any other sort of intermediates) have been reported, despite the close proximity of its close relatives north, west, and south of its limited range. In addition to its phenological separation from all other taxa, its interfertility is relatively low in experimental crosses with other members of the complex (Leuck, 1980). However, F_1 hybrids are easily produced in the greenhouse.

Synonym. *Echinocereus viridiflorus* Engelm. var. *davisii* (Houghton) W. T. Marshall. Marshall's recombination (in 1941) was valid even though he failed to cite the basionym, because full basionym citations were not required by the Code until after 1 January 1953 (Article 33.2).

Common Names. Davis' dwarf hedgehog cactus; Davis' green pitaya; Davis hedge cactus.

12. **Echinocereus viridiflorus** Engelm., Mem. Tour N. Mex. 91. 1848. GREEN-FLOWERED HEDGEHOG CACTUS and its allies. Plates 153–73. Mountain to desert habitats, sedimentary and igneous substrates, Trans-Pecos

TX and adjacent Mexico N through NM, TX Panhandle (also in Coke Co.), OK Panhandle, central and E CO, extreme W KS and NE, SE WY, and SW SD. 2,100–9,000 ft. Flowering (Feb–)Mar–Jun. $2n = 22$. Maps 34–36.

About 12 varieties of *E. viridiflorus* are known, 7 of them in the Trans-Pecos.

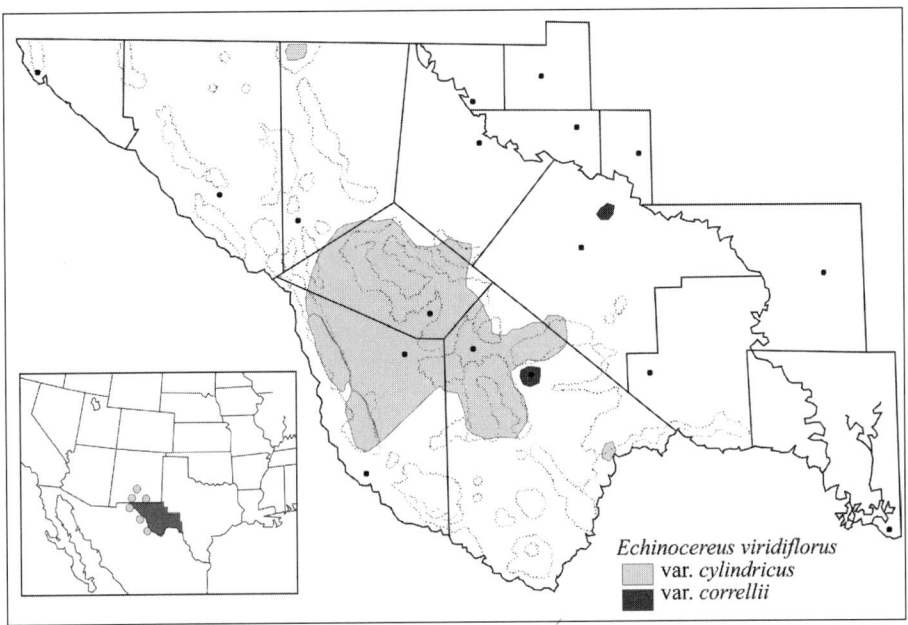

Map 34. Generalized distribution of *Echinocereus viridiflorus* var. *cylindricus* (small-flowered hedgehog cactus) and *E. viridiflorus* var. *correllii* (Correll's green-flowered hedgehog cactus).

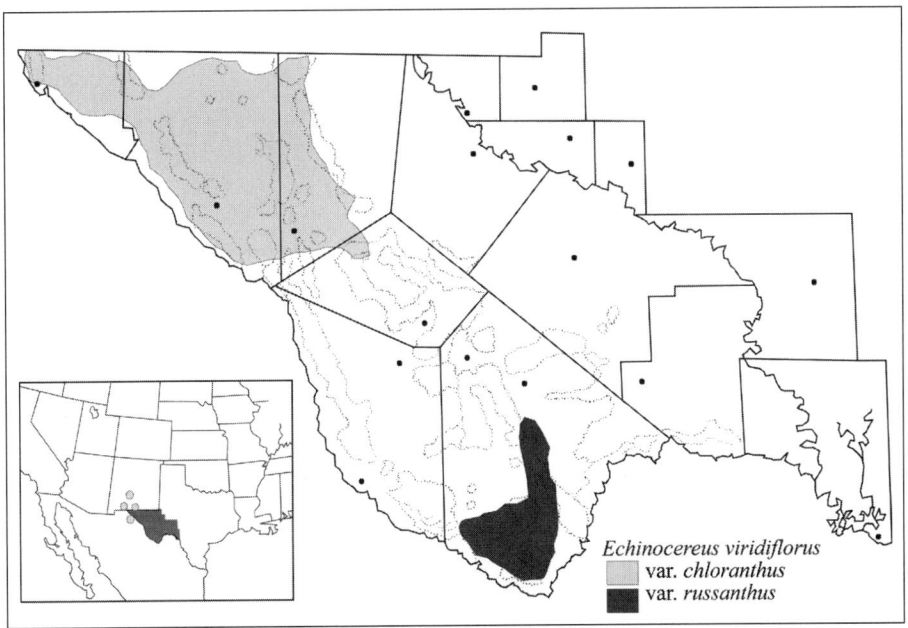

Map 35. Generalized distribution of *Echinocereus viridiflorus* var. *chloranthus* (western green-flowered hedgehog cactus) and *E. viridiflorus* var. *russanthus* (rusty hedgehog cactus).

Echinocereus 251

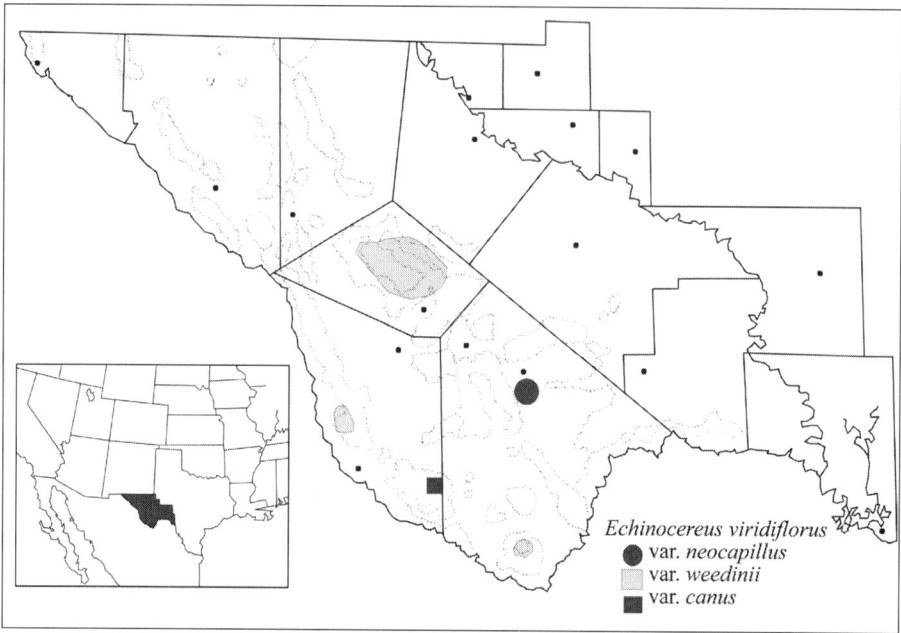

Map 36. Generalized distribution of *Echinocereus viridiflorus* var. *neocapillus* (Weniger's small-flowered hedgehog cactus), *E. viridiflorus* var. *weedinii* (Weedin's small-flowered hedgehog cactus), and *E. viridiflorus* var. *canus* (graybeard cactus).

Varieties not known to occur in the Trans-Pecos are listed in Appendix 2, including var. *viridiflorus* (Plate 153), which is found in the Texas Panhandle. The specific epithet is after the Latin *viridis*, "green," and *floris*, "flower," in reference to the green flowers of the typical variety.

Previous workers have referred to the taxa associated with *E. viridiflorus* as the *E. viridiflorus* complex (Leuck, 1980) or the small-flowered echinocerei (Weniger, 1969). One distinguished cactus specialist recently called the *E. viridiflorus* complex a "real can of worms," a tongue-in-cheek reference to the widely recognized morphological and populational variability, apparent intergradation, and resulting taxonomic uncertainty involving especially the Trans-Pecos members of the complex. Previous authors who dealt with the entire *E. viridiflorus* complex have recognized anywhere from one to six species. Leuck (1980) maintained *E. davisii* but lumped everything else into *E. viridiflorus*. Benson (1982) also visualized two species, but very different from those taxa recognized by Leuck: *E. viridiflorus* with four varieties (including *davisii*) and *E. chloranthus* with two varieties (*chloranthus* and *neocapillus*). Weniger (1969, 1984) advocated using four taxonomic species to organize the small-flowered echinocerei: (1) *E. viridiflorus* with three varieties (*viridiflorus*, *cylindricus*, and *standleyi* [including Benson's var. *correllii*]); (2) *E. davisii*; (3) *E. chloranthus* with two varieties (*chloranthus* and *neocapillus*); and (4) *E. russanthus*, a taxonomic segregate of *E. chloranthus*. Taylor (1985) organized the complex into two species: *E. chloranthus* with four varieties (*chloranthus*, *russanthus*, *cylindricus*, and *neo*-

capillus); and *E. viridiflorus* and three varieties (*viridiflorus, correllii,* and *davisii*). Among other characters, Taylor relied on the spurious long-wave UV absorption information provided by Leuck and Miller (1982), by which "target flowers" (inner tepal tips reflecting UV, flower centers absorbing UV) were inferred from photographs of *viridiflorus, correllii,* and *davisii,* and uniformly absorbing flowers were obvious from the much better photographs of *neocapillus, russanthus,* and *cylindricus*. Most recently, Blum et al. (1998) perceived six species (12 taxa in all, including some undiscovered until recently) in the *E. viridiflorus* complex. This arrangement included *E. davisii, E. viridiflorus* (subsp. *viridiflorus*; subsp. *correllii*), *E. chloranthus* (subspp. *chloranthus, cylindricus,* and *rhyolithensis* W. Blum & Mich. Lange), *E. russanthus* (subspp. *russanthus, fiehnii,* and *weedinii*), *E. neocapillus,* and *E. carmenensis*. Taxonomic novelties relative to previous classifications included *E. chloranthus* subsp. *rhyolithensis,* which was described as a phenotypic "mixture of *E. russanthus* and *E. chloranthus*" in southwest New Mexico; *E. russanthus* subsp. *fiehnii* in central Chihuahua, Mexico; *E. carmenensis* of northern Coahuila, Mexico; and an undescribed population in Coke County, Texas, known since the 1970s (Horst Kuenzler, pers. comm. to A. D. Zimmerman ca. 1972; Leuck, 1980) but never satisfactorily documented.

Artificial hybridization data demonstrate that all members of the *E. viridiflorus* complex are highly interfertile, except for *E. davisii*, which exhibits some degree of reproductive isolation (Leuck, 1980; A. M. Powell, unpub.). In the present treatment of the *E. viridiflorus* complex, two species are recognized, *E. viridiflorus* and *E. davisii*, in large part following Leuck (1980) and Zimmerman et al. (forthcoming). The seven varieties of *E. viridiflorus* recognized in our geographic area are vars. *correllii, cylindricus, chloranthus, russanthus, neocapillus, weedinii,* and (newly described) *canus*. Benson (1982) thought it feasible to recognize two groups as distinct species in the *E. viridiflorus* complex, based primarily upon spine characters, as was the preference of most previous authors. However, the spine characters of var. *chloranthus* bridge the phenetic "gap" that would otherwise exist between var. *cylindricus* (0–2 central spines, rarely three) and an informally recognizable "*russanthus* group," the taxa sharing high numbers of central spines.

Weniger (1969), Taylor (1985, 1988), Blum et al. (1998), and others have variously advocated or rejected the possibility that the *E. viridiflorus* complex is most closely related to the still poorly known *E. longisetus* (Engelm.) Lem. and its allies, of eastern Coahuila and Nuevo León, Mexico. *Echinocereus longisetus* and relatives are superficially somewhat similar to the *E. viridiflorus* complex, but they are much larger, have magenta flowers, and are unrelated. The known distribution of *E. longisetus* (Weniger, 1969) approaches the Rio Grande, and it might eventually be found in the Trans-Pecos, although there is no reliable report of *E. longisetus* in the United States. Taylor (1988) eventually advocated the much more strongly supported possibility that the true relationship of the *E. longisetus* group is with *E. stramineus* and not the *E. viridiflorus* complex, but Blum et al. (1998), without explanation, still mention *E. rayonesensis* N. P. Taylor, obviously part of the *E. longisetus* species-group, as possibly the closest ally of the *E. viridiflorus* complex.

Key to the Varieties

1. Central spines 0–3, stout or relatively stout, when present, with prominent bulbous bases; radial spines 14–23(–28), pectinately arranged, relatively stout, usually laterally compressed at their bases (2).
1. Central spines 6–12, relatively thin and flexible, with smallish-bulbous bases; radial spines (25–)30–45, appressed or spreading, slender, flexible, usually terete at their bases (4).

2(1). Central spines three or more per areole [will also key under second lead 1 above]; lower central spine usually white, mostly white or white on the upper surface, (1.5–)2.5–4.3 cm long, directed downward but usually curving upward; Culberson Co. W to El Paso Co. and S NM; (in part)
$\qquad$ 12c. *E. viridiflorus* var. *chloranthus*.

2. Central spines 0–3; lower central, when present, colored like the other central spines (yellow, reddish, or reddish-brown, rarely whitish), 0.7–1.4 cm long, directed downward or porrect, straight or slightly curved; mostly Jeff Davis and northern Brewster counties (3).

3(2). Plants all or mostly with a red or red-brown and white aspect; flowers usually unscented, typically opening only about 45°, of any color; radial spines 14–23; a familiar cactus of the Davis Mts and elsewhere
$\qquad$ 12a. *E. viridiflorus* var. *cylindricus*.

3. Plants all with a greenish-yellow and tan or ashy-white aspect; flowers lemon-scented, widely opening, all yellow-green; radial spines 19–28; mostly restricted in the northern Marathon Basin
$\qquad$ 12b. *E. viridiflorus* var. *correllii*.

4(1). Stems clothed in white spines; colored spine tips, when present, all red-purple (no yellow color-form present); seedlings with long, white, hair-like spines and not stiff spines in the areoles; flowers bright, light green; distribution Solitario Dome in SE Presidio Co.
$\qquad$ 12g. *E. viridiflorus* var. *canus*.

4. Not as above; either spines are fewer, or yellow color-forms present, or seedlings normal, or flowers varying to red/brown; distribution not in the Solitario (5).

5(4). Central spines 5–12 or fewer [will also key under first lead 1 above], relatively thin and flexible, acicular, with smallish-bulbous bases; radial spines (25–)30–45, appressed or spreading, slender, flexible, usually terete at the base; (in part when spines are numerous)
$\qquad$ 12c. *E. viridiflorus* var. *chloranthus*.

5. Not as above (6).

6(5). Flowers greenish-brown; seedlings with white, hairlike spines and not stiff spines in areoles; stems usually with sharply defined horizontal bands of darker/lighter spines; yellow spine-color phase present in every population; in Marathon Basin
           12e. *E. viridiflorus* var. *neocapillus*.

6. Not as above; horizontal bands of contrasting spine colors vague or none (7).

7(6). Flowers dull green or greenish-brown; stems with yellow, white, or red-tipped spines; higher elevations in Davis Mts and at Cattail Falls, Chisos Mts
           12f. *E. viridiflorus* var. *weedinii*.

7. Flowers reddish to reddish-brown; stems with reddish, brownish, white, or black spines, yellow spine-color phase absent except on W slope of Chisos Mts (where in contact with var. *weedinii*); widespread in southern Brewster Co.
           12d. *E. viridiflorus* var. *russanthus*.

12a. **Echinocereus viridiflorus** var. **cylindricus** (Engelm.) Engelm. ex Rümpler. SMALL-FLOWERED HEDGEHOG CACTUS. Plates 154, 155. [*Cereus viridiflorus* (Engelm.) Emgem. var. *cylindricus* Engelm., Proc. Amer. Acad. Arts 3: 22. 1856]. Widespread in both rocky and alluvial substrates of the Davis Mts (Jeff Davis, Brewster, and Presidio counties), SE to the Glass Mts, S through Alpine to the northern Del Norte Mts, W to the Chinati Mts and Sierra Vieja, with outliers at Reagan Canyon in Brewster Co. and the Guadalupe Mts, Culberson Co. (2,100–)4,000–5,800 ft. Flowering Apr–May. $2n = 22$. Also Eddy, Chavez, Lincoln, Otero, and Doña Ana counties, S NM. Mexico, NE Chihuahua. Map 34.

No modern authors have given full specific rank to this taxon. The taxonomic "splitters" who recognize *E. chloranthus* at the level of species have either placed var. *cylindricus* there (e.g., Taylor, 1985), or they have left *cylindricus* as a variety of *E. viridiflorus* instead (e.g., Benson, 1969b, 1982). The varietal epithet *cylindricus* is from the Greek *kylindros*, "cylinder," alluding to the characteristic stem shape in the taxon.

Identifying Characters. The stems of var. *cylindricus* usually are unbranched except in relatively old age (then up to 10 or more branches), to 20 cm long and 5–8 cm in diameter. The ribs number 12–17.

Like all other varieties of *E. viridiflorus*, but visibly different from *E. dasyacanthus*, the flowers (buds, fruits) are produced at midstem, or between midstem and the apex. During much of the year, traces of these flowers and fruits are visible (at least a few spine clusters). The flowers are funnelform, usually do not open fully, and may vary in color from amber or sulfur-yellow to greenish-brown, reddish-brown, or carmine. The flowers are 2–3.3 cm long. The inner tepals are 3–5 mm wide and rounded, apiculate, subacute, or acute-acuminate at their tips. The filaments are 7–8 mm long, and the yellow anthers 0.8–1.5 mm long. The style is green, 7–16 mm long, and it supports ca. nine green stigma lobes that are 1.5–3 mm long. The shallow nectar chamber is 1–1.5 mm deep and is mostly filled by the broad style base. The flowers of var. *cylindricus* are unscented (mostly if not always) in contrast with the lemony scent characteristic of *E. davisii*, *E. viridiflorus* var. *viridiflorus*, var. *correllii*, and some other taxa in this group.

The fruits of var. *cylindricus* mostly remain dark green, but those of some plants may turn dull red; they are somewhat fleshy with a sugary pulp (Leuck, 1980), oval to nearly round, 0.9–1.7 cm long, 0.8–1.2 cm in diameter, and indehiscent. The seeds are about 1–1.2 mm long.

Phenology. The usual flowering period for *E. viridiflorus* var. *cylindricus* is April to May. An individual plant may produce from one to more than 50 flowers in a season. Characteristically, flowers on the south side of a plant mature first. In a given population, the flowering period usually lasts 12–20 days, but the period may be extended by optimum conditions including precipitation. After extended dry periods plants do not flower or do not produce as many flowers as usual. Following rains in 1996, which finally arrived after a severe drought of five years, var. *cylindricus* was observed to flower during 8–30 June in populations from north of Fort Davis to south of Alpine. Late freezes may also extend the length of the spring flowering period. Fruits usually mature by May or June and may persist into late June. Under cultivation in Alpine in 1998, all of the varieties of *E. viridiflorus* developed mature green or reddish fruits by 21 May. Populations in the southern Davis Mountains have been observed to produce mostly green fruits, with relatively few ultimately turning dull red. Other populations throughout the range, at least in some years, may produce predominantly reddish (or dark purple) fruits (Leuck, 1980).

Sterile and Immature Specimens. Plants of var. *cylindricus* superficially resemble those of *E. dasyacanthus*, and these taxa may occur together at intermediate elevations in and near the Davis Mountains, where peripheral populations of var. *cylindricus* grow unusually near the relatively arid habitats of *E. dasyacanthus*. The more brightly red- and white-banded aspect of typical var. *cylindricus* usually allows easy distinction from a distance. The central spines are much more diagnostic: 0–3 (individually conspicuous) in a vertical row in var. *cylindricus* and 2–5 or more (individually inconspicuous, either very short or spreading in all directions) in *E. dasyacanthus*. The radial spines are about the same number in var. *cylindricus* and *E. dasyacanthus* (14–25), but in var. *cylindricus* the radials are more distinctly pectinate, 8–11 mm long, very variable in color but usually having relatively brighter, more distinctly purplish-red tips.

In var. *cylindricus*, areoles are elliptic to oblong or oval (rarely round), 3–4(–5) mm long, 2–2.5(–3) mm wide, with some wool at the stem apex but naked or nearly so downstem. Central spines are 0–2(–3) per areole (see below); when numbering two or more, they are disposed in a single row. Often there is a single prominent central spine, conspicuous to the naked eye, and a smaller second central. The largest central spine is borne in the middle to lower portion of the areole and is either porrect or deflexed; it is straight, or rarely slightly curved in any direction, usually is 0.6–2 cm long, stout, with a prominent bulbous base 1–1.5 mm in diameter. The base may be slightly laterally compressed. Central spine colors are gray or yellowish at their bases with reddish distal halves, or the main central may be entirely whitish to yellow. The maximum number of central spines (three per areole) has been observed in populations at the Nations Ranch northeast of Fort Davis, where intergradation between var. *cylindricus* and var. *weedinii* has been reported, and elsewhere in Jeff Davis County, for example, 3–4 miles south of Fort Davis. Radial spines are 14–23(–30) per areole, the longest 8–11 mm. The radial spine array is pectinate and tightly appressed to the stem (in all but the smallest plants). The longest radials are the ones positioned laterally, and the shortest are the upper ones in the areole. Lateral spines are rather stout, slightly bulbous-based, and laterally compressed on the proximal one-third. Radial spines are tan or yellowish, or reddish, or yellowish proximally and red on the distal half, or whitish with an ashy-white coating and red tips. Individual plants in the Davis and Guadalupe mountains have all-yellow radial spines and no central spines (see Biosystematics below for more detailed discussion).

In the areoles of juvenile plants, central spines often are absent, or one or two may be present. In seedlings the spines are shorter, slender, pectinately arranged, and usually appressed. In seedlings several months old, the radial spines may be angled away from the stem, perhaps as an artifact of vigorous growth.

Biosystematics. *Echinocereus viridiflorus* var. *cylindricus* has been universally treated as, at least, varietally distinct from var. *viridiflorus*. Leuck (1980) is the only author to have completely synonymized var. *chloranthus* with var. *cylindricus*. The spine characters of var. *cylindricus* are more like those of var. *viridiflorus* and var. *correllii* than those of any other members of the complex, but the flowers are more similar to those of var. *chloranthus*.

Variety *cylindricus* clearly appears to be closer to var. *chloranthus* than to varieties *russanthus, neocapillus, canus,* and *weedinii*. All of those share with each other unusually dense spination, with 6–12 or more slender centrals and 26–45 slender radials. Variety *chloranthus* is intermediate, with (0–)2–6 central spines, these not arranged in a single vertical row, and 17–23 radials. Leuck found that the phenotypes identifiable as vars. *cylindricus* and *chloranthus* had no external barriers to gene flow and that naturally occurring intermediate plants are numerous. Occasional plants with yellow radial spines, consequently having the general appearance of var. *correllii,* occur among otherwise typical var. *cylindricus* in the Davis Mountains north of Fort Davis. These are interpreted as a color-form within the population, not a separate taxon. A similar specimen from the Guadalupe Mountains ("5,500 ft., yucca enclosure, on top") also is interpreted as a yellow-spined form of var. *cylindricus*. According to Weniger (1984), *E. viridiflorus* var. *standleyi* (Britton & Rose) Orcutt, originally described as *E. standleyi* Britton & Rose, from near Cloudcroft in the Sacramento Mountains in south-central New Mexico, is another of these yellow-spined forms. Leuck (1980, p. 30) stated that var. *standleyi* was "not different from typical plants of *E. viridiflorus* var. *cylindricus* from southern New Mexico," and he treated them as synonymous, but then Leuck's concept of var. *cylindricus* was unusually broad, including all of var. *chloranthus,* which also occurs in southern New Mexico. Benson (1982) treated var. *standleyi* as a synonym of *E. viridiflorus* var. *viridiflorus,* without explanation. The taxonomic recognition of var. *standleyi* remains controversial, depending upon the understanding of yellow-spined color-forms wherever they occur within var. *cylindricus* in general.

Synonyms. Echinocereus standleyi Britton & Rose; *E. viridiflorus* Engelm. var. *standleyi* (Britton & Rose) Orcutt.; *E. chloranthus* (Engelm.) F. A. Haage var. *cylindricus* (Engelm.) N. P. Taylor; *E. chloranthus* (Engelm.) Hort. F. A. Haage subsp. *cylindricus* (Engelm.) W. Blum and Mich. Lange.

Common Names. Nylon cactus; New Mexico rainbow cactus; green-flowered cactus; cylinder bells; green-flowered torch cactus; green-flowered pitaya; brown-flowered pitaya; brown-spine hedgehog; green-flowered hedgehog-cactus.

12b. **Echinocereus viridiflorus var. correllii** L. D. Benson, Cact. Succ. J. (U.S.) 41: 128. 1969. CORRELL'S GREEN-FLOWERED HEDGEHOG CACTUS. Plates 156, 157. Endemic to the Trans-Pecos. Caballos Novaculite, rocky hills and slopes, and in alluvial grassland with novaculite rubble, near Marathon, NE Brewster Co., TX; also in limestone at a site N of Fort Stockton. 3,000–4,000 ft. Flowering Mar–May. $2n = 22$. Map 34.

The type locality of *E. viridiflorus* var. *correllii* is on private land near Fort Peña Colorado, ca. four miles southwest of Marathon. The varietal epithet honors D. S. Correll, a famous Texas botanist who accompanied L. D. Benson in collecting the holotype.

Identifying Characters. In habit *E. viridiflorus* var. *correllii* is very similar to var. *cylindricus*. In vegetative characters var. *correllii* is best distinguished from var. *cylindricus* by its greenish-yellow and ashy-white horizontal bands on the stems, or a yellowish aspect of the stems, and by its characteristically higher number of radial spines, (19–)20–27(–29), as opposed to 14–20(–25) radial spines in var. *cylindricus*. In var. *correllii* there are (0–)1–2(–3) central spines, and if more than one central is present, then one is much larger than the other(s), as in var. *cylindricus*.

The flowers (inner tepals) of var. *correllii* typically are greenish-yellow, but they may vary to (1) more greenish than yellow, (2) golden-bronze, (3) yellowish-brown, or (4) pale reddish-brown in a greenish-yellow background. Further studies will be needed to determine if the brownish and pale reddish colors of some specimens are present in every population or only in localized areas where there is other evidence of introgression from var. *cylindricus*. The campanulate corollas open widely compared to the flowers of var. *cylindricus;* the rounded, acute to attenuate-apiculate tepals often are reflexed at maximum expansion, as in *E. davisii*. Whole flowers are 3.5–3.8 cm long, and 2.5–3 cm across when fully open. First-day flowers are 1.5–2 cm across, then usually open wider during succeeding days. Leuck (1980) described the flowers as having a slight lemony scent, a trait verified by at least five investigators at Sul Ross. The scent is slightly stronger in flowers that have been open for more than one day. The outer tepals usually have a brownish midregion with yellowish margins. The anthers are light yellow. The style is greatly exserted from the stamens, with 6–10 green stigma lobes.

The fruits at maturity remain green or may turn dull red. The fruits split open along longitudinal lines and quickly dry. They are ovate, 1–1.3 cm long and 0.7–1 cm in diameter. The black seeds are ca. 1 mm long. Small ants have been observed to visit the dehiscent mature fruits, as if using a sugary pulp; possibly ants aid in seed dissemination.

Phenology. The usual blooming time for var. *correllii* is April into May, although in some seasons flowers may open in late March. Under outdoor cultivation in Alpine, the same plants bloomed on 14 April 1998 and then again on 24 May 1998. The flowers of var. *correllii* usually close at night, but some flowers may remain partially open, even on plants where other flowers are closed.

Sterile and Immature Specimens. Within the Marathon Basin near Marathon on novaculite substrates, var. *correllii* is recognizable by its combination of few central spines (or none), as in var. *cylindricus*, with the spines in alternating seasonal bands of whitish or yellowish color; sometimes all of the spines are yellowish. However, where novaculite is mixed with other substrates near the west, north, and east edges of the Marathon Basin, var. *correllii* can have red instead of yellow spines, apparently intergrading with var. *cylindricus* (Leuck, 1980). Outlying populations of var. *correllii* in the Glass Mountains and north of Fort Stockton (in limestone) have typical spine coloration.

The stems of var. *correllii* have 14–17(–19) ribs. The areoles are oblong, elliptic, or oval, 3–4 mm long, 2.5–3 mm in diameter, or circular, those near the apex with short wool. At least one central spine occurs in most populations, although plants with (0–)2(–3) centrals do occur. The larger "main" central spine is 0.7–1.3 cm long, yellowish or white, often with a reddish tip. The radial spines are pectinate and appressed. The longest radial spines are 8–9.5(–12) mm, in lateral position in the areole, and the upper ones are shortest. Basally, the radial spines are slightly bulbous and somewhat laterally compressed for nearly half their length. The radial spines are yellowish or ashy-white, seemingly with a thin whitish surface layer, with darker yellow or reddish tips.

In juvenile plants typically there are no central spines, and the shorter radial spines are clearly pectinate and appressed. In juveniles the yellow spine color seems to predominate. The spines of seedlings also are pectinate and appressed, or spreading in turgid specimens, and minutely pubescent.

Biosystematics. With one or two known exceptions, the entire distribution of var. *correllii* is restricted to novaculite habitats within a few miles of Marathon in northeastern Brewster County. A single collection from limestone hills north of Fort Stockton seems to be typical var. *correllii*, and a collection from near Hess Canyon in the Glass Mountains closely resembles var. *correllii*, although E. Leuck annotated the specimen as intermediate between var. *correllii* and var. *cylindricus*. Yellow-spined plants (without central spines) in the Davis and Guadalupe mountains with superficial resemblance to var. *correllii* are believed to be mere color-forms within populations of var. *cylindricus*. Weniger (1984) presumably interpreted all yellow-spined forms as a cohesive taxon when he regarded var. *correllii* to be a synonym in a geographically broad concept of *E. viridiflorus* var. *standleyi*.

On the basis of spine characters and intermediate-looking populations, var. *correllii* appears to be closely related to var. *cylindricus*. Leuck (1980) has shown that there are no apparent reproductive barriers between these taxa. Taylor (1985) contended that previous workers had placed too much emphasis on spine characters in relating var. *correllii* to var. *cylindricus* and that var. *correllii* instead belonged in an alliance with var. *viridiflorus* and *E. davisii* (see discussion under *E. davisii*). In addition to the floral characters emphasized by Taylor, the taxa *viridiflorus, correllii,* and *davisii* are supposed to have in common similar fruits that mostly remain green and are dehiscent (splitting and quickly drying), as in var. *neocapillus* (Leuck, 1980). The traits of the ripe fruits, in particular, are poorly documented and not necessarily reliable.

Synonym. *Echinocereus viridiflorus* Engelm. subsp. *correllii* (L. D. Benson) W. Blum & Mich. Lange.

Common Name. Correll hedgehog.

12c. **Echinocereus viridiflorus** var. **chloranthus** (Engelm.) Backeb. WESTERN GREEN-FLOWERED HEDGEHOG CACTUS. Plates 158, 159. [*Cereus chloranthus* Engelm., Proc. Amer. Acad. Arts 3: 278. 1856]. On both igneous and sedimentary substrates; in habitats that are more arid than is typical for the related var. *cylindricus;* El Paso Co. E through Hudspeth Co. and Culberson Co. 3,900–5,300 ft. Flowering late Mar–early May. $2n = 22$. Also southern NM, Doña Ana, Otero, Lincoln, and Eddy counties (Blum et al., 1998), but not in higher elevations of the Sacramento Mts. Barely documented from adjacent Chihuahua, Mexico, near the Rio Grande. Map 35.

Typical var. *chloranthus* usually is distinguished by a white, curved, lower central spine; otherwise, it is essentially a repository for plants intermediate between var. *cylindricus* (fewer central spines) and var. *russanthus* (more central spines). Such plants are abundant from the Franklin and Hueco mountains and adjacent New Mexico east around the Guadalupe Mountains, south through the Delaware Mountains and Rustler Hills to the Apache Mountains (and at least a few miles east of the Apache Mountains), where populations include individuals resembling var. *cylindricus,* and back west through the Baylor Mountains and the Sierra Diablo, and to near Sierra Blanca. Benson (1982) did not recognize either var. *russanthus* (which he reduced to synonymy) or var. *weedinii,* but instead identified all of those populations as *E. chloranthus* var. *chloranthus.* Old literature records (1980s and earlier) of *chloranthus* from farther northwest and southeast pertain to populations subsequently segregated as other taxa (*rhyolithensis* in New Mexico; *russanthus, weedinii,* and *neocapillus* in Texas). The epithet *chloranthus* is meant to be descriptive of the flowers, after the Greek, *chloros,* "green" or "greenish-yellow," and *anthos,* "flower," although the flower colors in this taxon vary from yellowish-green to brown or reddish-brown.

Identifying Characters. The var. *chloranthus* has the general habit of var.

cylindricus except that the stems often are relatively slender and the plants have a more bristly aspect, the effect of longer and/or more numerous central spines. At least one central spine is white, although often with a reddish tip, usually the one positioned lowest in the areole. The other centrals usually are reddish in color. The lower central usually is the longest, 2–4.3 cm, usually longer than centrals of var. *cylindricus*. In all there are (0–)2–6 central spines in var. *chloranthus*; when they number two or three, they are positioned in a single vertical row as they are in var. *cylindricus*; otherwise, they spread in all directions. The number of radial spines in var. *chloranthus* is 15–23, about the same number as in var. *cylindricus*, but the radials in var. *chloranthus* usually are longer, to 1.2–1.4(–1.7) cm long. In color aspect the stems may show horizontal bands of more white and more reddish spines, as is the case in var. *cylindricus*, except that this character in var. *chloranthus* is more obscured by the central spines. The radial spine color in var. *chloranthus* may be different in separate areoles, with all or most all of the radials ashy-white, most all of them reddish or reddish-black, or some of the spines in each areole may be reddish and some whitish. Or spines may be multicolored, cream, whitish, or gray at the base with reddish distal portions or tips, as may be the case in var. *cylindricus*. The var. *chloranthus* is easily distinguished from var. *russanthus* by its consistently fewer central spines and radial spines, by the larger caliber of the spines, and by the appressed angle of the radial spines.

The flowers of var. *chloranthus*, in overall aspect, are dark green to yellowish-green or greenish-brown. More specifically, inner tepals have been described as brownish-green or rusty distally and green proximally, or as having dark green midlines with the margins lighter green and suffused with brown. Outer tepals may have brownish midlines. Kolle (1978) described the tepal margins as olive-yellow. Almost chocolate-brown and reddish-brown flower colors have also been reported for var. *chloranthus*. The flower morphology in var. *chloranthus*, var. *cylindricus*, and var. *russanthus* is virtually identical. The flowers in all three taxa are funnelform, 2–3.4 cm long, with styles 1.2–2 cm long, usually 8(–10) green stigma lobes, pale yellow anthers on filaments 5–8 mm long, and shallow nectar chambers 1–1.5 mm deep. The inner tepals in all three taxa are 2.5–5 mm wide, with tips acute-apiculate, rounded, or even emarginate. The flowers of var. *chloranthus* are not detectably scented, at least in the several plants tested by the same investigators who could smell the lemon odor in certain related taxa such as var. *correllii*.

Mature fruits of var. *chloranthus* are green (or dull red, at least in taxonomically ambiguous populations in southern Culberson County), almost round, and with areoles that fall away, leaving a glistening fruit wall, at least in plants from the Franklin Mountains. The fruits are indehiscent. The black seeds are 1–1.3 mm long and broad, and papillate.

Phenology. The flowering period for var. *chloranthus* is from mid- to late March through April, and in some years into early May. Individual plants or stems may produce numerous flowers, up to about 38 (Worthington, 1986), over a period of about 15 days to one month.

Sterile and Immature Specimens. The stems of var. *chloranthus* are solitary or basally branched to form clumps of rarely more than 10 stems, these 7.5–20(–25) cm long, 5–7.5 cm in diameter. Stems have 11–16 ribs that are well defined and weakly undulate. The areoles are circular and ca. 3 mm long, or elliptic to oblong and 4 mm long, and 3–6 mm apart on the ribs. Only areoles near the stem apex have copious short white wool.

In var. *chloranthus* the central spine number, length, color, and disposition in the areole are different from those in var. *cylindricus*. In most populations of var. *cylindricus* there is no white central spine, although the main central may be gray in some plants, and the main central usually is straight, although occasionally it may be up-curved. In the Davis Mountains near Fort Davis, the presence of a long, white, up-curved, main central spine in a few populations of var. *cylindricus* suggests a history of introgression from var. *chloranthus*. In var. *chloranthus* the nonwhite central spines may be reddish to reddish-black, or even gray with a reddish tip. Some centrals may be dark on the upper surface and whitish on the lower surface. Rarely, the centrals are cream or yellowish.

The radial spine numbers, colors, and orientation in var. *chloranthus* are not significantly different from those of var. *cylindricus*. All the spines of var. *chloranthus* have the same basic morphology as those of var. *cylindricus* and its closest allies in the *E. viridiflorus* complex, except that the spines of var. *chloranthus* are longer, more slender, and less prominently bulbous-based. The spines of both var. *chloranthus* and var. *cylindricus* are clearly distinguishable from those of var. *russanthus*.

In juvenile plants of var. *chloranthus* the central spines may be absent or present. The spines are more slender in juveniles than in adults. The radial spines are short, pectinate, and appressed, except in turgid seedlings, where the radials may be spreading.

Biosystematics. The var. *chloranthus* is a morphologically identifiable but inconveniently variable and intergrading geographic race of the same species as var. *cylindricus*. After a biosystematic investigation, Leuck (1980) concluded that var. *chloranthus* was not distinguishable from var. *cylindricus*. Leuck's phenetic study graphically demonstrated that var. *chloranthus* and var. *cylindricus* intergrade to such an extent that there is a continuum of morphological variation between the extremes, and he observed that both morphological extremes may be found within some single populations. We, too, have seen this great variability (interpreted as a classic "hybrid swarm" arising from secondary contact between otherwise distinct taxa) in certain populations as far apart as the Franklin Mountains in El Paso County and near Kent in Culberson County, at least as far as spine morphology is concerned. Flower-color variation assorts independently of the spine morphotypes: In theory, "pure" var. *chloranthus* has green flowers, and "pure" var. *cylindricus* has red-brown flowers, but in practice this hoped-for taxonomic correlation is completely undermined by the great variability within each taxon.

Our review of the *E. viridiflorus* complex failed to reveal any "classic" hybrid zone linking clearly defined population systems of var. *chloranthus* and var. *cylindricus,* partly because both taxa are disjunctly distributed (every population is isolated geographically) and partly because "pure" *chloranthus* has been difficult to define. Each population could be interpreted as a unique link in a chain of intermediates. The spine characters of var. *chloranthus*, even near its type locality, vary from those of var. *cylindricus* to those resembling the red-spined color form of var. *weedinii*, or the northwestern populations recently segregated as subsp. *rhyolithensis* (Blum et al., 1998). Variety *chloranthus* could have originated through introgressive hybridization between these taxa. Variety *chloranthus* exhibits greater morphological variation than either pure *cylindricus* or any of the bristly-looking taxa such as *weedinii*, as expected of a hybrid product relative to its parental taxa.

The concept of *E. chloranthus* var. *chloranthus* embraced by Benson (1982) included the distinct taxon described by Weniger (1969) as *E. russanthus*. Benson evidently believed that there was a clinal gradation from the northwestern populations (var. *chloranthus*) with 2–6 central spines to the southwestern populations (var. *russanthus*) with 9–12 or more central spines. Specimens we have seen suggest that var. *chloranthus* has 3–4 central spines throughout most of its range, whereas populations with 5–6 central spines are found (at some sites up to 10 centrals) in the New Mexican populations recently segregated as "*E. chloranthus* subsp. *rhyolithensis*," which phenetically (but not geographically) bridge the gap between vars. *chloranthus* and *russanthus*. The present populations of var. *chloranthus* and var. *russanthus* are disjunct, so far as known, but the Mexican side of the river is practically unexplored. Variety *weedinii*, likewise, bridges the gap in spine counts between var. *chloranthus* and var. *russanthus*, but var. *weedinii* is different from either in its ecogeographic range and the presence (predominance) of a yellow spine-color form in its populations. Populations in the Rosillos Mountains and on Nine-Point Mesa have fewer, stouter spines than normal var. *russanthus* and so they, too, are technically intermediate between var. *cylindricus* and var. *russanthus*, although much more like var. *russanthus*.

In Benson (1982, fig. 713, p. 678), the photograph labeled "*Echinocereus chloranthus* var. *chloranthus*" is var. *russanthus*; figure 714 probably is "subsp. *rhyolithensis*."

Synonyms. *Echinocereus chloranthus* (Engelm.) F. A. Haage; *E. chloranthus* (Engelm.) F. A. Haage var. *chloranthus*; *E. viridiflorus* subsp. *chloranthus* (Engelm.) N. P. Taylor.

Common Names. Cylinder bells; green-flowered torch cactus; green-flowered pitaya; brown-spine hedgehog.

12d. **Echinocereus viridiflorus** var. **russanthus** (Weniger) A. D. Zimmerman, comb. nov. (forthcoming). RUSTY HEDGEHOG CACTUS. Plates 160–63. [*Echinocereus russanthus* Weniger, Cact. Succ. J. (U.S.) 41: 41. 1969]. Endemic to the Trans-Pecos, arid mountains, desert grasslands, among desertscrub, mostly igneous but also in limestone substrates, and on the southernmost novaculite hills of the Marathon Basin, extending upslope to mesic habitats in the Chisos Mts. 2,300–6,400 ft. Flowering (Feb–)Mar–Apr. $2n = 22$. Map 35.

As pointed out by Leuck (1980), it is surprising that this distinctive taxon remained submerged under *E. chloranthus* until it was described by Weniger (1969) as *E. russanthus*. This relatively abundant taxon consistently differs from everything else formerly included in var. *chloranthus* and is geographically isolated from them, in the southern Big Bend region. The epithet *russanthus* is after the Latin *russus*, "reddish," and Greek, *anthos*, "flower," alluding to the characteristic flower color of var. *russanthus*.

Identifying Characters. The stems of var. *russanthus* are cylindrical, often branched at the base, 5–25 cm long, and 4–9 cm in diameter. The central spines are flexible, almost as slender as the radial spines. There are (7–)9–12(–17) central spines and (18–)30–45 radial spines. The spines are diffusely spreading, making it difficult to count the radials without removing whole spine clusters.

The reddish flowers of var. *russanthus* are described by Weniger (1969) as having soft, thin inner tepals that have somewhat translucent, dull surfaces, as compared to the firm, opaque, glossy (almost waxy-looking) tepals of other taxa, but this alleged textural character is difficult to evaluate. Weniger also described the flowers of var. *russanthus* as the smallest he had seen on any *Echinocereus*, but we have observed considerable variation in the flower size (2–3.5 cm long) of var. *russanthus*, with some flowers easily as long as those of other members of the *E. viridiflorus* complex. The flowers of var. *russanthus* are rather narrowly funnelform, most like those of var. *chloranthus* or var. *cylindricus*.

Although the predominant flower color in var. *russanthus* appears to be rusty-red or russet-red, other flower colors have been observed by us or reported for the taxon: dark reddish-brown, beet-root red, and orange-brownish. We have not observed "bright red" flowers, reported by Taylor (1985), to occur in var. *russanthus*. The flower centers usually are greenish, which may account for one report of greenish-brown flowers in the variety. The greenish flower color commonly seen in several other varieties has not been observed in var. *russanthus*; inner tepals colored green, even bright green, are especially common in var. *neocapillus*, var. *canus*, and var. *weedinii*.

The inner tepals of var. *russanthus* are 2.5–3.5 mm wide; their tips are acute, acute-apiculate, rounded, or emarginate-mucronate. The filaments are 5–6(–10) mm long, with pale yellow anthers ca. 1 mm long. The yellowish style is 1.3–1.4(–1.8) cm long, supporting 8–10 green stigma lobes that are 2.5–3.5(–4) mm long. The shallow nectar chamber is ca. 1(–2) mm deep. The flowers of var. *russanthus* apparently lack scent, in contrast to most other taxa in this group.

The fruits are oval to almost spherical, 1–1.3 cm long, green or dull red or reddish-brown, succulent, not sugary, and indehiscent (Leuck, 1980). In var. *russanthus*, at least in some plants, fruits remain green at maturity and may not turn reddish, except that they develop a tinge of red or brown. The fruits are obscured by clusters of slender white spines, 10–20 in each areole. The seeds are ca. 1 mm long and broad.

Phenology. In var. *russanthus* flowering may begin in early February to March in desert populations at lower elevations in and near the Chisos Mountains. Flowering is progressively later, through April, at progressively higher elevations. Floral phenology is otherwise typical of the *E. viridiflorus* complex, as described under var. *cylindricus*.

Sterile and Immature Specimens. Short-spined specimens of var. *russanthus* appear practically identical to longer-spined forms of *E. dasyacanthus* until the flowers or the tiny fruits are seen. The stems of var. *russanthus* have 12–18 ribs. Areoles are at first circular and ca. 3 mm wide, becoming oval and 3–3.5 mm long. Young areoles at the stem apex have short white wool tufts or near-straight hairs. The stem surface is mostly obscured by the spines.

The longest (lowermost) centrals are 1–3.7 cm in var. *russanthus* and either straight or slightly curved upward. The centrals are bristlelike, with small-bulbous bases, these 0.27–0.35 mm thick. In var. *russanthus* the typical number of radial spines is 30–45, although fewer radials have been counted in some areoles. The longest radials are the lateral ones, 8–16 mm, and the shortest are uppermost.

The var. *russanthus* varies in spine length and color, particularly the central spines. The spines are reddish-brown to reddish-black, or white with grayish-purple (or reddish-brown) tips; rarely the spines are whitish-yellow, but these forms might reflect introgression from var. *weedinii* rather than intrinsic variation within var. *russanthus* itself.

Immature plants usually have fewer spines per areole but resemble adult plants in aspect. The spines of seedlings are relatively long, diffusely spreading as they are in adults, and minutely pubescent.

Biosystematics. The distribution of var. *russanthus* is essentially restricted to the southern Big Bend region of Trans-Pecos Texas. The taxon is most common on and near the lower slopes of the Chisos Mountains. It is also abundant in the Christmas and Rosillos mountains, north to Nine-Point Mesa, but those in the latter two sites are visibly different. Outlying populations extend north to about 17 miles south of Marathon, where they are geographically replaced by var. *neocapillus* (but mostly long-, red-spined plants of var. *neocapillus*, identifiable only by their remarkable hairy seedlings). One collection resembling typical var. *russanthus* from about four miles north of Fort Davis in the Davis Mountains (*Warnock 22681*) presumably is the red-spined color-form of var. *weedinii*, not really var. *russanthus*. Presumably, var. *russanthus* also grows in Coahuila, Mexico, adjacent to Big Bend National Park, but it has not been documented there. Reports from central Chihuahua, Mexico (Santa Clara Canyon in the Sierra del Nido), mentioned under *russanthus* by Taylor (1985), have now been taxonomically segregated as *E. russanthus* subsp. *fiehnii* (Trocha) W. Blum & Mich. Lange (Blum et al., 1998).

Within the range of var. *russanthus* the spine length and color vary geographically; other spine characters, such as number, position, diameter, and angles of divergence from the stem, are much less variable. Six or more unnamed subtaxa can be recognized: (1) Plants at some lower elevation sites in the Chisos Mountains, on Chilicotal Mountain, in the McKinney Hills just west of the Dead Horse Mountains, and in at least one site just north of the Rosillos Mountains, have relatively long, deflexed, central spines and relatively long radial spines that are reddish-brown to nearly black. (2) Plants in the Chisos Basin at 5,400 feet and in the outwash foothills on the north and west sides of the Chisos Mountains have relatively short centrals and radials. These plants are vegetatively almost identical to *E. dasyacanthus* (Leuck, 1980), although the particular form of *E. dasyacanthus* growing with them has unusually short spines for its species, superficially resembling *E. pectinatus*. (3) In the desert foothills southwest of the Chisos Mountains and north near Terlingua, in the southwest Christmas Mountains, and on the slopes of the Rosillos Mountains, the plants have rather long spines that are dark gray or whitish, with considerable reddish coloration. In these populations usually it is the lower, up-curving central spine that is whitish, resembling in habit, but not spine size or number, var. *chloranthus*. (4) Still other plants in the Christmas Mountains have relatively short, reddish-brown spines. (5) South of Marathon, the northernmost populations of var. *russanthus* have only 6–8 central spines and fewer radial spines (as in var. *neocapillus*); the centrals are long and the lowermost one deflexed; these grow in the novaculite substrate less than one mile from the southernmost population of var. *neocapillus*. Leuck (1980) and Kolle (1978) have already pointed out that numerous plants of var. *neocapillus* have reddish coloration in the spines and flowers, suggesting hybridization with var. *russanthus*. (6) Just north of the Rosillos Mountains in some igneous outcrops where discovered by Paul Whitefield, extending to the south slopes of Nine-Point Mesa, plants of var. *russanthus* have 1–3 lower central spines that are nearly black, and porrect, different from all other populations of var. *russanthus* known to us (Plate 163). These plants have 6–11 central spines that are slightly more stout than in other var. *russanthus*, but otherwise the spination seems to be rather typical of var. *russanthus*. Given a sufficiently narrow concept of taxa, any of these population systems could be interpreted as an undescribed taxon closely related to var. *russanthus*. The variation in spination of present-day var. *russanthus*, the most widespread variety of *E. viridiflorus* in desertic habitats in the southern Big Bend region, indicates a taxon "caught in the act" of ongoing evolutionary diversification below the species level.

Leuck (1980) reported that var. *russanthus* was occasionally sympatric with *E. viridiflorus* var. *cylindricus* in the southernmost Davis Mountains, presumably an error based on the single anomalous specimen mentioned above (*Warnock 22681*, SRSC). Hybridization between var. *weedinii* and var. *russanthus* on the west slope of the Chisos Mountains is inferred from the poor taxonomic separation between the yellow- and red-spined plants that grow intermingled there, scarcely better than what one would expect of a single dimorphic population of yellow/red plants. The higher elevation distribution of var. *weedinii* in the Davis, Chinati, and Chisos mountains is typical of Pleistocene relicts, taxa that were much more widespread prior to the climatic warming of the past 15,000 years. Today only the relatively xerophytic varieties *cylindricus*, *chloranthus*, and *russanthus* are common below 6,000 feet in the Trans-Pecos. For most cacti, 15,000 years is ecological time, not evolutionary time, but extensive hybridization could have taken place between varieties of *E. viridiflorus* as the ecogeographic ranges of all the plants shifted upward (or were stranded in favorable refugia such as novaculite outcrops) in response to the changing climate.

In habit *E. longisetus* var. *longisetus* and its allies, such as *E. rayonesensis* of Nuevo León, vaguely and superficially resemble the longest- and whitest-spined plants of *E. viridiflorus* var. *russanthus*. Reports of *E. longisetus* in the Trans-Pecos are errors, based upon (1) blatant misidentifications of long-spined var. *russanthus* and (2) in another case (the Harry Barwick collection) unlabeled horticultural specimens of typical *E. longisetus* that probably were imported from Mexico (as indicated by *Astrophytum* and other endemic Mexican plants in the same collection).

Synonym. *Echinocereus chloranthus* F. A. Haage var. *russanthus* (Weniger) Lamb ex G. Rowley.

Common Names. Brown-flowered cactus; foxtail cactus; rusty hedgehog.

12e. **Echinocereus viridiflorus** var. **neocapillus** (Weniger) Leuck ex A. D. Zimmerman, comb. nov. (forthcoming). WENIGER'S SMALL-FLOWERED HEDGEHOG CACTUS. Plates 164, 165. [*Echinocereus chloranthus* (Engelm.) F. A. Haage var. *neocapillus* Weniger, Cact. Succ. J. (U.S.) 41: 39, f. 4. 1969]. Endemic to the Trans-Pecos, on certain novaculite hills of the Marathon Basin, from ca. 10 to 17 mi S of Marathon. 3,700–4,500 ft. Flowering late Mar–early May. $2n = 22$. Map 36.

The var. *neocapillus* is one of five remarkable Marathon Basin endemic taxa (if one counts *Coryphantha hesteri*, which extends into the Del Norte Mountains and beyond), this one perhaps of more limited distribution than any of the others. All of the Marathon Basin endemics are either restricted to or most common upon Caballos Novaculite substrates. Perhaps the most remarkable feature of var. *neocapillus* is that the seedlings produce only long, soft, hairlike white spines in the areoles (Plate 166). The seedling "hairs" are the basis for the varietal epithet, after the Greek *neos*, "new" or "young," and the Latin *capillas*, "hair."

Identifying Characters. The cylindrical stems of var. *neocapillus* usually are single but occasionally branched, 8–25 cm long, and 3–7 cm in diameter. In aspect the stems are pale yellowish-green or with horizontal bands of white spines alternating with yellowish, tan, or light brown spines. In larger stems there may be 5–7 or more annual bands visible aboveground, and previous years' growth is compressed and weathered at the base of the plant. The white spines are mixed with some that are reddish or reddish and white. The annual growth markings of var. *russanthus* usually are much less prominent than the horizontal banding characteristic in populations of var. *neocapillus*.

In var. *neocapillus* there are 5–11 central spines, these slender and spreading in all directions, (0.4–)1.2–2.0(–2.7) cm long, yellow, often reddish distally, or all reddish. There are (26–)30–38(–45) slender radial spines, these usually crowded and interlocking with spines from adjacent areoles. The radials are 6–12 mm long, yellow or white, usually concolorous in the same areole.

The flowers of var. *neocapillus* are (2.5–)2.8–3 cm long, 1.5–2 cm in diameter, cylindrical-funnelform, campanulate, or rarely subrotate with reflexed tepals. There is some flower-color variation, with the inner tepals usually greenish-yellow, having been described by various workers as "yellowy," "greenish," "brown," or greenish-brown to bronze, occasionally with reddish margins and dark midstripes, and with greenish bases inside the flower with brownish distal portions of the inner tepals. The inner tepals are 2–3(–4) mm wide and subacute. The filaments are 7–12 mm long, and the light yellow anthers are ca. 1.2 mm long. The greenish-yellow style is ca. (1–)1.6 cm long, topped by 8–10(–12) green stigma lobes, these 2.5–4 mm long. The flowers of var. *neocapillus* resemble those of var. *correllii*, but they lack the lemon scent present in flowers of var. *correllii* and only rarely open as widely as those of var. *correllii*.

The fruits of var. *neocapillus* at maturity are oval to subglobose, 1.2–1.4 cm long, 0.7–0.9 cm in diameter, and covered with clusters of white spines, (10–)20–25 in each areole. Mature fruits remain green, or turn dull purplish-red; they split longitudinally and quickly dry out. The black seeds are 1–1.2 mm long and finely papillate.

Phenology. The known populations of var. *neocapillus* appear to have a flowering period of one month, or a little longer, often beginning in late March and extending through most of April or into early May. The flowers open again for 3–4 days after closing at night. When the fruits mature, they split open, and then small ants visit them as if using the sugary pulp and possibly disseminating seeds in the process.

Sterile and Immature Specimens. The seedlings and juvenile plants of var. *neocapillus* are remarkable for their production of long, hairlike white spines in place of normal spines in the areoles (Plate 166). In each areole there are about 40 flexuous "hairs," to 3 cm or more long. In the field, at first glance, seedlings have reminded us of a clump of milkweed (*Asclepias* L.) seeds among novaculite rocks or on mats of *Selaginella*. The tuft of flexible, white spines may be the only visible part of smaller seedlings, which are small enough to be completely obscured by the hairs. Older juveniles with stems 2–5 cm high (Plate 167) typically are still covered with white hairs, but at about this size, true spines are formed at the stem apex. The juvenile hairs may persist at the base of semimature plants (Plate 168). A healthy population of var. *neocapillus* should consist of plants in all age classes, from hairy seedlings to spiny, flowering adults without any remaining evidence of the white hairs that were present in juvenile stages.

After the seedling hairs are lost or hidden, the usually single stems of var. *neocapillus* are identifiable only by careful comparison with those of the other varieties. The stems have 12–16(–18) ribs. The areoles are nearly round, the apical ones with matted white hairs. The central spines have smallish-bulbous bases. The longest centrals are the ones positioned lowest in the areole, and the lower one to several centrals are angled down as in var. *chloranthus*. The longest radials are in lateral position in the areole, and uppermost radials are the shortest. The radials are yellowish, or in plants with horizontal bands, the radials are yellowish in the yellow bands and chalky white in the whitish bands.

Biosystematics. Blum et al. (1998) are unique in treating *neocapillus* as a separate species. Benson (1982), Weniger (1984), and Taylor (1985) included *neocapillus* as a variety of *E. chloranthus*, but all three authors had different concepts of *E. chloranthus*. Leuck (1980), in our opinion, correctly recognized the close relationship between var. *neocapillus*, var. *russanthus*, and var. *weedinii*. Variety *canus* and "*Echinocereus carmenensis*," with hairy seedling stages, were not discovered until after Leuck had completed his studies of the complex. The unnamed population in Coke County, Texas, with hairy seedlings, was known to Leuck but was ignored in his classification system for the group and barely mentioned in his dissertation.

The var. *neocapillus* either contacts or blends into var. *correllii* at the northern margin of its range, and it is abruptly replaced by a form of the closely related var. *russanthus* at the southern end of its range. The southernmost extension of var. *neocapillus*, so far as known, extends to about 17 miles south of Marathon, where it grows within less than one mile of var. *russanthus*. Some adult plants of var. *neocapillus* are practically identical to the northernmost plants of var. *russanthus* in spine color and general appearance (Leuck, 1980), identifiable only by their distinctive seedlings. Leuck examined one herbarium specimen (*Pierce 4164*, UNM) that he suspected "to be a hybrid of *E. viridiflorus* var. *neocapillus* with var. *correllii*" (p. 37). Zimmerman et al. (forthcoming) observed that var. *neocapillus* at its northernmost documented site exhibits the shortest central spines "as if introgressed" from the nearby vars. *correllii* and/or *cylindricus*.

Both var. *neocapillus* and var. *correllii* have yellowish spines and prominent horizontal yellowish and white annual growth banding of the stems, although in var. *neocapillus* the spines may be tan or light brown in the bands alternating with white spines. The stems of var. *neocapillus* are more slender than are those of vars. *correllii* and *cylindricus,* and var. *neocapillus* has slender and flexible central and radial spines in numbers greatly exceeding the stouter spines of vars. *correllii* and *cylindricus*. The yellow-green flowers of var. *neocapillus* typically are darker than those of var. *correllii*. When fully open, the flowers of var. *neocapillus* remain cylindrical or expand to funnelform, and lack a lemon scent. Geographic and certain morphological considerations suggest that var. *neocapillus* might have originated from hybridization between var. *correllii* and var. *russanthus*.

In Texas the only other taxa known to produce seedlings and juveniles like those of var. *neocapillus*, with white hairs in the areoles, are var. *canus* and the poorly known Coke County population. Taylor (1985) compared the seedlings of var. *neocapillus* to *E. delaetii* Guerke var. *delaetii* (now regarded as *E. longisetus* var. *delaetii*) from southern Coahuila, Mexico; these hairy *Echinocereus* plants superficially resemble seedlings of *Cephalocereus senilis* (Haw.) Pfeiff.

The unusual spine character, the white hairs of seedlings, was first discussed by Leding (1934). In his experimental work with the *E. viridiflorus* complex, Leuck (1980) grew var. *neocapillus* from seed and learned that the seedling hairs are produced for two years after germination under greenhouse conditions. Leuck speculated that the seedling hair state may last for 4–5 years in the field before normal spines are produced. The juvenile spines persist at the bases of mature stems. Juvenile-type spines may occur on new growth after injury, where according to Leuck they last for about one year before typical spines are produced.

Leuck (1980) speculated that in var. *neocapillus* the long white hairs of seedlings may be an adaptation for survival in the novaculite substrate. However, three other taxa are now known to have the hairy seedlings (including the Coke County population), and only one of them (var. *canus*) grows on novaculite.

Synonym. *Echinocereus neocapillus* (Weniger) W. Blum & Mich. Lange.

Common Names. Woolly hedgehog; goldspine hedgehog cactus.

12f. **Echinocereus viridiflorus** var. **weedinii** (Leuck ex W. Blum & Mich. Lange) A. D. Zimmerman, comb. nov. (forthcoming). WEEDIN'S SMALL-FLOWERED HEDGEHOG CACTUS. Plates 169, 170. [*Echinocereus russanthus* Weniger subsp. *weedinii* W. Blum & Mich. Lange, *Echinocereus* 216. 1998]. Trans-Pecos endemic. Exposed rocky slopes and ridge crests, higher elevations in the Davis Mts (Jeff Davis Co.), Chinati Peak (Presidio Co.) and western Chisos Mts (Brewster Co.). 5,300–8,382 ft. (extreme summit of Mt Livermore). Flowering from mid- and late Apr to near the end of May. $2n = 22$. Similar plants long known from a few high peaks and grasslands in SW NM (Catron, Luna, Sierra, Doña Ana, and Socorro counties) appear identical to the red-spined color-form in the type population of var. *weedinii*; these were treated as a disjunct northern portion of var. *weedinii* in the *Chihuahuan Desert Flora* manuscript (Zimmerman et al., forthcoming). They are widely separated from Mt Livermore across the range of var. *chloranthus*, with which they intergrade; recently they were named in a taxonomic split by W. Blum & Mich. Lange in Blum et al. (1998): "*Echinocereus chloranthus* subsp. *rhyolithensis*." Map 36.

Echinocereus viridiflorus var. *weedinii* was recognized as distinct by Kolle (1978), who discussed the taxon under a provisional name referring to Mount Livermore, the primary locality, and was reported but never formally published as var. *weedinii* by Leuck (1980). Both authors indicated that var. *weedinii* was geographically widespread, disjunct between higher elevations of the Davis, Chinati, and Chisos mountains. The epithet *weedinii* is after James F. Weedin, second author of this volume, who also recognized the taxon as distinct at the time Kolle was studying the *E. viridiflorus* complex. Apparently Benson (1982), Weniger (1984), and Taylor (1985) were not aware of the unpublished dissertation by Leuck (1980), where the existence of var. *weedinii* was recorded. Horst Kuenzler and Roland Wauer had each independently discovered the Chisos Mountains population a decade earlier.

Identifying Characters. The stems of var. *weedinii* are solitary or branched, cylindroid, tapering slightly at the apex, 7–20 cm long, and 3.5–6 cm in diameter. Characteristically, these plants are covered with golden-yellow spines (at the type locality), both centrals and radials, but individuals with red-tipped white spines are common within the population on Mount Livermore and elsewhere in the range of var. *weedinii*. Compared with var. *russanthus*, var. *weedinii* has fewer central spines (5–12), and its fewer radial spines (20–35) are more appressed; sometimes the spines are slightly larger in caliber than is typical of var. *russanthus*. In the Davis and Chinati mountains var. *weedinii* is mostly restricted to the highest peaks, where it tends to replace the much more common and widespread *E. viridiflorus* var. *cylindricus*. Typically, plants of vars. *weedinii* and *russanthus* are easily distinguished from var. *cylindricus* at a great distance by their dense covering of relatively long, slender, and numerous central spines, with the lower and longest one deflexed, giving the plants a rather shaggy appearance. Most plants of var. *weedinii* are yellow-spined, whereas most populations of var. *russanthus* are exclusively red-spined. The lower elevational extremes of var. *weedinii* and the upper elevational extremes of var. *cylindricus* have not been carefully mapped. On Mount Livermore they are altitudinally separated, or nearly so, by a zone of welded tuff at 8,000 feet, where neither taxon thrives. However, at their one site in the eastern Davis Mountains, all of the herbarium specimens are intermediate between typical var. *weedinii* and var. *cylindricus;* Leuck's (1980) concept of var. *weedinii* was heavily biased toward this introgressed population, and so "pure" var. *weedinii* (as at the top of Mount Livermore) is not reliably treated in his dissertation. For example, his description of "pectinate" spination for the seedlings of var. *weedinii* is suspect. In the Chisos Mountains, plants of var. *weedinii* at their lower elevational limits are completely surrounded by var. *russanthus,* and the two taxa are thought to hybridize (Leuck, 1980); there, the short- and red-spined plants are traditionally identified as var. *russanthus,* and only those with long yellow spines are safely identifiable as var. *weedinii.* Perhaps none of them is really var. *weedinii,* in the sense of the Mount Livermore taxon.

When in flower var. *weedinii* typically is distinguished from var. *russanthus* by its yellow-green inner tepals and presence of floral scent, whereas in var. *russanthus* the flowers are reddish. However, the flower color of var. *weedinii* varies widely from greenish to olive-yellow, yellow-orange, brownish-green, and even reddish-brown, as described by different workers. Potted specimens of var. *weedinii* from the Davis Mountains have been observed to produce yellow-green flowers on one plant, while a separate plant from the same locality formed greenish-brown to rusty-brown flowers. Flowers of var. *weedinii* also have been observed to change color over a period of several days, from greenish-brown or rusty-brown, when first opened, to yellow-green. The flowers are cylindroid to campanulate, 1.4–3 cm long, and 1.5–3 cm in diameter. The inner tepals are 1–1.8 cm long, 2–4 mm wide, and acute-apiculate or rounded-apiculate. The style is 1–2 cm long with 6–8 green stigma lobes. The filaments are 5–10 mm long, white-greenish, and support yellow anthers that are ca. 1 mm long. The nectar chamber is 1–2 mm deep. The flowers of var. *weedinii* are lemon-scented, at least at the type locality (Zimmerman et al., forthcoming), but this trait requires further evaluation.

The fruits at maturity are reddish (dark purple) or rarely green, globose to ovoid, 5–8 mm long, 5–7 mm in diameter, fleshy with a nonsugary pulp, and they are indehiscent. The black seeds are 1–1.2 mm long and finely papillate.

Phenology. The usual flowering period for var. *weedinii* is from mid- to late April, to near the end of May. Each stem may produce several flowers.

Sterile and Immature Specimens. The cylindroid stems of var. *weedinii* support 15–17(–20) ribs that are mostly obscured, along with the rest of the stem surface, by the spines. The areoles are oval to circular and ca. 3 mm long. The 5–12 central spines may be all-yellow, or in some plants the longest central is white or grayish-white. In some plants both yellow and red central spines occur in the same areole with a single white central, or the centrals in other populations are red or white. The grayish-purple spine tips characteristic of var. *russanthus* do not appear in var. *weedinii*. The percentage of red centrals seems highest in younger areoles at and near the stem apex. The longest, typically deflexed, central spine is in the lowest position in the areole. The central spines usually are 1–2.5(–2.7) cm long, but in some specimens from Mount Livermore, the longest centrals are nearly 4 cm. Specimens found among the short-spined populations of var. *russanthus* at Laguna Meadow in the Chisos Mountains have central spines at the shorter end of the size range. In general the central spines of var. *weedinii* are shorter, slightly larger in caliber, and have somewhat larger bulbous bases than is characteristic of var. *russanthus*.

The (20–)25–31(–35) radial spines in var. *weedinii* are 0.7–1.2 cm long, yellow, in some specimens with yellow bases and red tips, usually appressed and somewhat pectinate, or spreading in some plants. The radials are slightly larger in diameter, at least in many specimens, than in var. *russanthus*, which has (18–)30–45 spreading radials.

Seedlings of var. *weedinii* characteristically produce translucent yellow spines like those of adults, but fewer in number and shorter, as would be expected. Leuck (1980) reported that central spines were not found in seedlings but were gradually produced in apical areoles by maturing plants. We have observed 1 cm high seedlings of var. *weedinii* (from Jeff Davis County) with at least 1–2 centrals evident in each lateral areole. Other seedlings of var. *weedinii* propagated in the Sul Ross greenhouses lacked central spines, partially explaining Leuck's generalization that var. *weedinii* seedling spine clusters are "pectinate."

Biosystematics. The var. *weedinii* occurs as discrete populations of predominantly yellow-spined, yellow- and white-spined, or red- and white-spined plants at upper elevations in the Davis, Chinati, and Chisos mountains of Trans-Pecos Texas. Populations resembling the Mount Livermore type of var. *weedinii*, but completely lacking the yellow color-form, have long been known from southwestern New Mexico; these are the plants recently named *E. chloranthus* subsp. *rhyolithensis* (Blum et al., 1998). In the Trans-Pecos the normal range of var. *cylindricus* at middle to lower elevations in the Davis and Chinati mountains brings this taxon adjacent to the lower limits of var. *weedinii*. Plants individually identifiable as var. *chloranthus*, commonly found in sedimentary substrates, are known to occur near Kent in Culberson County (here mixed in taxonomically ambiguous populations otherwise resembling var. *cylindricus*) just a few miles northwest of the northern end of the volcanic Davis Mountains.

Leuck (1980) observed overlap in flowering time and no apparent reproductive incompatibility between var. *weedinii* and var. *cylindricus* in the Davis Mountains, and the same was true between var. *weedinii* and var. *russanthus* in the Chisos Mountains. He also reported evidence of hybridization (i.e., he found plants of intermediate appearance) between var. *weedinii* and var. *cylindricus* at a site below Timber Mountain in the eastern Davis Mountains and between var. *weedinii* and var. *russanthus* on a west-facing cliff in the Chisos Mountains.

Despite these (and other) taxonomically unclassifiable populations, Leuck somehow concluded that there was no evidence of significant introgression between var. *weedinii* and the other varieties; he speculated about ecological mechanisms that might be keeping respective genotypes segregated.

Specimens from Timber Mountain, in the Davis Mountains, vary in spine thickness from thick spines like those of var. *cylindricus* to slender centrals (and radials) like those of var. *russanthus* and typical (Mount Livermore) var. *weedinii*. The stout spines could be regarded as a taxonomic difference (yet another "microendemic" variety); a simpler explanation is that they have resulted from introgression from the surrounding population of var. *cylindricus*. If they did not have brilliant yellow spines, then they might be interpreted as a population of var. *chloranthus*, which itself may have arisen through genetic "swamping" of long-spined, higher-altitude populations as all but the highest peaks grew arid during the past 15,000 years.

Synonyms. None.

Common Names. Yellow-flowered cactus; Livermore golden spine hedgehog.

12g. **Echinocereus viridiflorus** var. **canus** A. M. Powell and J. F. Weedin, **var. nov.** GRAYBEARD CACTUS. Plates 171, 172. TYPE: UNITED STATES. Texas. Presidio Co., Solitario, S-facing crest of a novaculite ridge, ca. 4,847 ft, novaculite surfaces similar to those in the Marathon Basin, 23 Oct 1993, *A. M. Powell and S. A. Powell 6012*, with J. Hardy and P. Manning (HOLOTYPE and Isotypes, SRSC; Isotypes, TEX, MO).

Plantae plerumque caulibus solitariis, 6–15 cm longis, 3.5–6 cm diametro; costis 14–16. Spinae plerumque albae, una spina aut paucis spinis rubellis per quamque areolam; spinae centrales 8–15, graciles, flexiles plerumque appressae. Flores virides, 2.5–3.2 cm longi. Fructus virides, globosi aut ovoidei; semina atra.

Additional specimens examined: UNITED STATES. Texas. Presidio Co., Solitario. Sandstone ridge above Righthand Shutup, *J. J. Clark 143* (SRSC); steep slope near entry to Righthand Shutup, *J. J. Clark 1114* (SRSC); Caballos Novaculite ridge near upper drainage into Righthand Shutup, *J. E. Hardy 440* (SRSC); head of Righthand Shutup, *J. E. Hardy 683; 736; 749* (SRSC).

Endemic to the Trans-Pecos, exposed Caballos Novaculite ridge crests and mostly S-facing slopes with *Selaginella;* also on chert and igneous rock; W-facing aphyric rhyolite slope (Hardy, 1997); and on "sandstone"; inside the Solitario Dome, a rare form of laccolith and caldera (Hardy, 1997), in SE Presidio Co., TX. 4,400–4,800 ft. Flowering from early Mar through most of Apr, sometimes into May. $2n = 22$. Map 36.

This remarkable white-spined entity was discovered in 1984 by James Jeff Clark, who at the time was conducting a vegetative survey of the rugged Solitario Dome as part of a master of science thesis project. One of Clark's objectives was to examine the isolated novaculite exposures there, which had been little investigated botanically, to see if there were any novaculite endemics similar to those known to occur in the vastly more extensive novaculite hills of the Marathon Basin about 58 miles to the northeast. After examining the initial collections of an anomalous white-spined cactus, Clark and several colleagues recognized them as probably representing an undescribed taxon, but it was not until the early 1990s that the taxonomically crucial "hairy" seedlings were collected. The varietal epithet, *canus*, is a reference to the whitish aspect of the plants, after the Latin *canus*, "grayish-white."

Identifying Characters. In its natural habitat, the var. *canus* is easily distinguished from sympatric species by its usually solitary stems covered with mostly white spines and its shaggy appearance resulting from spreading central spines. With its white spines, var. *canus* bears slight resemblance to both *Coryphantha sneedii* var. *albicolumnaria* and *Mammillaria pottsii*, both of which occur in the Solitario, and to the whitest-spined individual plants of *E. dasyacanthus*. Even from a distance, each of the white-spined taxa is distinctive, however, and of course they would never be confused when in flower. No other member of the *E. viridiflorus* complex is known to occur in the Solitario Dome, although var. *russanthus* is common in igneous substrates of the nearby Christmas Mountains. Without the hairy seedlings, var. *canus* could be mistaken for the longest- and whitest-spined plants of vars. *russanthus* and *weedinii*.

The spine morphology of var. *canus* clearly indicates close relationship with vars. *russanthus*, *neocapillus*, and *weedinii*. In var. *canus* there are more spines per areole than in any other variety of *E. viridiflorus*, and it lacks the yellow color-form found in populations of var. *neocapillus* (the only other Trans-Pecos variety having the white hairy seedlings). In var. *canus* there are 8–15, usually 9–13, slender, flexible, very fragile central spines in each areole, the longest per areole 1.7–2.5 cm, and spreading in all directions. There are 30–48 very slender radial spines, pure white, a maximum of 5–8 mm long, and mostly appressed to the stem. Both central and radial spines are white, or one or more central spines in each areole may be reddish (usually on the distal half or only at the tip), particularly in upper and apical areoles. In mature areoles the white centrals under magnification may be black-speckled.

The flowers of var. *canus* vary in saturation from light green to bright golden-green, perhaps more nearly pure green than those of var. *viridiflorus* itself. The flowers are 2–3.4 cm or more long, at first funnelform, and then in a hot greenhouse usually opening widely. The inner tepals are 1.3–2 cm long, 3–5 mm across, and rounded-apiculate or acute-apiculate. The filaments are 5–8 mm long, and the pale yellow anthers are ca. 1 mm long. The style is slender, (1.1–)1.5–1.7 cm long, with usually 7–8 dark green stigma lobes, these 2.5–3(–5.8) mm long. The nectar chamber is ca. 1.5–2 mm deep. Every plant observed for long enough, in horticulture, has produced scented flowers; the flower odor of one cultivated plant of var. *canus* was perceived as lemony by five individuals at Sul Ross, but may be more complex.

The fruits of var. *canus* are not yet well characterized, but all of ours have remained permanently green; they are ovoid to oblong. Sometimes they split open longitudinally on at least one side and quickly dry. The fruits are obscured by slender white spines, 18–20 or more in each areole. The black seeds are subpyriform, 1–1.3 mm long, and finely papillate except at the truncate bases.

Phenology. The peak flowering period for var. *canus* appears to be from mid-March to mid-April, with flowering extending into May in some seasons. In the field, flowers have been observed to open as early as 5 March, and in some years anthesis may occur even earlier. As is characteristic in the *E. viridiflorus* complex, the flowers close at night and reopen for 3–4 days. The first-day flowers usually are funnelform, but in succeeding days usually they open wider until the inner tepals may recurve. The flowers often open as widely as those of var. *viridiflorus*, var. *correllii*, and *E. davisii*, a trait used by Taylor (1985) to divide the *E. viridiflorus* complex into two groups.

Sterile and Immature Specimens. The hairy seedlings are similar to those of var. *neocapillus*, but possibly the hairlike white spines are longer. Sexually mature stems of var. *canus* are ovoid-cylindroid, 6–15 cm long, and 3.5–6 cm in diameter. The rib number was counted as 14–16 in the relatively few specimens that have been available for examination. The areoles are oval and 2.5–3.5 mm long.

The central spines have slightly bulbous bases 0.5–1 mm in diameter. The lower, descending 1–3 centrals are the longest. The lateral radial spines are longest, usually interlocking with those of adjacent areoles.

Seedlings of var. *canus* less than 1–3 cm tall produce only flexible, hairlike white spines, these 0.5–1 cm or more long (Plate 173). The hairs are straight, wavy, or curved, and number ca. 40 per areole. They persist for several years as a dense "skirt" (1–2 cm wide) of white hairs on the lower portion of the stem. The hairlike spines become gray with age until the plants are 3–8 cm high, but usually they are lost on older plants. White and red normal spines are first produced in the apical areoles when juvenile stems are 1–3 cm long.

Biosystematics. The var. *canus* is narrowly endemic in novaculite on a few ridges and slopes in the west-central part of the Solitario, which is a circular eroded dome about nine miles in diameter, with severely tilted limestone strata forming an outer perimeter of concentric ridges that completely enclose the geologic structure, except for narrow gorges that drain water to the outside. Several novaculite exposures exist in the Solitario, but apparently only certain ones support var. *canus*. The var. *canus* is one of the most distinctive taxa of the *E. viridiflorus*, complex with its hairy seedlings, shaggy, white-spined stems, and bright golden-green flowers. Its total known range is the smallest of any others for members of the *E. viridiflorus* complex.

Among Trans-Pecos cacti, only var. *canus* and var. *neocapillus* have the hairy seedlings described for these taxa. Both var. *canus* and var. *neocapillus* are narrowly endemic on novaculite substrates; they are about 58 miles apart and differ from each other most conspicuously in spine and flower color. The intervening territory is occupied by var. *russanthus*, in many habitats, even those including novaculite. Although var. *canus* and var. *neocapillus* share two unusual features (substrate preference and the remarkable hairy seedlings), the hairy trait also occurs in widely disparate populations in Coke County and the Maderas del Carmen. The consistent occurrence of scented flowers in var. *canus,* and in at least the Mount Livermore population of var. *weedinii,* complicates Taylor's (1985) criteria for dividing the *viridiflorus* complex into two major groups.

Synonyms. None.

Common Name. "Graybeard cactus" is already in use in horticulture.

Appendix 2

Extralimital taxa of *Echinocereus* relevant to the Trans-Pecos flora and not discussed elsewhere in the Trans-Pecos treatment of the genus are summarized below.

Echinocereus fendleri var. **kuenzleri** (Castetter, P. Pierce, & K. H. Schwer.) L. D. Benson. KUENZLER'S HEDGEHOG CACTUS. [*Echinocereus kuenzleri* Castetter et al., Cact. Succ. J. (U.S.) 48: 77–78, figs. 1, 2. 1976]. Both Benson (1982) and Taylor (1985) recognized a third variety of *E. fendleri,* var. *kuenzleri* (provisionally including *E. hempelii* Fobe of north-central Chihuahua), a weakly defined taxon treated by Weniger (1984) as the distinct species *E. kuenzleri.* The var. *kuenzleri* is not known to occur in the Trans-Pecos, but taxonomic implications of its possible existence there was discussed by Taylor. Its restricted distribution just north of the Trans-Pecos in the Sacramento and Guadalupe mountains (Otero and Eddy counties) of southern New Mexico in woodlands at ca. 5,000 feet, suggests that var. *kuenzleri* eventually might be found in the Trans-Pecos. The var. *kuenzleri* and its Chihuahuan counterpart are distinguished by their unusually small spine clusters, with 2–6 radials that are up to 2.5 cm long, thick, angular, flattened, chalky-white, and often twisted, and no central spine (or rarely one central), thus revealing more of the dark green, soft, flabby stem surface and very tuberculate ribs, as compared with var. *fendleri.* According to Taylor, var. *kuenzleri* produces large flowers, up to 11 cm in length and diameter (based on Chihuahuan material?), the maximum flower size for the species, and it is therefore one of the most desirable ornamental types in the genus. The status of var. *kuenzleri* requires further study, especially with respect to its populational integrity. It is possibly an occasional spine form that is not distinct (Taylor, 1985) from var. *fendleri.* Weniger (1984) pointed out that var. *kuenzleri,* with its unusual spines, superficially resembles *E. triglochidiatus* var. *triglochidiatus,* and var. *kuenzleri* thus may be confused with young plants of *E. triglochidiatus* var. *triglochidiatus* in sterile condition.

Echinocereus viridiflorus Engelm. var. **viridiflorus,** Mem. Tour N. Mex. 91. 1848. GREEN-FLOWERED HEDGEHOG CACTUS. Plate 153. Distinguishing features of var. *viridiflorus* include relatively short, solitary or branching stems 3–9(–18) cm long; 10–15 ribs; spines white or reddish, frequently white at the base and reddish distally, the predominant spine colors often in horizontal bands encircling the stem, central spines 0–1(–3), radial spines 12–18, these pectinate and appressed; greenish-yellow flowers that are campanulate to rotate, having a strong lemony scent; and fruits that are green, dry, and dehiscent along longitudinal lines. South-central New Mexico north to South Dakota, and east to the Texas Panhandle in Oldham and Potter counties (see Leuck, 1980; Benson, 1982; Blum et al., 1998).

"Echinocereus carmenensis" W. Blum, Mich. Lange, & Scherer Foothills of the Sierra del Carmen, Coahuila, Mexico.

Echinocereus "russanthus subsp. fiehnii" (Trocha) W. Blum & Mich. Lange; endemic to the Sierra del Nido, central Chihuahua, Mexico.

Echinocereus "chloranthus subsp. rhyolithensis" W. Blum & Mich. Lange; the counterpart of *weedinii* in southwestern New Mexico but lacks the yellow-spined color-form.

Echinocereus viridiflorus var. nov.; an undescribed population in Coke County, Texas. Leuck (1980) barely mentioned the population (north of Robert Lee), which he evaluated as having features of both var. *correllii* (which it obviously is not) and var. *neocapillus* (which it obviously is not, but it has the "hairy" seedlings). The Coke County population is an unnamed taxon, but it has received more attention from horticulturists than from biologists (Blum et al., 1998).

Echinocereus pentalophus (DC.) Lem. ALICOCHE, LADY-FINGER CACTUS. Plates 174, 175. [*Cereus pentalophus* DC., Mém. Mus. Hist. Nat. 17: 117. 1828]. Plants forming low clumps to ca. 1 m in diameter. Stems branched, 20–65 cm long, 1–6 cm in diameter, erect or prostrate, reddish- to yellowish-green. Ribs 4–5. Central spine 0–1, less than 1–2 cm long, porrect to ascending, yellowish to brown; radial spines 3–7, 0.2–2 cm long, whitish to yellowish, upper ones very small. Flowers 8–10 cm long, 10–15 cm in diameter, bright pink to light magenta, the throat white to yellow; areoles on receptacular tube with brown bristlelike spines and cobwebby wool. Fruit ovoid, green, to 1.9 cm long, covered by brownish bristlelike spines and white wool, irregularly dehiscent. The blooming period for *E. pentalophus* reportedly is the last week in March through early April. Flowers open about noon, close in late afternoon, and sometimes open again for 2–3 days. The Texas entity of this species is *E. pentalophus* var. *procumbens* (Engelm.) P. Fourn., found in South Texas brushlands; more widespread in east to northeast Mexico, reaching elevations of ca. 4,300 feet. An endemic variety, *E. pentalophus* var. *leonensis* (Mathsson) N. P. Taylor, occurs in southeast Coahuila and Nuevo León, Mexico (Taylor, 1985). Typical *E. pentalophus* with larger stems, grows in tropical eastern Mexico. *Echinocereus pentalophus* from South Texas is not cold hardy enough to be grown outside in most of the Trans-Pecos. The epithet *pentalophus* refers to the five-ribbed stems.

Echinocereus berlandieri (Engelm.) Rümpler, BERLANDIER'S ALICOCHE. Plates 176, 177. [*Cereus berlandieri* Engelm., Proc. Amer. Acad. Arts 3: 286. 1856]. Plants forming low clumps to ca. 1 cm in diameter. Stems branched, 5–65(–200) cm long, 1.5–3 cm in diameter, dark green or purple-tinged, prostrate, apexes turned upward. Ribs 5–7, low, tubercles conspicuous. Central spines 1–3, 2.5–5.1 cm long, porrect to deflexed, yellowish to brown; radial spines 6–9, to 1 cm or more long, whitish, tips often brown. Flowers 6–9 cm long, 7–10 cm in diameter, purplish, the throat dark; anthers orange-yellow, filaments pale purplish; stigma lobes nine, green; areoles on receptacular tube with 12 or fewer spines, the upper spines sometimes curved. Fruit obovoid, green (turning reddish; Taylor, 1988), 2–3 cm long, 1.5–1.8 cm in diameter, the areoles with spine clusters similar to those on the stems, these deciduous. Found in brushland, commonly mesquite thickets, and grassland, South Texas, "near the Nueces River" and near the Rio Grande, Webb County southeast to Cameron County, reportedly also in adjacent Tamaulipas and Nuevo León, Mexico. Both *E. berlandieri* and *E. pentalophus* are similar in general appearance. When in flower, the species are easily distinguished from each other by the dark throat in *E. berlandieri* and the white to yellow throat in *E. pentalophus* (Taylor, 1985). *Echinocereus berlandieri* is named after the French explorer Jean Louis Berlandier, who traveled in Mexico and settled in Texas in the early 1830s. Taylor (1985, 1988) accurately observed that Weniger (1970, 1984) confused the nomenclature of *E. berlandieri* with a form of the common species *E. pentalophus*, and Weniger's concept of *E. blanckii* illustrates the actual *E. berlandieri*. *Echinocereus berlandieri* from South Texas is not cold hardy enough to be grown outside in most of the Trans-Pecos. *Echinocereus blanckii* (Poselger) Palmer, sensu stricto [*Cereus blanckii* Poselger, Allg. Gartenz. 21: 134. 1853], believed by Schulz and Runyon (1930) to be a distinct species, probably was based upon a plant specimen of *E. enneacanthus* var. *brevispinus*, but no type material was preserved or even illustrated. The cactus illustrated by Schulz and Runyon (p. 87) is a typical form of *E. berlandieri*.

Echinocereus papillosus Linke ex C. F. Först., Handb. Cacteenk. (ed. 2) 783. 1885. YELLOW-FLOWERED ALICOCHE. Plates 178–80. Plants forming loose clumps of 2–95 stems. Stems mostly erect or leaning, to 4–20 cm long, 2–5(–7) cm in diameter, deep green to brownish-green. Ribs (6–)7–10, mostly divided into conspicuous tubercles to 9 mm long. One central spine, ca. 2 cm long, porrect, brownish to yellow, or with a brown base, yellow middle, and brown tip; radial spines 7–11, whitish to yellowish-brown, usually brown-based. Flowers 5–7 cm long, 6–10 cm in diameter, pale straw-yellow with orange-red throats. Fruit subglobose, greenish, covered with short bristles. Typical *E. papillosus* var. *papillosus* (Plate 178), is distinguished by clumps of up to 10–12 stems, these to 20 cm long, and to 7 cm in diameter; Rio Grande plains, sandy loam, low elevations, from McMullen County south to near the Rio Grande. Some authors have attempted to recognize an endemic variety, *E. papillosus* var. *angusticeps* (Clover) W. T. Marshall (Plates 179, 180), distinguished by smaller plants with 5–95 stems, these 4–8 cm long, and 2–3 cm in diameter; restricted to Starr County and northern Hidalgo County in South Texas, growing under mesquite thickets. Taylor (1985) noted that *E. papillosus* seems to be related to *E. fendleri* in a different section of the genus than *E. berlandieri* with which it was assumed by some previous workers to be allied. Presumably, the specific epithet alludes to the conspicuous stem tubercles of *E. papillosus*. The varietal epithet *angusticeps* is a reference to the more slender stems of this taxon. *Echinocereus papillosus* is not cold hardy enough to be propagated outside in most of the Trans-Pecos.

Echinocereus poselgeri Lem., Cactees 57. 1868. DAHLIA CACTUS,

SACASIL, PENCIL CACTUS. Plates 181, 182. [*Wilcoxia poselgeri* (Lem.) Britton & Rose; *Cereus poselgeri* (Lem.) J. M. Coult.]. Plants clambering through shrubs, arising from fascicled (one to several), tuberous roots 5–10 cm long, 2.5–5 cm in diameter. Stems branched, 30–120 cm long, 0.7–1.5 cm in diameter, terete, more slender and woody near the base. Ribs 8–10, inconspicuous. Spines appressed except near stem apex; central spine one, to 7–9 mm long, turned against the upper radials, dark brown to black, bases lighter; radial spines 8–16, 2–4.5 mm long, flat against the stem, whitish or gray, dark-tipped. Flowers to 6 cm long and 7 cm in diameter, pinkish to magenta, midregions of inner tepals darker especially near the base, the margins sometimes white; pericarpel areoles with much loose wool and bristlelike brown spines. Fruit ovoid, dark green to brownish, fleshy and juicy but drying soon after ripening, to 2 cm long and 1 cm in diameter, wool and spines persistent. Distributed in South Texas, Cameron County and western Hidalgo County north along the Rio Grande to Laredo, sandy soils in brushlands; adjacent Mexico in Tamaulipas, Nuevo León, and Coahuila, extending into the eastern Chihuahuan Desert from eastern Coahuila to eastern San Luis Potosí, to 3,500 feet. Until recently, many workers preferred to place *E. poselgeri* in the separate genus *Wilcoxia* because of its distinctive stem (this taxon and *Opuntia leptocaulis* produce the most slender stems of any cacti in the United States except for *Peniocereus striatus*) and spine characters and its dahlialike tubers, but Taylor (1985) and others have shown that the taxon clearly belongs with *Echinocereus*. The specific epithet honors Dr. H. Poselger, the German cactus collector who conducted fieldwork in the southwestern United States and northern Mexico between 1849 and 1852 (Taylor, 1985). Outdoor plantings of *E. poselgeri* in 1996 at Alpine have survived the winter temperatures. Single large plants, properly supported, produce numerous large pink and white or magenta flowers. The species should be a favored ornamental when propagated according to the advice given by Taylor.

4. ECHINOCACTUS Link & Otto
Echinocactus

Plants unbranched (in our species). Stems subglobose, ovoid, becoming columnar or pyramidal, or hemispheroidal and depressed-flattened, 5–30 cm long, 10–30 cm wide. Ribs 7–27, rounded or rather sharp; pith and cortex firm, not mucilaginous. Ribs uninterrupted or with shallow horizontal furrows between the areoles. Areoles subcircular to elliptic. Spines prominently annulate or cross-ribbed, pale reddish to reddish-brown, or grayish-white to nearly black overlying reddish; central spines 1–3 per areole, 2.5–8 cm long, 2–9 mm wide at the base, straight or curving, thick-acicular or flattened but angular, tapering from base to apex, microscopically pubescent or not. Radial spines similar to centrals but smaller, 5–8 per areole, 2–5 cm long, 1–3 mm wide at the base, acicular or flattened but angular and subulate like the centrals. Flowers produced on new growth at the stem apex, each one in a woolly area connected with the spine-bearing portion of the areole; flowers 5–6.4 cm in diameter; floral tube above the ovary funnelform; inner tepals (in Texas species) pink to magenta; outer tepals aristate or attenuate-spinose; filaments (in Texas species) red to pink; style lobes pinkish to reddish. Fruit 1.4–3 cm long, 1–2.5 cm wide (in Texas species), with aristate scales and woolly hairs in the scale axils, pinkish or red at maturity but obscured or not by dense tufts of white-woolly hairs (in our two Texas species), later drying and splitting in some species, succulent and indehiscent in others. Seeds black, irregularly ovoid or obovoid, smooth or papillate, 2–3 mm long and wide; hilum basal, oblique, or appearing lateral. [*Homalocephala* Britton & Rose]. $x = 11$.

A genus of 6–7 species that range from Texas west to California and south into central Mexico. Four species occur in the United States. Four species occur in the Chihuahuan Desert Region, two of them in the Trans-Pecos. The genus name is derived in part from the Greek *echinos,* "hedgehog." Weniger (1984) included *Ferocactus, Glandulicactus, Ancistrocactus, Thelocactus, Echinomastus, Neolloydia, Astrophytum, Hamatocactus,* and *Sclerocactus* (not found in Texas) in a broad concept of *Echinocactus*. Benson (1982) excluded all of these genera from *Echinocactus* except for *Astrophytum*.

Key to the Species

1. Ribs usually eight, their crests rounded; mature plants rarely depressed-globose but usually subglobose, subpyramidal, or short-cylindroid; spines glabrous; flowers magenta; fruit pink, drying quickly after ripening, indehiscent or dehiscent basally, at maturity completely obscured by a dense wool covering

    1. *E. horizonthalonius.*

1. Ribs 13 or more, at least after reaching sexual maturity, their crests relatively sharp; mature plants depressed or flattened and hemispheroidal;

spines microscopically pubescent; flowers pink, rarely whitish; fruit bright red, succulent, drying slowly, not or only irregularly dehiscent, not obscured by wool

2. *E. texensis*.

1. **Echinocactus horizonthalonius** Lem., Cact. Gen. Sp. Nov. 19. 1839. EAGLE-CLAW CACTUS. Plates 183, 184. Common in desert mountains and flats, of both igneous and sedimentary origin, including clay and gypsum, every county of the Trans-Pecos. 2,500–5,500 ft. Flowering Apr–Jul, and opportunistic after rain, in Jul–Sep. $2n = 22$. East of the Trans-Pecos mostly in Val Verde Co.; one sight-record by Zimmerman in Crockett Co.; nineteenth-century specimens cited by Benson from Kinney and Duval counties; N to central NM and W to S-central AZ. Mexico, in Sonora (where disjunct and rare) and S through the Chihuahuan Desert to San Luis Potosí, and possibly even farther S, where displaying obvious geographic variation. Map 37.

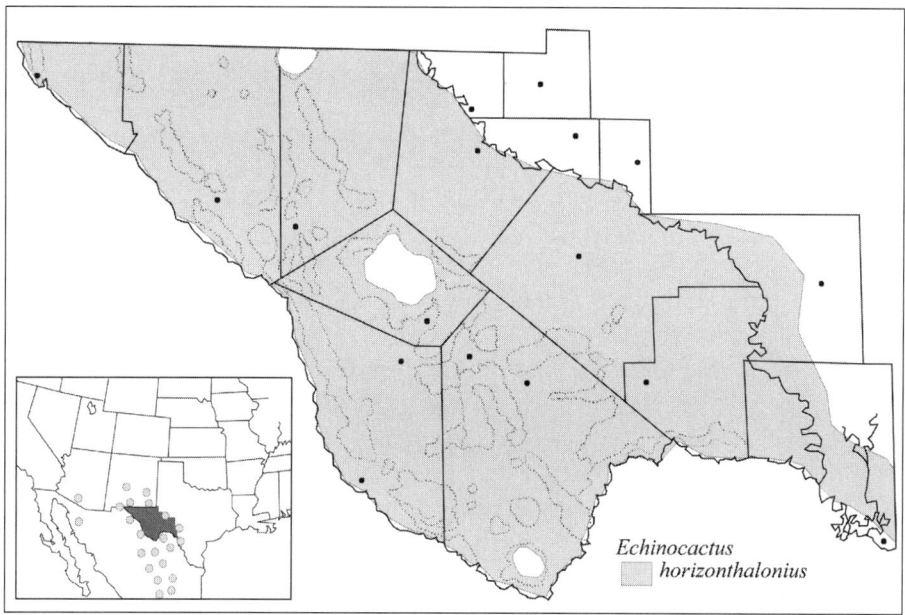

Map 37. Generalized distribution of *Echinocactus horizonthalonius* (eagle-claw cactus).

The flowers of *E. horizonthalonius* are visited by bees that undoubtedly serve as pollinators. The meaning of the specific epithet, frequently misspelled *horizontalonius,* is obscure unless one consults Lemaire's original diagnosis or the 1841 translation into German by Burghardt, quoted by Gottfried Unger: "*Areolen . . . waagrecht in die Quere stehend.*" (Passage provided by A. D. Zimmerman).

Identifying Characters. Echinocactus horizonthalonius is easily recognized by its solitary eight-ribbed stems that are gray-green to gray-blue and glaucous, subglobose, subpyramidal, or short-cylindroid (in older plants), and rigid, subulate spines that are prominently annulate and glabrous. The largest (lower) central spine, 1.8–4.3 cm long, curves downward, and two smaller upper centrals are compressed against or near the stem, along with the radial spines, and they are straight or curved. Typically, the radial spines are curved in toward the stem.

The magenta or bright rose-pink flowers are 5–7 cm long and 5–6.5(–9.5) cm in diameter when fully open, usually with a more intensely pigmented (darker) throat. The flowers are produced from among the dense, white wool-like trichomes that cover the stem apex; the flower buds are hidden by wool until near anthesis. The magenta or rose-pink inner tepals are ca. 3 cm long and 1.5 cm wide, with entire to serrate margins. The outer tepals are short, acute, and sometimes with a blackish spine at the tip. The funnelform receptacular tube and the pericarpel are furry with white "wool" from the axils of many small bracts with glabrous, blackish spinelike tips. The filaments are yellowish, and the anthers are yellow. The usually pinkish style supports 6–8 stigma lobes that are pinkish to olive.

The fruits remain completely or mostly hidden by the apical wool (long areolar and axillary trichomes). The fruits are pink to reddish, globose to ovoid-cylindroid, 1–3 cm long, and ca. 1 cm in diameter with some scales on the surface. At maturity the fruits quickly dry from the top downward, the relatively thin walls turning tan. The fruits are indehiscent or basally poricidal, in the latter case leaving a pile of seeds among the apical trichomes of the stem after the dry "shell" has been removed. The seeds are black or gray, subglobose to obovoid and angular, 2–3 mm long, the surface wrinkled and weakly papillate. The hilum appears to be basal or lateral.

Phenology. *Echinocactus horizonthalonius* is one of the few cactus species in the Trans-Pecos, other than *Opuntia*, for which the characteristic flowering time is delayed until late spring or early summer. The main flowering period for *E. horizonthalonius* appears to be in May and early June, although at least in some seasons and in some populations, major blooming episodes may occur as early as mid- to late April. The "main" or "major" flowering periods are ones in which most of the plants in a given population produce flowers. Clearly, *E. horizonthalonius* also is opportunistic in its flowering, with some (perhaps 20%; Worthington, 1986) of the same plants producing flowers again in response to adequate rainfall during the summer and perhaps in early fall, for up to three additional times during the year. Synchronous flowering of many populations, almost from one edge of the Chihuahuan Desert to the other, is at times surprisingly conspicuous during highway travel in the Trans-Pecos and northern Mexico. Informative observations by G. G. Raun (pers. comm.) include (1) early blooming on 16 April 1995 in the McKinney Hills (desert hills in Big Bend National Park) and on 19 April 1997 in the grasslands 4–5 miles south of Marathon; (2) a peak bloom (nearly 100% of the plants) on 5 June 1995 about two miles south of Marathon; and (3) complete failure to bloom, in a sample of about 50 plants, seven miles west of Marathon, after a fire burned away the apical wool.

Worthington (1986) observed that the duration between rainfall and flowering response in *E. horizonthalonius* (i.e., opportunistic flowering) was 3–12 days. His observations involved 18 plants under cultivation over a period of several years in El Paso, Texas, in some instances compared to plants in nearby natural populations. Individual flowers of *E. horizonthalonius* open at midday, close at night, and may or may not open again for 1–3 days. We have no records of the time required for fruit maturation.

Sterile and Immature Specimens. Mature specimens of *E. horizonthalonius* are orange- to grapefruit-size (rarely larger) with broadly eight-ribbed, depressed-globular, subpyramidal, or short-cylindroid stems and heavy, strongly annulate spines. The stems are 4–25(–30) cm tall and 8–15(–20) cm in diameter. The gray-green or gray-blue stem surface is partly obscured (in old plants) or easily seen (in young plants) through the vertical or (in very old age) spiraling rows or irregularly projecting or strongly decurved spines located on usually eight broad ribs. The (7–)8(–9) ribs are rounded and smooth or slightly constricted between areoles that are widely separated on immature plants and more closely spaced (1–2 cm apart) on adult plants. In adult plants there are (5–)8(–10) spines per areole, these stout, more or less subulate, strongly annulate, glabrous, gray to pink, tan, brownish, or nearly black, with the darker spine color sometimes evidently overlying a pale color. The only other Trans-Pecos species with prominently annulate spines are *E. texensis* and *Ferocactus wislizeni*. *Echinocactus horizonthalonius* is easily distinguished from *E. texensis* by rib number (more numerous in *E. texensis*) and rib shape (sharp, almost keeled in *E. texensis*), and from *F. wislizeni* by spine configurations.

The 1–3 central spines are 1.8–4.3 cm long and 1–2.5(–3) mm thick. In Trans-Pecos plants of *E. horizonthalonius*, usually there are three central spines: the main, longest (lower) central, which is usually descending and straight or decurved, and two shorter upper centrals that are more or less appressed, straight, and curved upward. The 5(–8) radial spines are similar to the central spines but typically curved toward the stem.

Immature specimens of *E. horizonthalonius* (e.g., those 2.5–8 cm in diameter) usually exhibit ca. eight spines per areole including the lower central, on green or blue-green, pincushion-like stems. The spines are already annulate on young stems 2.5 cm in diameter. Very rarely such small plants of *E. horizonthalonius* are spineless, in which case they may resemble mature plants of *Lophophora williamsii*. Seedlings of *E. horizonthalonius*, those less than 1 cm to 1.5 cm in diameter, are not ribbed, or else they are unevenly ribbed, with some of the ribs larger or smaller than the others. The ribs of seedlings resemble low, rounded tubercles, with a single areole, and 2–5 spines per areole (none of them distinctly "central"), these subulate, decurved, brownish or with some red or blackish-red, and the spines are not annulate. The seedlings of *E. horizonthalonius* resemble those of *Glandulicactus uncinatus*, but *G. uncinatus* can be distinguished by its hooked spines.

Biosystematics. The varieties of *E. horizonthalonius* recognized by Weniger (1984) to occur in the Trans-Pecos, var. *curvispinus* Salm-Dyck and var. *moelleri*

"Haage Jr.," do not appear to be morphologically and geographically distinct from each other and are not accepted here. Plants resembling the description of *E. horizonthalonius* var. *nicholii* L. D. Benson, a taxon said by Benson (1982) to be restricted to Arizona, occur in Big Bend National Park. Such plants in Texas appear to be older individuals in harsh desert habitats, and they are not given taxonomic recognition here.

No one has ever reported any evidence of hybridization between *E. horizonthalonius* and the regionally sympatric *E. texensis*. Chloroplast DNA data (Wallace, 1995) suggest that the two species are not closely related, a conclusion that also can be drawn through comparison of morphological characters.

Britton and Rose (1919–23) mistakenly included a distantly related Mexican species, *E. parryi* Engelm., in their concept of *E. horizonthalonius,* and they expanded their description of *E. horizonthalonius* to include *E. parryi* with 13 ribs. This apparently mistaken rib count has been copied in various subsequent descriptions of *E. horizonthalonius,* a species remarkably constant in its eight ribs.

Synonyms. The type material of *E. horizonthalonius* was conveyed to Europe by the botanical explorer Galeotti from deep in Mexico, possibly in San Luis Potosí, where these cacti are visibly different from our Trans-Pecos plants. Benson (1982) recognized two varieties of *E. horizonthalonius* north of the Mexican boundary: one he identified as *E. horizonthalonius* var. *horizonthalonius* (despite its very distant type locality); and the other he named as new to science, *E. horizonthalonius* var. *nicholii* L. D. Benson (supposedly a relictual endemic taxon, in Arizona and Sonora). Benson treated *E. horizonthalonius* var. *centrispinus* Engelm. as a synonym of *E. horizonthalonius* var. *horizonthalonius*. Weniger (1984) apparently recognized three varieties of the species: *E. horizonthalonius* var. *horizonthalonius*; *E. horizonthalonius* var. *curvispinus* Salm-Dyck; and *E. horizonthalonius* var. *moelleri* "Haage Jr.," ex Weniger, *nom. nud.*, in which Weniger included Benson's var. *nicholii* as a synonym. We suspect that several varieties of *E. horizonthalonius* should be recognized, but the nominate variety *horizonthalonius*, sensu stricto, is Mexican, not in our flora.

Common Names. Eaglesclaw cactus; eagle's-claw; eagle claw; eagle claws; turk's head; turk's head cactus; bisnagre; bisnaga de dulce; bisnaga meloncillo; manca mula; horse maimer; devil's head; devilshead; blue barrel; tepenexcomitl. "Turk's head" was used for *Melocactus communis* (Aiton) Link & Otto for hundreds of years and subsequently was misapplied to additional species, such as *E. horizonthalonius*. Eagle-claw cactus is vastly more appropriate for *Glandulicactus uncinatus.*

2. **Echinocactus texensis** Hopffer, Allg. Gartenz. 10: 297. 1842. HORSE-CRIPPLER. Plates 185, 186. Widespread in desert flats, substrates of igneous or sedimentary origin, desertscrub, grasslands, lower mountain slopes and basins, eastern two-thirds of the Trans-Pecos. 1,000–5,500 ft. Flowering Apr–May. $2n =$ 22. Widespread in Texas, from the S coastal plains at near sea level N, except in E Texas, to SW OK, across to SE NM. Mexico in the Chihuahuan Desert (mostly

in Coahuila and Nuevo León), and in Tamaulipas, S to vicinity of Tampico. Map 38.

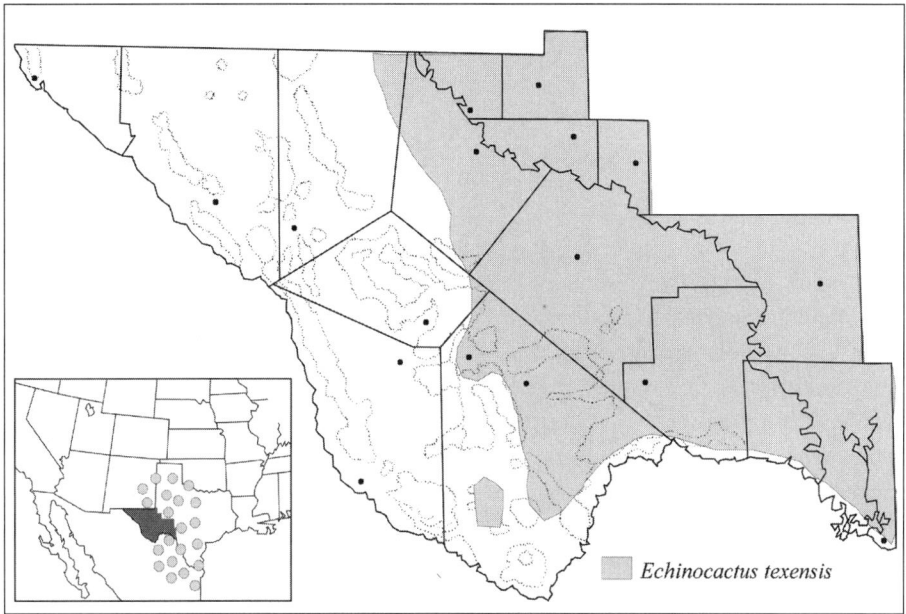

Map 38. Generalized distribution of *Echinocactus texensis* (horse-crippler).

Reportedly, *E. texensis* is more suitable to cultivation than many of the other barrel cactus types, because it is more cold hardy and capable of withstanding excessive ground moisture. Its wide distribution from sea level to desert floors and mountain slopes reveals its adaptability. In the Trans-Pecos *E. texensis* is particularly abundant in Pecos County and near Dog Canyon in Big Bend National Park. Numerous plants also have been located in alluvial basins near Alpine, Brewster County.

Identifying Characters. The depressed (flattened), solitary stems of *E. texensis,* deeply seated in the soil, provide one of the best traits for field identification. During dry periods the stems shrink to or slightly below ground level and may expand to several centimeters above the ground when moisture is plentiful. The stems are armed with long, rigid spines that have inspired the common name. Mature stems always have 13–27 ribs, and the ribs have sharper crests than those of *E. horizonthalonius.* The green to gray-green stems are as stiff as rubber tires and can be stepped on by heavy animals without harming the plant. Each areole has one long central spine (usually arching downward) and (5–)6–7 radials. The microscopically pubescent spines are flattened, light reddish-brown to pinkish, prominently annulate or cross-ribbed, broad at the base, and tapering to a point.

The flowers of *E. texensis* usually are pink but vary to rose-pink or salmon, rarely whitish, with prominent reddish centers. The flowers are 3–6 cm long and wide, usually smaller and paler those of *E. horizonthalonius*. The inner tepals are 2–3.2 cm long, (3–)6(–9) mm wide, with red bases, usually erose-margined, and apically aristate. The outer tepals are short, greenish-brown, partially covered with weblike trichomes, fimbriate or erose at the margins, and terminate in spinelike tips (these dark brown or blackish in herbarium specimens) that are microscopically pubescent. The filaments are pinkish to reddish, and the anthers are yellow. The style is pinkish to yellowish and supports 7–14(–17) stigma lobes that are ca. 3 mm long and pink to pinkish-white, sometimes red-striped underneath.

The bright red fruits are ovoid to globose, 1.5–5 cm long and 1.5–4 cm in diameter. The floral remnant is persistent. The crimson fruits are fully exposed, because *E. texensis* lacks the apical wool of *E. horizonthalonius*, but are inconspicuously ornamented by wool from the axils of bristlelike scales. Mature fruits persist on the plants for months, if not removed by animals. The fruits are mostly indehiscent, but some may rupture, forming longitudinal openings through which the bulging complement of black seeds is evident. The seeds are 2.5–3 mm long, weakly glossy, subreniform, or approaching globose or obovoid in shape, with a lateral, deeply concave hilum.

Phenology. In the Trans-Pecos *E. texensis* blooms in April and May. Unlike *E. horizonthalonius*, *E. texensis* does not appear to bloom opportunistically in response to summer rainfall. Under cultivation in Alpine, flowers open between 10:00 A.M. and 12:30 P.M. on warm, sunny days in late spring. Individual flowers close at night and may not open again the next day, or flowers may open again for 2–3(–4) days. The inner tepals are delicate, and flower opening may be influenced by wind, cool temperatures, or clouds. Fruit maturation requires several months. Under cultivation, where foraging animals are not a problem, red fruits may persist into the fall.

Sterile and Immature Specimens. The only species in the Trans-Pecos with which *E. texensis* might be confused in sterile condition is *E. horizonthalonius*. The two species both have heavy annulate or cross-ribbed spines, but the usually longer spines of *E. texensis* are microscopically densely pubescent, and the shorter spines of *E. horizonthalonius* are glabrous. The stems of *E. texensis* are flat-topped or hemispheric aboveground, whereas the stems of *E. horizonthalonius* grow taller and become short-cylindroid in old age. The stems of *E. texensis* are rarely branched, 10–30 cm long and 10–30 cm in diameter. The rib crests are without constrictions between the areoles. In desiccated plants the ribs may be sinuous. The stem surfaces are pale gray-green, blue-green, or brighter green. The apex of the stem is covered with dense, white wool, shorter than the apical wool of *E. horizonthalonius*.

The areoles are 3–6 cm apart on the rib crests, sometimes partly recessed, three-sided, and covered with white or gray wool. The spines, 7–8 per areole, are subterete to somewhat flattened, all prominently ringed or cross-ribbed, rigid, the surfaces tan, reddish to pinkish, or gray, and canescent. The spines are straight or decurved, menacing in appearance but not much obscuring the stem surface. In each areole there is a single central spine, the one that is responsible for the "horse-crippler" image of *E. texensis*. The central spine usually is deflexed but may be more or less porrect, straight or decurved, or even apically decurved into an open hook. The central spine may be (3–)4–6.5(–7.5) cm long and (1.5–)2–4(–8) mm wide. The upper surface of the central spine usually is flatter than the lower (underneath) surface, which may be longitudinally low-ridged. The radial spines are appressed or low-spreading, straight, or slightly decurved. Among the usually 6–7 radial spines, there are two diverging upper ones, two lateral, and two lower in position. The lateral radials usually are the longest at (1.2–)1.4–3.9(–5) cm, the lower ones the shortest, and the radials are about half as thick as the central spine. If a seventh radial is present, it is relatively short and directed upward between the upper pair.

Immature specimens of *E. texensis* that are 1–3 cm in diameter exhibit smaller and fewer (whitish) spines, about 5–6 per areole, than are found in areoles of older plants. A slender, brownish central spine is present in a few areoles in plants of this size grown from seed in the greenhouse. The spines in both juvenile and adult plants of *E. texensis* tend to be minutely and densely pubescent, unlike the glabrous spines of *E. horizonthalonius*. Long before sexual maturity, while still less than 1–3 cm in diameter, *E. texensis* often already has 13 narrow and sharply angled ribs, contrasting with the eight rounded ribs usually seen in mature *E. horizonthalonius*. Greenhouse-grown juveniles at 1–3 cm in diameter already have the hemispheric or flat-topped "land-mine" habit of adults, whereas juveniles of *E. horizonthalonius* may be either flat-topped or globular.

Biosystematics. Echinocactus texensis is the type species of a segregate genus, *Homalocephala* Britton & Rose (Taylor and Clark, 1983). Ferguson (1992) pointed out that the three taxa designated as the *E. polycephalus* Engelm. & Bigelow complex (Chamberland, 1997) also belong in *Homalocephala*, whereas *E. horizonthalonius* is less closely related. Chloroplast DNA data support the relationship of *E. texensis* with the *E. polycephalus* complex (Wallace, 1995) in the separate subgenus *Homalocephala*.

Synonym. Homalocephala texensis (Hopffer) Britton & Rose.

Common Names. Numerous popular names and variations of names have been applied to *E. texensis:* Horse crippler; devils pincushion; devil's head; devil's head cactus; candy cactus; visnaga; viznaga; manca caballo; devil's claw.

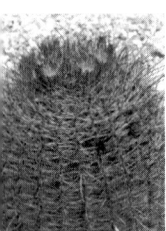

5. FEROCACTUS Britton & Rose
Barrel Cactus

Plants short-cylindroid, somewhat globose or ovoid in younger specimens. Roots diffuse. Stems unbranched or few-branched from the base, 15–100 cm long (1–3 m long in larger specimens of *F. wislizeni*), 10–45(–60) cm in diameter. Ribs (8–)12–28(–40), rib crests straight or crenate, without podaria except in *F. hamatacanthus*. Pith and cortex not mucilaginous. Areoles circular to elliptic; peglike secretory spines usually present, but sometimes short and inconspicuous. Central spines usually four (perhaps more) per areole, reddish to salmon overlain by light gray or yellowish to brown, strongly cross-ribbed or annulate, or only faintly so, or not annulate (in *F. hamatacanthus*), straight, curved, or hooked (the lower one largest and hooked), 3–11(–15) cm long, 0.5–3.5 mm broad, flattened, needlelike, or subulate; radial spines 8–13(–20) per areole, brown, reddish, or stramineous, or grayish, needlelike or slender and flexible, sinuate, straight or curving, annulate or not, 1.5–7 cm long; flower-bearing portion either confluent with the spine cluster or connected to the spine cluster by a felty groove. Flowers on new growth, borne in a ring near the apex but several centimeters away from the apical meristem; flowers 4.4–7.5 cm in diameter, floral tube obconic to funnelform; inner tepals yellow, orange-yellow, or red, narrowly lanceolate to oblanceolate, 2.5–4 cm long, 0.5–1 cm broad, acute, mucronate or mucronulate, entire or serrulate; outer tepals with greenish to reddish midribs and middle portions, margins yellow or orange-yellow; filaments yellow or orange-yellow, anthers yellow; styles yellow, stigma lobes long. Fruit fleshy with a thin wall, green to reddish-green, indehiscent (*F. hamatacanthus*), or with a thick, fleshy, yellow wall but dry interior, more or less dehiscent at the base (*F. wislizeni*), ovoid, globular, or short-cylindroid, 3–5 cm long, ca. 2.5 cm in diameter, with numerous conspicuous fimbriate or denticulate scales 3–4.5 mm long, 2.5–3.5 mm wide; floral remnant persistent. Seeds black, 1.4–1.6 mm long, 1–2.5 mm broad, 0.7–1 mm thick; hilum basal or subbasal. $x = 11$.

A genus of 23 (Taylor, 1984) or 25–30 (Lindsay, 1955) species, distributed from Texas west to California, but mostly in Mexico south to Oaxaca. About six taxa (four species) occur in the United States (Taylor, 1984). Two species occur in Trans-Pecos Texas. Benson (1969b, 1982) recognized three species of *Ferocactus* for Texas, with one of them, *F. setispinus* (Engelm.) L. D. Benson, mostly South Texas in distribution. In the current treatment *F. setispinus* is recognized as belonging to the separate genus *Hamatocactus*, as *H. bicolor*. The two true barrel cacti in Texas, *F. wislizeni* and *F. hamatacactus*, are actually quite distinct and placed in separate subgenera by Taylor and Clark (1983) and Taylor (1984). According to Benson (1982) the genus name *Ferocactus* means "wild or fierce cactus" after the Latin *ferus*, "wild." The only really large barrel cacti in the Trans-Pecos are larger specimens of *F. wislizeni* in the Franklin Mountains near El Paso, although reportedly few large plants remain near the roads after many years of them having been removed for landscape use. Weniger (1984) included *Ferocactus* in his megagenus concept of *Echinocactus*.

Key to the Species

1. Mature stems, 20–200 cm tall (in Texas), 19–60 cm in diameter; spines, at least some of them in addition to the hooked main central, strongly cross-ribbed or annulate; central spines rigid, 1.5–3.5 mm wide at base; fruit yellow at maturity, the rind thick, pulp dry

   1. *F. wislizeni.*

1. Mature stems 10–30(–45) cm tall, 7.5–17 cm in diameter; spines, even the hooked main central, not cross-ribbed but sometimes weakly annulate; central spines flexible, 1–2 mm wide at base; fruit green or reddish-tinged, the rind thin, pulp juicy

   2. *F. hamatacanthus.*

1. **Ferocactus wislizeni** (Engelm.) Britton & Rose. ARIZONA BARREL CACTUS. Plates 187–89. [*Echinocactus wislizeni* Engelm., Mem. Tour N. Mex. 96–97. 1848]. Rocky slopes and canyons, alluvial fans, and gravel soils, in Texas apparently restricted to the Franklin Mts. 3,500–5,300 ft. Flowering late summer. $2n = 22$. Reported from the Pecos River drainage in SE NM, W across southern NM from SW Lincoln Co. to Hidalgo Co.; southern AZ, mostly in Pima Co. and N to Yapavai Co., W to Yuma Co. Mexico, N Chihuahua and N Sonora. Map 39.

According to Benson (1982) *F. wislizeni* is the largest barrel cactus in the United States, with mature plants reaching 2.5–3 m or more high and ca. 0.6 m

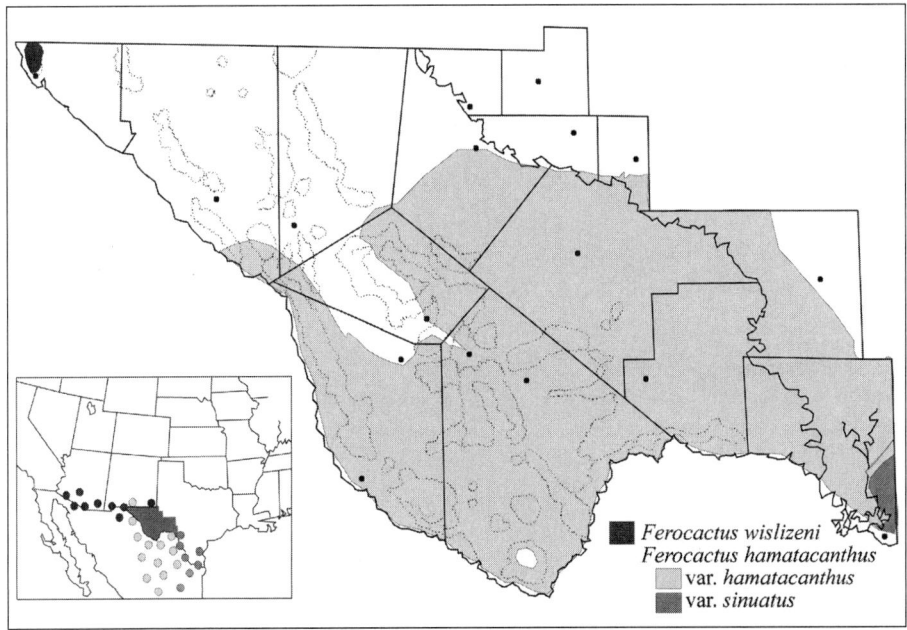

Map 39. Generalized distribution of *Ferocactus wislizeni* (Arizona barrel cactus), *F. hamatacanthus* var. *hamatacanthus* (giant fishhook-cactus), and *F. hamatacanthus* var. *sinuatus* (lower Rio Grande valley barrel cactus).

in diameter, particularly in southwestern Arizona. One plant measured in the Franklin Mountains in 1996 was 1.8 m tall and 2 m in circumference.

A few plants of Arizona barrel cactus have been found in Big Bend National Park, Brewster County, near old houses along the Rio Grande, but they surely were brought in by the early residents (Wauer, 1980). A report of two supposedly naturally occurring (in a mesquite thicket) plants of *F. wislizeni* near Glenn Spring in Big Bend National Park has been verified by photographs (A. D. Zimmerman, pers. comm.). Glenn Spring was the site of early ranching activity and a general store (Gomez, 1992; and see several historical record sources at Archives of the Big Bend, Bryan Wildenthal Memorial Library, Sul Ross State University, Alpine, Texas), and plants of *F. wislizeni* could have been transported there. In 2001 G. G. Raun observed a plant of *F. wislizeni* in natural surroundings ca. 33 miles south of Marfa, Presidio County. Naturally occurring *F. wislizeni* in Brewster and Presidio counties, if verified, would be remarkable.

The specific epithet honors early botanist-explorer Frederick Adolphus Wislizenus, M.D. The correct spelling of the epithet is *wislizeni* (Taylor, 1984), the genitive form of Wislizenus, and not *wislizenii*, as used in some treatments of the species (e.g., Benson, 1982).

Identifying Characters. Plants of *F. wislizeni* are hemispheroidal, barrel-shaped, or short-cylindroid single stems, often reaching 1 m high. Bowers (1998) studied *F. wislizeni* in the northern Sonoran Desert and determined that plants start to reproduce when they reach the size of ca. 19 cm in diameter. The stems are rarely branched and then probably only after injury. Usually over 20 ribs (up to 28) are shallowly notched above each areole and collectively support rather dense spines that partially obscure the stem. Each areole has (depending on interpretation) four central spines and (12–)16–20 radial spines. The central spines are arranged in a near cross although the spines are angled outward, and the lateral centrals are directed upward. The lower central is much larger and hooked downward or twisted to one side. The central spines are basically reddish or salmon-colored, but overlain with a gray surface, and strongly cross-ribbed. The hooked central is flattened and/or sulcate and 1.5–3.5 mm wide. The radial spines are ashy-gray, to 3.5–5.4 cm long, the lateral ones mostly appressed against the stem, flexible, and undulate, the upper and lower radials perhaps angled out from the stem and stiff, the lower three approaching the centrals in appearance but much smaller. The slender radials are cross-ribbed, or not.

Flowers are produced in adaxial prolongations of the spine-bearing areoles. Usually several flowers are produced at the stem apex, in good seasons ringing the apex. The showy flowers are 4–7 cm long and 4–5 cm in diameter. The inner tepals, outer tepals, and staminal filaments typically are orange-yellow but may range from pure yellow to red-striped or red. The anthers are yellow. The 18–20 or more stigma lobes are yellow to pale orange-yellow. The numerous, crowded pericarpel scales are green at first, turning yellow with the ripening fruit, the margins scarious, fimbrillate, and obtuse.

The fruit of *F. wislizeni* is yellow, barrel-shaped, with a thick fleshy wall, but dry interior, and with a persistent floral remnant. In good flowering years apical rings of yellow fruits each 4–5 cm long, ca. 2.5 cm in diameter, may persist for several months. The fruit walls support numerous semicircular, fimbriate scales. Mature fruits contain hundreds of edible black seeds. Some of the seeds escape the fruits through a perforate basal abscission scar during detachment of the fruit. The seeds are about 2–2.5 mm long, shiny or dull, microscopically reticulate and verrucose, and have a broad rim associated with the hilum.

Phenology. Usually, blooming follows significant rains during the summer. Two large individuals of *F. wislizeni* transplanted from the Franklin Mountains to the Sul Ross Cactus Garden in Alpine produced rings of flower buds in late June, and the first flower opened on 1 July 1996. The single flower was still open on 3 July when a second flower opened on the opposite side of the ring of buds. Flowers continued opening on this plant through 20 July. On second and third plants, by 18 July, less than half the buds had opened, and flowers lasted through all of July. Flowers open during late morning or about noon, at least partially close at night, and open again the next day for several days. One flower on one plant opened on 28 September in 1997. Grant and Grant (1979a) observed that plants of *F. wislizeni* in Arizona began to bloom in late August following summer rains and that the flowers were pollinated by several species of medium-size bees.

Sterile and Immature Specimens. Mature specimens of *F. wislizeni* in sterile condition usually are identifiable by their huge stems. No other cactus in the Franklin Mountains has stems approaching the size of those of *F. wislizeni*. Young plants of *F. wislizeni* and mature plants of *F. hamatacanthus* may be of similar size, but *F. hamatacanthus* rarely (if at all) occurs in the Franklin Mountains with *F. wislizeni*. Other useful vegetative characters include the ca. 20 or more ribs, mere notches in the rib crests in *F. wislizeni*, compared to ca. 12–17 ribs and strongly crenate ribs (almost divided into separate tubercles) in mature plants of *F. hamatacanthus*. In *F. wislizeni* the hooked spine is broader (to 3.5 mm) and strongly cross-ribbed.

Juvenile specimens of *F. wislizeni* have fewer ribs than mature plants and potentially may resemble *F. hamatacanthus* or even *Echinocactus horizonthalonius* or *Glandulicactus uncinatus*, which are common in the Franklin Mountains. Unlike *F. wislizeni*, *G. uncinatus* always has hooked lower radial spines. Juvenile plants of *F. wislizeni* that are 2–4 cm in diameter have indistinct ribs or some distinct ribs with short tubercles and circular areoles with whitish wool. Their central spines are solitary, relatively rigid, reddish-brown proximally and yellowish distally, straight, curved apically, or somewhat hooked at the apex. Usually there are 6–8 radial spines that are reddish-brown, straight, and appressed. Both centrals and radials are densely minutely pubescent. Actively growing new spines in apical areoles of juvenile plants are beet-red at the bases. In immature plants of *F. wislizeni*, usually those larger than 4 cm in diameter, the spine number in each areole is 1–3 centrals and 8–10 radials, as it is in *F. hamatacanthus*. As soon as plants are old enough to produce hooked central spines, *F. wislizeni* can be distinguished from *F. hamatacanthus* by its prominent cross-ribbing or annulations of the spines.

Even immature plants of *F. wislizeni* usually have more ribs than the typical eight ribs in adults of *E. horizonthalonius*. Healthy juveniles of *F. wislizeni* have green stems that are darker than the blue-green, often glaucous stems of *E. horizonthalonius*.

Biosystematics. Three sharply defined taxa were interpreted as varieties of a widely ranging *F. wislizeni* by Taylor (1984). Bravo-Hollis and Sánchez-Mejorada (1991a) accepted the supposed conspecificity of "var." *tiburonensis* Linds., endemic on Tiburon Island off Sonora in the Gulf of California, but not "var." *herrerae* (J. G. Ortega) N. P. Taylor, a species from south Sonora and Sinaloa.

Synonym. See basionym under *F. wislizeni*.

Common Names. Southwestern barrel cactus; Southwest barrel-cactus; Southwest barrelcactus; barrel cactus; visnaga; biznaga; biznaga de agua; compass barrel; candy barrel; fishhook barrel-cactus; fishhook barrel; fish-hook barrel; fish-hook cactus. All barrel cacti in the United States are southwestern, and most have hooked spines.

2. **Ferocactus hamatacanthus** (Muehlenpf.) Britton & Rose. GIANT FISHHOOK-CACTUS. [*Echinocactus hamatacanthus* Muehlenpf., Allg. Gartenz. 14: 371. 1846].

Two varieties of *F. hamatacanthus* are recognized in Texas, extreme southern New Mexico, and Mexico.

Key to the Varieties

1. Ribs relatively thick, in younger plants with rounded tubercles; hooked central spine angled on one side, or terete, relatively stiff but still flexible, microscopically merely scabrous or sometimes short-pubescent; stigma lobes 11–14; fruits green at first, turning reddish-green with age
    2a. *F. hamatacanthus* var. *hamatacanthus*.

1. Ribs compressed and narrow, even in younger plants; hooked central spine flattened, very flexible, microscopically pubescent; stigma lobes 8–10; fruits remaining green
    2b. *F. hamatacanthus* var. *sinuatus*.

2a. **Ferocactus hamatacanthus var. hamatacanthus.** GIANT FISHHOOK-CACTUS. Plates 190–92. Rock crevices and various soil types, igneous and limestone, wooded mountains, desert mountains, alluvial deposits in basins and desert. Widely distributed in the Trans-Pecos, except rare northward and westward; (no modern records N or W of the Davis Mts; Zimmerman et al., forthcoming), SE to the Devils River in Val Verde Co. (Weniger, 1984). 800–7,500 ft. Flowering usually in the summer, Jun–Aug, occasionally earlier or later. $2n = 22$. South-central NM probably in Doña Ana and Otero counties. Northern Mexico, W of the Sierra Madre Oriental, S to San Luis Potosí and Durango, W to Chihuahua, in the CDR. Map 39.

The epithet *hamatacanthus* is derived from the Latin *hamatus,* "hooked," and Greek *akantha,* "thorn," in reference to the hooked main central spine. *Ferocactus hamatacanthus* var. *hamatacanthus* is one of the most widely distributed cacti, but not most abundant, in the Big Bend and east-central regions of the Trans-Pecos. Plants occur from the lowest desert up to the wooded mountains (rarely 6,000 feet and higher), frequently under shrubs and in crevices of rocks. They are locally common in the limestone mesas and alluvial valleys of the eroded Stockton Plateau in Pecos County and the desert mountains and desert basins in the Big Bend region.

Identifying Characters. In Texas, the stems of var. *hamatacanthus* are usually solitary, but Weniger (1984) remarked that plants of var. *hamatacanthus* might form larger clusters of stems (more than 2–3) after injury, a phenomenon verified by our observations. We have observed several instances where seemingly uninjured plants had more than three stems, particularly plants growing in rock crevices.

Plants are already sexually mature when stems reach 15 cm high or less. Plants with stems 10–30 cm high and 10–25 cm in diameter are relatively common in the Trans-Pecos. Larger plants 40–45 cm high are rarely encountered north of Mexico, although experienced observer Barry Hughes reported seeing a single plant near 90 cm high in Boquillas Canyon. Stems to 60 cm are reported for var. *hamatacanthus* by Benson (1982) and Weniger (1984). In overall aspect the stem tips of mature plants have a covering of red spines or a mixture of red spines and yellowish hooked central spines. The 10–13(–17) ribs are partially obscured, mostly by interlacing radial spines. The ribs are large, 2.5–5 cm high and thick, and partially divided into rounded tubercles.

The flowers of *F. hamatacanthus* are yellow (Taylor, 1987), funnelform, 6–8(–10) cm long, 6.5–8.5 cm in diameter, with yellow anthers, and with 11–14 yellow stigma lobes 4.5–6 mm long. In the Trans-Pecos the inner tepals usually are totally lemon-yellow, or pale green at the base, glossy, 3–4 cm long, ca. 1 cm wide. The flowers are fragrant. The pericarpel surface exhibits 20–40 small, triangular scales with greenish-yellow, fimbriate margins.

The fruits are obovoid to oblong, 3–5 cm long, ca. 2.5 cm in diameter, usually with a persistent floral remnant, and with 20–40 widely spaced scales on the greenish surface, these with fimbriate-scarious margins. The fruits are thin-walled, with white pulp, juicy, and sweet. Taylor and Clark (1983) reported that when fruits are mature, seed dispersal follows the rupturing of the fruits near the apex and the extruding of seeds in a liquid, but in our experience the fruits are indehiscent. The juicy fruit interior is markedly different from the dry fruit of *F. wislizeni.* The seeds of *F. hamatacanthus* are black, shiny, ca. 1.5 mm long, ca. 1 mm broad, ovate, microscopically pitted on the surface, and with a basal-lateral hilum that is marked by a sharp, narrow hilum-micropylar rim (Taylor and Clark, 1983).

Phenology. Flowers of *F. hamatacanthus* usually are produced in the summer, from June through August. The fruits mature slowly, persist through the summer and fall, apparently reaching full maturity in late fall and winter as they develop reddish or reddish-brown coloration, especially around the scales. It is not known if flowering in *F. hamatacanthus* is stimulated by rainfall. Individual flowers open about midday, partially close at night, and open again for several days. Fruits possibly require about six months to reach maturation, but this aspect remains to be thoroughly studied.

Sterile and Immature Specimens. Even in sterile condition adults of *F. hamatacanthus* var. *hamatacanthus* are not likely to be confused with any other species in Texas. The natural populations of *F. hamatacanthus* are not sympatric with the other large barrel cactus species in Texas, *F. wislizeni*. Smaller plants of *F. hamatacanthus* are superficially similar to *Glandulicactus uncinatus* var. *wrightii*, but the two species are readily distinguished by the all-straight radial spines in *F. hamatacanthus*, whereas the three lower radial spines in each areole are hooked in *Glandulicactus*. The Texas distribution of *F. hamatacanthus* var. *hamatacanthus* may overlap and even intergrade with *F. hamatacanthus* var. *sinuatus*, particularly in the vicinity of the Devils River. In Mexico *F. hamatacanthus* var. *hamatacanthus* and *F. hamatacanthus* var. *sinuatus* replace each other on opposite slopes of the Sierra Madre Oriental (Zimmerman et al., forthcoming).

The globular to short-cylindroid stems of var. *hamatacanthus* usually are dark green but may be gray-green or even reddish-tinged. The stems are predominantly tuberculate, particularly in smaller plants, or they may have 10–13 or more poorly defined or obvious ribs. Even on ribbed stems rounded tubercles usually dominate the plant aspect. Areoles in mature plants of *F. hamatacanthus* are circular to elliptic. The circular areole diameter is about 5–7 mm, and the elliptic areoles reach 10–13 mm long, including mostly the spine-bearing portion. On older tubercles the areoles narrow on the adaxial end (toward the stem). White to gray wool occurs in younger areoles.

The spine pattern of *F. hamatacanthus* is characterized by four centrals and (8–)10–14(–20) radials. Of the four central spines, three are regarded as upper centrals in position in the areoles, and they are angled upward as well. The upper centrals are subterete or somewhat compressed or angled, perhaps flattened on the adaxial surface, they range to ca. 8 cm long, and they are straight (not hooked) at the tip. In most areoles one central spine originates in a lower central position. This lower central also might be described as the "main central" because it is potentially the longest of the four central spines, typically reaching 9–11(–15) cm, and the lower central is also marked by its hooked tip. The hooked lower central is yellow or stramineous, especially in younger plants, and the flexible shaft is basically straight (not loosely spiraling), although usually it is twisted one or two turns. Characteristically, the shaft of the hooked lower central is 1.5–2 mm wide, flattened adaxially, typically with a narrow groove in the middle of the flat surface, and usually otherwise rounded or angled in cross section. Two hooked central spines occur in the areoles of some plants, but areoles with one hooked lower central are characteristic of *F. hamatacanthus*. The central spines (and radials), particularly the younger ones, may be microscopically scabrous. Up to eight central spines have been reported for var. *hamatacanthus* (Benson, 1982).

The radial spines in var. *hamatacanthus* are somewhat variable in number, diameter, and position. In general the (8–)10–14(–20) radial spines are up to (1.5–)3.5–7 cm long, terete, somewhat compressed or angled, and the radials may be relatively stout and stiff or somewhat thin and flexible, or even sinuate. The longest radials are the upper ones that project upward and away from the stem and the lateral ones that are appressed against the horizontal plane of the stem, often interlacing with the lateral, appressed spines of adjacent areoles that are 1–3 cm apart. The lower radials usually project downward. The spines in mature areoles may vary in color from gray to reddish. In some plants of var. *hamatacanthus*, particularly on the upper one-half or one-third of the stems, the collective dominant color of the spines is red. The spines closest to the stem, radials and upper centrals, may be predominantly red on upper parts of the plant and gray to reddish down the stem.

Immature specimens of var. *hamatacanthus* tend to have larger, swollen, or at least more prominent tubercles than in some adult plants. The ribs are usually less than 13, and the ribs may be poorly defined. The areoles of immature plants have fewer and shorter central and radial spines than is typical in adult areoles. The number of central spines is 1–3 in immature plants, with the hooked main central appearing early, usually before all three upper centrals appear. The hooked main central in immature plants tends to be more terete or angled and not necessarily flat on the adaxial surface. In immature areoles 8–10 radial spines are usual, and the radials often are slender and gray. Greenhouse-grown juvenile plants of var. *hamatacanthus* that are 1–4 cm in diameter have prominent, conical tubercles 3–8 mm long. We have observed hundreds of these seedlings, and all have been tuberculate. The circular areoles have white wool. The juveniles have one (apparent) central spine that is much larger than the radials, and it is slender, very flexible, reddish with a yellow distal portion or yellowish throughout, straight, curved apically, or somewhat hooked at the apex. Usually there are 6–9 radial spines that are white and more or less appressed. In most juveniles examined, both centrals and radials are glabrous or scabrous as seen under a 10× lens, although especially in some smaller plants (1–2 cm in diameter), a portion of at least some spines are as prominently short-pubescent as they are in *F. wislizeni* at this stage of growth. The spines are bright red in apical areoles of juvenile var. *hamatacanthus* plants.

Biosystematics. Relatively comprehensive studies of *Ferocactus* (Taylor and Clark, 1983; Taylor, 1984) place *F. hamatacanthus* in the *Recurvus* group of section *Bisnaga*. This is the section of *Ferocactus* that is characterized by juicy fruits that do not dehisce via a basal pore but (very rarely) may rupture near the apex, with the seeds ultimately being released in a liquid, and by seeds less than 2 mm long with a sharp, narrow hilum-micropylar rim. The other barrel cactus species in Texas, *F. wislizeni*, is in a different subgeneric group, section *Ferocactus*, which is characterized by fruits with a dry interior, dehiscence via a basal pore, and seeds to 3 mm long and with a broad hilum-micropylar rim. *Ferocactus hamatacanthus* and *F. wislizeni* are allopatric in distribution and so distantly related that they probably would not hybridize if they did occur together.

Two varieties of *F. hamatacanthus* are recognized to occur in Texas (Benson, 1982; Taylor, 1984; Zimmerman et al., forthcoming): var. *hamatacanthus* and var. *sinuatus*. Most workers (e.g., Weniger, 1984; Taylor, 1984; Zimmerman et al., forthcoming) have interpreted as separate taxa the western mountain and Chihuahuan Desert entity and the southeastern entity occurring northwest to the Devils River in Texas. Weniger (1984) treated var. *sinuatus* as a distinct species (*Echinocactus sinuatus* A. Dietr.). Benson's concept of *F. hamatacanthus* var. *hamatacanthus* included specimens from the humid shrublands in South Texas and adjacent Tamaulipas, Mexico, at elevations of about 30 feet. Benson (1982) treated var. *sinuatus* as a rare entity restricted to San Patricio County in South Texas.

Synonyms. Hamatocactus hamatacanthus (Muehlenpf.) F. M. Knuth; *Echinocactus longihamatus* Galeotti; *E. longihamatus* var. *crassispinus* Engelm.; *Ferocactus hamatacanthus* var. *crassispinus* (Engelm.) L. D. Benson.

Common Names. Whiskered barrel cactus; turkshead echinocactus; turks head; visnaga; bisnaga costillona; biznaga espinosa; biznaga ganchuda; biznaga limilla; biznaga de tuna; biznaga de limilla; Texas longhorn; turk's head.

2b. **Ferocactus hamatacanthus** var. **sinuatus** (A. Dietr.) L. D. Benson. LOWER RIO GRANDE VALLEY BARREL CACTUS. Plate 193. [*Echinocactus sinuatus* A. Dietr., Allg. Gartenz. 19: 345. 1851]. Not known to occur in the Trans-Pecos. This small eastern geographic race replaces var. *hamatacanthus* E of the desert, from the Devils River eastward; eventually some populations of *F. hamatacanthus* W of the Pecos River might be interpreted as var. *sinuatus*. The distributional range for var. *sinuatus*, as the taxon was interpreted by Taylor (1984) and Bravo-Hollis and Sánchez-Mejorada (1991a), is in S TX, mostly in the brushlands from near Brownsville to Eagle Pass, NE to the edge of the Hill Country near Camp Wood, and on W past the Devils River, and in adjacent Mexico E of the Sierra Madre Oriental. $2n = 22$. Map 39. Varietal identification of *F. hamatacanthus* specimens is difficult near the Devils River and along the eastern margin of the Chihuahuan Desert in Coahuila and Nuevo León, where taxonomically ambiguous intermediates occur.

According to Taylor (1984), var. *sinuatus* is distinguished by its smaller stems to 30 cm long and 20 cm in diameter; ribs ca. 13, these narrow and more acute at the crest; radial spines 8–12, with at least some of them markedly flattened; central spines four, with the lowermost one strongly flattened, pubescent, and somewhat flexuous; stigmas 8–10; fruits sometimes globose and to 2.5 cm long, maturing green (Weniger, 1984) to dark brownish-red (Taylor, 1985); and seeds ca. 1 mm in diameter. By contrast var. *hamatacanthus* typically has larger stems; ribs 12–17, rounded, and strongly tuberculate; radial spines 8–20, terete; central spines 4(–8), terete or somewhat flattened; stigmas 11–14; fruits ovoid to oblong, to 5 cm long; and seeds to 1.6 mm in diameter.

6. HAMATOCACTUS Britton & Rose
Twisted-Rib Cactus

A genus of one species, as interpreted here, distributed from Central Texas to South Texas and adjacent northeastern Mexico. The taxon has been aligned by various workers with at least four different genera. The genus name *Hamatocactus* is rooted in the Latin *hamatus,* "hooked," and the Greek *acantha,* "thorn," a reference to the hooked central spines.

1. **Hamatocactus bicolor** (Terán & Berland.) I. M. Johnst., non *Thelocactus bicolor* (Galeotti) Britton & Rose. TWISTED-RIB CACTUS. Plates 194, 195. [*Cactus bicolor* Terán & Berland., Memorias de la Comision de Limites 4. 1832]. Plants unbranched or branched at the base. Roots diffuse. Stems green, hemispheric when young, then becoming ovoid to cylindroid, 3.6–12(–20) cm long, 4.5–12 cm in diameter. Ribs 13, spiraling or straight, slender, sinuate, sharp at the crests; narrow tubercles raised on the ribs. Areoles 2.5–3 mm in diameter, becoming elliptic or ovate, 4.5–5 mm long; after sexual maturation a felted groove expands adaxially for a few millimeters; usually one or a few golden (dark with age), cylindrical, glands adjacent to the spines. Spines not obscuring the stem; central spine one, porrect, hooked, 1–3.8 cm long, acicular or rarely flattened, minutely scabrous, yellowish, becoming ashy-gray or reddish-brown; radial spines 10–19, yellowish, whitish, or reddish-brown, acicular, slightly diffuse, straight or slightly curved toward the stem, the longest (upper ones) 1.2–3.2 cm. Flower buds produced in a felty adaxially extension groove between the stem and the gland-/spine-bearing portion of the areole, leaving a circular pit after flower or fruit abscission; flowers yellow with red centers, 3.7–7 cm long, 4–7 cm across; inner tepals yellow (to cream or ivory) with red bases, oblanceolate, 2–2.5 cm long, 0.6–0.9 cm wide, margins entire toothed, or lacerate, apex acute and cuspidate; outer tepals to 2 cm long, mostly green but with reddish margins, the lower ones obtuse, or with broad, somewhat auriculate distal portions, the upper ones acute and finely fringed; stamens with weak, swirled filaments, these reddish, pale yellow, or whitish, supporting pale yellow anthers; style yellow to greenish-yellow, with 5–11 stigma lobes, these 3–7 mm long and pale yellow to orangish; ovary ca. 4.5 mm long. Fruit bright red, fleshy, spherical or nearly so, 8–13 mm wide, dehiscent or indehiscent, with a maximum of about 10–15 scales, usually fewer, these whitish, with fringed margins; floral remnant persistent. Seeds black, minutely papillate, obovoid, usually 1–1.4 mm long, 0.8–1 mm wide, expanded around the micropyle, hilum basal. $x = 11$.

Hamatocactus bicolor occurs in heavy soils in grassland and shrubland habitats, particularly in mesquite thickets, from Brown and San Saba counties of Central TX SE to Travis Co., S to Cameron Co. at the S tip of TX, and W to Val Verde Co. in limestone habitats near the mouth of the Devils River. Reported from S Terrell Co. From near sea level to ca. 1,000 ft. Flower spring–summer. $2n = 22$. Adjacent Mexico, Tamaulipas, to be expected in Nuevo León, and Coahuila. Map 40.

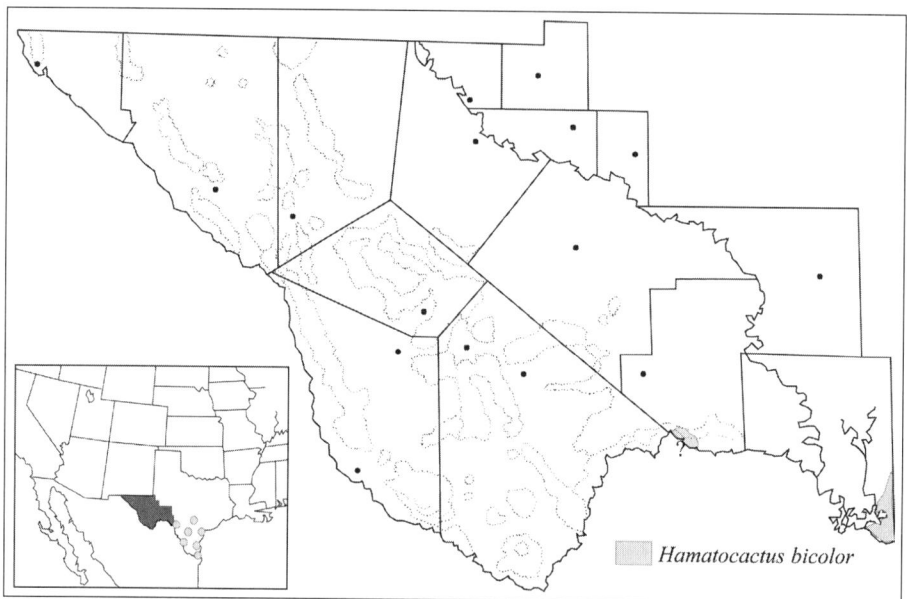

Map 40. Generalized distribution of *Hamatocactus bicolor* (twisted-rib cactus).

Hamatocactus bicolor is not documented to occur in the Trans-Pecos, but on 23 October 1997, Jim Talbot photographed a specimen said by the collector to have been taken from near the Rio Grande in the vicinity of the Brewster-Terrell county line, generally south of Sanderson. The known occurrence of *H. bicolor* near the mouth of the Devils River suggests that it could exist in sheltered microhabitats up the Rio Grande at least as far as Terrell County. The specific epithet, *bicolor*, presumably refers to the two colors in the flowers, yellow with red centers.

Identifying Characters. In the field *H. bicolor* is most likely to be confused with *Ferocactus hamatacanthus*, *Ancistrocactus* spp., and also possibly with *Glandulicactus uncinatus*. *Hamatocactus bicolor* is readily distinguished by its stem shape, 13 narrow ribs, and spine pattern. The flowers are borne near the stem apex and are 5–7 cm in diameter, smaller than the pure yellow flowers of *F. hamatacanthus*. The flowers of *Glandulicactus* are relatively small, red-maroon, and usually exist in clusters at the stem apex. The flowers of *Thelocactus bicolor* are magenta. In flowers of *H. bicolor* the red centers are the collective manifestation of red midregions coloring the basal-central portion of each yellow inner tepal, with the basal midregions tapering to an attenuated point from the base to about midtepal. The filaments are reddish and collectively swirled around the greenish-yellow style. The anthers are yellow to cream-colored. The stigma lobes are pale yellow to yellow, and curving in different directions.

The scarlet fleshy fruits of *H. bicolor* are spherical or slightly elongate and ca. 1 cm long, usually with fringed scales on the shiny surface. The floral remnant is persistent but usually abscises before the very persistent fruits fall from the stem. By the time the fruits ripen and turn red, often they are positioned on the upper stem some distance away from the apex. The fruits are indehiscent for a long period but eventually may form vertical slits. The small black seeds are somewhat obovoid but with a larger somewhat globose, minutely papillate end and a smaller somewhat cylindroid, smooth end. The hilum is recessed at the small end of the seed.

The fruits of *F. hamatacanthus* are larger than those of *H. bicolor* and either remain green or develop a reddish or brownish tinge. The fruits of *Glandulicactus* and *Hamatocactus* are fleshy and red, 2–2.5 cm long, ellipsoid, and exhibit numerous auriculate, deltoid scales ca 4.5 mm long. The fruits of *T. bicolor* are greenish-brown, ovate, and to 1.7 cm long.

Phenology. In south-central Texas *H. bicolor* is reported to bloom all summer and from April to October (Weniger, 1984). Under cultivation in Alpine, the plants bloom in the late spring (25 May in 1998) and early summer (June–July), but they appear to be equally prolific in flowering during August and in the early fall. Flowers are produced at the stem apex. During heavy blooming several flowers may open simultaneously on the same stem. Individual flowers open by midday (as early as 10:00 A.M.), close at night, and may not open again after the first day. The flowers are aromatic and attract several types of bees, which probably serve as pollinators. In Alpine, ripening fruits turn red by mid- to late fall. Such red fruits contain black, presumably viable seeds. We assume that fruits mature in 2–3 months, but they may persist for additional months through the winter. In some cases the fruits may shrivel, turn black, and persist for years. Jim Talbot has observed such shriveled black fruits still attached at near midstem on older plants of *H. bicolor*. Germination experiments have demonstrated that the seeds from persistent black fruits were viable (A. M. Powell, unpub.). The age of these fruits and seeds was estimated at about five years.

Sterile and Immature Specimens. *Glandulicactus uncinatus* apparently is not common in the South Texas range of *H. bicolor*, but it exhibits a prominent hooked central spine and might be confused with *Hamatocactus* on that basis. The principal central spine of *G. uncinatus* is directed upward, usually straw-colored except reddish-brown or blue-gray at the hooked tip, and 4–9.5 cm long, whereas the hooked central spine of *H. bicolor* is usually much shorter (1–3.8 cm long), directed outward, and may be ashy-gray, reddish-brown, or yellowish.

Adult plants of *H. bicolor* not in flower usually are easily recognized by the 13 ribs that may be sinuate, undulate, and spiral around the hemispheric to columnar stems. The ribs are slender and with rather sharp crests. The ribs have shallow to rather deep sinuses between the areoles. In some plants of *H. bicolor*, the ribs are straight, for example, in the population recognized by Weniger (1984) as the var. *setaceus* Engelm. (as *Echinocactus setispinus* var. *setaceus*). The spines do not obscure the dark green stems. The single, hooked central spine is not as long as in other relatively large cacti with hooked central spines. The central spine is rather stiff and enlarged at the base. The radial spines are thin and flexible. The radial spine number appears to be lower in northern and eastern parts of the range, with 10–15 spines being the norm, whereas in Cameron, Starr, and Hidalgo counties of deep South Texas (Weniger, 1984), 12–19 radial spines are common. The upper three radial spines, spreading in plane with the other radials, are obviously larger in diameter and might be interpreted as appressed upper central spines. The spines of *H. bicolor* appear smooth to the naked eye and under hand-lens magnification, but the spines, at least in young areoles, are minutely scabrous, not rough to the touch, but with short hairlike projections as seen under higher magnification.

In immature plants of *H. bicolor* the slender, sharp ribs and needlelike, flexible spines are reliable distinguishing traits. Even in juvenile plants, *H. bicolor* has a hooked, single central spine and slender, needlelike radials. *Ancistrocactus scheeri*, found within the range of *H. bicolor*, has one central spine that is hooked and 1–2 straight upper centrals.

Biosystematics. The taxon treated here as *H. bicolor* has been positioned in different genera by modern workers in Cactaceae. Benson (1982) included *H. bicolor* in *Ferocactus* (the barrel cactus genus) as *F. setispinus*, where it is anomalous in flower and fruit characters. Weniger (1984) accepted *H. bicolor* into the traditional broad genus *Echinocactus* (as *E. setispinus* Engelm.) and recognized two varieties of the species, *E. setispinus* var. *hamatus* Engelm. and *E. setispinus* var. *setaceus* Engelm., to occur in Texas. Bravo-Hollis and Sánchez-Mejorada (1991a) included *H. bicolor* in subgenus *Hamatocactus* of the genus *Hamatocactus*. These authors also placed *Glandulicactus uncinatus* in a second subgenus, *Glandulicactus*, of the genus *Hamatocactus*, indicating their perception of close relationship between *Hamatocactus* and *Glandulicactus*. *Glandulicactus* is treated in the current work as a distinct genus. After analysis of numerous characters, Anderson (1987) concluded that *H. setispinus* belonged with *Thelocactus*, sensu lato, yet *Glandulicactus* was not included in the study.

We recognize *Hamatocactus* as monotypic, with the single species *H. bicolor* having its major distribution in south-central Texas and extending across the Rio Grande into Tamaulipas, and perhaps extending as well into the adjacent Mexican states of Nuevo León and Coahuila. The two varieties of *H. bicolor* accepted by Weniger (1984) are treated as synonyms of *H. bicolor* by Benson (1982) and Bravo-Hollis and Sánchez-Mejorada (1991).

Synonyms. *Echinocactus setispinus* Engelm.; *Hamatocactus setispinus* (Engelm.) Britton & Rose; *Ferocactus setispinus* (Engelm.) L. D. Benson; *Thelocactus setispinus* (Engelm.) E. F. Anderson.

Common Names. Fishhook cactus; hedgehog cactus.

7. THELOCACTUS (K. Schum.) Britton & Rose
Glory of Texas

A genus of 10 or 11 species (Anderson, 1987; Zimmerman et al., forthcoming) located mostly in the CDR of Texas and northeastern Mexico but also occurring as far south as Tamaulipas, Querétaro, and Hidalgo. Only one species of *Thelocactus* occurs in Texas. The genus name is taken from the Greek *thele,* "nipple," in reference to the tubercles.

Thelocactus has been recognized as separate genus by all twentieth-century cactus specialists (e.g., Anderson, 1986, 1987; Anderson and Ralston, 1978; Glass and Foster, 1977; Benson, 1982; Bravo-Hollis and Sánchez-Mejorada, 1991a) except for Weniger (1984), who combined this and many other genera with *Echinocactus.* Our species, *T. bicolor,* may be more closely related to *Hamatocactus* (Anderson, 1987) than to the type species of *Thelocactus.* Herein we have treated *Hamatocactus* as a monotypic genus.

1. **Thelocactus bicolor** (Galeotti) Britton & Rose. GLORY OF TEXAS, MARATHON BASIN THELOCACTUS. Plates 196–99. [*Echinocactus bicolor* Galeotti, Abbild. Beschr. Cact. 2: t. 25. 1848, non *Hamatocactus bicolor* (Terán & Berland.) I. M. Johnst.]. Plants deep-seated or tall, typically with solitary stems, sometimes branched from the base. Stem surface green to gray-green or yellow-green, ovoid or elongate-ovoid to subconical, (3–)5–18(–28) cm long, 3.5–11(–17) cm in diameter. Ribs 8–13, either low and poorly developed (stems mostly tuberculate) or prominent, and often helically slanted in old age. Tubercles 9–12 mm long, 11–20 mm broad, round, subconical, either prominent or fused basally and forming low ribs, crowded and separated only by narrow spaces; pith and cortex firm, not mucilaginous; pith including vascular strands at least in some plants; druses present in hypodermis (Anderson, 1987). Areoles 5–16 cm apart, circular, ca. 3 mm wide, spine-bearing and woolly, with a short, wide, woolly areolar groove (most apparent on older plants) on the adaxial surface of each tubercle; areolar glands conspicuous or not. Spines completely or only partly obscuring stem surface, most of the centrals and radials multicolored, either stramineous at the tips and bases, or the bases gray or whitish, or pinkish to reddish in the middle zone, or reddish to the base; most colorful on new growth; spines (9–)12–30 per areole; central spines 1–4, the upper three, if present, like the radials except slightly larger, the lower one, the one most centrally positioned in the areole, porrect or deflexed, straight, reddish or gray, (1.3–)2–4.5(–6) cm long, 0.4–1.5 mm wide, flattened on one side (in var. *bicolor*) or not, keeled or not; other spines 11–19(–25), of two types: the upper 1–2(–4) are uniquely specialized, either subcentrals or radials, depending upon terminology, and are flattened, 2–7.5(–10) cm long, to 1.5 mm wide, straight, erect, yellow, becoming gray with age; the lower 10–14 obvious radials are terete, 1–1.5 cm long, appressed or slightly recurving toward stem, varicolored, reddish in the central portion, gray at the base, yellowish at the tips. Flowers 4–8 cm long and 4–8 cm in diameter, arising from woolly adaxial prolations of the areoles; inner

tepals magenta or rose-pink, basally darker and forming a darker red center of the flower, oblanceolate, 2.8–5.4 cm long, ca. 0.5–1.2 cm broad, acute to mucronate, minutely denticulate; outer tepals oblanceolate to 4 cm long, ca. 7.5 mm broad, with greenish-rose midribs, the margins pale rose, apex rounded or acute, minutely fringed; filaments cream to whitish at the base, reddish distally, ca. 1.0 cm long, anthers yellow, 1.5 mm long; style whitish, 2–3 cm long, ca. 1.5 mm in diameter; stigma lobes 7–13, ca. 4.5 mm long, reddish to orange or yellowish; ovary in anthesis with short-fimbriate scales. Fruits green to brownish-red, 0.7–1.8 cm long, 0.6–1.2 cm in diameter; pericarpel thin, not juicy, the fruit rapidly drying after ripening, dehiscent through a large basal pore, the pericarpel wall with 15–20 fimbriate scales. Seeds black, obovoid, truncate at the base, 1.5–2.5 mm long, ca. 1.75 mm broad, finely reticulate-papillate. $x = 11$.

Chihuahuan desertscrub, semidesert grassland, and Tamaulipan scrub. 2,300–4,000 ft. Flowering (Feb–)May–Sep. Texas, Starr Co. and the Trans-Pecos. Mexico in adjacent states of Chihuahua, Coahuila, Nuevo León, and Tamaulipas, and S to San Luis Potosí. Map 41.

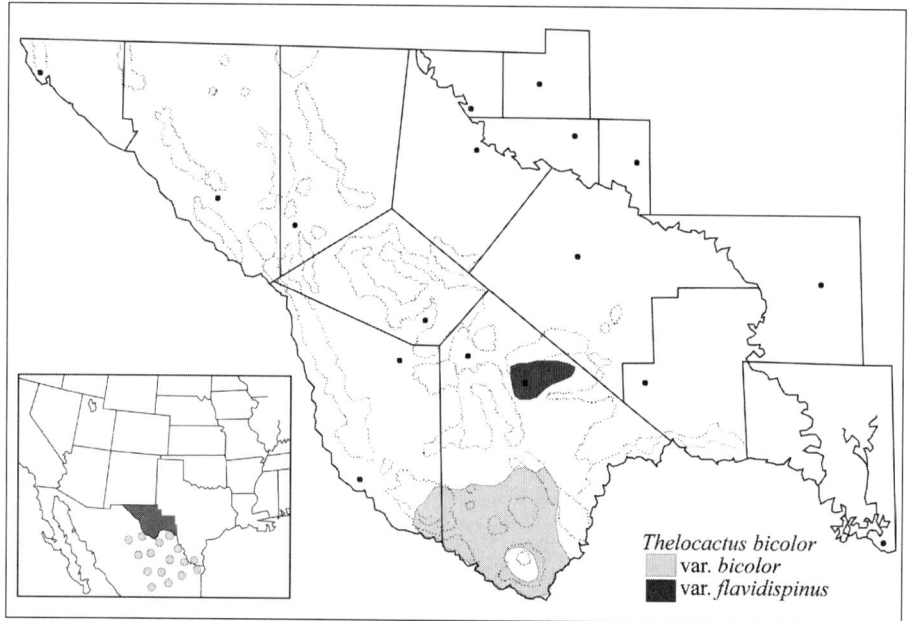

Map 41. Generalized distribution of *Thelocactus bicolor* var. *bicolor* (glory of Texas) and *T. bicolor* var. *flavidispinus* (Marathon Basin thelocactus).

Three Chihuahuan Desert varieties of *T. bicolor* are recognized by Glass and Foster (1977), Benson (1982), Bravo-Hollis and Sánchez-Mejorada (1991a), and Anderson (1987). Two of those varieties occur in Trans-Pecos. One or more additional varieties of *T. bicolor* occur in Mexico. Anderson recognized *T. bicolor* var. *schwarzii* (Backeb.) E. F. Anderson in Tamaulipas. Zimmerman et al. (forthcoming) recognized three varieties of *T. bicolor* in the Chihuahuan Desert

Region, including var. *bolaensis* (Ruenge) A. Berger, which is restricted to southern Coahuila. Anderson considered var. *bolaensis* to be a "form" of *T. bicolor* var. *bicolor*. *Thelocactus bicolor* primarily is a species of northeastern Mexico that extends into Texas in the lower Big Bend region of the Trans-Pecos and on the Rio Grande plains in deep South Texas. The specific epithet presumably alludes either to the color-banded (stramineous and reddish) spines or the red-centered "target" pattern of the flowers.

Key to the Varieties

1. Stems 7.5–18(–25) cm long, 5–9(–13) cm in diameter; ribs notched by well-defined tubercles; central spines four (upper three like the radials but slightly larger), tinged with pink, longest one to ca. 4.5 cm, keeled; radial spines 15–17, mostly pinkish-tinged; flattened upper spines 5–10 cm long, the main one ca. 1.5 mm broad
   1a. *T. bicolor* var. *bicolor*.

1. Stems (3–)5–9 cm long, 3.5–6 cm in diameter; ribs almost completely divided into tubercles; central spines 1(–3) (the upper two like the radials), stramineous (or slightly pink-tinged), ca. 2 cm long, not keeled; radial spines 12–14(–19), stramineous or slightly pinkish-tinged; flattened upper spines 2.5–3.8 cm long, the main one 0.5–1.5 mm broad
   1b. *T. bicolor* var. *flavidispinus*.

1a. **Thelocactus bicolor** var. **bicolor**. GLORY OF TEXAS. Plates 196, 197. Infrequent and/or localized, in silty alluvial and gravelly soils of both sedimentary and igneous origin, desert flats, hills, and *bajadas* with *Larrea* Cav., *Agave lechuguilla* Torr., and *Prosopis* L., southern Presidio and Brewster counties. 2,300–4,000 ft. Flowering (Feb–May–)Jul–Sep, opportunistic. $2n = 22, 44$. Disjunct to S TX, on Rio Grande plains in Starr Co., ca. 50–300 ft, in Tamaulipan scrub. Mexico, Chihuahua and Coahuila S to San Luis Potosí, E through Nuevo León to northern Tamaulipas. Map 41.

Thelocactus bicolor var. *bicolor* is infrequent almost throughout its range in southern Presidio and Brewster counties, but plants are locally common in certain sites. Collectors highly prize the plants, mostly because of their attractive flowers.

Identifying Characters. Mature stems of *T. bicolor* var. *bicolor* are ovoid to short-cylindroid or subconical, 8–18(–38) cm long and 6–11(–17) cm in diameter, mostly tuberculate when young, or with the rows of tubercles either vertical or spiraling on 8–13 ribs when older. The most prominent spine characters are the wood-shaving-like upper spines in each areole and the bi- or tricoloration of the central and certain radial spines. The flattened upper spines usually are whitish throughout, not bicolored. The bi- or tricolored spines are pinkish to reddish in the central part of each spine, if not all the way or nearly all the way to the base, and with gray to tan bases and stramineous or tan tips.

The flowers of *T. bicolor* var. *bicolor* are among the most spectacular of the Texas cacti. The flowers are bright magenta to rose-pink, 4–8 cm in length and diameter, with dark red centers. When fully open, the tepals are reflexed. The inner tepals are 2.8–5.4 cm long, 0.5–1.2 cm wide, magenta or rose-pink (sometimes with whitish margins or centers), deep red at the bases, and attenuate at the tips. The outer tepals are oblong, and they have greenish midregions and whitish, fimbriate margins. The filaments are reddish or yellowish, and the anthers are yellow. The pinkish-red style supports 7–13 stigma lobes that are reddish to red-orange or pale reddish-brown, or yellowish. The pericarpel surface exhibits scales.

The fruits of *T. bicolor* var. *bicolor* are ovoid to nearly globular, 0.7–1.8 cm long, and 0.6–1.2 cm in diameter. At maturity the fruits are dry, green to brownish-red, and dehiscent at the base, forming a prominent basal pore. The floral remnant is persistent. The seeds are black, 1.5–2.5 mm long, obovoid-pyriform, with the surfaces finely papillate.

Phenology. Probably the typical flowering period for var. *bicolor* is in the spring, followed by multiple flowering episodes after rains during summer or early fall (September or October), about 7–10 days after precipitation. The first flowering period, regardless of previous precipitation, for populations north of Big Bend National Park, may occur in late March to June or July. A report by Wauer (1997) that he observed several large plants of var. *bicolor* in full bloom on 24 February 1968, near the River Road in Big Bend National Park, suggests that the initial flowering period for var. *bicolor* may be in late winter or early spring in lower-elevation sites, such as those near the Rio Grande south of the Chisos Mountains. Flowers open about noon on sunny days, close at night, and may open again for a second day. The length of time required for fruit maturation appears to be 1–1.5 months.

Sterile and Immature Specimens. No other Texas cactus species has spine characters identical to those of *T. bicolor*. In *T. bicolor* var. *bicolor* there are 1–4 central spines, only one of which is centrally placed in the areole and porrect. This porrect "main" central spine may be reddish to gray in color, (1.3–)2–4.5(–6) cm long, and either subterete or flattened on the upper (adaxial) side, and rounded or keeled on the underneath side. In older plants the main central spine may be flattened or grooved on the upper surface. Thus, a partially flattened main central spine is typical of *T. bicolor* var. *bicolor*, but the main central may be terete in some areoles and plants. Typically in mature areoles there are three upper central spines borne near the upper edge of the areole near the origin of the upper radial spines. The upper centrals are angled upward or somewhat appressed and under magnification are noticeably smaller in diameter and slightly shorter in length than the main central spine. The upper centrals are terete.

The radial spines of *T. bicolor* var. *bicolor* are of two types, at least by our interpretation. There is a circle usually of 13–15 "ordinary" radial spines that are terete or laterally compressed toward the base, ca. 1.5 cm long, and bicolored with a reddish hue in the central region of each spine. These ordinary radial spines are appressed against the stem and straight, or particularly the lower ones, slightly recurved. The two or three uppermost spines, "flat upper radial spines," are whitish or grayish, flattened, and thin, sometimes slightly curled like stiff wood shavings, with the main (usually median) one to 5–10 cm long, ca. 1.5 mm wide, typically longer and slightly wider than the other one or two. The flat upper radial spines are either erect or appressed against the stem or the spines in other areoles for about half their lengths. The flat upper spines provide the best means for identifying sterile specimens of *T. bicolor*.

In juvenile plants of *T. bicolor* the spine characters described above are already evident. In the areoles of immature plants, however, the main central spine is more commonly terete, and there are only 1–2 upper central spines. The main central may be absent in some areoles. The number of ordinary radial spines may be only 11–13, and the 1–3 flat upper central spines are shorter than in adults, the longest sometimes reaching 3–4 cm. Spine coloration is basically the same in juveniles and adults, except that in juveniles the ordinary radials may be uniformly pigmented rather than bicolored, and the flat upper radials typically are yellowish in areoles high on the stem. The flat upper spines become ashy-gray with age.

Biosystematics. The widely disjunct distribution of *T. bicolor* var. *bicolor* in Texas seems unusual until one considers that this taxon also is common and widely distributed in northeastern Mexico. In Texas *T. bicolor* var. *bicolor* thrives under relatively humid conditions near Rio Grande City in Starr County, and under arid, desert conditions mostly in silty or gravel soils in the southern Big Bend region. All of the varieties are allopatric with respect to each other.

Variety *bicolor* is distinguished from var. *flavidispinus* most conspicuously by its lack of a yellow spine-color phase and by its larger ovoid to cylindrical or subconical stems usually 8–35 cm long. In var. *bicolor* there are four central spines, three of them borne in the upper part of the areole near the radials and one positioned near the center of the areole. The lower central is longer, to 4.5–6 cm, porrect, and flat on the upper side. The upper flattened spines are longer (5–10 cm), very flat, and ashy-gray in mature areoles. The two varieties also are separated geographically and ecologically. The var. *bicolor* occurs in miscellaneous substrates in southern Brewster and Presidio counties, whereas var. *flavidispinus* is mostly restricted to novaculite exposures in the Marathon Basin of northeast Brewster County. Plants of var. *flavidispinus* are spheroidal to short-cylindroid, typically 3–8(–13) cm high, with 1–3 central spines, one in the areole center and two arising in the upper areole margin almost among the radials. The porrect central spine in var. *flavidispinus* is ca. 2 cm long and rarely, if at all, flattened on the upper surface. The longer (to 3.8 cm) flat upper spine is yellow, even in mature areoles, but does turn whitish with age, as in var. *bicolor*.

Collections of *T. bicolor* from certain novaculite substrates in the Solitario, a geologic formation in southern Presidio and Brewster counties, are small like those of var. *flavidispinus* and consequently appear intermediate between var. *bicolor* and var. *flavidispinus*. Some of these plants growing side by side resemble either var. *bicolor* or var. *flavidispinus*. After careful consideration, we have concluded tentatively that these are stunted plants of var. *bicolor*, the smaller size perhaps being influenced by factors similar to those that cause many cactus species growing in novaculite in the Marathon Basin to be smaller.

Synonyms. *Echinocactus bicolor* Galeotti var. *schottii* Engelm.; *E. bicolor* var. *tricolor* (K. Schum.) Knuth; *Thelocactus pottsii* (Salm-Dyck) Britton & Rose sensu Bravo; *T. bicolor* (Galeotti) Britton & Rose var. *schottii* (Engelm.) Davis *nom. nud. fide* Backeb.; *T. bicolor* (Galeotti) Britton & Rose var. *schottii* (Engelm.) Krainz; *T. bicolor* (Galeotti) Britton & Rose var. *schottii* (Engelm.) L. D. Benson, *comb. superfl.*; *Ferocactus bicolor* (Galeotti ex Pfeiff.) N. P. Taylor.

Common Names. Texas pride; bicolor cactus. We believe that not any of the popular names is truly appropriate for this primarily Mexican taxon, but when in flower it is appropriate for Texas.

1b. **Thelocactus bicolor** var. **flavidispinus** Backeb., Sukkulentenkunde 1941: 6.1941. MARATHON BASIN THELOCACTUS. Plates 198, 199. Common in soils in and near novaculite exposures, hillsides, and basins, often in alluvium associated with or derived at least in part from novaculite, occasionally in soils of other derivation, Marathon Basin, Brewster Co. 4,000–4,500 ft. Flowering late Mar–May, also Jun–Sep, opportunistic after rains. $2n = 22$. The reports of *T. bicolor* var. *flavidispinus* in Starr Co. of S TX (Benson, 1982) are based upon misidentified *T. bicolor* var. *bicolor* (Zimmerman et al., forthcoming). Map 41.

The varietal epithet *flavidispinus* is derived from the Latin *flavidus*, "somewhat yellow," and *spina*, "spine." Plants of *T. bicolor* var. *flavidispinus* are abundant at some sites in the Marathon Basin.

Identifying Characters. Any *Thelocactus* located in the Marathon Basin probably is var. *flavidispinus*. The var. *bicolor* is not verified to occur in the Marathon Basin, although it would not be unexpected, particularly in southern portions of the area. Sexually mature plants of var. *flavidispinus* typically are smaller in height and in diameter than those of var. *bicolor*. Plants of var. *flavidispinus* may produce flowers when the stems reach only ca. 3 cm high. Other morphological traits useful in distinguishing var. *flavidispinus* from var. *bicolor* include the porrect central spine that is rather short (1.5–2 cm) and terete and the shorter (2.8–3.5 cm) flat upper spines. In the areoles of mature plants of var. *flavidispinus*, there may be only one central spine, the one that is porrect or sometimes turned downward. Typically, however, there are what appear to be 1–2 additional central spines at the upper margin of the areole, their position essentially in line with the radials, but they are slightly larger at the base and slightly longer than the radials. One or two flattened upper spines occur typically in areoles of mature plants, and these always are borne among or under the normal radials at the upper edge of the usually circular areoles; they typically are yellowish in upper areoles, but weathering to grayish in areoles down the stem.

Radial spines number (12–)15–17(–20). The majority of the radial spines around the usually circular areoles are (0.9–)1.5–2(–2.4) cm long, basically needlelike except typically laterally compressed near the base, and many of them are curved back toward the stem. In color the "typical" radials are yellowish (rarely reddish) in upper areoles and rose to grayish or bicolored with reddish central zones, gray bases, and yellowish tips in areoles down the stem. The 1–2 flat upper spines at the top of the areole may be poorly defined on young plants or flat on only one side. On older plants they are more like pointed wood shavings and become grayish, as in var. *bicolor*. In var. *flavidispinus* the largest flat upper spine may reach 3.8(–4) cm long and 1(–1.5) mm broad.

The flowers of var. *flavidispinus* are like those of var. *bicolor*. The inner tepals are up to 4 cm long, ca. 1 cm wide, with entire margins and attenuate apexes. The tepals may be reflexed distally in fully open flowers. The stamen filaments are yellowish or pale red, and the anthers are yellow. The ca. 11 stigma lobes are reddish or yellowish, especially toward the tips.

The fruits and seeds of var. *flavidispinus*, so far as known, are like those of var. *bicolor*.

Phenology. The common flowering time for var. *flavidispinus* is in late spring, from late March to late May. Additional flowering periods in the summer, from June to September, appear to be stimulated by sufficient rains. Flowers open 7–10 days after rain (G. G. Raun, pers. comm.). On 3 May 1994 G. G. Raun observed the "second or third" flowering episode that spring in the same population of var. *flavidispinus* south of Marathon. Individual flowers open about midday under warm, sunny conditions, close at night, and may or may not open again the next day. Fruit maturation in cultivated plants requires about one month.

Sterile and Immature Specimens. Mature stems of var. *flavidispinus* are unbranched, spheroidal or short-cylindroid, 3–8(–13) cm long, 3–6(–7) cm in diameter, with ca. 13 ribs, these defined by tubercles united only at their bases. Immature plants often are flat-topped and strictly tuberculate. The stem surface is green and obscured by spines in areoles 0.6–1.5 cm apart. Mature sterile specimens are most reliably identified by locality, stem size, and the 1–3 central spines, only one of which is clearly a central spine. The porrect, essentially needlelike, "main" spine, located in the center of the areole, is (0.9–)1.5–2(–2.4) cm long in var. *flavidispinus*. Usually there are two upper central spines, but these are located essentially in line with the upper radials. One or two of the upper flat spines in the upper areoles typically are yellowish, and one is noticeably longer and broader than the other. This "main" upper flat spine may be erect or curved along the plane of the stem and is 1.8–4 cm long. The typical radial spines are bicolored or tricolored with reddish central zones, or in younger or older areoles, respectively, the radials may be either yellowish or grayish throughout. At least some of the radial spines in var. *flavidispinus* typically are curved back toward the stem.

In juvenile specimens 1–1.5 cm high and 2.5–3 cm in diameter, the central spine is absent in all areoles or in all but some areoles toward the stem apex. The areoles are circular or elliptic and are surrounded by about 16 radial spines. In the youngest plants the flat, yellow, upper radials are also absent, but as plants grow older, a single upper radial appears and is longer than the other radials, usually is yellow, and in youngest areoles is needlelike or flat on the upper side, before appearing more flattened in older areoles. A single, porrect central spine appears in upper areoles of slightly older plants. The ordinary radial spines are laterally compressed at the base but otherwise are acicular, and at least some of them are curved toward the stems. In color the radials of juveniles are most often yellowish, or yellow or reddish at the base and yellow distally, or bicolored with red in the middle or distal regions of the otherwise yellow spines. Completely red radial spines are present in apical areoles of some plants in the Marathon Basin, particularly in a population at the eastern margin of the Basin.

Biosystematics. Backeberg (1951) elevated var. *flavidispinus* to species level after originally describing the taxon as a variety. Subsequent workers have preferred to maintain these taxa as varieties (Glass and Foster, 1977; Benson, 1982; Bravo-Hollis and Sánchez-Mejorada, 1991a; Anderson, 1987; Zimmerman et al., forthcoming). Weniger (1984) treated var. *flavidispinus* as distinct (*Echinocactus flavidispinus*), citing allopatric ranges and distinctive morphological features as reasons for specific status.

Perhaps the most distinctive features of var. *flavidispinus,* besides its restricted ecogeographic distribution on novaculite hills in the Marathon Basin, is the consistently smaller size of the plants along with correspondingly smaller tubercle and spine characters and the few distinctive spine characters (shorter, typically terete "main" central spine, fewer [1–3] central spines, and recurved radial spines). We regard the apparent edaphic endemism and the smaller plant size of var. *flavidispinus* as being of particular taxonomic significance. Those knowledgeable about Texas cacti are aware that there are several distinctive taxa of diminutive cacti in the Marathon Basin, all of them either restricted to or associated with the localized and geologically peculiar novaculite substrates.

Synonyms. Thelocactus flavidispinus (Backeb.) Backeb.; *Echinocactus flavidispinus* (Backeb.) Weniger; *Ferocactus bicolor* (Galeotti. ex Pfeiff.) N. P. Taylor var. *flavidispinus* (Backeb.) N. P. Taylor.

Common Name. A published common name is flatspine thelocactus (Warnock, 1977).

8. LOPHOPHORA J. M. Coult.
Peyote

Lophophora has been treated in detail by Anderson (1980, 1996a), who recognized two closely related species for the genus. One species is *L. williamsii*, the wide-ranging entity that occurs in Mexico and in Texas, and the other is *L. diffusa* (Croizat) Bravo, which is restricted to near Vizarrón in the state of Querétaro, Mexico. Historically, *Lophophora* has been placed in several different genera, *Echinocactus*, *Anhalonium* Lem., *Mammillaria*, and *Ariocarpus* (Boke and Anderson, 1970; Anderson, 1980; Benson, 1982), but among the leading cactologists of today there appears to be virtual unanimity in accepting *Lophophora*. All populations of *L. williamsii* in Mexico and Texas produce essentially the same alkaloids. *Lophophora diffusa* has a slightly different alkaloid content. Both *Lophophora* species have received extensive cultural attention (Anderson, 1996a), and the common name, peyote, is applied to both taxa of *Lophophora*. The origin of the genus name, *Lophophora*, is from the Greek *lophos*, "the crest," and *phoreus*, "a bearer," in reference to the tufts of hairs borne in each areole. The common name, peyote, originated from *peyotl*, the Aztec name for the taxon.

1. **Lophophora williamsii** (Lem.) J. M. Coult. PEYOTE. Plates 200–202. [*Anhalonium williamsii* Lem., Handb. Cact. 233. 1885]. Plants spineless, low, flat to the ground or rounded, often gray, blue-green to darker green, rarely reddish-green; taproot fleshy, broadly carrot-shaped, 6–12 cm long. Stems solitary to numerous (up to 50), flesh soft, extending well below the ground as fleshy basal stems or roots, depressed-globose or flattened and depressed in the centers, 2–7.5 cm high, (4–)5–10(–12) cm in diameter. Ribs (5–)8(–13), low and demarked by narrow grooves, straight or spiraled, each rib composed of usually low, fused tubercles, these more distinct at the stem apex, the tubercles imperfectly hexagonal and perhaps with slight wrinkles when desiccated, to 2.5 cm in diameter. Areoles essentially round, 0.5–1.5 cm apart at the tips of the humplike tubercles, each bearing a compact, cylindroid tuft of soft whitish or yellowish trichomes, the trichomes 7–10 mm long, with age turning grayish and eventually broken or worn away; stem surface dull, microscopically papillate; pith and cortex more or less flaccid, not mucilaginous. Spines absent in the areoles except in small seedlings (too tiny to be found in the wild), where spines are rudimentary and weak. Flowers (1–)1.5–2.5 cm in diameter, 1–3 cm long, produced from within areoles on young tubercles, in the woolly depressed centers of stems; inner tepals mostly pink in our specimens, also pinkish-red to nearly white, usually darker near midribs, rarely yellowish-white, the largest ones elliptical, 0.8–1.4(–2.2) cm long, 2–5 mm wide, mucronate or rarely attenuate, margins ciliate or entire; outer tepals with midribs greenish, margins greenish-pink or whitish, the largest ones narrowly elliptical or oblanceolate, 3–13 mm long, 1–3 mm wide, acute, mucronate, margins minutely ciliate distally; stamens with filaments white, rarely magenta, 1–2 cm long, anthers yellow, to 1.4 mm long; style white, rarely pinkish, 0.5–1.4 cm long, stigma lobes (3–)4–8, 1–3 mm long,

white, rarely pinkish; pericarpel naked, in anthesis turbinate, 3–4.5 mm long, closely subtended by areolar hairs of the stem apex to 1 cm long. Fruit naked, clavate or nearly cylindroid, 1.1–2.5 cm long, 2–4.5 mm in diameter distally, whitish, pinkish in most of our specimens, or pinkish-red, weakly succulent, quickly becoming dry, translucent, and brownish-white or membranous-white after ripening, having first grown rapidly and exserted above the apical hairs. Seeds black, 1–1.5 mm long, 1–1.2 mm wide, somewhat pear-shaped, cells of testa strongly convex; hilum basal, large and flattened. $x = 11$.

Mostly in rocky limestone substrates, or in alluvium derived from calcareous parent material, among and under shrubs and other Chihuahuan Desert vegetation, or in rock crevices. Presidio Co., at least two localities in southern part of county, nearly extirpated from a significant portion of its range in one population; Brewster Co., one documented locality, on an igneous mountain; Val Verde Co., at least two known sites near the Pecos River. 1,000–4,500 ft. Flowering Mar–Sep (Anderson, 1980) but most commonly Jun–Aug in the Trans-Pecos. $2n = 22$. South TX in Hidalgo, Starr, Jim Hogg, Zapata, and Webb counties and then NE near the Rio Grande to the Pecos River. Mexico from Tamaulipas and Nuevo León SW to San Luis Potosí and NE Zacatecas, and N through eastern Chihuahua and much of Coahuila to TX (Anderson, 1980). Map 42.

Weniger (1984) recognized two varieties of *L. williamsii* for Texas, the var. *williamsii* in South Texas and *L. williamsii* var. *echinata* (Croizat) Bravo-Hollis in the Trans-Pecos. Weniger distinguished the two varieties primarily by stem number of individual plants ("clump size") and cold hardiness. The South Texas variety was said to have smaller stems forming larger stem clusters in older plants, with the plants being more tender, growing under partial shade, and being

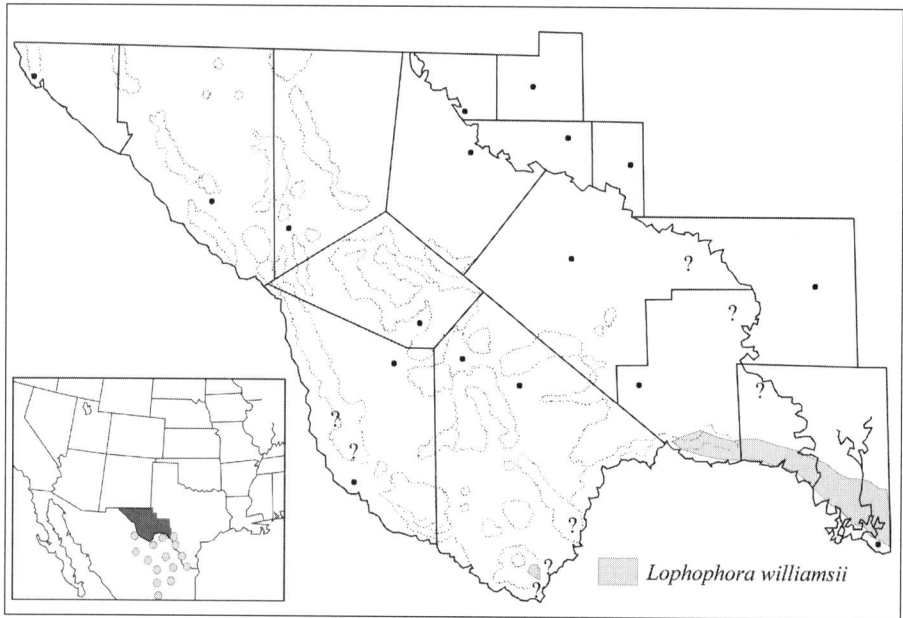

Map 42. Generalized distribution of *Lophophora williamsii* (peyote).

easily damaged by frost. The Trans-Pecos variety reportedly has larger stems that remain solitary or form clusters of 2–3 stems in older plants, and the plants often grow on dry hillsides, where they survive greater heat and cold. We follow Anderson (1980, 1996a) in not recognizing varieties for the Texas *Lophophora*. Although Weniger's observations are probably mostly correct for the South Texas and Trans-Pecos populations (i.e., the distributional extremes), other workers have found that these character differences are not evident in populations of intermediate distribution. In our own field investigations in southern Presidio County, we have found numerous multistemmed plants (5–12+ stems) among the predominantly single-stemmed plants.

Identifying Characters. The spineless stems of *L. williamsii* in the Trans-Pecos commonly are solitary, in clusters of 2–3, or less commonly in clusters of 12 or more stems. The stems are deep-seated with fleshy taproots, usually are flat-topped, and barely or not at all extending above the soil surface. The stems are gray-green, usually soft, and inconspicuously ribbed. The podaria usually are imperfectly low-conical, each with a circular areole densely filled with erect or bent, dirty-white trichomes to 1 cm long.

The usually pinkish flowers are morphologically similar to those of a *Mammillaria*, a *Coryphantha*, or *Ariocarpus fissuratus*. The flowers are nearly apical in position. The inner tepals may vary in color from rose-pink to pale pink with whitish margins. The filaments are whitish, and the anthers are yellow. According to Patty Manning, the stamens are sensitive (i.e., contracting rapidly against the style when touched). The style is ca. 8 mm long and supports short white or pinkish stigma lobes.

The fruits are weakly succulent, usually pinkish, and partially hidden by the long trichomes ("wool") of the stem apex. After maturity the fruits become dry with thin walls, contract in size, and turn brownish. Usually only the upper half of the fruits contains seeds (Anderson, 1996a). The black seeds are small and with strongly convex testa cells. The large, basal hilum, almost as wide as the seed, is nearly rounded or V-shaped. The seeds of *L. williamsii* are almost identical to those of *Ariocarpus*. After the thin fruit walls rupture, not by a patterned dehiscence, seeds are disseminated by rainwater (Anderson, 1996a).

Phenology. The principal flowering time for *Lophophora* seems to be in March to May. In southern Presidio County, however, it is not uncommon for individual plants or stems to produce flowers from June through September, and we suspect that the same is true for other Trans-Pecos populations. Cultivated plants have been observed by us to bloom in June, July, and August. One plant propagated from seed grew to ca. 4 cm in diameter and produced a single flower in less than two years, in contrast to the relatively slow growth pattern described by Anderson (1996a). Twelve other plants grown from seed planted at the same time did not bloom after two years. In one cultivated plant, a full-size fruit (its seeds not examined for viability) was produced in five weeks after the flower withered. Anderson reported that fruits of *L. williamsii* are obscured by the apical trichomes for about a year, before the seeds mature and the fruits elongate rapidly and extend above the apical wool.

Sterile and Immature Specimens. Mature stems of *L. williamsii* are hemispheric or depressed-globular, in the Trans-Pecos usually extending barely above ground level during most of the year. In the Trans-Pecos, stems usually are solitary, but older plants with clustered stems are flat and not mounded as is typical in more southerly portions of the range, from central Coahuila, Mexico, southward and into South Texas. The stems are 5–12 cm in diameter, weakly ribbed with low podaria, or they are merely weakly tuberculate. Stem surfaces throughout the range are blue-gray or blue-green, but in the Trans-Pecos mostly pale gray-green, and the stems are soft and perhaps flaccid when compressed. The circular areoles bear dense tufts of trichomes. Multistemmed plants form either as a result of natural lateral shoot formation or as a consequence of apical injury, such as careful peyote harvesting, where stems are decapitated at ground level, conserving the ancient root system and promoting the formation of new stems at the periphery of the old ones. Anderson (1996a) noted that not all populations of *L. williamsii* have the same tendency to develop cespitose plants. In his experience, peyote stems rarely rotted, and stem tops or even pieces of stems readily formed adventitious roots (presumably under laboratory conditions, not field conditions).

In the Trans-Pecos *L. williamsii* is easily distinguished from all other species, but it has been confused by some inexperienced observers with *Ariocarpus fissuratus* and *Echinocactus horizonthalonius*, both of which are widely distributed in the same type of limestone habitats that appear to be capable of supporting *L. williamsii*. Like *L. williamsii*, *A. fissuratus* is completely spineless and characterized by flattened stems extending just above ground level, but *A. fissuratus* is easily recognized by its rigid, copiously fissured, and wrinkled tubercles. In aspect the flat or weakly hemispheric stems of *A. fissuratus* support a rosette of roughened, usually pointed tubercles, often a starlike habit. Immature stems of *E. horizonthalonius* often are depressed-globular and have relatively broad, smooth tubercles and may resemble *L. williamsii* in these traits, but *E. horizonthalonius* has stout spines in the areoles. Juvenile plants of *E. horizonthalonius* with few spines or, rarely, with almost no spines, are the ones that are most likely to be confused with *L. williamsii*. Much additional morphological detail regarding *L. williamsii* is available in Boke and Anderson (1970) and Anderson (1996a).

Immature stems of *L. williamsii* are miniature versions of the adults, presenting no special problems in identification. The areolar trichomes may be yellowish or white, becoming dirty-white or gray with age.

Biosystematics. The documented distribution of *L. williamsii* in the Trans-Pecos includes one modest-size population near Shafter in southern Presidio County and one very small population in Big Bend National Park in extreme southern Brewster County. Other small populations, not documented by herbarium specimens, persist near the Rio Grande valley from Presidio County downstream through Terrell and Val Verde counties, and disjunctly along and near the Rio Grande south to Hidalgo County in the Rio Grande valley, where the size of historical populations greatly exceeded those in the Trans-Pecos (Weniger, 1984; Anderson, 1996a). In Texas the main region in which *L. williamsii* has been harvested commercially includes parts of Starr, Jim Hogg, Webb, and Zapata counties (Anderson, 1995). Most of the total range of *L. williamsii* is in northeastern Mexico, extending south of the Rio Grande to just north of Ciudad San Luis Potosí and occupying habitats in Chihuahuan desertscrub or Tamaulipan scrub in the Mexican states of Chihuahua, Coahuila, Zacatecas, San Luis Potosí, Nuevo León, and Tamaulipas.

Lophophora williamsii usually occurs in limestone hills or in alluvium derived from limestone (Anderson, 1996a). Most plants are found under shrubs (e.g., *Prosopis, Acacia* Mill., *Larrea*) or in the "protection" of other vegetation (e.g., *Agave lechuguilla, Hechtia* Klotzsch). In the Trans-Pecos the known distribution of *L. williamsii* is in limestone except at the one site in southern Brewster County, where the cited locality is on a small mountain of igneous origin. The small Brewster County population might or might not have been introduced (Anderson, 1996a). In Presidio County, plants of *L. williamsii* occur in thin rocky soil on rounded limestone hills, in alluvium on and between the hills, and rarely in soil-filled crevices of otherwise solid rock.

A second species of *Lophophora*, *L. diffusa*, occurs in Querétaro north of Ciudad Querétaro. By comparison *L. williamsii* is exceedingly widespread in distribution, occupying habitats at elevations of 2,000–6,000 feet in the CDR to 150 feet in the Tamaulipan scrub of South Texas.

At present the nearest generic relatives of *Lophophora* are not well understood. Preliminary DNA studies support the morphological and biogeographical evidence that *Lophophora* is most closely related to *Obregonia* Frić ex A. Berger. Other data also suggest relationships with *Aztekium* Boed., *Strombocactus* Britton & Rose, *Turbinicarpus* Buxb. & Backeb., *Ariocarpus*, *Pelecyphora* Ehrenb., *Echinocactus*, and *Thelocactus* (Zimmerman, 1985; Anderson, 1996a).

One special chemical feature of *Lophophora* is its high concentration of alkaloids. Some of the alkaloids are psychoactive compounds, most notably mescaline (3,4,5-trimethoxy-beta-phenethylamine) which is primarily responsible for the hallucinogenic (psychotomimetic) properties of *L. williamsii* (peyote). A total of 57 alkaloids (and related compounds) have been isolated from *L. williamsii* (Anderson, 1996a). Some of the alkaloids are unique to *L. williamsii* and *L. diffusa*, but others also are found in different cactus genera and even different plant families. Mescaline is not one of the alkaloids that is unique to *Lophophora*. Mescaline is known to occur in the North American genera *Opuntia*, *Pachycereus* (A. Berger) Britton & Rose, *Pelecyphora*, *Polaskia* Backeb., and *Stenocereus* (A. Berger) Riccob., and in the South American genera *Echinopsis* Zucc. (*Trichocereus*), *Gymnocalycium* Pfeiff., and *Stetsonia* Britton & Rose. In all, mescaline and/or other *Lophophora* alkaloids are known to occur in 12 genera of cacti (Gibson and Nobel, 1986; Nobel, 1994; Anderson, 1996a), including species of *Ariocarpus* (*A. fissuratus* of the Trans-Pecos) and *Carnegiea gigantea*.

Information about alkaloid chemistry has been useful in evaluating intergeneric relationships with *Lophophora*. Not all of the genera suspected of being related to *Lophophora* have been investigated thoroughly for their alkaloids, but so far *Obregonia*, *Turbinicarpus*, *Ariocarpus*, and *Pelecyphora* are known to produce peyote alkaloids (Cheatham and Johnston, 1995; Anderson, 1996a). Alkaloid chemistry also has been useful systematically in supporting specific distinction between *L. williamsii* and the morphologically similar *L. diffusa* (Anderson, 1996a). Studies have shown that *L. diffusa* may completely lack the alkaloids anhalinine, anhalonine, and hordenine, "essentially" may lack mescaline, and produces smaller amounts of four other *L. williamsii* alkaloids. The two peyote species differ significantly in these and other aspects of chemical content.

Uses and Other Ethnobotany. Historical evidence suggests that peyote always has been used in association with sacred religious ritual and as medicine (Anderson, 1995). Possibly no North American plant species has been more intensely investigated culturally than *L. williamsii* (Anderson, 1980, 1996a; Stewart, 1989; Morgan, 1976; Morgan and Stewart, 1984; Newcomb, 1956).

There is archeological evidence that hallucinogenic plants were associated with human cultures in the New World as early as 8500 B.C., suggesting that knowledge about mind-altering plants was present among the earliest inhabitants of the New World (Anderson, 1996a). Early humans used hallucinogenic plants as a means to stay in contact with the spirit world. Such concepts are retained in some cultures in the twentieth century, where peyote allows communication with spirit forces and provides an avenue for dealing with those spirits that cause evil or sickness in the body.

Apparently the first manuscript reference to peyote cactus was about 1560 in Mexico. After that time there are numerous references through Spanish writings in Mexico regarding ritualistic use of peyote. In Mexico today traditional use of peyote persists among Tarahumara, Huichol, Cora, and Tepecano Native American tribes (Anderson, 1996a).

A new type of peyotism originated in the United States during the latter part of the twentieth century, rapidly spreading throughout Indian reservations that were created during that time. The new form of peyotism combined elements of the traditional peyote culture with aspects of Christian symbolism (Anderson, 1996a). At the beginning of the twenty-first century, more than 250,000 Native Americans were using peyote as a sacrament in a religion called the Native American Church (Anderson, 1995).

It is a violation of federal law (Public Law 91–513) for non-Indians to possess or use either peyote or its alkaloid mescaline. Under this law, peyote is considered to be a Schedule I controlled substance. Thus, federal law prohibits even simple possession by cactus collectors, horticulturists, and scientists. Scientists can apply for a permit through the U.S. Attorney General to use peyote in teaching or research. Conviction for illegal possession of peyote can lead to a fine of up to $5,000 or imprisonment for up to one year, or both.

The primary source of peyote cactus for ceremonial use by the Native American Church of the United States and Canada is from the "peyote gardens" in Starr, Jim Hogg, Webb, and Zapata counties in South Texas (Anderson, 1995). The South Texas natural populations of peyote occur almost entirely on private lands. Federal and Texas laws permit the collecting of peyote and its use by Native Americans in their religious ceremonies. Collectors of peyote must be licensed by the Federal Drug Administration and by the Texas Department of

Public Safety. Peyote collectors, mostly Hispanic, are called *peyoteros*. The licensed *peyoteros* secure access to certain natural populations of *L. williamsii* and harvest the plants by cutting off the tops of the stems. The stem tops, known as peyote "buttons," are sold to the Native American Church in fresh "green" or dried condition. Continual harvesting of peyote for more than 100 years in the South Texas region has led to diminished natural populations of *L. williamsii* and concern about conservation of the "peyote gardens."

During Native American religious ceremonies, peyote often is ingested by swallowing fresh or dried buttons. The buttons are briefly chewed and softened by saliva, spit out in the hands, rolled into a ball, and then quickly swallowed, at least in part to avoid the characteristic bitter taste of the buttons. Areolar trichomes, particularly the dense apical "wool," is pulled from the buttons before they are first placed in the mouth. Reportedly, most Native Americans prefer fresh buttons, but these are not always available because of the time interval between collection and ritual use. During a ceremony, individuals may eat 2–15 or more buttons, usually fewer than 20 but perhaps as many as 50. Different forms of peyote consumption are practiced by the various Native American tribes. Some prepare a tea by boiling buttons and allowing them to steep in warm water, then drink the tea, or combine drinking the tea and eating whole buttons or pieces of the stems. Details regarding the ceremonial use of peyote, effects on users of the material, and other aspects of the peyote experience are addressed by Anderson (1996a).

Peyote is considered to have significant medicinal value, in addition to its religious use. Native Americans, including the Tarahumara of Mexico, use peyote tea or powder, either ingested or rubbed on the skin, as treatment for specific ailments or for general therapeutic purposes. There is some experimental evidence that the peyote alkaloids have antibiotic activity (Anderson, 1996a).

The concentration of mescaline in peyote is estimated to be about 1–8% of the dry weight of the plant (Bruhn et al., 1978; Anderson, 1996a). Total alkaloid concentration essentially is the same in roots and stems, except for hordenine, which mostly is in the roots of *L. williamsii*. The practice of harvesting and using stem tops is maintained traditionally so that plants will be left in place to regenerate new stems for future use. The potency of the alkaloids in peyote lasts for many years. The persistence of compounds in plant tissues is attested to by the analysis of alkaloids in peyote buttons excavated from a burial cave in central Coahuila, Mexico, dated A.D. 810 to 1070 (Bruhn et al., 1978).

"Peyote is not a dangerous drug that has victimized Native Americans as alcohol has done. Rather, it is a sacred plant having a history of use of more than 6,000 years. It is used only ceremonially and medicinally. It is not addicting, nor does it cause harmful effects. It is one of the most important medicines to Native Americans. Their religion, in which peyote is used as the sacrament, is highly moral and serious. For anyone who has experienced the night-long ceremony of singing, praying, and meditating, there can be only respect and admiration. Serious efforts must be made to assure the continued supply of peyote for members of the Native American Church" (Anderson, 1995, p. 73).

Synonyms. *Echinocactus williamsii* Lem.; *Mammillaria williamsii* J. M.

Coult.; *Ariocarpus williamsii* Voss; *E. rapa* Fischer & C. Meyer ex Regel; *Anhalonium lewinii* Henn. ex Lewin; *E. lewinii* Henn.; *Lophophora williamsii* var. *williamsii* Lem. ex Salm-Dyck; *L. williamsii* var. *lewinii* J. M. Coult.; *L. lewinii* Rusby; *M. lewinii* Karst.; *E. williamsii* var. *lutea* Rouhier; *L. echinata* var. *lutea* Croizat; *L. williamsii* var. *lutea* Soulaire; *L. echinata* Croizat; *L. williamsii* var. *echinata* (Croizat) Bravo-Hollis; *L. williamsii* var. *decipiens* Croizat; *L. williamsii* var. *pentagonia* Croizat; *L. williamsii* var. *pluricostata* Croizat; *E. williamsii*, "hylaeid," *anhaloninica* (= *E. lewinii*) Schum.; *E. williamsii* var. *anhaloninica*; *L. fricii* Haberm.; *L. jourdaniana* Haberm.; *A. jourdanianum* (Rebut ex Maass) Lewin, *nom. nud.*; *E. jourdanianus* Rebut ex Maass, *nom. nud.*; *L. jourdaniana* Kreuz. See Anderson (1996, p. 213) concerning the misidentification of *L. williamsii* as *E. rapa*.

Common Names. Dry whisky; drywhisky; whisky cactus; mescal buttons; mescal button; mescal; divine cactus; divine herb; medicine of God; devil's root; diabolic root; raiz diabolica; dumpling cactus; cactus pudding; turnip cactus; white mule; Indian dope; moon "P"; the bad seed; tuna de tierra; piote; piotl; peyotl; peyori; pezote; pejote; peyot; pellote; piule; peote; challote; Native American names in their own languages (see Anderson, 1996a).

9. ARIOCARPUS Scheidw.
Living Rock Cactus

As monographed by Anderson (1960, 1962, 1963, 1964, 1965), *Ariocarpus* is a mostly Chihuahuan Desert genus of six species in two subgenera. All six of the species occur in northeastern Mexico, extending south to San Luis Potosí, disjunct to one taxon in Querétaro. Four species occur in the Chihuahuan Desert Region where three species and two varieties are endemic (Zimmerman et al., forthcoming). Only one of the species, *A. fissuratus*, reaches the United States, where its primary distribution is in limestone habitats of the Big Bend region of the Trans-Pecos along and near the Rio Grande. The genus name is derived (Anderson, 1965) from the intent by Scheidweiler, who described the genus, to depict the fruit as shaped like a small pear (the Latin *aria*, usually a suffix denoting similarity, but in this case used as a prefix meaning "pear" or *pyrus*, and the Greek *karpos*, "fruit"). [*Anhalonium* Lem.; *Neogomesia* Castañeda; *Roseocactus* A. Berger]. $x = 11$.

1. **Ariocarpus fissuratus** (Engelm.) K. Schum. var. **fissuratus**. LIVING ROCK CACTUS. Plates 203–5. [*Mammillaria fissurata* Engelm., Proc. Amer. Acad. Arts 3: 270. 1856]. Plants spineless, usually unbranched, deeply seated, taprooted, usually barely extending aboveground, often drawn below ground level. Stems turnip-shaped, flattened, concave, or weakly hemispheric, 0–2(–10) cm high (aboveground), 5–10(–13) cm in diameter, gray-green, becoming yellowish and horny with age. Tubercles crowded and overlapping, flattened or slightly convex, the exposed part deltoid (pointed or slightly rounded at the tip), 1–2 cm long, 1.5–2.5 cm wide, tough-skinned and rigid, prominently fissured and tuberculate. Areoles spineless, 10–15 mm long, 3–4 mm wide, evident as densely woolly (matted hairs or trichomes) short grooves to 3 mm wide in the centers of the flattened tubercles (the tubercles with naked lateral grooves as well), the areolar groove rarely absent. Flowers axillary, arising from the adaxial extremities of areoles at the base of young tubercles near the stem center; flowers 1.5–3.5 cm long, 2.5–4.5(–5) cm wide, inner tepals dark to light magenta or pink, 1.3–3.4 cm long, 0.4–1 cm wide, margins pink or whitish; outer tepals brownish or greenish, perhaps with a tinge of pink, 1.2–3.5 cm long, 0.5–0.9 cm wide, acute or rounded; filaments whitish; anthers deep yellow, ca. 0.7 mm long; style whitish, 1.5–1.9 cm long, ca. 1 mm thick, stigma lobes white, 5–10, 1.2–5 mm long, exserted 1–3 mm above the anthers; ovary naked, 3–4.5 mm long. Fruit naked, oval to clavate or cylindroid, whitish to greenish at maturity, 1–2.4 cm long, 0.5–1 cm wide, fleshy at first, drying brownish, remaining embedded in the wool, ultimately disintegrating, leaving numerous seeds in the woolly center. Seeds black, 1–2 mm long, shiny, globose to obovoid, cells of the testa strongly convex, with a large, pale, basal hilum. $2n = 22$.

Rocky limestone habitats, rarely in gypseous soils, Hudspeth, Presidio, Brewster, and Pecos counties, S and E to Terrell and Val Verde counties, usually with lechuguilla and other desertscrub; rarely with *Pinus remota* in the Del Norte Mts

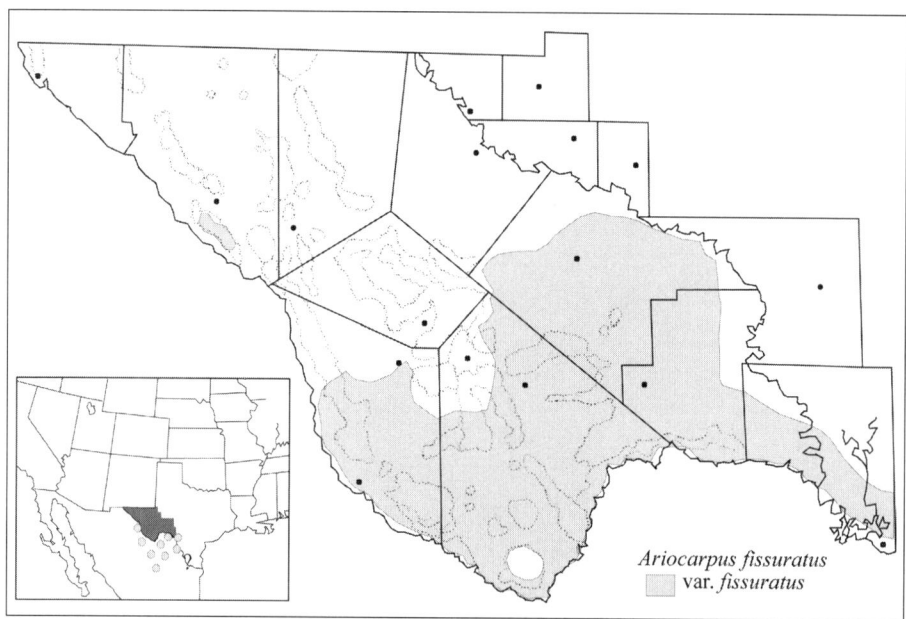

Map 43. Generalized distribution of *Ariocarpus fissuratus* var. *fissuratus* (living rock cactus).

ca. 10 mi SE of Alpine. 1,500–4,500 ft. Flowering Sep–Nov, peaking in early to mid-Oct. Fruit maturing in the summer or fall. $2n = 22$. Mexico, NE Chihuahua, N Coahuila (from Cuatro Ciénegas northward; Zimmerman et al., forthcoming), NE Durango. Map 43.

Ariocarpus fissuratus is unusual among Trans-Pecos cacti in that it flowers only in the fall. The epithet *fissuratus* refers to the fissured upper surfaces of the tubercles.

Identifying Characters. *Ariocarpus* is easily distinguished by its spineless, usually solitary, flattened stems that in aspect may resemble a multipointed star. Basically, the stems are depressed-globose with the concave, flat, or weakly hemispheric tops at ground level or only a few centimeters aboveground. The complex starlike pattern is the result of overlapping, flattened, usually pointed tubercles projecting horizontally. In younger plants the tubercles may be suberect, or the peripheral tubercles may extend beyond the sphere of the stem and yield a star-shaped habit, the inspiration for one of the popular names (star cactus) of the species. The closely packed tubercles also are described (Zimmerman et al., forthcoming) as forming a mosaic that blends with the background of limestone gravel, although the stems do not always blend with limestone in color. The stems may be gray-green, or pale green to deep green, or with age yellow-green to tan or brownish, particularly in outer rows of tubercles. In *A. fissuratus* the imbricate, apically pointed (or rounded), deeply fissured and roughened, rigid tubercles allow easy distinction from all other cacti in the Trans-Pecos.

The usually rose-pink to magenta flowers of *A. fissuratus* var. *fissuratus* are about 3–4 cm in diameter when widely open. Less commonly flowers may be pale pink or rarely nearly white. The filaments are white, and the anthers are yellow to orangish. Both the style and 5–10 stigma lobes are white.

The fruits are oval to clavate or cylindroid and 1–2.4 cm long. Freshly ripened fruits are succulent and white to greenish-white but within a few days become dry and shriveled. Mature dry fruits with thin walls remain firmly attached to the stem and do not dehisce or abscise. Such fruits often remain hidden in the apical wool. Dissemination of seeds apparently depends upon disintegration of the fruit walls or mechanical tearing of the walls. The small black seeds have rough surfaces formed (as seen under magnification) by "strongly convex" protuberances of the seed coat.

Phenology. All species of *Ariocarpus* are fall bloomers, usually in October. The peak flowering time for *A. fissuratus* var. *fissuratus* in the Trans-Pecos is October. In some seasons at least a few plants in some populations may bloom as early as September or as late as November. Under cultivation in Alpine, the first *Ariocarpus* flower of the 1998 season was on 9 September. Flowers open on bright, sunny days and close at night. Fruit maturation requires 8–9 months or more, with ripe fruits being present usually in June or July. In some cases fruit maturation may require more than a year. Apparently, there is no natural abscission process for ripened fruits. Bisected stems of *A. fissuratus* have revealed the presence of seed-filled old fruits that were buried in tissue formed through successive seasons of growth.

Sterile and Immature Specimens. *Ariocarpus fissuratus* is common in many desert limestone habitats in the southern Big Bend region, but individual plants typically are difficult to locate unless one is standing almost directly over the plants. The plants are not apparent in side view because the stems usually do not extend far, only one or a few centimeters, if at all, above ground level. During the winter and drought periods, gradually desiccated internal tissues cause the stems to contract slightly below ground level, often allowing the surrounding limestone gravel to fall on the stem tops, further obscuring the plants. In some populations it is possible to find plants that can be seen in side view (i.e., hemispheric stems that project more than a few centimeters above ground level). The largest plant of *A. fissuratus* var. *fissuratus* we have observed is one from southern Terrell County, shown to us by Jim Talbot, which was 10 cm high and 13 cm in diameter. The stem of this unusually large-size plant was seven rows of tubercles above ground level.

Fortunately, the identifiable features of *A. fissuratus* are best seen when looking directly down on top of the stem. In Texas the spineless, rigid, flattened, strongly fissured and wrinkled, usually pointed tubercles arranged in an imbricated starlike configuration allow for the distinction of *A. fissuratus* from any other cactus species, including *Lophophora williamsii,* which is also spineless but very different in stem morphology and highly localized in distribution. The exposed distal portions of the tubercles usually are deltoid, but in some plants the tubercles have rounded apexes and are more hemispheric. In *A. fissuratus* var. *fissuratus* there is a central areolar groove, usually filled with wool, extending much of the length of the tubercles. The exposed portions of the tubercles are prominently fissured and coarsely rugose on either side of the areolar groove, including prominent fissures that are parallel to and near the margins. The exposed (distal) portions of the horizontally or vertically oriented tubercles are 0.8–2 cm long, and 1.1–2.5 cm wide toward the bases. The areoles are elongate, occupying the central groove on the exposed upper surface of the tubercles. In older tubercles the areolar groove may be short, extending only to near the middle, or longer, reaching essentially to the tip.

In some presumably older plants of the Trans-Pecos, the outer rows of tubercles are yellowish or brownish. In such plants, often the inner rows of tubercles are the normal gray-green color. The yellowish color is the result of an unusually thick, noncellular cuticle overlying the epidermis, which yellows with age and persists intact long after the underlying tissue has died and disappeared. Many larger, presumably older plants in the Trans-Pecos do not exhibit any yellowish tubercles but instead have only normal-looking, gray-green tubercles visible above the soil surface.

Seedlings to 1 cm high usually produce 3–4 bristlelike spines at the stem apex. Juvenile plants to 1 cm in diameter may retain some of the spines. Spines in slightly older plants are sporadic or rudimentary.

Biosystematics. *Ariocarpus fissuratus* is a Mexican species that extends into the United States only in Trans-Pecos Texas. Two varieties were recognized (Anderson, 1965; Benson, 1982; Weniger, 1984; Zimmerman et al., forthcoming): var. *fissuratus* distributed from northeastern Durango, Mexico, north through eastern Chihuahua and northern Coahuila into Texas; and var. *lloydii* (Rose) W. T. Marshall, which occurs only deep in Mexico. Variety *lloydii* appears to intergrade with var. *fissuratus* near Cuatro Ciénegas and at Sierra de la Paila in Coahuila (Zimmerman et al., forthcoming). The var. *lloydii* is traditionally distinguished from var. *fissuratus* by longer, more convex stems that on average are larger in size; more important, its tubercles are more rounded at the apex, weakly fissured, and only finely rugose. The larger plants examined in the Trans-Pecos have prominently fissured tubercles, like those of smaller, typical-size plants of var. *fissuratus*. The larger plants within the northern range of var. *fissuratus* do not appear to represent isolated plants of var. *lloydii*. Anderson and Fitz Maurice (1997) concluded that infraspecific taxa, such as var. *fissuratus* and var. *lloydii*, should not be recognized.

The distribution map for *A. fissuratus* var. *fissuratus* in Benson (1982) showed several localities for var. *fissuratus* near the Rio Grande in Texas south of Val Verde County to Cameron County. No other author, before or since, has mentioned *Ariocarpus* in Texas south of Val Verde County. In Warnock (1970, p. 86), *Ariocarpus fissuratus* is miscaptioned as *Coryphantha vivipara*.

Uses and Other Ethnobotany. *Ariocarpus fissuratus* contains several alkaloids, some hallucinogenic, probably the basis of medicinal and ceremonial uses reported by Cheatham and Johnston (1995). Dried slices of *A. fissuratus* have been used medicinally in Mexico for the treatment of fever. More specifically, the Mexican Kickapoo used juice of the plants to treat tuberculosis, and the Tarahumara used chewed green plant poultices on superficial wounds and bruises. The

Tarahumara also used *A. fissuratus* in ceremony, believing the plant they called *sunamí* to be more powerful than peyote (*Lophophora williamsii*), which they also used.

The ceremonial and medicinal use of *A. fissuratus* by the Tarahumara and the Huichol long predates the attention by white settlers. Ethnobotanical and historical interest in the hallucinogenic properties of this species was evident by the late 1880s (Cheatham and Johnston, 1995). Apparently, contemporary drug cultists are widely aware of the reputation of *A. fissuratus*, although there are multiple instances of the more widely distributed *A. fissuratus* having been confused with the more localized and more famous *L. williamsii*. During the late 1970s a relative "newcomer" to the region led one of us to a "new population of peyote" in southern Brewster County. Upon arrival at the site, the newcomer proudly pointed out a healthy population of *A. fissuratus*, which contained several decapitated plants. Apparently, the newcomer had misidentified the species on the basis of some physiological effects he had expected from *L. williamsii*. *Ariocarpus fissuratus* does not contain mescaline, the well-known hallucinogenic alkaloid present in *L. williamsii*, but the effects of its alkaloids are sufficient to have inspired the common name "dry whiskey" (assuming that it was accurately used in reference to *A. fissuratus*). According to Nobel (1994) the alkaloids in *A. fissuratus* have the effect of increasing the endurance and visual perception of long-distance runners, and *Ariocarpus* is used in this manner by Indians, along with other Trans-Pecos cacti, *Echinocereus coccineus*, *Epithelantha micromeris*, and *Mammillaria heyderi*.

Synonyms. *Anhalonium fissuratum* (Engelm.) Engelm.; *A. engelmannii* Lem.; *Roseocactus fissuratus* (Engelm.) A. Berger.

Common Names. Crack star; living rock; dry whiskey; ariocarpus; peyote cimarrón; star cactus; star rock; chaute; chautle; chautl; sunamí.

10. NEOLLOYDIA Britton & Rose
Cone Cactus

According to Anderson (1986) *Neolloydia* is a genus of 15 species distributed mostly in east-central and northeast Mexico with only one species, *N. conoidea*, extending into the United States. Benson (1982) included *Echinomastus* with *Neolloydia*, thereby recognizing seven species of *Neolloydia* for the United States, most of which are treated herein as *Echinomastus*. Weniger (1984) treated *N. conoidea* as a member of a broadly conceived *Echinocactus*. A more narrow concept of *Neolloydia* also was favored by Bravo-Hollis and Sánchez-Mejorada (1991a), who placed some of the *Neolloydia* species recognized by Anderson in *Thelocactus* and *Normanbokea* Kladiwa & Buxb. They treated *Neolloydia* as consisting of three species in two subgenera, and they delineated three varieties of *N. conoidea*, two of them strictly Mexican in distribution, and *N. conoidea* var. *conoidea* for Mexico and Texas. Zimmerman et al. (forthcoming) accepted two species of *Neolloydia*, both occurring in the CDR, and two varieties of *N. conoidea*, these mostly Mexican in distribution. *Neolloydia* has been associated systematically with a number of other genera, including *Turbinicarpus*, *Rapicactus* Buxb. & Oehme ex Buxb., *Pediocactus* Britton & Rose, *Sclerocactus* Britton & Rose, and *Gymnocactus* Backeb. Anderson compiled a taxonomic comparison of *Neolloydia*, *Thelocactus*, *Normanbokea*, *Gymnocactus*, and *Turbinicarpus* involving 26 diagnostic characters. In the Trans-Pecos *N. conoidea* is easily distinguished from *Thelocactus*, *Echinomastus*, and all other taxa by numerous vegetative and floral characters.

Among the many distinctive features of *N. conoidea*, compared with *Echinomastus* (mostly reviewed by Anderson, 1986), are the many-stemmed growth habit, lack of ribs, prominent tubercles, copious white-woolly hairs at the stem apex, lack of mucilage cells in the stem, and one hypodermal layer. In *Echinomastus* the growth habit is single-stemmed, ribs are present, tubercles are not as prominent, and there are woolly hairs mostly associated with young areoles at the stem apex, but not as prominently as in *N. conoidea*. Also, the stems of *Echinomastus* have mucilage cells in outer cell layers, and there are 1–3 hypodermal layers.

Distinctive reproductive traits in *N. conoidea*, compared with *Echinomastus*, include magenta flowers, white stigma lobes, spherical pollen 60–65 µm in diameter with 12–15 apertures and reticulate exine, round fruit with dry, papery walls (at maturity) without appendages (scales), and dehiscence by vertical slits. *Neolloydia* also has pyriform seeds, 1.3–1.6 mm in diameter, and with a large basal hilum featuring a prominent lip on one side.

In Texas, *Echinomastus* flowers are white to very pale pinkish, the stigma lobes are green or red, the pollen is elongate and equatorially 45–53 µm in diameter and with a very finely punctate exine, the fruit bears scales (bracteoles) and a strongly persistent floral remnant, and the seeds are ovoid-reniform, 1.8–2.2 mm in diameter, and with a large "lateral" hilum.

The genus name is taken from the Greek *neos*, "new," and *lloydia*, after Francis Ernest Lloyd (1868–1947), professor of botany at McGill University in Montreal. $x = 11$.

1. **Neolloydia conoidea** (DC.) Britton & Rose var. **conoidea**. TEXAS CONE CACTUS. Plates 206, 207. [*Mammillaria conoidea* DC., Mém. Mus. Hist. Nat. Paris 17: 112. 1828]. Plants branched or not. Roots diffuse. Stems globose to cylindroid, 5–10(–12) cm high, 2.5–6.5(–7) cm in diameter, gray-green to somewhat yellowish-green, typically with a white-woolly apex; hypodermis one-layered; pith and cortex not mucilaginous, rarely with druses; a yellow viscid layer present in the bark beneath old epidermal tissue, unique to this genus; pith relatively small, without vascular strands. Ribs none, but tubercles in 8–13 spiral rows. Tubercles prominent, 7–12 mm long, 8–18 mm broad, conical-compressed, each with a felted areolar groove from the apex to the axil. Areoles circular, 2–5 mm in diameter, 8–12 mm apart, the youngest ones with white wool. Spines rather dense, usually only partially obscuring the stem; central spines stout with bulbous bases, black or dark brown, mostly 3–4 per areole (but 0–6), straight, the lower one longest, 1.7–2.4(–2.8) cm, 0.6–1 mm thick, porrect or angled slightly downward, the upper 1–3 angled upward, 0.9–1.7 cm long, ca. 0.5 mm thick; radial spines (9–)14–17(–25), appressed, whitish, often with a dark tip, 0.6–1.2(–1.7) cm long, straight, needlelike, 0.3 mm in diameter, with bulbous bases. Flowers 2.5–3.2 cm long, 3–5.5 cm in diameter; inner tepals magenta or bright rose-pink, oblanceolate, 1.5–3.2 cm long, 0.5–1.1 cm wide, entire, mucronate; outer tepals whitish with a greenish midvein, elongate, 1–1.8 cm long, 4–8 mm wide, margins entire, apexes rounded-mucronate; filaments whitish, 3–6 mm long, anthers deep yellow, 1 mm long; style whitish, 0.7–1.1 cm long, ca. 1 mm or less wide, stigma lobes 4–7, white to cream, 2–3 mm long, slender; pericarpels naked (rarely with 1–2 scales). Fruit arising at base of apical tubercle, hidden in early development, later exposed by deciduous floral remnant, green then greenish-brown or tan at maturity, spheroidal, 4–8(–10) mm long, 4–8 mm in diameter, dry and papery at maturity, dehiscing by vertical slits, the fruits persistent. Seeds black to gray, strongly papillate (testa cells strongly convex), 1–1.6 mm long, 0.8–1.2 mm in diameter, pyriform, hilum large and basal, with a lip over part of hilum. $2n = 22$.

Stable rocky limestone habitats (substrate at one site said by the collector to be igneous), mostly in desert mountains, S, central, and E Brewster Co., SE Presidio Co., E to S Pecos Co., Terrell Co. into Val Verde Co. near Del Rio. Outliers in El Paso and Culberson counties (Benson, 1969b). 1,500–4,000 ft. Flowering Mar–Jul, usually May–Jun. $2n = 22$. Edwards Co., TX. Northeastern Mexico in Coahuila, Nuevo León, Durango, Zacatecas, San Luis Potosí, Querétaro, S to Hidalgo, and E to Tamaulipas. Map 44.

The specific epithet is derived from the Greek *konos*, "cone," and *oideos*, suffix meaning "a form or type of," presumably in reference to the somewhat cone-shaped stems or tubercles. Weniger (1984, p. 136) remarked that *N. conoidea* is "never very common in Texas," but our field observations have revealed that this species is exceedingly abundant at certain sites, for example, west of Sanderson,

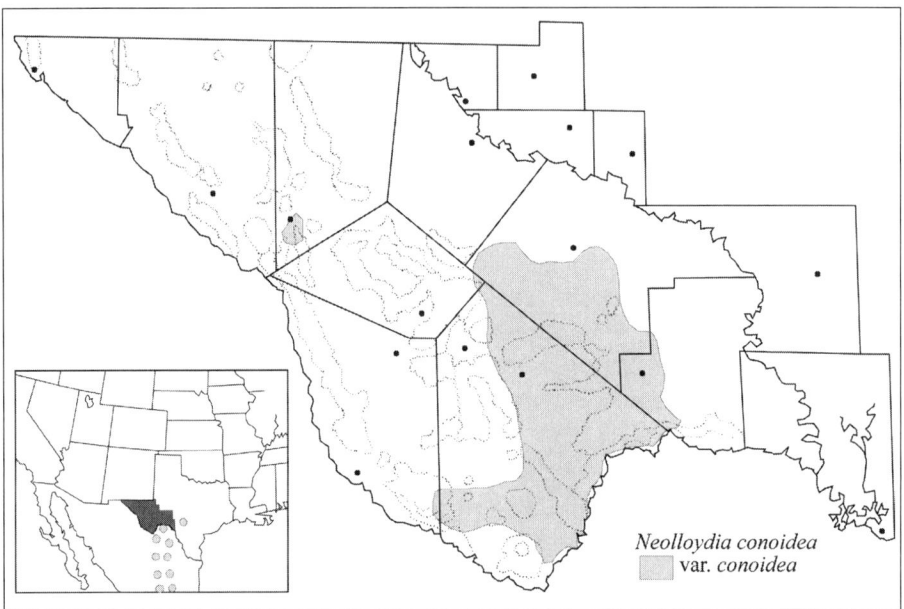

Map 44. Generalized distribution of *Neolloydia conoidea* var. *conoidea* (Texas cone cactus).

in the Dead Horse Mountains, and on certain stable limestone formations elsewhere, particularly in central and southern Brewster County.

Identifying Characters. Full-size plants of *N. conoidea* in the Trans-Pecos usually are branched with several stems, the largest stems 5–10(–12) cm long, 2.5–6.5(–7) cm in diameter. Plants that appear single-stemmed often have small basal branches. The green or gray-green stem surfaces usually are visible through the spines. The stem apex characteristically exhibits a prominent vestiture of white-woolly hairs.

Both *Coryphantha macromeris* and *C. ramillosa* are regionally sympatric with *N. conoidea* var. *conoidea* in the southeastern Trans-Pecos, and all share grooved tubercles with pink/magenta flowers and comparable numbers of radial and central spines. These coryphanthas usually are much larger than *N. conoidea*, and they have juicy fruits with smooth brown seeds. The central spines of *C. macromeris* and *C. ramillosa* often are black, and the radial spines often are gray, as in *N. conoidea*, but in the two coryphanthas both spine types typically are longer (3–4 centrals, the main one 3.5–6.5 cm long; ca. 10–12 radials, 1.5–5 cm long), and they are somewhat flattened or angular, or both, as compared to the shorter, terete spines of *N. conoidea*.

The magenta or bright rose-pink flowers of *N. conoidea* allow easy distinction from *Echinomastus*, which has white, tan, greenish, or pale pink-striped flowers. The flowers of *N. conoidea* are about 3–5.5 cm in diameter. The style and stigma lobes are white, or occasionally the lobes are cream-white to pale yellowish. The flowers of *N. conoidea* var. *conoidea* are similar to the magenta flowers of *C. macromeris* and *C. ramillosa*, except that the *Coryphantha* flowers are larger at 5.5–6 cm long and 4.5–6 cm across; *C. macromeris* has fringed, not entire, outer tepals.

The fruits of *N. conoidea* var. *conoidea* are inconspicuous even when fully ripe, being relatively small and hidden in the copious white wool at the stem apex. As the spherical fruits enlarge to 4–8 mm in diameter, they become evident, as the apical portion extends slightly above the wool. Then dry, greenish-brown or tan fruits remain indehiscent, or ultimately rupture, forming one or more longitudinal slits near the apex.

Phenology. The principal blooming period for *N. conoidea* var. *conoidea* in the Trans-Pecos is from May to June, but flowering occurs as early as March and may extend into August. In 1998 at the Sul Ross Cactus Garden, following a severe drought and then about one-half inch of rain ca. two weeks earlier, there was a major bloom on 3 August, with one plant in flower on 17 August and three plants in bloom on 1 September, following additional rains. The flowers of *N. conoidea* typically are fully open by mid- to late morning, and in at least some instances may begin closing by 1:00 P.M. They become more completely closed as the afternoon goes on, as though affected by hot sun. Each flower may open again for 1–3 days. One to three flowers may be produced at the same time at the apex of a single stem.

Sterile and Immature Specimens. Even in single-stemmed individuals of *N. conoidea*, the best vegetative distinguishing characters are the prominent white-woolly hairs on and between the tubercles at the stem apex, a porrect black main central spine, particularly in upper areoles, and gray-white radial spines outlining circular areoles and appressed against the plane of the stem. The main central spine in middle or lower areoles may be gray or gray with a dark distal half. On mature stems the woolly hairs are not present much below the apex, but short woolly hairs may persist in some of the areoles on tubercles near the stem apex.

Two to four central spines are typical in the circular areoles of *N. conoidea*. The main central spine, the one that extends perpendicular, or is erect or deflexed, and that is actually positioned as the lower central spine, is usually dark to black in color, at least on the distal half, is terete with a bulbous base, and is larger in diameter than the 1–3 upper central spines. The main central (lower central) spine usually is 2–2.4 cm long. The 1–3 upper central spines are angled upward. The upper centrals are terete, to 1.7 cm long, and gray to dark throughout, or with a dark tip. The needlelike radial spines of *N. conoidea* are evenly spaced around the circular areole, and they are appressed against the plane of the stem. The usually 14–17 radial spines are ashy-white, sometimes with dark tips, and about 1 cm or more long.

Immature specimens of *N. conoidea*, and small stem branches, exhibit much the same characters as adults, except that the somewhat translucent radial spines are shorter and often tend to be oriented horizontally instead of radially. In juvenile areoles there may be only one or sometimes two central spines, with the main central spine porrect and with only the distal half dark in color. Stem apexes are covered with white-woolly hairs even in juveniles. In seedlings to 2 cm tall, the stems are cylindroid-clavate. The nearly black central spines (at least distally) contrast with the translucent or whitish radials. The spines are glabrous, except that the centrals may be scabrous.

Biosystematics. *Neolloydia conoidea* var. *conoidea* in Texas is mostly restricted to limestone substrates that are stable, such as the Boquillas formation, and less commonly in alluvial mixtures of limestone gravel and smaller soil particles. One Brewster County collection (*Warnock 11064,* in bloom on 10 June 1952, SRSC) is anomalous: "NNE of Alpine: Kokernot Ranch. In igneous soil," 4,500 feet.

In addition to the widely distributed var. *conoidea,* one or two other varieties of *Neolloydia* are of more limited range and are restricted to northeastern Mexico (Bravo-Hollis and Sánchez-Mejorada, 1991a; Zimmerman et al., forthcoming). *Neolloydia conoidea* as a species is most closely related to *N. matehualensis* Backeb., which occurs in Nuevo León and San Luis Potosí, Mexico. *Neolloydia gautii* L. D. Benson, described as a rare species in Hardin County of East Texas, is based on a cultivated plant of a common Mexican species (A. Zimmerman, unpub.).

Synonyms. *Echinocactus conoideus* (DC.) Poselger; *Coryphantha conoidea* (DC.) Orcutt ex A. Berger; *Neolloydia texensis* Britton & Rose; *N. conoidea* var. *texensis* (Britton & Rose) Kladiwa & Fittkau.

Common Name. Texas cactus. Neither Texas cactus nor Texas cone cactus is ideally appropriate because *N. conoidea* var. *conoidea* is mostly Mexican in distribution, and the stems do not much resemble a pine cone.

11. GLANDULICACTUS Backeb.
Eagle-Claw Cactus

A genus of three taxa (Ferguson, 1991), one of which, *Glandulicactus uncinatus,* occurs in Texas, New Mexico, and Mexico, and another of which, *G. crassihamatus* (Weber) Backeb., is restricted to Querétaro, Guanajuato, and San Luis Potosí, Mexico. The genus name is derived from Latin *glandula,* "a gland," and *cactus.* Presumably, the glands alluded to are the several yellowish glands in the narrow areolar groove. $x = 11$.

1. **Glandulicactus uncinatus** (Galeotti ex Pfeiff.) Backeb. var. **wrightii** (Engelm.) Backeb. EAGLE-CLAW CACTUS. Plates 208–10. [*Echinocactus uncinatus* Galeotti ex Pfeiff. var. *wrightii* Engelm., Proc. Amer. Acad. Arts 3: 272. 1856]. Plants typically with solitary stems, occasionally branched at the base. Roots diffuse. Stems globose, short-cylindroid, or ovoid, green to bluish-green with a gray glaucescence, (5–)7.5–15(–30) cm long, 5–7.5(–10) cm wide. Ribs 9–13, somewhat prominent but deeply notched, 6–9 mm broad, protruding 9–15 mm. Areoles 2–2.5 cm apart, ca. 4.5 mm in diameter, younger ones especially with gray or yellowish wool, with several yellowish glands in a narrow area of the short groove connecting upper spine-bearing portion of areole and lower flower-bearing portion. Spines abundant but not totally obscuring the stem; central spines tannish-white to stramineous and pinkish, 1–4 per areole (on mature stems), the principal central spine turned upward, prominently hooked, 5–9(–12) cm long, 1–1.5 mm wide at the base, somewhat flattened; radial spines 5–8 per areole, including 2–3 upper ones in a quasi-central spine position; three lower radials subterete, reddish or reddish-tan, hooked, slightly flattened; one straight lower radial may be present underneath the hooked ones; lateral and upper radials subulate, tan, stramineous, or cyanic, not hooked, somewhat flattened. Flowers 2–4 cm long, 2–3 cm wide, borne near stem apex at adaxial ends of short areolar grooves; inner tepals brick-red (or maroon or garnet), largest oblanceolate, 1.2–2.2 cm long, 3.5–6 mm wide, obtuse to acute, minutely toothed; outer tepals with brownish midribs but otherwise brownish or brick-red, broadly oblanceolate, 1.2–2.1 cm long, to 6 mm wide, the margins scarious, upper ones obtuse, and minutely fringed-denticulate; filaments yellow (sometimes brownish or maroon), ca. 6 mm long, anthers yellow; style reddish, to 1.2 cm long, stigma lobes 10–14, 5–6 mm long, pale yellow to dull orange; pericarpel with minutely toothed scales to 6 mm long. Fruit indehiscent, fleshy, red, ovate or globose, 1.5–2.5 cm long, 1–2 cm thick, with numerous (13–35) conspicuous scales, these deltoid, auriculate, ca. 4.5 mm long, scarious-margined with naked axils; fruit pulp white (red?); floral remnant persistent. Seeds black, basically oblong-obovoid and curving, 1.3–1.5 mm long, ca. 1 mm broad, ca. 0.8 mm thick, upper portion minutely papillate, base smooth; hilum basal, with a conspicuous rim. $2n = 22$.

Desert hills, flats, in desertscrub and grasslands, igneous and limestone derivation, commonly growing in or adjacent to grass clumps. In TX mostly restricted to the Chihuahuan Desert, almost throughout the Trans-Pecos, with a few outliers in Crockett and Val Verde counties. 1,000–5,000 ft. Flowering Mar–May(–Aug). 2n = 22. Rare in Starr and Victoria counties of S TX. Chihuahuan Desert of southern NM, Sierra, Doña Ana, Otero, and Eddy counties. Mexico, E Chihuahua, Coahuila, Nuevo León, Zacatecas, Durango, and NW Sonora. Map 45.

Glandulicactus uncinatus var. *wrightii* is one of the most widespread and common cacti in the Trans-Pecos, and it extends like *Thelocactus bicolor* dis-

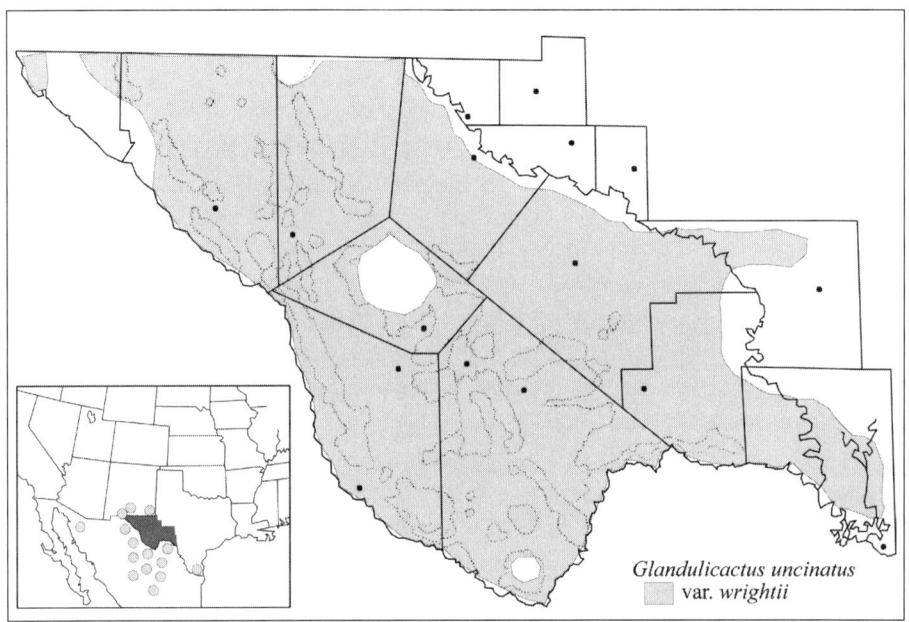

Map 45. Generalized distribution of *Glandulicactus uncinatus* var. *wrightii* (eagle-claw cactus).

junctly into South Texas. The descriptive specific epithet is taken after the Latin *uncus*, "hook," in reference to the hooked spines of this species. The varietal epithet honors the early plant collector Charles Wright (see *Mammillaria wrightii*).

Identifying Characters. Plants of *G. uncinatus* var. *wrightii* usually have unbranched stems that are cylindroid-ovoid, blue-green or grayish in color, with rather prominent ribs, deeply notched and thus almost divided into tubercles. The stems are dominated by the presence of a rather long, dull yellow, hooked central spine in each areole. Mature stems usually are 7–15 cm long and not more than 8 cm in diameter. Stems to 15 cm long or more are rare. In the field *G. uncinatus* var. *wrightii* is most likely to be confused with *Ancistrocactus brevihamatus* and young plants of *Ferocactus hamatacanthus*, which occupy much the same range and share the presence of a single, long, hooked central spine in each areole. Older plants of *F. hamatacanthus*, a barrel cactus, are not likely to be confused with *G. uncinatus* var. *wrightii* because of much larger stems and tubercles and a very different spine combination.

The flowers of *G. uncinatus* var. *wrightii* are very distinctive. Their inner and outer tepals are brick-red, dark brownish-purple, maroon, or dark red in color, much the color of *Mammillaria pottsii* flowers. Several flowers usually develop together at the stem apex. They are relatively small, 2–4 cm long and 2–3 cm in diameter, usually cylindrical-funnelform in shape. Crowded yellow anthers contrast with the dark red tepals and pale yellow to dull orange stigma lobes. In *F. hamatacanthus* the flowers are considerably larger and yellow in color.

The fleshy fruits of *G. uncinatus* var. *wrightii* are rather bright red, with numerous conspicuous white-fringed scales. It is common to see several fruits developing simultaneously at a stem apex. In *F. hamatacanthus* the fleshy fruits are much larger, to 4.5 cm long and 3.5 cm in diameter, green to reddish-green, and with numerous conspicuous scales.

Phenology. The principal blooming period for *G. uncinatus* in the Trans-Pecos is from March to May, with April being the most active month. Populations in the southern and southeastern Trans-Pecos generally are the earliest to produce blooms. Flowering episodes are sporadic, perhaps rare, in the summer as late as August. Healthy plants produce numerous buds in a spiral at the stem apex, with usually 3–5 or more flowers opening at the same time. Flowers sometimes open by 9:00 A.M., but usually opening is stimulated by warm temperatures later in the morning (L. Coleman and A. Kirk, pers. comm.). Individual flowers close partially (two-thirds of the way; Worthington, 1986) at night and open again for 2–3 or more days. Fruit maturation apparently requires only a month or two. Brilliant red fruits appear in May or June or sometimes later in the summer, depending upon sporadic flowering.

Sterile and Immature Specimens. Sterile specimens of *G. uncinatus* var. *wrightii* usually are easy to identify because of a distinctive spine combination. In mature plants each areole supports usually 11–12 spines, with different sizes, types, and colors arranged in a certain pattern in each areole. The dominant spine is relatively long, erect, and hooked, borne near the center of the areole. This "main" central spine is 5.5–10 cm long, to 1.5 mm wide, typically dull yellow, microscopically scabrous, usually flattened or flat on the adaxial side and angled on the other (younger spines often subterete), with the shaft straight, bent, or twisted in a loose spiral. Older main central spines down the stem may lose the yellow color and become gray or blue-gray, and reddish-brown at the tip. The remaining 10–11 spines in each areole are interpreted here as three central spines and 7–8 radial spines. All spines are suberect or appressed against the plane of the stem except for the porrect or erect hooked central spine. In areoles of immature plants (discussed below), the central and radial spine positions are relatively clear, but in older areoles of mature plants, the position of all spines except the main central spine could be interpreted as radial.

In mature plants most areoles exhibit the following spine configuration, beginning at the lower margin of the areole and proceeding upward: three primary lower radial spines, 2–3 cm long, ca. 1 mm wide, one at the bottom, and two lateral ones slanting down, all three of these subterete, dark reddish, and hooked at the darker tips; one smaller radial spine, not hooked but perhaps curved at the tip, borne underneath the lowermost hooked radial spine; two lateral radial spines, one on each side, 2–3.5 cm long, ca. 1 mm wide, angled upward, somewhat flattened, stramineous throughout or gray at base and cyanic distally, curved but not hooked at the tip; two upper central spines, 3–3.6(–4.2) cm long, ca. 1 mm wide, borne above the main central in a near-radial spine position, one on each side angled upward, these stramineous to reddish and not hooked; one upper central spine positioned near the upper edge, 3–4.5 cm long, ca. 1.5 mm wide at the base, flattened, or flat adaxially and angled abaxially, stramineous, approaching wood-shaving-flat and as wide as the main central spine, tapering to a point; two upper radials, 3–3.5 cm long, these flat, stramineous or becoming gray. There is a total of 12 spines in a "typical" mature areole, including one hooked main central spine. In each areole only the main central and the three main lower radial spines are hooked.

Juvenile plants display about eight spines in each areole, with the total increasing to 12 spines in older plants. The hooked main central spine may be absent in all, or at least some, areoles of immature plants, and the hooked spines, if present in juveniles, are located in areoles toward the stem apex. The hooked central spine typically is yellow even in juvenile plants. In the eight- or nine-spine areole of immature plants, the spines typically present, from bottom to top, are the three hooked lower radials, the two lateral radials, the one hooked main central, two upper radials (or centrals, depending upon interpretation of position at this point), and one (flat, broad) upper central (or radial). One of the upper radials may be missing in an eight-spine areole. In somewhat older areoles one flat upper radial spine is added to form a 10-spine areole. An upper central or upper radial and the lowermost nonhooked radial are added to form the 12-spine areole of mature plants.

Seedlings of *Glandulicactus* are strictly tuberculate. In seedlings 1–2 cm in diameter, there are 4–7 rather thick spines per areole. The spines are yellow, red proximally, or brown, densely microscopically short-pubescent, and all of them in an areole are hooked. Even in immature plants to 5–7 cm tall, all or most of the spines are hooked, or at least curved at the tip, particularly in upper areoles of the stem.

Sterile mature and immature specimens of *Ferocactus hamatacanthus*, with which *G. uncinatus* var. *wrightii* might be confused, have very different spine configurations in each areole, although a hooked main central spine is produced in both taxa. In general, *F. hamatacanthus* produces typically 14–19 spines in each areole; four centrals, three upper and one lower, and only the obviously lower central is hooked. The hooked lower central is the longest spine in the areole. In *F. hamatacanthus* not any of the 10–15 radial spines is hooked.

Biosystematics. In Trans-Pecos Texas the widespread and common *G. uncinatus* var. *wrightii* is most frequently found in desert or semidesert habitats growing within tufts of grass such as chino grama (*Bouteloua ramosa* Scribn. ex Vasey). Throughout its range *G. uncinatus* var. *wrightii* is also associated with many other plant species and may be found on bare ground, but its close association with grasses suggests that grasses are the most important nurse plants for this taxon.

There appear not to be any close relatives of *G. uncinatus* var. *wrightii* in Texas. The typical variety, var. *uncinatus,* with green fruits, is restricted to San Luis Potosí, Mexico (Ferguson, 1991; Zimmerman et al., forthcoming). The third taxon, *G. crassihamatus,* from even farther south in Mexico, is very closely related. The true generic position of *G. uncinatus* requires further evaluation. *Glandulicactus* has been aligned with *Echinocactus, Ferocactus, Hamatocactus, Thelocactus, Ancistrocactus,* and *Sclerocactus.* Benson (1969b, 1982), without any explanation, placed *G. uncinatus* in *Ancistrocactus,* a genus that clearly has different stem anatomy and spine, flower, fruit, and seed characters. Weniger (1984) immersed *G. uncinatus* in the broad generic concept of *Echinocactus.* Bravo-Hollis and Sánchez-Mejorada (1991a) followed Buxbaum, who placed *Glandulicactus* into the genus *Hamatocactus* as a separate subgenus. *Glandulicactus* and *Hamatocactus,* sensu stricto (i.e., monotypic *H. bicolor*), are similar in habit, hooked central spines, and fruit and seed types. The brick-red flowers of *G. uncinatus* are very different from the spreading yellow flowers with red centers in *H. bicolor. Glandulicactus* was combined with *Sclerocactus* by Hunt and Taylor (1986) but excluded from *Sclerocactus* by Heil and Porter (1994). *Glandulicactus* is maintained as distinct by Zimmerman et al. (forthcoming).

Synonyms. *Echinocactus wrightii* J. M. Coult.; *Hamatocactus wrightii* (Engelm.) Orcutt; *Echinomastus uncinatus* (Galeotti) F. M. Knuth var. *wrightii* Engelm. ex F. M. Knuth; *H. uncinatus* Galeotti ex Borg var. *wrightii* Engelm. ex Borg; *Ancistrocactus uncinatus* (Galeotti ex Pfeiff.) L. D. Benson var. *wrightii* (Engelm.) L. D. Benson; *H. uncinatus* (Galeotti) Buxb. var. *wrightii* (Engelm.) Bravo-Hollis; *Sclerocactus uncinatus* (Galeotti) N. P. Taylor var. *wrightii* (Engelm.) N. P. Taylor; *Glandulicactus wrightii* D. J. Ferguson.

Common Names. Catclaw cactus; catsclaw Echinocactus; turk's head; catclaw cactus; brown-flowered hedgehog; mountain fishhook cactus; Texas hedgehog; vaca. The four hooked spines per areole are like a set of talons, inspiring the name eagle-claw cactus.

12. ANCISTROCACTUS Britton & Rose
Fishhook Cactus

Plants typically with solitary stems, rarely branching at the base, usually with only a broad, rounded apex visible aboveground, thus deeply seated. Roots diffuse except in *A. scheeri*, which has fleshy taproots with bulbous swellings or bulbous secondary roots. Stems ellipsoid to subglobose or obovoid to elongate-obovoid, or obconic-turbinate, nearly flattened at apex, (1.5–)2–10(–15) cm long, 2.5–8(–10) cm in diameter; stem surface dark green to pale blue-green; pith and cortex strongly mucilaginous. Ribs low and indistinct in some species (*A. scheeri*) but with prominent tubercles, absent (in young plants), or indistinct with prominent tubercles in others (*A. tobuschii*; *A. brevihamatus*). Tubercles 5–12 mm long, 5–9 mm broad, protruding 5–12 mm; areoles usually 6–9 mm apart. Areoles subcircular, becoming elongated with age, 1.5–4.5 mm in diameter, a felted areolar groove present, often with hemispheric glands evident. Spines dense, in mature plants sometimes obscuring the stem, minutely canescent; central spines grayish or whitish, tending toward black on lower portions and curve of the hook, light yellow with reddish tips and maturing grayish, or mostly yellowish and often maturing grayish-white; central spines differing in juvenile and mature plants, (3–)4 per areole, upper centrals 0–2(–3), straight, turned upward, principal central spine porrect and hooked (absent on immature plants and young adults of *A. brevihamatus*), 0.9–2.5(–4.5) cm long, needlelike or somewhat flattened, 0.4–0.8 mm wide at the base; radial spines like centrals or more tannish to whitish, 7–22 per areole, straight, the longest 6–12 mm, needlelike, radiating parallel to the stem, or somewhat irregularly. Flowers borne at adaxial end of areole, adjacent to spine clusters on new growth near stem apex in most mature plants, the flowers 2.5–4 cm long, 1.5–4 cm in diameter; inner tepals 1.5–3 cm long, 2–4.5 mm broad, entire or minutely laciniate; bright green, light yellow, cream-colored, pinkish, brown to white, the midrib area perhaps brownish or purplish; outer tepals 1–1.5(–2) cm long, 2–6 mm broad, with greenish, brownish-red, or purplish midribs, and pale yellow to greenish-white margins, apically rounded and entire or denticulate; nectar chamber funnelform, 1–2 mm long; filaments pale or reddish, anthers yellow; style green to greenish-white, stigma lobes 4–10, 1.5–2.5 mm long, green to yellowish; pericarpel with 0–20 scale bracts. Fruit indehiscent, (0.9–)1.5–2.5(–3.1) cm long, (3–)6–9 mm across, fleshy but thin-walled, ultimately disintegrating with 1–13 scales, these broad and membranous, green at maturity, drying slightly with a rosy tinge, or yellowish or tan. Seeds dark reddish brown to blackish, finely glossy, 1.5–2 mm in maximum dimension, ca. 1 mm thick, roundish to subreniform; hilum appearing lateral, relatively large, concave. $x = 11$.

A genus of two or three poorly defined species distributed from the southern Trans-Pecos to South Texas mostly along and near the Rio Grande and into northern Mexico. All three of the "species" occur in Texas. One species with three varieties occurs in the CDR. Two of the taxa, both varieties of *A. brevihamatus*, occur in or near the Trans-Pecos. The genus name is derived from the

Greek *ankistron,* "fishhook," in reference to the hooked central spines and hence "fishhook cactus."

1. **Ancistrocactus brevihamatus** (Engelm.) Britton & Rose. SHORT-SPINED FISHHOOK CACTUS, SNIPE CACTUS. Plates 211–19. [*Echinocactus brevihamatus* Engelm., Proc. Amer. Acad. Arts 3: 271. 1856]. On or near various limestone habitats except usually not on steep slopes, also gypsum in Mexico, Tamaulipas scrub or Chihuahuan desertscrub, degraded grassland, live-oak savanna E of Devils River. Brewster Co. E to Val Verde and Kinney counties. 1,000–4,100 ft. Flowering winter–early spring. $2n = 22$. Mexico, Coahuila, Nuevo León, and Tamaulipas, mostly E of the Sierra Madre Occidental. Map 46.

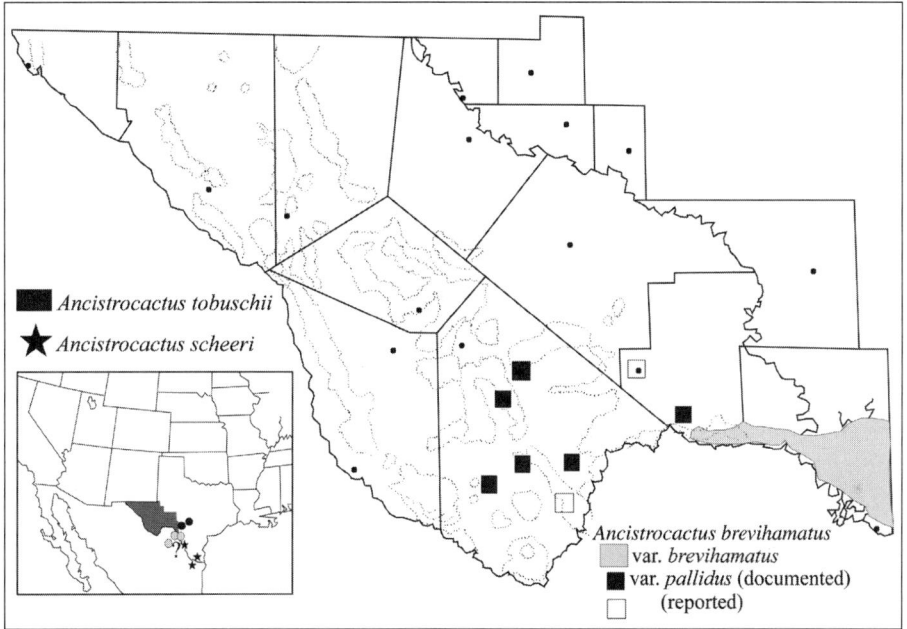

Map 46. Generalized distribution of *Ancistrocactus brevihamatus* var. *brevihamatus* (short-spined fishhook cactus), *A. brevihamatus* var. *pallidus* (snipe cactus), and *A. tobuschii* (Tobusch fishhook cactus), and *A. Scheeri* (Scheer's fishhook cactus).

Ancistrocactus brevihamatus was treated by Benson (1982) as a synonym of *A. scheeri* (Salm-Dyck) Britton & Rose. Weniger (1984) and Zimmerman et al. (forthcoming) recognized *A. brevihamatus* as a distinct species. Benson (1982) included *Glandulicactus uncinatus* in *Ancistrocactus*. The only one of the three species of *Ancistrocactus* that extends into the Trans-Pecos is *A. brevihamatus*. The report (Wauer, 1980) of *A. tobuschii* (W. T. Marshall) W. T. Marshall ex Backeb. from the Dead Horse Mountains in Big Bend National Park (Brewster County) is erroneous. The plant from the Dead Horse Mountains is either *A. brevihamatus* var. *pallidus* (Zimmerman et al., forthcoming) or *G. uncinatus* (S. Brack, pers. comm. to D. Miller, 1996), although the description of "yellow-green" flower color given by Wauer does not match either var. *pallidus* (white flowers) or *G. uncinatus* (maroon flowers). Of the four varieties of *A. brevihamatus* (Zimmerman et al., forthcoming), three occur in the CDR, two of them in Texas, and both of these occur in the southeastern Trans-Pecos. The specific epithet is derived from the Latin *brevis*, "short," and *hamatus*, "hooked," a reference to the relatively short, hooked central spine.

Key to the Varieties

1. Hooked central spine 0.8–1.5 mm in diameter; flowers brown, pink and brown, green and brown, or rose-pink with pale margins

   1a. *A. brevihamatus* var. *brevihamatus*.

1. Hooked central spine 0.3–0.6(–0.8) mm in diameter; flowers off-white or cream-colored

   1b. *A. brevihamatus* var. *pallidus*.

1a. Ancistrocactus brevihamatus var. **brevihamatus.** SHORT-SPINED FISHHOOK CACTUS. Plates 211, 212. Various limestone habitats, rocky or alluvial, mostly in Tamaulipan scrub. Val Verde Co., Devils River W to near Comstock. Terrell Co., near Rio Grande. 1,000–2,500 ft. Flowering Jan–Mar. $2n = 22$. Kinney and Uvalde counties. Mexico, Coahuila W to near Cuatro Ciénegas, and Nuevo León. Map 46.

The distribution of var. *brevihamatus* is mostly east of the Pecos River in Val Verde County, east to Kinney and Uvalde counties, but it has been documented as far west as southern Terrell County. At the northeast edge of its range, it intergrades with the parapatric and very weakly distinguished *A. tobuschii*. *Ancistrocactus tobuschii* is federally and state listed as an Endangered species. *Ancistrocactus brevihamatus* is more sharply distinct from *A. scheeri*, which in general has a more southerly distribution in Texas (and adjacent Mexico), extending north to southern Kinney County where it is sympatric with *A. brevihamatus* var. *brevihamatus* (Zimmerman et al., forthcoming).

Identifying Characters. The stems of var. *brevihamatus* usually are globular and unbranched. In very old specimens the stems may be short-cylindroid. The stems vary in size, 3–6 cm in length and diameter in most sexually mature plants, to ca. 13 cm long and 8–9 cm in diameter in old plants. The stems are tuberculate in younger plants, and older specimens have 8–13 ribs. Vegetatively, the most conspicuous feature of *Ancistrocactus* is the single, porrect, hooked, lower central spine, 1.5–4.3 cm long. Usually there are three additional, straight central spines, which are erect, conspicuously dorsiventrally flattened, may be slightly curved but never hooked. Typically, there are 7–14 appressed radial spines. Both central and radial spines usually are whitish-gray to gray-white in color, or the centrals are gray and the radials whitish.

The flowers of var. *brevihamatus* are inconspicuous. The inner tepals usually are pinkish-brown to olive-green with reddish-brown midregions, may be brownish, pink and brown, green and brown, or dull rose-pink with pale margins (Zimmerman et al., forthcoming), varying to "green to rose or salmon-orange," and "darkened or muddy looking." The flowers are funnelform, 2–4 cm or more long, and (1–)2–3(–4) cm in diameter, like those of all other *Ancistrocactus* taxa, differing only in color. The apical spines typically restrict the opening of flowers. Flowers of *A. tobuschii* are bright yellow, turning golden-yellow with age (rarely cream-yellow or yellowish-green), and those of *A. scheeri* are green (even bright green) or greenish-yellow. In var. *brevihamatus* the inner tepals are 0.9–1.7(–2.4) cm long, 2.5–5(–7) mm wide, sublinear, and acute at the apex. The stamens have rose-colored filaments and yellow anthers. The relatively short style supports 4–10 stigma lobes that are 1.5–2.5 mm long and pale reddish-purple or green to yellow in color. The pericarpel wall bears 0–20 scale bracts. The outer tepals are entire to denticulate.

The fruits of var. *brevihamatus* are cylindroid-ovoid, (0.9–)1.5–2.5(–3) cm long, 0.6–1.3 cm in diameter, green at maturity, or often with a tinge of pink or rose when very ripe. The green fruits of the closely related *A. tobuschii* also may develop a rose tinge when fully mature, and when fully ripe the fruit walls become papery thin and turn tan to papery, at least in cultivated specimens. The fruits of var. *brevihamatus* are indehiscent and weakly succulent until after maturity, when the pericarpel wall dries. The floral remnant is persistent. The helmet-shaped or reniform seeds are 1.7–2 mm across in maximum dimension, glossy, and dark reddish-brown to nearly black. Fresh seeds sometimes are golden-brown, but usually they are reddish-brown, and drying darker.

Phenology. The flowering period for var. *brevihamatus* is from late January to mid-March, with peak flowering usually in late February. In the Sul Ross greenhouse, plants from Val Verde County have produced open flowers as early as 20 January, and the flowers have lasted for as long as nine days, closing at night and opening again the next day. This flowering period is shared by the parapatric *A. tobuschii* (Sutton et al., 1997). The closely related South Texas species, *A. scheeri,* has a potential blooming period from late November to early March. At a site in Val Verde County mature fruits were found to be present on var. *brevihamatus* in mid-April, and the fruits were mostly gone from all the plants by late May (Miller, 1995). In greenhouse specimens referred to above, fruits of var. *brevihamatus* remained green for about 3–4 months, about one month longer than in var. *pallidus* and *A. tobuschii*. Lockwood (1995) and Emmett (1995) reported some phenological and other information for *A. tobuschii*. Numerous seed-propagated plants of *A. tobuschii* from Kinney County produced flowers in Alpine during late December and early January 1997, at the age of two years and two months. The flowers were greenish-yellow with pale brownish midregions of the inner tepals.

Sterile and Immature Specimens. Other species with hooked spines that may be sympatric and potentially confused with *Ancistrocactus* in vegetative condition are *Glandulicactus uncinatus* var. *wrightii, Ferocactus hamatacanthus,* and *Hamatocactus bicolor. Glandulicactus* is by far the most similar to *Ancistrocactus,* but it is easily distinguished by its longer, yellowish, often upturned hooked central spine, and especially by its hooked lower radial spines. In *Ancistrocactus* the radial spines are never hooked. Mature plants of both *Ferocactus* and *Hamatocactus* are larger than those of *Ancistrocactus,* and even in immature plants, the ribs and spine patterns are different from those of *Ancistrocactus*.

At immaturity the globular, often flat-topped stems of var. *brevihamatus* are different from the cylindroid, relatively tall stems of *A. scheeri* but identical to those of *A. tobuschii*. Very old plants of var. *brevihamatus* may be cylindroid. The single-stemmed plants of var. *brevihamatus* often are deep-seated. The roots of var. *brevihamatus* are diffuse and without tuberous enlargements, or there may be a short conical taproot. In contrast, the root system of *A. scheeri* features a rather long, fleshy taproot, ca. 1 cm in diameter, sometimes with bulbous swellings, and tuberous secondary roots as well. Both var. *brevihamatus* and *A. tobuschii* have similar root systems.

In var. *brevihamatus* immature stems are tuberculate, but with advancing age, 8–13 ribs are formed, these deeply notched between the areoles. The podaria are more or less conical and protrude 1 cm or less. The stem surface is dark green, gray-green, to pale blue-green. The pith and cortex are strongly mucilaginous. The pith occupies one-fourth to one-third of the inner stem diameter and does not contain vascular bundles. The areoles are nearly circular and 3–4 mm in diameter at sexual maturity but become progressively more elongate with age, ultimately extending as a felty groove to the axils. One or more hemispheric glands often are conspicuous in the felty areolar grooves. In younger plants flower buds are borne in the areolar groove adjacent to the spine clusters, but with advancing age buds are formed in an axillary position in areoles near the stem apex. Whitish wool is present in circular areoles near the stem apex.

One of the most identifiable vegetative features of *Ancistrocactus* is the presence of a single porrect, hooked central spine. In adults of var. *brevihamatus* the hooked central is 0.8–1(–1.2–1.5) mm wide, usually gray, subterete or dorsiventrally compressed, and bulbous-based. In cross section the hooked central is either elliptic, flat on the upper surface and round on the lower, or angled on one or both surfaces. The hooked central seems to be gray, or brown on the upper surface, in most plants of average age, except in the upper areoles where the hooked central may be yellowish and only ca. 0.7 mm wide. In very old plants the hooked central may be even more flattened, dark brown, and to 1.5 mm wide. The distal portion of the hooked central may be pale gray just below and into the hook, this being especially apparent on spines with a dark shaft. The upper centrals are flexible, 2.3–3.5(–5) cm long, 0.8–1.5 mm wide, enlarged basally, and usually gray in color. If not actually flat on both surfaces, the upper centrals in cross section are narrowly elliptic or angled, or with a groove extending most of the length on the lower surface. When only two upper centrals are present, they are arranged in a "V." When three upper centrals are present, the pattern is a "bird's foot," with the middle spine shorter than the two lateral ones. All of the spines, including the radials, are red-brown at the tip. In var. *brevihamatus,* most commonly there are 10–14 radial spines, at least in northwestern populations. The radial spines are mostly appressed and terete except that some of the upper ones are flattened, resembling the upper centrals in shape. In fact, 1–2 of these flattened radials are intermediate in position between the upper centrals and the radials and might be interpreted as subcentrals. The flattened upper radials are the longest, 1.5–2(–2.9) cm, and the lateral ones are 0.8–1.2(–2.2) cm long. Radial spines are 0.2–0.5 mm wide. In color the radials usually are pale gray (whitish) or tan but may be yellowish, particularly in fresh areoles.

Radial spine characters have been cited in the literature as somewhat distinctive in *A. brevihamatus*, *A. tobuschii*, and *A. scheeri*. Reportedly characteristic of *A. brevihamatus* are 12–14 radials that are tan to pale gray. *Ancistrocactus tobuschii* has been reported to have 7–9 or 7–12 yellowish radials, whereas *A. scheeri* is characterized by 15–22(–28) radials that are translucent yellow. Recent observations of these three taxa, albeit as yet no comprehensive populational studies, have tended to verify the larger number of radial spines in *A. scheeri*. In *A. brevihamatus* and *A. tobuschii*, however, radial spine numbers overlap significantly, although *A. brevihamatus* on average seems to have more radials, and radial spine color differences for these taxa may prove to be a useful distinguishing feature.

In juveniles of var. *brevihamatus* the hooked central spine may be absent in all areoles or present in just the upper areoles (Zimmerman et al., forthcoming). In juveniles propagated from seed and grown in a Sul Ross greenhouse, stems as small as 1 cm long and 1.5–2 cm in diameter produced the hooked central in upper areoles. Apparently in nature larger juvenile stems and even some young adults may lack the hooked central. Greenhouse-grown juveniles may produce 9–14 radials with none of the radials in an upper position (i.e., all of the radials in upper lateral, lateral, or lower positions in the areole).

Greenhouse-grown seedlings of var. *brevihamatus* smaller than ca. 1.5 cm in diameter lack the hooked central and upper centrals. The seedlings have (5–)8(–10) radial spines that are ca. 5 mm long, translucent-white, and copiously short-pubescent. In seedlings all of the spines are slender and more or less terete. In most seedlings one or two of the spines are in the position of upper radials, where they might be expected to develop as upper central spines. The hooked central spine in upper areoles of juvenile stems, those 1–1.5 cm long and 1.5 cm wide, is merely scabrous among pubescent radials. In young adult and adult plants, upper central spines are microscopically scabrous but never pubescent. The hooked central and radials are glabrous or sparingly scabrous. One or two red glands are produced in the areolar grooves on juvenile stems.

Seed-propagated seedlings and juveniles of the closely related *A. tobuschii* from Kinney County also have been observed during the current study. Immature stems of *A. tobuschii* produce 8–12 radials and usually a hooked central spine in the upper areoles. The hooked central, if present, is yellowish or golden-yellow, and the radials are yellow or cream-colored and short-pubescent in at least some individuals. Red or colorless glands are produced in areolar grooves on juvenile stems.

Biosystematics. In the Trans-Pecos var. *brevihamatus* apparently exists as scattered individuals or small populations mostly near the Rio Grande, but the extent of its distribution west of the Pecos River is not known. Morphological differences between older and younger plants of *A. brevihamatus* and *A. scheeri* are more conspicuous than the taxonomically significant differences between the species (Zimmerman et al., forthcoming.; Benson, 1982), not including the very distinctive fleshy-tuberous roots of *A. scheeri*. Valid interpopulational comparisons in *Ancistrocactus* require the use of specimens of comparable ages (Zimmerman et al., forthcoming). Older plants are more or less ribbed, with elongate areoles on the tubercles, and have a shaggy, "wild-spined" appearance. Younger plants are strictly tuberculate (not ribbed), with short areoles, and have a "neat-spined" (shorter spine) aspect.

Two other species of *Ancistrocactus* are found in Texas, but they are not known to occur in the Trans-Pecos. *Ancistrocactus tobuschii* is most similar morphologically and perhaps most closely related to var. *brevihamatus*. Both taxa have been reported in Val Verde County near the Devils River north of Lake Amistad and in Kinney County, but dimorphism with respect to flower color should not be mistaken for sympatry. East of Kinney County, all of the populations are the yellow-flowered *A. tobuschii*.

Ancistrocactus tobuschii (Plates 213, 214) was described (Marshall, 1952) from a single locality near Vanderpool in Bandera County. Very few additional populations were known when it was federally listed as Endangered in 1983. The rare-plant status of *A. tobuschii* generated considerable interest among certain workers (U.S. Fish and Wildlife Service, 1987), and by 1987 (Poole and Riskind, 1987) this yellow-flowered taxon was known from five Hill Country counties: Bandera, Kerr, Kimble, Real, and Uvalde. An early assumption regarding the ecological requirements of *A. tobuschii* was that it was most common in (if not restricted to) alluvial soils near drainages in oak-juniper woodland. Subsequent fieldwork (Emmett, 1995; Poole and Janssen, 1995; Lockwood, 1995; Texas Parks and Wildlife, unpub. reports; Sutton et al., 1997) revealed that the usual habitat for *A. tobuschii* is on the rocky hills and plateaus. In fact, in the western portion of its range, where *A. tobuschii* occurs in proximity to *A. brevihamatus*, the two taxa appear to be mostly ecologically distinct. Field-workers from the Texas Parks and Wildlife Department have observed that all of those fishhook cactus plants that they saw in bloom on mesa tops in thin soil had yellow flowers (i.e., *A. tobuschii*), whereas those seen in bloom in alluvial draws below the same mesa tops had pinkish-brown or pinkish-green flowers (i.e., *A. brevihamatus*). Several workers (Weniger, 1984; Poole and Janssen, 1995) have pointed out vegetative and floral differences between *A. tobuschii* and *A. brevihamatus*. Some other workers who have participated recently in documenting westerly limits in the distribution of *A. tobuschii* (unpub. reports and pers. comm.) have noted possible hybridization, introgression, and morphological similarity between *A. tobuschii* and *A. brevihamatus*, particularly in Val Verde and Kinney counties. At present it is not clear whether *A. tobuschii* and *A. brevihamatus* should be regarded as distinct species, two intergrading varieties, or merely integrating flower color morphs of the same taxon.

Ancistrocactus scheeri (Plates 215, 216), commonly known as Scheer's fishhook cactus, is regarded by most workers (e.g., Weniger, 1984; Zimmerman et al., forthcoming) as a distinct species distributed in South Texas and adjacent Mexico. This Tamaulipan taxon extends northwest in Texas to the Anacacho Mountains in Kinney County, where it is sympatric with *A. brevihamatus*. *Ancistrocactus scheeri* also approaches *A. brevihamatus* in distribution and spine number in eastern Coahuila (Zimmerman et al., forthcoming). The fleshy-tuberous root system of *A. scheeri* is not found in any other species of *Ancistrocactus*. Cultivated plants of *A. scheeri* produced fruits that were green, ovate-elongate to oblong-ovate, 2.7–4 cm long, 0.9–1.5 cm in diameter, with membranous scales mostly on the upper half (a few smaller ones below the middle). A floral remnant 7–8 mm long persists at the subtruncate top of each fruit.

Benson (1982) mistakenly neotypified *A. brevihamatus* from *A. scheeri* in South Texas. Benson (pp. 801–2) mixed together plants of *A. brevihamatus* and *A. scheeri* in one photograph (fig. 840) and identified another photograph (fig. 842) of *A. brevihamatus* as *A. scheeri*, but he did distinguish a "*brevihamatus* type" habit in these photographs. Benson considered *A. brevihamatus* to be as a synonym of *A. scheeri*.

Synonyms. *Echinocactus scheeri* Salm-Dyck var. *brevihamatus* (Engelm.) F. A. C. Weber; *Ancistrocactus scheeri* auct. non *A. scheeri* (Salm-Dyck) Britton & Rose, in part (excluding type).

Common Name. Fishhook cactus.

1b. **Ancistrocactus brevihamatus** var. **pallidus** A. D. Zimmerman ex A. M. Powell, var. nov. SNIPE CACTUS. Plates 217–19. TYPE: UNITED STATES. Texas. Brewster Co., near Marathon, 20 Mar 1993, *A. M. Powell and S. A. Powell 5924* (HOLOTYPE, SRSC!).

Caules non ramosi, globulares aut plani ad apicem, 3–6 cm diametro, tuberculati. Spinae albae, canae, luteolae aut fulvae; spinae centrales 4; spina infima uncata, 2–2.5 cm longa, 0.3–0.6 cm diametro. Flores albi aut cremei. Fructus virides, 1–2 cm longi; semina atra.

Additional specimens examined: UNITED STATES. Texas. Terrell Co., 24 km S of Sanderson, *J. Talbot* s.n. of Oct 1997 (SRSC). Brewster Co., near Marathon, *P. R. Manning 1016* (SRSC); Del Norte Mts, 10 mi N of Santiago Peak, *A. M. Powell and S. A. Powell 6215* (SRSC); near Agua Fria, *S. A. Powell 6218* (SRSC).

Rounded limestone hills with *Parthenium argentatum* and other desertscrub, mixed novaculite alluvium in degraded grassland, bentonite in desertscrub, or Del Rio clay exposed in and among limestone mesas. Central and SW Brewster Co. and Terrell Co. S of Sanderson. 1,400–2,000 ft. Flowering Feb–Mar. $2n = 22$. Endemic. Map 46.

The varietal epithet is after the Latin *pallidus*, "pale," descriptive of the white to cream-colored flowers.

Identifying Characters. Thus far only a few specimens of var. *pallidus* have been available for study, but the stems appear always to be unbranched and globular or flat-topped. The largest stems we have observed are 6–7 cm long and 6.5 cm in diameter. One large stem extended only 3 cm aboveground. Most plants are smaller and seem to be sexually mature when 1.5–2.5 cm in diameter. Plants of var. *pallidus* closely resemble those of *A. brevihamatus* var. *brevihamatus* except that the hooked central spines in most plants are more slender, 0.3–0.4(–0.8) mm wide in Brewster County populations and 0.4–0.6 mm wide in the one Terrell County population examined. In var. *brevihamatus* the central spines are wider, to 1–1.5 mm, and usually gray or sometimes brown in color. In addition, in var. *pallidus*, the central spines are yellowish or tan, especially in mid- and upper areoles, in Brewster County populations, and light brown to varnish-brown on the upper spine surfaces in the Terrell County population. In lower areoles the hooked spines usually are gray.

The flowers of var. *pallidus* are like those of var. *brevihamatus* except that the inner tepals are white (off-white) or cream-colored, sometimes with pale brown midregions. The outer tepals have broad midregions that are greenish to greenish-red or greenish-brown, and they have greenish-white margins. The tepals are 3–8 mm wide and rounded, obtuse, or acute at the apex. The flowers are 2.5–3.8 cm long, basically tubular, opening wider, to 2.5 cm, only where the spines permit. The stamens are sensitive, their filaments are colorless or pale green, and the anthers are yellow. The pale green style is ca. 2 cm long. The 5–9 stigma lobes are cream-colored or pale green, and ca. 2 mm long.

The fruits of var. *pallidus* are green in development and in maturity. Mature fruits are 0.8–1.4 cm long, 0.6–1 cm wide, and may turn light brown or tannish. Upon drying, the pericarpel wall is tan or paper-white in color, thin and easily torn, with the seeds barely visible through the wall. The seeds are black, glossy, helmet-shaped or subreniform, and 1.7–2 mm long. The floral remnant is persistent.

Phenology. The principal flowering period for var. *pallidus* seems to be in mid- to late February, extending into mid-March. Under sunny conditions flowers open partially by about 10:00 A.M., and they are more fully open by noon or shortly thereafter. Individual flowers close at night and open again for 2–4 days. Under greenhouse conditions flowers may open again for 4–6 days. In one large plant (5.5 cm in diameter) from Brewster County, six buds were produced at the same time, and later four flowers were open at the same time. Fruit maturation requires one month and three weeks to 2–3 months. Careful observation of fruit maturation has been accomplished with only six plants. Plants (from a Brewster County population) grown from seed in the Chihuahuan Desert Research Institute greenhouse reached sexual maturity the first year and produced flowers the same year (D. Miller, pers. comm.). A single plant from Terrell County, maintained in the Sul Ross greenhouse, produced three buds, and the first open flower on 15 January. The flower was closed until about noon, partially open (ca. 1.5 cm across) at 1:00 P.M., and fully open by 2:00 P.M. The flower closed at night and opened again for five days before it was collected. The flower was protandrous, with the anthers dehiscing on the first day open. Another five plants from

Brewster County, maintained in the greenhouse, produced among them 14 flowers that opened at about noon, closed at night, and opened again for about 5–6 days.

In greenhouse-cultivated plants of *A. brevihamatus* var. *pallidus* and *A. tobuschii,* fruits ripened to the thin-walled stage at about the same time, in mid- to late April. The tan or papery thin-walled fruits of var. *pallidus* and *A. tobuschii* also are similar in morphology. The greenhouse-grown plants of *A. brevihamatus* var. *brevihamatus* produced fruits that remained green for about a month longer than those of var. *pallidus* and *A. tobuschii,* and the fruits of var. *brevihamatus* never did turn thin-walled and tan. The fruits of *A. scheeri,* as seen in cultivated specimens, are similar to those of *A. brevihamatus* var. *brevihamatus* in phenology and morphology.

Ancistrocactus brevihamatus var. *pallidus* often is sympatric with *Glandulicactus uncinatus* var. *wrightii,* and small plants of *G. uncinatus* are similar to those of var. *pallidus.* The two taxa are easily distinguished in vegetative condition by the straight radial spines in var. *pallidus* and hooked lower radials in *Glandulicactus.*

Sterile and Immature Specimens. Except for certain characters, var. *pallidus* closely resembles var. *brevihamatus* in vegetative features. All of the plants we have observed so far are tuberculate. In var. *brevihamatus* younger plants are tuberculate, and older plants have 8–13 ribs. The subconical tubercles in var. *pallidus* are 0.5–0.9 cm long and just as wide at the base, or to 1–1.4 cm broad. The spine-bearing portion of the areole is circular and 2.5–3 mm in diameter. The felty areolar groove has 1–2 red to reddish-orange glands that under certain conditions exude a clear, spherical, viscous droplet (ca. 2 mm in diameter) that is mildly sweet.

In var. *pallidus* usually there are four somewhat flexible central spines as there are in var. *brevihamatus,* one porrect, hooked lower central, and three erect, straight or curved, flattened, upper centrals. The main vegetative difference between var. *pallidus* and var. *brevihamatus* is the size and color of the central spines, alluded to above. In var. *pallidus* the central spines are more slender, 0.3–0.4(–0.8) mm wide in the Brewster County populations, than in the Terrell County collection. In Brewster County plants, the younger centrals are yellowish or tan, but these become gray with age. In the Terrell County plants the centrals are brown on the upper surface and pale gray elsewhere, including distally, except at the red-brown tips. In Brewster County plants the hooked central is 1.8–2.5(–4.5) cm long, flat on the upper surface, and rounded underneath. Another character of the hooked central spine, seemingly unique in var. *pallidus,* is that the hook is smaller on nearly straight shafts, compared to a larger hook on curved shafts in var. *brevihamatus.* In var. *pallidus* the upper centrals are thinner than the hooked central, with a flattened upper surface, but still with a shallow-rounded lower surface. The two lateral upper centrals are 2.5–4(–5.5) cm long, longer than the middle upper central at 1.8–2.5(–3) cm. Microscopically, the centrals are antrorsely scabrous, a feature that is particularly evident at the margins of the more flattened upper central spines. In the Terrell County collection, the hooked central spine is 1–3 cm long, longer in upper areoles, and the upper centrals are 2–4.3 cm long. A smaller, younger plant from Terrell County exhibited only gray spines, as in some var. *brevihamatus*, but they were slender as in other var. *pallidus* so far examined.

Radial spines are (7–)8(–10) in number in plants of var. *pallidus* from Brewster County. The radials are pale gray, appressed, usually acicular, and 1–2(–2.5) cm long. In the Terrell County collection, there are 10–12 radial spines per areole, these pale gray except at the dark tip, and they are up to 1.3 cm long.

Greenhouse-grown seedlings of var. *pallidus* usually have seven spines per areole. These are all radial spines, except sometimes for one upper spine that often is erect and may be a central. The spines are pubescent. No hooked spines are present in seedlings. Small and apparently immature plants observed in the field had 7–8 radial spines.

Biosystematics. The eight reported populations of var. *pallidus*, most of them separated by 15 miles or more, suggest that the western distributional limits are just west of the Del Norte and Santiago mountain axis in central and southwestern Brewster County, extending west to near Agua Fria, and that the taxon extends east to south-central Terrell County. In spring 2001, Bonnie McKinney (pers. comm., with specimen in hand) located a small population of var. *pallidus* in Black Gap Wildlife Management Area. All eight documented populations are small, possibly consisting of 5–100 plants. The taxon also has been reported, but not documented by specimens, from the Dead Horse Mountains in southern Brewster County and from a site in south Terrell County. The var. *pallidus* has not been widely collected probably because it is rare and also because individual plants are difficult to see in their habitats. Repeated expeditions over a period of several years in search for the exceedingly cryptic plants of var. *pallidus* were labeled "snipe hunts" by one of the regular participants. We extrapolate, however, that the taxon probably is present in low-density populations at additional localities in Brewster and Terrell counties.

In the western part of its range, var. *pallidus* is characterized by eight radial spines. The collection from Terrell County has 10–12 radials, approaching the typical number of radial spines for var. *brevihamatus*, which is known to occur just nine miles away. The Terrell County collection of var. *pallidus* also approaches var. *brevihamatus* in central spine morphology (i.e., in slightly wider centrals, to 0.6 mm, and brown color on the flattened upper surface of some centrals). Varietal recognition of var. *pallidus*, rather than distinction as a species, is supported by the tentative evidence of intergradation with var. *brevihamatus* in Terrell County.

The seemingly close relationship of A. *brevihamatus* var. *pallidus* and A. *brevihamatus* var. *brevihamatus* discussed above is brought into question by the dissimilar fruits of these taxa. The fruit morphology and development of var. *pallidus* seems identical to that of A. *tobuschii*. The closest present geographic relationship of var. *pallidus* is with var. *brevihamatus*, but based upon fruit characters, it is possible that the two true evolutionary taxon pairs in Texas *Ancistrocactus* are var. *pallidus* with A. *tobuschii* and var. *brevihamatus* with A. *scheeri*.

Most of our detailed observation of fruits have been with cultivated plants kept in the greenhouse.

A third variety of *A. brevihamatus,* as yet undescribed (Zimmerman et al., forthcoming), occurs near Cuatro Ciénegas in south-central Coahuila, Mexico. The undescribed variety has white, rose-pink, or magenta flowers and is known from only one or two populations restricted to nearly pure gypsum. The brownish-flowered var. *brevihamatus* occurs just a few miles away in south-central Coahuila on substrates other than gypsum.

Synonyms. None known.

Common Names. White-flowered fishhook cactus seems appropriate, but we prefer the shorter, humorous appellation, snipe cactus.

13. TOUMEYA Britton & Rose
Grama-Grass Cactus

A genus of one species distributed in east-central Arizona, western New Mexico, and about 18 miles into Trans-Pecos Texas in northeastern Hudspeth County, possibly in adjacent Culberson County. *Toumeya papyracantha* has an unsettled taxonomic history, having been placed in several different genera, *Mammillaria*, *Pediocactus*, *Sclerocactus*, and *Toumeya*. The genus name was proposed by Britton and Rose in honor of Dean James W. Toumey, whose work with certain cacti was significant. $x = 11$.

1. **Toumeya papyracantha** (Engelm.) Britton & Rose. GRAMA-GRASS CACTUS. Plates 220, 221. [*Mammillaria papyracantha* Engelm., Pl. Fendl. Novi-Mexicanae 4: 49. 1849]. Plants with solitary stems, rarely branched from the base in old age. Stems cylindroid or slightly obconic, 2.5–7.5(–10) cm long, 1.2–2(–4) cm thick. Ribs none. Tubercles elongate, 1.5–5 mm long, dark gray-green, rather soft; glands pale reddish to tan. Areoles with relatively dense spines obscuring surface of stem. Central spines 1–3(–4) per areole, whitish or pale brown to gray, erect or ascending, often curled, flattened and thin, 1.2–2.7 cm long (to 4–5 cm long in cultivation), 0.4–2 mm wide, centrals (longer one especially) resembling dried grass leaves or wood shavings; radial spines 6–9 per areole, silvery-white to grayish, flexible, flattened, straight, appressed, 2–5 mm long, 0.3–0.6 mm wide. Flowers 2–2.5 cm wide, ca. 2.5 cm long, funnelform-campanulate; inner tepals white, ca. 2 cm long, ca. 4 mm in diameter; outer tepals with reddish-brown midribs and whitish margins, to 2 cm long, 3 mm wide; filaments whitish; anthers cream-colored; style greenish to cream-colored, to 2 cm long; stigma lobes 4–5, cream-colored or pale green, 1.5 mm long; pericarpel with or without scale bracts, spines also occasionally present in pericarpel areoles (Reeves, 1994). Fruit usually indehiscent, green, drying to tan, subglobose, 4.5–7 mm long, to ca. 4.5 mm wide; floral remnant persistent. Seeds black, ca. 3 mm in largest dimension, ca. 1 mm thick, broadly obovoid, finely papillate-checkered; hilum relatively small. $2n = 22$.

Salt flats and gypsum habitats, northeastern Hudspeth Co., associated with *Sporobolus airoides* (Torr.) Torr. (alkali sacaton), according to Reeves (1994), and probably *Bouteloua gracilis* (Kunth) Lag. ex Griffiths (blue grama), as it is in NM and AZ. Northeastern Hudspeth Co., near the NM state line N of Dell City and about 18 mi or more S into TX. 3,600–4,000 ft. Flowering Apr–May. $2n = 22$? Arizona, in S Navajo Co., western and central NM (at 5,000–7,200 ft). Map 47.

Benson (1982) remarked about how plants of *T. papyracantha* are small and commonly overlooked, particularly because they typically grow in or near clumps of grama grass, and the cactus plants themselves have central spines that resemble dried leaf blades. In Hudspeth County where *T. papyracantha* has been located by Reeves (1994), plants perhaps 10 cm or more tall were found associ-

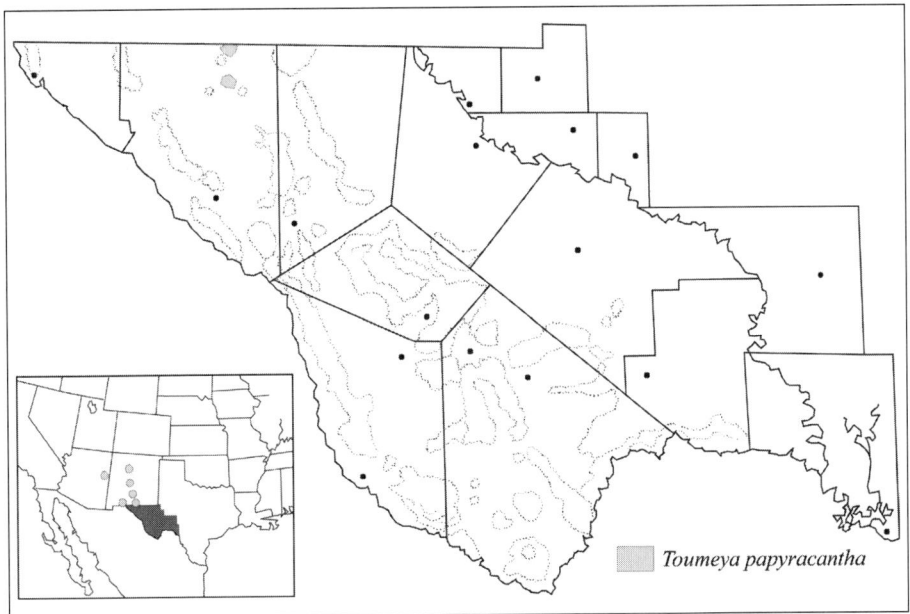

Map 47. Generalized distribution of *Toumeya papyracantha* (grama-grass cactus).

ated with a rather large, clumped grass species, alkali sacaton, but still well camouflaged within or beside the grass.

Identifying Characters. The most identifiable feature of *T. papyracantha* is the flattened, papery, twisted, and curved central spines that resemble dried blades of grass. In upper areoles the central spines tend to stand or curve upward, collectively overarching the stem in a disorganized fashion. Centrals in low areoles extend in any direction. The central spines usually are pale brown, often mottled, fading to pale gray with age.

The white flowers are produced at the stem apex. The flowers are narrowly bell-shaped or tubular-funnelform and usually not widely opening. The inner tepals are essentially white or white with brownish midribs, and they are acute to acute-attenuate at the apex. The outer tepals are white with reddish-tan, or brown, to dark brownish midribs. Outer bracts have greenish-brown midregions, whitish margins, and rounded apexes. The outermost bracts have broader, greenish-brown midregions.

The subglobose fruits of *T. papyracantha* are less than 1 cm in diameter, green, and usually indehiscent. At full maturity the fruits become tan and dry. The seeds are black and shiny, somewhat flattened, and ca. 3 mm long, actually surprisingly large for a small cactus species. Under magnification the glossy seed coat is seen to have a fine raised pattern.

Phenology. The flowering period for *T. papyracantha* throughout its range in New Mexico and Arizona is in the spring, most commonly in April and May. We assume that the Trans-Pecos populations also bloom in April and May. Under greenhouse conditions one plant produced four buds in February. Anthesis of the most advanced bud occurred on 23 February, two others opened on 26 February, and the fourth opened a day later. The flowers opened relatively late in the afternoon, between 3:00 and 4:00 P.M., closed at night, and opened again for 5–6 days.

Sterile and Immature Specimens. Throughout most of its range the usually unbranched, cylindroid stems of *T. papyracantha* extend aboveground for only a few centimeters. In the Texas populations associated with alkali sacaton grass near the salt lakes in northeastern Hudspeth County, the stems appear to extend 10 cm or more above the ground, as estimated from the photographs provided by Reeves (1994). The elongate-conical, soft tubercles protrude to 5 mm. The stem surface is dark gray-green but usually is obscured by the spines. The areoles are elongated, to 5 mm, with short gray wool on mature stems, and areoles are spherical on small stems. Areoles at the stem apex produce yellowish wool. One to several pinkish to tan glands are present on upper edges of the areoles.

In each areole there are 1–3(–4) central spines. The main central spine, the longest and widest, is the lower central. The main central is flat, thin, flexible, curved, and often twisted. The longest central has an evident midrib on the ventral surface. The margins of the largest central often are involute and minutely pubescent. Two or three upper central spines usually are present in each areole. The upper centrals are flattened and curved but shorter and not as wide as the lower central. The 6–9 radial spines are silvery-white or gray, flattened, thin, straight, appressed, and contrast distinctly with the centrals. The radial spines are upper lateral (not upper), lateral, and lower in position in the areole. The longest radials, to 5 mm, are the lowermost ones.

Immature stems of *T. papyracantha* throughout most of its range are very slender, sometimes remaining less than 6 mm in diameter when 2–3 cm long. The spines of juveniles are similar to those in adults except that the central spines are relatively short.

Biosystematics. *Toumeya papyracantha* is distributed throughout much of the western half of New Mexico and into northeastern Arizona. This primarily New Mexican species was only recently discovered in Texas, first in 1991 just across the state line (by Phillip Clayton) north of Dell City in Hudspeth County and then about 18 miles into Texas southeast of Dell City (Reeves, 1994). The typical habitat for *T. papyracantha* usually is described as grassy hills with grama grasses (usually blue grama), but the grama-grass cactus actually occurs in a number of soil types and habitats in addition to the igneous alluvium of grama grasslands. In Doña Ana and Otero counties in southern New Mexico and in Hudspeth County, Texas, *T. papyracantha* is found in certain gypseous substrates. The species is found in saline habitats near the margins of salt lakes associated with alkali sacaton in northeastern Hudspeth County and in more or less sandy soil at sites in New Mexico.

The plants often are described as being exceedingly cryptic, growing in or near "fairy rings" of blue grama, in that the spines resemble dried grass leaves (Benson, 1982). Weniger (1984) contended that *T. papyracantha* was among the rarest of all cacti in the Southwest. In Texas the taxon surely ranks with *Ancistrocactus brevihamatus* var. *pallidus* as being either rare or so well camouflaged that it appears to be rare.

New molecular data (Porter et al., 2000) have supported previous morphological evidence that *T. papyracantha* might be a neotonous derivative of *Sclerocactus*. This supports the recommendation by the IOS Working Party (Hunt and Taylor, 1986) that *Toumeya* be treated as a synonym of *Sclerocactus*. For purposes of the present treatment we have elected to recognize *Toumeya* as a distinct genus. Benson (1982) and Weniger (1984) included *T. papyracanthus* in *Pediocactus*.

Synonyms. *Pediocactus papyracanthus* (Engelm.) L. D. Benson; *Sclerocactus papyracanthus* (Engelm.) N. P. Taylor.

Common Names. Paperspine cactus; paper-spined cactus; Toumeya; grama grass cactus; paper-spine plains cactus.

14. ECHINOMASTUS Britton & Rose
Pineapple Cactus

Plants with solitary stems, 2.5–15(–22) cm high, 2–8 cm in diameter. Roots diffuse. Stems cylindroid, ovoid, obovoid-cylindroid, to nearly globular, blue-green, cortex mucilaginous; pith lacking vascular bundles; druses lacking. Ribs (8–)13–21, usually 4–9 mm high, usually evident beneath the tubercles or appearing as rows of tubercles. Tubercles 3–10 mm long, with a short and/or wide adaxial felted groove extending from the spine-bearing areole to the base, flowers borne in a groove away from the spines. Areoles elliptic to nearly circular, 2.2–3 mm long or wide, usually 6–10 mm apart on tubercle apexes, areolar glands absent. Spines dense, obscuring the stem or mostly so, except in some *E. warnockii*; central spines tan or stramineous, with tips chalky-blue or pinkish, gray, or stramineous, (2–)4 per areole, straight, porrect, or pointing upward, or the lower central slightly curving downward, acicular, 1.2–2(–2.4) cm long, usually larger or swollen at the base; radial spines similar to centrals or whitish, 12–32 per areole, the longest 0.7–1.8(–2.4) cm, diffuse, appressed, or slightly curved back toward the stem, usually larger or swollen at the base. Flowers borne on new growth at the stem apex, 2–3(–3.5) cm in diameter and about as long; inner tepals whitish (*E. warnockii* and *E. mariposensis*) to pinkish-white (*E. intertextus*), the largest lanceolate to oblanceolate, 1.2–2 cm long; outer tepals with green to reddish or reddish-brown midribs, margins whitish to pink (or scarious), oblanceolate to lanceolate, 0.9–2.0 mm long; filaments pale yellow, anthers yellow or cream-yellow; style 1–1.2 cm long, green to yellow-green; stigma lobes 5–10, 1–2 mm long, green or reddish (rarely white); pericarpel bracts 1–20, these scalelike and appressed, with scarious margins, axils naked. Fruit barely succulent, green, at maturity drying to tan, brownish, or dull reddish, elongate, 5–10 mm long, with a few membranous scales, dehiscent longitudinally or basally (in *E. intertextus*), the floral remnant persistent. Seeds black, ovate-reniform, 1.5–2.2 mm in greatest dimension, papillate-reticulate (verrucose); hilum lateral, large and deep. $x = 11$.

A genus of 5–9 species distributed from the southwestern United States to central Mexico. Four species occur in the Chihuahuan Desert Region. About five or six species occur in the United States. Three or four species are found in Texas, with three of these in the Trans-Pecos.

The genus *Echinomastus* was named by Britton and Rose (1919–23) and since then has been the subject of considerable taxonomic repositioning. Benson (1982) included *Echinomastus* in *Neolloydia*, Weniger (1984) included *Echinomastus* as part of *Echinocactus*, and Bravo-Hollis and Sánchez-Mejorada (1991a) chose to recognize *Echinomastus* as a distinct genus. Anderson (1986) reviewed the extensive evidence that *Echinomastus* should be separated from *Neolloydia* (see *Neolloydia*). Hunt and Taylor (1986) included *Echinomastus* in a broad concept of *Sclerocactus*, but Heil and Porter (1994), in their revision of *Sclerocactus*, concluded that *Echinomastus* should be maintained as a separate genus. At this time we are persuaded by Glass and Foster (1975), Anderson (1986), Heil and Porter (1994), Zimmerman (1985), and Zimmerman et al.

(forthcoming) that *Echinomastus* is best treated as a distinct genus. The genus name is derived from the Greek *echinos*, "hedgehog" (for "spiny"), and *mastos*, "breast," in reference to the spiny tubercle apexes.

Key to the Species

1. Stems green, usually visible through the spines; central spines stramineous, grayish, or reddish, usually the distal part if not the whole spine, pink or dark reddish; anthers cream-colored; stigma red

   1. *E. intertextus.*

1. Stems blue-green or blue-gray, tinted by powdered wax on the surface, visible through the spines or not; central spines stramineous or tan, the distal parts, if not nearly the whole spines, chalky-blue, blue-gray, or blue-brown; anthers yellow; stigma green (2).

2(1). Stems usually visible through spreading radial spines; radial spines tan or stramineous and chalky-blue, at least at the tips, 11–15(–17) per areole, 1.5–2.4 cm long; central spines straight or slightly curved, similar to radials but slightly larger in diameter, upper three (usually) angled upward, blue-gray distally, and slightly longer (1.8–2 cm) than the lower central one that is angled upward (or porrect)

   2. *E. warnockii.*

2. Stems usually not visible through somewhat appressed and overlapping radial spines; radial spines ashy-white, (20–)26–32 per areole, (0.3–)0.4–0.7(–1.1) cm long; central spines (upper ones) markedly larger in diameter (0.4–0.9 mm basally) and longer than radials (upper centrals to 1.5–2.1 cm long), the upper three upswept, curved, and chalky-blue for most of the length, particularly those of the upper areoles in mature plants, lower central usually much shorter, (0.3–0.4–)0.5–0.7(–1.1) cm long, slightly turned downward or curving downward

   3. *E. mariposensis.*

1. **Echinomastus intertextus** (Engelm.) Britton & Rose. WOVEN-SPINE PINEAPPLE CACTUS, LONGCENTRAL WOVEN-SPINE PINEAPPLE CACTUS. Plates 222–26. [*Echinocactus intertextus* Engelm., Proc. Amer. Acad. Arts 3: 277. 1856]. Common in the grasslands and grassy areas of woodlands, Big Bend region of the Trans-Pecos, Brewster, Presidio, and Jeff Davis counties, primarily igneous-derived substrates, in the central mountains and occasional in the more desertic mountains, reported on lower to midslopes of the Chisos Mts, W to El Paso Co. where common in the Franklin Mts, with but few reports from mountains in Culberson and Hudspeth counties. (3,000–)4,000–5,800 ft. Flowering Feb–Apr. $2n = 22$. Southeast AZ, SW and central NM. Mexico, Chihuahua, possibly in N Coahuila and NE Sonora. Map 48.

Two varieties of *E. intertextus* have been recognized by most authors who

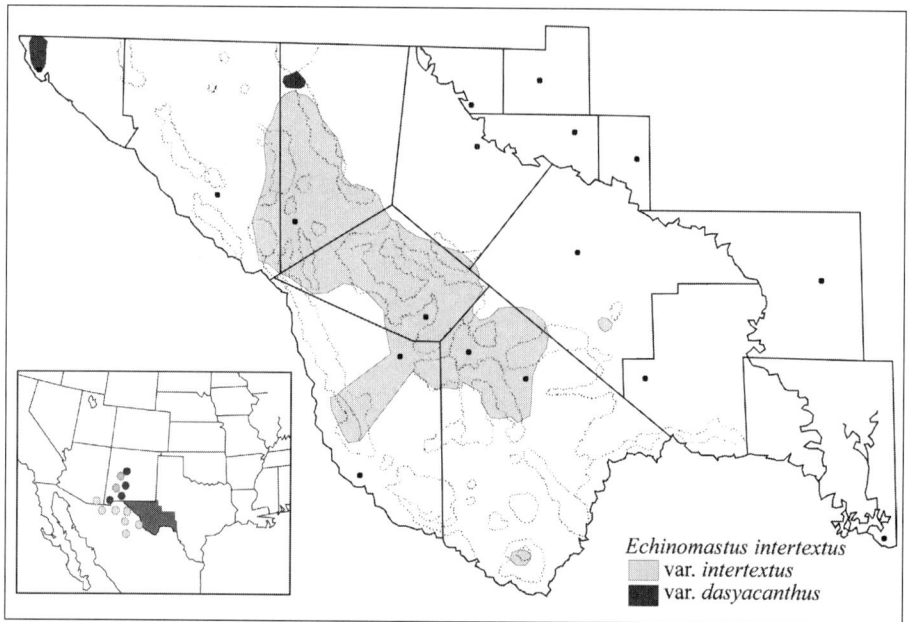

Map 48. Generalized distribution of *Echinomastus intertextus* var. *intertextus* (woven-spine pineapple cactus) and *E. intertextus* var. *dasyacanthus* (longcentral woven-spine pineapple cactus).

have dealt with cacti of Texas. The specific epithet *intertextus* presumably refers to the interwoven spines, particularly at the plant apex, and is from the Latin *inter*, "between," and *textilis*, "woven."

Key to the Varieties

1. Stems spherical in young plants, becoming ovoid to ovoid-cylindroid in age, 5–12(–20) cm high; radial spines and upper centrals markedly appressed, the longest 0.9–1.9(–2.1) cm; lower central spine 0.05–0.4(–0.5) cm long

    1a. *E. intertextus* var. *intertextus*.

1. Stems ovoid to cylindroid, 7–15 cm long; radial spines and upper centrals not appressed, somewhat diffuse, the longest 1.2–2.0(–2.4) cm; lower central spine 0.4–1.5(–2) cm long

    1b. *E. intertextus* var. *dasyacanthus*.

1a. **Echinomastus intertextus** var. **intertextus**. WOVEN-SPINE PINEAPPLE CACTUS. Plates 222, 223. Mostly in the central mountains, Jeff Davis, N Brewster, N Presidio counties, also Chinati Mts, Del Norte Mts, E to SW Pecos Co., W to E Hudspeth Co. and El Paso Co. (Franklin Mts). 4,000–5,800 ft. Flowering Feb–Apr. $2n = 22$. West to SW NM (Hidalgo, Luna, and Socorro counties) and SE AZ. Mexico, NE Sonora, Chihuahua, possibly N Coahuila. Map 48.

Echinomastus intertextus var. *intertextus* is by far the most common variety of *Echinomastus* in the Trans-Pecos, being most abundant in the grama grasslands and woodlands of the central mountain region (the Davis Mountains). *Echinomastus intertextus* var. *dasyacanthus* in the Trans-Pecos seemingly is most common in the Franklin Mountains of El Paso County. Most published records of var. *intertextus* from central New Mexico, and many reports from southern New Mexico, probably pertain to *E. intertextus* var. *dasyacanthus* or intermediates between the two varieties. In general the two varieties of *E. intertextus* replace each other geographically in New Mexico (Zimmerman et al., forthcoming). Benson (1982) implied that there is an ecological distinction between var. *intertextus* and var. *dasyacanthus*, in that the former occurs mostly at slightly higher elevations in semidesert grassland near the edge of the Chihuahuan Desert, whereas var. *dasyacanthus* is found mostly in the Chihuahuan Desert (arid mountains and lower slopes of mountains) in desert grassland. Our observations of the *E. intertextus* varieties in the Trans-Pecos basically support Benson's conclusions, except that we would not describe the high grasslands (ca. 5,200 feet) in the Davis Mountains, where var. *intertextus* is common, as desert grasslands, but instead as plains grasslands (Powell, 2000). And we suspect that the *E. intertextus* var. *dasyacanthus* specimens in Brewster County, as alluded to by Benson (1982), are actually hybrids between *E. intertextus* var. *intertextus* and *E. warnockii* (see below).

Identifying Characters. In the field var. *intertextus* is most easily distinguished from var. *dasyacanthus* by its appressed spines (radials and upper centrals), lying flat against the stem, and by the short lower central spine. In aspect the plants of var. *intertextus* appear as if they could be safely handled. The porrect lower central spine in the Davis Mountains populations usually is only 1–3 mm long. The plants of var. *intertextus* often are globular to ovoid in shape, compared to the more ovoid to cylindroid stems of var. *dasyacanthus*. In var. *dasyacanthus* the radial and upper central spines are somewhat diffuse, obviously not flattened against the stem as in var. *intertextus*, and the porrect lower central spine is noticeably longer.

Among Trans-Pecos cacti *E. intertextus* might be confused in the field mostly with *E. warnockii*, although both of these taxa in their pure forms are ecologically separated. Putative hybrids involving *E. intertextus* and *E. warnockii* may be found in adjacent or mixed populations with the parental species in some localities at or near the desert margin, such as the southeastern Davis Mountains and Chisos Mountains (discussed below under var. *dasyacanthus*). *Echinomastus intertextus* var. *dasyacanthus* and *E. warnockii* both have diffuse radial and upper central spines, but the spines in each species are different in color. The spines of *E. intertextus* typically are stramineous with pink or dark reddish tips, or the whole spines are reddish. The spines of *E. warnockii* typically are tan to stramineous or gray at the base, and chalky-blue in color on the distal portions or just the tips, or blue-gray essentially throughout. Most often the stem is visible through the low-spreading spines, and in *E. intertextus* the stem is green, whereas in *E. warnockii* the stem is somewhat glaucous. *Echinomastus intertextus* var. *intertextus*, with its appressed spines, is not readily confused with *E. warnockii*, which has diffuse spines, and these two ecologically separated taxa rarely occur in close proximity in the Trans-Pecos. *Echinomastus warnockii* has fewer radial spines (11–17) per areole than does *E. intertextus* (13–25), and the lower central spine in *E. warnockii* is as long as the radial spines.

In a field setting *E. intertextus* may be similar to *E. mariposensis*, *Neolloydia conoidea*, *Echinocereus viridiflorus*, or perhaps *Coryphantha echinus* but should not be confused with these taxa. *Echinomastus mariposensis* has slender, whitish radials that overlap those in adjacent areoles and obscure the stem, and *E. mariposensis* is not known to occur with other *Echinomastus* species. *Neolloydia conoidea* usually is multistemmed, the radials are whitish, and the longer central spine is black or dark brown. *Echinocereus viridiflorus* is an extremely variable species, with several varieties distributed throughout the Trans-Pecos, some of them in habitats with *E. intertextus*. *Echinocereus*, however, is identifiable by its purely ribbed stems, spine clusters different from those of *E. intertextus*, and flowers borne on the sides of the stems. *Coryphantha echinus* has ashy-gray to whitish, appressed radial spines, and in adult plants, a prominent, whitish porrect central spine.

The white flowers of *E. intertextus* var. *intertextus* are 2.5–3(–3.8) cm long and 2.3–3 cm in diameter. The closely spaced or interlocking apical spines often interfere with full opening of the flowers. The inner tepals are 1.3–2.5 cm long, 2–5.5 mm wide, bright white to pale pink, or with a pale pink midstripe, the apexes acute and sometimes mucronate. The outer tepals are dull white to pale pinkish, with pink midstripes, and to ca. 2 cm long. The filaments are pale greenish and the anthers are cream-yellow. The greenish style supports 6–12 stigma lobes that are bright red, pink, or rarely white.

When in flower both varieties of *E. intertextus* are easily distinguished from *E. warnockii* and *E. mariposensis*. Typically, the stigma is red in *E. intertextus* and green in *E. warnockii* and *E. mariposensis*. Also, the tepals in *E. intertextus* are white, often with a pinkish tinge and pink midstripes, whereas the tepals of *E. warnockii* typically are white with greenish, greenish-brown, or tan midstripes. The tepal midstripes in flowers of some *E. warnockii* are pinkish to pinkish-brown. Other vegetatively similar taxa are easily distinguished when in flower. *Neolloydia conoidea* produces magenta flowers, *Echinocereus viridiflorus* bears red-brown, greenish-brown, or green flowers, and *Coryphantha echinus* has yellow flowers.

The fruits of *Echinomastus intertextus* are green, globose to ovoid, and 0.8–1.5 cm long, with a few surface scales. At maturity the scarcely succulent green fruits become dry and may turn tan, or dull reddish, or pink. The fruits are circumscissily dehiscent at the base in a manner similar to that of *Thelocactus*. After dehiscence the fruit wall is easily dislodged, leaving a small pile of black seeds on the basal fruit remains and the stem surface. The seeds themselves are readily dry and not at all adherent. Seed dissemination probably follows any gentle mechanical disturbance, and most of the seeds apparently fall between the spines to the base of the plants. Experimental removal of seeds even with slender forceps is difficult because of closely spaced spines. The seeds are reniform to nearly spherical, somewhat glossy, but microscopically papillose, as a result of strongly convex cells of the testa.

Phenology. All three species of *Echinomastus* are among the earliest of all Trans-Pecos cacti to bloom in the spring, and all three species produce white to pinkish-white flowers at the stem apex. The peak blooming time for var. *intertextus* in the central mountain region is March, although early spring flowering may extend from late February to April. The spring blooming season extends into May, at least in some years, probably following a very dry winter and early spring and then late spring rains. In 1995 a population of var. *intertextus*, probably introgressed with *E. warnockii* (Raun, 1997), bloomed in May. Zimmerman et al. (forthcoming) reported blooming in var. *intertextus* as late as July, probably in response to "exceptional rains." Sexually mature plants characteristically produce numerous buds, up to about 20 (Worthington, 1986), crowded at the stem apex. Several flowers, sometimes 4–7, may open at the same time. Individual flowers open about 11:00 A.M., close at night, and may open again for about three days. The flowers are protandrous, with dehiscent anthers developing 1–2 days before erect stigma lobes spread from a near-erect position. Developing fruits, often 8–10 or more, are crowded into a low, conelike cluster under restrictive apical spines. Fruit maturation requires only 1–2 months. In a 1996 experiment with var. *intertextus* and *E. warnockii* under greenhouse conditions, artificial cross-pollination in late March resulted in the collection of viable seeds of var. *intertextus* in early May.

Sterile and Immature Specimens. The largest plants of var. *intertextus* we have seen in the Trans-Pecos were near Fort Davis, where stems of some plants were 17–20 cm long. Sterile specimens of *E. intertextus* var. *intertextus* can be distinguished by their appressed radial spines, 8–20 mm long, and short lower central spine. The lower central is positioned in the center of the areole and typically is 0.5–4 mm long, frequently less than 2 mm long, or merely rudimentary.

In juvenile plants of var. *intertextus* the radial spines are appressed as they are in adults, but shorter. The lower central may be absent in the areoles of juveniles. Immature specimens of *E. intertextus* may resemble those of *Coryphantha echinus*. Adult plants of *C. echinus* have a prominent porrect lower central spine, about 1.5–2 cm long, along with usually three upper centrals that are appressed with the radials, and grayish to ivory-white radial spines (16–26) appressed to the stem. Juvenile plants of *C. echinus* lack the porrect central spine that is typical in adults, or other central spines, and thus resemble *E. intertextus* juveniles that also lack central spines. The radial spines in *E. intertextus* juveniles, however, are cyanic, at least on the distil portions, and the radial spines of *C. echinus* are white. In *C. echinus* the radial spines in juvenile plants are not only appressed to the stem, but they may be curved back toward the stem. The appressed radial spines of *E. intertextus* var. *intertextus* usually are not curved back toward the stem. In juvenile plants of *E. intertextus* var. *dasyacanthus* the radial spines are not appressed but extend out from the stem at a low angle. Furthermore, in *C. echinus* the elliptic, usually woolly, areoles are larger, at ca. 2 mm long, than are those of *E. intertextus*, at ca. 1 mm long.

The distributions of *E. intertextus* var. *intertextus* and *C. echinus* slightly overlap in the Trans-Pecos. *Coryphantha echinus* typically occurs in more desertic habitats than does var. *intertextus*, but the two taxa occasionally occur together in the central mountain region. *Coryphantha echinus* also may occur with *E. intertextus* var. *dasyacanthus*, but even in juvenile plants the cyanic radial spines of var. *dasyacanthus* are not appressed against the stem as are the white radials of *C. echinus*.

Biosystematics. *Echinomastus intertextus* var. *intertextus* and *E. intertextus* var. *dasyacanthus* are distinct enough in their extreme morphological and distributional forms that they could be recognized as separate species. Some Trans-Pecos collections of these taxa reveal what appears to be intergradation of morphological characters. The lower central spine length (ca. 5 mm) in some Davis Mountains populations of var. *intertextus* approaches the usual lower central spine length of var. *dasyacanthus* (7–15 mm). A 1959 collection from the desertic Fresno Canyon in southern Presidio County (*Warnock 17348*, SRSC) has the short lower central of var. *intertextus* (1–2.5 mm long) but somewhat diffusely spreading radial spines similar to those of var. *dasyacanthus* (or of *E. warnockii*). The Fresno Canyon population, which has never been re-located, possibly is of hybrid origin involving *E. intertextus* var. *intertextus* and *E. warnockii*. There are similar intergradations in the morphological and ecological traits of var. *intertextus* and var. *dasyacanthus* in the Franklin Mountains of El Paso County and at sites in southwestern New Mexico.

The color photograph (no. 154) in Benson (1982) labeled "*Neolloydia intertexta* var. *intertexta*" appears to portray a putative hybrid, *Echinomastus intertextus* var. *intertextus* × *E. warnockii*, presumably from the Chisos Mountains. The photograph was by Roland Wauer, well-known chief park naturalist in Big Bend National Park, where perfectly typical *E. intertextus* var. *intertextus* apparently does not occur.

Synonyms. *Neolloydia intertexta* (Engelm.) L. D. Benson; *Sclerocactus intertextus* (Engelm.) N. P. Taylor.

Common Names. Previously published common names applied to both var. *intertextus* and var. *dasyacanthus* are woven-spine pineapple cactus; white biznagita; blanco viznagita; early bloomer; white-flowered visnagita.

1b. **Echinomastus intertextus** var. **dasyacanthus** (Engelm.) Backeb. LONG-CENTRAL WOVEN-SPINE PINEAPPLE CACTUS. Plates 224, 225. [*Echinocactus intertextus* Engelm. var. *dasyacanthus* Engelm., Proc. Amer. Acad. Arts 3: 277. 1856]. Desert mountains, slopes, and grasslands. In the Trans-Pecos restricted to the Franklin Mts, El Paso Co.; the isolated population keying to var. *dasyacantha* in Big Bend National Park may owe its long spines to introgressive hybridization from *E. warnockii* into ordinary short-spined var. *intertextus* and may not be a genuine var. *dasyacanthus*. 3,800–5,600 ft. Flowering Feb–Mar. $2n = 22$. New Mexico, Otero Co. (Jarilla Mts) N to Bernalillo Co., W to Luna Co. (Florida Mts) and Hidalgo Co. (Little Hatchet Mts). Map 48.

The var. *dasyacanthus* is considered here to be a taxon of central and southwestern New Mexico that barely enters the Trans-Pecos in El Paso County. The varietal epithet *dasyacanthus* refers to the shaggy aspect created by the relatively long, protruding spines and is from the Greek *dasy*, "shaggy," and *acantha*, "spines," which contrasts with the smooth aspect of typical *E. intertextus*.

Identifying Characters. The stems of var. *dasyacanthus* are ovoid to cylindroid and 7–17 cm or more long when sexually mature. In habit the plants look bristly and uncomfortable to handle, compared to the smooth innocuous-looking var. *intertextus*. The "bristly" or "shaggy" look of var. *dasyacanthus* results from slightly appressed to spreading radial and upper central spines, along with a porrect lower central spine that is usually 0.7–1.5(–2) cm long. The 16–24 radial spines of var. *dasyacanthus* usually are longer (0.8–2.4 cm) than those of var. *intertextus*. The dominant spine color in both varieties varies from gray or brown to dull reddish; especially on distal portions of the spines, reddish inner layers and spine tips may lend whole stems a dull and dark reddish hue, especially when the spines are wet.

The flowers, fruits, and seeds of var. *dasyacanthus* appear to be identical to those of var. *intertextus*. The white flowers, often with pinkish midregions and tinge, are marked by red to pink stigma lobes and cream-yellow anthers. The fruits are green but upon ripening may turn tan to dull or pale reddish throughout or at the apex. In some specimens of var. *dasyacanthus* we have observed from El Paso County, the fruits tend not to be as crowded at the stem apex as they are in var. *intertextus*, where they are physically compressed into a low cone by restricting appressed spines. The fruits in var. *dasyacanthus* are not as tightly trapped because the spines are more divergent.

Phenology. The characteristic flowering period for var. *dasyacanthus* in the Franklin Mountains and adjacent New Mexico is from late February to March, and sometimes into early April (Worthington, 1986). Individual plants usually produce numerous flower buds at the apex, up to about 20, and these open in succession, several (up to about six) at a time over a period of 30 days, more or less. Individual flowers open again for about three days after closing at night. Flowers remain closed under cool or cloudy conditions. Fruit maturation probably requires only 1–2 months as in var. *intertextus*.

Sterile and Immature Specimens. Fertile material does not help for identification to the varietal level in *E. intertextus*, because an adequate sample size shows the same range of variations in flowers, fruits, and seeds of both taxa. Adults of the two taxa differ *conspicuously*, but not always *consistently*, in the spine characters discussed above. In the Franklin Mountains, plants matching the spine characters of both var. *dasyacanthus* and var. *intertextus* (and intermediates) occur together. In one populational sample from the southwest side of the mountains, 20 plants had the characters of var. *dasyacanthus*, and only one plant exhibited the appressed radial spines and short lower central spines of var. *intertextus*. The stems usually are unbranched, but two plants of var. *dasyacanthus* from the Franklin Mountains had one or two small stems at the base, possibly formed as the result of injury. The central and radial spines of "pure" var. *dasyacanthus* populations in New Mexico are 0.8–2.4 cm long and 0.2–0.5 mm thick. The centrals are slightly thicker than the radials.

In immature plants of var. *dasyacanthus* the spreading radial spine character is expressed to some degree, although in juveniles the radials are somewhat appressed. Side-by-side comparison of seedlings (1–2 cm tall) of var. *dasyacanthus* and var. *intertextus* allows distinction between the "somewhat appressed" radials of var. *dasyacanthus* and the truly appressed radials of var. *intertextus*. In greenhouse-grown seedlings of var. *dasyacanthus*, the porrect lower central spine was absent in all of the areoles, and there were 13–18 glabrous radials in each areole.

Biosystematics. In its extreme form with elongated stems, spreading radial spines, and long lower central spines, var. *dasyacanthus* is easily distinguished from var. *intertextus*, although intergradation between the two varieties apparently occurs throughout much of their ranges in El Paso County and in New Mexico (Zimmerman et al., forthcoming). Collections from southern Brewster County cited by Benson (1982) as var. *dasyacanthus* (see his distribution map, p. 789) are in fact morphologically different from plants of var. *dasyacanthus* in El Paso County and New Mexico. Specimens from Brewster County identified by Benson as var. *dasyacanthus* have somewhat diffuse radial spines and porrect central spine 3–5 mm long. In those particular Brewster County populations, far from the main range of var. *dasyacanthus*, we believe that the spine characters are inherited from *E. warnockii* through interspecific hybridization (Plate 226).

Specimens of *Echinomastus* from the arid western slopes of the Chisos Mountains, the grasslands near Marathon, and several arid sites between Marathon and the Chisos Mountains were identified by Benson (1982) as *E. intertextus* var. *dasyacanthus*. These specimens are similar, but not identical, to var. *dasyacanthus* in habit, except that their porrect central spines are only 3–5 mm long in specimens examined, shorter than those that are typical for var. *dasyacanthus* in populations more than 200 miles to the west in El Paso County. The stigma lobes are consistently reddish in flowers of the Chisos Mountains population, a feature that is characteristic of both varieties of *E. intertextus*. Plants in the Marathon population produce flowers with both red and green stigma lobes. Green stigma lobes are characteristic of *E. warnockii*. Raun (1997) searched for but did not locate any populations of *E. intertextus* at accessible sites between Marathon and the Chisos Mountains.

Echinomastus warnockii is a common species in desert habitats of the lower Big Bend region, extending north to the lower slopes of the central mountains northwest along and near the Rio Grande, and in desert habitats to Culberson and southern Hudspeth counties. In the Trans-Pecos, *E. intertextus* var. *intertextus* occupies mountain habitats, mostly at mid-elevations. Successful crosses in horticulture demonstrate that *E. warnockii* and *E. intertextus* var. *intertextus* are interfertile, at least as far as the production of F$_1$ hybrids that resemble in spine characters the putative var. *dasyacanthus* in southern Brewster County (Powell, 2002).

The Chisos Mountains population of *E. intertextus* occurs on an arid *bajada* of the western slopes at ca. 4,500 feet, near Lower Oak Canyon. The population consists of several hundred individuals that exhibit considerable variation in stem and spine morphology. The stems of many older plants are slender and cylindroid and sometimes stipitate at the base. The radial spines (16–19) are either appressed, as in var. *intertextus*, or somewhat diffuse, as in var. *dasyacantha* and *E. warnockii*. The lower central spine in some plants slightly exceeds a length (3–5 mm) that typifies most var. *intertextus* (1–3 mm). Smaller plants of the Lower Oak Canyon population are virtually indistinguishable in habit from *E. warnockii*. In this population we have examined flowers of about 100 plants, and all of them have exhibited the red stigma lobes unique to *E. intertextus*. All of the fruits, too, are typical of *E. intertextus,* with the gaping basal pore that leaves most of the seeds behind in the spines when the fruit is pulled from its seat. *Echinomastus warnockii,* with its usual indehiscent or minutely and ineffectively poricidal fruits, is common in desert habitats less than one-half mile below the Lower Oak Canyon population. One plant in this parapatric population of *E. warnockii* was observed to have pinkish outer tepals, a trait that is characteristic of *E. intertextus,* but green stigma lobes like those of all other *E. warnockii*. This one flower character is the only evidence suggesting introgression into the vast regional population system of *E. warnockii*.

To our knowledge, typical var. *intertextus* has not been documented anywhere in Big Bend National Park. A search of the west slopes above the Lower Oak Canyon population has resulted in the discovery of only a few small plants of *Echinomastus,* none of which appeared to be var. *intertextus*. Reports of this taxon by Wauer (1980) from near Ward Spring and on Burro Mesa are possibly additional plants of the putative hybrids. Even with the apparent present-day absence of var. *intertextus* in the Chisos Mountains, we suggest that the best explanation for the occurrence of long-spined var. *dasyacanthus*-like plants in southern Brewster County is introgressive hybridization from *E. warnockii* into var. *intertextus*.

Benson (1982) included a color photograph (no. 155) of a flowering cactus labeled "*Neolloydia intertexta* var. *dasyacantha*" that is clearly *Echinomastus warnockii,* displaying perfectly typical spine characters, stem-surface color, and the green stigma lobes that never occur in *E. intertextus,* as well as the darker yellow anthers (clearly visible in the photograph).

Synonyms. *Neolloydia intertexta* (Engelm.) L. D. Benson var. *dasyacantha* (Engelm.) L. D. Benson; *Sclerocactus intertextus* (Engelm.) N. P. Taylor var. *dasyacanthus* (Engelm.) N. P. Taylor; *Echinomastus dasyacanthus* (Engelm.) Britton & Rose; *Thelocactus dasyacanthus* W. T. Marshall ex Kelsey & Dayton.

Common Names. The common names listed under var. *intertextus* (above) are also applied to var. *dasyacanthus*.

2. **Echinomastus warnockii** (L. D. Benson) Glass & Foster. WARNOCK'S CACTUS. Plates 227, 228. [*Neolloydia warnockii* L. D. Benson, Cact. Succ. J. (U.S.) 41: 186. 1969]. Common on desert gravel hills and benches, limestone hills, alluvial flats, occasional in gypsum flats with desertscrub vegetation, most abundant in southern Presidio and Brewster counties, but also in desert habitats NW to Jeff Davis, Culberson, and southern Hudspeth counties. 1,900–4,500 ft. Flowering usually Feb–Mar, Big Bend National Park and vicinity. $2n = 22$. Mexico, adjacent to the Big Bend region in NE Chihuahua (W of Ojinaga) and expected in NE Coahuila. Map 49.

Echinomastus warnockii is one of the most common cactus species in the

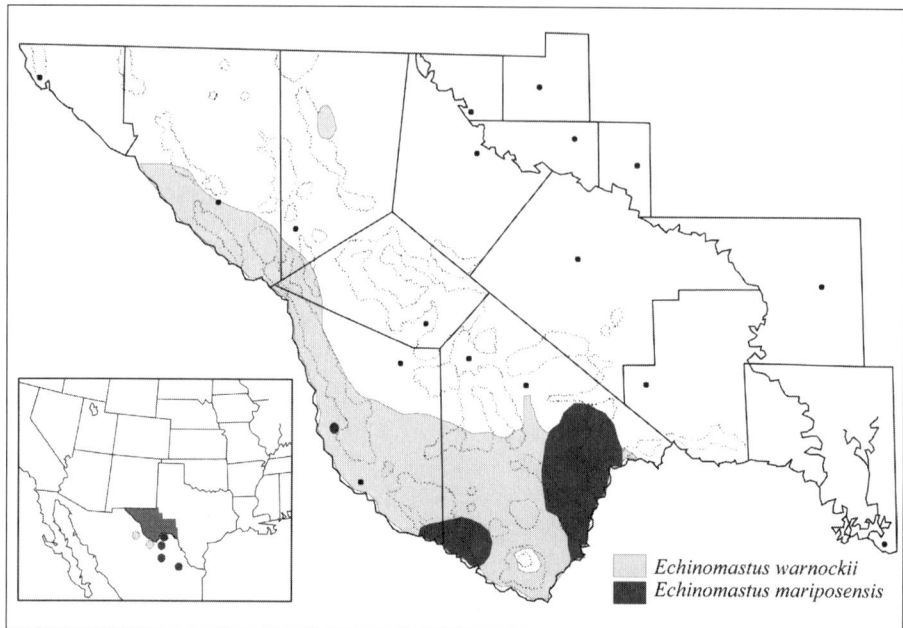

Map 49. Generalized distribution of *Echinomastus warnockii* (Warnock's cactus) and *E. mariposensis* (Mariposa cactus).

desert southern Big Bend region and adjacent areas. At one time, before the late 1980s, it was regarded by some workers as a relatively rare species. Recent field excursions in Big Bend National Park have revealed abundant plants of *E. warnockii* of all sizes and presumably all ages. Juvenile plants the size of a quarter are abundant in many habitats, and we assume that the species is thriving. The largest plants we have observed, and presumably the oldest, were about 20 cm high. The specific epithet honors Barton H. Warnock, longtime professor of biology at Sul Ross State University, who collected the species several times before it was named by Benson (1969a).

Identifying Characters. The solitary blue-green stems of *E. warnockii* have tubercles that are coalescent into low ribs, and spreading, divergent spines that do not hide the stem in turgid, healthy plants. The 11–15(–17) radial spines are sufficiently divergent, stiff, and overlapping that it is difficult to measure their length with an ordinary ruler, although some of the radial spines may be somewhat appressed. The radial spines are acicular and of slightly smaller caliber than the usually three (in adult plants) upper central spines that merge upward with the radials. In adult plants a single porrect or upward-directed lower central spine is evident. Radial spines are tan to stramineous, gray at the base and "chalky-blue" or blue-gray (rarely reddish-purple) on the distal portion, or nearly the whole spines may be cyanic. The color of the central spines is similar to that of the radials, or the centrals, particularly the lower central, may be blue-gray (rarely dark reddish-purple) for more that half of their length. *Echinomastus warnockii* is so different from the related species *E. mariposensis* that the two taxa are not likely to be confused in the field.

The white flowers of *E. warnockii* are 2.5–3(–3.8) cm long and 2.3–3 cm in diameter when widely open with reflexed tepals. In size and general appearance the flowers of *E. warnockii* are the same as in *E. intertextus* and *E. mariposensis*. The inner tepals of *E. warnockii* are 1.3–2.6 cm long, 3–5.5 mm wide, and pure white or with thin greenish midstripes. The outer tepals usually have greenish or greenish-brown midregions and whitish margins. Rarely, the tepal midstripes or midregions are pinkish or pinkish-brown in specimens matching *E. warnockii* in other characters. It is not known if the seemingly rare pink coloration in flowers of *E. warnockii* is part of flower-color variation in this species or the result of introgression from *E. intertextus*. The stamen filaments of *E. warnockii* are pale green or whitish, and the anthers are bright yellow. The style is green, as are the 6–10 stigma lobes. The green stigma lobes and relatively dark yellow anthers are reliable floral differences between *E. warnockii* and *E. intertextus*, which has red stigma lobes and cream-yellow anthers.

The light green fruits of *E. warnockii* are globular, 0.7–1 cm in diameter, thin-walled, and mostly indehiscent. The thin-walled fruits become dry and fragile after ripening, and in some cases may split longitudinally (Weniger, 1984; Zimmerman et al., forthcoming). Ripe fruits usually are green but may develop a pinkish tinge on exposed surfaces. The floral remnant is persistent.

Phenology. *Echinomastus warnockii* is one of the first cactus species to bloom each winter and spring in the lower Big Bend country, along with *Mammillaria lasiacantha*, with which it shares much the same distribution, and along with the relatively obscure *Coryphantha duncanii* and *E. mariposensis*. The usual blooming period of *E. warnockii* is from late February to late March. Plants south of the Chisos Mountains near the Rio Grande may bloom in early February. Near Marathon in northern Brewster County, plants resembling *E. warnockii* have bloomed as late as May, but these plants near Marathon are suspected to be introgressed from *E. intertextus*. Several flower buds and ultimately several flowers are produced near the stem apex, as they are in other species of *Echinomastus*. One to several flowers may be open at the same time, and a single plant may produce flowers for several days to several weeks. Individual flowers open about 1:00 P.M. on warm, sunny days, close at night, and open again for 2–4(–5) days. The flowers have remained open for 1–2 days after experimental cross-pollination. The protandrous flowers are visited by carpenter bees, honeybees, various beetles, and ants (Raun, 1997). Fruit maturation in *E. warnockii* requires 1–2 months.

Sterile and Immature Specimens. The usually unbranched stems of *E. warnockii* typically are globose or ovoid in younger plants, and with age and size they become larger ovoid, oblong, or short-cylindroid. Most plants in the Big Bend region are approximately 3–8 cm tall. Much larger stems, to 20 cm long, are occasionally encountered in populations near the Rio Grande. The largest plants we have seen were in the Black Gap Wildlife Management Area. Very old stems may produce one or more short basal branches. About 13 more or less spiraling ribs are found in younger plants of *E. warnockii*, whereas 20 or more ribs are characteristic of older stems. Definitive short tubercles are evident on the ribs.

Juvenile plants of *E. warnockii* are easily distinguished from *E. intertextus*, and usually from *Echinocereus dasyacanthus* and *Coryphantha echinus*, by the spine and stem characters discussed above. Juveniles of *Echinomastus warnockii* have circular areoles in which there is a central region ca. 1 mm across that is devoid of spines, and 11–15(–17) radial spines with blue-gray tips, and often one or two central spines with cyanic tips. Our observations suggest that the youngest plants of *E. warnockii* produce 0–1 central spine per areole and that somewhat older and larger juveniles produce 1–2 spines per areole. Still slightly older plants nearing sexual maturity may develop the full complement of four central spines, with the upper three sharply angled upward in position with the radials, and with the single lower central more or less set apart at a more outward angle. As discussed above, close examination of areoles and spine clusters in juvenile plants of *C. echinus* always allows easy field distinction from *Echinomastus*. In seedlings of *E. warnockii* that are less than 1 cm long, at least some of the radial spines are pubescent.

Biosystematics. Morphologically and ecologically, *Echinomastus warnockii* clearly is distinct from what appears to be its most closely related species, *E. mariposensis*. The two species overlap in distribution throughout most of their ranges in the southern Big Bend region, but we have not observed these species growing together in any part of this range. There is an undocumented report that the two species occur together in at least one site in the Black Gap Wildlife Management Area in southern Brewster County. In Texas, *E. warnockii* is by far the most widespread and abundant of the two taxa, apparently being adapted to a wider range of desert soil types and other ecological conditions. *Echinomastus mariposensis* is also abundant in the southern Big Bend region, but in what appears to be a more restrictive limestone habitat.

Preliminary artificial hybridization studies involving *E. warnockii* and *E. mariposensis* (Powell, unpub.) have so far failed to show the extent of any reproductive isolation between the species. Preliminary tests have demonstrated that *E. warnockii* is self-incompatible, but they have not yet resulted in successful crosses between the species.

In Benson (1969a) the two photographs accompanying the original description of *E. warnockii* and titled as *E. warnockii* in the figure legend clearly are photographs of *E. mariposensis*. The photographs were provided by R. H. Wauer and represented plants that Wauer collected in the Dead Horse Mountains (Glass and Foster, 1975; pers. comm., from Wauer to D. Zimmerman). In Heil and Brack (1988) the names and figure legends for *E. warnockii* and *E. mariposensis* (figs. 32, 33) were switched.

Synonyms. *Echinomastus pallidus* Backeb., *nom. nud.*; *Echinocactus erectocentrus* J. M. Coult. var. *pallidus* (Backeb.) Weniger, *nom. nud.*; *Sclerocactus warnockii* (L. D. Benson) N. P. Taylor; *Echinocactus warnockii* (L. D. Benson) Weniger, *nom. nud.*

Common Names. Warnock biznagita; white-flowered cactus.

3. **Echinomastus mariposensis** Hester, Des. Pl. Life 17: 59. 1945. MARIPOSA CACTUS. Plates 229, 230. Locally common in thin Boquillas limestone soil, fre-

quently on mesa and ridgetops, but also in saddles and on slopes, often in gravel-like rubble or in relatively stable rocky substrates, southern Presidio and Brewster counties, most common from S and NE of Lajitas E beyond Black Gap Wildlife Management Area; one collection from near Ruidosa in W-central Presidio Co. 2,500–3,700 ft. Flowering Feb–Mar. $2n = 22$. Mexico, well documented in eastern Coahuila. Map 49.

Echinomastus mariposensis is one of the early-blooming cactus species in the southern Big Bend region. The relatively small plants, covered with ashy-white spines and with white flowers, are difficult to locate against the pale background of limestone rock. *Echinomastus mariposensis* was listed federally as a Threatened species in 1979, and in 1983 was accorded the same status in Texas. Our field studies, in concurrence with those of Raun (1997), have shown that *E. mariposensis* is restricted to specific habitats within its geographic range but that it is widespread and abundant across its range. The species is much more widespread than was suggested by Benson (1982) or Weniger (1984). We have observed, as has Raun, that many populations of *E. mariposensis* are healthy and thriving, and at some sites, young adults and juveniles are so abundant that it is difficult to walk without stepping on the plants. Sexually mature plants of *E. mariposensis* usually are 2.5–7 cm tall. Presumably older plants that are 10–12 cm or more tall and 6 cm or more wide have been observed at some sites in Big Bend National Park and in Mexico (Smith, 1973). The specific epithet is taken from the name of a cinnabar mine and mining town near the type locality (Hester, 1945) on private land northwest of Terlingua in Brewster County.

Identifying Characters. Sexually mature plants of *Echinomastus mariposensis* are ashy-white in aspect and are the general size and shape of golf balls or tennis balls. Thus, the unbranched stems typically are subglobose, broadly ellipsoid, or cylindroid. Larger plants approaching 10–12 cm tall are not common in Texas. The ashy-white habit is the result of overlapping white radial spines. Plants often exhibit a slightly darker hue because central spines are cyanic on the distal portions. The blue-green stem surfaces are obscured by the dense spine cover. The stems are strictly tuberculate, or they may exhibit ca. 21 ribs. The areoles are 6–12 mm apart.

The radial spines of *E. mariposensis* are slender, (20–)25–32(–36) in number, and are 0.3–1.1(–1.5) cm long. In older areoles most of the radial spines are oriented horizontally. Usually there are four central spines per areole (rarely six) in mature plants: one lower central, and three upper centrals. The central spines are markedly heavier than the slender radials, and they have small-bulbous bases. The porrect or down-curved lower central spine is (0.3–)0.6–1.1(–1.5) cm long, and 0.3–0.6(–0.9) mm thick at the base. The upper central spines are ascending-appressed and sometimes curved upward (especially in upper areoles), and they are (0.9–)1.3–2(–2.1) cm long and 0.2–0.6 mm thick. Typically, the central spines are chalky-blue or blue-black on the distal portions. Tentative field identification of *E. mariposensis* comes from observing subglobose, small, ashy-white plants with upswept cyanic central spines, these often forming a diffuse tuft of spines at the apex.

The white flowers of *E. mariposensis* are 2–3 cm long and 2–3(–3.5) cm in diameter, and they are borne at the stem apex. The inner tepals are 1–2 cm long, 1.5–4 mm wide, somewhat spatulate, and usually rounded or obtuse and sometimes notched or toothed at the apex. The inner tepals are white or very pale pink, except for flesh-pink, pale yellowish-tan, or pale green midregions. The outer tepals are either white or pale pink at the margins and have greenish or brownish midregions. The filaments are colorless to pinkish, and the anthers are cream-yellow to slightly darker yellow. The style is pale green and supports 5–8 green to yellow-green stigma lobes.

The green to yellowish-green fruits are globose to oblong, ca. 1 cm long, and ca. 0.8 cm thick. After ripening the fruits remain green, usually are indehiscent, but may split longitudinally on one side. The thin, dry fruit walls display a few scales. The seeds are black, subreniform or subglobular, 1.6–2 mm long, and the surfaces are finely papillate.

Phenology. Apparently the only flowering time for *E. mariposensis* is in late February to mid-March. Plants of this species have not been observed to respond opportunistically to rains later in the season. Numerous flower buds are produced at the plant apex. Several flowers, often four or five, may open at the same time. Individual flowers apparently open after noon, about 1:00 P.M., close at night, and open again under warm, sunny conditions for 3–4(–5) days. Fruit maturation requires only 1–2 months. In the field we have observed apparently mature fruits as early as 17 March. Fruit maturation may continue in some seasons into late April. Indehiscent ripened fruits may persist under apical spines until May.

Sterile and Immature Specimens. *Echinomastus mariposensis* is easily distinguished from the closely related *E. warnockii* by the spine characters discussed above. In the field some nonflowering plants of *E. mariposensis* resemble single-stemmed plants of *Mammillaria pottsii*. Plants of both species are whitish in overall appearance; both have an abundance of appressed, slender, ashy-white radial spines; and they display up-curved, cyanic central spines in upper areoles. In addition, both species occur in similar limestone habitats in the western portion of the range of *E. mariposensis*. Plants of *M. pottsii*, however, typically are multistemmed, and at least in larger plants, the stems are cylindroid. *Mammillaria pottsii* has a very different spine pattern with eight centrals and usually over 40 radials. When in flower in late spring, *M. pottsii* is distinctive with a row of small reddish-maroon flowers circling the stem below the apex.

In the field *E. mariposensis* also could be confused with *Coryphantha sneedii* var. *albicolumnaria*, *C. tuberculosa* var. *tuberculosa*, *C. tuberculosa* var. *varicolor*, or *C. dasyacantha*. All of these taxa can be distinguished from *E. mariposensis* by their central spine morphology, which includes usually several, spreading, central spines, with these oriented as inner and outer centrals. Also, mature plants of *C. sneedii* var. *albicolumnaria*, *C. tuberculosa* var. *tuberculosa*, and *C. dasyacantha* characteristically are multistemmed, although sexually mature, single-stemmed plants of these taxa are not uncommon. *Coryphantha tuberculosa* var. *varicolor* typically is single-stemmed, as is *E. mariposensis*. In the eastern portion of its range, *E. mariposensis* might be confused with *Neolloy-*

dia conoidea. The latter taxon, however, is easily distinguished by its branching stems forming clumps and spine characters. The flowers of *N. conoidea* are magenta in color.

Microscopic examination of immature plants of *E. mariposensis* shows that each areole toward the base of the plant exhibits 20–30(–32) appressed white radial spines that are ca. 2 mm long. The radial spines of adjacent areoles overlap but not as much as in adult plants. In juveniles typically only one central spine is evident in areoles positioned near the stem base, and two centrals are characteristic in areoles positioned near the apex of the stem. When one central spine is present, it is borne near the center or a little above center in the areole. Central spines in areoles of immature plants are much like those in adult plants in shape and color, but they are shorter at only 3–5 mm long. And the one upper central in the upper areoles of immature plants is curved upward, as it is in older plants. Observations of juvenile plants suggest that the first prominent central spine to appear in areoles may be interpreted as the upper central, and the second central to develop may be equivalent to the lower central of adult areoles. Observation of immature plants also suggests that the two central spines may develop simultaneously, at least in certain individuals. At least some of the spines are pubescent in seedlings that are ca. 1 cm tall.

Biosystematics. *Echinomastus mariposensis* is believed to be most closely related to *E. warnockii*. These two species occupy much the same geographic range in southern Presidio and Brewster counties, although the taxa appear to be ecologically isolated. The authors have walked a significant portion of the Texas range of *E. mariposensis* without encountering *E. warnockii* in those habitats. No suspected hybrids between *E. mariposensis* and *E. warnockii* have been detected. In general, *E. mariposensis* occupies more stable Boquillas limestone substrates at ridge and mesa crests, while *E. warnockii* occurs in a variety of habitats, including those apparently derived from both igneous and sedimentary parent materials, including gypsum.

Synonyms. *Neolloydia mariposensis* (Hester) L. D. Benson; *Echinocactus mariposensis* (Hester) Weniger, *nom. nud.*; *Sclerocactus mariposensis* (Hester) N. P. Taylor.

Common Names. Silver column cactus; Mariposa viszagita; Mariposa Lloyd's cactus; Lloyd's Mariposa cactus; Lloyd's Mariposa; golfball cactus. The U.S. Fish and Wildlife Service and the Texas Parks and Wildlife Department list this taxon as Threatened under the common name Lloyd's Mariposa cactus. The application of "Lloyd's" to a common name for *Echinomastus mariposensis* is enigmatic because Lloyd (presumably F. E. Lloyd) apparently had nothing to do with this species.

15. EPITHELANTHA
F. A. C. Weber ex Britton & Rose
Button-Cactus

Plants small, whitish. Roots diffuse (in Texas taxa). Stems unbranched or cespitose, cylindroid or subglobose, somewhat acute, truncate, or concave at the apex, 1–4(–6) cm long, 2–4(–6) cm in diameter, when in clumps to 15 cm wide; pith and cortex not mucilaginous; hypodermis one-layered, with few spheroidal druses. Ribs none. Tubercles very small, numerous, not visible through the spines. Areoles subcircular. Spines innocuous, white to ashy-gray, 20–90 per areole, in 1–5 series, slender, straight, 2–6(–7) mm long, ca. 0.1 mm wide, smooth or microscopically scabrous, those in lateral areoles flattened against the stem, some of those in areoles lining the apical depression standing erect or inclined and collectively forming a tuft of spines among clear, hairlike spines and trichomes (wool) at the stem apex, the erect spines ultimately breaking near the middle and contributing, as the stem grows, to the shorter-spined areoles on the sides of the stem. Flowers produced on new growth at or near the stem apex, each formed at a tubercle apex in a felted area adjacent to and merging with the spine-bearing part of the areole, the area leaving a near-circular scar after fruit fall; flowers usually small and inconspicuous, pink to pale pink (rarely nearly colorless, silvery-white, or pale yellow), 0.6–1.7 cm long, 0.3–1.2(–1.7) cm wide; inner tepals 3–13(–21) per flower, usually pinkish or whitish; outer tepals entire or fimbriate, with pink midribs and pale pink margins; receptacular tube funnelform, naked, deciduous during fruit development; stamens 10–40, filaments reddish or pale yellow, anthers white, yellow, or pink; stigma lobes white, usually 3–4; pericarpel in anthesis about 1–2 mm long. Fruit bright red, clavate or cylindroid, 3–20 mm long, 2–3(–5) mm thick, thin-fleshy at maturity, usually with 5–11 seeds, or fruits smaller, gray to colorless, and with fewer seeds; floral remnant caducous. Seeds black, glossy, obovoid, 1.2–1.5 mm long, impressed-reticulate; hilum "lateral," elongate. $x = 11$.

A genus of two species distributed from western Texas to Arizona, and in northeastern Mexico. Two taxa occur in the United States, and in Texas. Five or six taxa of *Epithelantha micromeris* are recognized for Mexico (Bravo-Hollis and Sánchez-Mejorada, 1991a; Glass and Foster, 1978). The genus name is derived from the Greek *epi*, "on"; *thel*, "nipple"; and *anthos*, "flower," in reference to the position of flowers near the apex of tubercles.

Boke (1955) carefully documented the areole morphology of *Epithelantha*. This was a formidable task, because *Epithelantha* areoles are tiny and their development is obscured by their spines and wool. The stem apex must be destroyed, or at least damaged severely, before the areole structure can be seen. Boke stated in essence that *Epithelantha* areoles are dimorphic and may give rise to spines in one portion of the areole and a flower in another part of the same areole. This was a misuse of the word *dimorphic*. The word normally would imply two different kinds of areoles, not specialization of parts within "the same areole." Boke merely documented that *Epithelantha* has the same areole structure as that in most other genera of cacti. Subsequent authors have been confused by Boke's comparison between *Epithelantha* and *Mammillaria* (i.e., "Dimorphic areoles are produced in *Mammillaria*"). However, Boke's own definition of "dimorphic areoles" does not imply *Mammillaria*-like areoles (with the two parts physically disjunct from each other). "Dimorphism" in Boke's sense is the standard form of any cactus areole: the flower grows from the spineless adaxial extremity of the areole rather than from the center of the spine cluster. Boke's frequently cited paper provides no evidence for a close relationship between *Epithelantha* and *Mammillaria*.

Key to the Species

1. Stems whitish-gray, relatively rough in appearance; areoles and their spines 4–5(–7) mm across (on sides of stems); spines of each areole in 1–3 series, 20–30(–40) in all; flowers 6–8.5 mm long, 3–4.6 mm wide, only the tips exposed above the spines and wool of the stem apex; stamens 10–16
    1. *E. micromeris* var. *micromeris*.

1. Stems whitish (or cream-colored), smooth and shiny in appearance; areoles and their spines 2–2.5(–4) mm across (on sides of stems); spines of each areole in many series, 33–40(–50–90) in all; flowers 10–17 mm long and wide, conspicuously exserted above the spines at stem apex; stamens 20–40
    2. *E. bokei*.

1. **Epithelantha micromeris** (Engelm.) F. A. C. Weber ex Britton & Rose var. **micromeris**. COMMON BUTTON CACTUS. Plates 231, 232. [*Mammillaria micromeris* Engelm., Proc. Amer. Acad. Arts 3: 260. 1856]. Infrequent to locally common in rocky substrates of limestone and igneous origin, hills and ridges in the desert and grasslands, to be expected in every county, but currently known in the eastern Trans-Pecos in eastern Presidio Co., Brewster Co., and Pecos Co. E to Val Verde Co., and in western Culberson Co. and El Paso Co. 2,000–5,000 ft. Flowering late winter to early spring. $2n = 22$. East in Texas to Medina, Real, and Bandera counties, and to Upton Co., N to Howard Co., W across S-central NM (rare except in Otero and Eddy counties), N to near Belen on limestone (sight records, S. Brack and D. Ferguson) to AZ, rare in S Cochise Co. Presumably northern Mexico, but identifiable Mexican specimens thus far are *E. micromeris* var. *greggii* (Engelm.) Borg. Map 50.

In the Trans-Pecos the plants of *E. micromeris* var. *micromeris* usually are unbranched, and they have been described aptly (Benson, 1982) as "fuzzy white balls." The whitish-gray spines completely cover the wartlike greenish tubercles. Small, usually pinkish flowers are borne at the stem apex in an upswept apical tuft of spines. In late spring slender, bright red fruits stand prominently on the "button-size" whitish or gray plants. The specific epithet is taken after the Greek *mikros*, "small," and *meros*, "part," for the small characters of the species.

Identifying Characters. In the field *E. micromeris* is potentially confused with

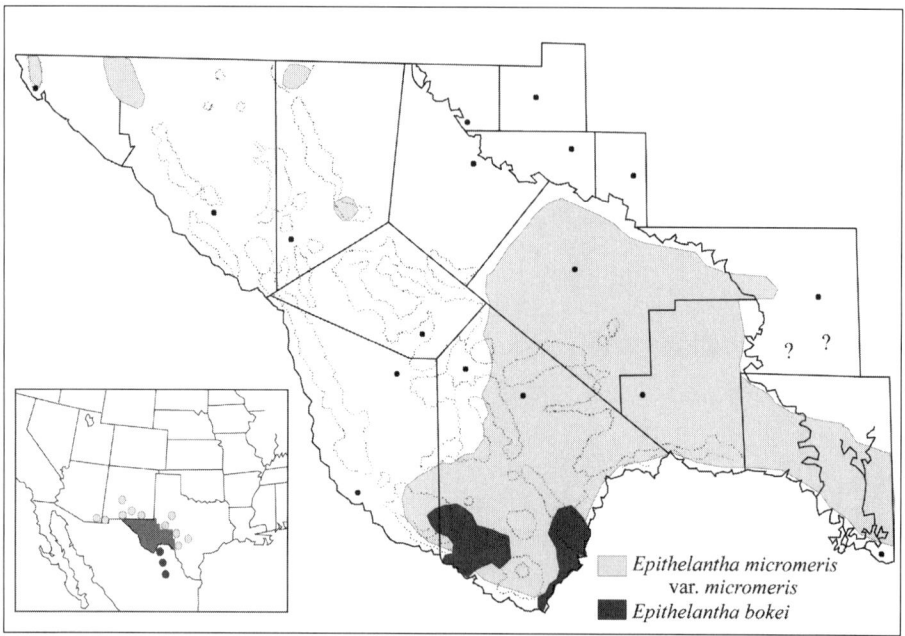

Map 50. Generalized distribution of *Epithelantha micromeris* var. *micromeris* (common button cactus) and *E. bokei* (Boke's button-cactus).

the closely related *E. bokei, Mammillaria lasiacantha,* and seedlings of *Coryphantha vivipara.* All of these have small stems hidden by whitish spines. In sexually mature *Epithelantha* the spines at the stem apex are about twice as long as the spines in areoles on the sides of the stem. The reason is that during the course of one season or more the distal halves of the longer spines break away, leaving shorter spine clusters that are appressed against the stem in older, lateral areoles. In *M. lasiacantha* also the spines mostly are appressed against the stem, but spines do not disarticulate in the middle and become shorter in the process of forming the spine clusters down the stem.

In flowers of *E. micromeris* there are 5–8 pink or pale yellow inner tepals per flower, with the largest of these only 2.5–3.5 mm long and 1–2 mm wide. Each flower has only 10–16 stamens, with pink to cream-yellow anthers. The 3–4 stigma lobes are white. The reduced flowers of *E. micromeris* are the smallest of any cactus species in the Trans-Pecos.

The red fruits of *E. micromeris* are naked, cylindroid, usually 1–2 cm long, 2–3 mm thick, hollow, few-seeded, with thin but modestly juicy walls. The black, glossy seeds are 1–1.4 mm long, somewhat comma- or helmet-shaped, with a deep cavity associated with the hilum.

When in flower, *Epithelantha* and *M. lasiacantha* are easily distinguished. The flowers of *Epithelantha* are somewhat funnelform (the largest 1–1.7 cm across) and pink (rarely white or pale yellow). The flowers of *M. lasiacantha* are on the shoulders of the stem, and both inner and outer tepals usually are "candy-striped" with sharply defined red to brownish midregions or midstripes and prominent whitish margins. The bright red ripe fruits of *Epithelantha* and *M. lasiacantha* are not readily distinguished, but the seeds inside them are very different in shape and surface sculpture.

Phenology. The peak blooming time for *E. micromeris* in the Trans-Pecos is not well understood. Potentially, flowering occurs from late winter, in February, to early spring, in March and April. Individual flowers open on warm days, close at night, and may open again for one or two days. The flowers often do not open widely. Several flowers may be open at the same time, all tightly clustered at the stem apex. Red fruits appear in late spring or early summer (April–June). As many as 10–12 or more red fruits may develop almost simultaneously at the plant apex.

Sterile and Immature Specimens. In *Epithelantha*, areoles have a circular (radial) arrangement of spines appressed against the stem, but in *M. lasiacantha* most of the spines in lower lateral areoles are oriented to the sides, forming somewhat bilateral spine clusters. The upper areoles in *M. lasiacantha* have more spines radiating in all directions, but still they are basically bilateral, and all the spines are not appressed against the stem as they are in *Epithelantha*. Spines in adjacent areoles in *Epithelantha* are slightly overlapping if at all, whereas those in *M. lasiacantha* are significantly overlapping. Microscopically, the longer spines in *E. micromeris* are minutely antrorsely barbed and sometimes dark-colored on much of the distal half. The all-white spines of *E. bokei* are minutely fringed. The white spines of *M. lasiacantha* are either glabrous or pubescent.

Sterile specimens of *E. micromeris* and *E. bokei* usually can be distinguished without difficulty. In aspect the plants of *E. micromeris* are ashy-gray or purplish-gray (in Mexico there is a reddish-brown color-phase in one variety) and rougher in texture than the bright white or creamy-white plants of *E. bokei*, with a satinlike texture. "Slump rings" are more often evident on the smooth, shining surface of mature plants of *E. bokei* in flaccid, desiccated condition. The radial spines are longer in *E. micromeris* than in *E. bokei*, resulting in spine clusters that are 4–5(–7) mm across in *E. micromeris* and 2–2.5(–4) mm across in *E. bokei*. The spine clusters in *M. lasiacantha*, by comparison, are (6–)9(–10) mm in diameter, although it is more difficult to get an exact measurement because the spines are more overlapping. The spine bases of *E. micromeris* (and in *M. lasiacantha*) vary from white to pale yellowish or light brown, collectively creating a darker spot in the center of each spine cluster, whereas in *E. bokei* the spine bases usually are white and not brown.

Immature specimens of *Epithelantha* and *M. lasiacantha* reveal most of the spine characters discussed above. The distinguishing features can be seen with the naked eye or with a hand lens.

Biosystematics. About eight taxa of *Epithelantha* have been recognized, but all of these have been treated by Bravo-Hollis and Sánchez-Mejorada (1991a) or other authors as varieties of *E. micromeris*. Six taxa are recognized by Zimmerman et al. (forthcoming), the species *E. micromeris* and *E. bokei*, and five varieties of *E. micromeris*, four of them in the CDR. Clearly, the understanding of the *Epithelantha* complex could benefit from a thorough systematic study, particularly in Mexico where most of the enigmatic variation exists. In the United States no one has ever claimed the presence of more than two taxa of *Epithelantha*. The taxon of long standing is *E. micromeris,* having been described in 1856, whereas *E. bokei* was not described until 1969 when Benson was preparing his treatment of Texas cacti.

Epithelantha micromeris var. *micromeris* is geographically relatively widespread in the southwestern United States, but *E. bokei* barely enters the Big Bend region of Trans-Pecos Texas from a wide range in eastern Coahuila. The two taxa usually are physically separated in the Trans-Pecos, but they are known to occur side by side (maybe 30 cm apart) in at least one site in the Solitario Dome in Presidio County (Hardy, 1997), without any evidence of hybridization. The two species also grow intermingled in Coahuila, Mexico, apparently without hybridization (Zimmerman, et al., forthcoming). At present the recognition of *E. micromeris* and *E. bokei* seems to be firmly supported by compelling morphological, reproductive, and ecological differences between the taxa. Through transposition of captions in Benson's (1969, p. 186) original diagnosis of *E. bokei*, confusion was created with (ironically) some of the finest published photographs of the diagnostic differences between areoles of these two sharply differentiated taxa.

Synonym. See basionym under *M. micromeris.*

Common Names. Several popular names have been applied to *Epithelantha* in general: Button cactus; mulato; tapon.

2. **Epithelantha bokei** L. D. Benson, Cact. Succ. J. (U.S.) 41: 185. 1969. BOKE'S BUTTON-CACTUS. Plates 233–35. Rocky limestone soils, ridges, tops, and slopes of desert mountains, near the Rio Grande in southern Brewster and Presidio counties. 2,300–4,400 ft. Flowering May–Jun. $2n = 22$. Coahuila, Mexico, SE to near Saltillo. Map 50.

Epithelantha bokei often appears as one of the more common small cacti in the limestone hills of southern Brewster County, particularly southeast of the Chisos Mountains. In this area *E. bokei* is much more common than *E. micromeris.* The two small cactus species are easily distinguished in the field, even from a standing position. The plants of *E. bokei* are usually seen as short-cylindroid stems, solitary or in clusters, flat or concave at the apex, and a shining, smooth coat of close-set spines covering the sides of the stems. Characteristically, the stems of *E. micromeris* are more rounded at the apex (but immatures of both species have concave apexes), the close-set coat of spines is ashy-white or gray, and the texture is more coarse in appearance and feel (but immatures of both species are smooth and shiny). The species *E. bokei* is named after Norman H. Boke, plant anatomist, who at the University of Oklahoma studied this species and other cacti. Boke also supervised several students whose investigations contributed to knowledge about the Cactaceae.

Identifying Characters. Floral and vegetative features by which *E. bokei* can be distinguished from *E. micromeris* are presented above under *E. micromeris.* In summary, the larger flowers (when they are present) and the smooth, white stems (with practice) of *E. bokei* are dependable macroscopic identifying traits. Microscopically, in *E. bokei* the individual spines in lateral areoles are only ca. 1 mm long, with the whole appressed spine cluster usually only 2–2.5 mm across, as opposed to spine clusters that are 4–5(–7) mm across in *E. micromeris*. The spines are arranged in indefinitely many series in *E. bokei,* and in only 1–3 series (except for an adaxial bundle) in *E. micromeris*. Adjacent spine clusters are merely touching or barely overlapping in *E. bokei,* whereas spine tips from adjacent clusters in *E. micromeris* usually are overlapping. The tiny spines themselves in *E. bokei* have margins that appear to be more white than the central portion of the spines. The ashy-white spines of *E. micromeris* do not have margins that are distinctive in color.

The pale pink to silvery-white flowers of *E. bokei* are always larger at 1–1.7 cm in diameter than those of *E. micromeris*. In each flower there are 13–21 inner tepals, the largest ones 5–6(–9) mm long and 1.5–3 mm wide. There are 8–13 outer tepals, these often fringed. The inner tepals are translucently pale, usually with very diffuse midstripes that are either pink, pinkish-tan, or yellow-green. There are 20–40 stamens in each flower, with pinkish to cream-yellow anthers. The stigma lobes are white. Although the flowers of *E. bokei* are larger than those of *E. micromeris* and tend to open wider, they are still small by comparison with other cacti. The flowers of *E. bokei* are obligately outcrossing.

The red fruits of *E. bokei* are naked, cylindroid when elongated, 0.8–1.3 cm long, and 1.5–2.3 mm thick. Fruits elongate and turn red only under favorable conditions. Under unfavorable conditions they remain small and green. The seeds are black, glossy, 1–1.4 mm long, comma- or helmet-shaped, and somewhat bowl-like with a deep cavity associated with the hilum.

Phenology. The peak blooming period for *E. bokei* is not well documented, but it appears to be in May–June, later than for *E. micromeris*. Early- or late-season sporadic blooming in *E. bokei* needs to be investigated. Under cultivation in Alpine, numerous plants were in flower on 21–26 June 1999 after rains two weeks earlier, and a single plant was in flower on 25 September 1998. On 23 February 1987 a single plant of *E. bokei* was found in fruit along the River Road in Big Bend National Park. Individual flowers open on warm days, close at night, and may open again for one or two more days. The flowers usually open widely. Several flowers may be open at the same time, these clustered at the center (apex) of truncate or concave stem tops. One to several red fruits may elongate and develop at the same time, probably in late summer and fall.

Sterile and Immature Specimens. The vegetative characters discussed above are obviously not totally sufficient for distinguishing both adult and juvenile specimens of *E. bokei*. Immature specimens of *E. bokei* are similar to the adults and not likely to be confused with any other Trans-Pecos cactus species, but immature and adult *E. micromeris* have "smooth" spinaton and often are misidentified as *E. bokei*.

Biosystematics. Glass and Foster (1978) reduced *E. bokei* (Benson, 1969a) to a variety of *E. micromeris*. Weniger (1984) did not give taxonomic recognition to the taxon. Weniger commented without documentation that the two entities intergrade and that taxonomic recognition as a species was not justified. No one who knows how to identify *E. bokei* and *E. micromeris*, however, has ever seriously claimed that they intergrade. Benson (1969b) noted that in horticulture *E. bokei* has been known as *E. greggii* Engelm., having been confused with the Mexican taxon *E. micromeris* var. *greggii* (Glass and Foster, 1978).

Synonym. Epithelantha micromeris var. bokei (L. D. Benson) Glass & R. A. Foster.

Common Names. Boke button cactus; Boke's cactus; Boquillas button cactus.

16. MAMMILLARIA Haw.
Pincushion Cactus

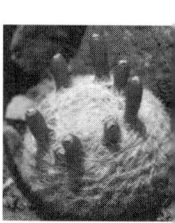

Plants with stems cylindroid, globose, or turbinate, in which case, the tops are flat to the ground if not contracted below ground level during dry periods. Stems single or branched, 1–15(–25) cm long, 1.8–12(–20) cm wide; pith and cortex often containing latex, mucilaginous or not. Ribs none. Tubercles separate and usually prominent, cylindrical or conical. Areoles circular or subcircular, located at tips of tubercles. Spines smooth, whitish to gray or blue-gray to nearly black, or yellowish, or brown with some reddish color; central spines none to several, straight, curved, or hooked at the end, 0.3–2(–2.5) cm long, slender, needlelike, usually only 0.1–0.6 mm thick at the base; central and radial spines perhaps poorly differentiated; radial spines usually smaller than centrals and lighter in color, (6–)10–80 per areole, needlelike, not hooked. Flowers produced on old growth of preceding seasons and thus usually appear below the apex and lateral on the stems, typically formed between tubercles, not apparently connected with areoles or tubercles directly, the fertile areoles between tubercles bristly, woolly, or nearly naked; flowers 0.6–3.5(–7.5) cm in diameter; pericarpel naked or with only rudimentary scale bracts; floral tube funnelform, campanulate, or flower rotate. Fruit typically red or pink, fleshy at maturity, ripening 1–12 months after pollination; usually thin-walled, cylindroid, clavate, to obovoid, 0.9–2(–4) cm long, (0.2–)0.4–0.9(–2.6) cm in diameter; floral remnant deciduous. Seeds maturing usually before fruit ripens, the seeds black, glossy, brown, reddish, or yellowish, microscopically either pitted, raised-reticulate, foveolate, or wrinkled, longer (1–2 mm) than broad; hilum usually basal, sometimes appearing oblique. *Neomammillaria* Britton & Rose; *Dolichothele* (K. Schum.) Britton & Rose; *Chilita* Orcutt; *Ebnerella* Buxb.; *Leptocladia* J. Agardh; *Leptocladodia* Buxb.; *Porfiria* Boed.; *Krainzia* Backeb.; *Mammilloydia* Buxb.; *Haagea* Frič; *Pseudomammillaria* Buxb.; *Solisia* Britton & Rose; *Oehmia* Buxb.; *Phellosperma* Britton & Rose; *Bartschella* Britton & Rose; *Cochemiea* (K. Brandegee) Walton; *Mamillopsis* (Morren) Weber ex Britton & Rose; *Cactus* L., in part. $x = 11$.

A genus of 150–316 species occurring in the southwestern United States and Mexico, with a few species in South America (Colombia and Venezuela), Central America, and the West Indies. Eight species occur in Texas, all but *M. sphaerica* in the Trans-Pecos. All twentieth-century authors except for Weniger (1970, 1984) have maintained *Mammillaria* as distinct from *Coryphantha* while pointing out that the most consistent difference between the two genera is the lack of an adaxial areolar groove on the tubercles in *Mammillaria*. *Mammillaria* usually also differs by producing flowers on older growth in a ring 1 cm or more below the apex, whereas in *Coryphantha* flowers usually are produced on current growth at the stem apex, although some coryphanthas exhibit intermediate flower position. In general, the Trans-Pecos mammillarias are a group of dissimilar species. *Mammillaria grahamii*, *M. wrightii*, *M. lasiacantha*, *M. pottsii*, and *M. prolifera* are all distinctive, but *M. meiacantha*–*M. heyderi* are a species-pair.

Key to the Species

1. Hooked central spine(s) present; flowers rose-pink, rose-purple, or magenta (2).
1. Hooked central spine(s) absent; flowers white to pink with prominent-colored midstripes, or flowers maroon-red or yellow (3).
2(1). Central spines (2–)3–4, only one (the lower one) prominent, porrect, hooked, dark red-brown or blackish, the others inconspicuous, appressed-erect, straight, relatively slender; radial spines 19–35; pith and cortex not mucilaginous; outer tepals minutely fimbriate; fruit when fully ripe bright red, clavate; Franklin Mts, El Paso Co., to NW Presidio Co.

                                    1. *M. grahamii* var. *grahamii*.

2. Central spines 1–4(–7), one or usually all hooked, all protruding, brownish; radial spines usually ca. 13 (rarely 8–30); pith and cortex mucilaginous; outer tepals prominently fimbriate; fruit when fully ripe green to dull purplish, globose to ovoid; Franklin Mts, El Paso Co.

                                    2. *M. wrightii* var. *wrightii*.

3(1). Plants when mature in low, matlike clumps, with 10–20 or more subglobose to short columnar stems of different sizes; radial spines numerous, the ones closest to the stem white, hairlike, flexible, usually curved and twisted; distributed in Val Verde Co.

                                      7. *M. prolifera* var. *texana*.

3. Plants when mature with solitary stems, these flat-topped, hemispheric, to depressed-globose, or with usually branched stems, these cylindroid or clavate; radial spines few to numerous, white, tan, yellowish, brown, reddish-brown or gray, not hairlike, all straight; distributed throughout Trans-Pecos (4).
4(3). Stems narrowly cylindroid or clavate, 6–15 cm or more tall, usually branched at the base; flowers maroon-red, the inner tepals usually with darker midregions

                                      4. *M. pottsii*.

4. Stems depressed globose, flat-topped or hemispheric (aboveground), unbranched; flowers white to pink, the inner tepals usually with conspicuous midstripes of pink or purple to brown or tan (5).
5(4). Stems depressed globose to short cylindroid, usually 1.5–4 cm in diameter, stem surface completely hidden by very numerous, thin, interlaced whitish spines

                                      3. *M. lasiacantha*.

5. Stems flat-topped or hemispheric (the part aboveground), usually 7–15 cm or more in diameter, stem surface and its prominent tubercles evident through slender or thick, spreading, stramineous, gray, to red-brown radial spines (6).

6(5). Radial spines usually 5–7 per areole, to 0.35–0.7 mm thick; fruits dull red to purplish-pink

5. *M. meiacantha.*

6. Radial spines usually 13–17, to 0.15–0.45 mm thick; fruits bright red

6. *M. heyderi.*

1. **Mammillaria grahamii** Engelm. var. **grahamii**, Proc. Amer. Acad. Arts 3: 262. 1856. GRAHAM'S FISHHOOK CACTUS. Plates 236, 237. Desert mountains, igneous or limestone, desertscrub or grasslands. Northwest Presidio Co. near Candelaria, below the rim of the Sierra Vieja, to El Paso Co., Franklin Mts. 3,200–4,600 ft. Flowering May–Jun, sporadically in Sep. $2n$ = 22. Southern NM, S and central AZ, and extreme E CA. Mexico, Sonora and Chihuahua. Map 51.

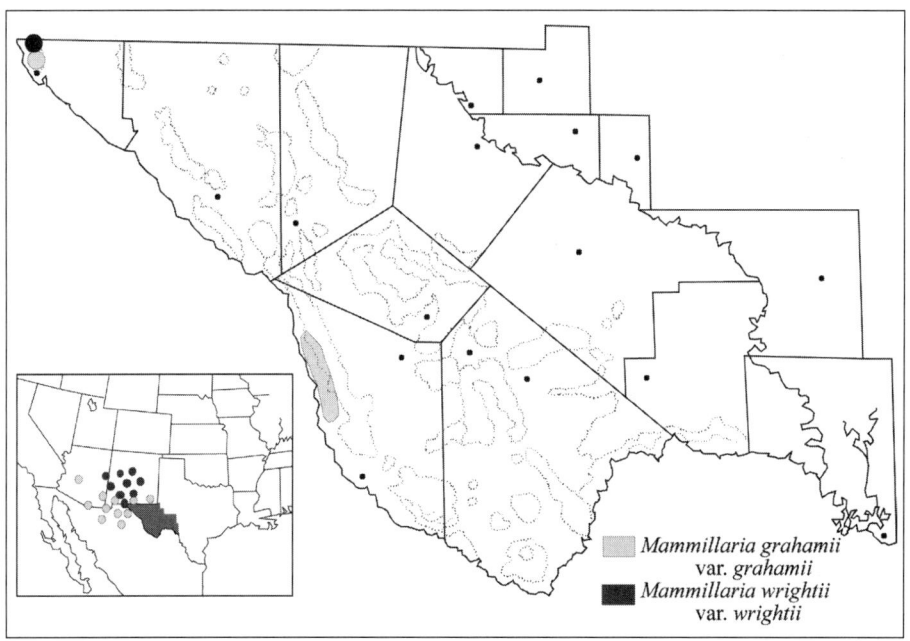

Map 51. Generalized distribution of *Mammillaria grahamii* (Graham's fishhook cactus) and *M. wrightii* var. *wrightii* (Wright's fishhook cactus).

The occurrence of *M. grahamii* in northwest Presidio County marks the southeastern distributional limits of hooked-spined mammillarias, so far as known. Benson (1982) recognized another variety of this species, *M. grahamii* var. *oliviae* (Orcutt) L. D. Benson, from southeast Arizona, with shorter, straight central spines. The specific epithet is after Colonel James Duncan Graham (1799–1865), topographical engineer, astronomer, and surveyor with the eastern portion of the U.S.-Mexican border survey. Mount Graham in Arizona was named after him.

Identifying Characters. The plants of *M. grahamii* are unbranched or branched at the base. Plants in Presidio County often are cespitose with 5–10 or more stems of different sizes. The stems are globose to short-cylindroid, 5–10(–15) cm long, 4–6.5(–8) cm in diameter, and whitish to gray in aspect because of a rather dense covering of radial spines that number 19–25(–35) in each areole. Upon close examination the single, porrect, hooked central spine is evident. The slender hooked spine is 1.1–1.7(–2.3) cm long, and its dark red-brown to blackish color is in contrast to the whitish radial spines.

The flowers of *M. grahamii* are bright rose-pink or magenta, 2–3.5(–4.3) cm in diameter when fully open with reflexed tepals. The colorful inner tepals are 1–1.6 cm long and 0.4–0.8 cm wide. The outer tepals are minutely fimbriate, with this feature best observed under magnification. The stamens have pinkish filaments and yellow to pale orange anthers. The relatively long style extends ca. 6 mm or more above the anthers and supports 6–10 slender, green, stigma lobes, these 3–7 mm long.

The fruits are bright red at maturity. Before maturation the fruits are green and spheroidal but usually gradually elongate, becoming somewhat club-shaped, 1.2–2.7 cm long, and 5–8 mm in diameter as they turn red. The seeds are black, globose to pear-shaped, and ca. 1 mm long.

Phenology. In Presidio County the early peak flowering time for *M. grahamii* seems to be in June. Flowers appear on some plants as early as May, at least in some years. In El Paso County, Worthington (1986) observed over three years that in *M. grahamii* flowering occurred from late May to late June and perhaps opportunistically thereafter in July and September. His observations of several plants revealed that each episode of flowering lasted for 2–4 days and that individual flowers last for one day and are closed at night. Further studies are needed to determine if the sporadic flowering is in response to rainfall and if in some years flowers appear earlier than May, particularly in the relatively protected habitat below the Sierra Vieja rim in northwest Presidio County. Fruit maturation requires several months, after which green fruits abruptly elongate and turn bright red. Plants cultivated in Alpine have produced mature fruits in September. Flowering patterns and reproductive ecology for *M. grahamii* have been studied in a population near Tucson, Arizona (Bowers, 2002).

Sterile and Immature Specimens. Because of the hooked central spine in each areole, *M. grahamii* is not likely to be confused with any other species, except perhaps with *M. wrightii* in the Franklin Mountains. The best identifying vegetative features of *M. grahamii* are the single hooked central spine in each areole, the 20–30 or more radial spines per areole, and the nonmucilaginous pith and cortex. In the Franklin Mountains the similar species *M. wrightii* usually has more than one hooked central spine, usually 15 or fewer radial spines, and mucilaginous pith and cortex. Both species lack latex in the stems.

The tubercles of *M. grahamii* are 0.6–1.2(–1.5) cm long and 5–7 mm wide at the base. The axils between the tubercles are naked (i.e., lacking hairs or bristles). The areoles are circular and 1.5–2 mm in diameter. In each areole usually there are 2–3 central spines. The lower central is the "main" central, in that it is porrect and hooked. The hooked central is only 0.2–0.4 mm thick, with a slightly enlarged base, and often is darker toward the distal half. The lower main central spine, although positioned below the other 1–3 central spines, actually emerges from the center of the areole surrounded by short, matted hairs. The other 1–2(–3) centrals, located in the upper position just below the upper radials, are erect-appressed, straight (not hooked), relatively slender, and inconspicuous. Usually there are two of these erect "subcentrals" in a "V" pattern. The upper subcentrals are inconspicuous compared to the hooked central, but they are whitish on the lower half, reddish- or yellowish-brown on the distal half, and thus easily distinguishable from the whitish radial spines. The radial spines in each areole are slender, straight, 0.7–1 cm long, appressed, and overlapping with the tips of spines from adjacent areoles. The radial spines usually are white, although in dried specimens under magnification sometimes they appear to be gray, stramineous, or pale tan.

Seedlings and juvenile plants have been observed under greenhouse conditions. Juveniles are unbranched, depressed-globose, and whitish in overall aspect and produce the hooked central spine at early age. Plants less than 0.5 cm tall and 1 cm in diameter already have a hooked central in essentially every areole, along with 18–20 white radial spines appressed to the stem. The central spines in juveniles are reddish-brown, lighter in color than is characteristic of adult plants, and with a slightly darker color toward the apex.

Biosystematics. The Texas populations of *M. grahamii*, all in desert habitats near the Rio Grande, represent the extreme eastern range of the species. Plants are common in certain habitats, particularly among fields of very dark volcanic rock in northwest Presidio County, where multistemmed, presumably older plants are common.

Synonyms. *Mammillaria microcarpa* Engelm.; *M. milleri* (Britton & Rose) Boed.

Common Names. Graham fishhook; fishhook cactus; fishhook nipple-cactus; pincushion cactus; sunset cactus; sunset fishhook cactus; lizard-catcher; tangled fishhook; fishhook mammillaria; fishhook pincushion.

2. **Mammillaria wrightii** Engelm. var. **wrightii**, Proc. Amer. Acad. Arts 3: 262. 1856. WRIGHT'S FISHHOOK CACTUS. Plates 238, 239. Rare in the Franklin Mts, El Paso Co. ca. 5,000 ft? Flowering after summer rains. 2n = 22. Grassy slopes, Great Plains grasslands, pinyon-juniper woodland, most of NM except the E plains and the extreme N part, into E-central AZ. Mexico, reportedly Chihuahua (Laguna de Santa Maria). Map 51.

Mammillaria wrightii barely enters Texas in the Franklin Mountains of western El Paso County. It was collected from the Franklin Mountains in 1909, but since then it has not been rediscovered (Worthington, 1995). Its major distribution is in New Mexico, where it is part of a complex of hooked-spined mammillarias. Another variety of this species, *M. wrightii* var. *wilcoxii* (Toumey ex K. Schum.) W. T. Marshall, occurs in New Mexico, Arizona, and Sonora and Chihuahua, Mexico. The specific epithet honors Charles Wright (1811–85), early plant collector in Texas and a plant collector with the early U.S.-Mexico border survey.

Identifying Characters. In habit *M. wrightii* usually is unbranched, with the tapering base of the stem deeply seated in the ground. During long dry periods and in winter, the whole stem may retract to near ground level. The stems are flat-topped or globose and 4–8 cm in diameter, with tubercles (0.7–)1.2(–2) cm long. The tubercle axils are naked. The pith and cortex are mucilaginous. Both *M. wrightii* and the superficially similar *M. grahamii* occur in the Franklin Mountains, if *M. wrightii* still is extant there. *Mammillaria wrightii* generally is easily distinguished by most of the characters listed above under *M. grahamii* and usually by 2–3 (less often 1–4 or more) hooked central spines. *Mammillaria wrightii* also differs vegetatively in having usually 8–15 radial spines, compared to 20 or more radials in *M. grahamii*.

The flowers of *M. wrightii* are rose-purple or magenta, typically 2.5–4 cm long and 2.5–7.5 cm in diameter. The outer tepals are prominently fringed. The anthers are bright yellow. The style is greenish, pinkish above, supporting 7–11 stigma lobes that are yellow or pale green, or rarely reddish.

The globose or ovoid fruits, previously described as "grapelike," usually are 1.3–2 cm long and 1.2–2.5 cm in diameter. The fruits are green or dull purple, very juicy, and differ conspicuously from the elongate, bright red mature fruits of *M. grahamii*. The seeds of *M. wrightii* are black, pitted, and 1.3–1.5 mm long.

Phenology. Specific information is lacking regarding flowering times for *M. wrightii* in the Franklin Mountains, the only known locality for this species in Texas. In general *M. wrightii* flowers in the summer following rains (Zimmerman and Zimmerman, 1977) and fruits during the autumn.

Sterile and Immature Specimens. The usually unbranched, globose stems of *M. wrightii* are 4–8 cm long aboveground when expanded during moist periods, but when desiccated the stems become flat-topped and shrink to ground level, in part the result of a prolonged underground base. The flaccid tubercles are tipped by circular areoles with short white hairs. There are 1–4(–7) slender central spines, typically 2–3, and usually all of them are hooked, except sometimes the upper one. The centrals are stiffly spreading, (0.6–)1.2–1.4(–2) cm long, and reddish-brown to blackish-and-white. The radial spines, 0.5–1 cm long, the lateral ones longest, are appressed along the plane of the stem surface. The radials are white or brown-tipped, or the upper radials are brownish throughout.

Very small juvenile plants of *M. wrightii* may have only one hooked central spine per areole and may be confused with *M. grahamii*, except that even in juveniles usually there are fewer radial spines in juveniles of *M. wrightii* than in *M. grahamii*. At least two hooked centrals per areole usually are present in juveniles of *M. wrightii* with a stem diameter of more than 1 cm.

Biosystematics. *Mammillaria wrightii* var. *wrightii* is not closely related to *M. grahamii,* the only other hooked-spined *Mammillaria* in Texas. An extensive review of the *M. wrightii* complex in New Mexico, Arizona, and northern Mexico was published by Zimmerman and Zimmerman (1977).

Synonyms. *Neomammillaria wrightii* (Engelm.) Britton & Rose; *Chilita wrightii* (Engelm.) Orcutt; *Ebnerella wrightii* (Engelm.) Buxb.

Common Names. Wright's fishhook; Wright fishhook.

3. **Mammillaria lasiacantha** Engelm., Proc. Amer. Acad. Arts 3: 261. 1856. GOLF BALL CACTUS. Plates 240–42. Desert mountains, hills, and alluvial flats, usually limestone but also igneous substrates. Often with lechuguilla scrub or other desertscrub. Throughout much of the Trans-Pecos except the central Davis Mts. 1,500–4,500(–5,500) ft. Flowering Feb–Mar(–Jun). $2n = 22$. Southeast and S-central NM. Mexico, NE portion S to San Luis Potosí, NW Sonora. Map 52.

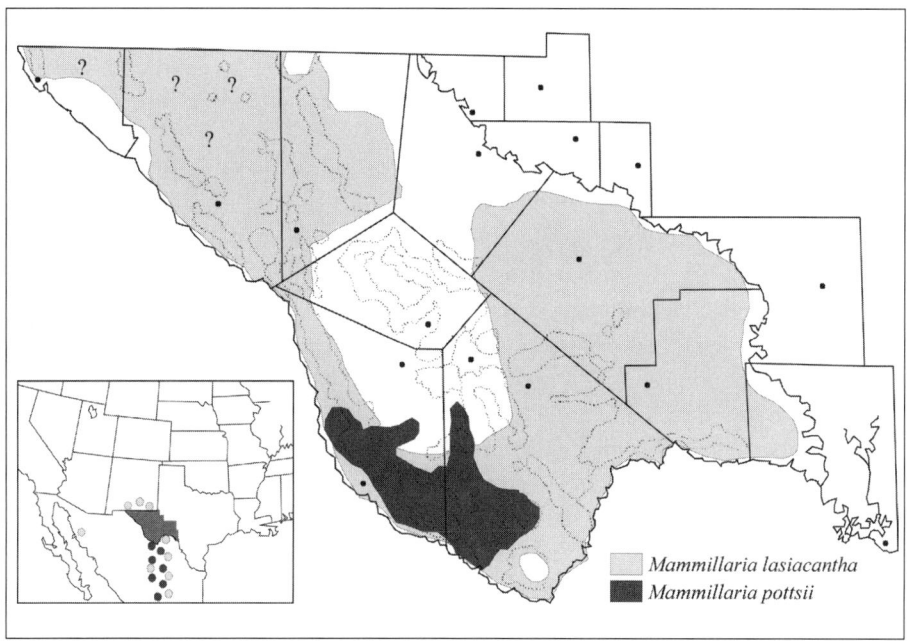

Map 52. Generalized distribution of *Mammillaria lasiacantha* (golf ball cactus) and *M. pottsii* (Potts' mammillaria).

Mammillaria lasiacantha is one of the most widespread small desert cacti in the Trans-Pecos. Usually its common name, golf ball cactus, is an accurate reference to size, but the plants are mature at the size of ping-pong balls or smaller, and plants the size of tennis balls are found occasionally, especially near Candelaria, Presidio County, and near the Old Ore Road, Big Bend National Park. In the Trans-Pecos *M. lasiacantha* is most common in rocky limestone hills and in alluvium derived from limestone rather than in igneous substrates. Some of the largest plants of this species occur in igneous-derived soils in northwest Presidio County, below the Sierra Vieja rim. The specific epithet is after the Greek *lasios*, "hairy," "woolly," or "shaggy," and *acantha*, "thorn" or "spine," a reference to the whitish spines that clothe the plants.

Identifying Characters. The unbranched stems of *M. lasiacantha* usually are depressed-globose, 1–3 cm tall, 1.5–3.5 cm in diameter, deeply seated in the substrate and inconspicuous, with the stem completely obscured by whitish spines. The larger plants found at some sites are subglobose to short-cylindroid, to

7–8(–13) cm long, and 6–7 cm in diameter. Usually there are 40–60 white or ashy-gray radial spines per areole, these 2.5–5(–6) mm long, thin, mostly appressed, and those of adjacent areoles often are overlapping. All of the spines are interpreted as radial spines, although 1–6 short spines, 0.6–2 mm long, resembling the true radials, occupy the central part of the areole. The spines are "innocuous" in that plants can be handled without spine pricks. The spines are glabrous or plumose, with the short, fine hairs easily detected on the spines with the aid of a hand lens. *Mammillaria lasiacantha* in habit resembles *Epithelantha*, particularly *E. micromeris*. Spine clusters in *M. lasiacantha* usually are 9–10 mm across, but in *Epithelantha* the spine clusters generally are 2–5(–7) mm in diameter. Both *M. lasiacantha* and *E. micromeris* exhibit a relatively coarse stem texture, whereas *E. bokei* has a satinlike surface texture.

The flowers of *M. lasiacantha* are white or cream-colored with conspicuous midstripes of pink, purplish, reddish-brown, salmon-red, or greenish-tan colors. The flowers are 0.9–1.5(–2) cm long, 0.8–1.0(–1.8) cm in diameter, and rotate when fully open. The inner tepals are 4.5–8 mm long, 1.5–2.7 mm wide, and obtuse at the apex. The outer tepal margins are either entire or irregular. The filaments are pale yellow, and the anthers are yellow. The style is greenish, slightly longer than the stamens, and supports usually 4–5 green or yellow-green stigma lobes, these 0.3–1 mm long.

The fruits of *M. lasiacantha* are bright red, cylindroid or clavate, 1–2(–2.5) cm long, and (3–)4–5 mm in diameter. The fruit surface is naked. A floral remnant is persistent at the fruit apex. The seeds are black, pitted, globose or somewhat comma-shaped, and 1–1.2 mm long. Stem bisections have revealed the presence of seed clusters embedded in the cortex below the stem apex, often 1.5–4 cm below the apex. Apparently the seeds are embedded by new growth before fruit abscission. This phenomenon is known to occur in other cacti, including *Ariocarpus fissuratus*. The elongated red fruits of *M. lasiacantha* are similar in appearance to those of *Epithelantha*.

Phenology. *Mammillaria lasiacantha* is one of the first cacti to flower each year in the Trans-Pecos. Earliest flowering is in February in the southern Big Bend area, particularly south of the Chisos Mountains in Big Bend National Park. The flowers open again for several days and close at night. Flowers usually do not open during cold or cloudy days. Fruit maturation requires several months, perhaps five or more, for ripening.

Sterile and Immature Specimens. The plants of *M. lasiacantha* are deep-seated with fibrous roots. Typically, rounded stems ca. 2–3 cm in diameter protrude ca. 1–2 cm above ground level. When desiccated, the stems may withdraw to ground level. A translucent latex is sporadically present in the outer part of the cortex. Tubercles are 3–6(–8) mm long, 2–3 mm in diameter, cylindroid, and obscured by the whitish spines. The axils between the tubercles are naked, or in young plants with some woolly hairs. Areoles are circular and ca. 1.5 mm in diameter or elliptic-oval and 1.5 mm long and 1 mm wide, with some short hairs disappearing with age. Areoles are 3–5 mm apart. To the naked eye the areoles are marked by a light brown spot, the collective result of brown spine bases, evident on the stem in a field of otherwise whitish spines. The usually white spines in some plants may be pale pink with tiny pinkish-brown tips. The radial spines vaguely appear to lie in several appressed series or layers, with spines often overlapping with those of adjacent areoles. Radial spine numbers are difficult to count precisely, but (26–)40–60(–90) spines per areole have been reported. The spines are slender, the longest ones in upper and lateral positions in the areole. Some plants have an apical ring of dirty-yellow or pale tan spines. Spine surfaces are glabrous or plumose.

Juvenile plants manifest fewer spines in each areole. Juveniles commonly have plumose spines, a condition that may or may not be retained in areoles of adult plants. *Mammillaria lasiacantha* can be distinguished from *Epithelantha* even in juveniles by microscopic examination of the spine clusters or with the naked eye after some experience in viewing the superficial appearance of stems.

Biosystematics. Weniger (1984) recognized two varieties of *M. lasiacantha* based mostly upon plumose versus glabrous spines: *M. lasiacantha* var. *lasiacantha* with microscopically plumose spines; *M. lasiacantha* var. *denudata* Engelm. with glabrous spines, or some spines with only a few hairs. Both Benson (1982) and Zimmerman et al. (forthcoming) concluded that in some populations of *M. lasiacantha* only one spine form may be evident, but in other widely separated populations both spine forms are present. In the Trans-Pecos, adult plants with plumose spines are common in Pecos County, including at the type locality near Fort Stockton and at several other known sites in Culberson, Brewster, and Val Verde counties. The two spine forms do not appear to be segregated in distinct populations, although Weniger (1984) also correlated plant and flower size with spine forms.

Synonyms. *Mammillaria lasiacantha* var. *denudata* Engelm.; *M. lasiacantha* var. *minor* Engelm.; *M. denudata* Engelm. ex Berger.

Common Names. Golf ball pincushion; lacespine cactus; lace-spine cactus; lacyspine cactus; fuzzy mammillaria.

4. **Mammillaria pottsii** Scheer ex Salm-Dyck, Cact. Hort. Dyck. [ed. 1844] 104. 1844. POTTS' MAMMILLARIA. Plates 243, 244. Mostly if not entirely limestone hills, slopes, mesas, flats, and desert habitats with lechuguilla and other desertscrub. South Presidio Co., N to near Shafter; SW Brewster Co., N to near Nine-Point Mesa. 2,800–4,000 ft. Flowering late Feb–Mar, rarely later. $2n = 22$. Northeast Mexico S to Durango and Zacatecas. Map 52.

Mammillaria pottsii is distinctive among other mammillarias of the Trans-Pecos. It is mostly restricted to the southern Big Bend region, in assorted sedimentary habitats such as those in the vicinity of Terlingua. The major distribution of *M. pottsii* is in Mexico, where its relatives are located. Presumably, the specific epithet honors F. H. Potts (1824–88), a mining engineer in the Sierra Madre, Mexico. Reportedly, Potts was a source of cactus plants for the Kew Royal Botanic Gardens in London and ultimately for the foremost German cactologist, Salm-Dyck. Instead, the epithet may be for John Potts (see *Opuntia pottsii*, p. 147).

Identifying Characters. Plants of *M. pottsii* usually are branched from the base, although single-stemmed, presumably young plants are not uncommon. The stems are narrowly cylindroid or clavate, 6–15(–20) cm long, (1.5–)2–3.5(–4) cm in diameter, and completely obscured by whitish radial spines. The central spines are pale reddish to gray or nearly white, with darker, usually dark gray, reddish, brown or chalky-blue distal portions. The relatively heavy central spines contrast sharply with the slender, numerous, whitish radial spines. The upper central spine, particularly in areoles near the stem apex, characteristically is curved upward and blue-gray on the distal half. In this feature young single-stemmed plants of *M. pottsii* may resemble *Echinomastus mariposensis* with which *M. pottsii* is sympatric in the vicinity of Terlingua. Otherwise, *M. pottsii* is similar to some plants of *Coryphantha tuberculosa* and *C. sneedii* var. *albicolumnaria,* both of which occur in the Terlingua area, but these species are readily distinguished by stem, spine, flower, and fruit characters.

The flowers of *M. pottsii* are maroon-red, deep red, or rusty-red, or reddish-purple. The small campanulate flowers are 0.9–1.3 cm long and 0.6–1.3 cm in diameter. Typically, a ring of flowers may encircle the stem one to several centimeters below the apex. The flowers usually do not open widely, but the inner tepals are reflexed near the tips. The inner tepals usually have darker midregions and paler margins, with acute apexes. The stamens are cream-colored to pale yellow. The style is reddish, with 4–5 narrow stigma lobes, these reddish-purple to orange-yellow.

The fruits of *M. pottsii* are bright red at maturity, clavate, and ca. 1.5 cm long. The seeds are dark reddish-brown or brownish-black, pitted, nearly oval, and ca. 1 mm long.

Phenology. The typical peak flowering time for *M. pottsii* appears to be in March. In some years flowering may begin in early February, with sometimes an occasional plant or two blooming in late January, and flowering bursts may extend into April. The small flowers of *M. pottsii* open about noon or before for at least three days, often 5–6 days, and close at night. Fruit development and ripening under natural conditions appear to require one to several months. In plants under cultivation in Alpine, fruits matured in 2–2.5 months.

Sterile and Immature Specimens. The stems of *M. pottsii* do not produce latex. The tubercles, usually obscured by spines, are closely aggregated, conical to ovate in shape, 3–5 mm long, blue-green, and tough in texture. Axils between the tubercles exhibit abundant and persistent white wool. In the flowering zone the longest axillary wool is longer than the tubercles. In *M. pottsii* throughout its range, central spine number reportedly varies from 6–12, but most of the plants in the Trans-Pecos have eight central spines. The centrals usually are packed together in an oval configuration, with their bulbous bases somewhat laterally compressed. The shaft of the centrals is terete, 0.2–0.3 mm thick, and thus relatively heavy compared to the radials. Most of the centrals are 5–8 mm long, diffuse, straight or weakly curved, tan, reddish, to nearly white basally, and distally usually dark brown, reddish, or bluish. A single upper central is 1–1.5 cm long and visually could be regarded as the "main" central. The upper central usually is up-curved, conspicuously so especially in the upper areoles, where typically the distal portion of each up-curved central is chalky-blue in color. The appressed radial spines are (27–)36–49 in number, 3–4.5(–6) mm long, slender (0.05–0.10 mm thick), and white or pale tan.

Juvenile specimens typically have fewer central and radial spines. In lower (older) areoles of juveniles there are perhaps 2–3 central spines, and the number of centrals increases to about five in areoles at midstem. Even in juveniles the centrals are bulbous-based and diffusely spreading. Juvenile areoles of *M. pottsii* support 28–33 radial spines, these mostly appressed, slender, and white.

Biosystematics. *Mammillaria pottsii* is a Mexican species whose distribution barely enters in the United States in the southwestern Big Bend region of Trans-Pecos Texas. There do not appear to be any systematic problems in Texas populations of the species, although more information should be obtained regarding its ecology, flower-color variation, and phenology.

Synonyms. *Mammillaria leona* Poselger; *Cactus pottsii* (Scheer ex Salm-Dyck) Kuntze; *Neomammillaria pottsii* (Scheer ex Salm-Dyck) Britton & Rose; *Chilita pottsii* (Scheer ex Salm-Dyck) Orcutt; *Coryphantha pottsii* (Scheer ex Salm-Dyck) A. Berger; *Leptocladodia leona* (Poselger) Buxb.

Common Names. Foxtail cactus; Potts' nipple cactus; rat-tail cactus; rattail cactus.

5. **Mammillaria meiacantha** Engelm., Proc. Amer. Acad. Arts 3: 263. 1856. NIPPLE CACTUS. Plates 245, 246. Mountains to desert, oak-juniper-pinyon woodland, grasslands, desertscrub. Widespread in the Trans-Pecos, most common in Culberson, Jeff Davis, Presidio, and Brewster counties, also Terrell Co., perhaps in adjacent counties as well. 3,000–7,300 ft. Flowering Mar–May. $2n = 22$. Central and S-central NM, probably to SE AZ. Mexico, mostly Coahuila, probably in E Chihuahua and N Zacatecas. Map 53.

These flat-topped or hemispheric, circular plants with prominent, rather closely spaced tubercles are abundant in the central mountains of the Trans-Pecos and in desert habitats of the southwestern Big Bend. Only a few other cacti in the Trans-Pecos have such broad ecological tolerances. The specific epithet denotes the Greek *meion*, "fewer," and *acantha*, "spine," in reference to the smaller number of spines per areole, compared to related taxa.

Identifying Characters. *Mammillaria meiacantha* is similar in habit and several other characters to the regionally sympatric *M. heyderi*. Both species are unbranched, deep-seated (with a thick, but basal stem), circular in outline, and flat-topped or hemispheric. The aboveground portion of the stem when turgid is 2.5–5 cm high. When desiccated, the stem may shrink to near ground level. Hemispheric stems (i.e., with rounded tops) are usually larger and presumably of older plants or those growing in shaded conditions. *Mammillaria meiacantha* is readily distinguished from *M. heyderi* by its 5–7(–9) relatively thick radial spines. *Mammillaria heyderi* usually has 13–17 radial spines.

The flowers of *M. meiacantha* are white to pale pink, often with pink or reddish-brown tepal midstripes. Typically, there is a ring of blooms outside the

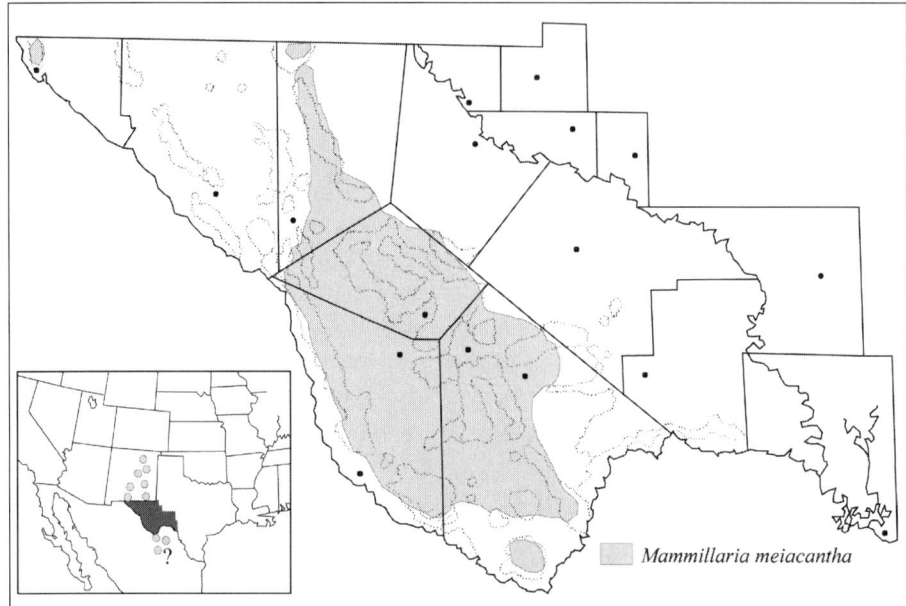

Map 53. Generalized distribution of *Mammillaria meiacantha* (nipple cactus).

newer growth in the center of the stem. The flowers are 2–3 cm long and 2–3 cm or more in diameter. When fully open, the tepals usually are reflexed. The inner tepals are 0.8–1.6 cm long. The stamen filaments, ca. 8 mm long, are pinkish to nearly white, and the anthers are cream-colored to yellowish. The style is 1.2–1.6 cm long, longer than the stamens, and supports 6–9 light green stigma lobes, these 3–5 mm long.

The fruits of *M. meiacantha* are broadly clavate, 2–3.2 cm long, and rose-pink to dull red at maturity. The seeds are reddish-brown, pitted, and 1.1–1.2 mm long.

Phenology. The peak flowering time for *M. meiacantha* appears to be controlled by prevailing temperatures of the spring season, which also involves altitude. Flowering occurs earlier in the desert, or on south slopes, than in the mesic mountains. The main flowering period is March through May. Individual flowers open about noon, last 3–4 days, and usually close at night. Some of the flowers may remain open, or partially open, during the night and may attract crepuscular pollinators, but this remains to be studied. In the central mountains of the Trans-Pecos it is not unusual to see plants in full flower during mid- to late May, and occasionally flowering persists into June. Fruit maturation requires several months, and ripe fruits may persist for a year or more.

Sterile and Immature Specimens. The circular, flat-topped stems of *M. meiacantha* are 7–16(–30) cm in diameter. In the Trans-Pecos, plants to 20 cm in diameter are uncommon, and plants to 30 cm across are rare. Latex is abundant throughout the stem cortex and in the tubercles. The latex is white and very sticky. The tubercles are subpyramidal, dark green or blue-green, 0.8–1.7 cm long, and 0.4–0.9 cm in diameter above the base. The tubercle axils seasonally support tufts of white wool with hairs 3–5 mm long. Areoles are circular, ca. 2 mm across, with short matted hairs when young but become naked with age. In each areole usually there is one central spine, as interpreted by most authors, but the "central spine" actually is positioned at the upper (toward the stem apex) periphery of the areole, almost in line with the radial spines. Technically, *M. meiacantha* has 0–1 central spine. The central spine, when present, is porrect or ascending, (0.3–)0.5–1.2 cm long, stout, tapering to a bulbous base, 0.5–0.7 mm thick above the base, terete or, rarely, slightly compressed. The central spine is red-brown to nearly black, sometimes glaucous near the apex. In each areole there are (4–)5–7(–9) stout radial spines, these more or less appressed or spreading at a low angle, the longest 0.9–1.3 cm. The lower (outer) three radials are the longest, and all are stout, except that the 2–3 upper (inner) ones are conspicuously smaller. The largest radials have blackish or dark brown tips but otherwise are reddish-brown, gray, or yellowish.

Juvenile plants have smaller tubercles and fewer, light-colored spines than adults. Very young plants of *M. meiacantha* are recognizable because of their habit and prominent tubercles.

Biosystematics. *Mammillaria meiacantha* appears to be closely related to *M. heyderi*. The two taxa are regionally sympatric, but they are not known to hybridize, and they are distinguished by both spine and fruit characters. Benson (1982) and Bravo-Hollis and Sánchez-Mejorada (1991b) recognized *M. meiacantha* as a variety of *M. heyderi*. Weniger (1984) and Zimmerman et al. (forthcoming) treated *M. meiacantha* as a separate species.

Synonyms. *Cactus meiacanthus* (Engelm.) Kuntze; *Neomammillaria runyonii* Britton & Rose; *N. meiacantha* (Engelm.) Britton & Rose; *Mammillaria runyonii* (Britton & Rose) Cory; *M. melanocentra* Poselger var. *meiacantha* (Engelm.) Craig; *M. melanocentra* Poselger var. *runyonii* (Britton & Rose) Craig; *M. gummifera* Engelm. var. *meiacantha* (Engelm.) L. D. Benson; *M. heyderi* Muehlenpf. var. *meiacantha* (Engelm.) L. D. Benson.

Common Names. Heyder nipple-cactus; little chilis or bisnaga de chilitos; small-spined cream pincushion.

6. **Mammillaria heyderi** Muehlenpf., Allg. Gartenz. 16: 20. 1848. HEYDER'S PINCUSHION CACTUS, WESTERN HEYDER PINCUSHION CACTUS. Plates 247–49. A relatively widespread species from S TX N to SW OK, W to S NM and SE AZ. 1,000–5,000 ft. Flowering Mar–Apr? $2n = 22$. In all states of Mexico along the TX border, W to Sonora, S to Durango, possibly San Luis Potosí and Zacatecas, possibly N Yucatán, if *Mammillaria gaumeri* (Britton & Rose) Orcutt is conspecific (Zimmerman et al., forthcoming). Map 54.

What might be termed the *M. heyderi* complex has been interpreted by previous workers as either a group of closely related species (Britton and Rose, 1919–23) or as a species with several varieties (Benson, 1982; Weniger, 1984; Bravo-Hollis and Sánchez-Mejorada, 1991b), in some cases (e.g., Benson, 1982) including *M. meiacantha*. Zimmerman et al. (forthcoming) regarded *M. heyderi*

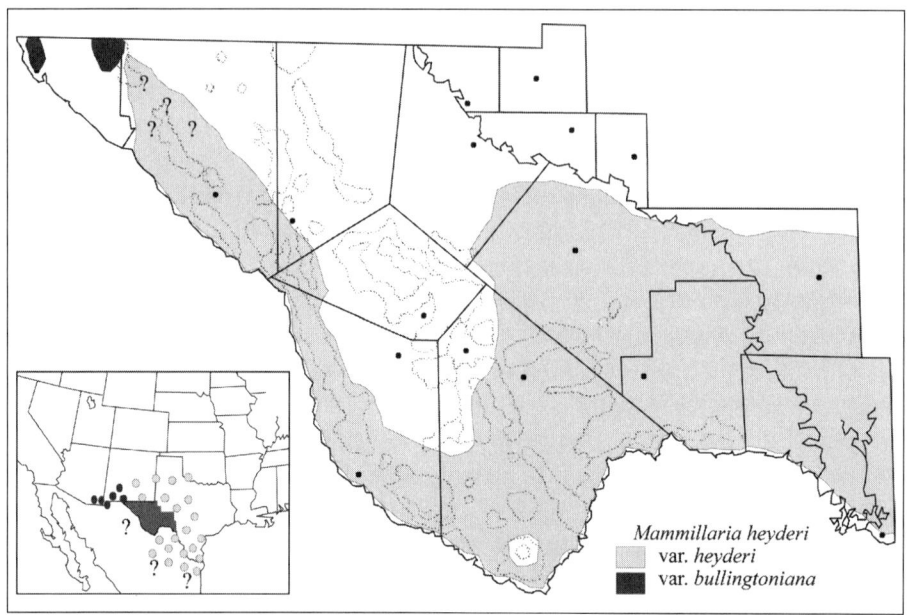

Map 54. Generalized distribution of *Mammillaria heyderi* var. *heyderi* (Heyder's pincushion cactus) and *M. heyderi* var. *bullingtoniana* (western Heyder pincushion cactus).

as a species of two varieties (possibly three), both of which occur in the Trans-Pecos. The specific epithet honors Edward Heyder (1808–84).

Key to the Varieties

1. Radial spines usually 13–17 or more, the longest ones usually 6–11 mm; central spines usually 0.35 mm in diameter or less; distribution widespread in the southeastern Trans-Pecos

   6a. *M. heyderi* var. *heyderi*.

1. Radial spines usually 10–14, the longest ones 9–15 mm; central spines usually 0.35–0.45 mm in diameter; distribution in El Paso Co.

   6b. *M. heyderi* var. *bullingtoniana*.

6a. Mammillaria heyderi var. heyderi. HEYDER'S PINCUSHION CACTUS. Plates 247, 248. Mostly in limestone substrates, desertscrub or semidesert habitats. Sporadic in occurrence from near Sierra Blanca to near Indian Hot Springs, Hudspeth Co., SE to where more common in SE Brewster, Pecos, Terrell, and Val Verde counties. 1,000–4,600 ft. Flowering Mar–Apr. $2n = 22$. South TX and S-central TX N to SE OK and SE NM, excluding the TX Panhandle where absent or rare; according to Zimmerman et al. (forthcoming), reports from SE AZ and SW NM are based on *M. heyderi* var. *bullingtoniana*. Mexico, northern Tamaulipas, Nuevo León, and probably Coahuila. Map 54.

Mammillaria heyderi var. *heyderi* is by far the more common of the two varieties of *M. heyderi* in the Trans-Pecos, particularly in the southeastern portion. The var. *heyderi* extends southeast along and near the Rio Grande to deep South Texas in Cameron County, where it occurs at near sea level, up the coast to near Corpus Christi, and inland northwest to near Austin and beyond to the Edwards Plateau.

Identifying Characters. The plants of var. *heyderi* are similar in appearance to those of *M. meiacantha* with unbranched, circular, flat-topped or hemispheric stems, with deep-seated, thick, but not much elongated bases. The flattened hemispheres are (4–)7–15 cm in diameter and extend 2–5 cm above the ground. Prominent tubercles, 9–15 mm long and 3–4 mm in diameter above the bases and arranged in spiral rows, are evident through the relatively thin spines. Usually there is one central spine and 13–17 or more radial spines. In *M. meiacantha* the spines are fewer and heavier in appearance, distinguishing features that in the field are evident at a glance.

The flowers of var. *heyderi* are white to cream-colored or slightly pinkish, usually with prominent pink or greenish-brown midregions. The flowers typically appear in a circle outside new growth at the flattened apex. The flowers are 2–3.8 cm in length and diameter. When fully open, the tepals are reflexed. The stamen filaments are whitish to pink, with yellow anthers. The 5–10 stigma lobes are light green, cream-colored, or pinkish-tan, and elevated slightly above the anthers.

The fruits are bright red when ripe, broadly clavate, and 1–3.5(–3.8) cm long. The bright red fruits usually are easily distinguishable from the pinkish to dull red fruits of *M. meiacantha*. The seeds are reddish-brown, pitted, and 1–1.2 mm long. Microscopically, the raised margins of the deeply concave testa (seed coat) cells are conspicuous.

Phenology. Flowering typically occurs in March and April. Individual flowers open about noon, last for 3–4 days, and close at night. Persistent fruits often are present along with flowers, having involved about a year in the maturation process. It appears that fruits do not always require a year to ripen, but this phenological aspect needs investigation.

Sterile and Immature Specimens. Presumably, older plants and shade forms often are hemispheric, but typically the single stems of var. *heyderi* are aptly described as flattened hemispheres. The dark green tubercles are weakly angled and firm, or with the ventral side sharply keeled. A white, very sticky latex is abundant in the stem cortex and tubercles. The axils between the tubercles have short wool that tends to fall away with age. Areoles are circular, ca. 2 mm in diameter, with white wool, this diminishing with age. Usually there is one central spine, (0.5–)2–7(–8) mm long, tapering to a point and rather stout. The central spine is centrally located in the areole, unlike in *M. meiacantha*, where the central spine is in a subradial position. In younger areoles the central position of the single central spine is accentuated by the presence of short white hairs in the areole. The porrect central spine is dark reddish-brown or yellow-brown. Rarely, two central spines are present in var. *heyderi*, but possibly not in Trans-Pecos populations. Across the range of var. *heyderi* there is considerable variation in number of radial spines, (7–)13–17(–26) or more. In the Trans-Pecos usually there are 13–17 radial spines, these slender and relatively weak, with the lower 3–5 longer and stouter (approaching the central in appearance) than the upper ones. The lower radials are 0.6–1.1(–1.6) cm long, in Trans-Pecos populations usually 0.6–0.9 cm long. The most appressed radial spines are whitish or brown to yellowish-brown, often with darker red-brown tips.

Juvenile plants are miniature versions of the adults but with smaller tubercles and fewer and smaller spines. In the field, juveniles are most frequently observed under grasses, shrubs, or other nurse plants, as is the case with most, if not all, Trans-Pecos cactus species.

Biosystematics. Weniger (1984) accepted three varieties of *M. heyderi* in Texas, with at least two of them, var. *heyderi* and var. *applanata* (Engelm.) Engelm., occurring in the Trans-Pecos. The third variety recognized by Weniger, var. *hemispherica* Engelm., had its major distribution in southern Texas and adjacent Mexico but also was interpreted by Weniger to occur in Oklahoma and New Mexico. These varieties and others have been regarded by various authors as belonging with the *M. heyderi* complex. Zimmerman et al. (forthcoming) treated var. *applanata* and var. *hemispherica* as synonyms of *M. heyderi* var. *heyderi* while recognizing two varieties for the Trans-Pecos, var. *heyderi* and var. *bullingtoniana*. Zimmerman et al. noted that plants with 9–13 radial spines, those segregated by Weniger as var. *hemispherica*, occur sporadically throughout much of the range of otherwise typical var. *heyderi*. Both Weniger and Zimmerman et al. considered *M. meiacantha* to be a species distinct from *M. heyderi*, contrary to the taxonomy favored by some other authors, including Benson (1982). We have adopted the taxonomic disposition of Zimmerman et al. largely because *M. heyderi* and all of its "varieties" are readily distinguishable from *M. meiacantha*. The two taxa exhibit consistent morphological differences, and they also appear to be ecologically distinct. Although *M. heyderi* and *M. meiacantha* are regionally sympatric, they do not appear to hybridize or even to occur together. Comprehensive field, morphological, or experimental analyses of the *M. heyderi* complex, however, are lacking.

Synonyms. *Mammillaria applanata* Engelm.; *M. heyderi* var. *applanata* (Engelm.) Engelm.; *M. gummifera* Engelm.; *M. gummifera* var. *applanata* (Engelm.) L. D. Benson; *M. hemispherica* Engelm.; *M. heyderi* var. *hemispherica* Engelm.; *M. heyderi* var. *gummifera* (Engelm.) L. D. Benson; *M. gummifera* var. *hemispherica* (Engelm.) L. D. Benson; *M. declivis* A. Dietr.; *M. texensis* Labour.; *M. buchheimeana* Quehl; ? *M. waltheri* Boed.; ? *M. hemispherica* var. *waltheri* (Boed.) Craig; *Cactus heyderi* (Muehlenpf.) Kuntze; *C. texensis* (Labour.) Kuntze; *C. gummifera* Kuntze; *C. gummiferus* (Engelm.) Kuntze; *C. heyderi* var. *hemisphericus* (Engelm.) J. M. Coult.; *C. hemisphericus* Small; *Neomammillaria heyderi* (Muehlenpf.) Britton & Rose; *N. applanata* (Engelm.) Britton & Rose; *N. gummifera* (Engelm.) Britton & Rose; *N. hemispherica* (Engelm.) Britton & Rose.

Common Names. Pancake cactus; Heyder mammillaria; nipple cactus; biznaga de chilitos, little chilis; pancake pincushion.

6b. Mammillaria heyderi var. **bullingtoniana** Castetter, P. Pierce, & K. H. Schwer., Cact. Succ. J. (U.S.) 48: 138. 1976. WESTERN HEYDER PINCUSHION CACTUS. Plate 249. Reportedly in El Paso Co. (Zimmerman et al., forthcoming), desert grassland, and desertscrub. 3,800–4,800 ft. Flowering Mar–Apr. $2n = 22$. Southwestern NM from Doña Ana Co. W to SE AZ. Probably Chihuahua, Mexico. Map 54.

Zimmerman et al. (forthcoming) recognized var. *bullingtoniana* as one of two varieties of *M. heyderi* that occur in the CDR. The distribution of this taxon in Arizona and New Mexico is reasonably well established, but its status in El Paso County remains to be determined. Worthington (1995) has collected plants identified as *M. heyderi* in the Franklin and Hueco mountains. Zimmerman et al. pointed out that erroneous previous reports of the Arizona species *M. macdougalii* Rose in southwest New Mexico stem from the misidentification of sterile or poorly prepared specimens of *M. heyderi* var. *bullingtoniana* (Martin and Hutchins, 1981).

Identifying Characters. The var. *bullingtoniana* is similar to var. *heyderi* (Zimmerman et al., forthcoming), except that the stem and tubercle measurements are about one-fourth larger in var. *bullingtoniana*. Also, in var. *bullingtoniana* there are 10–14 radial spines, fewer than is typical for var. *heyderi*, and the longest (lower) radials are 0.9–1.5 cm and heavier than the radials that are 0.6–1.1 cm long in var. *heyderi*. The central spine of var. *bullingtoniana* is slightly larger in diameter, usually 0.35–0.45 mm, than in var. *heyderi*.

Presumably, the flowers and fruits of var. *bullingtoniana* are like those of var. *heyderi*.

Phenology. Presumably, the flowering time for var. *bullingtoniana* in El Paso County is March to April.

Biosystematics. Systematic studies of var. *bullingtoniana* are lacking.

Synonyms. None known.

Common Names. Flat cream pincushion; cream cactus; the original "cream cactus" is *M. macdougalii* of Arizona (A. D. Zimmerman, pers. comm.).

7. Mammillaria prolifera (Mill.) Haw. var. **texana** (Poselger) Borg. HAIR-COVERED CACTUS. Plates 250, 251. [*Cactus prolifera* Mill., Gard. Dict. ed. 8, No. 6. 1768; *Mammillaria prolifera* (Mill.) Haw., Syn. Pl. Succ. 177. 1812; *M. multiceps* Salm-Dyck, Cact. Hort. Dyck. 1849. 81. 1850; *M. pusilla* DC. var. *texana* Engelm., Proc. Amer. Acad. Arts 3: 216. 1856; *M. prolifera* var. *multiceps* Borg, Cacti 316. 1937]. Base of limestone cliffs near the Rio Grande, Langtry, Val Verde Co. ca. 1,500 ft. Flowering in Mar. $2n = 22, 44, 66$? East to Edwards and Bandera counties, S to near the Rio Grande valley in Hidalgo and Brooks counties, in deep soil with shrubs and grasses, crests of rocky hills, and limestone ledges, E to Bexar Co. and Aransas Co. Adjacent Mexico in Coahuila, Nuevo León, and Tamaulipas. Also West Indies. Map 55.

In North America, *M. prolifera* essentially is a species of south and south-central Texas and adjacent Mexico. It extends west of the Pecos River only a few miles to its only known locality near Langtry in Val Verde County. Weniger

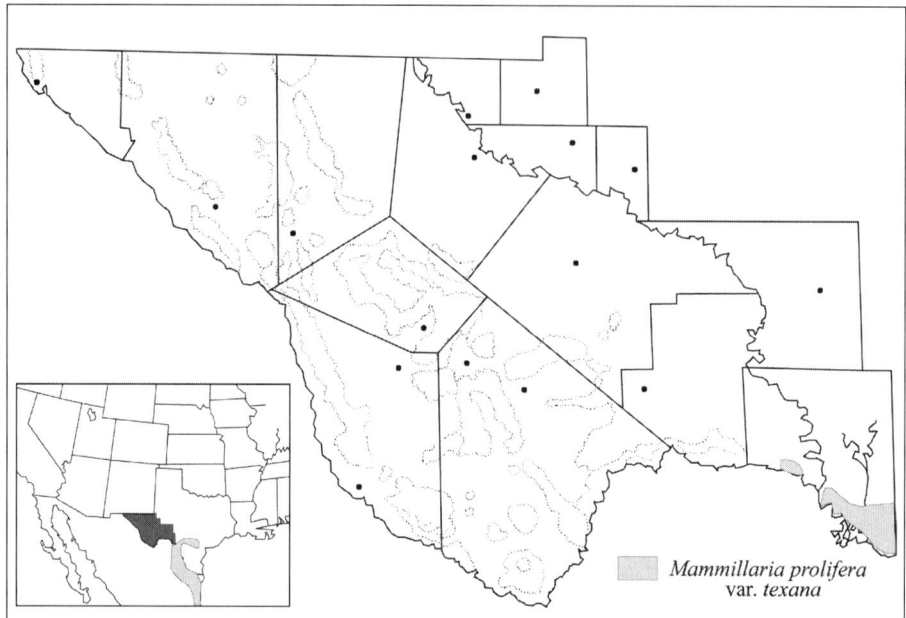

Map 55. Generalized distribution of *Mammillaria prolifera* var. *texana* (hair-covered cactus).

(1984) remarked that *M. prolifera* is susceptible to freezing, a possible limiting factor for the species in the Trans-Pecos, except that cultivated *M. prolifera* has survived (–12°C [10°F]) in a cactus garden in Alpine. The epithets "prolifera" and "multiceps" (from the Latin *ceps*, "head") presumably are in reference to the prolific, many-stemmed plants.

Identifying Characters. Mature plants of *M. prolifera* usually form dense, low mounds or mats, with the several to many stems representing a range of sizes. Clumps of 12–20 or more stems are not unusual, forming mats to near 30 cm in diameter. Flowering stems are up to 5 cm or more long, 1.2–5 cm in diameter, and spherical to ovate or short-columnar. Perhaps the most identifiable vegetative characters involve the spines. In each areole there are 8–11(–12) relatively slender, spreading, central spines with bulbous bases. There are numerous slender radial spines in several series, grading from flexible spines inside near the centrals to longer, undulating, hairlike spines at the outside of the areole. The central spines are translucent to pale yellow at the base and reddish-brown to nearly black on the distal half. The radial spines are white to pale yellow. The twisting, curved, hairlike radial spines overlapping and mostly obscuring the stem provide the most unusual and most distinctive identifiable feature for the "hair-covered cactus."

The flowers of *M. prolifera* usually are dirty-yellow. The color of the inner tepals also may be described as tannish-yellow or light yellow, with pale rose to rose-purple midregions. The cylindrical to funnelform flowers are 1.5–2 cm long and 1–1.5 cm in open diameter. The inner tepals may be reflexed apically. The stamens are 4.5–7 mm long with pale yellow or whitish filaments and yellow anthers. The style is ca. 1 cm long, cream-colored, and supports 4–8 yellowish stigma lobes, these 1.5–4 mm long. Flowers are subtended by some white wool and several hairlike bristles.

The mature fruits are bright red, juicy, clavate or narrowly obovoid, 1.3–2(–2.3) cm long, and 4.5–5 mm in diameter. The fruit surface is smooth, lacking scales, and the floral remnant is persistent. The seeds are black, pitted, asymmetrically obovate to nearly round, and 1–1.3 mm long.

Phenology. Not enough is known about the flowering and fruiting phenology of *M. prolifera* in the Trans-Pecos. In natural populations peak flowering appears to be in March, although reportedly sporadic flowering may occur through late spring or even into the summer. Outdoor cultivated plants in the Trans-Pecos do exhibit sporadic flowering into the early summer, with flowers and fruits often appearing together on the same stems. Plants cultivated in the greenhouse seemingly may produce flowers almost any time of the year, beginning in January. The time required for fruit maturation appears to be a couple of months or more.

Sterile and Immature Specimens. Single-stemmed plants are seldom observed because stems multiply by branching, in one or a few seasons forming prolific mats of stems of different sizes. Woolly hairs and hairlike bristles persist between the tubercles at sites where flowers and fruits were produced. The tubercles are cylindrical to slightly conical, 4.5–7 mm long. Areoles at the end of the tubercles, ca. 3 mm apart, are circular, ca. 1.5 mm in diameter, the younger ones with white wool. A tan circle in the center of each areole is seen under magnification to be the tan base of the central spines. The spreading central spines are slender by most comparisons, the longest ones 7–8(–9) mm. The peripheral centrals are slightly smaller in caliber and shorter than the innermost 2–3 in the areole. The peripheral centrals are more or less appressed, and at least one "main" central is porrect-ascending. The central spines usually are translucent to pale yellow on the basal half and light reddish-brown to nearly black on the distal half. The centrals are microscopically pubescent. The 30–60 white radial spines are arranged in several series, all slender and flexible, with the innermost ones straight and more or less intergrading with the outermost centrals. Outside, toward the stem, the radials are progressively more slender, longer, and more hairlike. The outermost radials are the longest ones, 5–10(–15) mm, but difficult to measure because they are twisted and curved, intermingled with crinkled, hairlike spines of adjacent areoles. Some of the shorter radial spines, those closest to the centrals, are microscopically pubescent. The longer, crinkled, hairlike radials are smooth.

The characteristic hairlike radial spines are present in the areoles of juvenile stems and provide one of the best means of identifying sterile specimens of any age. In juvenile plants there are fewer central and radial spines than in adult plants. In Texas, the plants of *M. prolifera* often are difficult to locate, even though they form mats of many stems, because the low plants frequently are obscured by surrounding vegetation. Plants of *M. prolifera* with their "little mounds of furry, egglike stems" (Benson, 1982, p. 877) are not likely to be confused with any other cactus species in the Trans-Pecos.

Biosystematics. Benson (1982) recognized two varieties of *M. prolifera*: the var. *texana* (Poselger) Borg. in Texas and adjacent Mexico; and var. *prolifera* outside North America, mainly in the West Indies. Weniger (1984) commented about the existence of two spine forms of this species in Texas and Mexico, one distinguished by central spines that are dark red-brown or black on the distal portions and one with centrals whitish or yellowish throughout. As previously distinguished, var. *texana* has dark central spines, and var. *prolifera* has light-colored centrals. So far as known, the Trans-Pecos population of *M. prolifera* produces dark central spines and should be included with var. *texana*. Numerous geographic variants have been described or otherwise observed from throughout much of the broad distribution of *M. prolifera*, some in addition to the synonymy cited below (Bravo-Hollis and Sánchez-Mejorada, 1991b; Pilbeam, 1981). The complete synonymy for *M. prolifera* is extensive. Only partial synonymy is included below. If truly conspecific with the Greater Antillean cacti, then the name *M. prolifera* is correctly applied to this species; at specific rank, as employed by Weniger (1984), *M. multiceps* is the correct name for our mainland taxon.

Synonyms. *Mammillaria multiceps* Salm-Dyck; *M. multiceps* Salm-Dyck var. *texana* Engelm. ex F. M. Knuth; *M. multiceps grisea* Meinsh.; *M. pusilla* (DC.) Sweet f. *texana* Schelle; *M. texana* Young; *M. prolifera* (Mill.) Haw. var. *texana* (Poselger) Borg.; *M. prolifera* (Mill.) Haw. f. *multiceps* Schelle; *M. prolifera* (Mill.) Haw. var. *multiceps* (Salm-Dyck) Schum. ex Borg; *M. prolifera* (Mill.) Haw. f. *texana* (Engelm.) Krainz.

Common Name. Grape cactus.

Appendix 3

Another Texas species of *Mammillaria* not discussed elsewhere in the Trans-Pecos treatment of the genus is summarized below.

Mammillaria sphaerica A. Dietr., Allg. Gartenz. 21: 94. 1853. PALE MAMMILLARIA. Plate 252. [*Cactus sphaericus* (A. Dietr.) Kuntze; *Dolichothele sphaerica* (A. Dietr.) Britton & Rose; *Mammillaria longimamma* DC. var. *sphaerica* (A. Dietr.) L. D. Benson]. Plants with numerous stems, forming mounds or clumps to ca. 5 cm high and ca. 30 cm in diameter. Roots fleshy, to 2.5 cm thick. Stems light green, spherical, or depressed-spherical, ca. 5 cm long, 2.5–6 cm in diameter; tubercles soft, turgid, spreading, mammiform-cylindroid, 1.2–2.5 cm long, usually tapering toward apex, when desiccated becoming much smaller, firm, and closely arranged. Areoles circular, 1.5 mm in diameter. Spines not obscuring the stem; central spine usually one (reported 1–4), 1–1.3 cm long, slightly thicker than radials; radial spines 12–15, slender, the longest (lateral and upper) ones 1–1.6 cm, smooth, brownish, yellow to gray distally. Flowers fragrant, lemon-yellow, 5–6.5 cm in diameter, funnelform; inner tepals yellow, oblanceolate, 2–3 cm long, 4.5–6 mm wide, acute or acuminate, the margins entire; outer tepals with greenish-brown midregions and yellow margins; filaments and anthers yellow, the stamens swirled around the yellow style; stigma lobes eight, yellow, 6–7.5 mm long. Fruit green, ovoid, ca. 1.3 cm long, persisting, ultimately turning reddish. Seeds black, pitted. Distribution in South Texas in counties located below a line between Laredo to Corpus Christi, where it is rather common under shade in the brush country, and into Tamaulipas and Nuevo León, Mexico. Benson (1982) treated *M. sphaerica* as a variety of *M. longimamma,* a species otherwise restricted to central Mexico and often segregated from *Mammillaria* as the genus *Dolichothele* (Bravo-Hollis and Sánchez-Mejorada, 1991b). According to Weniger (1984) plants of *M. sphaerica* are readily distinguished from other mammillarias by their light green, almost yellow-green, stems. The plants bloom in spring and summer. The specific epithet alludes to the characteristic spherical stems.

17. CORYPHANTHA (Engelm.) Lem.
Pincushion Cactus, Beehive Cactus, Cory Cactus

Plants of solitary or of a few branching stems, or forming mounds of 50–200 stems. Roots diffuse or with a short taproot. Stems cylindroid to subglobose, 1.5–10(–12) cm long, 1–7 cm in diameter. Ribs none. Tubercles prominent and separate. Tissues sometimes with spheroidal or lenticular druses; mucilaginous pith and cortex present in some species, merely watery or mealy in others. Areoles circular to elliptic. Spines 3–90 per areole, smooth, white, gray, yellow, brown, pinkish to reddish, or black; spines sometimes modified into glands (extrafloral nectaries), red, orange, yellow, green, or brown in color; central spines 1–10 or more per areole, straight or curved, less often twisted, 0.3–2.3 cm long, 0.2–1 mm wide at the base, needlelike or awl-shaped, somewhat elliptic in cross section, these in most species grading into the radials; radial spines usually similar to centrals in color (or lighter), shape, and size, 5–40 per areole, typically straight, needlelike, 0.3–2.3 cm long, 0.1–0.5 mm wide. Flowers produced on new growth at stem apex; flowers formed in special region at adaxial base of tubercle, away from the spine-bearing part of areole and connected on the tubercle by a felted groove; flowers 1–6.5 cm long, 1–9 cm in diameter; floral tube funnelform; inner tepals 0.4–4 cm long, 0.1–1.5 cm wide, often glossy, nearly white with pink or brownish midribs, pink, salmon, magenta, brownish, or yellow; filaments white to pinkish, anthers yellowish; styles and stigma lobes white, to pink, reddish, or green. Fruit fleshy, often juicy, thin-walled, green or red, cylindroid, clavate, ellipsoid, or suborbicular, 2–6 cm long, 0.2–2.5 cm in diameter, surface appendages absent or few and rudimentary, never spinose-tipped with naked axils; floral remnants persistent or caducous. Seeds comma-shaped, reniform, obovoid, or spheroidal, 1–3 mm in longest dimension, reddish, yellowish, tan, brown, or black, smooth and shining, or the coats dotted or reticulate; hilum appearing lateral, basal, or oblique. [*Mammillaria* subgenus *Coryphantha* Engelm.; *Lepidocoryphantha* Backeb.; *Escobaria* Britton & Rose; *Neobesseya* Britton & Rose; *Ecobesseya* Hester; *Cochiseia* W. Earle; *Cumarinia* Buxb.]. $x = 11$.

A genus of about 70 species distributed from western Canada through the western United States and into Mexico, south to Oaxaca, and in Cuba. A majority of the species occur in Mexico, with about 33 taxa in the United States. About 41 species occur in the Chihuahuan Desert Region. About 17 species of *Coryphantha* are reported for Texas, and 14 of these, many of them endemic, occur in the Trans-Pecos. The size of the genus depends in part upon whether closely related taxa, such as the subgenus *Escobaria* and *Cochiseia* are combined with or segregated from *Coryphantha* (Taylor, 1986; Zimmerman, 1985; Zimmerman et al., forthcoming). The genus name is derived from the Greek *coryph*, "summit," and *anthos*, "flower," in reference to the characteristic position of flowers on new growth at the stem apex.

It is widely recognized that *Coryphantha* and *Mammillaria* are closely related, and some workers (e.g., Weniger, 1984) prefer to combine them as a single genus. The most consistent distinction of *Coryphantha* from *Mammillaria* is the presence in *Coryphantha* of an areolar groove extending from the spine clusters down the tubercles to the tubercle axils. Another frequently cited difference between the two taxa is the production of flowers on new growth at the stem apex in *Coryphantha*, whereas in most *Mammillaria* species flowers are produced in a ring at least 1 cm below the stem apex. Intermediate flower positions are known in species of both genera. In one subgenus of *Coryphantha* (subgenus *Coryphantha*), areolar glands are present, but areolar glands are completely absent in subgenus *Escobaria* and the related genus *Mammillaria*. When present, the areolar glands occur within, or at either end of, the areolar grooves. Areolar glands in this genus are domelike structures ca. 1 mm in diameter (unlike the longer, peg-shaped glands of *Ferocactus* and *Hamatocactus*), and they may occur singly or in groups. Although the glands are ephemeral (Zimmerman, 1985), sometimes they leave detectable circular impressions in the areoles. *Coryphantha* is also related to the clade containing *Echinomastus*, but only superficially similar to *Neolloydia*.

Recognition of *Escobaria* as a distinct genus is favored by Europeans (Hunt, 1978; Taylor, 1978, 1979, 1986). In the present work we have treated *Escobaria* (species 1–9) as a subgenus of *Coryphantha*, for the reasons reviewed by Zimmerman (1985). *Escobaria*, described as a genus by Britton and Rose (1919–23), commemorates the work of brothers Rómulo and Numa Escobar, of Mexico City and Juárez.

Key to the Species

1. Stems 1–2 cm long, 0.7–1.7 cm in diameter; spines appressed, relatively short and thick, the largest ones less than 6.5 mm long, with abruptly pointed or obtuse tips; flowers magenta; endemic in certain novaculite exposures, Marathon Basin, Brewster Co.
   4. *C. minima*.
1. Stems usually considerably larger than 1–2 cm long and wide (individual stems sometimes this small); spines appressed or not, thin or thick, tapering gradually to a point; flowers yellow, magenta, or other colors; in various substrates including novaculite, but not restricted to the Marathon Basin (2).
2(1). Stems flat-topped or hemispheroidal, deep-seated, usually with a fleshy base below ground, when desiccated the stem top drawn to near ground level or below, 1.5–4(–4.7) cm in diameter; plants often in flat-topped or depressed-hemispheric clumps 5–20 cm wide; flowers magenta; in or near the Marathon Basin, Brewster Co.
   5. *C. hesteri*.
2. Stems typically cylindroid to short-cylindroid, with a fleshy base or not, not drawn below ground level when desiccated, usually 2 cm or more in diameter; flowers yellow, magenta, or other colors; Marathon Basin or elsewhere in the Trans-Pecos (3).

3(2). Central spines hooked (at least some of them)

                                          12b. *C. scheeri* var. *uncinata*.

3. Central spines straight or gently curved (4).

4(3). Radial spines 20–40 or more, often fewer than 20 in *C. vivipara* var. *vivipara* and *C. tuberculosa* var. *varicolor*, often poorly differentiated from outer centrals; spines overlapping and mostly to completely obscuring the stem (except in var. *vivipara* and var. *varicolor*; species of subgenus *Escobaria* (5).

4. Radial spines 9–15, rarely 15–20, except 16–27 in *C. echinus*; centrals (if present) usually clearly differentiated from radials, often a main, porrect, stout central; spines usually not obscuring the stems (except in some *C. echinus*); species of subgenus *Coryphantha* (11).

5(4). Flowers 2.1–6.5 cm in diameter (6).

5. Flowers 1.3–2 cm in diameter (7).

6(5). Flowers magenta; spine clusters persisting on tubercles near base of stems

                                          6. *C. vivipara*.

6. Flowers pink, sometimes pinkish-white or white; spine clusters abscising from tubercles near base of stems, leaving corncoblike bases of older stems

                                          9. *C. tuberculosa*.

7(5). Seeds black; stigmas green (8).

7. Seeds red-brown or orange-brown; stigmas white (except green in *C. pottsiana*) (10).

8(7). Plants with a thick taproot; floral remnant deciduous; spines snowy-white; usually in limestone crevices, S Brewster and Presidio counties, near the Rio Grande

                                          3. *C. duncanii*.

8. Plants with diffuse roots; floral remnant persistent; spines snowy-white to brown; commonly other substrates and habitats (9).

9(8). Flowers 1.3–1.9 cm in diameter, the inner tepals 0.8–1.2 cm long; spines relatively few, white to brown; commonly in gravel or loam, *Larrea* and *Prosopis* desert, or among igneous rocks in Jeff Davis and Brewster counties

                                          1. *C. dasyacantha*.

9. Flowers 0.6–1.6 cm in diameter, the inner tepals 4.5–9 mm long; spines relatively numerous, snowy-white; in the Trans-Pecos restricted to the upper Chisos Mts, igneous and limestone rocks

                                          2. *C. chaffeyi*.

10(7). Stigma white; seeds larger (see text); pith and cortex containing lenticular druses 0.5–1 mm in diameter, these evident to the naked eye (also present in *C. vivipara*); S Brewster and Presidio counties, rare in SW Pecos Co., Franklin Mts and Guadalupe Mts

                    7. *C. sneedii.*

10. Stigma green; seeds smaller (see text); pith and cortex containing only spheroidal druses 0.05–0.3 mm in diameter, these not evident to the naked eye; Val Verde Co.

                    8. *C. pottsiana.*

11(4). Plants cespitose, forming mats or mounds 20–100 cm in diameter; pith and cortex mucilaginous; areolar grooves extending only one-half, sometimes three-fourths, the distance from spine clusters toward tubercle axils; central spines, at least the lowermost one, (2.5–)3.5–7 cm long, usually striate or angular, gray to black, often flexible and slightly curved; flowers rose-pink or magenta

              10. *C. macromeris* var. *macromeris.*

11. Plants (in the Trans-Pecos) of solitary or few stems (except for *C. echinus* at certain sites including S of the Chisos Mts); pith and cortex non-mucilaginous; areolar grooves extending from spine clusters to tubercle axils; central spines various but usually less than 4 cm long; flowers yellow, rose-pink to magenta (12).

12(11). Flowers rose-pink to magenta, the inner tepals sometimes whitish at bases; resembling *C. macromeris,* but stems solitary or in small clumps (in Texas), central spines shorter, the porrect central terete, dark brown or black, and radial spines acicular (dorsiventrally flattened in *C. macromeris*); Brewster and Terrell counties, along and near the Rio Grande, and in adjacent Mexico

                    11. *C. ramillosa.*

12. Flowers yellow; not much resembling *C. macromeris;* wide distribution in the Trans-Pecos, or E Trans-Pecos (13).

13(12). Tubercles 1.5–3(–4) cm long; radial spines 6–16(–20); flowers golden, pale greenish to dull yellow

                    12. *C. scheeri.*

13. Tubercles 0.8–1.5 cm long; radial spines 16–27; flowers bright to golden-yellow (14).

14(13). Radial spines 16–27; plants mostly of solitary stems throughout most of the Trans-Pecos, multistemmed with clumps or mounds to 50–80 cm across in S Brewster and S Presidio counties

                    13. *C. echinus.*

14. Radial spines 8–15; plants usually multistemmed, in clumps 30–50 cm or more across; E Trans-Pecos, Pecos Co., mostly S-central Texas

                    14. *C. sulcata.*

1. **Coryphantha dasyacantha** (Engelm.) Orcutt. DESERT PINCUSHION CACTUS. Plates 253–56. [*Mammillaria dasyacantha* Engelm., Proc. Amer. Acad. Arts 3: 269. 1856]. *Larrea, Prosopis*, lechuguilla and other desertscrub, gravel slopes and loamy flats, or oak-juniper woodland among rocks, igneous and limestone substrates, usually rare but locally common at some sites. El Paso, Hudspeth, Jeff Davis, Presidio, Brewster, and Pecos counties, mostly desert habitats, except in the Davis Mts and eastern Glass Mts. 2,500–5,500 ft. Flowering Mar–May (Jun–Jul). $2n = 22$. To be expected in NE Chihuahua and N Coahuila, Mexico. Map 56.

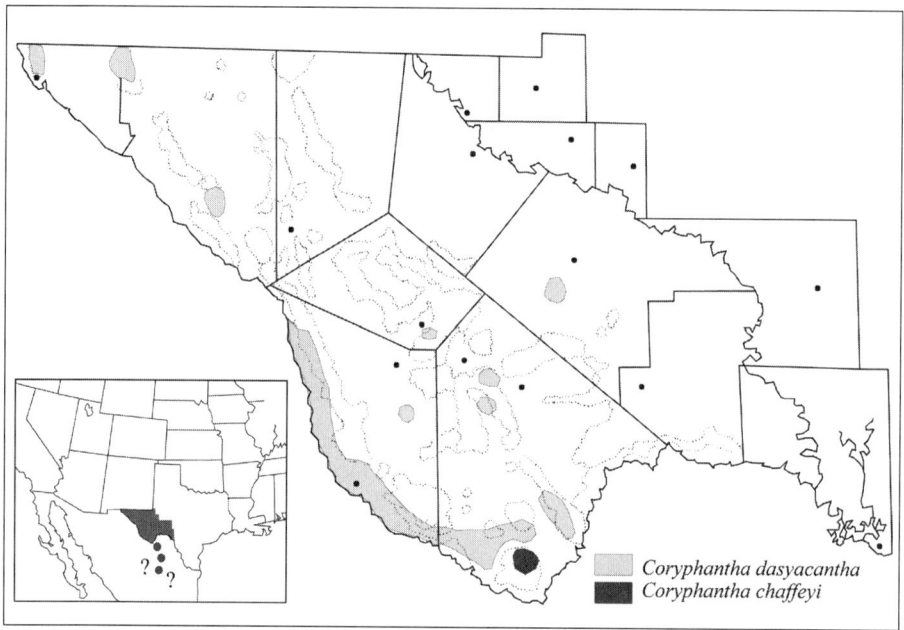

Map 56. Generalized distribution of *Coryphantha dasyacantha* (desert pincushion cactus) and *C. chaffeyi* (Chaffey's pincushion cactus).

Coryphantha dasyacantha is a Chihuahuan Desert endemic species essentially restricted to parts of Trans-Pecos Texas and probably adjacent Mexico. At least two morphotypes occur in the Trans-Pecos, one in the eastern Glass Mountains and southern Big Bend desert country, north along the Rio Grande to near El Paso, and the other in the Davis Mountains south of Fort Davis. There is one report of the species from the southern Davis Mountains about 20 miles south of Alpine and one record from about 30 miles south of Alpine; localities plotted by Benson (1982) are almost half based on misidentified specimens of other species.

Our field investigations suggest that *C. dasyacantha* usually is rare, contrary to the pronouncement by Weniger (1984, p. 195) that the taxon "is the commonest species of group often called Escobarias." Heil and Brack (1988) reported that *C. dasyacantha* is rare in Big Bend National Park. In our experience the taxon is rare throughout most of its range in the desert, although at a few sites near the Rio Grande in Presidio County, it is locally common. At the Jeff Davis County and northern Brewster County sites, the taxon is not uncommon, but it is cryptic.

Coryphantha dasyacantha is very closely related to *C. chaffeyi* and less so to *C. duncanii*; all share black, pitted seeds, red fruits, green stigma lobes, and other characters. Other Trans-Pecos coryphanthas, except *C. minima*, have brown or reddish-brown seeds. The specific epithet is after the Greek *dasy*, meaning "hairy" or "shaggy," and *akantha*, in reference to spines.

Identifying Characters. Plants of *C. dasyacantha* usually are single-stemmed, although older plants may exhibit several stems. Young stems are globular, but become elongate to cylindroid with age. Stems are 4.5–10(–17) cm long and (2–)3–4.5(–6) cm in diameter. Plants growing in the shade of small trees or shrubs in the desert may form the most elongated stems, 15–17 cm tall. In *C. dasyacantha* there are (3–)4–9(–11) central spines and 21–31 (sometimes more) radial spines. Usually there are 5–6 spreading centrals and several appressed subcentrals. The central spines are slightly larger in diameter than the radial spines.

The morphotype of *C. dasyacantha* in Jeff Davis County and northern Brewster County also usually is single-stemmed, but clusters of 2–5 stems are occasional. In the field the Davis Mountain plants are depressed-globose or globose, usually 1–4 cm tall, and 1.5–3(–5.5) cm in diameter. The central spines usually are darker (often reddish-brown) at least on distal portions, although the radials are white, and the Davis Mountain plants of *C. dasyacantha* are not as white in aspect as the desert plants. Under greenhouse cultivation the Davis Mountains morphotype has formed stems to ca. 10 cm long and has developed the white cloak of spines, closely resembling in habit the desert morphotype.

The flowers of *C. dasyacantha* and closely related species are among the smallest in subgenus *Escobaria*, which is noted for its smallish flowers, compared to subgenus *Coryphantha*. The flowers of *C. dasyacantha* are 1.5–2.5 cm long and 0.9–1.5 cm in diameter. The inner tepals are (0.7–)0.8–1(–1.2) cm long and 1.5–3 mm wide, with the apexes acute-attenuate and mucronate. The outer tepals are conspicuously fringed. The flowers of the desert morphotype are slightly larger than those of the Davis Mountains morphotype, and there are differences in color. Flowers of the desert form basically are white or cream-colored. Close examination shows that the inner tepals have white or cream-colored margins and sharply defined, usually brown, midregions. The midregion color may be pinkish or brownish-green in flowers of some plants. In size, shape, and color the flowers of the desert morphotype of *C. dasyacantha* are essentially indistinguishable from those of *C. duncanii*. The stamens show yellow anthers, ca. 0.5 mm long, and white or colorless filaments 5–8 mm long. The style is greenish, 1–1.3 cm long, with 4–6 green stigma lobes, these 1–2.5 mm long. Flowers of the Davis Mountains form are smaller, and the inner tepals are light-pink with salmon-pink (darker) midregions.

The fruits of *C. dasyacantha* are bright red, usually clavate, sometimes cylindroid to narrowly ellipsoid, (0.7–)1.3–2.7(–3.5) cm long, and (3.5–)4–6(–7) mm in diameter. The longest fruits produced by *C. dasyacantha*, measured at 3.5 cm, were observed on plants from near Shafter in Presidio County. The floral remnant is persistent. The seeds are black, pitted, subspherical, and 1–1.2 mm in longest diameter. The red fruits of the Davis Mountains morphotype are 1–1.7 cm long and broadly elliptic, much like those of *C. chaffeyi*.

Phenology. The flowering period for *C. dasyacantha* usually is March to April in the desert morphotype but may extend into May and June or even July in some years, presumably in response to moist, cool conditions. The Davis Mountains morphotype typically produces flowers in April, but earlier or later in some years. Flowers close at night but open again for 2–3 days. Mature fruits develop after about 1.5–2.5 months, usually in June–August, but sometimes as early as May.

Sterile and Immature Specimens. In the southern part of its range, *C. dasyacantha* is regionally sympatric with *C. tuberculosa* and is most likely to be confused with either var. *tuberculosa* or var. *varicolor* of this species. *Coryphantha dasyacantha* also sometimes is regionally sympatric with *C. sneedii* and *C. duncanii*, which are similar in some characters. *Coryphantha dasyacantha* has stems larger than those of *C. duncanii* and lacks the tuberous taproots of *C. duncanii*. *Coryphantha dasyacantha* is easily distinguished from any of these species, except *C. duncanii*, when in flower. In *C. dasyacantha* the pith and cortex are not mucilaginous. Tubercles, each with a felty groove, are (4–)7–8(–12) mm long and 3–5(–7) mm in diameter. Areolar glands are not present in the grooves, and the grooves appear to lose trichomes with age. The areoles are circular, 2–2.5 mm in diameter, with some white wool in younger areoles.

In *C. dasyacantha* the central spines are slightly larger in diameter than the radial spines, which are appressed against the plane of the stem. The centrals have larger bulbous bases than the radials, and the inner ones are angled diffusely away from the stem. Usually there are 5–6 inner diffuse centrals, 0.2–0.3(–0.4) mm thick, grading outward into subcentrals and transitional with the slender, white, appressed radial spines. The longest centrals are 1.2–1.7 cm. In some populations there is one centrally located, main, porrect central spine in older areoles, surrounded by 8–9 peripheral centrals that are diffuse to somewhat appressed. Even the radials are transitional in size, slightly larger away from the stem and more slender toward the stem. In desert forms of *C. dasyacantha*, the central and radial spines are white, although the centrals and sometimes the radials have red-brown to nearly black tips. The radial spines are 6–9(–10) mm long. In the Davis Mountains morphotype the central spines are more heavily pigmented. Most of the 4–6 main centrals are brown to reddish-brown or nearly black at the tips, or for more than half their length, or to the base. In older areoles the centrals tend to be white. Plants in the Davis Mountains are smaller, usually 1–3 cm high and 2–5 cm in diameter, than those in the desert.

Juvenile plants of *C. dasyacantha* usually are spherical and have fewer central (sometimes two) and radial (often 12–22) spines. In seedlings the central spines may appear to be absent. The spines are white and pubescent in seedlings and juveniles.

Biosystematics. *Coryphantha dasyacantha* is the most widespread of three closely related black-seeded species in the Trans-Pecos. All three species, including *C. chaffeyi* and *C. duncanii*, seem to be weakly separated ecologically and by certain vegetative and floral characters. The flowers of *C. dasyacantha* and *C. duncanii* are almost indistinguishable. Desert and Davis Mountains (Plates 255, 256) morphotypes of *C. dasyacantha* are discussed above. The two forms differ mainly in ecology, plant size, central spine color, and flower color. Further studies are needed in order to clarify the relationship between the desert and mountain forms. The ecological and morphological differences suggest that the Davis Mountains population might be recognized as a distinct taxon.

Another morphotype of *C. dasyacantha* is known from one collection at 4,125 feet in a canyon east of the desertic Dead Horse Mountains in southern Brewster County. The single stem examined from this collection was smaller than most desert *C. dasyacantha*, ca. 1.5 cm long and 3 cm in diameter, and had shorter main central spines, 4–6 mm long, and smaller, pinkish-salmon flowers. The flowers of this form were ca. 9 mm long with inner tepals 6–7 mm long, even smaller than the flowers of *C. chaffeyi*. In fact, the form approaches *C. chaffeyi* in some spine and flower characters. Zimmerman et al. (forthcoming) noted that *C. dasyacantha* and *C. chaffeyi* might be conspecific. Taylor (1986) recognized *C. chaffeyi* as a var. of *C. dasyacantha*.

The superficial vegetative similarities of *C. dasyacantha* and *C. tuberculosa* were highlighted through the taxonomic disposition of these taxa by Benson (1982). Benson recognized two varieties of *C. dasyacantha*, the typical variety and the var. *varicolor* (Tiegel) L. D. Benson. His variety *varicolor* is clearly *C. tuberculosa* (Zimmerman, 1985) and is included in the present treatment as *C. tuberculosa* var. *varicolor* (Tiegel) A. D. Zimmerman (forthcoming). *Coryphantha dasyacantha* and *C. tuberculosa* are easily distinguished by floral characters.

Weniger (1984) remarked that *C. dasyacantha* is common on mountain slopes, ledges, and summits and on exposed rock surfaces. Weniger did not recognize *C. chaffeyi* to occur in Texas, and it is possible that he was including this taxon, which is common at some sites in oak-juniper-pinyon woodland in the Chisos Mountains, in his description of the occurrence of *C. dasyacantha*. In our field experience, *C. dasyacantha* is rare throughout most of its range, and it usually is found on gravel slopes and in silty flats associated with creosotebush, mesquite, and other desertscrub.

In Benson (1982) the color plate 176 labeled "*Coryphantha dasyacantha* var. *dasyacantha*" is *C. sneedii* var. *albicolumnaria*. Taylor (1986) noted that Benson's plate 176 is mislabeled, but identified it as *Escobaria orcuttii*. Taylor also pointed out correctly that figure 889 in Benson is *Coryphantha vivipara* (as *Escobaria vivipara*) and not *C. dasyacantha* var. *dasyacantha*.

Synonyms. *Cactus dasyacanthus* (Engelm.) Kuntze; *Escobaria dasyacantha* (Engelm.) Britton & Rose; *Escobesseya dasyacantha* (Engelm.) Hester.

Common Names. Dense mammillaria; Big Bend eggs; desert pincushion.

2. **Coryphantha chaffeyi** (Britton & Rose) Fosberg. CHAFFEY'S PINCUSHION CACTUS. Plates 257, 258. [*Escobaria chaffeyi* Britton & Rose, Cact. 4: 56. 1923]. Rock crevices, sometimes with *Selaginella*, open slopes, among rocks in grassy areas, oak-juniper-pinyon woodland, igneous rock and limestone sub-

strates, upper Chisos Mts, Brewster Co. 4,700–7,300 ft. Flowering Mar–May(–Jun). $2n = 22$. Coahuila (Sierra de la Madera; Sierra de la Paila), N Zacatecas, N San Luis Potosí, NE Durango. Map 56.

Coryphantha chaffeyi primarily is a Mexican species, known to the United States only from the Chisos Mountains of Big Bend National Park. The species was first recognized as part of the United States flora by Zimmerman et al. (forthcoming) but first reported by Heil and Brack (1988). Benson (1982) listed *C. chaffeyi* as a species from outside of the United States (Zacatecas, Mexico), and Weniger (1984) made no mention of the taxon. The specific epithet honors Dr. Elswood Chaffey, who collected the species near Cedros, Zacatecas, Mexico, in June 1910 (Britton and Rose, 1919–23).

Identifying Characters. The stems of *C. chaffeyi* are usually unbranched, although plants with clusters of 2–3 stems are not uncommon in the Chisos Mountains. Stems are globular to short cylindroid, 3–7(–11) cm long, 3–4.5(–6) cm in diameter, and densely covered by white spines. In habit *C. chaffeyi* closely resembles *C. dasyacantha*.

The flowers of *C. chaffeyi* are ca. 1.2 cm long and 0.7–0.9(–1.1) cm in diameter, smaller than those of *C. dasyacantha*, and provide one of the most reliable means for distinguishing the taxa. The inner tepals are pinkish to cream-white with prominent, broad, pinkish-salmon to light brown midregions and narrow margins. The tepals are acute and mucronate. The anthers are yellow and 0.4–0.5 mm long. The filaments are 4–6 mm long and pink. The style is 7–8 mm long, greenish, and supports green stigma lobes that are only ca. 1 mm long.

The brilliant red ripe fruits are oval, broadly ellipsoid, or broadly clavate, 0.8–1.1 cm long, and 3–5(–6) mm in diameter. The seeds are black to dark brown, pitted, 1–1.2 mm long, and somewhat comma-shaped.

Phenology. Flowers of *C. chaffeyi* usually appear in early to mid-April but may open as early as mid- to late March. Flowering may persist into May or even June, at least in some years. Individual flowers open about noon, close at night, and open again for 2–4(–6) days. The bright red fruits develop in 1–3 months, at least in cultivated plants, and are present on plants in June, July, and August. A few ripe fruits have been observed on a few plants in mid-September near Boot Spring in the Chisos Mountains.

Sterile and Immature Specimens. In the Chisos Mountains at upper-mid to higher elevations, cacti with relatively short, globose to short-cylindroid stems densely covered by white spines should be *C. chaffeyi*, although *C. tuberculosa* var. *varicolor* does occur rarely in the upper Chisos. The pith and cortex are not mucilaginous. The tubercles are 3–6 mm long. There are no areolar glands. Areoles are circular and 2–2.5 mm in diameter. In each areole there are 30–50 or more spines, usually more than in *C. dasyacantha*. Also, the spines of *C. chaffeyi* noticeably are more slender than in *C. dasyacantha*. In *C. chaffeyi* usually there are 8–10 central spines, these diffuse and (0.4–)0.7–1.5 cm long. The slender white centrals have bulbous bases. The centrals may be yellowish or reddish at the tips. There are 22–40 or more slender, appressed, radial spines. The radials are 5–9 mm long, white to gray, and sometimes yellowish at the tips.

Seedlings and juveniles of *C. chaffeyi* have fewer and shorter spines per areole, centrals and radials, than do young adults or sexually mature plants. The spines of juveniles are white and conspicuously pubescent. Cultivated seedlings of C. chaffeyi are more white in aspect than cultivated seedlings of C. duncanii, a collective result of white, pubescent spines in C. chaffeyi.

Biosystematics. Taylor (1986) recognized C. chaffeyi and C. duncanii as varieties of *Escobaria dasyacantha*. All three taxa are closely related, but we have elected to follow Zimmerman et al. (forthcoming), at present, in treating them as distinct species. *Coryphantha chaffeyi* differs from C. dasyacantha by spines that are more slender, more numerous per areole, and always snowy-white except at the tips. The flowers of C. chaffeyi are smaller, and average smaller in all their parts, than those of C. dasyacantha. The fruits of C. chaffeyi, at least those observed in cultivated specimens, are broadly ellipsoid or broadly clavate and shorter, compared to the cylindroid-clavate and longer fruits of the desert forms of C. dasyacantha. In addition, C. chaffeyi appears to be confined to higher elevations, in the Chisos Mountains and also in Mexico, whereas both C. dasyacantha (except for the Davis Mountains plants) and C. duncanii are desert entities.

The Dead Horse Mountains population (at present, known from one collection), discussed above as a form of C. dasyacantha, is similar in certain traits to C. chaffeyi. The small flowers of this form are similar in size and color to those of C. chaffeyi and smaller than the flowers of any other C. dasyacantha, particularly the desert morphotype. In addition, the short central spines (4–6 mm long) of the Dead Horse Mountains form are similar in size to those in some plants of C. chaffeyi. The Dead Horse Mountains form was collected in a deep canyon east of the mountains, growing at an elevation of 4,125 feet. This site in stable limestone is decidedly different ecologically from the habitats of C. chaffeyi in the Chisos Mountains, although C. chaffeyi does occur in limestone in at least one site west of Laguna Meadow (ca. 6,800 feet) in the Chisos Mountains. Tentatively, we have interpreted the Dead Horse Mountains collection as a small-flowered form of C. dasyacantha, but its closest relationship could be with C. chaffeyi.

Consideration has been given to the possibility that C. chaffeyi and the Davis Mountains form of C. dasyacantha are related. Both populations occur in igneous mountain systems of similar age about 70–115 miles apart. In 1995 Jack Brady conducted successful reciprocal artificial hybridization between plants from the two populations. Numerous juveniles resulting from the cross were 1.5–4.5 cm tall in 1997. All of the hybrid plants are characterized by red-brown central spines, a spine color that is typical of the Davis Mountains population of C. dasyacantha. The flowers and fruits in natural populations of the Davis Mountains morphotype are similar to those of C. chaffeyi, even more so (in smaller size) in the population 30 miles south of Alpine, but the spine characters of the morphotype are like those of C. dasyacantha. Further study may indicate that the Davis Mountains population should be described as a distinct taxon or merged with C. chaffeyi.

Synonym. *Escobaria dasyacantha* (Engelm.) Britton & Rose var. *chaffeyi* (Britton & Rose) N. P. Taylor.

Common Name. Bisquit cactus.

3. **Coryphantha duncanii** (Hester) L. D. Benson. DUNCAN'S PINCUSHION CACTUS. Plates 259, 260. [*Escobesseya duncanii* Hester, Des. Pl. Life 13: 192. 1941]. Locally common to rare, desert limestone hills. Near the Rio Grande (within 25 mi), Big Bend National park NW to SE Presidio Co. 2,000–3,700 ft. Flowering Feb–Mar(–May). $2n = 22$. Also Sierra Co., NM (rare, 4,500–5,000 ft); mistakenly reported by Benson (1982) from Doña Ana Co., NM. Not yet documented from Mexico (specimens from Coahuila are intermediate with a related species, not typical *C. duncanii*). Map 57.

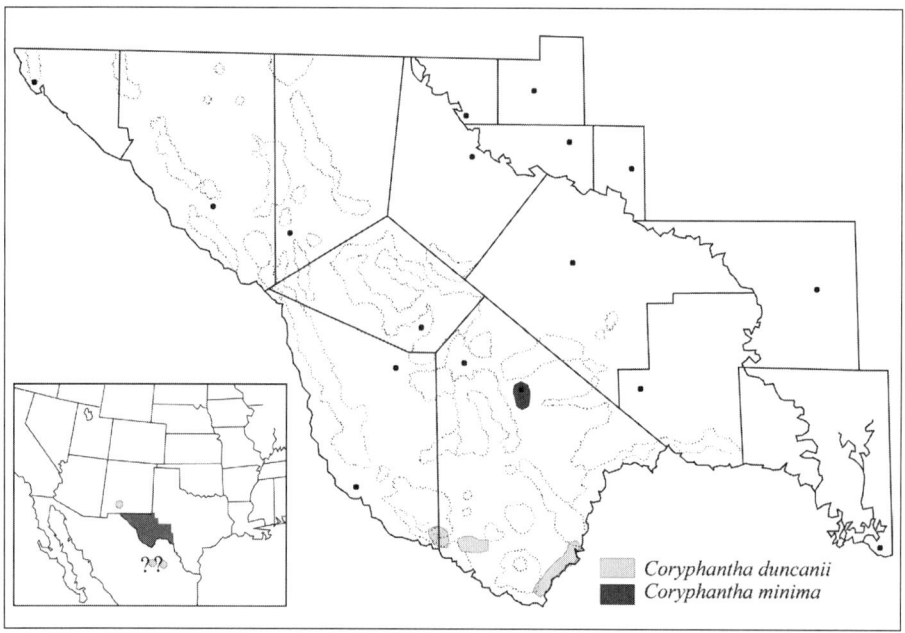

Map 57. Generalized distribution of *Coryphantha duncanii* (Duncan's pincushion cactus) and *C. minima* (Nellie's pincushion cactus).

Coryphantha duncanii is a Chihuahuan Desert endemic species found primarily in crevices of layered limestone in the hottest parts of the Big Bend desert region. The small plants, densely covered with white spines, are difficult to notice against a background of white rocks. The species is named after Capt. Frank Duncan, a miner who owned a claim near the type locality.

Identifying Characters. *Coryphantha duncanii* closely resembles the desert morphotype of *C. dasyacantha,* especially in flower morphology, but the stems of *C. duncanii* are much smaller. The stems of *C. duncanii* usually are unbranched, although plants with 2–3 stems are not uncommon. Plants with 5–8 stems are rare. Plants with smaller, immature branches are seldom found. The plants are deep-seated in crevices, or less often in soil, the root system being dominated by a succulent taproot, sometimes carrotlike, and up to 20–30 cm long. The usually inconspicuous stems are ovoid, spherical, to somewhat conical, 1.5–3.5(–6) cm tall, 1.1–3(–3.4) cm in diameter. The spines are snowy-white, or some of the larger spines, particularly near the apex, are tan to reddish-brown. In *C. duncanii* the inner central spines, those most easily interpreted as central, are absent, or rarely there is one obvious central. There are about 23–40 radial spines. Most likely to be confused with *C. duncanii* in the Trans-Pecos are smaller or immature plants of *C. tuberculosa* var. *tuberculosa*, particularly the more white-spined individuals of var. *tuberculosa*, which characteristically occur in rock crevices in habitats that appear suited for *C. duncanii*. *Coryphantha duncanii* is most easily distinguished from *C. dasyacantha* by its smaller stems, fleshy taproot, and 0(–1) central spine. Within the known range of *C. duncanii*, *C. dasyacantha* is most likely to be found in alluvial substrates, not the rock crevices favored by *C. duncanii*.

The flowers of *C. duncanii* are similar to those of *C. dasyacantha* (larger than those of *C. chaffeyi*). They are 1.5–3 cm long, 1.3–1.9 cm wide, with inner tepals 0.8–1.2 cm long and 1.5–3 mm wide. The inner tepal margins are white to cream-colored, rarely pinkish, and usually there are prominent pinkish, purplish, brown, or brownish-green midstripes. The outer tepals are conspicuously fringed. The anthers are yellow, and the filaments whitish or pinkish. The style is greenish, and the usually four stigma lobes are green, sometimes described as "yellow."

The bright red ripe fruits are cylindroid, ellipsoid, or clavate, 1.1–2 cm long, and not juicy. In *C. duncanii* the floral remnant is caducous (cleanly abscising, leaving a shallow umbilicus), a diagnostic trait for distinguishing the species and its two or more Mexican relatives from *C. dasyacantha* and all other similar species. The seeds are black, pitted, globose, and ca. 1.2 mm in diameter.

Phenology. *Coryphantha duncanii* is one of the early-blooming Trans-Pecos cacti, with flowers appearing in mid- to late February and in March. In some years flowers may form as late as early May. Under cultivation in Alpine, single plants produced flowers on 22 August and 15 September 1998 following rain a week or more earlier. The flowers do not open on cool or cloudy days. Flowers close at night and may open again for 2–3 days. Red fruits usually appear in May.

Sterile and Immature Specimens. The deep-seated, inconspicuous stems of *C. duncanii* usually are densely covered by the white spines. The pith and cortex are not mucilaginous. The tubercles are 3–6 mm long. The areoles are circular, ca. 1.5 mm in diameter, with the light brown bases of the spines forming what is to the naked eye a small brown circle. These brown areoles are also present in *C. dasyacantha* and *C. chaffeyi*. Areolar glands are not present in the areolar grooves.

During the microscopic examination of specimens of *C. duncanii* for the present study, only a single inner central was found in one subapical areole. Commonly, there are (1–)3–9 or more spines, 1–1.4(–1.9) cm long, which we have interpreted as outer centrals. These are slightly heavier than the slender radials and have small-bulbous bases that are laterally compressed and light brown at the base. The radials are slender, white, rarely brown-tipped, and to ca. 9 mm long. In general aspect the stems are bristly.

Seedlings of *C. duncanii* have fewer and shorter spines, often 13–15 radial spines, and usually no central spines. The spines are visibly pubescent in seedlings but become glabrous in adults. In the field immature plants of *C. duncanii* are difficult to distinguish from juvenile stems of *C. tuberculosa*.

Biosystematics. Taylor (1986) is the only author to have treated *C. duncanii* as a variety of *C. dasyacantha* (under the generic name *Escobaria*). Among Trans-Pecos cacti, *Coryphantha duncanii* most closely resembles "a stunted, crevice-limited growth form" (A. D. Zimmerman, pers. comm.) of *C. dasyacantha* or an anomalous lowland occurrence of *C. chaffeyi*. *Coryphantha duncanii* is distinguished mostly by its habitat, smaller stems, tuberous taproot, lack of inner central spines, and the caducous floral remnant that characterizes the fruit in a whole group of Mexican taxa. The flowers and seeds of *C. duncanii* and *C. dasyacantha* are similar, but *C. duncanii* is more likely conspecific with *C. zilziana* Boed. of Mexico. The desert-dwelling *C. duncanii*, like its Mexican relative *C. zilziana*, is segregated ecologically from the mountain-dwelling *C. chaffeyi*.

Synonyms. *Escobaria duncanii* (Hester) Backeb.; "*Mammillaria duncanii* (Hester)" Weniger, *nom. nud.*; *E. dasyacantha* (Engelm.) Britton & Rose var. *duncanii* (Hester) N. P. Taylor.

Common Names. Duncan's pincushion; Duncan's cactus; Duncan snowball.

4. **Coryphantha minima** Baird, Amer. Bot. 37: 150. 1931. NELLIE'S PINCUSHION CACTUS. Plates 261, 262. Restricted to outcrops on Caballos Novaculite and related chert-rich soils, associated with *Selaginella*, on rocky, grassy hills in the central Marathon Basin, Brewster Co. 3,700–4,300 ft. Flowering (Mar–)Apr–May(–Jun). $2n = 22$. Endemic to the Trans-Pecos. Map 57.

Coryphantha minima is one of the most exceptional cactus species in the Trans-Pecos. It is remarkable for its marble-size stems, unusual peglike spines, and substrate specificity. *Coryphantha minima* was federally listed by the U.S. Fish and Wildlife Service as Endangered in 1979 and listed by the Texas Parks and Wildlife Department as Endangered in 1983. The specific epithet is derived from the Latin *minimus*, "smallest," in reference to the small stems.

Identifying Characters. In the wild (but not in horticulture) the stems of *C. minima* usually are unbranched, rarely with 2–3 stems, these spherical to cylindroid, 1–2(–2.7) cm tall, and 0.7–1.7(–2.5) cm in diameter. The spines are ca. 15–28 per areole, all appressed against the stem and innocuous. The most identifiable aspect of the spines is that they are relatively short and thick and cylindroid or weakly clavate. Most cactus spines are needlelike and taper to a point, but the "peglike" spines of *C. minima* are acuminate or obtuse. In *C. minima* there are no projecting central spines.

The magenta flowers of *C. minima* when fully open are easily as large as or larger than most of its stems. The flowers are 1.3–1.6 cm long and 1.5–2.7 cm wide. The inner tepals are 0.7–1.2 cm long, 2.5–4 mm wide, and acute-obtuse. These inner tepals characteristically are magenta, but sometimes they are rose-pink to light pink, perhaps even white basally, with weakly defined midstripes, or distally with darker midlines. The outer tepals are greenish-brown to pinkish-brown with weakly defined midstripes and prominently fringed margins. The filaments are greenish, and the anthers are bright yellow or orange-yellow. The style is greenish and rather short, with 4–8 green stigma lobes, these 0.5–1.5 mm long.

The fruits are spherical, ovoid, or obconic, 1.5–6 mm long, 1.5–4 mm in diameter, and they either are green at maturity or may develop a yellowish tinge. The floral remnant is persistent, but fragile and easily lost by breakage, but not cleanly falling from an abscission zone. Old fruits are persistent among the spines, quickly drying and turning nearly white with thin, fragile pericarpels that ultimately disintegrate without ever dispersing. The black, pitted seeds are visible through the thin, dry pericarpels before fruit disintegration. The seeds are obovoid to weakly pyriform and 0.8–1 mm long.

Phenology. The usual flowering time for *C. minima* seems to be from about mid-April to mid-May. On 19 April 1997 only a few plants were in flower in one population that was available for examination that year. Seemingly peak flowering has been observed by us in different years to be in early to mid-May. In some years, flowers may open as early as late March or as late as June. Under cultivation in Alpine, a few plants have flowered as late as 20 July. The fruits appear to develop within about one month or less.

Sterile and Immature Specimens. *Coryphantha minima* should not be confused with any other cactus species in or near its range, because of its unique peg-like spines. There are several tuberculate cactus species of several genera in the Marathon Basin, but only a few have adult stems that are small enough to compare with those of *C. minima*. For example, the tiny stems of *Echinocereus davisii* and the much more common *C. hesteri* are immediately distinguishable from those of *C. minima* after properly identified plants of *C. minima* have been seen; otherwise, novices may try to make every *C. hesteri* plant into the scarcer species. The adults of these and seedlings of several other tuberculate species are comparable in small size to *C. minima*, but there is no particular history of misidentifications.

The pith and cortex of *C. minima* are not mucilaginous. The dark green tubercles are 2–5 mm long, 2–4 mm in diameter, and conical to ovoid. The areoles supporting spine clusters are spherical and ca. 2 mm in diameter. Areolar glands are absent. Of the 15–27 spines per areole, most are radial spines. There are no inner central spines, and in aspect it appears as if there are no central spines at all. In the upper parts of the spine clusters, however, (1–)3(–4) of the larger spines, 4–6 mm long, are interpreted as outer centrals, these being appressed in plane with the pectinate radials. There are 13–24 radial spines, the upper and lateral ones to 3.5–5 mm long. All of the spines, at least the larger ones, are glabrous, cylindroid, 0.3–0.5 mm thick, abruptly pointed, and laterally compressed at the base. The spines are pale tan, to pinkish-gray, weathering to gray, and sometimes with darker tips on some of the larger spines. In most populations of *C. minima*, the predominant spine color seems to be tan in upper areoles and gray in older areoles.

The seedlings of *C. minima* are miniatures of the tiny adults, with shorter and fewer spines per areole. In very young seedlings the spines are pubescent. The peglike spine morphology is evident even in the smallest seedlings and serves to distinguish young plants of *C. minima* from other cactus species.

Biosystematics. *Coryphantha minima* is now included in the subgenus or segregate genus *Escobaria* (Taylor, 1986) by most workers. Interspecific relationships with *C. minima* are not clear; *C. hesteri* and *C. vivipara* have similar flowers, but *C. minima* differs sharply by its black seeds and green stigma lobes.

Synonyms. *Escobaria minima* (Baird) D. R. Hunt; *Coryphantha nellieae* Croizat; *Mammillaria nellieae* (Croizat) Croizat; *E. nellieae* (Croizat) Backeb. The synonymous epithet *nellieae* and source of the common name were after Mrs. Nellie Davis of Marathon, Texas.

Common Names. Dwarf cory cactus; least cory cactus; Nellie's cory cactus; Nellie cory cactus; Nellie corycactus; birdfoot cactus. The "Nellie" part of names associated with this species was in use for a long time and is well established, even though it is based upon the synonym *Mammillaria nellieae* and perhaps should be maintained in use.

5. **Coryphantha hesteri** Y. Wright, Cact. Succ. J. (U.S.) 4: 274. f. 273. 1932. HESTER'S PINCUSHION CACTUS. Plates 263–65. Crevices of rocks or rocky soil, grassland on Caballos Novaculite or limestone substrate, oak-juniper-pinyon woodland and grassland in igneous substrate. Mostly Brewster Co., Marathon Basin; localized in and near Del Norte Mts, S and SE of Alpine.; Pecos Co., extreme S portion; Terrell Co., W of Sanderson. 3,600–5,300 ft. Flowering Apr–June(–Nov). 2n = 22. Endemic to the Trans-Pecos. Map 58.

Coryphantha hesteri was once thought to be a rare species restricted to nova-

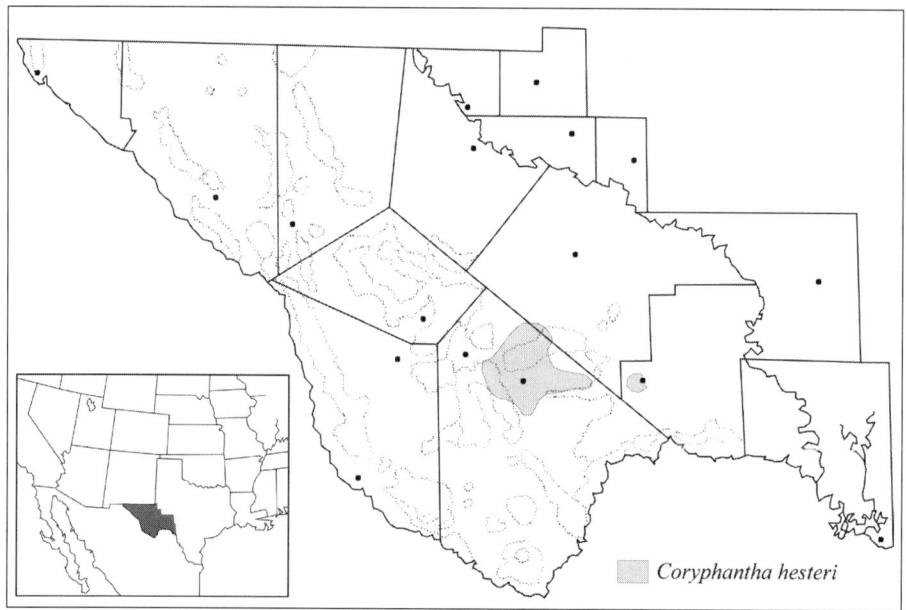

Map 58. Generalized distribution of *Coryphantha hesteri* (Hester's pincushion cactus).

culite substrates in the Marathon Basin of northern Brewster County and was listed as a class 3C species by the U.S. Fish and Wildlife Service. A status survey by Heil and Anderson (1982) found it to be abundant and widespread in the Marathon Basin in both novaculite and limestone habitats. More recent studies, both our own and those of Raun (1996), have documented a wider distribution in novaculite, limestone, and igneous substrates and a greater population density than was previously known. *Coryphantha hesteri* is considered to be one of the several special diminutive cacti of the Trans-Pecos, even though its stems in some individuals may be larger than those of the other tiny cacti; and the stems frequently occur in clumps. The type locality of *C. hesteri* is near Mount Ord ca. 10 miles southeast of Alpine, Brewster County (Wright, 1932). The species is named after J. P. Hester, who collected the species in 1930 and made it available to Ysabel Wright.

Identifying Characters. The stems are solitary or the plants cespitose, with few- or many-stemmed clumps being common. Clumps 0.5–2 cm across are not uncommon, and large stem clusters to 30 cm or more across are present in the Marathon Basin. The stems are deep-seated, usually with a fleshy, enlarged taproot, and individually and collectively are flat-topped or depressed-hemispheric, or the stems may be hemispheroidal, spherical, to egg-shaped, 1–6.5(–9) cm long aboveground, and 1.5–4.7(–5.5) cm in diameter. During dry periods, when stems are desiccated, they tend to withdraw below the ground or into crevices with little more than spine clusters being visible. The plants often are further obscured by grasses and other plants. *Coryphantha hesteri* is one of the most cryptic cactus species in the Trans-Pecos, especially in grassland habitats, where many stems are 1–1.5 cm aboveground and 1.5–3 cm in diameter.

The glabrous, straight spines of *C. hesteri* number (12–)13–18(–23) per areole, with no manifest central spines. The spines are white with red or brown tips, weathering to gray, or the centrals, when present, are reddish to brown-gray for one-half or more of their length. The centrals, when present, are 0.9–1.3(–1.5) cm long, and 0.2–0.3 mm thick. The longest radial spines, upper and upper lateral in position, are 0.7–1.3 cm long and (0.1–)0.2–0.3 mm thick.

The magenta to rose-pink flowers are 1.8–2.5 cm long and 1.3–2(–3.4) cm in diameter. The inner tepals are 1–1.7 cm long, 2–4 mm wide, with apexes acute or obtuse to acuminate. The outer tepals are densely fringed. The stamens are 3–5 mm long with yellow anthers ca. 0.5 mm long and colorless to pinkish filaments. The style is 6–10 mm long, greenish to greenish-yellow, often reddish above. The (4–)5–6(–7) stigma lobes, 1–3 mm long, are white to cream or pale pinkish.

The fruits are green, sometimes turning greenish-red, globose to ovoid, (3.5–)5–8(–10) mm long, 3–6 mm in diameter, and quickly drying (Zimmerman, 1972). The floral remnant usually is persistent on the fruit. The seeds are dark brown, pitted, globose, and 0.9–1.1(–1.2) mm in diameter.

Phenology. The primary flowering period for *C. hesteri* appears to be April–June, but flowering may continue periodically until the first freeze in November. Since 1991 the first author (pers. obs.) has been studying the flowering phenology of *C. hesteri* on a regular basis in a population about nine miles southeast of Alpine. After the initial spring flowering period in April and May, *C. hesteri* is opportunistic, with some plants or many plants in the population producing flowers about eight days after a significant rain. This flowering response about eight days after rain and continuing for several days has also been observed by G. G. Raun (pers. comm.) in a population of *C. hesteri* west of Marathon. Individual flowers open about noon, close in the late afternoon, and usually do not open again. Flowers may not open on cool or cloudy days. The flowers are visited by bees of 2–3 sizes. One of the types of bees has been identified by entomologist J. V. Richerson as a plasterer bee (Colletidae). The identified bee before capture visited, in succession, flowers on several plants of *C. hesteri*, and the bee was "loaded" with pollen identified microscopically as that of *C. hesteri*.

Sterile and Immature Specimens. Plants of *C. hesteri* with hydrated (swollen) stems are much more easily located in the field than are desiccated plants with stems at near ground or rock level. In hydrated plants particularly, the deep green or gray-green stems are clearly visible through the spines. The pith and cortex usually are not mucilaginous. Relatively large lenticular druses, the largest 0.3–0.4(–0.5) mm in diameter, about half or less than half the size of those in *C. vivipara*, are present in the cortex. The tubercles are 5–9(–12) mm long and 4–6(–7) mm in diameter. The areoles are circular, ca. 2 mm in diameter, with white trichomes in younger areoles. Areolar glands are absent.

It often appears as if there are no central spines in the spine clusters of *C. hesteri*. Close examination may reveal 1–3(–4) upper spines that are larger in diameter and longer than the radials. These larger upper spines, with their bases slightly displaced inward, are interpreted as central spines. They are appressed in the same plane as the true radials. During the current study plants were examined from throughout the known distribution of *C. hesteri*. It was determined that many more specimens did not produce central spines than those that exhibited them, although where populational samples were available, it was found that plants in the same population may have 0–1 or 1–3 centrals. Even when there appeared to be no central spines macroscopically, in some cases microscopic examination showed that one spine in the normal position of centrals was slightly larger and might be "interpreted" as a central. The usually acicular central spines have tan bulbous bases, are white, pale yellow, or colorless for the proximal half, and are reddish or brown on the distal half. The acicular radial spines have small, tan, laterally compressed, bulbous bases. The pectinate (appressed) radials are whitish except for red tips, or sometimes reddish distal halves.

Coryphantha hesteri is not likely to be confused with adult specimens of any of the several species of tuberculate cacti (several genera) found in and near the Marathon Basin. Seedlings of *C. tuberculosa*, *Echinocereus viridiflorus*, and perhaps other species may resemble *C. hesteri*, but all of these taxa can be distinguished by close examination of spine characters.

Seedlings of *C. hesteri* usually lack central spines, and the radial spines are shorter and fewer in number. The spines of seedlings are glabrous.

Biosystematics. *Coryphantha hesteri* belongs with the group of species recognized herein as subgenus *Escobaria* (genus *Escobaria*; Taylor, 1986). Interspecific relationships of *C. hesteri* are not clear, but it shares some characters with *C. vivipara*, including conspicuous lenticular druses of the pith and cortex.

The known distribution of *C. hesteri* in the Marathon Basin extends in all directions from the town of Marathon, occupying several types of sedimentary substrates (Raun, 1996). The species extends from Marathon as far south as 17 miles and as far east as 15 miles. The extent of the range north and west of Marathon is not as well known, but it is suspected that *C. hesteri* occurs in the Glass Mountains to the north and west, and in the Del Norte Mountains to the west. The species occurs in limestone west of Sanderson (Terrell County). No significant differences have been detected in plants from the different substrates.

Synonyms. *Escobaria hesteri* (Y. Wright) Buxb.; *Mammillaria hesteri* (Y. Wright) Weniger, *nom. nud.*

Common Names. Hester's dwarf cactus; Hester cory cactus.

6. **Coryphantha vivipara** (Nutt.) Britton & Rose. BEEHIVE CACTUS. Plates 266–71. [*Cactus viviparus* Nutt., in Fraser's Cac. No. 22 (Cat. Pl. Upper Louisiana). 1813]. Throughout most of the Trans-Pecos, mostly excluding central and southern Presidio and Brewster counties. 3,000–5,000 ft. Flowering late spring, early summer. $2n = 22, 44$. Central TX to Canada, W to SE CA, Great

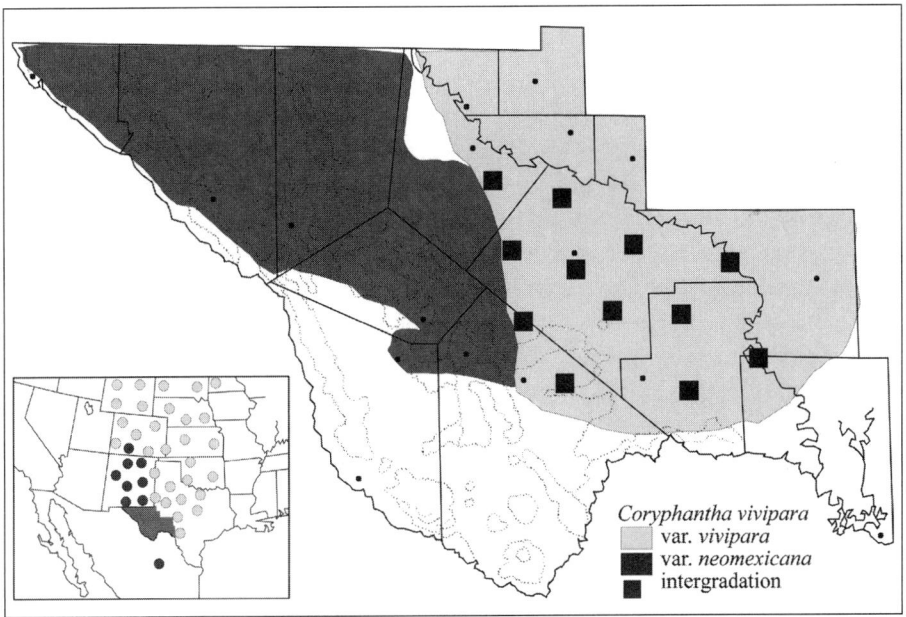

Map 59. Generalized distribution of *Coryphantha vivipara* var. *vivipara* (eastern beehive cactus), *C. vivipara* var. *neomexicana* (New Mexico beehive cactus), and and intergrades *C. vivipara* var. *radiosa*.

Plains and S Rocky Mts. Coahuila, Chihuahua, and Sonora, Mexico, on diverse substrates in habitats from desertscrub to conifer forest. Map 59.

Coryphantha vivipara is the most widespread and abundant species of the genus, extending from Texas through the Great Plains to Canada, and west through the southern Rocky Mountains and desert mountains to southeast California. The plants are locally common and have been frequently collected. Seven varieties are recognized throughout its broad range, according to Zimmerman (1985), the treatment that is followed here. Three varieties of *C. vivipara* occur in Texas: var. *vivipara*, var. *neomexicana,* and var. *radiosa*. Two varieties are evident in the Trans-Pecos, as well as intermediates, with the third variety (var. *radiosa*) only east of the Trans-Pecos. Benson (1982) recognized seven varieties of *C. vivipara*, two in Texas and one of them in the Trans-Pecos. Weniger (1984) treated six varieties of *C. vivipara,* four in Texas and two of them in the Trans-Pecos. The taxonomic interpretations of Zimmerman differ considerably from those of Benson and Weniger. The specific epithet includes the Latin *vivus*, "alive," and probably *pareo*, "to bring forth," presumably from the branching stems of the type collection.

Key to the Varieties

1. Spines 13–29 per areole, not obscuring the tubercles; central spines lustrous, reddish-brown or orange; stigmatic surfaces pale pink to bright magenta

    6a. *C. vivipara* var. *vivipara*.

1. Spines 24–55 per areole, obscuring the tubercles; central spines white, or if brown then opaque and drab; stigmatic surfaces usually white (but underlying parts variable, white to reddish)

    6b. *C. vivipara* var. *neomexicana*.

6a. **Coryphantha vivipara** var. **vivipara**. EASTERN BEEHIVE CACTUS. Plates 266, 267. Various alluvial substrates, including gypsum, often with grasses, mesquite, and prickly pear. Mostly eastern Trans-Pecos, one collection in N Culberson Co., others in NE Brewster Co., Pecos Co., and Ward Co. just E of the Pecos River. 3,000–4,500 ft. Flowering May–Jun. $2n = 22, 44$. From northern Edwards Plateau in TX, N through the Great Plains into S Canada, W to NM, CO, WY, and MT. Map 59.

Coryphantha vivipara var. *vivipara* is the most widespread of the seven varieties of the species. It is common in north-central Texas but not often collected in the Trans-Pecos. It is to be expected in the eastern Trans-Pecos, where it may intergrade with *C. vivipara* var. *radiosa* (Plates 268, 269), a variety with its major distribution on the Texas Edwards Plateau.

Identifying Characters. The globose to ovoid stems of var. *vivipara* are unbranched or sometimes profusely branched. The stems are 2.5–10(–20) cm tall and 3–10.5 cm in diameter. The spines are about 13–29 per areole, straight, and glabrous. Usually there are four central spines. The four centrals include (0–)1 inner central and 3(–7) outer centrals. The usually four centrals are often in a "bird's-foot" arrangement, one descending-porrect, and three ascending, the longest 2–2.3 cm. The central spines usually are bright reddish-brown, at least on the distal half, contrasting with gleaming white radial spines and bright green tubercles that are visible between the spines. The centrals may be horn-colored, or rarely yellowish, whitish, or banded. In some plants a few of the radials may be colored like the centrals, or have dark tips. There are 10–26 radial spines, these 0.7–2 cm long. The radials are appressed.

The relatively large flowers of var. *vivipara* usually are intense magenta but may be rose-pink or pale rose. The flowers are (2–)2.5–5(–5.7) cm long, (2.5–)3–6.7(–8) cm in diameter. They are similar to those of *C. macromeris* and *Mammillaria wrightii*. The inner tepals of var. *vivipara* are 1.5–3.5 cm long, (1.3–)2–6 mm wide, unicolored or sometimes with a darker midregion, and whitish or greenish at the base. The outer tepals are greenish with conspicuously fringed margins. The stamen filaments are magenta to whitish or greenish-white. The anthers are ca. 1 mm long and yellow. The 5–13 stigma lobes are magenta to pale rose-pink (rarely white), and apiculate.

The green, juicy fruits are ovoid to obovoid, or ellipsoid, 1.2–2.8 cm long, and 0.7–2 cm in diameter. Mature fruits may turn dull brownish-red, or greenish-brown, or even tan on some exposed portions. The floral remnant is persistent. The seeds are bright red-brown, pitted, 1–2.4 mm long, usually comma-shaped to subreniform, or the smallest seeds obovoid.

Phenology. In the Trans-Pecos plants of var. *vivipara* usually bloom in May or early June, but sometimes as late as July and August. Regionally, the blooming season lasts 1–2 months (Zimmerman, 1985), but a given population may exhibit synchronous blooming and may display open flowers for only a few days each year. Individual flowers open about 1:00 P.M. on warm, sunny days and stay open until mid- to late afternoon. The flowers close at night and may or may not open again. Our observations suggest that flowers usually open only one day, if the day is warm and sunny, but each stem often supports several mature flower buds, many of which open the next or another day. There is some indication that *C. vivipara* might be opportunistic with respect to flowering, (Zimmerman, 1985; Worthington, 1986) and that some plants may bloom again in the summer (e.g., August) following adequate rainfall. The aspect of sporadic late flowering in var. *vivipara* requires further investigation.

Fruit maturation in var. *vivipara* requires about three months, possibly as few as two months, or as many as five months (Zimmerman, 1985). Even after maturation, with fully developed seeds, the green, juicy fruits may persist for one or more months before withering. Thus, fruits may remain on the plants for 4–6 months. Under cultivation in Alpine, fruits have been observed on plants as late as in October. The length of time required for actual fruit maturation is not well understood.

Sterile and Immature Specimens. In the Trans-Pecos most plants consist of unbranched globose to ovoid stems with diffuse roots. In northeastern and west-central Texas multistemmed plants of var. *vivipara* are so large that they superficially resemble *C. macromeris,* but these large plants apparently are not present in the Trans-Pecos. The pith and cortex are not mucilaginous. The tubercles are 0.8–2.5 cm long, 0.3–0.8 cm in diameter, somewhat flaccid, and rather widely separated when stems are turgid. Large lenticular druses, to 0.75–1 mm in diameter, are always present in older parts of the pith and cortex. In Trans-Pecos plants the central spines usually are not at all appressed, although this may be the case in populations elsewhere. Usually the inner central is porrect, and the three (or more) ascending outer central spines are noticeably thicker (0.2–0.6 mm) than the radials and have prominent bulbous bases. The inner (shortest), porrect central spine is 1–1.8 cm long, and the outer centrals are 1.5–2.3 cm long. The mostly white, appressed radial spines typically are slender and easily distinguished from the usually colored centrals.

In juveniles and seedlings, there are fewer and shorter spines per areole. Central spines, both the porrect one and the ascending upper one or two, are evident in seedlings less than 1 cm long. The spines, both centrals and radials, are conspicuously pubescent in seedlings.

Biosystematics. There is considerable populational variation in *C. vivipara* var. *vivipara*, as might be expected in a taxon that ranges from Texas to Canada and occupies various substrates at elevations of 600–8,200 feet. The taxon in its pure form does not appear to be widely distributed or abundant in the Trans-Pecos. It is restricted mostly to the eastern portion of the region, where it intergrades with var. *neomexicana,* which is more widespread in the Trans-Pecos, and var. *radiosa,* which is centered in the Edwards Plateau area of Central Texas. According to Zimmerman (1985), var. *vivipara* most closely resembles *C. vivipara* var. *arizonica* (Engelm.) W. T. Marshall of Arizona and neighboring states. Extensive information regarding the systematics of var. *vivipara* is available (Zimmerman, 1985; Weniger, 1984; Benson, 1982; Fischer, 1971, 1980), where the polymorphic nature of the taxon is reflected in differing taxonomic interpretations.

Coryphantha vivipara belongs to section *Pseudocoryphantha* of the genus *Coryphantha* along with three other species, including the polytypic *C. sneedii* (Zimmerman, 1985). Within the section, *C. vivipara* is closely related to *C. chlorantha* (Engelm.) Britton & Rose (northwestern Arizona, southern Nevada, southern California) which was treated as a variety of *C. vivipara* [under the name *deserti* (Engelm.) W. T. Marshall] by Benson (1982), and *C. alversonii* (J. M. Coult.) Orcutt (southern California), also regarded by Benson (1982) as a variety of *C. vivipara*. Zimmerman (1985) used the common occurrence of the giant lenticular druses, and similar exomorphology, in proposing a close relationship between *C. vivipara* and *C. sneedii*. The nearly identical flowers of *C. vivipara, C. macromeris,* and *Mammillaria wrightii* are not regarded (Zimmerman, 1985) as evidence of close affinity because the taxa differ significantly in most other characters, and each species has its own set of obviously closer relatives.

Synonyms. *Mammillaria vivipara* (Nutt.) Haw.; *Echinocactus viviparus* (Nutt.) Poselger; *M. vivipara* (Nutt.) Haw. var. *vera* Engelm.; *M. radiosa* K. Schum. f. *vivipara* (Nutt.) Schelle; *Neomammillaria vivipara* (Nutt.) Britton &

Rose ex Rydb.; *Escobaria vivipara* (Nutt.) Buxb.; ? *M. montana* Blanc; ? *M. hirschtiana* Haage F. ex Hirscht; *C. columnaris* Lahman; *C. oklahomensis* Lahman; *E. oklahomensis* (Lahman) Buxb.

Common Name. The name eastern beehive cactus is coined because this variety has the widest distribution at the eastern range of the species.

6b. **Coryphantha vivipara** var. **neomexicana** (Engelm.) Backeb. NEW MEXICO BEEHIVE CACTUS. Plates 270, 271. [*Mammillaria vivipara* (Nutt.) Haw. var. *radiosa* (Engelm.) Engelm. subvar. *neomexicana* Engelm., Proc. Amer. Acad. Arts 3: 269. 1856]. Desertscrub to mountain woodlands and basin grasslands, rocky exposures or alluvium, sedimentary, and igneous substrates. Every county of the Trans-Pecos except perhaps Val Verde Co., rare or nonexistent in S Presidio and Brewster counties. 2,300–5,200 ft. Flowering (Apr–)May–Jun(–Aug). $2n = 22$. Across most of NM except extreme eastern portions, into SW CO, W-central AZ. Mexico, N Chihuahua, central Coahuila, gypsum. Map 59.

Benson (1982) included *C. vivipara* var. *neomexicana* under *C. vivipara* var. *radiosa*. Weniger (1984) recognized *C. vivipara* var. *neomexicana,* but his interpretation of the taxon was not the same as that by Zimmerman (1985), who included Weniger's *Mammillaria fragrans* (Hester) Weniger in *C. vivipara* var. *neomexicana*. According to Zimmerman, var. *neomexicana* is the most widely distributed taxon of *C. vivipara* in the Trans-Pecos, from El Paso County east to Pecos County. Intermediates between var. *neomexicana* and var. *radiosa* occur in Pecos and in northeastern Brewster and Terrell counties. The varietal epithet *neomexicana* reflects the distributional center of the taxon in New Mexico.

Identifying Characters. The stems of var. *neomexicana* are globose, ovoid, to cylindroid, and covered with spines to the extent that the tubercles are mostly or completely obscured. The spines characteristically are snowy-white, although the centrals may be tan, reddish-brown, or dark reddish-brown, and the radials may have dark tips. There are 25–54 spines per areole, usually including 4–10 or more centrals and 30 or more radials, these all straight and glabrous.

The magenta flowers of var. *neomexicana* are like those of var. *vivipara* described above except that the stigma lobes are white. The underlying nonstigmatic tissues are variable in color, often white or reddish (Zimmerman, 1985).

The fruits are green, ovoid to obovoid, 1.2–2.8 cm long, and are like those of var. *vivipara*. The bright red-brown, pitted seeds of var. *neomexicana* are 1.3–2(–3) mm long, significantly larger on average than those of var. *vivipara*.

Phenology. The flowering phenology of var. *neomexicana* has been evaluated by Zimmerman (1985) and Worthington (1986). It appears that the typical flowering time in the Trans-Pecos is May and early to mid-June, although the first flowers may appear in some populations in late April, and sporadic flowering occurs in July and August. It is speculated that prevailing, local climatic conditions of the season, particularly temperature and precipitation, influence the timing of peak flowering. Sporadic flowering in July to mid-August possibly is a response to adequate rainfall, as is known to be the case with *C. hesteri*. Flowers open about midday under warm, sunny conditions and close in late afternoon. Worthington observed that flowers of var. *neomexicana* (reported as var. *radiosa*) last one day, usually not opening again the next day, and that all of the flower buds on one plant usually are expended in 1–3 days. Zimmerman stated that individual flowers may last for two days, particularly if the first day is cloudy. Tentatively, Zimmerman concluded that fruit maturation is concentrated in August–October, about three months after flowering. Pending definitive studies on the subject, fruit maturation is difficult to determine from circumstantial data because juicy fruits may persist on the plants in the winter, at least into January and February, well after full development of the seeds.

Sterile and Immature Specimens. The stems of var. *neomexicana* usually are unbranched, but presumably older plants may form clusters of six or more stems. The stems are (5–)6–20 cm long, (3–)3.5–7.5 cm in diameter, and are nearly covered or completely covered with white or drab-tan spines, some of them perhaps with colored tips or shafts. The tubercles are 0.9–2 cm long, 0.4–0.6 cm in diameter, and usually are mostly or completely obscured by spines. The areoles are circular and ca. 3 mm in diameter. Usually there are 5–11 central spines, (0–)1(–4) of them inner centrals, and (3–)4–10(–14) of them outer centrals. The central spines in areoles of adult plants usually are diffusely arranged if there are more than six centrals, but in a "bird's-foot" arrangement if less than six centrals. The inner central spine usually is the largest in diameter (0.3–0.7 mm) of all the central spines and is shorter (0.8–1.4 cm long) than the longest outer centrals. When in a "bird's-foot" arrangement, the usually 1–4 outer centrals are ascending-appressed, these heavier by far than the subcentrals or radials, and the longest outer central is 1.4–2 cm. The (19–)25–34(–40) radial spines are relatively slender (0.08–0.3 mm wide), white, appressed, and 0.8–1.4(–2) cm long.

In and near the Marathon Basin, var. *neomexicana* may be confused with *C. tuberculosa* var. *varicolor* or smaller, turgid plants of var. *neomexicana* may resemble larger plants of *C. hesteri*. Elsewhere in the Davis Mountains, var. *neomexicana* and *C. tuberculosa* var. *varicolor* may be sympatric. Dense-spined (white) forms of var. *neomexicana* near the Guadalupe Mountains (e.g., in the Patterson Hills) resemble the largest-stemmed forms of *C. sneedii*. *Coryphantha vivipara* var. *neomexicana* and *C. sneedii* have very different flowers.

In seedlings and juveniles of var. *neomexicana*, the spines are conspicuously pubescent. Subadults have shorter and fewer spines. Central spines are present even in seedlings, where they are often ascending-appressed and light brown. Reportedly (Zimmerman, 1985), juveniles of var. *neomexicana* have been misidentified as adults of *Mammillaria lasiacantha*. Most of the El Paso County reports of *M. lasiacantha* (rare in El Paso County) are based on immature plants of *C. vivipara* var. *neomexicana*. The two taxa are easily distinguished by spine characters.

Biosystematics. *Coryphantha vivipara* var. *neomexicana* as tentatively circumscribed by Zimmerman (1985, p. 263) "is a group of contiguous, weakly defined geographical races left over as a taxonomic unit after all of the better-defined surrounding races have been duly circumscribed." Typical var. *neomexi-*

cana occurs in the Rio Grande valley in southern New Mexico, for example, in and near the Organ Mountains. The plants have long, spreading brown to grayish, central spines and large rose-pink flowers. In Texas the populations of var. *neomexicana* in and near the Davis Mountains have relatively short white spines and few central spines, very much like plants geographically separated in northern New Mexico (Zimmerman, 1985) and unlike the more "typical" var. *neomexicana* to the west in southern New Mexico. Further studies involving detailed populational investigations will be necessary before it is apparent if the Davis Mountains population of var. *neomexicana,* or other geographic races of var. *neomexicana,* should be formally described as separate taxa. Meanwhile, Zimmerman's approach to dealing with the widespread and variable *C. vivipara* complex is embraced herein.

Elsewhere in Texas *C. vivipara* var. *radiosa,* mostly of the Edwards Plateau, intergrades with var. *neomexicana* in the eastern Trans-Pecos region. This includes, sensu lato, *C. fragrans* Hester, a taxon recognized by Weniger (1970) as a separate species (as *Mammillaria fragrans* of Terrell County and the Big Bend region). The review by Zimmerman is convincing that *C. fragrans* is not distinguishable from the intermediates between *C. vivipara* var. *neomexicana* and var. *radiosa.*

To the north and northeast of the Edwards Plateau, var. *radiosa* intergrades with var. *vivipara* (Zimmerman, 1985). Intergradants of var. *radiosa* and var. *vivipara* are also found in the southeastern Trans-Pecos (e.g., in Terrell County), these being regionally sympatric with var. *radiosa* and var. *neomexicana* intermediates. Supposedly pure forms of var. *vivipara* also occur in the eastern Trans-Pecos. The var. *radiosa* can be distinguished from var. *vivipara* by its more numerous white-looking spines, 27–33 per areole, and from var. *neomexicana* by its fewer, glassy-looking spines and pink to magenta stigma lobes.

The *C. vivipara* var. *neomexicana* reported from the Chisos Mountains (as var. *radiosa*) by Benson (1969b, 1982) and Wauer (1980) is *C. chaffeyi* (Zimmerman, 1985). The existence of *C. vivipara* in Big Bend National Park in the foothills of the Chisos Mountains and on gravelly banks near the Rio Grande (Wauer, 1980; as *Mammillaria fragrans*) remains to be documented. The photograph identified in Warnock (1970) as *C. vivipara* is of *C. pottsiana.*

Synonyms. *Mammillaria vivipara* (Nutt.) Haw. var. *neomexicana* (Engelm.) Engelm; *M. vivipara* subsp. *radiosa* (Engelm.) Engelm. var. *neomexicana* (Engelm.) Engelm.; *M. radiosa* Engelm. var. *neomexicana* (Engelm.) Engelm.; *Cactus radiosus* (Engelm.) J. M. Coult. var. *neomexicanus* (Engelm.) J. M. Coult.; *C. neomexicanus* (Engelm.) Small; *M. radiosa* Engelm. f. *neomexicana* (Engelm.) Schelle; *M. neomexicana* A. Nelson ex J. M. Coult. & A. Nelson; *Coryphantha neomexicana* (Engelm.) Britton & Rose; *C. radiosa* (Engelm.) Rydb. var. *neomexicana* (Engelm.) Schelle; *Escobaria neomexicana* (Engelm.) Buxb.; *E. vivipara* (Nutt.) Buxb. var. *neomexicana* (Engelm.) Buxb.; *M. vivipara* (Nutt.) Haw. var. *radiosa* (Engelm.) Engelm. subvar. *borealis* Engelm.; *M. vivipara* (Nutt.) Haw. subsp. *radiosa* (Engelm.) Engelm.; *M. radiosa* Engelm. var. *borealis* (Engelm.) Engelm.; *M. borealis* (Engelm.) Engelm. ex Britton & Rose;

M. vivipara var. *borealis* Engelm. [sic] Weniger; *C. fragrans* Hester; *M. fragrans* (Hester) Weniger *comb. nud.*

Common Names. Fragrant cactus; spiny-star cactus; *estria del tarde;* New Mexico beehive; sour cactus.

7. **Coryphantha sneedii** (Britton & Rose) A. Berger. SNEED'S PINCUSHION CACTUS, GUADALUPE PINCUSHION CACTUS, LEE'S PINCUSHION CACTUS, SILVERLACE CACTUS. Plates 272–83. [*Escobaria sneedii* Britton & Rose, Cact. 4: 56, f. 54. 1923]. Desertscrub to conifer woodlands, limestone rock outcrops (in the Trans-Pecos), rarely in rocky alluvium. 2,000–8,700 ft. Flowering Mar–May, opportunistically after summer rains. $2n$ = 22. Southern NM, SE AZ. Chihuahua and probably N Coahuila, Mexico. Map 60.

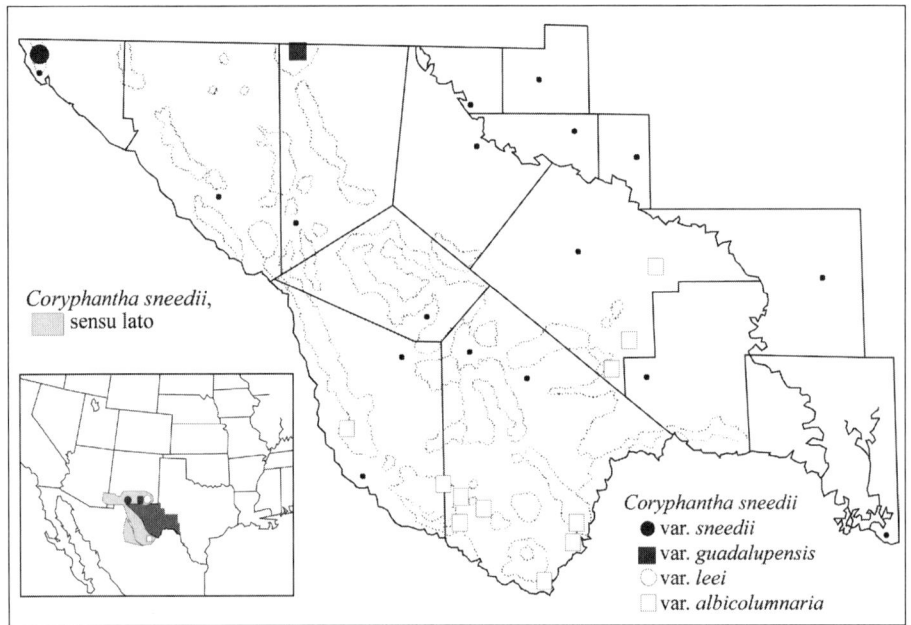

Map 60. Generalized distribution of four varieties of *Coryphantha sneedii* (Sneed's pincushion cactus).

Coryphantha sneedii is a polytypic species consisting of 9–10 geographic races, each of which is treated as a variety by Zimmerman (1985). Three of these are found within the political boundaries of Trans-Pecos Texas, and others are located on mountains visible across the boundaries in Chihuahua and New Mexico: in the Organ Mountains, Doña Ana County; the San Andres Mountains, Doña Ana County; the Sacramento Mountains, Otero County, New Mexico; just across the Rio Grande in Chihuahua, Mexico. The varieties of *C. sneedii* are seldom encountered, usually in the isolated rock outcrops high on mountainsides except for the tall Big Bend variety, which is locally common at several frequently visited sites in and around Big Bend National Park. The specific epithet is taken after the original collector, J. R. Sneed, who collected the type specimen in the Franklin Mountains, El Paso County.

Coryphantha sneedii is sometimes sympatric with *C. vivipara* in the Trans-Pecos, most notably in the Guadalupe Mountains. To the experienced eye, plants of the two species usually can be distinguished by habit. More specific distinguishing traits have been outlined by Zimmerman (1985). *Coryphantha sneedii* plants are smaller in all parts than *C. vivipara:* mature stems are 1.3–7 cm in diameter; the largest spines are 0.15–0.60 mm thick; radial spines are 0.01–0.25 mm thick; flowers 0.7–2.5 cm wide; inner bracts 0.5–1.4 cm long; fruits 2.5–7 mm in diameter; and seeds 1.1–1.5 mm long. Also in var. *sneedii* the central spines are always in a radiating arrangement, and the total spines per areole, 31–95, are relatively numerous. The smaller flowers usually are pale pink to whitish with conspicuous (darker) midregions. In anatomical consideration, there is no medullary vascular system in *C. sneedii*. In *C. vivipara* mature stems are 2.9–11 cm in diameter, the largest spines are 0.2–0.9 mm thick, radial spines 0.7–0.6 mm thick, flowers 0.8–6.7 cm wide, inner bracts 1.1–3.5 cm long, fruits 5–20 mm in diameter, and seeds 1.3–2.4 mm long. In *C. vivipara* the central spines either are in a "bird's-foot" arrangement or radiating (as in *C. sneedii*), and the spines are fewer, 11–55, per areole. The larger flowers consistently are magenta to rose-pink in the Trans-Pecos. A medullary vascular system (vascular bundles in the pith) is present in *C. vivipara*.

In distinguishing *C. sneedii* from *C. tuberculosa*, radiating central spines are characteristic of all the varieties of *C. sneedii*, whereas in *C. tuberculosa* the "bird's-foot" arrangement is typical. The central spine arrangement represents a consistent difference between the taxa in addition to the usually snowy-white spines of *C. sneedii* and the pale gray to ashy-white spines of *C. tuberculosa*. *Coryphantha sneedii* also differs from *C. tuberculosa* in characters of the fruits (green to brownish-red in *C. sneedii*, red in *C. tuberculosa*), seeds, flowers (color and size), anthers, and phenology (typical flowering time and anthesis).

Key to the Varieties

1. Plants often profusely branched, dominated by immature stems, up to 200 or more, 1.3–3 cm in diameter; spines 62–95 per areole, all mostly appressed and conspicuously pubescent; inner central spines, the shortest ones, 1–8 mm long; NE end of Guadalupe Mts, in NM

   7c. *C. sneedii* var. *leei*.

1. Plants less profusely branched, sometimes but not often dominated by immature stems, 1–100 or more stems, 1.4–6.5 cm in diameter; spines 31–68 per areole, not or only slightly appressed, glabrous or inconspicuously pubescent; inner central spines, the shortest ones, 2–22 mm long; SW end of Guadalupe Mts, mostly in TX (2).

2(1). Plants averaging 14–24 stems, up to 100 or more; spines shorter and more numerous (see text); Franklin Mts, El Paso Co., TX, and Franklin Mts and Bishop's Cap Mt, Doña Ana Co., NM

                                                            7a. *C. sneedii* var. *sneedii*.

2. Plants averaging 1–5 stems, up to 26; spines longer and fewer (see text); Guadalupe Mts, and eastern Pecos Co. to lower Big Bend region (3).

3(2). Stems mostly 3–6 cm long; Guadalupe Mts, high altitudes

                                                            7b. *C. sneedii* var. *guadalupensis*.

3. Stems mostly 7–25 cm long; lower Big Bend region, low altitudes

                                                            7d. *C. sneedii* var. *albicolumnaria*.

7a. **Coryphantha sneedii** var. **sneedii**. SNEED'S PINCUSHION CACTUS. Plates 272–74. [*Escobaria sneedii* Britton & Rose, Cact. 4: 56, f. 54. 1923]. Limestone crevices, usually steep S-facing slopes, with desertscrub and *C. tuberculosa*. El Paso Co., Franklin Mts, TX (type locality), and Doña Ana Co., NM, Franklin Mts and Bishop's Cap Mt (at least in Doña Ana Co. always on or near Silurian-Ordovician-Cambrian limestone; Zimmerman, 1985). 3,900–6,000 ft. Flowering Mar–May, perhaps blooming again in summer after rains. $2n = 22$. Map 60.

Coryphantha sneedii var. *sneedii* is endemic to the Chihuahuan Desert Region. As interpreted here, this rare taxon occurs only in the western tip of the Trans-Pecos, in the Franklin Mountains of El Paso County and adjacent Doña Ana County, New Mexico; also on adjacent Bishop's Cap Mountain, a few miles north of the Franklin Mountains. The var. *sneedii* is listed as Endangered by the U.S. Fish and Wildlife Service (November 1979) and by the Texas Parks and Wildlife Department (April 1983).

Identifying Characters. Plants of var. *sneedii* are cespitose, often forming mounded clumps of less than 25 densely white-spined stems, but in some clumps there are up to 100 or more stems. In most clumps only 3–5 stems obviously are full-size, and the rest are immature stems of various sizes. The stems are globose when very young and become cylindroid or clavate with age, 3–13 cm long, and 1.3–3(–4.5) cm in diameter. In aspect the stems are bristly, although the radial spines are essentially appressed, and the spines are snowy-white.

The flowers basically are whitish or pinkish, 1.2–2.5 cm long, and do not open widely. The inner tepals are white with prominent midregions, these broad or narrow, and pink, magenta, dull lavender, brown, or greenish. The distal portions of the inner tepals exhibit the deepest color. The inner tepals are 0.5–1.4 cm long and 0.8–4 mm wide. The outer tepals are fimbriate in all varieties of *C. sneedii*. The filaments are whitish to pale green, pink, or magenta. The 3–7 stigma lobes are white, rarely pink or cream-colored, and 1–3 mm long.

The semisucculent green fruits are obconic to cylindroid, 0.6–1.5 cm long, and 0.25–0.6 cm in diameter. The fruits are green when mature but with age may turn brownish-pink on exposed portions and become nearly dry. The floral remnant is persistent. The pitted seeds are red-brown or orange-brown, becoming darker when dry, and (0.9–)1.1–1.5 mm long.

Phenology. Zimmerman (1985) concluded that the whole species *C. sneedii*, including var. *sneedii*, is spring-blooming. Flowering may commence in early March at the lowest elevations and in early May at the highest elevations. The main flowering time for var. *sneedii* is in April, sometimes extending into May (Worthington, 1986). Summer blooming is unpredictable but may be opportunistic following substantial rainfall, as is the case with some other predictably spring-blooming coryphanthas. Under cultivation in Alpine, two plants of var. *sneedii* were in flower on 17 September 1998, following rain about two weeks earlier. The flowers open about noon, or sometimes during late morning, and close in the afternoon. Flowers usually open again for several days. Individual plants with only a few flowering stems have been observed to produce flowers for 3–14 days (Worthington, 1986). The spring flowering season of var. *sneedii* precedes the initial flowering period of the vegetatively similar (but not closely related) and sympatric *C. tuberculosa*, whose flowers open in mid-afternoon and stay open until evening. Mature fruits are present on plants of var. *sneedii* from August to November. The fruit maturation time appears to be about 3–4(–5) months.

Sterile and Immature Specimens. In vegetative condition one of the most diagnostic distinguishing features of *C. sneedii* (separating it from all related Texas species except *C. vivipara*) is the giant lenticular druses, 0.5–1 mm in diameter, that are present in mature pith and cortical tissue. The stem tissue when exposed has a granular texture owing to the large druses, which are evident to the naked eye in fresh or dried tissue. All other species of *Coryphantha*, except *C. vivipara* and its allies (*C. chlorantha* and *C. alversonii*), lack lenticular druses of that size. Other species of *Coryphantha* have relatively small spheroidal druses usually less than 0.4 mm in diameter, except for *C. hesteri*, which has lenticular druses to 0.4(–0.5) mm in diameter.

Plants of var. *sneedii* on average have numerous small stems that are snowy-white. This habit feature allows tentative identification but was greatly overemphasized in the early literature; the stems of many plants of var. *sneedii* are unbranched, and not every plant of var. *sneedii* is "snowy" white. Several taxa of *Coryphantha* in Texas, not just *C. sneedii* var. *sneedii*, sometimes have a habit of dense clumps of small stems.

The roots of all varieties of *C. sneedii* are diffuse, or they may develop as short taproots. The pith and cortex of the stems are not mucilaginous. The tubercles of var. *sneedii* are 4–9 mm long and 2.5–4 mm in diameter. Of the 31–68 spines per areole, 0–6 (usually 2–3 in Texas) are inner central spines, and 5–18 (usually 13–14 in Texas) are outer centrals. The shortest inner centrals are 2–13 mm long, but average ca. 5 mm long, and the other inner centrals are 3–15.5 mm long. The outer central spines usually are 4–16.5 mm long. The inner centrals are porrect (usually one) or variously spreading or appressed. The largest central spines obviously are heavier, 0.25–0.4 mm thick, than the radial spines and have bulbous bases. Fresh centrals are snowy-white, but the larger centrals usually develop darker tips, these tan, to brownish-red, or reddish-brown. The 24–46 (averaging 37 in Texas) radial spines are snowy-white, appressed, 3–12 mm long (averaging 5.5 mm long in Texas), and slender. The outer radials, those closest to the stem, are the most slender at 0.05–0.12 mm wide. By comparison, the appressed radials in var. *sneedii* are not as appressed as those of var. *leei*. Additional details in spine morphology are available in Zimmerman (1985).

Juveniles of var. *sneedii* have fewer central (0–4) and radial spines (15–25), and the spines are shorter, usually 1–1.5 mm long, in very young stems. The spines are pubescent in juveniles but glabrous in adults.

Biosystematics. As implied above, *C. sneedii* is most closely related to the *C. vivipara* complex, a species-group with which it shares many characters and the only other taxon in *Coryphantha* that possesses the giant lenticular druses. *Coryphantha sneedii* is not closely related to *C. tuberculosa*, although some previous workers (Taylor, 1979; Benson, 1982) evidently misinterpreted superficial similarities between the taxa. Benson reported that *C. sneedii* var. *sneedii* and *C. tuberculosa* (as *C. strobiliformis*) were "known" to intergrade, but instead the taxa appear to be reproductively isolated and do not hybridize (Zimmerman, 1985). Benson even included two varieties of *C. sneedii*, as did Taylor (1979), with *C. tuberculosa* (as "var. *durispina*" and "var. *orcuttii*"). Taylor (1986) later corrected this mistake. See Zimmerman (1985) for misidentifications involving var. *sneedii*, many of them published.

Synonyms. *Mammillaria sneedii* (Britton & Rose) Cory; *C. pygmaea* Frič.

Common Names. Sneed pincushion cactus; Sneed cory cactus; Sneed's carpet escobaria.

7b. **Coryphantha sneedii** var. **guadalupensis** (S. Brack & K. D. Heil) A. D. Zimmerman, comb. nov (forthcoming). GUADALUPE PINCUSHION CACTUS. Plates 275, 276. [*Escobaria guadalupensis* S. Brack & K. D. Heil, Cact. Succ. J. (U.S.) 58: 165. 1986]. Mostly steep limestone slopes, frequently S-facing exposures, SW Guadalupe Mts, Culberson Co., mostly higher elevations in with oak-juniper-pinyon associations. 4,500–8,700 ft. Flowering Apr–May (opportunistically in summer). $2n = 22$. Endemic to the Guadalupe Mts, extending a few miles into NM. Map 60.

This is the most recently described of the three varieties of *C. sneedii* in Texas, as "*Escobaria guadalupensis*," named after the mountain range in which it occurs. The type locality is from the highest peak in Texas in the Guadalupe Mountains. The Guadalupe Mountains lie mostly in southeastern New Mexico, extending just into Texas and harboring at least six of the highest peaks in Texas, in Guadalupe Mountains National Park.

Identifying Characters. The plants high in the Guadalupe Mountains are similar in most characters to the largest adult stems of var. *sneedii*. They mostly are solitary or in small clumps of 2–3(–8) robust stems. A few plants exhibiting tight, rounded clumps of stems have been observed in these populations (B. Wauer, pers. comm.).

The flowers of var. *guadalupensis* are described by Heil and Brack (1986) as having inner tepals that are pale yellow or cream-colored to pink, with pink midregions. The flower color has been described as "orangish-pink" in most of the Guadalupe Mountain plants by Brent Wauer (pers. comm.), who has observed numerous populations of var. *guadalupensis*.

The flower, fruit, and seed characters all appear to fall well within the range of variation described for var. *sneedii* in El Paso County and adjacent New Mexico (Zimmerman, 1985). The fruits of var. *guadalupensis* are described (Heil and Brack, 1986) as green when ripe.

Phenology. It is expected but not documented that phenological characteristics of var. *guadalupensis* correspond with those described for var. *sneedii*. For var. *guadalupensis* at high elevations in the Guadalupe Mountains, average blooming time probably is later in the year than is typical for var. *sneedii* in the Franklin Mountains.

Sterile and Immature Specimens. Juvenile specimens examined from Dog Canyon in the Guadalupe Mountains show 1–8 clustered stems. The spines are shorter, all appressed, with no protruding central spines, and the spines in the juveniles are conspicuously pubescent. Two other species of *Coryphantha* with stems densely white-spined occur in the Guadalupe Mountains; *C. tuberculosa* and *C. vivipara* var. *neomexicana*. Characters for distinction of these species from *C. sneedii* are discussed under the species *C. sneedii* as a whole.

Biosystematics. The Zimmerman (1985) monograph of section *Pseudocoryphantha* (including *C. sneedii*) was completed a year before var. *guadalupensis* was described (Brack and Heil, 1986). In his treatment, Zimmerman studied specimens from the southwest end of the Guadalupe Mountains, in what is now Guadalupe Mountains National Park, and concluded that they belonged with var. *sneedii*. Other specimens in the northeastern Guadalupe Mountains in Eddy County, New Mexico, were evaluated by Zimmerman as intermediate between these and var. *leei*. The geographic race in the southwestern Guadalupe Mountains probably deserves recognition as a variety because of its seeming isolation and differences in habit from var. *sneedii* and var. *leei*. Because all the varieties of *C. sneedii* are geographic races (Zimmerman, 1985) with some degree of morphological differentiation, all of them can be identified geographically, except for those in the central mid-altitude part of the Guadalupe Mountains, connecting "*Escobaria guadalupensis*" with *C. sneedii* var. *leei* of Carlsbad Caverns National Park.

Brent Wauer, former ranger in Guadalupe Mountains National Park, avid explorer of the Guadalupe Mountains, and astute observer of the plants there, noted that *C. sneedii* was "all over the upper Guadalupes." Wauer also observed that *C. sneedii* in the southwestern Guadalupe Mountains were "sloppy" in habit, compared to var. *sneedii* and var. *leei*, with taller stems in less concentrated clumps, usually 2–3 stems together, or a few more, and that the flowers were orangish-pink and somewhat translucent. He believed that var. *sneedii*, the morphotype in El Paso County, may not occur at all in the Guadalupes. His concept of var. *sneedii* taken from the Franklin Mountains was of plants with many more stems in tighter, more rounded clumps, and plants with pinkish, more opaque flowers compared to those in the Guadalupes. His field experience matches that of all other explorers, finding var. *leei* only in the northeastern Guadalupes near Carlsbad Caverns. Wauer described the var. *leei* morphotype as "tight clumps of little balls" in reference to the numerous, small stems with tightly appressed spines.

Synonym. See basionym under var. *guadalupensis*.

Common Name. We recommend using Guadalupe pincushion cactus.

7c. **Coryphantha sneedii** var. **leei** (Rose ex Boed.) L. D. Benson. LEE'S PIN-CUSHION CACTUS. Plates 277, 278. [*Escobaria leei* Rose ex Boed., Mamm.-Vergleichs-Schlussel 17. 1933]. Limestone outcrops (ledges on the N-facing slopes at the type locality), desertscrub, often dominated by lechuguilla, NE end of Guadalupe Mts, Carlsbad Caverns National Park, Eddy Co., NM. 3,750–5,500 ft. Flowering Apr–May. $2n = 22$. Map 60.

The Guadalupe Mountains are a relatively small, narrow range located mostly in southeastern New Mexico, and extending barely into Texas. The main mountain mass trends north and northwest in Otero and Eddy counties, but a smaller arm extends to the northeast from the Texas portion of the mountains into New Mexico, gradually terminating about 10 miles north of the New Mexico–Texas state line. The end of the northeast arm is the site of the famous Carlsbad Caverns and the general locality of *C. sneedii* var. *leei*. The varietal epithet honors W. T. Lee, who collected the taxon in Rattlesnake Canyon.

Identifying Characters. Plants of var. *leei* have been described as profusely branching cushions of up to 200–250 stems tightly packed together and of various sizes. These are the types of plants that are familiar in horticulture and like the type specimen (Zimmerman, 1985). The cushion plants are dominated by small, immature stems and represent an extreme form of the species.

The relatively few mature stems are 3.5–10 cm long, 1.3–3 cm in diameter, and are most similar to the stems of *C. sneedii* var. *sneedii*. The stems are densely white-spined, with 62–95 spines per areole, and all the spines are tightly appressed (with the exception of some inner centrals), even reflexed toward the stem, resulting in a knobby surface of the stem. The stem surface is visible between the tubercles.

The flowers of var. *leei* are 1.1–1.5(–2) cm long and are often described as pinkish-brown or pinkish-green, with deep pink midstripes. Inner tepal color variation, according to Zimmerman (1985), may include pale rose-pink, reddish-purple to dull brownish-lavender, or pale tan with yellowish-pink midregions. The inner tepals may be greenish-white and dull pink distally. The inner tepals often have prominent whitish or cream-colored margins and sharply defined midstripes, but the midstripes are not evident in flowers of some plants. The inner tepals usually are linear-lanceolate, but in some plants they are broadly ovate or oblong. The flowers have yellow anthers, ca. 0.5 mm long, on filaments 3–4 mm long. The whitish style is 9–10 mm long, supporting 3–6 white stigma lobes.

The fruits are ellipsoid to oblong, or narrowly obovoid, 0.5–1.2 cm long,

3.5–5 mm in diameter, green or pale brownish-green at maturity, and develop a pinkish or dull brownish-red tint on exposed surfaces. The seeds are reddish-brown, drying darker, 1.2–1.5 mm long, and more or less comma-shaped.

Phenology. The principal flowering time for var. *leei* is mid- or late April until early or mid-May (Zimmerman, 1985). The flowers open about noon and close at night, and individual flowers open again for 3–4 days (Castetter and Pierce, 1966; Heil, 1984). Mature fruits are formed at least by mid- to late August, into September, and rarely later in the year. Fruit maturation requires 4–5 months.

Sterile and Immature Specimens. The roots of young plants are fibrous, but older plants form relatively slender taproots. The often profusely branched stems are innocuous because the spines are tightly appressed. The pith and cortex are not mucilaginous. The tubercles are 3.5–7.5 mm long, 2.5–4 mm in diameter, and cylindroid. The areoles are white-woolly when young, with the wool remaining in the areolar groove. The center of the spine cluster often protrudes, with spines reflexed, the spines sometimes almost against the surfaces of the tubercles. On average there are about five (ranging from 1–12) inner central spines, usually one of them porrect. There are 10–22 outer centrals, usually about 17, these 3–10 mm long. The central spines clearly are 0.25–0.6 mm thick. The central spines are snowy-white when fresh, rarely stramineous, with some of the largest spines dark-tipped. The white radial spines are 38–74 in number, commonly about 50, 3–8 mm long, and 0.03–0.15 mm thick.

In seedlings and on immature stems, the spines are fewer and shorter than on mature stems. The spines of seedlings and small branches are conspicuously pubescent. They are glabrous or microscopically puberulent on older stems. In many plants of var. *leei,* most of the stems are spherical, short-clavate, or cylindrical, and most of the stems are immature.

Biosystematics. Most of the measurements given above were taken by Zimmerman (1985) from plants of var. *leei* in Rattlesnake Canyon, the type locality. Zimmerman has concluded that var. *leei* exhibits phenotypic plasticity and polymorphism within its populations and that a "perfectly intermediate" population between var. *leei* and var. *sneedii* (sensu lato, including specimens now classified as var. *guadalupensis*) occurs in West Slaughter Canyon, 7.5 miles to the southwest of the type locality.

According to Zimmerman, the profusely branched "cushion plants" are the "extreme" form of var. *leei,* the one that has been most perpetuated in horticulture and the one that is most familiar to cactus enthusiasts. At the type locality it is especially the senescent older clumps that consist of mostly small, immature stems and perhaps plants that have been injured. Plants consisting largely of adult stems more closely resemble typical *C. sneedii* but with the more densely short-spined aspect of var. *leei.* The "cushion plant" forms of var. *leei* essentially are monstrosities. It is the relatively ordinary-looking old adult stems of var. *leei* that should be used for purposes of taxonomic comparison with other taxa.

The distinctive vegetative features of var. *leei,* as compared to those of all other populations of the species, including var. *guadalupensis,* are the smaller and more numerous stems in greater numbers, shorter and more numerous spines that are more closely appressed, and more strongly pubescent spines on immature stems. The population in West Slaughter Canyon is morphologically and ecogeographically intermediate between var. *leei* and the larger, more nearly "normal" varieties such as, closest at hand, var. *guadalupensis*. West Slaughter Canyon is geographically intermediate between the type locality of var. *leei* at Rattlesnake Canyon and the stronghold of var. *guadalupensis* among the high peaks at the southwest end of the Guadalupe Mountains.

Populations of var. *leei* in and near Rattlesnake, Slaughter, Midnight, Yucca, and upper Walnut canyons were reported by Burgess and Northington (1981), who did not attempt to evaluate the taxonomy of each population. Apparently var. *leei* is rare or occasional on certain limestone outcrops (possibly the Tansil Limestone formation; Heil, 1984) associated with desertscrub at the lowest elevations (ca. 3,750 feet in Rattlesnake Canyon) or with mountain shrubs at the highest elevations (5,500 feet near West Slaughter Canyon). At lower elevations the plants tend to occur on north-facing slopes. The status of var. *leei* as a rare taxon in the Guadalupe Mountains recently has received statistical evaluation in a populational study (Baker and Johnson, 2000).

Synonyms. *Escobaria sneedii* Britton & Rose var. *leei* (Rose ex Boed.) D. R. Hunt; *E. sneedii* subsp. *leei* (Rose ex Boed.) D. R. Hunt.

Common Names. Lee pincushion cactus; Lee's carpet escobaria.

7d. **Coryphantha sneedii** var. **albicolumnaria** (Hester) A. D. Zimmerman, comb. nov. (forthcoming). SILVERLACE CACTUS. Plates 279–83. [*Escobaria albicolumnaria* Hester, Des. Pl. Life 13: 129. 1941]. Mostly if not entirely limestone, rocky-gravelly slopes, rocky outcrops, and alluvium; desertscrub, often associated with *Euphorbia antisyphilitica* Zucc., *Hechtia texensis,* and *Jatropha dioica* Cerv. Mostly lower Big Bend near the Rio Grande, rare in E Pecos Co. 1,850–4,450 ft. Flowering Mar–May (Jun in cult.). $2n = 22$. Chihuahua, Mexico; expected in Coahuila across from Big Bend National Park. Map 60.

Coryphantha sneedii var. *albicolumnaria*, erroneously included with *C. tuberculosa* (as *C. strobiliformis* var. *durispina*) by Benson (1982), is by far the most widespread variety of *C. sneedii* in the Trans-Pecos. The varietal epithet includes the Latin *albi,* "white," *colum,* "pillar" (or "column"), and *aria,* a suffix forming an adjective ("connected with"), in reference to the cylindroid stems cloaked in white spines.

Identifying Characters. Plants of var. *albicolumnaria* are characterized by unbranched or few-branched, cylindroid stems densely covered by bristly white spines. Older plants may have 2–5 stems, these blunt apically, and presumably very old plants form loose clumps of up to 26 mature stems. The stems are mostly erect, rarely decumbent, 7–15(–25) cm long, and 2.5–5.5(–6.5) cm in diameter. There are 35–55 spines per areole, with the central spines spreading in all directions. There are 24–35 radial spines, these mostly appressed. All the spines are snowy-white when fresh, or the central spines may be chalky-pink, and distally the larger centrals usually are chalky-pink, tan, reddish-brown, or blackish-purple.

The usually pink flowers of var. *albicolumnaria* are 1.5–2(–2.8) cm long and do not open widely (tubular-funnelform). The inner tepals may be pale rose-pink, to deep pink, or even bright magenta or glossy milk-white. Sharply defined midstripes may or may not be present. The inner tepals are linear and 1.5–3 mm wide. The filaments usually are white and support yellow anthers. The style is relatively short and pinkish, with the 3–7 short, white, or pale pink stigma lobes not much exserted above the anthers.

Mature fruits of var. *albicolumnaria* are either pale green or red. The fruits are cylindroid, or narrowly obovoid, 1–1.7 cm long, and 0.5–0.7 cm in diameter. This is the only variety of *C. sneedii* in the Trans-Pecos that has brilliant red fruits in at least some of the populations. The green fruits usually are yellow or apricot-tinged at maturity, and the red fruits are either bright red or dull pinkish-red. The seeds are pitted, red-brown, and 1–1.5 mm long.

Phenology. Tentative information suggests that the usual flowering period for var. *albicolumnaria* is March through May, and into early June. Opportunistic blooming in late summer apparently has not been documented for var. *albicolumnaria,* as it has for var. *sneedii.* Reportedly, the flowers open about noon and close in late afternoon. Individual flowers may open again for 2–3 successive days. It is not unusual to see 3–5 flowers open at the same time on a single stem, among buds that also will open in due course. Zimmerman (1985) estimated that the time required for fruit maturation may be as little as two months, or sometimes much longer.

Sterile and Immature Specimens. Characteristically, the plants are of solitary or few-branched stems, usually all mature stems in adult plants. The stems glisten with white, spreading spines, or some of the larger spines are merely dark-tipped. In some populations of var. *albicolumnaria* the betacyanic coloration of the central spines involves the whole distal half or more, and the overall coloration of the plants has a cyanic hue. Such plants, found in the lower Big Bend area, still can be identified from a great distance by the stem shape and diffuse central spines. The tubercles are 6–11 mm long and 4–6 mm in diameter. The areoles are circular and ca. 3 mm in diameter. In each areole there are about 45 spines. There are 1–5, usually 1–3, spreading, inner central spines, the shortest one 0.9–2.2 cm long, and the longest ones 1.3–2.5 cm. There are 9–16 outer centrals, usually about 13, with the longest ones 0.9–2.6 cm. All of the central spines are spreading, with the innermost ones angled more in plane with the stem. The longest radial spines, usually the upper, rarely the lateral ones, are 0.5–1.4 cm and usually about 1 cm. The central spines are 0.3–0.5 mm thick, and the slender white radial spines 0.05–0.2 mm thick.

There are similar species that occur throughout some or all of the range of var. *albicolumnaria*. The one closely related species is *C. vivipara* (apparently absent from the southern Big Bend area). All varieties of *C. sneedii* have giant lenticular druses in mature stem tissues. The flattened druses, 0.5–1 mm wide, can be seen with the naked eye but are best identified under magnification, in fresh or dried older stem tissue. The morphotype of *C. vivipara* var. *neomexicana* most likely to be found with var. *albicolumnaria* usually has conspicuously brown central spines in a "bird's-foot" arrangement, very different from the radiating centrals in var. *albicolumnaria*. However, other morphotypes of *C. vivipara* var. *neomexicana*, particularly the one in and near the Guadalupe Mountains, typically have white central spines in a radiating arrangement, like *C. sneedii*. Floral and fruit differences between *C. vivipara* and *C. sneedii* are more profound than the vegetative differences.

Both *C. tuberculosa* and *C. dasyacantha* lack the characteristic lenticular druses of var. *albicolumnaria*. The stems of *C. tuberculosa* may be solitary but usually are clumped and characteristically taper at the apex, unlike the subtruncate stems of var. *albicolumnaria*. Also, the spines of *C. tuberculosa* may be white, and in color resemble those of var. *albicolumnaria*, but typically the spines of *C. tuberculosa* are pale gray or pale tan, giving an ashy-gray or pale gray aspect even from a distance, and at least not the gleaming translucent-white that characterizes most populations of *C. sneedii*. The floral and fruit differences between *C. tuberculosa* and *C. sneedii* are more striking than the vegetative differences.

Superficially, var. *albicolumnaria* is most difficult to distinguish from *C. dasyacantha*, which occurs basically in the same area near the Rio Grande in the southern Big Bend and into eastern Pecos County. In specific habitats var. *albicolumnaria* usually is found in rocky or gravelly substrates associated with stable limestone formations, whereas *C. dasyacantha* tends to favor igneous or limestone gravel or silty alluvium on *bajadas*, in desert flats, or on gravel benches near the Rio Grande. Both *C. dasyacantha* and the large-stemmed varieties of *C. sneedii* have numerous radiating central spines that are snowy-white in color. In addition to the absence of large lenticular druses in its tissues, *C. dasyacantha* is reliably distinguished by its flower (with green stigma lobes), seed shape, and seed color (black).

From a distance, some plants of *Mammillaria pottsii* might be confused with var. *albicolumnaria*. Typically, the columnar, characteristically branching stems of *M. pottsii* are noticeably more slender than those of var. *albicolumnaria*. More specifically, the central spines of *M. pottsii* usually are curved, and those in upper areoles are purplish-gray at least near the tips. Areolar grooves, of course, are absent in *M. pottsii*, and in *M. pottsii* the tubercle axils are densely woolly in the zone below the stem apex where flowers are produced. Flowers and fruits of *M. pottsii* are very different from those of var. *albicolumnaria*.

In seedlings of var. *albicolumnaria* as compared to adults, the spines are shorter, fewer in number, and conspicuously pubescent. Immature branches may have only 4–8 central spines and 14–23 radial spines. The spines are glabrous on immature branches and in adults. The spines are translucent-white on both seedlings and immature stems.

Biosystematics. According to Zimmerman (1985), var. *albicolumnaria* closely resembles and probably is most closely related to *C. sneedii* var. *villardii* (Castetter, P. Pierce, & K. H. Schwer.) A. D. Zimmerman (forthcoming) and *C. sneedii* var. *sandbergii* (Castetter, P. Pierce, & Schwer.) A. D. Zimmerman (forthcoming), both of which are restricted to southern New Mexico. A closer geographic relationship also exists with an undescribed, red-fruited variety that occurs in limestone substrates in Chihuahua, Mexico, from near Ojinaga northwest to near

Samalayuca, distinguished vegetatively by much shorter, thinner, and more numerous spines.

Actually, two undescribed varieties of C. *sneedii* in Chihuahua produce only red fruits (so far as known), and all the other varieties of the species are green-fruited, except in some plants of var. *albicolumnaria* (Zimmerman, 1985). The var. *albicolumnaria* exhibits a color dimorphism in its fruits, at least in some plants in the central part of its range. Zimmerman suggested the possibility that the red fruits in plants of var. *albicolumnaria* might be the result of introgression. In 1996 apparently bona fide var. *albicolumnaria* was discovered near San Carlos in Chihuahua (across from Lajitas, Texas), possibly its first record in Mexico, but nothing is known about its fruit color. In any case the plants near San Carlos demonstrate closer proximity of var. *albicolumnaria* to the red-fruited variety than previously had been documented.

Some populations of var. *albicolumnaria* occur at the lowest elevations of any C. *sneedii* variety, just above 1,800 feet in the southeastern part of its range. These low-elevation plants exhibit the longest stems and the longest and heaviest spines in any population of C. *sneedii*; these robust stem and spine features are correlated with hot, dry habitats (Zimmerman, 1985). Relatively deep-pink flowers, produced by certain plants within various populations of var. *albicolumnaria*, seem to predominate there.

Although var. *albicolumnaria* is one of the most widely distributed of the races of C. *sneedii*, along with one red-fruited variety in Mexico, Zimmerman considered the range to be relictual; he noted that the Pecos County populations seemed to be "relicts on the verge of extinction" (p. 392), disjunct and composed only of widely separated individuals. Other populations in the main part of the range were dominated by relatively old (large) plants, as if reproducing poorly. He recommended that var. *albicolumnaria* should be evaluated with respect to density, age-class-distributions, and other factors. A significant part of the range of var. *albicolumnaria* lies within the traditional collecting grounds for the wholesale cactus trade in Texas. Fieldwork in connection with the current study, encouragingly, has revealed several populations that were dominated by small plants, including one locality where var. *albicolumnaria* was exceedingly abundant, probably the dominant cactus in the area.

Benson (1982) treated var. *albicolumnaria* as a variety of C. *tuberculosa*, under the name C. *strobiliformis* var. *durispina* (Quehl) L. D. Benson, while recognizing only two varieties of C. *sneedii*, var. *sneedii* and var. *leei*. According to Zimmerman (1985), Benson perpetuated a taxonomic misconception initiated by Marshall (in Marshall and Bock, 1941) and followed by Backeberg (1961). Marshall considered *albicolumnaria* to be conspecific with C. *tuberculosa*. Subsequently, Backeberg listed the name *albicolumnaria* in synonymy of *Escobaria tuberculosa* var. *tuberculosa* while at the same time including cultivated *albicolumnaria* plants of dubious origin under the misapplied name *durispina*. An invalid neotype for the epithet *durispina*, from a Texas population of *albicolumnaria*, was designated by Benson (1969b, 1982). Use of the epithet *albicolumnaria*, and not *durispina*, for var. *albicolumnaria*, was explained by Zimmerman. Bravo-Hollis and Sánchez-Mejorada (1991b) also treated var. *albicolumnaria* as a variety of *C. tuberculosa* (as *Escobaria strobiliformis* var. *durispina*). In Benson (1982, fig. 176) the color photograph labeled "*Coryphantha dasyacantha* var. *dasyacantha*" (by R. Wauer) is *C. sneedii* var. *albicolumnaria*.

Synonyms. *Escobesseya albicolumnaria* (as *albocolumnaria*) Hester ex L. D. Benson, *nom. nud., pro. syn.*; "*Mammillaria albicolumnaria* (Hester)" Weniger, *comb. nud.*; *Coryphantha albicolumnaria* (Hester) Zimmerman; *Escobaria tuberculosa* (Engelm.) Britton & Rose var. *durispina* auct. non Quehl; *C. strobiliformis* (Poselger) Moran var. *durispina* auct. non Quehl; *M. strobiliformis* Scheer var. *durispina* Quehl; *E. strobiliformis* (Poselger) Scheer ex Boed. var. *durispina* auct. non Quehl; *E. durispina* Hort., *nom. nud.*

Common Names. White column; white-spine cob cactus; snowcone nipple cactus.

8. **Coryphantha pottsiana** (Poselger) A. D. Zimmerman. RUNYON'S PINCUSHION CACTUS. Plates 284, 285. [*Echinocactus pottsianus* Poselger, Allg. Gartenz. 21: 107. 1853]. Rocky limestone habitats, often in dense scrub, Val Verde Co., reported near the mouth of the Pecos River, relatively common near the W bank of the lower Devils River. 900–1,000 ft. Flowering Feb–Mar. 2n = 22. Deep S TX, mostly in the Rio Grande valley, Webb and Duval counties S to Cameron Co. Adjacent Mexico, NE Coahuila (N to Monclova), N Nuevo León, and Tamaulipas. Map 61.

To our knowledge *C. pottsiana* has not been documented in the Trans-Pecos

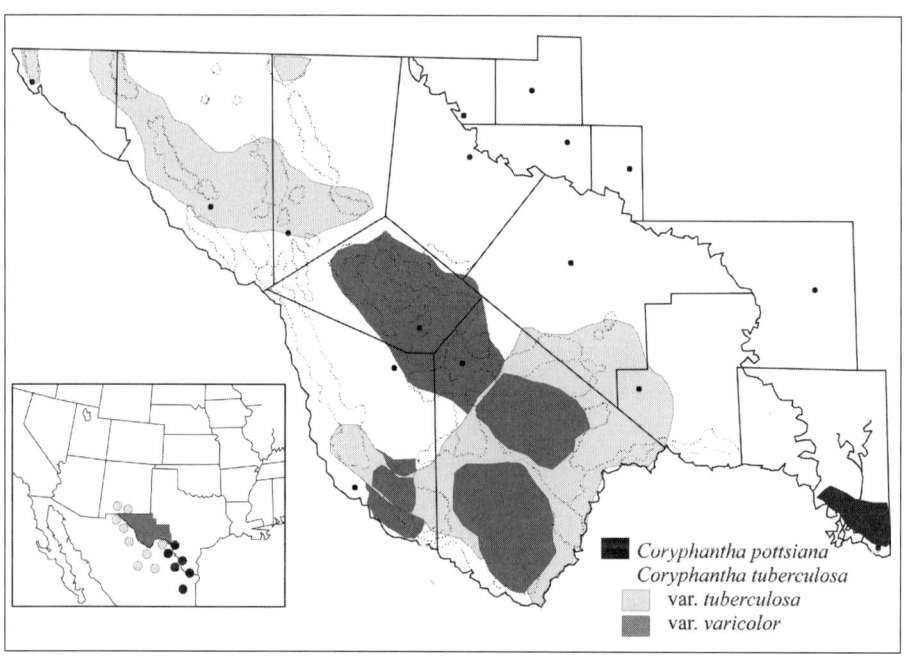

Map 61. Generalized distribution of *Coryphantha pottsiana* (Runyon's pincushion cactus), *C. tuberculosa* var. *tuberculosa* (cob cactus), and *C. tuberculosa* var. *varicolor* (varicolor cob cactus).

(i.e., west of the Pecos River), although reportedly it occurs near the mouth of the Pecos River (Weniger, 1984). The taxon has been verified in similar habitats not more than 25 miles away on the west bank of the lower Devils River, now Amistad Reservoir. Both Benson (1982) and Weniger (1984) used the epithet *robertii* for this taxon before Zimmerman (forthcoming) determined that the oldest name is *pottsiana*. The name *pottsiana* presumably is after F. H. Potts (see *Mammillaria pottsii*, p. xxx) or John Potts (see *Opuntia pottsii*, p. xxx).

Identifying Characters. The plants of *C. pottsiana* are in clumps of a few to several or numerous stems. The clumps may be 10–30 cm or more wide in Val Verde County and reportedly are up to 1 m wide in the lower Rio Grande valley, but usually they are small and inconspicuous. The individual stems are ovoid to cylindroid, at maturity 3–6 cm long and 1.5–3 cm in diameter, densely covered with spines and bristly in aspect. In Val Verde County plants we have evaluated, there are 25–38 spines per areole, including (5–)7–8 inner centrals and 20–30 outer centrals and radials. The central spines spread in all directions, some of them porrect, either white with dark tips or dominated by reddish-brown or nearly black distal portions. The radial spines are white, or off-white, sometimes with brown tips.

The flowers are pale purplish-brown, 1.7–2.5 cm long, and 1–1.5 cm in diameter. Actually, there appears to be considerable flower color variation in *C. pottsiana*, with this being evident in the descriptions of flower color by different authors. In flowers we have examined from Val Verde County populations, the inner tepals have broad midregions, these purplish (maroon) or lavender to pale purple, or faintly so, with pale brownish margins. The midregions have been described by other authors as reddish-brown or maroon, and the margins as tan, buff, or greenish-yellow. The inner tepals are linear to lanceolate, 9–11 mm long, 2–3 mm wide, attenuate, acute, or apiculate at the apex, and with entire margins. The outer tepals have greenish-purple midregions and pinkish margins. The outer tepals basically are lanceolate, to 1.5 cm long, ca. 3 mm wide, acute, and with conspicuously fringed margins. Yellow anthers 0.5–1 mm long are supported by pink, purplish, magenta, or colorless filaments that are 5–6 mm long. The stamens are sensitive, in that they twist around the style after being touched. The style is ca. 1 cm long, and green to reddish-brown. The 5–7 stigma lobes are green and 1.5–2 mm long.

The fruits of *C. pottsiana* are red in the Val Verde County populations, so far as known, and they are red in plants of South Texas and adjacent Mexico. The fruits are spherical to ovoid, 6–9 mm in diameter, juicy, the juice reddish, and moderately sweet. The floral remnant is persistent. Some populations in Mexico have green fruits (Zimmerman et al., forthcoming). The seeds are comma-shaped, 1–1.2 mm long, pitted, brown or reddish-brown, at least when fresh, perhaps drying darker. The hilum is clearly basal.

Phenology. In Val Verde County flowers of *C. pottsiana* usually are open from mid-February to mid-March. In some seasons at least, flowers may appear in early February. In greenhouse cultivation flowers close at night and open again before noon for up to 6–8 days. One seven-stemmed greenhouse plant (from Val

Verde County) produced a total of 32 flowers over a period of 48 days, from 1 February to 20 March. Under greenhouse cultivation, plants from Val Verde County produced fruits (after cross-pollination) that remained green for 2–3 months before turning red. The medium to dull red color appeared rather rapidly over a period of about 3–6 days, developing from the fruit base to the apex.

Sterile and Immature Specimens. The whitish clumps of *C. pottsiana* are decorated with reddish-brown or reddish-black diffuse central spines. These branched clumps resemble *C. sneedii* in habit and in most other vegetative and floral features, except for the smaller seeds, different internal anatomy, and green stigma lobes of *C. pottsiana*. *Coryphantha sneedii* is not sympatric with *C. pottsiana* in Texas. In Val Verde County, *C. pottsiana* might be confused at first glance with *Mammillaria prolifera* or *Neolloydia conoidea*, because of their clustering habits, but these taxa are obviously distinct in most vegetative and floral characters. *Coryphantha tuberculosa* var. *tuberculosa* likewise resembles *C. pottsiana* in its clumped habit and many other details, including the fruit and seed, but *C. tuberculosa* is not clearly documented to occur in Val Verde County.

The stem surfaces of *C. pottsiana* are barely visible, if at all, through the dense covering of spines. Relatively closely spaced tubercles, ca. 6 mm long, display circular, spine-bearing areoles 1.5–2 mm in diameter. The areolar groove, extending down the tubercle to the stem axis, is densely woolly on fresh growth. The inner central spines are acicular, projecting, with at least some of them more or less porrect. The innermost centrals are 1.3–1.5 cm long. The longest inner central is the uppermost one that is appressed-ascending and 1.8–2.3 cm long. The inner central spines clearly are more robust than the true radial spines and slightly larger in diameter than the outer radials that grade into the main radials. The central spines are pale yellow or tan at the base with reddish-brown, dark brown, or nearly black distal portions (one-half the length or more). The 20–30 outer central and true radial spines are slender, acicular, and white or gray throughout, or sometimes with pale red-brown tips. The longest radials are the upper and lateral ones at ca. 1 cm, and most all of the radials are appressed.

Immature stems are similar to those of adult plants except that they are smaller and have fewer spines per areole. The central spines are darkly pigmented only at the tips, or on the distal half in some centrals. Seedlings of *C. pottsiana* are even whiter in aspect than juveniles, with fewer and shorter spines per areole.

Biosystematics. *Coryphantha pottsiana* clearly belongs in the subgenus *Escobaria*, perhaps most closely related to *C. dasyacantha* and *C. chaffeyi*. *Coryphantha pottsiana* has many vegetative and floral characters in common with *C. sneedii*, *C. dasyacantha*, and *C. chaffeyi* (Zimmerman, 1985). In Texas the distribution of *C. pottsiana* is allopatric with respect to all of the other escobarias, restricted to the Rio Grande corridor, from the mouth of Devils River, where it is well documented on the west shores of Lake Amistad, to the Gulf of Mexico.

Certain Mexican populations of *C. pottsiana* have green fruits, fewer and larger stems, slightly longer and more numerous spines, and unusual hairlike upper radial spines present in unweathered young areoles (mostly near the stem apex). In Val Verde County populations of *C. pottsiana,* plants so far examined often exhibit the hairlike upper radial spines along with the normal straight spines.

In Warnock (1970) the photograph labeled "no. 4" on page 86 is of *Coryphantha pottsiana,* and the discussion under *Coryphantha vivipara* on page 87 is appropriate for *C. pottsiana* and not for *C. vivipara.* On page 86 of Warnock (1970), obvious transposition of captions with an adjacent photograph of *Ariocarpus fissuratus* (labeled "no. 5") is merely a printing error. Only misidentification can explain Warnock's use of the name *C. vivipara* associated with the photograph (p. 86) and the associated discussion (p. 87).

Synonyms. *Mammillaria emskoetteriana* Quehl; *Escobaria runyonii* Britton & Rose (*non Coryphantha runyonii* Britton & Rose); *C. emskoetteriana* (Quehl) A. Berger; *C. robertii* A. Berger; *C. muehlbaueriana* Boed.; *C. piercei* Fosberg; *E. emskoetteriana* (Quehl) Borg; *Neobesseya muehlbaueriana* (Boed.) Boed.; *E. muehlbaueriana* (Boed.) F. M. Knuth; *M. escobaria* Cory; *E. emskoetterana* (Quehl) Backeb. (comb. superl.); *E. bella* Britton & Rose.

Common Names. Runyon's escobaria; junior Tom Thumb cactus.

9. **Coryphantha tuberculosa** (Engelm.) A. Berger. COB CACTUS, VARICOLOR COB CACTUS. Plates 286–89. [*Mammillaria tuberculosa* Engelm., Proc. Amer. Acad. Arts 3: 268. 1856]. Rocky exposures, sedimentary and igneous, desert and mountain grassland or juniper-oak woodland. Throughout most of the Trans-Pecos, except extreme NE portion. 1,700–6,700 ft. Flowering Apr–Aug. 2*n* = 22. In TX restricted to the Trans-Pecos, except perhaps in Val Verde Co. Southern NM. Mexico, to S Coahuila. Map 61.

Four varieties of *C. tuberculosa* are recognized by Zimmerman et al. (forthcoming), two of them restricted to Mexico in parts of Coahuila, Durango, and Chihuahua. Two varieties occur in the Trans-Pecos. Benson (1982) erroneously treated *C. tuberculosa* var. *varicolor* as a variety of the distantly related species *C. dasyacantha.* Benson treated *C. tuberculosa* as *C. strobiliformis* (Poselger) Moran while recognizing three varieties, two of which clearly belong with *C. sneedii.* Bravo-Hollis and Sánchez-Mejorada (1991b), like Benson, used the epithet *strobiliformis* and erroneously treated part of *C. sneedii* (under the misapplied name "*durispina*") as a variety of this species. Unlike Benson, however, Bravo-Hollis and Sánchez-Mejorada accurately included the var. *varicolor* with *C.* "*strobiliformis*" (= *C. tuberculosa*). Weniger (1984) used the correct specific epithet for the species (as *Mammillaria tuberculosa*), but he treated var. *varicolor* as a separate species. The specific epithet, *tuberculosa,* is intended to be descriptive, after the Latin *tuberculum,* "knob," and *osa,* as a termination meaning "full of," apparently in reference to the bare tubercles collectively lending a corncob aspect at the base of old stems in many plants.

Key to the Varieties

1. Stems almost always branched, forming clumps of 3–50 or more stems; mature stems ovoid or cylindroid, often pointed at the apex, (2.2–)3–4(–5) cm in diameter; plants usually restricted to limestone habitats, or at least to near limestone
                                                9a. *C. tuberculosa* var. **tuberculosa**.

1. Stems almost always solitary, rarely branched after injury, mature stems oblate or globose, in old age cylindroid, (2.9–)3.6–5.6(–6.5) cm in diameter; plants usually restricted to igneous or novaculite habitats
                                                9b. *C. tuberculosa* var. *varicolor*.

9a. **Coryphantha tuberculosa** (Engelm.) A. Berger var. **tuberculosa**. COB CACTUS. Plates 286, 287. Limestone habitats, desert mountains, ridges, eroded outcrops, rarely extending into adjacent igneous habitats, often associated with lechuguilla. Throughout most of the Trans-Pecos, from El Paso Co. E to W Terrell Co., absent or uncommon in Jeff Davis and Reeves counties, in Presidio and Pecos counties mostly in S portions, reported but not reliably documented in Val Verde Co. at the mouth of the Devils River. 1,700–5,500 ft. Flowering Apr–Aug. $2n = 22$. Southern NM, Doña Ana, Otero, and Eddy counties, possibly Luna and Sierra counties. Mexico, S to central Coahuila and NE Durango, E Chihuahua. Map 61.

In the Trans-Pecos *C. tuberculosa* var. *tuberculosa* is a common species in desert limestone habitats, where potentially it might be confused in vegetative condition with several other densely spined species. When in flower, *C. tuberculosa* is readily distinguished from any other species in the Trans-Pecos, provided that either (1) one knows to check the anther color, the stigma color, and the internal structure of the receptacular tube, or (2) one has prior familiarity with the particularly wide, pale aspect and almost vespertine timing of the fully open flowers in late afternoon.

Identifying Characters. Plants of var. *tuberculosa* frequently grow in crevices of limestone rock and almost always form clumps of branched stems. Commonly, each clump is represented by 3–20 stems, but large, presumably older plants may have 50 or more stems. Probably less than 1% of adult plants are unbranched (Zimmerman et al., forthcoming). The stems are ovoid to cylindroid and sometimes slightly pointed at the apex. The stems are 5–14(–16) cm long and (2.2–)3–4(–5) cm in diameter. Spine color is a subtle but often useful trait in distinguishing *C. tuberculosa* from several other taxa. The general aspect of the spines of Trans-Pecos varieties of *C. tuberculosa* are white, pale gray, or pale tan, not the gleaming translucent-white typical of certain other species (Zimmerman et al., forthcoming), including *C. sneedii*. The tips of larger spines are pinkish-gray or tan, or darker gray-brown or reddish-brown. Typically in var. *tuberculosa* there are only 1–2 inner central spines, usually two in areoles of adult stems. If there is only one, it is in a lower position and usually 0.5–1 cm long, but perhaps to 1.5 cm long. If there are two inner centrals, the one in the lower position is always the shortest, stoutest, and either porrect or deflexed, and the longer upper one is 1.3–1.8 cm and either ascending or porrect. In some plants, presumably with age, there appear to be four inner centrals (the upper ones may be interpreted as outer centrals), still with the lower one always the shortest and stoutest. In each areole usually there is only one porrect central. There are 17–25(–36) light gray radial spines.

The flowers of C. tuberculosa are unique in the genus, but the most important features are best observed in fresh condition. In particular, (1) the sterile distal portion of the receptacular tube is unusually long, almost tubular, between the stamens and the inner tepals, and (2) the anthers and pollen in var. *tuberculosa* are very pale yellow, almost white, probably lighter in color than those of any other species of *Coryphantha*. The 4–6(–8) stigma lobes are white and (1.8–)2–4 mm long. The flowers are (1.8–)2–3(–3.2) cm long and 2.1–3.5(–4) cm in diameter. The flower shape is funnelform to salverform when fully expanded. The inner tepals are (0.9–)1.1–1.6(–1.9) cm long, 1.5–2.5(–3.5) mm wide, and nearly white to pale pink or pale rose-pink in color. Typically, the inner tepals are more intensely pigmented in their central regions, but a prominent midstripe seldom is evident. The outer tepals are conspicuously fringed.

The dull to bright red fruits of var. *tuberculosa* are broadly ellipsoid to ovoid as exposed above the spines but seem to be broadly clavate as seen when separated from the plants. The fruits are (0.8–)1.3–2.5 cm long and 3.5–6.5(–7.5) mm in diameter. A variety of C. tuberculosa in southeastern Chihuahua has green, maroon, or, at best, only rarely red fruits (Zimmerman et al., forthcoming). The fruits of both the Trans-Pecos varieties always are red when fully ripe. In the Big Bend area, we have observed nearly full-size green fruits positioned just inside (toward the apex) bright red or dull red fruits on the same stem, as though nearly mature green fruits turn red abruptly at full maturity. The seeds of var. *tuberculosa* are reddish-brown, pitted, and 0.95–1 mm long, always smaller than the seeds of C. sneedii (0.9–1.6 mm long.)

Phenology. The typical flowering period for var. *tuberculosa* is in the spring. The flowering period throughout most of its range may be May to August (Worthington, 1986), in general later in the spring than C. sneedii. In the hottest parts of the lower Big Bend, near the Rio Grande, var. *tuberculosa* may be in full flower by late March or the first of April. Flower development appears to be progressively later at slightly higher elevations away from the Rio Grande. In one season on 18 April, var. *tuberculosa* was observed to be in near full flower at the tunnel in the south part of Big Bend National Park and in mature bud stage at Persimmon Gap at the north edge of the park. Apparently blooming in June–August is sporadic, possibly in response to ample rainfall (Worthington, 1986). The sporadic flowering periods may last 1–10 days. In August both flowers and red fruits have been observed on the same stems of var. *tuberculosa*. Under cultivation in Alpine, following rains, flowers have been observed as early as April and as late as early September on stems with red fruits. Individual flowers close at night and open again the next day for more than one day. Flowers may be partially open as early as noon, but they characteristically open relatively late in the afternoon, about 3:00–4:00 P.M., and remain open until dusk or into early evening. Fruit maturation appears to require 3–4 months. Red fruits may persist through the winter, on some plants at least until April.

Sterile and Immature Specimens. The roots of *C. tuberculosa* are fibrous and diffuse. The pith and cortical tissues are firm and nonmucilaginous, with very small, mostly spheroidal druses. The rather slender and numerous tubercles are conoid-cylindroid, (6–)8–11 mm long, 3–6 mm in diameter. Tubercles on the lower stem at ground level or below are naked because the spine clusters and often the whole tubercle tips have abscised, revealing in many older stems a characteristic corncob aspect. In plants with older stems, the "warty," spineless tubercle bases often are evident at or above ground level, or the spineless bases may not be visible until the stems are excavated. The areoles are circular, 2.5–3 mm in diameter, with the tan spine bases forming a tan inner circle. The gray-green to green tubercles and stem surfaces are more or less obscured by the spines. The spines are straight, glabrous, and brittle. Individually and in overall aspect the spines may be white, but not snowy-white, or may be gray to grayish-tan. The spines do not exceed 2 cm long. Usually there are two inner central spines in upper and lower positions. The inner centrals are stout, 0.25–0.5 mm thick, with bulbous bases. Often there are about five outer centrals (4–9 in all), the longest three of these usually in an upper position, appressed with the radials, but having the stouter appearance of central spines and being to 1.3–1.8 cm long. The central spines are whitish (but not as snowy-white or translucent as those of *C. sneedii*) or grayish-tan with the tips pinkish-tan to reddish-brown or reddish-black. Usually there are 19–25 radial spines, these whitish to light gray, (0.5–)0.7–1(–1.3) cm long, with laterally compressed and bulbous bases. The largest of the slender radials are only 0.1–0.2 mm thick.

Juvenile plants may have single stems, although in the field small and presumably young plants seem to branch early. Plants grown from seed in the greenhouse may remain unbranched at least for several months. Stems 1–2 cm tall and to 2 cm in diameter may be unbranched in cultivation. The spines of juvenile plants and seedlings are glabrous.

Biosystematics. *Coryphantha tuberculosa* is the type species of *Escobaria*, regarded here as a subgenus of *Coryphantha* (Zimmerman, 1985). In some previous cactus literature (e.g., Benson, 1982) *C. tuberculosa* was known as *C. strobiliformis*. The epithet *strobiliformis* is incorrectly applied to *C. tuberculosa*. *Coryphantha strobiliformis* (Poselger) Moran is a Mexican (Chihuahuan) taxon related to *C. henricksonii* (Glass & Foster) Glass & Foster.

Among Trans-Pecos cacti *C. tuberculosa* is vegetatively most similar to *C. sneedii* and *C. dasyacantha*. Many previous workers have misinterpreted the vegetative similarities as evidence of close relationship. Taylor (1979) placed *C. tuberculosa* in his "*Escobaria strobiliformis* group," an assemblage that included *C. sneedii*. Taylor (1986) later segregated *C. tuberculosa* from *C. sneedii* in recognizing an "*Escobaria sneedii* group" from which *C. tuberculosa* (as *C. strobiliformis*) was separated. In his concept of *C. tuberculosa* (likewise under the misapplied epithet *strobiliformis*), Benson (1969, 1982) included two varieties ("var. *durispina*" [misapplied to var. *albicolumnaria*] and var. *orcuttii*) that belong with *C. sneedii*.

There is little question that previous taxonomic treatments asserting close relationship between *C. tuberculosa*, *C. sneedii,* and *C. dasyacantha* were based upon superficially similar spine and habit features, at a time when reproductive characters were not well known or appreciated by the authors concerned. Actually, the practiced observer can easily distinguish field or pressed specimens of *C. tuberculosa*. The spines of *C. tuberculosa* are ashy-white or pale gray (with pinkish-tan or purplish tips). The central spines are in a modified "bird's-foot" arrangement, with the upper porrect inner central, when present, much longer than the lower central or other centrals and radial spines. The central spines seldom radiate evenly in *C. tuberculosa*. They always radiate, spokelike, in *C. sneedii* and usually in *C. dasyacantha*. Also, in *C. sneedii* the spines typically are relatively bright and snowy-white. On average, *C. tuberculosa* has fewer spines per areole than does *C. sneedii* or *C. dasyacantha*. Field experience and familiarity with the overall aspect of *C. tuberculosa,* particularly the typical appearance that results from spine color and central spine position, allow tentative identification at a considerable distance from the plants. Internally, *C. tuberculosa* lacks the giant lenticular druses that are characteristic of *C. sneedii* and *C. vivipara*.

Coryphantha tuberculosa blooms in late afternoon and evening, unlike *C. sneedii* or any other species of *Coryphantha*. The flowers of *C. tuberculosa* are pale pink, pale lavender-pink, or pure white; they open widely with reflexed tepals, and they are larger than the small, pale, often striped flowers that do not open very widely in *C. sneedii* and *C. dasyacantha*. The very pale yellow, nearly white anthers in *C. tuberculosa* are unique in *Coryphantha*. The stigma lobes in *C. tuberculosa* and *C. sneedii* are white, and the stigma lobes are green in *C. dasyacantha*. Other floral differences are discussed by Zimmerman (1985). The fruits of *C. tuberculosa* are red, in populations north of Mexico, like those of *C. dasyacantha*, in contrast with the green fruits of *C. sneedii* except in *C. sneedii* var. *albicolumnaria*, which has red and green fruit morphs. The reddish-brown seeds of *C. tuberculosa* consistently are smaller than the reddish-brown seeds of *C. sneedii* and also slightly smaller than the black seeds of *C. dasyacantha*.

The precise interspecific relationship of *C. tuberculosa* is not understood. At present *C. tuberculosa* seems to stand alone, with its unique flowers, but it shares confusing vegetative resemblance with many other species, such as *C. sneedii* and *C. pottsiana*. As stated above, however, *C. tuberculosa* does not seem to be closely related to *C. sneedii* or any of the other Trans-Pecos coryphanthas. Genetic isolation between *C. tuberculosa* and *C. sneedii* is supported by the apparent lack of hybridization between these species or any other coryphanthas with which *C. tuberculosa* is sympatric. Benson (1969, 1982) reported intergradation between *C. tuberculosa* and *C. sneedii* in the Franklin Mountains, but the study by Zimmerman (1985) found no evidence of any hybridization between the species. In fact, *C. tuberculosa* consistently blooms later in the spring than does *C. sneedii*. The flowers of *C. tuberculosa* open in late afternoon, whereas those of *C. sneedii* open around noon. In certain years there may be some overlap in spring flowering periods, and some overlap in diurnal opening times. Sporadic summer flowering, commonplace in *C. tuberculosa*, but rare in *C. sneedii*, might occasionally bring *C. tuberculosa* and *C. sneedii* into bloom simultaneously, thus providing additional opportunities for cross-pollination. In southeastern Brewster County we have once (May 1981) observed *C. tuberculosa* var. *tuberculosa* and *C. sneedii* var. *albicolumnaria* growing side by side and in full flower at the same time. Thus, there are some opportunities for pollen transfer between the species but still no evidence of hybridization.

Curiously, Weniger (1984, p. 195) remarked that *C. tuberculosa* was "much less common than usually thought." Our own field experience has revealed that *C. tuberculosa* is widespread and common in certain habitats in the Trans-Pecos.

According to Zimmerman (1985) the lectotype of *Mammillaria tuberculosa* is a mixed collection consisting of *C. tuberculosa* var. *tuberculosa* and an undescribed variety of *C. sneedii* (A. D. Zimmerman, forthcoming). The illustration published by Engelmann (1856, p. 268) is of *C. tuberculosa* and not *C. sneedii*. Zimmerman (1985, p. 390) discovered that Benson (1969b, 1982) "treated most specimens of *C. sneedii* var. *albicolumnaria* as *C. tuberculosa* var. *tuberculosa*," resulting in subsequent partially inaccurate descriptions and a poorly defined concept of *C. tuberculosa*.

Synonyms. *Mammillaria strobiliformis* Scheer ex Salm-Dyck; *Escobaria tuberculosa* (Engelm.) Britton & Rose; *Coryphantha tuberculosa* (Engelm.) Orcutt.

Common Names. Corncob escobaria; cob cory cactus.

9b. **Coryphantha tuberculosa** (Engelm.) Orcutt var. **varicolor** (Tiegel) A. D. Zimmerman, comb. nov. (forthcoming). VARICOLOR COB CACTUS. Plates 288, 289. [*Coryphantha varicolor* Tiegel, Monatsschr. Deutsch. Kakt. Gesellsch. 3: 278. 1932]. Usually in volcanic mountains or other igneous rock habitats, oak-juniper woodlands and rocky sites in grasslands to lower elevations in desertscrub, and in novaculite exposures in the Marathon Basin and Solitario, associated with grassland or desertscrub, rarely extending onto limestone substrates. Mostly Presidio, Jeff Davis, and Brewster counties, rare in S Pecos Co. 2,300–5,500(–6,500) ft. Flowering May–Aug. $2n = 22$. Known only from Trans-Pecos TX, but surely not endemic, considering its abundance on the peaks overlooking adjacent Coahuila and Chihuahua, Mexico. Map 61.

The var. *varicolor* predominantly is a taxon of the igneous central mountains of the Trans-Pecos, but it extends south into the Chinati and Chisos mountains and into novaculite substrates. The varietal epithet is derived from the Latin prefix *vario*, "variable," and *color*, presumably in reference to the different colors in central and radial spines.

Identifying Characters. The var. *varicolor* is distinguished by its single-stemmed plants that are similar in aspect to the stems of var. *tuberculosa* except that they are larger in diameter. The globose, oblate, or cylindroid stems are 4–13(–16) cm long and (3–)3.6–6(–6.5) cm in diameter. Plants growing in novaculite substrates tend to be smaller in size. In var. *varicolor* the green to gray-green stem surfaces are more visible through the spines than in var. *tuberculosa*, evidently because there are fewer, shorter, and more slender spines. Also, the tubercles are slightly farther apart on slightly broader stems than in var. *tuberculosa*. In var. *varicolor* there are 1–2 inner central spines, usually two, with the shortest one porrect or descending and in a near-central position, and an upper one (or three) that is longer and erect or ascending. The upper inner central spines also may be interpreted as outer centrals. There are 15–21(–24) radial spines appressed against the plane of the stem. In var. *varicolor* the spines are slightly fewer (radials), shorter, more slender, and the centrals with more color, than in var. *tuberculosa*, although spine clusters in the absence of stems do not provide for reliable identification of the varieties.

The pale pink to white flowers of var. *varicolor* are virtually identical to those of var. *tuberculosa*. They are 2–3(–3.2) cm long and 2–3.5(–4) cm wide when fully open with reflexed tepals. The inner tepals are 1–1.9 cm long, 1.5–2.5(–3.5) mm wide, and acute to attenuate at the apex. The outer tepals are conspicuously fringed. The anthers are very pale yellow. The stamen filaments are ca. 8 mm long and cream-colored. The style is 1–1.3 cm long, and it supports 4–6(–8) white stigma lobes that are ca. 2.5 mm long. The fruits and seeds of var. *varicolor* are identical to those of var. *tuberculosa*.

Phenology. The general flowering period for var. *varicolor* appears to be from May to August. Preliminary observations suggest that populations at lower elevations flower earliest in the season, in May as expected, and that they may blossom again in June or later in the summer. A July blooming period for var. *varicolor* has been observed in the Davis Mountains. These were casual observations over a period of several years, and it is not known if the flowering episodes in July represent a main blooming period or if they reflect sporadic or opportunistic late-season flowering. Opportunistic blooming may be the rule in var. *varicolor*. One plant four miles south of Marathon was observed blooming on 16 September 1991. Cultivated plants in Alpine have bloomed as early as 13 April and as late as early September. It is suspected that fruit maturation requires 3–4 months. We have observed red fruits on plants in October. Specific phenological studies in var. *varicolor* are needed.

Sterile and Immature Specimens. Living specimens of var. *varicolor* are most reliably distinguished from var. *tuberculosa* by their broader, single stems and more evident stem surfaces. In pressed specimens, where it may not be apparent if the source plant was single-stemmed, tentative distinction between var. *varicolor* and var. *tuberculosa* can be achieved by noting stem size and the density of spines covering the stems. The var. *varicolor* might be confused with *C. sneedii* var. *albicolumnaria* in the southern Big Bend region, although var. *varicolor* usually occurs in igneous substrates, whereas var. *C. sneedii* occurs on or near limestone. Both var. *varicolor* and *C. dasyacantha* are vegetatively similar and sometimes cannot be identified without their strikingly different flowers or seeds.

The details of spine characters for var. *varicolor* are about the same as those discussed under var. *tuberculosa* (above), although there are subtle differences. In var. *varicolor* there are (0–)1–2 porrect inner central spines. The lowest inner central is the shortest, 0.4–0.6 mm, and the upper central is the longest, 1.1–1.7 cm. The lower inner central is porrect or descending, and the upper inner central is porrect or ascending and stands inside 1–2 (commonly two) upper lateral inner centrals of about the same size. Usually there are four obvious central spines (one lower and short, one upper and longest, and two laterals). The central spines in general appear to be shorter and more slender than the equivalent spines in var. *tuberculosa*. In var. *varicolor*, the central spines are tan to pinkish-tan, are pale reddish distally, and often have darker reddish-brown tips. The centrals in upper stem areoles are the most colorful and usually more colorful than in var. *tuberculosa*. The radial spines are gray and appressed. The radials are fewer in number than in var. *tuberculosa* and somewhat shorter at 7–8 mm long in many central mountain populations, although some populations of var. *varicolor*, mostly in south Brewster and Presidio counties, have radials 8–12 mm long.

The seedlings and juvenile plants of var. *varicolor* have fewer, shorter spines than do adults. Juvenile stems ca. 1 cm long produce the characteristic 1–2 inner central spines, one of them porrect, at least under cultivated conditions. Particularly in novaculite outcrops of the Marathon Basin, immature plants of var. *varicolor* are characterized by depressed-globose or flat-topped stems with "smooth" surface features. Here the porrect central spine is short, and the outer central spines are more or less appressed with the radial spines.

Biosystematics. Plants that appear to be intermediate between var. *varicolor* and var. *tuberculosa* occur in some populations in the lower Big Bend region, for example, near Shafter in Presidio County, in Brewster County on the south slopes of Elephant Mountain, and in the Paint Gap Hills in Big Bend National Park. Seemingly intermediate specimens mostly are in the form of solitary, relatively slender stems lacking the dense spination of var. *tuberculosa*, or solitary, relatively thick stems with the dense spination of var. *tuberculosa*.

Plants of *C. tuberculosa* in novaculite substrates of the Marathon Basin are different in vegetative appearance from var. *varicolor* throughout most of its range. In novaculite the stems are usually flat-topped and short, and the spine covering is relatively smooth and innocuous. The smooth appearance is the result of shorter, appressed radial and outer central spines and shorter porrect central spines. Superficially, the spines appear more slender as well, but microscopic examination suggests that the smaller-spined form is shorter-spined but not thinner-spined. Preliminary observations also suggest that the smaller-spined form tends to include more multistemmed plants than in some other populations of var. *varicolor*. At present we regard the small-spined form as an edaphic variant, because plants with similar appearance are of sporadic occurrence elsewhere in the distribution of var. *varicolor*, where they often are interpreted simply as juveniles, and because the typical form of var. *varicolor* with longer spines occurs sporadically in novaculite habitats in the Marathon Basin. Remarkable edaphic endemism in Cactaceae (e.g., *C. minima* and *E. davisii*) is associated with novaculite in the Trans-Pecos, inviting further investigation of *C. tuberculosa* in this substrate.

In Benson (1982), figure 892 is a photograph of *C. tuberculosa* var. *varicolor*, which is labeled "*C. dasyacantha* var. *varicolor*."

Synonyms. *Escobaria varicolor* Tiegel; *E. dasyacantha* (Engelm.) Britton & Rose var. *varicolor* (Tiegel) D. R. Hunt.; *Coryphantha dasyacantha* (Engelm.) Orcutt var. *varicolor* (Tiegel) L. D. Benson; "*Mammillaria varicolor*" (Tiegel) Weniger, *comb. nud.*

Common Names. Cob cactus; varicolor cory cactus; varicolor cactus; mountain cob cactus.

10. **Coryphantha macromeris** (Engelm.) Lem. var. **macromeris** BIG-NEEDLE PINCUSHION CACTUS. Plates 290, 291. [*Mammillaria macromeris* Engelm., Mem. Tour N. Mex. 97. 1848]. Most common in sandy alluvium, gravel benches, or clay, in the open or under shrubs, various substrates but rarely on steep rocky slopes or in crevices. Throughout the Trans-Pecos, perhaps most common in southern Presidio and Brewster counties. 1,500–4,000(–6,000) ft. Flowering sporadically May–Sep, mostly Jun–Aug. $2n = 22$. In TX E and NE to Glasscock and Ector counties. Southern NM, Chaves, Eddy, Lea, Doña Ana, Sierra, and Luna (near Columbus) counties. Mexico, NE Chihuahua, Coahuila, NE Durango, and N Zacatecas to Tamaulipas. Map 62.

Coryphantha macromeris is one of the most widespread cactus species in the Trans-Pecos, occupying nearly all substrate types, including gypsum. In Texas two varieties of *C. macromeris*, var. *macromeris* and var. *runyonii* (Britton &

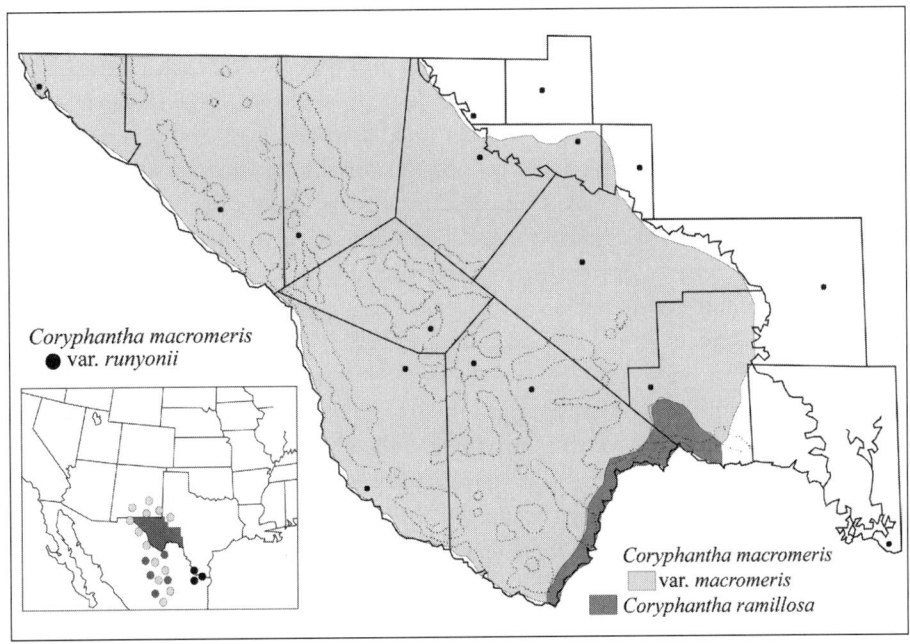

Map 62. Generalized distribution of *Coryphantha macromeris* var. *macromeris* (big-needle pincushion cactus), *C. ramillosa* (whiskerbrush pincushion cactus), and *C. macromeris* var. *runyonii* (Runyon's coryphantha).

Rose) L. D. Benson, were recognized by Benson (1982), but Weniger (1984) treated var. *runyonii* as a distinct species (as *Mammillaria runyonii* Britton & Rose). The var. *runyonii* (Plates 292, 293) is restricted to the lower Rio Grande valley, on gravelly hillsides and in silt often under shrubs near the Rio Grande, from about Roma in Starr County to near Brownsville in Cameron County, and in adjacent Tamaulipas, Mexico. The South Texas variety forms particularly untidy clumps, with relatively short tubercles usually 1–2 cm long, the result of relatively rapid proliferation of axillary branches. The epithet *macromeris* is after the Greek *makros,* "large," and *meros,* "a part," probably a reference to the elongated tubercles that characterize the taxon. The epithet *runyonii* is after Robert Runyon, an early (ca. 1920) botanist in the Rio Grande valley, who studied the taxon and provided specimens for Britton and Rose, who described it as a distinct species. *Coryphantha macromeris* and the remaining coryphanthas in the current treatment (species 10–14) belong to subgenus *Coryphantha,* unless *C. macromeris* is taxonomically segregated as the monotypic *Lepidocoryphantha* Backeb.

Identifying Characters. Mature plants of var. *macromeris* are cespitose, forming many-stemmed low mats or hemispheric mounds 5–25 cm high and up to 1 m wide. The stems are hemispheroidal to short-cylindroid, 5–23 cm long and 4–8(–13) cm in diameter. The pith and cortex are mucilaginous. The tubercles are conspicuous, somewhat flaccid and soft-skinned, narrowly conoid-cylindroid, and (1–)1.5–3(–4.5) cm long. *Coryphantha macromeris* is similar to *C. ramillosa* in the characters mentioned above except that *C. ramillosa* usually is single- or few-stemmed, the pith and cortex are not mucilaginous, and the tubercles are shorter. The presence or absence of mucilage in living plants usually can be determined simply by removing and examining a tubercle, even from a relatively small immature stem.

In *C. macromeris* there are 7–21 spines per areole, (1–)3–7 central, and (6–)9–15(–17) radial. In Trans-Pecos populations usually there are four central spines, these usually 3–7 cm long. The lowermost central spine is porrect or descending, and the other centrals are diffuse or weakly appressed. The central spines usually are angular in cross section, typically flattened and grooved on one side and rounded on the other side. The centrals commonly are dark gray to black, but in some plants they are pale gray to tan or stramineous. The radial spines are 1.5–2.5(–3.4) cm long (the longest ones), tan to stramineous or brown, and sometimes gray to whitish in age.

The bright rose-pink to magenta flowers are 3–5(–6) cm long, and 4–7 cm in diameter. The inner tepals are 3–4 cm long, 4.5–6 mm wide, and long-acuminate or acute and toothed or erose at the apex. In the Trans-Pecos the inner tepals are more commonly rose-pink than magenta, often with a darker midline and pale margins. The outer tepals are brownish to green with prominently fringed margins. The filaments are 1–1.7 cm long, greenish-white or purplish-pink distally, contrasting with the yellow anthers that are 1–1.5 mm long. The style is 2–2.7 cm long and supports 7–13 stigma lobes that are white or cream-colored. The stigma lobes are 4–6 mm long, often with slender, soft points at the tips.

The fruits of *C. macromeris* are ovoid to obpyriform, 1.4–2.5(–3) cm long, 1.2–1.8 cm in diameter, and dark green at maturity. The fruit pulp is whitish to pink. The floral remnant is persistent. Mature fruits have a few scales at the top (the basis for the generic name *Lepidocoryphantha*), near the base of the floral remnant. The reddish-brown seeds are finely and weakly raised-reticulate, as seen under magnification, but otherwise appear to be nearly smooth. The seeds are 1.2–1.5 mm long and more or less comma-shaped to globose.

Phenology. In the Trans-Pecos *C. macromeris* most commonly blooms in June, July, and August, but plants may produce flowers as early as May and at least as late as September. We have observed numerous plants in the same population all in full flower at the same time (large plants, each with 20–30 or more open flowers), while on many other occasions, only a few, scattered, multi-stemmed plants in a given population produced just one or two open flowers. Summer flowering in *C. macromeris* appears to follow adequate rainfall. This opportunistic pattern, with 4–6 flowering periods in the same season, has been documented by Worthington (1986) in the Franklin Mountains and has been reported for plants cultivated in Alpine by G. G. Raun (pers. comm.). Worthington noted that flowers of *C. macromeris* close at night and do not open again after the first day. Usually flowers are fully open, with reflexed tepals, by late morning or early afternoon. Fruits appear on plants from August to December. We suspect that fruits mature in 3–4 months, but we have no documentation for this. Ripe fruits have a pleasant, sweet aroma, sweeter in smell than a ripe banana, but the taste is not correspondingly sweet.

Sterile and Immature Specimens. Plants of *C. macromeris* with single or only a few stems are not as commonly encountered in the field as are plants with several to numerous stems. Branching begins near the time of sexual maturation, or even earlier, with new stems developing from areolar grooves of lower tubercles, ultimately forming flat or mounded clusters of 20–50 stems in some plants that commonly are 20–100 cm across. The stems are deep-seated but not flat-topped, above fleshy taproots. The stems are dark green to gray-green or blue-green, globular to cylindroid, with conspicuous tubercles and dark central spines. Features of the internal stem also are useful in identifying sterile specimens. In addition to the mucilaginous pith and cortex of *C. macromeris*, the slender pith (10% of stem diameter) lacks a medullary vascular system. The vegetatively similar and nonmucilaginous *C. ramillosa* has a wide pith (ca. 50% of stem diameter) with a conspicuous medullary vascular system.

The relatively large, soft-skinned tubercles, usually 1.5–3 cm or more long and 0.6–1.5 cm wide at the base, are an important feature in tentative identification of sterile specimens. Even more important are the areolar grooves that extend only one-half to three-fourths the length of the tubercles, from the spine clusters toward the tubercle axils. In mature plants of all other *Coryphantha* species, the areolar grooves extend the full length of the tubercles. The areoles at the tip of each tubercle are circular, ca. 4 mm in diameter, with white wool, the wool diminishing with age. Areolar glands are sporadically present. The usually four central spines are pale gray to black and radiate in all directions, and are

3–7 cm long. The stem and plant aspect is dominated by the dark, projecting central spines. The centrals often are slightly flexible, sometimes slightly curved, characteristically angular to straight in cross section, with large-bulbous bases that are a golden hue. Although black or very dark central spines are characteristic in *C. macromeris* throughout most of the Trans-Pecos, some plants examined have pale gray, tan, or stramineous centrals (e.g., in the Franklin Mountains). In Trans-Pecos populations of *C. macromeris,* usually there are 2–3 upper subcentrals appressed in near–radial spine position. These subcentrals are most like the main centrals in their angular and grooved appearance, except that they usually are gray and smaller than the main centrals. Usually there are 12–14 radial spines in Trans-Pecos plants of *C. macromeris,* but the number may vary from 6 to 17. The radials are dorsiventrally flattened with prominent, golden bulbous bases, these ca. 2 mm thick. The radials are gray to nearly white or tan to stramineous in color.

There are several spine differences between *C. macromeris* and the superficially similar *C. ramillosa*. Differences in central spines include the usually four, angular, radiating centrals. Actually in *C. macromeris,* typically one porrect or descending lowermost central can be distinguished, with the other three radiating at all angles. In *C. ramillosa* there is a single main (inner) central that is terete and porrect and usually three upper (outer) centrals that are somewhat appressed and ascending, either terete or angled. Radial spine differences include the dorsiventrally flattened radials of *C. macromeris* and the laterally compressed or terete radials of *C. ramillosa*. At present these spine differences are based upon the examination of a limited number of specimens of *C. ramillosa*.

Immature specimens of *C. macromeris* are variable in many characters, including spines reduced in number and length. In some juveniles the central spines may be absent, or the number may be reduced to 1–3, and in immature plants the centrals often are stramineous to gray. Often there are only 5–9 radial spines in immature plants of var. *macromeris*. The spines are glabrous in juveniles as they are in adults. The strongly mucilaginous cortex of *C. macromeris* is useful in tentative identification of juvenile or stunted individuals.

Biosystematics. The widespread *C. macromeris* is remarkably facultative in its tolerance of ecological conditions in the Trans-Pecos. The species is found in many habitats and most substrate types where one might expect to find cacti in the region, except in high mountains, wooded areas, steep slopes, and rock crevices. The species is most common in desert habitats, particularly in desertscrub vegetation. Populations may consist of scattered plants over large areas or numerous plants in restricted habitats. The densest populations we have observed are in the clay, silt, or gravel soils near the Rio Grande in Presidio and Brewster counties. The largest plants we have observed were spreading mats in Cretaceous clays south of the Chisos Mountains and mounded clusters in Cretaceous clays near Study Butte.

Coryphantha macromeris var. *runyonii* (Runyon's coryphantha), restricted to the lower Rio Grande valley near sea level (Map 62), is geographically isolated from other parts of *C. macromeris*. Some authors have recognized var. *runyonii* as a distinct species. In habit the plants are small matted clumps, with gray-green irregular stems 5–7 cm long and 1.5–3.8 cm in diameter. The stems display tubercles 1–1.8 cm long with very broad bases, 1–4 central spines to 5 cm long, and 3–11 radial spines. The spines are terete, radiating, and gray and brown in color.

Synonyms. Mammillaria heteromorpha Scheer; *Echinocactus macromeris* (Engelm.) Poselger; *Lepidocoryphantha macromeris* (Engelm.) Backeb.; *Coryphantha macromeris* (Engelm.) Orcutt.

Common Names. Nipple beehive; big needle cactus; long mamma; big nipple cory-cactus.

11. **Coryphantha ramillosa** Cutak, Cact. Succ. J. (U.S.) 14: 164. 1942. WHISKERBRUSH PINCUSHION CACTUS. Plates 294, 295. Desert limestone hills, ridges, benches, and mesa slopes, associated with *Agave lechuguilla*. Brewster and Terrell counties, near the Rio Grande. 1,350–3,000 ft. Flowering Aug–Sep. $2n = 22$. Mexico, N Coahuila, adjacent to Big Bend National Park and southeastern Brewster Co., and probably adjacent to Terrell Co., S to near Cuatro Ciénegas. Map 62.

In Texas *C. ramillosa* is restricted to certain limestone habitats near the Rio Grande in Brewster and Terrell counties. The species was federally listed as Threatened in 1979 and listed in Texas as Threatened in 1983. The specific epithet apparently was derived from the Latin *ramulosus*, "full of branches," but with reference to its superficial appearance in the field rather than its actual architecture.

Identifying Characters. One of the most conspicuous features reportedly distinguishing *C. ramillosa* from the regionally sympatric and superficially similar *C. macromeris* (Benson, 1982; Weniger, 1984) is the single or sparingly branched stems of *C. ramillosa* and the multibranched mats or mounds of *C. macromeris*. In Trans-Pecos populations, at sites where it is possible to observe several plants, the branching habit is a good distinguishing character for *C. ramillosa* and *C. macromeris*. Populations of *C. ramillosa* in Brewster County are predominated by plants with single stems, although some plants with 8–10 stems also are present in these populations. The same sparsely branching habit has been reported for populations in Terrell County, with about 60% of the plants unbranched, although plants with 22 or more stems are found south of Sanderson (J. Talbot, pers. comm.). In limestone hills across the Rio Grande from Big Bend National Park, we have observed a preponderance of low-mounded, multibranched (20–25 stems) plants in one population, where single-stemmed plants were also present.

The stems of *C. ramillosa* are more compact and tidy in appearance than the stems of *C. macromeris*. Part of this stem aspect is the result of subtly distinctive spination. In *C. ramillosa* there are 14–27 spines per areole, with one main (inner), porrect central, usually three upper (outer) centrals, and usually 13–16 radial spines. The single porrect, terete inner central is different from the several (typically four) radiating, angular main centrals of *C. macromeris*.

The flowers of *C. ramillosa* basically are the same color as those of *C.*

macromeris but a little smaller, 3.8–6.5 cm long and 3–5 cm in diameter The glossy inner tepals usually are rose-pink but may be pale pink to deep rose-purple (nearly magenta); they are 1.7–2.5 cm long, 3–4.5 mm wide, and attenuate-apiculate. The outer tepals are entire (unlike the long-fimbriate outer tepals of *C. macromeris*. The filaments are whitish and 6–8 mm long, supporting bright yellow to pale orange anthers 1–1.3 mm long. The style is 1.5–1.8 cm long. The 6–7 stigma lobes are white and 3.5–7 mm long.

The fruits of *C. ramillosa* are dark green to gray-green, obovoid, globose, or elliptic, 1.6–2.5 cm long, and 1.2–1.6 cm in diameter. The fruits are juicy, with colorless or greenish pulp. The floral remnant is persistent. There are no scales on the fruit surface, even at the top, in contrast with the fruits of *C. macromeris*. The reddish-brown to golden-brown seeds are globose to comma-shaped and 1.0–1.5 mm long. In specimens so far examined from Terrell County, the seeds of *C. ramillosa* were ca. 1 mm long, smaller than those of *C. macromeris* from the same region (at ca. 1.3 mm long). The seed coat is shiny but microscopically finely raised-reticulate. The seeds often are pointed at the proximal end where they are keeled. The keel ridge is lined with firmly attached funicular mucilage cells (Zimmerman et al., forthcoming). The seeds often have flattened sides.

Phenology. The flowering period for *C. ramillosa* is in summer and early fall. Numerous field reports suggest that plants in Brewster County tend to produce flowers in August and in September, and as late as 22 September. Potted plants in Alpine have bloomed from mid-July to late September, usually following significant precipitation. Single-stemmed cultivated plants tend to produce several flowers (1–7) that open on the same day or only a few days apart, and then usually do not produce flowers again until the next year, whereas branched cultivated plants may produce flowers more than once in the same season. From this information we extrapolate that in natural populations *C. ramillosa* flowers opportunistically, midsummer through September (perhaps into October), depending upon seasonal rainfall patterns (also noted by G. Raun, pers. comm.). Individual flowers in cultivation usually close permanently before sunset. We estimate that fruit maturation requires 3–4 months. Ripe fruits appear on cultivated plants in October through November. In natural populations ripe fruits may persist into December. Ripe fruits have a sweet aroma, similar to that of *C. macromeris,* but the juice is not sweet.

Sterile and Immature Specimens. The plants of *C. ramillosa* are strongly tap-rooted, or some plants have diffuse roots. The stems are deep-seated in rocky soil or rock crevices, with the aerial part hemispheric or flat-topped. Sexually mature stems are 4–9 cm long and 4–7(–9.5) cm in diameter. Cross section of the stems reveals a pith of large diameter, ca. 50% of lesser stem diameter, compared to ca. 10% of lesser stem diameter in *C. macromeris*. In *C. ramillosa* the pith contains a conspicuous medullary vascular system (this absent in *C. macromeris*). The pith and cortex are "dry" (i.e., nonmucilaginous) in *C. ramillosa* (copiously mucilaginous in *C. macromeris*). In *C. ramillosa* dark green or medium gray-green tubercles are 0.8–2 cm long, 0.6–0.9 cm in diameter (smaller than in *C. macromeris*), conical, and ascending. The areolar grooves of mature stems extend the full length of their tubercles, from the spines to the tubercle axils. Areolar glands usually are absent in *C. ramillosa*. The areoles are circular, ca. 3 mm in diameter, the young ones with white wool.

The spines in *C. ramillosa* are glabrous or microscopically pubescent. The spines usually are straight, but the longest ones may be slightly curved. In *C. ramillosa* there are as many or more spines per areole as in *C. macromeris*, but the central spine configuration usually is distinctive. Usually in *C. ramillosa* the single (inner) main, porrect, central spine is terete, and the usually three (2–3) upper centrals are angular or flattened in cross section, at least in the basal one-third, and they are appressed and directed upward in a "bird's-foot" pattern. In some plants there are 2–5 outer centrals, some of them often radiating at different angles from the single inner central. The inner central is 2.2–3.8(–4.3) cm long, 0.5–0.7 mm thick, with a golden bulbous base 1–1.3 mm in diameter. The inner central usually is whitish, but all of the centrals may be gray to dark gray with black distal portions, reddish-brown with black tips, or dark brown. One or more of the upper, appressed centrals may resemble the radial spines in color. The longest (upper) of the (9–)13–16(–20) radial spines is 1–2(–3) cm. The glaucous-white radials, sometimes black-speckled, are slightly or prominently laterally compressed and sometimes twisted in their appressed positions. Collectively, at the areole rim, the radial spines resemble spokes from the hub of a wagon wheel, the effect of their coalescent bulbous bases forming what appears to be a callus or circular rim in which the radial spines are inserted. In *C. ramillosa* the radial spines are laterally compressed near the base, but in *C. macromeris* the radials are dorsiventrally flattened.

Juveniles of *C. ramillosa* usually have shorter and fewer spines per areole than in adults. The spines are glabrous or microscopically pubescent.

Biosystematics. *Coryphantha ramillosa* is infrequent to locally common in desertic limestone habitats, near the Rio Grande in southeastern Brewster County. The species also is common at certain limestone sites in Terrell County generally south of Sanderson. The distribution of *C. ramillosa* in Mexico has not been carefully investigated, but relatively large multistemmed plants are known to be abundant at one locality in Coahuila adjacent to Big Bend National Park.

Current distributional evidence suggests that *C. ramillosa* primarily is a Mexican species, with populations extending from central Coahuila, and perhaps eastern Chihuahua, northeast to the Rio Grande, and barely into the United States in extreme southern Brewster and Terrell counties. In the United States *C. ramillosa* is most easily mistaken for the widespread *C. macromeris* because of its pinkish-magenta flowers and similar spine clusters. Much more closely related to *C. ramillosa*, however, is the Mexican taxon *C. delaetiana* (Quehl) A. Berger, which extends from Durango north to Coahuila and Chihuahua and may occur within about 10 miles of the Rio Grande in Chihuahua west of Ojinaga. Zimmerman et al. (forthcoming) noted that *C. delaetiana* is almost identical to *C. ramillosa* and may prove to be conspecific; *C. delaetiana* has pale yellow flowers and differs from *C. ramillosa* in quantitative vegetative characters. Both taxa are also related to the pale yellow-flowered Mexican species *C. nickelsae* (K. Brandegee) Britton & Rose, which replaces them in relatively mesic Coahuila across northern Nuevo León to northern Tamaulipas (Zimmerman, 1985).

Synonym. *Mammillaria ramillosa* (Cutak) Weniger.

Common Names. Cory cactus; bunched cory cactus; whiskerbrush; Big Bend mammillaria.

12. **Coryphantha scheeri** (Muehlenpf.) L. D. Benson. LONG-TUBERCLED CORYPHANTHA. Plates 296–99. [*Mammillaria scheeri* Muehlenpf., Allg. Gartenz. 15: 97–98. 1847]. Open grassy areas in oak-juniper-pinyon woodlands in the central mountains, to *Larrea* and other desertscrub or brushland, or degraded grasslands, alluvium derived from igneous rock and limestone, often in silty or sandy soils. Found in most Trans-Pecos counties; no specimens from Terrell or Val Verde counties; E to Upton, Crane, Ward counties. 2,700–5,200 ft. Flowering Apr–Jul (Aug–Sep). $2n = 22$. North to S NM, W to SE AZ. Mexico, Chihuahua and N Sonora. Map 63.

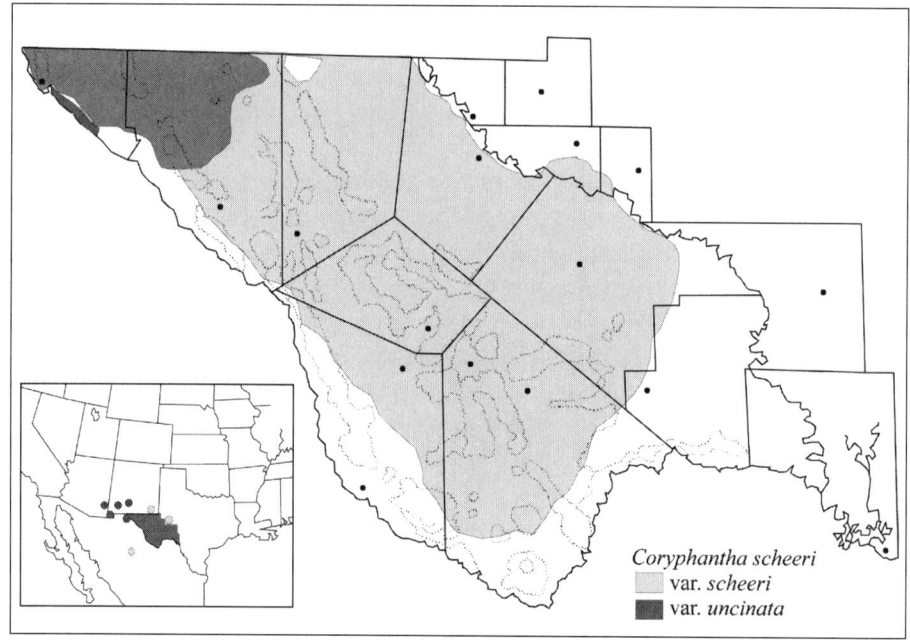

Map 63. Generalized distribution of *Coryphantha scheeri* var. *scheeri* (long-tubercled coryphantha) and *C. scheeri* var. *uncinata* (El Paso long-tubercled coryphantha).

Populations of *C. scheeri* in the Trans-Pecos always consist of widely spaced or scattered individuals. Rarely are more than one or a few plants found within several meters of each other. This widely spaced distributional pattern results from infestation by cactus-specialist beetle larvae (A. D. Zimmerman, unpub.). In Texas, *C. scheeri* is practically restricted to the Trans-Pecos, where it is widely distributed, but poorly collected and is not well understood in terms of habit, morphological variability, and taxonomy. Benson (1969b, 1982) accepted four varieties of *C. scheeri* from throughout its range, with at least three of them supposedly occurring in the Trans-Pecos, a taxonomic concept that was adopted by Bravo-Hollis and Sánchez-Mejorada (1991b). Weniger (1984) recognized only one variety of *C. scheeri* in Texas but apparently accepted the var. *robustispina* (Schott ex Engelm.) L. D. Benson in Arizona and Sonora, Mexico. After reevaluating this species for the *Chihuahuan Desert Flora*, Zimmerman concluded that three varieties were recognizable: two in the Chihuahuan Desert and var. *robustispina* in the Sonoran Desert. Both of the weakly differentiated Chihuahuan Desert varieties occur in the Trans-Pecos. The specific epithet honors Frederick Scheer (1792–1868), who evaluated and described the cacti for B. C. Seemann (1825–71) in *The Botany of the Voyage of H.M.S. Herald Under the Command of Captain Henry Kellett . . . during the years 1845–51* (1852–57).

Key to the Varieties

1. Central spines always straight; central Trans-Pecos and SE NM
   12a. *C. scheeri* var. *scheeri*.

1. Central spines of immature and young adult plants always hooked downward (often straight in older plants); El Paso Co., SW NM, SE AZ
   12b. *C. scheeri* var. *uncinata*.

12a. Coryphantha scheeri var. **scheeri.** LONG-TUBERCLED CORYPHANTHA. Plates 296, 297. Woodland and grassland in the central mountains, and degraded grassland-brushland, or less often in desertscrub, usually in alluvium derived from igneous or limestone. Pecos, Brewster, Presidio, Jeff Davis, Reeves, Culberson, and Hudspeth counties. 2,700–5,200 ft. Flowering Apr–Jul. $2n = 22$. In TX, E in Ward Co. near Royalty. Southeast NM in Eddy Co., near the Pecos River. Mexico, Chihuahua, near Ciudad Chihuahua (type locality of *scheeri*). Map 63.

The variety *scheeri* as treated here corresponds most closely with Benson's (1982) concept of var. *scheeri* and var. *valida*. By our definition of the taxon, only one variety of *C. scheeri* is present in the central mountain region of the Trans-Pecos and north along the Pecos River to southeastern New Mexico.

Identifying Characters. The stems of both Texas varieties of *C. scheeri* usually are single. Rarely, there are a few basal branches. The stems are globose to ovoid or cylindroid in older plants or flat-topped and not much protruding above the ground when young. Low, flat-topped stems prevail in mountain woodland habitats. The stems reach 5–10 cm or more tall and 5.5–8.5 cm in diameter. The tallest plant we have seen in the field was estimated at 25 cm. It was growing in brushland near Tunis Spring in eastern Pecos County. Like those of *C. macromeris*, the stems are dominated by the prominent tubercles, which are 1–2.5 cm long and 0.8–1.3 cm thick, ovoid to subcylindroid, and dull gray-green, and also by stout spines that are few enough in number and spread far enough apart so that the stem surface and tubercles are not much obscured.

In the eastern var. *scheeri* the one main central spine is always straight, porrect or slightly ascending, 2–2.5 cm long, ca. 1 mm wide, terete with a large-bulbous base, and mostly pale stramineous, gray, or pinkish-gray, with dark red to red-brown at the tip. In some specimens there are 1–2 upper, smaller subcentrals. The radial spines of all varieties are stout and have large-bulbous bases, the lower radials being the largest. The radials are terete or dorsiventrally flattened. In the eastern variety, they number 8–12, the largest ones 1.5–2 cm long. The radials are appressed and spokelike, pale gray to nearly white to stramineous (in younger areoles), with dark tips.

The yellow flowers, with or without a reddish throat, are 4–5.5 cm long and 3.5–5 cm in diameter. The specific shades of yellow or yellowish coloration of the flowers are different from the "clear" bright yellow of *C. echinus,* more nearly the color of *C. sulcata* flowers. The inner tepals are dark golden-yellow, pale greenish-yellow, to translucent dull yellow, sometimes bronze-tinted basally or in a midstripe, sometimes turning pinkish-yellow or bronze before wilting. The inner tepals are 2.3–3.5 cm long and 5–8 mm wide. The outer tepals are minutely fringed, unlike the entire outer tepals of *C. echinus, C. sulcata,* and *C. ramillosa.* The filaments are reddish to reddish-purple or pale reddish and support yellow anthers. There are 6–11 stigma lobes, these 3–7 mm long on a pale reddish style. The stigma lobes are yellow, or sometimes cream-colored, cream-pink, or slightly orange-yellow.

The fruit is light green at maturity, fusiform-cylindroid, 4–5 cm long, and ca. 1 cm in diameter. The fruit surface is microscopically papillate. The floral remnant is deciduous, leaving bare the truncate-concave fruit apex. The seeds are bright reddish-brown, 2.3–3.5 mm long, and smooth.

Phenology. The main flowering period for var. *scheeri* appears to be in May and June, at least in the Trans-Pecos central mountain region. Cultivated plants in Alpine have bloomed as early as 12 April, and it is suspected that early flowering, perhaps in April, may occur in plants at lower elevations. Blooming in July may actually represent a secondary blooming episode in opportunistic response to adequate rainfall. A plant from a site near the Brewster-Pecos county line had several mature fruits and three flower buds in early July 1997. This indicates the occurrence of multiple flowering periods, probably in opportunistic fashion after rains. Periodic later blooming, in August or even September, is suspected but not documented. Late flowering with mature fruits persisting into October has been observed under cultivation in Alpine. Individual flowers usually last one day and close at night. We have observed as many as 7–10 flowers open at the same time at the apex of one stem, but the flower buds do not always mature synchronously; a single plant may support one or several open flowers on each of several days in a row. We estimate that fruit maturation in var. *scheeri* requires at least 2–3 months. Ripe fruits are exceedingly aromatic, with the aroma of tropical fruits, something like the smell of ripe banana, but richer.

Sterile and Immature Specimens. The roots are diffuse in most plants examined, but flat-topped stems that protrude barely above the ground are deep-seated. Anatomically, the stem pith occupies one-third to nearly one-half of the lesser stem diameter and contains a medullary vascular system. The pith and cortex are nonmucilaginous. In adult plants areolar glands often are present in areolar grooves that extend the full length of tubercles. Hairs are conspicuous in young areoles but not in areolar grooves. The prominent tubercles and spine pattern (described above) usually allow for easy identification. *Coryphantha scheeri* is regionally sympatric with *C. echinus*, but these taxa have distinctive spine configurations; *C. macromeris* has much more similar spines but has mucilaginous parenchyma, tuberous roots, very thin "skin" (essentially lacking a hypodermis), and a very narrow pith that lacks a medullary system.

Juvenile plants of all the varieties have shorter, fewer spines per areole than adults, and the spines are glabrous or microscopically scabrous, as they are in adults. Cultivated juveniles in Alpine, 1 cm high or less, had (4–)5(–6) radial spines and no centrals. The spines were whitish, sometimes stramineous in the middle or distally, and had brown or reddish-brown tips. The areoles were pale yellow or whitish with short trichomes of the same color.

Biosystematics. Weniger (1984) did not recognize either var. *valida* or var. *uncinata* as being distinct from var. *scheeri*. The current treatment follows Zimmerman et al. (forthcoming), who recognized the populations of *C. scheeri* east of Hudspeth County in the Trans-Pecos and in southeastern New Mexico as one taxon, and the populations in El Paso County (extending west into southwestern New Mexico and adjacent Arizona) as a second taxon, var. *uncinata*. Populations of *C. scheeri* so far examined in Hudspeth County are geographically and morphologically intermediate between var. *scheeri* and var. *uncinata*. A specimen from Eddy County, New Mexico, has curved spines, suggesting intermediates with var. *uncinata* there, too.

Benson (1969b, 1982) reviewed the nomenclature for *C. scheeri*, with different conclusions each time. *Coryphantha scheeri* Lem. (Cactées 35. 1868) provides the basionym (i.e., the epithet *scheeri* was not validly published until 1868, by Lemaire, and then only by accident; it was intended as a recombination, not as the formal introduction of a new species). Benson failed to recognize that by 1868 *C. robustispina* had already been named, independently, by Engelmann, based on the Arizona populations; thus, *robustispina* is the older name (Taylor, 1998) and must be taken up for the species. Taylor published the new combinations as subspecies of *C. robustispina* (i.e., subsp. *scheeri*, subsp. *uncinata*). Because we recognize varieties of *C. robustispina*, instead of subspecies, and because the varietal combinations were not published at the time this manuscript went to press, we have retained use of the *C. scheeri* names. The basionym of *C. robustispina* (Schott ex Engelm.) Britton & Rose is *Mammillaria robustispina* Schott ex Engelm. [Proc. Amer. Acad. Arts 3: 265–266, 1856].

Synonyms. *Mammillaria robustispina* Schott ex Engelm.; *Echinocactus muehlenpfordtii* Poselger of 1853(non *Mammillaria muehlenpfordtii* Förster of 1847); *Cactus scheeri* (Muehlenpf.) Kuntze; *Coryphantha muehlenpfordtii* (Poselger) Britton & Rose; *C. scheeri* Lem.; *C. scheeri* Lem. var. *valida* (Engelm.) L. D. Benson, in part, as to lectotype; *M. engelmannii* Cory; *M. engelmannii* L. D. Benson ex W. T. Marshall; *Coryphantha neoscheeri* Backeb.; *C. scheeri* (Kuntze) L. D. Benson; *C. engelmannii* Lem., *nom. nud.*; *C. engelmannii* (Cory) Backeb., *nom. nud.*

Common Names. Needle mulee; Scheer cory cactus.

12b. **Coryphantha scheeri** var. **uncinata** L. D. Benson, Cact. Succ. J. (U.S.) 41: 234. 1969. EL PASO LONG-TUBERCLED CORYPHANTHA. Plates 298, 299. Grasslands and degraded grasslands in *Larrea*-dominated desertscrub. El Paso Co. 3,300–4,000 ft. Flowering May–Sep. 2*n* = 22. Southwest NM, Doña Ana, Sierra, Luna, Grant, and Hidalgo counties; adjacent SE AZ, Cochise and Graham counties. Mexico, sight records in extreme northern Chihuahua; expected in extreme NE Sonora. Map 63.

In Benson (1982) the specimens mapped as *C. scheeri* var. *valida* were from populations of all three currently recognized varieties of the species. Benson erroneously considered var. *uncinata* as a rare hooked-spined entity restricted to El Paso County. The varietal name *uncinata* is derived from the Latin *uncus*, "hook," in reference to hooked tips of central spines in some plants.

Identifying Characters. The var. *uncinata*, as recognized herein, always has strongly curved and/or hooked central spines on immature or young adult plants. Old adult plants usually have straight central spines. The direction of any curving or hooking in central spines is downward. According to Zimmerman et al. (forthcoming), all the plants of *C. scheeri* in El Paso County are one taxon, var. *uncinata*. The var. *uncinata* has 1–4 central spines, usually 14–16 radial spines, and all parts average slightly larger and more numerous than those of the eastern variety.

The flowers and fruits are similar in all varieties (despite reports to the contrary), except that they may be slightly larger (correlated with plant size in general) in the relatively robust western varieties. The inner tepals are golden-yellow to dull yellow and sometimes are bronze-tinged basally. The inner tepal bases also may or may not be reddish, or with red streaks, at the base, forming a reddish throat. They redden with age (see Phenology, below).

Phenology. Worthington (1986) observed that in the Franklin Mountains of El Paso County, this species (reported as *C. scheeri* var. *valida*) might produce flowers opportunistically multiple times each year after rains from August through September. Individual plants may produce several flowers that open on the same day, or anthesis may not be synchronous. Flowers usually last one day and close at night; those few that reopen usually are dull bronze-pink during their second day. Fruits mature during the fall or winter. Fruit maturation probably requires 2–4 months, although this has not been carefully studied; they remain succulent, firmly attached to the plant, for many months after ripening.

Sterile and Immature Specimens. Immature and young adult plants from El Paso County westward always have hooked central spines (seedlings lack central spines completely), superficially resembling seedlings of *Ferocactus wislizeni* (which grows in the same region) or *Ancistrocactus brevihamatus*.

Biosystematics. Zimmerman et al. (forthcoming) reported that plants in Hudspeth County may be intermediate between var. *uncinata* and the small, entirely straight-spined eastern variety. In the Trans-Pecos the hook-spined form probably is restricted to El Paso County, but it may extend into western Hudspeth County. One specimen from a stable gypsum substrate in northeastern Hudspeth County closely resembles El Paso County plants. Thus far, the documented range of exclusively straight-spined populations (one must evaluate the young adults and immatures) is entirely east of Hudspeth County. Not many specimens of *C. scheeri* from Hudspeth County have been available for study.

Since it was originally described by Benson (1969a), on the basis of a single old herbarium specimen, var. *uncinata* has been regarded as rare or even extinct. In fact, the type specimen is a slightly atypical of the variety that is geographically widespread from El Paso County northward and westward. Benson (1982) misidentified other specimens of the widespread variety (and straight-spined specimens of var. *robustispina*) as var. *valida*, but the lectotype of var. *valida* is a specimen of the small eastern variety from the central Trans-Pecos. The epithet *uncinata*, not yet formally available as a *robustispina* varietal combination, is the oldest name available for the widespread western variety. The photographs in Benson (figs. 854, 855, and 856) labeled "var. *valida*," are of var. *uncinata*.

Benson tabulated a few additional differences between the varieties, including areolar trichomes, flower color, and outer tepal margins, but these range from inconstant to simply inaccurate. The var. *robustispina*, from the upper edge of the Sonoran Desert in south-central Arizona, often is branched in habit, often has curved and hooked central spines, like var. *uncinata*, and is somewhat more robust in vegetative traits than any of the Texas plants.

Synonyms. *Mammillaria scheeri* Muehlenpf. var. *valida* Engelm. (in part, excluding lectotype); *Coryphantha scheeri* Lem. var. *valida*, sensu L. D. Benson (in part, excluding lectotype and straight-spined specimens of var. *robustispina*); *C. robustispina* subsp. *uncinata* (L. D. Benson) N. P. Taylor.

Common Names. Needle "mulee" beehive; pineapple cactus.

13. **Coryphantha echinus** (Engelm.) Britton & Rose. SEA-URCHIN CACTUS. Plates 300–5. [*Mammillaria echinus* Engelm., Proc. Amer. Acad. Arts 3: 267. 1856]. Degraded grasslands and desertscrub, limestone and (rarely) igneous substrates, throughout the southern and eastern parts of the Trans-Pecos; formerly reported W to El Paso Co., but without satisfactory documentation. 1,000–4,800 ft. Flowering in late spring and summer. $2n = 22$. In TX sporadic NE to Howard and Coke counties. Mexico, Coahuila and NE Chihuahua. Map 64.

All three varieties of *C. echinus* are endemic in the Chihuahuan Desert Region, two of them in Texas mostly in the Trans-Pecos, but also elsewhere in Texas. The common name "sea-urchin cactus" is particularly appropriate for these plants with globose stems, appressed radial spines, and single protruding central spine in each areole. The specific epithet is equally descriptive, as the Greek root *echinos* means "hedgehog" or "sea urchin."

Key to the Varieties

1. Plants usually remaining single-stemmed (neotenous?); older stems remaining globose to subglobose, 3–10 cm long, 3–5.5 cm wide; spines with a whitish aspect (central spines later in life, sometimes absent); northern and/or higher altitudes, widespread in the Trans-Pecos
    13a. *C. echinus* var. *echinus*.

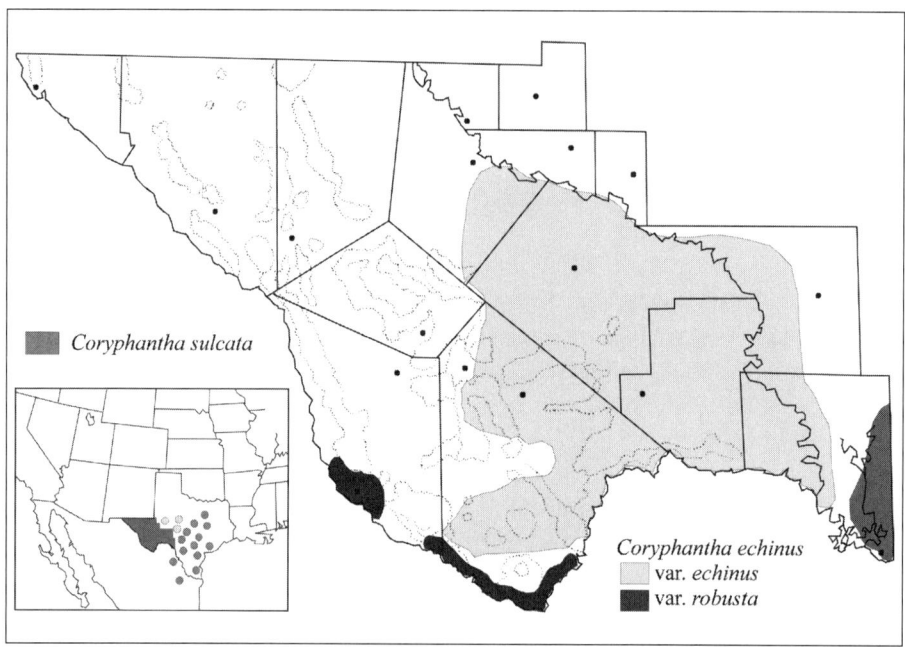

Map 64. Generalized distribution of *Coryphantha echinus* var. *echinus* (sea-urchin cactus), *C. echinus* var. *robusta* (multistemmed sea-urchin cactus), and *C. sulcata* (grooved nipple cactus).

1. Plants usually becoming multistemmed; older stems ovoid to cylindroid, rarely remaining subglobose, 7–15 cm or more long, 5–7 cm wide; spines with a dark aspect (central spines appearing early in life); along and near the Rio Grande in southern Brewster and Presidio counties
13b. *C. echinus* var. *robusta*.

13a. **Coryphantha echinus** var. **echinus**. SEA-URCHIN CACTUS. Plates 300, 301. Throughout much of the southern and eastern Trans-Pecos, from the Chisos Mts and Terlingua area N and NE to Pecos Co., E to Terrell and Val Verde counties (rarely W to El Paso Co.?). 1,000–4,800 ft. Flowering (spring) summer. $2n = 22$. Texas, sporadically NE to Howard and Coke counties. Mexico, formerly expected in N Coahuila, but now thought to be replaced by other varieties from the Rio Grande southward. Map 64.

The var. *echinus* is by far the most widely distributed of the two varieties of *C. echinus* in Texas. Perhaps it is most common on and near limestone soils of the eroded Stockton Plateau in Pecos and Terrell counties, and in eastern Brewster County. It also occurs in the limestone mountains of the southern Big Bend. The type locality is listed as "on the Pecos River."

Identifying Characters. The var. *echinus* characteristically is unbranched, and the globose or subglobose solitary stems usually are recognizable at a glance, with their whitish appressed radial spines and with a single porrect central spine in each areole. The plants can be viewed, with a little imagination, as having the aspect of a white sea urchin. Older plants may form a few branches at the base of a main stem, or what appears to be branches are at times juvenile plants growing from seed at the base of their parent (J. Talbot, pers. comm.).

The flowers of var. *echinus* are 3.5–5 cm long and wide, yellow inside but greenish or dull reddish externally, and often are reddish in the throat (filaments and inner lining of the receptacular tube). In completely open flowers the tepals are reflexed. The inner tepals are glossy yellow, 2.2–3.2 cm long, 0.4–1 cm wide, and attenuate or caudate apically. The outer tepals are entire. At times the whole throat of the flower, including the base, may appear reddish, largely because the red color of stamen filaments, especially the distal portions, is reflected by the mirrorlike surface of the (actually yellow-pigmented) inner tepal bases and the sterile distal portion of the receptacular tube. The anthers are bright yellow. The style and stigma lobes are pale yellow. Usually there are 10–12 stigma lobes, these 3–4 mm long.

The green to light green fruit of var. *echinus* is 1.2–2.5 cm long, 1–1.9 cm in diameter, and generally ovoid. The floral remnant is persistent. The seeds are reddish-brown, 1.7–1.9 mm long, and with smooth shiny surfaces.

Phenology. The main flowering time for var. *echinus* seems to be in June and July, probably following rains (Wauer, 1980), but any correlation of blooming with precipitation has not been studied. In some years flowering occurs as early as April, as has been documented photographically in eastern Pecos County, but it is not known if spring blooming is a regular phenomenon often associated with rainfall. Flowers open about midday, close surprisingly early in the afternoon, and usually do not open again the next day. One or several flowers may be open at the same time on a single stem. Fruit maturation is believed to require 3–4 months. In cultivation at least, ripe fruits may persist on plants for considerable periods after maturation. During the ripening process, fruits expand to the extent that they may press tightly against the radial spines of surrounding areoles.

Sterile and Immature Specimens. The plants of var. *echinus* have diffuse roots. The globose stems may become ovoid or conical with age. The stems are 3–10 cm long and 3–5.5 cm in diameter. Inside stems, the pith constitutes in diameter less than one-third of the lesser stem diameter, and there is no medullary vascular system. The pith and cortex are nonmucilaginous. The coniccylindroid tubercles are porrect or slightly ascending, 0.8–1.2 cm long, ca. 6–7 mm in diameter, and gray-green. The areoles when young are circular and 2.5–3 mm in diameter, often becoming elongated or oval with age, and ca. 4 mm long. Whitish or brownish wool occurs in the areolar grooves on tubercles near the stem apex. A tuft of wool may occur at the base of areoles.

In var. *echinus* there are 16–31 spines per areole. All the spines usually are drab white with brown tips. The spine number includes 0–4 centrals and 16–24(–27) radials. Usually by the age of sexual maturity there is one porrect, main (lower) central and 2–3(–4) upper centrals (or subcentrals). The upper centrals clearly are centrals in length and stout morphology (with bulbous bases), but they are appressed in position with the radials. The main, porrect central usually is straight or slightly decurved but is strongly decurved in some populations. The main central is 0.8–1.5 cm long, 0.6–1 mm thick, bulbous- or buttressed-based, ashy- (drab-) white, and brown-tipped. The appressed upper centrals are slightly longer and thinner than the porrect central, dorsiventrally compressed, and arranged in a "V" when two are present or a "bird's foot" when three are present. The radial spines, unlike the central spines, are laterally compressed, especially at the enlarged bases, but distally the radials usually are terete. The radials are appressed, pectinately arranged, and often curved toward the stem surface. The longest radials are 0.8–1.2 cm. In var. *echinus* the color of the radials, like that of the centrals, seems to be drab-white, but under magnification the radials appear to be ashy-white, gray-white, or even pale yellowish-tan, these colors overlying yellowish inner layers.

Immature plants of var. *echinus* are characterized by the absence of the porrect (lower, main) central spine. One or two of the appressed upper centrals (subcentrals) may be present in areoles of immature plants, along with shorter and fewer radial spines. The porrect central spine also may be absent in what appears to be young adult stems to 5 cm long and ca. 5 cm in diameter and in some adult plants. The spines of juveniles are glabrous, as they are in adults. At least some of the spines of seedlings may be microscopically pubescent-scabrous at the distal ends.

Biosystematics. Plants of var. *echinus* are common at some sites, but usually infrequent, and at other localities seemingly absent from large areas of apparently typical habitat on or near limestone in the southeastern Trans-Pecos. The few reports of var. *echinus* in igneous substrates need further documentation. The synonym *C. pectinata*, with its type locality "on the Pecos River," is based on a form of var. *echinus*, the "pectinate form," that lacks the lower (porrect) central spine in each areole. Essentially all immature plants of var. *echinus* lack porrect centrals, and occasional young adult and sometimes even sexually mature plants as well may lack the protruding centrals. Fully adult plants without porrect centrals, the "pectinate" form, seem to be otherwise identical to the common form of var. *echinus*. The pectinate form probably appears throughout much of the population of var. *echinus*. We have seen specimens of the pectinate form from both 10 miles north and 20 miles south of Fort Stockton in Pecos County, and from 25 miles northwest of Comstock in Val Verde County, near the Pecos River.

Benson (1982), without explanation, included *C. echinus* in *C. cornifera* (DC.) Lem. *Coryphantha cornifera* is a species common in central Mexico, mostly in the states of Hidalgo and Querétaro, far from the range of *C. echinus*; they are not closest relatives. Zimmerman et al. (forthcoming) recognized two varieties of *C. echinus*, both of them endemic to the CDR, and one of them, var. *echinus*, in the United States. The multistemmed populations along and near the Rio Grande in southern Brewster and Presidio counties are a third variety. These large plants essentially replace the typical form in their limited ecogeographic range. Benson seemed to be aware of the multistemmed *C. echinus*, in that his description of *C.* "*cornifera*" included characters of the taxon. Furthermore, one of Benson's photographs, figure 879, is clearly var. *robusta*. Weniger (1984) seems not to have included the multistemmed populations found near the Rio Grande in his descriptions of either *echinus* or "*Mammillaria scolymoides* Scheidw."

The Texas plants misidentified by Weniger as "*Mammillaria scolymoides*," a poorly known Mexican species, may prove to be either an unnamed species or yet another part of *C. echinus*. Plants matching Weniger's description and illustration of "*M. scolymoides*" are inferred to be sympatric at some sites with var. *echinus*, based upon vague data and undocumented sight records, but we have not yet found them together. They are large plants with decurved central spines.

Synonyms. Mammillaria pectinata Engelm.; *Coryphantha pectinata* (Engelm.) Britton & Rose; *C. cornifera* (DC.) Lem. var. *echinus* (Engelm.) L. D. Benson. *Mammillaria scolymoides* Scheidw. and *C. scolymoides* (Scheidw.) A. Berger are based upon a Mexican species; reports from Texas are misidentifications.

Common Names. Hedgehog cory-cactus; rhinoceros cactus.

13b. **Coryphantha echinus** var. **robusta** A. M. Powell, **var. nov.** MULTI-STEMMED SEA-URCHIN CACTUS. Plates 302–4. TYPE: UNITED STATES. Texas. Brewster Co., Big Bend National Park, gravel benches and arroyos, ca. 1 mi above the River Road (east) turnoff, Chihuahuan desertscrub, 25 Apr 1992, A. M. Powell 5829 (HOLOTYPE, SRSC!)

Plantae cespitosae, tumuli aut tegetes 30–70 cm diametro. Caules plerumque 15–50; caules maturi 7–15 cm longi, 5–7 cm diametro. Spinae plerumque 19–33 per quamque areolam; spinae centrales 4, spina inferiore porrecta; spinae radiales plerumque (15–)20–24(–29), appressae. Flores vivide lutei, sine faucibus rubris. Fructus virides, ovoidei, usque ad 2.8 cm longi; semina rufa.

Additional specimens examined: UNITED STATES. Texas. Brewster Co., Big Bend National Park (BBNP), gravel benches and arroyos ca. 0.5 mi above the River Road (east) turnoff, *A. M. Powell* and *S. A. Powell 5825* (SRSC); BBNP, ca. 1 mi above River Road (east) turnoff, *A. M. Powell* and *S. A. Powell 5827* (SRSC); BBNP, *D. Smith 2214* (SRSC).

Gravel hills and benches, and other alluvial substrates of igneous and limestone origin, desertscrub and degraded desert grassland. Along and near the Rio Grande, southern Presidio and Brewster counties. 2,000–3,500 ft. Flowering Apr–May (summer). $2n = 22$. Mexico, expected across the Rio Grande in Chihuahua (from Ojinaga) and Coahuila (from Boquillas). Map 64.

The var. *robusta* is best exemplified by the relatively large cespitose plants, in mounds and mats, found in the gravel hills south of the Chisos Mountains and north of Lower Tornillo Creek. The varietal epithet, taken from the Latin *robustus*, "strong," alludes to the robust multistemmed plants.

Identifying Characters. The presumably older plants of var. *robusta* have 15–50 or more stems of various sizes, forming mounds or mats 15–30 cm or more high and 30–70 cm in diameter. Numerous small stems, 1–3 cm high and wide on these larger plants, originate near the bases of the larger stems in the mound or mat. Sexually mature stems are subglobose or ovoid to cylindroid and are 7–15 cm or more long and 5–7 cm or more in diameter. In aspect the stems of var. *robusta* usually are darker than the stems of var. *echinus*, because of darker gray spine color. Older stem bases often are sheathed by nearly black spines. The porrect central spines are gray to brownish-red or nearly black, or gray with nearly black tips. The radial spines usually are gray.

In populations south of the Chisos Mountains, many older plants of var. *robusta* have senescent or dead stems in interior portions of the clumps. At the bases of senescent and dead stems, spine clusters tend to abscise *en masse*, with all of the spines turning very dark, remaining adherent, and forming a loose sheath around part of the stem bases. In some plants masses of decomposed centrally located stems contribute to a sprawling, matlike habit, with apparently healthy stems at the periphery of original tightly cespitose mounds or mats.

The flowers of var. *robusta* are bright yellow like those of var. *echinus*, usually 5.5–6.5 cm long and wide. The inner tepals are 2.7–3.4 cm long, usually 7–8 mm wide, and pure yellow at the base (i.e., in the flowers so far examined, without any reddish tinge that would contribute to a reddish throat). The inner tepal apexes often are emarginate-aristate. The stamens have reddish to reddish-orange filaments and yellow anthers ca. 1.5 mm long. The pale yellow style is ca. 2.4 cm long and supports 10–13 yellow to pale yellow stigma lobes. In flowers of var. *robusta* so far examined experimentally, the filaments are sensitive in response to mechanical stimulation. When swept with a cotton-tipped applicator, the filaments exhibit immediate and rapid movement in unison, twisting inward around the style. In specimens so far examined, the flower buds of var. *robusta* are smaller than those of var. *echinus*, and they are ovoid with an abrupt point at the apex. The buds of var. *echinus* are broad-based, subspherical in shape, and blunt-tipped.

The fruits of var. *robusta* are green to light green, 1.5–2.8 cm long, 1.3–1.9 cm in diameter, and generally ovoid, closely resembling those of var. *echinus*, except slightly larger. The floral remnant is persistent. The reddish-brown seeds virtually are identical to those of var. *echinus*.

Phenology. For two seasons in the early 1990s, populations of var. *robusta* south of the Chisos Mountains produced flowers in April, with some plants continuing to bloom in early May. It is not known if spring is the typical blooming period for the taxon or if plants also produce flowers opportunistically following rains in the summer. Under cultivation in Alpine, plants were observed to bloom on 5, 12, and 26 August after modest rains about two weeks earlier. Flowers usually open about midday, close at night, and do not open again the next day. It is suspected that fruits require 2–4 months to mature. In cultivation fruits may remain green on the plant for about one year.

Sterile and Immature Specimens. Mature stems of var. *robusta* resemble those of var. *echinus* except that usually they are slightly larger and more cylindroid in shape, and their spine-cover is darker gray. Usually there are more spines per areole, and the spines are longer. In areoles of mature var. *robusta*, typically there are four central spines and (15–)20–24(–29) radial spines. The lower (main) central spine is porrect, straight or sometimes slightly decurved, 1.5–2.5 cm long, ca. 1 mm thick, and bulbous-based. The main central usually is gray, or gray with black speckles as seen under magnification, or on some stems is brownish-red to nearly black, particularly on the distal portion, or dark only at the tip. The three upper central spines, in a "bird's-foot" arrangement, are appressed in plane with the radial spines. The upper centrals are more slender than the lower central, are bulbous-based, and are 1.7–2.5 cm long. The usually gray radial spines, (18–)20–24(–29) in number, are appressed, usually with the distal portions curved toward the stem. The upper radial spines are 1.7–2.4 cm long, longer than the lateral (1.2–2.2 cm) and lower (1–1.6 cm) radials. The bases of radial spines are enlarged but not bulbous, and they are laterally compressed.

Immature stems are basically spheroidal, and the spines are pale gray to nearly white. The areoles of the smallest stems in clumps of var. *robusta* may produce either the typical porrect central spine, or especially in apical areoles, the lower central may be absent. Stem branches just a few centimeters long presumably are immature, but in many cases at this size, they already have protruding centrals, unlike var. *echinus,* where areoles in stems this size usually lack the porrect central.

Biosystematics. The known distribution of var. *robusta* is near the Rio Grande from near the Chisos Mountains in southern Brewster County northwest to the south-central part of Presidio County below the Sierra Vieja and Chinati Mountains, and into adjacent Chihuahua. Large cespitose plants are relatively common on certain gravel hills south of the Chisos Mountains. Smaller plants of this taxon appear to be sporadic throughout the range described above, north at least as far as Dugout Wells in Big Bend National Park, and near Terlingua west of the Chisos Mountains. The var. *robusta* may intergrade with var. *echinus* in the northern part of its range, but this aspect has not yet received adequate study. It is possible that var. *robusta* extends farther north; "near Marathon" is indicated for the locality of a photographed specimen of what appears to be var. *robusta* (Benson, 1982, fig. 879). Possibly Benson's figure 879 was imprecisely mislabeled as to locality (the desert pavement visible in the photograph is a typical substrate farther south, but not "near" Marathon). We have observed numerous specimens of the typical single-stemmed var. *echinus* from near Marathon, but we have not located any cespitose individuals there.

The var. *robusta* in its cespitose habit is conspicuously distinct from var. *echinus,* which predominantly is single-stemmed throughout its range. The prominent habit difference, along with the several other vegetative and floral distinctions discussed above, suggest that var. *robusta* should be separated from var. *echinus*. At present we have elected to raise the lower Big Bend populations to varietal rank within *C. echinus,* their obvious closest relative, because intergradation is not yet documented.

In Big Bend National Park var. *robusta* is a typical associate of *Opuntia agge-*

ria and *Echinocereus chisoensis*. The var. *robusta* is more widely distributed than is *E. chisoensis* but more restricted than *O. aggeria*. The large cespitose habit of var. *robusta* may be the ecological manifestation of very old plants growing in relatively porous, deep substrates that are located in the hottest winter and summer habitats in the Chihuahuan Desert Region of Texas.

Plants in a population south of the Sierra Vieja in Presidio County, included here with var. *robusta*, are slightly different from the rest of *C. echinus*. The plants eventually branch (3–4, perhaps more, stems) with strongly decurved central spines. Single-stemmed plants also have been observed. This depauperate population (Plate 305) presumably has been depleted over the years by commercial cactus harvesters. At present we regard the poorly known Sierra Vieja population as part of var. *robusta*.

Synonyms. None known.

Common Name. We propose the use of multistemmed sea-urchin cactus for this cespitose taxon.

14. **Coryphantha sulcata** (Engelm.) Britton & Rose. GROOVED NIPPLE CACTUS. Plates 306, 307. [*Mammillaria sulcata* Engelm., Boston J. Nat. Hist. 5: 246. 1845]. Reported in limestone habitats near the mouth of the Pecos River, Val Verde Co., E Pecos Co., and N Brewster Co. 1,000–3,500 ft. Flowering May, summer. $2n = 22$. Edwards Plateau E to Austin Co., and N to Somervell, Tarrant, and Denton counties, S to Duval Co. (Benson, 1982). Map 64.

As far as known, *C. sulcata* is not yet documented for the Trans-Pecos. Weniger (1984) reported that *C. sulcata* extended west to the mouth of the Pecos River. Our own field investigations have not revealed the positive existence of *C. sulcata* in or near the eastern periphery of the Trans-Pecos. The species is included in the current work because we have attempted to review all of the cacti of Texas, some in neighboring areas, and to provide more detailed treatment for those taxa with distributions within or approaching the political boundaries of the Trans-Pecos. *Coryphantha sulcata* is traditionally endemic to Texas, but certain taxa from northeast Mexico are obviously related and nearly identical (e.g., *C. roederiana* Boed. and *C. obscura* Boed). The specific epithet is after the Latin *sulcus*, meaning "a furrow," probably a reference to the grooved tubercles.

Identifying Characters. Eastern plants of *C. sulcata* are cespitose, forming clumps to 30 cm or more in diameter, potentially with dozens of stems (of different sizes). The larger stems are spherical or obovoid, often compressed at the apex, 4–8(–12) cm long, and 6–8 cm in diameter. Usually there are 9–16(–18) spines per areole, with one, porrect, decurving, lower (main) central spine. The lower central is 1.1–1.7 cm long, typically 0.7–0.8 mm thick. Usually there are 1–2(–3) erect upper centrals. There are 8–13(–17) appressed radial spines, all of similar length at 1–1.5 cm. Rarely, there are up to 24 spines per areole, including 3–4 subcentrals. The spines are yellowish in upper (younger) areoles and gray (with age) in lower areoles. The green tubercles and stem surface are not much obscured by the spines. In general aspect, individual stems of *C. sulcata* resemble expanded, heavily watered stems of *C. echinus*. Distributions of the two taxa are not yet documented to overlap; mostly they are on opposite sides of the Pecos River. The taxa are easily distinguished vegetatively when the multistemmed, adventitious rooting habit of eastern, mesomorphic populations of *C. sulcata* is compared to the usually solitary-stemmed plants of *C. echinus* var. *echinus,* or when normal spine clusters of mature specimens are compared. However, western plants of *C. sulcata* and its Mexican allies branch much less; and *C. echinus* plants branch eventually, if they live long enough.

The flowers of *C. sulcata* usually are golden-yellow or dark golden-yellow compared to the bright glossy yellow in flowers of *C. echinus*. Rarely, the flowers of *C. sulcata* are greenish-yellow. These color differences are most easily appreciated when the two species are observed together in cultivation. The flowers of both species usually have a red throat, a coloration that may be in part the result of red pigment at the base of the inner tepals and in part the reflection of color from the dense complement of red filaments. The flowers are 4–6 cm long and 3.5–8 cm in diameter. The inner tepals are 3–4 cm long and ca. 7 mm wide, acute-aristate at the apex, and distally often slightly darker golden-yellow. The exposed upper filaments are 7–10 mm long (lower filaments up to 18 mm long), reddish distally, or rarely greenish. The anthers are ca. 1 mm long and yellow. The yellowish or greenish style is ca. 1.7 cm long, usually longer than the stamens, supporting 7–11 cream-colored or greenish-yellow stigma lobes, these ca. 3.5 mm long. The flower buds of *C. sulcata,* at least in plants so far available for study, are relatively long and pointed, sagittate in outline, and project above the spines before anthesis. The floral remnant is persistent. In *C. echinus,* the flower buds are shorter, and in *C. echinus* var. *echinus* much broader than those of *C. sulcata,* and they do not project above the spines.

The ellipsoid to fusiform fruits of *C. sulcata* usually are green at maturity, 1.5–3.5 cm or more long, and 1–1.5 cm in diameter. In one cultivated plant observed, fruits left on a plant for several months (apparently after maturation) became enlarged in diameter and developed a broadly ovoid shape, and the fruit surface turned from green to dull red. The seeds are dark reddish-brown, ca. 2 mm long, smooth and shiny, and somewhat comma-shaped.

Phenology. Plants flowering in May have been reported from Somervell County, just southeast of Fort Worth, near the northern limits of the range of *C. sulcata*. Plants in cultivation at Alpine have bloomed as late as July. This suggests that *C. sulcata* blooms in late spring, if conditions are right, and then it may flower sporadically in the summer. Although Benson (1982) reported that individual flowers of *C. sulcata* open each day for 2–3 days, in hot weather they last only one afternoon. Evidence from a single specimen suggests that fruit maturation requires 3–4 months.

Sterile and Immature Specimens. The plants of *C. sulcata* are readily branching and clustering, forming new stems from areolar grooves at the bases of old stems, except at the western/xeric edge of their range. New stems, like older stems, tend to be subglobose. The tubercles of mature stems are 0.8–1.5 cm long, nearly cylindrical or obliquely conoidal (the adaxial faces are shorter), with broad bases ca. 1 cm or more wide. The soft tubercles often curve upward on turgid stems or may sag under desiccated conditions. The areoles are circular, ca. 3.5 mm in diameter, with copious short white wool, the wool diminishing with age. Trichomes may be pale yellowish in areoles closest to the stem apex.

The stout, porrect, usually decurved lower central spine is terete (ca. 1 mm thick) with a bulbous base ca. 1.5 mm wide. In mature areoles the porrect (main) central spine usually is gray to nearly white, or gray with some black speckling and reddish-brown or black tips. In younger areoles the main central may be yellowish or pinkish with a brown tip. The 1–2 upper central spines, appressed in plane with the radials, are more slender than the main central, about the same length as the main central, and similar in color to the radials. The radial spines usually are slightly curved toward the stem. The radials are terete, but their bases are slightly laterally compressed. The radials are easily distinguished from the appressed upper centrals, which have bulbous bases. The yellowish to gray or gray-white radials have reddish-brown or black tips. The typical number of radial spines in *C. sulcata* (8–15) is fewer than in *C. echinus* var. *echinus*. The radials are all about the same length in *C. sulcata* and young plants of *C. echinus*, but in *C. echinus* the upper radials of old plants are obviously longer than the lower radials.

In immature plants of *C. sulcata* the porrect central spine is absent, usually as are the upper centrals. The areoles are circular with yellowish or white wool. Immature stems 1–3 cm long already have (8–)11–13 radial spines, these glabrous, translucent, gray-white, or yellowish, and with red-brown tips. In seedlings there are 6–8 radial spines. The radial spines are slightly curved inward in both seedlings and larger juvenile stems.

Biosystematics. Benson (1982), without explanation, extended his taxonomic circumscription of *C. sulcata* in Texas to include a second taxon, var. *nickelsae* (K. Brandegee) L. D. Benson, primarily of Mexican distribution. Weniger (1984) did not mention the taxon *nickelsae* in his treatment of the cacti of Texas and neighboring areas. Every other author except Benson has considered *nickelsae* to be a distinct Mexican species, *C. nickelsae*, that barely (if at all) enters Texas. *Coryphantha nickelsae* occurs in limestone and associated habitats in eastern Coahuila, northern Nuevo León, northern Tamaulipas, and (according to Benson) just across the Rio Grande into Texas. Benson mapped *C. nickelsae* in Maverick County, but in text gave the locality as Laredo, Webb County; the only documentation cited by Benson is vague. *Coryphantha nickelsae* intergrades with the *C. delaetiana* species-group (which includes *C. ramillosa*), in clean sympatry with the *C. sulcata* species-group (which includes *C. echinus*). *Coryphantha sulcata* has numerous close relatives in Mexico, as well as *C. echinus*, but *C. nickelsae* is not among them.

According to rumor, *C. sulcata* occurs in the Trans-Pecos. One knowledgeable cactus enthusiast, who seems to recognize all the common species, told us that *C. sulcata* is rare in vacant lots of Fort Stockton and elsewhere in the eastern part of Pecos County. One specimen tentatively identified as *C. sulcata* was collected by P. Manning in north-central Brewster County.

Synonyms. *Mammillaria strobiliformis* Muehlenpf. (non Engelm.); *M. cal-*

carata Engelm.; *Coryphantha calcarata* Lem.; *Cactus calcaratus* (Engelm.) Kuntze; *C. scolymoides* (Scheidw.) Kuntze var. *sulcatus* (Engelm.) J. M. Coult.; *M. radians* DC. var. *sulcata* (Engelm.) K. Schum.; *C. sulcatus* (Engelm.) Small.

Common Names. Nipple cactus; finger cactus; pineapple cactus.

Appendix 4

Another Texas species of *Coryphantha*, and other Texas genera not discussed elsewhere in the Trans-Pecos treatments, are summarized below.

Coryphantha missouriensis (Sweet) Britton & Rose var. **caespitosa** (Engelm.) L. D. Benson. NIPPLE CACTUS. Plate 308. [*Mammillaria similis* Engelm. var. *caespitosa* Engelm., Boston J. Nat. Hist. 6: 200. 1850]. Plants usually branching, occasionally solitary, forming clumps 5–10 cm high, 15–30 cm or more in diameter. Roots fibrous, but stem bases usually deeply seated in soil. Stems dark green, the larger ones usually depressed-globose, 2.5–5 cm long, 3.5–5(–10) cm in diameter. Tubercles 12–15 mm long. Areoles round or oval, 1.5–2 mm in diameter, white-woolly when young. Spines (10–)12–15(–17) per areole; central spines usually absent, rarely one, when present only slightly heavier and longer than the radials; radial spines 0.8–1.4 cm long, spreading, straight, slender, pubescent when young, gray to white, yellowish when young, often with brown tips. Flowers 2.5–5 cm in length and diameter when fully open, the inner tepals golden-yellow or greenish-bronze, or gold streaked with pink, or pale yellow, narrowly lanceolate, attenuate-aristate; filaments whitish to yellowish or light green; anthers yellow; style much longer than stamens, green to yellowish; stigma lobes 4–6, 3–5 mm long, green or yellowish. Fruits globose to oval, 1–2 cm long, remaining green into the winter (Weniger, 1984), then turning bright red, persisting into next flowering season at stem periphery. Seeds black, obovoid or subglobose, 1.5–2 mm in greatest diameter, pitted. $2n = 44$.

Distributed in south-central and north Texas, mostly east of the Edwards Plateau, as far south as Bexar County, and east to Walker County, north through Oklahoma; Arkansas (near the Red River, based on one old, vague specimen label); Kansas, where present in most of the counties but seldom seen; and, according to Weniger (1984), also west to eastern Colorado. Weniger maintained that var. *caespitosa*, as *Mammillaria similis* Engelm., was "rather common near Austin, Texas," and "occasionally found in a band about 100 miles wide" from Waco to the Dallas–Fort Worth area. Benson (1982) attempted to recognize *C. missouriensis* var. *robustior* (Engelm.) L. D. Benson, geographically intermingled with var. *caespitosa*. In all Benson recognized four varieties of *C. missouriensis* in the United States.

In general *C. missouriensis* is cold hardy but often short-lived in cultivation. The species tends to be very inconspicuous in cover of other vegetation, especially during the winter and in dry periods when desiccated stems are drawn to near ground level. The plants are most obvious during the late spring when the usually yellow flowers are produced in abundance (beginning 13 April and extending into July, cultivated in Alpine). The specific epithet is after the Missouri River, not the state of Missouri (where *C. missouriensis* is rare or absent). The varietal epithet describes the cespitose or densely clumped low habit of var. *caespitosa*, but the northwestern variety also forms clumps, at least in old age.

Acanthocereus tetragonus (L.) Hummelinck TRIANGLE CACTUS. Plates 309, 310. [*Cereus tetragonus* L., Sp. Pl. 1: 466. 1753; *Cactus pentagonus* L., *Acanthocereus pentagonus* (L.) Britton & Rose; *Cereus pentagonus* (L.) Haw.]. Plants with stems erect or sprawling, often clambering in other vegetation, ultimately arching over unless supported. Roots fibrous. Stems 3(-4-5)-angled, branching once or twice or more, individual growth increments 30-200 cm long, 2.5-5 cm in diameter, entire branched stems to 6-7 m long. Ribs sharply angled, 1.2-5 cm high, the stem axis slender. Areoles round, 1.5-3 mm in diameter, typically 2-5 cm apart, with short, whitish wool. Spines usually 7-8 per areole, light brown, aging gray, acicular or slightly flattened, with bulbous bases; central spines 1-3(-4), the porrect or slightly deflexed lower central 1.8-4 cm long, with 0-2 lateral centrals and 0-1 upper central; radial spines usually 5-7, light brown to gray, acicular, mostly appressed. Flowers white, nocturnal, ca. 10 cm in diameter, 17-25 cm long; inner tepals white, lanceolate to oblanceolate, to 5 cm long, ca. 1 cm wide, attenuate at the apex; outer tepals lanceolate, greenish, with darker midregions, these perhaps reddish-purple; filaments white, shorter than the tepals, anthers light yellow, ca. 3 mm long; style 1.7-2 cm long, stigma lobes 10-12, white; pericarpel greenish to greenish-blue, ca. 2.5 cm long, ca. 1.5 cm in diameter, extending into a slender, greenish floral tube, with areoles closely spaced on the lower (ovary) part and widely spaced on the tube. Fruit bright red, shiny, broadly elliptic to ovate, 3-7(-10) cm long, 2.5-5(-7) cm in diameter, with 1-4 spines per areole on low tubercles, the fruit pulp red, edible, moderately sweet. Seeds black, shiny, obovate, 2.5-3 mm long.

South Texas in Kenedy, Willacy, Hidalgo, Cameron, and Webb counties. Also southern Florida and doubtfully Louisiana. Down the east coast of Mexico through Central America, and east along and near the coast in South America to Venezuela, and in both the Greater and Lesser Antilles.

Acanthocereus is a genus of 12 or fewer species in tropical and subtropical lowlands. The plants are remarkable for their sprawling, clambering, and branching (especially of prostrate stems) habit and fast growth. In South Texas stems may grow 5-6 feet under optimum conditions in a summer season (Weniger, 1984). Plants of *A. tetragonus* are sensitive to frost, and aboveground parts may be killed at 0°C (32F°) or slightly below freezing. In the Trans-Pecos the plants are easily grown through the winter in heated greenhouses, where they occasionally produce flowers (one instance in September). The white flowers are large and showy, with minimal fragrance or a pineapple aroma (Cheatham and Johnston, 1995). Texas plants of *A. tetragonus* usually have three-angled stems (hence the common name triangle cactus), although plants in Mexico and Central America may have (4-)5-angled stems, the inspiration for the epithets *tetragonus* and *pentagonus*. Other common names used in Texas are barbwire cactus, night-blooming cereus, organo, and pitahaya. The economic uses of *A. tetragonus* were discussed by Cheatham and Johnston, and wildlife uses were outlined by Everitt and Drawe (1993).

Astrophytum asterias (Zucc.) Lem. SEA-URCHIN CACTUS, STAR CACTUS. Plate 311. [*Echinocactus asterias* Zucc., Abh. Bayer. Akad. Wiss. München 4: 13. 1845]. Plants unbranched. Roots diffuse. Stems green, depressed globose to globular, at maturity 5–15 cm in diameter, 2.5–6 cm high, retracting into the ground when desiccated. Ribs usually eight, very low (their crests nearly flat) separated by narrow but distinct vertical grooves. Stem surface dotted by ca. 1 mm clusters of short white hairs. Areoles circular, 3–5 mm in diameter, in straight rows, densely filled with yellow or gray wool. Spines absent in adults. Flowers yellow with orange throats, 4.5–5.4 cm long, 3.8–5.2 cm in diameter, opening widely; inner tepals lanceolate-acuminate, ca. 2.5 cm long, margins entire; outer tepals lanceolate-attenuate, covered with hairs; filaments yellowish, anthers yellow; style yellowish, with 10–12 yellow stigma lobes ca. 4 mm long; pericarpel covered with "scales" (bracteoles), these having blackish spinelike tips, and with wool in the axils. Fruit green or pinkish, fleshy, oval, 1.5–2 cm long, ca. 1.2 cm in diameter, densely covered with areolar wool and spinescent bracteoles, drying after maturity, ultimately abscising. Seeds blackish or dark brown, shiny, 2–3 mm in broadest diameter.

A rare species, reported by one source (Endangered and Threatened Wildlife and Plants, 17.12, 1996) to survive at only two sites, one in Starr County of the lower Rio Grande valley and the other in Tamaulipas, Mexico, with an estimated total population of ca. 2,000 plants, but not reliably surveyed; Weniger listed the range as including parts of Starr and Hidalgo counties. The range in Mexico, greatly reduced by conversion to farmland, is poorly documented.

The spineless plants of *A. asterias,* also known as star peyote, superficially resemble those of *Lophophora williamsii* (peyote) and reportedly possess some of the same alkaloids (Anderson, 1996a). Both Benson (1982) and Weniger (1984) included *A. asterias* in *Echinocactus,* a relationship that is supported by DNA studies (Wallace, 1995). As treated by most authors during the past century, *Astrophytum* Lem. is a genus of four species, all of them of Mexican distribution, two of them in the CDR; only the rare *A. asterias* barely extends into Texas. The genus name and specific epithet allude to either the myriad white dots on the stem surface (epidermal trichome tufts unique to this genus) or to the five-pointed "starlike" cross-sectional outline of the stem of *A. myriostigma* Lem., the type species, after the Greek *asteros,* "star," and *phyton,* "plant."

Pereskia aculeata Mill. LEMONVINE. Plates 312, 313. [*Cactus pereskia* L., Sp. Pl. 1: 469. 1753]. Plants shrubs or lianas, forming vines 3–10 m long, the trunk 2–3 cm or more in diameter. Stems straggling to clambering; long shoots scandent to winding; distal twigs to ca. 4 mm thick, green to reddish. Areoles on twigs ca. 2 mm in diameter, on the trunk cushionlike and to 15 mm in diameter. Leaves flat, pinnately veined, weakly succulent, long-persistent, short-petiolate, broadly lanceolate to oblong or ovate, 4.5–7 cm long, 1.5–5 cm broad or smaller, green or purplish, discolorous in some cultivars; petioles 3–7 mm long. Spines paired and clawlike on young shoots and twigs, later supplemented by long, straight spines (to 25 per areole) in older areoles of the main shoots, 1–3.5 cm long, abruptly thickened at the base, brown to black, gray with age. Inflorescences terminal, racemose or paniculate. Flowers whitish, heavily fragrant, nocturnal, 2.5–5 cm in diameter; receptacles cup-shaped or turbinate, 5–6 mm in diameter, on slender pedicels, bracteate, with trichomes and spines in areoles in axils of the bracts; outer tepals green, 2–5, ovate, 4–11 mm long; inner tepals white or greenish-white or with yellowish or pink tinge, obovate to spatulate, 1.5–2.5 cm long; stamens numerous, 5–10 mm long, filaments whitish proximally, white or yellow to reddish distally, anthers yellow; ovary superior at anthesis but in fruit enclosed by the perigynous rim of the receptacular cup and filament bases; style 10–12 mm long, white, stigma lobes 4–7, white, 3–5 mm long, erect or suberect. Fruit yellow to orange at maturity, globular to subglobular, fleshy, 1.5–2.5 cm in diameter, bearing 6–15 areoles with spreading spines 3–8 mm long. Seeds only 2–5 (solitary in each carpel, unusual for cacti), black, smooth, flat or weakly concave on one side (bilaterally asymmetrical, unlike all other cactus seeds), almost circular in general outline, 4.5–5 mm in diameter.

This tropical leafy cactus genus was unknown in Texas until 11 May 1996 (Ideker, 1996) in Willacy County, lower Rio Grande valley. The plants were located on a wooded tract of land surrounded by cotton fields, with stems climbing and spreading like vines in trees and hanging within 2 m of the ground. Previously the species, also known as Barbados gooseberry, was known to occur in Florida, southern Mexico, Central America, and the West Indies, where probably it is naturalized after cultivation (also escaped in South Africa), and in lowland tropical South America, its natural territory (Leuenberger, 1986). The plants are assumed to be naturalized in Willacy County but probably not introduced directly at the site where they were discovered. The genus name commemorated the French scholar Nicolas Claude Fabri de Peiresc (1580–1637), although the nomenclaturally correct spelling of the genus name turned out to be *Pereskia*, probably reflecting the French pronunciation of Peiresc. The specific epithet is derived from the Latin *aculeus*, "needle," obviously a reference to the spines of this broad-leafed cactus.

Selenicereus spinulosus (DC.) Britton & Rose QUEEN OF THE NIGHT. [*Cereus spinulosus* DC., Mém. Mus. Hist. Nat. 17: 117. 1828; *Selenicereus pseudospinulosus* Weing.] Plants clambering. Roots aerial. Stems many, usually 2–4 m long, light green, 1–2 cm in diameter. Ribs 4–6, acute. Areoles 1.5–2.5 cm apart. Spines brown, 1 mm long; central spine 1; radial spines 6–7. Flowers nocturnal, white to pinkish, 10–12 cm long, to 8.5 cm or more in diameter. Reported for South Texas. Main distribution in Tamaulipas, Veracruz, Hidalgo, San Luis Potosí, Oaxaca, and Chiapas, Mexico. A photograph is available in Anderson (2001).

Selected Glossary

ABAXIAL. The side of an organ away from the main axis; for example, the lower side or edge of a tubercle (opposite of adaxial).

ABSCISSION. Separation of a plant part through disintegration of a special layer(s) of cells.

ACICULAR. Needlelike; slender, elongate, circular in cross section, and tapering to a pointed apex.

ACUMINATE. A shape abruptly narrowed into a long, pointed apex.

ACUTE. Tapering to an apex, at less than 45°.

ADAXIAL. Adjacent to the axis (facing the stem); said of the upper side or edge of a leaf or tubercle.

ADVENTITIOUS. Formed in an unusual place; said especially of roots that sprout directly from stems or fruits instead of from preexisting roots.

ALLOGAMY. Cross-fertilization; outcrossing (opposite of autogamy).

ALLUVIUM. A geologic substrate deposited by running water; clay, silt, sand, gravel, or loose rocks.

ANNULAR. In the form of a ring, such as annular thickenings encircling stems or spines.

ANNULATE. Having ringlike bands; said of spines that show swollen daily growth increments, as in *Ferocactus wislizeni*.

ANTHER. The upper pollen-producing part of the stamen, consisting mostly of the pollen sacs.

ANTHESIS. The time of opening of a flower.

ANTHOCYANIN. A water-soluble flavonoid pigment (red, purple, or blue) occurring in cell vacuoles of many types of plants except cacti and their relatives.

ANTHROPOGENIC. Engendered by humans.

APEX. The uppermost point; the tip (pl., apexes or apices).

APICAL. At the apex.

APICULATE. A shape ending abruptly in a short, protruding point (an apiculation), but not a hard prickly point (see Mucronate).

APOMIXIS. Reproduction without fertilization; asexual reproduction; in plants either vegetative reproduction or some other form of bypassing sexual reproduction, in which meiosis and fusion of gametes are partially or totally suppressed.

APPRESSED. Lying flat against the underlying surface; said of radial spines spreading horizontally instead of pointing outward.

ARACHNOID. "Spidery"; said of certain plants having sticky or tangled cobwebby hairs; certain plant parts may be arachnid-like or arachnoid.

ARBORESCENT. Treelike in habit.

AREOLAR GROOVE. On a cactus podarium (only in certain genera), a narrow continuation of the areole, extending adaxially from the spine cluster.

AREOLE. A modified nonphotosynthetic short shoot, borne at the upper end or protruding summit of each podarium; an axillary bud, although its subtending leaf is obsolete or rudimentary; see Areoles in Introduction.

ARIL. A structure attached to a seed, which develops as an outgrowth in the region of the hilum, from the funiculus, and partially or entirely envelops the seed; used with reference to the funicular covering (see Funiculus) of *Opuntia* seeds.

ARILLATE. Bearing an aril.

ASCENDING. Angled upward.

ASEXUAL. Reproducing without sex, as in vegetative reproduction or some other form of apomixis.

ATTENUATE. Gradually tapering toward the tip into a point, involving a more or less flattened structure.

AUTOGAMY. The process of self-pollination and self-fertilization; autogamous is the opposite of allogamous.

AWL-SHAPED. Tapering from the base to a sharp point.

AXIL. The upper angle formed by the attachment of a leaf (or a similar structure) against its underlying surface.

AXILLARY. Developed in an axil; subtended by a leaf or similar structure.

AXIS. The main stem or main line of development of a plant part.

BARBELLATE. Minutely barbed.

BEAKED. Having a firm, elongate, slender, projecting structure.

BERRY. A fleshy or pulpy fruit usually with several to many seeds embedded in the pulp; in most cases a cactus fruit is interpreted as a modified berry derived from an inferior ovary.

BETACYANIN. One of the two color groups of betalains: red, magenta, pink, and so on.

BETALAINS. A class of water-soluble vacuolar pigments with almost the same colors as anthocyanins and anthoxanthins but a very different nitrogen-containing chemical structure; found only in Caryophyllales and certain fungi.

BETAXANTHIN. One of the two color groups of betalains: yellows and orange-reds.

BISEXUAL. When in reference to a flower, both male and female reproductive organs present; hermaphroditic.

BRACT. A reduced, modified leaf, often subtending a reproductive structure.

BRISTLE. A nearly rigid, slender, hair- or spinelike structure.

BULBOUS. Swollen and bulblike.

CADUCOUS. Falling away early, even prematurely.

CALCIUM OXALATE. Calcium salts deposited in crystal form usually in cell vacuoles.

CALLUS. Hard, differentiated tissue often lighter in color, usually raised above normal tissue.

CAM. The acronym for Crassulacean acid metabolism.

CANESCENT. Gray-pubescent with a dense covering of hairs.

CAPITATE. In a dense cluster or head.

CENTRAL. When in reference to spines, the individual(s) that originates anywhere inside the radials.

CESPITOSE, CAESPITOSE. Having numerous stems that form a dense clump.

CHLORENCHYMA. Green photosynthetic cells in the outer cactus stem, inside the collenchyma and outside the cortex.

CHOLLA. Type of cactus plant characterized by erect or suberect habit and cylindroid stem joints.

CILIATE. Having marginal hairs.

CIRCUMSCISSILE. Opening along a circular horizontal line.

CLADODE. A leaflike stem, such as the flattened green stem of a prickly pear.

CLAVATE. Club-shaped.

CLINE. A character gradient, as between two populations or a series of contiguous populations; clinal.

CLONE. A group of individual plants propagated asexually from a single individual.

COLLENCHYMA. Cell type characterized by thick nonlignified cell walls, such as in the hypodermis of a cactus stem, just inside the epidermis and outside the cortex and chlorenchyma.

COLUMN. A floral term, used in reference to the thin expanse of tissue extending from the top of the ovary up to the floor of nectar chamber.

COLUMNAR. Having the appearance of a column; said of cacti with large, columnlike stems, as in the saguaro of the Sonoran Desert.

COMPRESSED. Flattened, usually laterally.

CONCOLOROUS. Of uniform color.

CONICAL. Cone-shaped, with the point of attachment at the broad base of the cone.

CORTEX. Thick region of colorless, water-storing parenchyma cells in cactus stems inside the chlorenchyma and outside the vascular cylinder.

COTYLEDONS. In cacti the two first leaves of the embryo, apparent in most opuntioid seedlings, perhaps not evident in cactoid seedlings (not true leaves).

CRASSULACEAN ACID METABOLISM (CAM). A form of C_4 photosynthesis where CO_2 fixation occurs at night when stomates are open and where the light and dark reactions of C_3 photosynthesis occur during the day; most cacti undergo CAM metabolism.

CRESTATION. Crescent-shaped.

CRESTED. A cactus stem or stems with abnormal fanlike apical growth.

CRYPTIC. Hidden, not easily detected, perhaps as a result of being small or camouflaged as in some cacti.

CULTIVAR. A horticultural form maintained through cultivation by humans.

CUSP. A sharp rigid point.

CUSPIDATE. Ending in a cusp.
CUTICLE. A waxy, noncellular layer on the outer surface of a plant organ.
CYANIC. Dark blue but also used here in reference to red, blue, purple, violet, and similar colors, as opposed to yellow, green, or white.
CYLINDROID. In the form of a cylinder.
DECIDUOUS. Falling away in senescence.
DECLINED. Turned downward.
DECUMBENT. Lying on the ground, with the apical end ascending.
DECURRENT. Extending down and adnate to the stem, as the bases of some podaria.
DECURVED. A curved axis with tip pointing downward.
DEEP-SEATED. With the stem base or most of the stem below ground level.
DEFLEXED. Abruptly bent downward from near the base.
DEHISCENT. Opening by splitting along regular lines; seeds may be released after dehiscence of the fruit.
DELTOID. Having a triangular shape.
DENTICULATE. With sharp teeth extending perpendicular to the margin.
DEPRESSED. Pressed down endwise; more or less flattened.
DESCENDING. Extending gradually downward.
DETERMINATE GROWTH. Growing to a particular, predetermined size or form, then stopping.
DIMORPHIC. Occurring normally in two dissimilar forms.
DIOECIOUS. Having either male or female reproductive organs, or flowers, on different individuals; dioecy.
DIPLOID. Having two sets of chromosomes per nucleus; $2n$.
DISCOID. Circular and flat; disklike.
DISCOLOROUS. Used when the upper and lower surfaces of a leaf are of different colors.
DISTAL. Away from the point of attachment; toward the apex.
DISTINCT. Separate.
DIURNAL. Occurring during the daytime.
DIVERGENT. Spreading away from each other.
DOG CHOLLA. Type of cactus plant characterized by ground-hugging, clumped or mounded habit and cylindroid stem joints.
DORSAL. Back; the back or outer surface of a part or organ; the opposite of ventral; abaxial surface.
DORSIVENTRAL. Flattened and with definite dorsal and ventral surfaces.
DRUSES. Compound crystalline structures of calcium oxalate, oblong, spheroidal, or lens-shaped in most cacti.
EDAPHIC. Of or relating to the soil.
ELLIPSOID. Elliptic in three dimensions.
ELLIPTIC. In the form of an ellipse; a flattened circle, about twice as long as wide, widest in the middle, narrowed to rounded or pointed ends.
EMARGINATE. With a shallow notch at the apex.
ENDEMIC. Occurring naturally only in a certain geographic area.

ENTIRE. Surface with a smooth margin.
ENTOMOPHILOUS. Pollinated by insects or dispersed by the agency of insects; entomophily.
EPIDERMIS. The cells forming the surface layer of a plant organ.
EPIPHYTE. A nonparasitic plant growing upon another plant above the ground.
EPITHET. An adjective used as a noun; in nomenclature the second part of the binomial representing the specific epithet in a species or intraspecific name (e.g., *aggeria* in *Opuntia aggeria*).
ERECT. Standing straight up or nearly so.
EROSE. With an irregular margin.
ESCAPE. An introduced taxon that spreads and persists away from cultivation, without the aid of humans.
FASCIATION. Abnormal fanlike or undulate growth of the stem apex, as a result of growth from an altered apical meristem.
FELTED. With intertwining, matted hairs.
FIBROUS. Said of a root system in which there are several major roots of about equal caliber arising from the same area.
FILAMENT. Of the stamen, the slender stalk that bears the anther.
FIMBRIATE. With a fringed margin.
FIMBRILLATE. Finely fimbriate.
FISSURE. A natural or perhaps unnatural cleft, narrow opening, or crack in tissue or between organs or parts of an organism.
FLACCID. Limp, perhaps from loss of water in tissues; floppy.
FLEXUOUS. Curving in and out; wavy.
FLORAL REMNANT. The shriveled remains at the apex of a cactus fruit. These may be caducous or persistent.
FLORAL TUBE. In Cactaceae, the funnelform section of the flower.
FLOWER. In Cactaceae, the specialized tip of the fertile long shoot, from the base of the ovary to the tepals, including stamens and pistil(s).
FOLIACEOUS. Leaflike.
FOVEOLATE. Pitted, the pits numerous or solitary.
FRUIT. The ripened ovary of a plant (whether juicy or dry), containing the mature seeds. In Cactaceae, the berrylike fruit, sensu stricto, is enclosed by surrounding "accessory tissues" (see Pericarpel), but the whole structure falls as one unit and is called the "fruit" for all practical purposes.
FUNICULUS. Attachment stalk of ovule to placenta.
FUNNELFORM. Shaped like a funnel, with a tube gradually widening upward.
FUSIFORM. Spindle-shaped; narrowed to a point at both ends of a swollen middle.
GEITONOGAMY. Pollen transfer and fertilization between different flowers on the same plant.
GERMINATION. The initiation of growth, as occurs during the emergence of the embryonic organs from a seed.
GLABRATE. Hairy at first, but later the hair falling away.
GLABROUS. Not hairy.
GLAND. A tissue that secretes.

GLAUCOUS. With a powdered bluish, white, or gray wax on the surface (easily rubbed off), as on the surfaces of unweathered cladodes or certain opuntias.

GLOBOSE, GLOBULAR. Essentially spherical; spheroidal.

GLOCHID. One of the small, retrorsely barbed, easily dislodged modified spines occurring in the areoles of *Opuntia;* glochids usually are manifestly smaller and otherwise different from spines.

GYNODIOECIOUS. Having pistillate and perfect flowers on separate plants.

HABIT. The general aspect, shape, or appearance of a whole plant, in nature.

HABITAT. The locality, site, and particular local environment occupied by an organism.

HAIR. A slender projection of cells.

HEMISPHERIC. Shaped like half a sphere; hemispheroidal; hemispherical.

HERKOGAMOUS. Having the stamens and stigma positioned so that self-pollination is prevented.

HEXAPLOID. Having six sets of chromosomes per nucleus; $6n$.

HILUM. A depression or scar on the seed where the funiculus was attached; by definition the hilum is the basal point of the cactus seed, although it may appear to be lateral.

HOLOTYPE. The particular permanently preserved specimen upon which a taxon has been based and with which its scientific name is permanently associated; the type specimen, deposited in an herbarium.

HOOKED. With a hook at the end, otherwise with a straight or slightly curved shaft.

HYPODERMIS. A layer of cells just beneath the epidermis.

IDIOBLAST. Cells markedly differing from other cells in the same tissue, in cacti often the cells containing crystals, thus crystal idioblasts.

IMBRICATE. Overlapping like shingles.

INCISED. Having the margin indented by irregular, sharp, deep incisions.

INDEHISCENT. Not regularly splitting open, as in some fruits.

INDETERMINATE GROWTH. Not growing to a predetermined size or form and then stopping; growth continues as long as conditions are favorable.

INDURATE. Hardened.

INFERIOR OVARY. An ovary enclosed by and adnate with the floral tube, in most cacti including the receptacle and forming the pericarpel.

INNER TEPALS. The innermost or petaloid tepals in cactus flowers (see Tepal).

INNOCUOUS. Harmless, as in cactus stems that are spineless or easily handled owing to spines closely appressed against the stem, or spines too short, blunt, or soft to prick the skin.

INTRODUCED. Native to another area and imported, often becoming established (naturalized) and spreading.

INTROGRESSION. Hybridization followed by the hybrids and their descendants crossing back to one or both parents.

JOINT. One segment of a stem composed of several "joints" as used by some authors; "stem segment" should be used in place of "joint."

LACERATE. Irregularly cut, as if slashed.

LACINIATE. Lacerate into narrow, usually pointed segments.

LANCEOLATE. Having the shape of a lance (i.e., 4–6 times as long as broad), broadest near attachment end, tapering to the apex.
LATERAL. On the side; extending to the side.
LATEX. Viscid, usually whitish liquid substance produced in specialized cells and canals in certain cactus species, such as in some *Mammillaria*.
LECTOTYPE. A type specimen selected from one or more specimens cited in the original publication, if no holotype was indicated in the original published description, when the holotype is found to belong to more than one taxon, or if the holotype is missing (or otherwise selected according to the *International Code of Botanical Nomenclature*).
LENTICULAR. Lens-shaped, like a biconvex lens.
LINEAR. Narrow, considerably longer than wide, with parallel sides.
LOBE. A segment of an organ; a lobe may be almost any shape.
LONG SHOOT (FERTILE AND STERILE). Shoot or portion of a shoot with relatively long internodes, when on a plant with long-shoot–short-shoot organization.
MATURATION. The process of becoming mature; ripening, as with a fruit exhibiting mature color and texture, and containing seeds.
MATURE. Having completed natural growth and development; having undergone maturation.
MEDULLARY BUNDLES. Vascular bundles that extend through the pith in certain cacti, or in most cacti in the cortex and usually outward to ribs, tubercles, and areoles.
MERISTEM. A region of tissue where the cells are actively dividing or have the potential to divide and produce new cells.
MERISTEMATIC. Having the potential or the function of cell division.
MERISTEMATIC GROOVE. The linear, sulcate, or recessed groove on the adaxial surface of the tubercle; same as tubercle groove and areolar groove, also called meristematic groove because of meristematic activity there.
MICROPYLE. The opening in the integument through which the pollen tube enters the ovule.
MIDREGION. The middle area of a tepal or other structure, along and usually near the midrib.
MIDRIB. The major vein along the middle of a tepal or other structure.
MONOECIOUS. Having staminate and pistillate flowers on the same plant.
MORPHOTYPE. An individual plant or population distinguishable from another by one or more morphological features.
MUCILAGE. A viscid or slippery and viscous complex carbohydrate substance in cactus tissues, mostly stems, produced in specialized cells, reservoirs, or canals.
MUCILAGINOUS. Secreting mucilage.
MUCRONATE. Bearing a mucro, which is a short, sharp, terminal projection of tissue similar in texture to the rest of the structure.
MUCRONULATE. Minutely mucronate.
NATIVE. Naturally occurring in an area.
NECTAR. A sugary fluid that attracts animals to plants.

NECTAR CHAMBER. Floral chamber around the base of the style and below the lowermost stamen insertion, often containing nectar that is secreted from special surrounding tissues or the nectary.

NECTARY. A gland or tissue producing nectar.

NERVE. A visible vein.

NOCTURNAL. Occurring during the hours of darkness; said of flowers that open at night and close during the day.

Nomen nudum. An invalid name, published without sufficient information to satisfy criteria of valid publication, according to the *International Code of Botanical Nomenclature;* nom. nud.

NOTHOSPECIES. Any and all members of a hybrid species, including later-generation progeny and backcrosses; nothotaxon.

Ob-. A Latin prefix usually denoting inversion.

OBCONIC. Conic, but attached at the pointed end of a cone.

OBLANCEOLATE. Lanceolate, but attached at the narrow end.

OBLIQUE. Diagonal.

OBLONG. Rounded structure with more or less parallel sides and about two or three times as long as wide.

OBOVATE. Ovate, but attached at the narrow end.

OBOVOID. Ovoid in three dimensions, but attached at the narrow end.

OBTUSE. Blunt or rounded at the apex or base; less than 45° angle at apex.

OCTOPLOID. Having eight sets of chromosomes per nucleus; *8n,* as in *Opuntia ficus-indica.*

ORBICULAR. Approximately circular in outline; orbiculate.

OUTCROSSING. Pollination between different individuals of a population; open pollination, crossbreeding, outbreeding.

OUTER TEPALS. The outer or sepaloid tepals in cactus flowers, transitional with inner (petaloid) tepals (see Tepal).

OVAL. Having the shape of an egg; according to *Merriam-Webster's Collegiate Dictionary* (10th edition), broadly elliptical.

OVARY. The lower, expanded portion of the pistil, in cacti (except for some species of *Pereskia*) surrounded by pericarpel tissue and inferior in position; containing the ovules.

OVATE. Generally egg-shaped in outline: like elliptical, but wider at the proximal end.

OVOID. Ovate, but in three dimensions.

OVULE. An immature potential seed, still in the ovary prior to fertilization and/or ripening; anatomically, an integument surrounding a megasporangium and female gametophyte.

PAD. A term used in common reference to the strongly flattened stem segment of a prickly pear; cladode or phylloclade.

PANICLE. A branching inflorescence, composed of pedicels and peduncles.

PAPILLAE. Low, usually rounded projections.

PAPILLATE. Having papillae.

PAPILLOSE. Minutely papillate.

PAR. Photosynthetically active radiation (i.e., the red and blue light spectra and the photons in those spectra).

PARENCHYMA. Physiologically active cells with thin walls; collectively, a soft tissue composed of these cells.

PARIETAL. Borne at the periphery of the single locule.

PECTINATE. Like the teeth of a comb.

PEDICEL. The stalk of an individual flower; not evident in most cacti.

PERIANTH. The sepals and petals collectively, or the tepals collectively.

PERICARPEL. The stem tissue precisely surrounding the ovary, the external surface of which bears areoles in cacti that have areolate pericarpels.

PERIDERM. Bark on the outer surface of older stems in some cacti, usually developing in patches.

PERIPHERAL. On the margin.

PERISPERM. Reserve food in the cactus seed, derived from diploid nucellar tissue, unlike the triploid endosperm of most angiosperms.

PERSISTENT. Remaining attached, perhaps beyond the usual period, before falling away, or tending not to fall away.

PETAL. One of the separate parts of the corolla (the inner perianth organ), usually pigmented and showy; in Cactaceae the showy parts of flowers usually are inner tepals.

PETALOID. Petal-like in appearance or position, as in a petaloid perianth part, a tepal that is similar to a petal.

PHELLOGEN. Cork cambium.

PHENOLOGY. Study of temporal aspects of recurrent natural phenomena, and their relation to climate, season, and weather (e.g., flowering phenology).

PHOTOSYNTHESIS C_3; C_4; CAM. Conversion of light energy to chemical energy in chloroplasts; in the C_3 type CO_2 is "fixed" initially through the formation of a 3-carbon compound; in the C_4 type CO_2 is "fixed" initially through the formation of a 4-carbon compound.

PHYLLOCLAD. The leaflike flattened stem segment (pad) of a prickly pear.

PHYTOLITHS. Crystals in fossils or subfossils.

PILE. A reference to trichomes closely standing together like the pile of a carpet.

PILOSE. Having soft, slender hairs.

PISTIL. The female reproductive structure of the flower, consisting of the stigma (in cacti with several stigma lobes), which is connected by a tubular style to the expanded ovary.

PISTILLATE. Flower with one (in Cactaceae) or more pistils, but no functional stamens.

PITH. The central tissue of a stem, usually composed of parenchyma cells and surrounded by a vascular cylinder.

PLACENTA. The ovary tissue that gives rise to ovules and serves in attaching the funiculus (ovule stalk) to the ovary.

PLACENTATION. Arrangement of placentae and therefore ovules in the ovary.

PLOIDY. General reference to number of sets of chromosomes in a nucleus.

PLUMOSE. Featherlike, with fine hairs projecting from the sides of a shaft.

PODARIUM. Collective term for stem-surface protuberances in cacti (pl., podaria).

POLLEN. The microscopic, often spheroidal microgametophytes produced, developed, and usually disseminated from the anther of a stamen.

POLYPLOID. Having three or more sets of chromosomes per nucleus.

PORRECT. Perpendicular to the surface.

PPF. Photosynthetic photon flux (i.e., only those photons with wavelengths in the visible spectrum that are absorbed by chlorophyll and used in the light reaction of photosynthesis).

PRICKLY PEAR. Type of cactus plant characterized by strongly flattened stem joints; *nopal*.

PROJECT. Protrude outward, as in projecting spines.

PROLIFEROUS. In cacti the formation of buds from fruit areoles; reproducing by special buds or shoots.

PROSTRATE. Flat on the ground.

PROTANDROUS. Floral condition in which anthers mature and release pollen before the stigma of the same flower is receptive.

PROXIMAL. Close to the point of attachment; toward the base.

PUBERULENT. Minutely pubescent.

PUBESCENT. Hairy; most precisely meaning fine, soft hairs. The noun form, pubescence, is commonly used as a general term to indicate the presence of hair of any kind.

PULP. The pulpy tissue in the fruit, or other organ.

PUNCTATE. Covered with dots (glands) or pits.

RADIAL. When in reference to spines, the individuals or series that radiate from the periphery of the spine cluster in a plane that is more or less parallel with the stem surface.

RAPHIDES. Elongated crystalline structures of calcium oxalate, often rodlike and aggregated into bundles.

RECEPTACLE. The stem tip that supports the floral organs.

RECEPTACULAR TUBE. Tubular, funnelform, or cuplike part of the cactus flower that extends from just above the pericarpel and column to the outer tepals; essentially the "floral tube."

RECURVED. Curving backward and downward.

REFLEXED. Bent or extending abruptly downward or backward.

RELIC. Persisting in only a part of its previous range; relict.

RENIFORM. Kidney-shaped.

RETICULATE. Netlike; reticulum.

RETRORSE. Extending downward or backward.

REVOLUTE. Rolled backward.

RHIZOME. A horizontal underground stem.

RHOMBIC. Diamond-shaped.

RIBS. Confluent podaria forming longitudinal files on the stem surface in cacti, arising from stem tissue.

RIND. The wall of the cactus fruit.

ROOT. Underground anchoring and absorptive organ of a plant, anatomically differ-

ent from the stem, with no nodes or internodes; a storage root is enlarged with reserve food.
ROTATE. Saucerlike.
RUDIMENT. A vestige of an organ or structure.
RUFOUS. Reddish-brown.
RUGOSE. Wrinkled.
SCABROUS. Rough to the touch; bearing minute projections such as short, stiff hairs.
SCARIOUS. Membranous (thickish) and translucent.
SCLERIFIED. Having a thick cell wall that is impregnated with lignin.
SECTION. A subgeneric taxonomic category.
SEED. A mature ovule, with a hardened seed coat and an embryo inside.
SEED COAT. The hardened outer wall of a seed, derived from the integument.
SELF-COMPATIBILITY. The condition whereby a plant can be self-fertilized.
SELF-INCOMPATIBILITY. The condition whereby plants with perfect flowers cannot produce offspring through self-pollination.
SELF-POLLINATION. Pollen transfer from anther to stigma of the same flower or to another flower on the same plant.
SENSU LATO. In a broad sense.
SENSU STRICTO. In a narrow sense.
SEPAL. One of the separate parts of the corolla (the outer perianth organ), usually not as pigmented and as showy as petals, in Cactaceae the showy parts of flowers are inner tepals, and nonshowy parts are outer tepals.
SEPALOID. Sepal-like in appearance or position as in a sepaloid perianth part; a tepal that is similar to a sepal.
SERIES. A subgeneric taxonomic category (below section level).
SERPENTINE. Winding or turning, resembling a serpent in form.
SERRATE. With marginal teeth, forward projecting and acute, resembling those of a saw; saw-toothed.
SERRULATE. Minutely serrate.
SESSILE. Not stalked; directly attached.
SHEATH. On spines the paper-thin epidermal layer that separates from the lignified spine core.
SHOOT. The aboveground vegetative part of a plant; including apical meristems, stems, and leaves.
SHORT SHOOT. The highly contracted lateral branch of a shoot, when on a plant with long-shoot and short-shoot organization.
SHRUB. A woody plant with several stems arising from ground level; usually shrubs are smaller than trees.
SINUATE. Sinuous; wavy.
SINUS. The space or recess between two lobes or lobelike structures.
SKIN. Thin outer layer of the cactus stem, consisting of the cuticle, epidermis, and hypodermis.
SPATULATE, SPATHULATE. Spatula-shaped.
SPECIES. A living population or population system of genetically closely related

individuals; the most commonly used intraspecific categories are subspecies and varieties.

S PHEROIDAL . Sphere-shaped.

S PINE . A hard, sharp-pointed structure derived from leaf tissue; the spines of cacti are specialized leaves that develop from the bud in the areole or from secondary buds derived from the areole.

S PINE C LUSTER . The characteristic spine configuration in a single areole, this remaining intact upon abscission.

S PINIFEROUS . Bearing spines. Spinescent.

S PINOSE . Spinelike or ending in a spine; used by some authors to mean bearing spines.

S PREADING . Extending in several directions.

S TAMEN . The male reproductive structure of a flower, consisting of a pollen-producing anther with pollen sacs and a slender supporting filament.

S TAMINATE . Flower with stamens but no functional pistil.

S TEM . The plant axis aboveground, or usually so, with nodes and internodes; anatomically different from the root; the stems of opuntioid cacti are jointed.

S TIGMA . The terminal part of the pistil of a flower, usually supported by the style; functioning in pollen reception; in cacti comprising 3–20 cylindroid lobes or branches.

S TIPE . The stalk of an organ.

S TRAMINEOUS . Straw-colored.

S TRIATE . With longitudinal lines.

S TROPHIOLE . An outgrowth at or near the hilum of some seeds.

S TYLE . The tubular organ of the pistil, connecting the stigma and ovary.

Sub-. As a prefix, meaning somewhat or slightly.

S UBERIN . Waxy substance produced by outer stem cells in some cacti; the substance impregnating walls of cork cells.

S UBGENUS . A subgeneric taxonomic category; a group of related species, series, or sections; the category is not used formally unless there are at least two subgenera recognized.

S UBSPECIES . An intraspecific taxonomic category with rank between that of species and section or variety.

S UBTENDING . Positioned below or to the outside.

S UBULATE . Awl-shaped; flattened in cross section and tapering to an apical point.

S UCCULENCE . The condition in plants where relatively abundant water-storing tissues allow drought tolerance; succulent.

S ULCATE . Grooved or furrowed lengthwise.

S UPERIOR O VARY . An ovary above the origin of the floral tube or perianth.

S YNONYM . A name or a name combination that has been applied to a taxon but is not the correct name for the taxon in a given taxonomic treatment; a discarded name is said to be in synonymy.

T APROOT . The primary root usually of larger size than its branch roots.

T AXON . A taxonomic unit of any rank.

T EPAL . Perianth segment not clearly differentiated as calyx (sepals) or corolla (petals).

TERETE. Circular in transverse section.
TETRAPLOID. Having four sets of chromosomes per nucleus; $4n$.
THERMOPERIODISM. Response of an organism to fluctuating temperatures.
THIGMOTROPISM. An orientation response to touch or contact; thigmotropic.
TOMENTOSE. Woolly-pubescent; usually covered with matted, soft, wool-like hairs that are not straight.
TRANSVERSE. Across, at right angles to the axis.
TREE. A relatively large, usually branched and woody plant with a single main trunk.
TRICHOME. A usually multicellular plant hair; a trichome arises from the epidermis of a plant organ.
TRUNCATE. Having an abrupt ending, as in an apex at right angles to the axis.
TUBER. A thickened underground stem, functioning as a storage organ.
TUBERCLE. Individually protruding podarium on stem surface in cacti, arising from both leaf and stem tissue, bearing an areole at the apex; podarium.
TUBERCLE GROOVE. A narrow sulcus or indentation on the adaxial surface of a tubercle.
TUBEROUS. Having the aspect of a tuber but not necessarily derived from a stem.
TURBINATE. Top-shaped; inversely conical.
TURGID. Swollen; inflated with water.
TYPE. See Holotype.
TYPE LOCALITY. The locality at which the type specimen of a taxon was collected.
TYPE SPECIES. The species of a genus with which the generic name is permanently associated.
TYPE SPECIMEN. The holotype, or designated type according to the *International Code of Botanical Nomenclature*.
UMBILICUS. A cuplike depression at the apical end of fruits in *Opuntia*.
UNCINATE. Hooked.
UNISEXUAL. An individual with either staminate or pistillate reproductive structures but not both.
UPRIGHT. Standing vertical or nearly so.
URCEOLATE. Urn-shaped.
VARIETY. A subspecific taxonomic category, the lowest-ranking taxon commonly recognized.
VASCULAR BUNDLE. A strand composed of xylem and phloem and often a vascular cambium; a vascular trace is a vascular bundle that extends into a leaf, branch, or flower; a vein.
VASCULAR CYLINDER. The cylinder of vascular bundles formed in a stem or root, best observed in cross section.
VASCULAR TISSUE. Tissues of a plant that transport water in xylem and sugars in phloem.
VEGETATIVE. Pertaining to any part of a plant except the flower and fruit.
VEGETATIVE REPRODUCTION. Asexual reproduction from any plant organ but not from seed after sexual reproduction.
VEIN. Threadlike conducting tissue in a leaf or leaflike structure; vascular bundle.

VELUTINOUS. Velvety vestiture; clothed with dense, straight, erect hairs.

VENATION. The arrangements of veins as seen in a leaf or leaflike structure.

VENTRAL. Front; the inner surface of a part or organ; the opposite of dorsal; adaxial surface.

VERRUCOSE. With a wartlike surface.

VERTICIL. A whorl.

VERTICILLATE. Arranged in whorls; a group of three or more organs at the same level.

VESTIGE. A rudiment.

VESTIGIAL. Rudimentary; poorly developed.

VICARIANT. Closely related taxa isolated geographically.

VILLOUS. With soft, long, closely spaced or interlaced hairs.

VISCID. Sticky.

WEED. A usually introduced plant capable of becoming established and often widely distributed in a variety of habitats, perhaps becoming a nuisance.

WOOD SKELETON. A dried cylinder of wood, as in some chollas, exhibiting spaces where soft tissues were present in the living stem.

WOOLLY. With long, strongly interlaced hairs, collectively matted.

XERIC. Arid; dry.

XEROPHYTE. A plant that is adapted to dry or xerophytic conditions.

Literature Cited

Alston, R. E., and B. L. Turner. 1963. *Biochemical systematics.* Englewood Cliffs, NJ: Prentice-Hall.

Anderson, E. F. 1960. A revision of *Ariocarpus* (Cactaceae). I. The status of the proposed genus *Roseocactus. Amer. J. Bot.* 47:582–89.

———. 1962. A revision of *Ariocarpus* (Cactaceae). II. The status of the proposed genus *Neogomesia. Amer. J. Bot.* 49:615–22.

———. 1963. A revision of *Ariocarpus* (Cactaceae). III. Formal taxonomy of the subgenus *Roseocactus. Amer. J. Bot.* 50:724–32.

———. 1964. A revision of *Ariocarpus* (Cactaceae). IV. Formal taxonomy of the subgenus *Ariocarpus. Amer. J. Bot.* 51:144–51.

———. 1965. A taxonomic revision of *Ariocarpus* (Cactaceae). *Cact. Succ. J. (U.S.)* 37:39–49.

———. 1980. *Peyote the divine cactus.* Tucson: University of Arizona Press.

———. 1986. A revision of the genus *Neolloydia* B. & R. (Cactaceae). *Bradleya* 4:1–28.

———. 1987. A revision of the genus *Thelocactus* B. & R. (Cactaceae). *Bradleya* 5:49–76.

———. 1995. The "peyote gardens" of South Texas: A conservation crisis? *Cact. Succ. J. (U.S.)* 67:67–73.

———. 1996a. *Peyote the divine cactus.* 2nd ed. Tucson: University of Arizona Press.

———. 1996b. The genus *Opuntia* in the Galápagos Islands. *Cact. Succ. J. (U.S.)* 68:298–305.

———. 1999a. *Ariocarpus:* Some reminiscences. *Cact. Succ. J. (U.S.)* 71:180–90.

———. 1999b. Some nomenclatural changes in the Cactaceae, subfamily Opuntioideae. *Cact. Succ. J. (U.S.)* 71:324–25.

———. 2001. *The cactus family.* Portland, OR: Timber Press.

Anderson, E. F., and W. A. Fitz Maurice. 1997. *Ariocarpus* revisited. *Haseltonia* 5:1–20.

Anderson, E. F., and M. E. Ralston. 1978. A study of *Thelocactus* (Cactaceae). I. The status of the proposed genus *Gymnocactus. Cact. Succ. J. (U.S.)* 50:216–24.

Anthony, M. 1954. Ecology of the *Opuntiae* in the Big Bend region of Texas. *Ecology* 35:334–47.

———. 1956. The Opuntiae of the Big Bend region of Texas. *Amer. Mid. Nat.* 55:225–56.

Bach, D. 1998. Conservation—A common-sense approach. *Cact. Succ. J. (U.S.)* 70:297–99.

Backeberg, C. 1951. Some results of twenty years of cactus research. *Cact. Succ. J. (U.S.)* 23:150–51.

———. 1961. *Die Cactaceae*. Vol. 5. Jena, Germany: Gustav Fischer.

Backeberg, C., and F. M. Knuth. 1935. *Kaktus ABC*. Kopenhagen: Gyldendalske Boghandel Nordisk Forlag.

Baker, M. A., and R. A. Johnson. 2000. Morphometric analysis of *Escobaria sneedii* var. *sneedii*, *E. sneedii* var. *leei*, and *E. guadalupensis* (Cactaceae). *Syst. Bot.* 25:577–87.

Beesley, F. 1982. The barbed beauty. *Texas Highways,* May, 3–9.

Behnke, H. D. 1976. Ultrastructure of sieve-element plastids in Caryophyllales (Centrospermae), evidence for delimitation and classification of the order. *Pl. Syst. Evol.* 126:31–54.

Benson, L. 1969a. The cacti of the United States and Canada—New names and nomenclatural combinations. *Cact. Succ. J. (U.S.)* 41:124–28, 185–90, 233–34.

———. 1969b. Cactaceae. Pp. 221–317 in *Flora of Texas,* vol. 2., pt. 2, edited by C. L. Lundell et al. Renner: Texas Research Foundation.

———. 1970. Cactaceae. Pp. 1087–113 in *Manual of the vascular plants of Texas,* edited by D. S. Correll and M. C. Johnston. Renner: Texas Research Foundation.

———. 1977. Our daily bread and preserving the ecosystem. *Cact. Succ. J. (U.S)* 49:257–60.

———. 1982. *The cacti of the United States and Canada*. Stanford, CA: Stanford University Press.

Betancourt, J., T. R. Van Devender, and P. Martin, eds. 1990. *Packrat middens: The last 40,000 years of biotic change*. Tucson: University of Arizona Press.

Blum, W., M. Lange, W. Rischer, and J. Rutow. 1998. *Echinocereus*. Monograph. Belgium: n.p.

Bobich, E. G., and P. S. Nobel. 2001. Biomechanics and anatomy of cladode junctions for two *Opuntia* (Cactaceae) species and their hybrid. *Amer. J. Bot.* 88:391–400.

Boke, N. H. 1955. Dimorphic areoles of *Epithelantha*. *Amer. J. Bot.* 42:725–33.

———. 1979. Root glochids and root spurs of *Opuntia arenaria* (Cactaceae). *Amer. J. Bot.* 66:1085–92.

———. 1980. Developmental morphology and anatomy in Cactaceae. *BioScience* 30:605–10.

Boke, N. H., and E. F. Anderson. 1970. Structure, development and taxonomy in the genus *Lophophora*. *Amer. J. Bot.* 57:569–78.

Boke, N. H., and R. G. Ross. 1978. Fasciation and dichotomous branching in *Echinocereus* (Cactaceae). *Amer. J. Bot.* 65:522–30.

Bowers, J. E. 1996. Environmental determinants of flowering date in the columnar cactus *Carnegiea gigantea* in the northern Sonoran Desert. *Madroño* 43:69–84.

———. 1997. The effect of drought on Engelmann prickly pear (Cactaceae: *Opuntia engelmannii*) fruit and seed production. *Southwest. Nat.* 42:240–42.

———. 1998. Reproductive potential and minimum reproductive size of *Ferocactus wislizeni* (Cactaceae). *Desert Plants* 14:3–6.

———. 2002. Flowering patterns and reproductive ecology of *Mammillaria grahamii* (Cactaceae), a common, small cactus in the Sonoran Desert. *Madroño* 49:201–6.

Boyle, T. H., R. Karle, and S. S. Han. 1995. Pollen germination, pollen tube growth, fruit set, and seed development in *Schlumbergera truncata* and *S.* × *buckleyi* (Cactaceae). *J. Amer. Soc. Hort. Sci.* 120:313–17.

Bravo-Hollis, H. 1978. *Las Cactáceas de México*. Vol. 1. México, D.F.: Universidad Nacional Autónoma de México.

Bravo-Hollis, Sánchez-Mejorada, R. 1991a. *Las Cactáceas de México*. Vol. 2. México, D.F.: Universidad Nacional Autónoma de México, México.

———. 1991b. *Las Cactáceas de México*. Vol. 3. México, D.F.: Universidad Nacional Autónoma de México.

Breckenridge, B. 1999. Conservation (from cacti-etc.). *Cact. Succ. J. (U.S.)* 71:104.

Breckenridge III, F. G. 1981. A systematic study of the *Echinocereus enneacanthus* complex (Cactaceae). Master's thesis, Sul Ross State University, Alpine, Texas.

Breckenridge, F. G., III, and J. M. Miller. 1982. Pollination biology, distribution, and chemotaxonomy of the *Echinocereus enneacanthus* complex (Cactaceae). *System. Bot.* 7:365–78.

Britton, N. L., and J. N. Rose. 1919–23. *The Cactaceae*. 4 vols. Carnegie Institute Washington Publication 248. Washington, DC: Carnegie Institute.

———. [1937] 1963. *The Cactaceae*. Vol. 3. Reprint, New York: Dover.

Bruhn, J. G., J. E. Lindgren, and B. Holmsdedt. 1978. Peyote alkaloids: Identification in a prehistoric specimen of *Lophophora* from Coahuila, Mexico. *Science* 199:1437–38.

Brummitt, R. K., and C. E. Powell. 1992. *Authors of plant names*. Kew, England: Royal Botanic Gardens.

Bryant Jr., V. M. 1974. Prehistoric diet in southwest Texas: The coprolite evidence. *American Antiquity* 39:407–20.

Burbank, L. 1907. *The new agricultural-horticultural opuntias*. Los Angeles: Kruckeberg Press.

Burger, J. C., and S. M. Louda. 1995. Interaction of diffuse competition and insect herbivory in limiting brittle prickly pear cactus, *Opuntia fragilis* (Cactaceae). *Amer. J. Bot.* 82:1558–66.

Burgess, T. L., and D. K. Northington. 1981. *Plants of the Guadalupe Mountains and Carlsbad Caverns National Parks: An annotated checklist*. Contrib. no. 107. Alpine, TX: Chihuahuan Desert Research Institute.

Carey, R. H. 1980. Safeguarding the cacti. *Garden*, Sep.–Oct. 1980, 4–7, 31.

Casas, A., A. Valiente-Banuet, A. Rojas-Martínez, and P. Davila. 1999. Reproductive biology and the process of domestication of the columnar cactus *Stenocereus stellatus* in Central Mexico. *Amer. J. Bot.* 86:534–42.

Castetter, E. F., and P. Pierce. 1966. *Escobaria leei* Bödecker rediscovered in New Mexico. *Madroño* 18:137–40.

Chamberland, M. 1996. Cultivation and conservation. *Cactus and Succulent Society of America, Newsletter* 68:33–34.

———. 1997. Systematics of the *Echinocactus polycephalus* complex (Cactaceae). *Syst. Bot.* 22:303–13.

Cheatham, S., and M. C. Johnston. 1995. *The useful wild plants of Texas.* Vol. 1. Austin, TX: Useful Wild Plants.

Clark, W. D., G. K. Brown, and R. L. Mays. 1980. Flower flavonoids of *Opuntia* series *Cylindropuntia*. *Phytochemistry* 19:2042–43.

Clark, W. D., and B. D. Parfitt. 1980. The flower flavonoids of *Opuntia* series *Opuntiae*. *Phytochemistry* 19:1856–57.

Correll, D. S., and M. C. Johnston. 1970. *Manual of the vascular plants of Texas.* Renner: Texas Research Foundation.

Cory, V. L., and H. B. Parks. 1937. *Catalogue of the flora of the state of Texas.* Bull. no. 550. College Station: Texas Agricultural Experiment Station.

Cota, J. H. 1993. Pollination syndromes in the genus *Echinocereus:* A review. *Cact. Succ. J. (U.S.)* 65:19–26.

Cota, J. H., and C. T. Philbrick. 1994. Chromosome number variation and polyploidy in the genus *Echinocereus* (Cactaceae). *Amer. J. Bot.* 81:1054–62.

Cronquist, A. 1988. The evolution and classification of flowering plants. New York: New York Botanical Garden, Bronx.

Crook, R., and R. Mottram. 1995. *Opuntia* index: Part 1: Introduction and A–B. *Bradleya* 13:89–118.

———. 1996. *Opuntia* index: Part 2: Nomenclatural note and C–E. *Bradleya* 14:99–144.

———. 1997. *Opuntia* index: Part 3: Nomenclatural note and F. *Bradleya* 15:98–112.

———. 1998. *Opuntia* index: Part 4: G–H. *Bradleya* 16:119–36.

———. 1999. *Opuntia* index: Part 5: Nomenclatural note and I–L. *Bradleya* 17:109–31.

———. 2000. *Opuntia* index: Part 6: M–O. *Bradleya* 18:113–40.

Cullmann, W., E. Götz, and G. Gröner. 1984. The encyclopedia of cacti. Portland, OR: Timber Press.

Demarest, M. 1981. Prickly but imperiled species. *Time* 117:78.

Deno, N. C. 1994. The critical role of gibberellins in germination and survival of certain cacti. *Cact. Succ. J. (U.S.)* 66:28–30.

Diamond, D. D., D. H. Riskind, and S. L. Orsell. 1988. A framework for plant community classification and conservation in Texas. *Texas J. Sci.* 39:203–21.

Diggs Jr., G. M., B. L. Lipscomb, and R. J. O'Kennon. 1999. Shinners & Mahler's illustrated flora of north central Texas. Sida, Botanical Miscellany, no. 16. Fort Worth: Botanical Research Institute of Texas; Sherman, TX: Austin College.

DiMartino, L. 1996. *Echinocereus.* Baveno, Italy: Cactus & Co.

Dubrovsky, J. G. 1996. Seed hydration memory in Sonoran Desert cacti and its ecological implication. *Amer. J. Bot.* 83:624–32.

Earle, W. H. 1963. *Cacti of the southwest.* Tempe, AZ: Daily News.

———. 1980. Cacti of the southwest. 2nd ed. Phoenix, AZ: Desert Botanical Garden.

Edwards, J. R. 1994. Nibble a nopalito. *Texas Parks and Wildlife,* Aug., 29–35.

Eisner, T., and S. Nowicki. 1980. Red cochineal dye (carminic acid): Its role in nature. *Science* 208:1039–42.

Emmett, R. T. 1995. Reproduction, mortality, and temporal changes in plant size for the endemic Tobusch fishhook cactus (*Ancistrocactus tobuschii*). Ph.D. diss., University of Texas at Austin.

Emory, W. H. 1859. *Report of the United States and Mexican boundary survey: Cactaceae by George Engelmann.* Vol. 2, 1–78. Washington, DC.

Endangered and Threatened Wildlife and Plants. 1996. 17.12. Endangered and threatened plants. Washington, DC: U.S. Government Printing Office.

Engelmann, G. 1856 [1857]. Synopsis of the Cactaceae of the territory of the United States and adjacent regions. *Proc. Amer. Acad. Arts* 3:259–346.

———. 1859. Cactaceae of the boundary, from the report of the U.S. and Mexican Boundary Survey and the order of Lt. Col. W. H. Emory, Maj. 1st Cav. and U.S. Commissioner. Vol. 2, pt. 1, 1–78. Washington, DC.

Everitt, J. H., and M. A. Alaniz. 1981. Nutrient content of cactus and woody plant fruits eaten by birds and mammals in South Texas. *Southwest. Nat.* 26:301–5.

Everitt, J. H., and D. L. Drawe. 1993. Trees, shrubs, and cacti of South Texas. Lubbock: Texas Tech University Press.

Felger, R. S., and M. B. Moser. 1985. People of the desert and sea. Tucson: University of Arizona Press.

Felker, P., and J. Moss, eds. 1995. Proceedings, Professional Association for Cactus Development, first annual conference, San Antonio, Texas, 1–92. Dallas: Prof. Assoc. for Cactus Development.

Ferguson, D. J. 1986. *Opuntia chisosensis* (Anthony) comb. nov. *Cact. Succ. J. (U.S.)* 58:124–27.

———. 1987. *Opuntia cymochila* Eng. & Big: A species lost in the shuffle. *Cact. Succ. J. (U.S.)* 59:256–60.

———. 1988. *Opuntia macrocentra* Eng. and *Opuntia chlorotica* Eng. & Big. *Cact. Succ. J. (U.S.)* 60:155–60.

———. 1989. Revision of the U.S. members of the *Echinocereus triglochidiatus* group. *Cact. Succ. J. (U.S.)* 61:217–24.

———. 1991. In defense of the genus *Glandulicactus* Backeb. *Cact. Succ. J. (U.S.)* 63:87–91.

———. 1992. The genus *Echinocactus* Link & Otto, subgenus *Homalocephala* (Britton & Rose) stat. nov. *Cact. Succ. J. (U.S.)* 64:169–72.

Fischer, P. C. 1962. Taxonomic relationship of *Opuntia kleiniae* de Candolle and *Opuntia tetracantha* Toumey. Master's thesis, University of Arizona, Tucson.

———. 1971. Taxonomical and ecological relationship of the *Coryphantha vivipara* complex in the Cactaceae. Ph.D. diss., University of California, Berkeley.

———. 1980. The varieties of *Coryphantha vivipara*. *Cact. Succ. J. (U.S.)* 52:186–91.

———. 1989. *70 common cacti of the southwest*. Tucson, AZ: Southwest Parks and Monuments Association.

Fleming, T. H., and J. N. Holland. 1998. The evolution of obligate pollination mutualisms: Senita cactus and senita moth. *Oecologia* 114:368–75.

Fleming, T. H., S. Maurice, S. L. Buchmann, and M. D. Tuttle. 1994. Reproductive biology and male and female fitness in a trioecious cactus, *Pachycereus pringlei* (Cactaceae). *Amer. J. Bot.* 81:858–67.

Fleming, T. H., M. D. Tuttle, and M. A. Horner. 1996. Pollination biology and the relative importance of nocturnal and diurnal pollinators in three species of Sonoran Desert columnar cacti. *Southwest. Nat.* 41:257–69.

Flora of North America. Forthcoming. New York: Oxford University Press.

Forstner, P. I. 1996. Naturalized succulents in the Australian flora. *Haseltonia* 4:57–65.

Fraser, J. G., and R. D. Pieper. 1972. Growth characteristics of *Opuntia imbricata* [Haw.] DC. in New Mexico. *Southwest. Nat.* 17:229–37.

Ganders, F. R. 1976. Self-compatibility in the Cactaceae. *Cact. Succ. J. (Gr. Brit.)* 38:39–40.

Garcia-Aguilar, M., and E. Pimienta-Barrios. 1996. Cytological evidences of agamospermy in *Opuntia* (Cactaceae). *Haseltonia* 4:39–42.

Geissman, T. A. 1962. The chemistry of flavonoid compounds. New York: Macmillan.

Geller, G. N., and P. S. Nobel. 1987. Comparative cactus architecture and PAR interception. *Amer. J. Bot.* 74:998–1005.

Gibson, A. C. 1998. Photosynthetic organs of desert plants. *BioScience* 48:911–20.

Gibson, A. C., and K. E. Horak. 1978. Systematic anatomy and phylogeny of Mexican columnar cacti. *Ann. Missouri Bot. Gard.* 65:999–1057.

Gibson, A. C., and P. S. Nobel. 1986. *The cactus primer.* Cambridge, MA: Harvard University Press.

Glass, C., and R. Foster. 1975. The genus *Echinomastus* in the Chihuahuan Desert. *Cact. Succ. J. (U.S.)* 47:218–23.

———. 1977. The genus *Thelocactus* in the Chihuahuan Desert. *Cact. Succ. J. (U.S.)* 49:213–20, 244–51.

———. 1978. A revision of the genus *Epithelantha*. *Cact. Succ. J. (U.S.)* 50:184–87.

Gomez, A. R. 1992. The Glenn Springs–Boquillas raid reconsidered: Diplomatic intrigue on the Rio Grande. *J. of Big Bend Studies* 4:97–113.

Gould, F. W. 1962. Texas plants—A checklist and ecological summary. College Station: Texas Agricultural Experiment Station.

Graham, V. 1987. *Growing succulent plants.* Portland, OR: Timber Press.

Grant, V., and W. A. Connell. 1979. The association between *Carpophilus* beetles and cactus flowers. *Pl. Syst. Evol.* 133:99–102.

Grant, K. A., and V. Grant. 1967. Records of hummingbird pollination in the western American flora. III. Arizona records. *Aliso* 6:107–10.

Grant, V., and K. A. Grant. 1979a. Pollination of *Echinocereus fasciculatus* and *Ferocactus wislizenii*. *Pl. Syst. Evol.* 132:85–90.

———. 1979b. Systematics of the *Opuntia phaeacantha* group in Texas. *Bot. Gaz.* 140:199–207.

———. 1979c. Pollination of *Opuntia basilaris* and *O. littoralis*. *Pl. Syst. Evol.* 132:321–25.

———. 1979d. The pollination spectrum in the southwestern American cactus flora. *Pl. Syst. Evol.* 133:29–37.

———. 1982. Natural pentaploids in the *Opuntia lindheimeri–phaeacantha* group in Texas. *Bot. Gaz.* 143:117–20.

Grant, V., K. A. Grant, and P. D. Hurd Jr. 1979. Pollination of *Opuntia lindheimeri* and related species. *Pl. Syst. Evol.* 132:313–20.

Grant, V., and P. D. Hurd. 1979. Pollination of the southwestern opuntias. *Pl. Syst. Evol.* 133:15–28.

Griffith, M. P. 2000. Breeding systems and natural interspecific hybridization in *Opuntia* of the northern Chihuahuan Desert Region. Master's thesis, Sul Ross State University, Alpine, Texas.

———. 2001. A new Chihuahuan Desert hybrid prickly pear, *Opuntia* × *rooneyi* (Cactaceae). *Cact. Succ. J. (U.S.)* 73:307–10.

———. 2002. *Grusonia pulchella* classification and its impacts on the genus *Grusonia*: Morphological and molecular evidence. *Haseltonia* 9:86–93.

Griffiths, D. 1905. The prickly pear and other cacti as stock food. *USDA Bur. Plt. Indus. Bull.* 74. Washington, DC.

———. 1908–11. Illustrated studies in the genus *Opuntia*. Vols. 1–4. Missouri Botanical Garden annual reports 19–22: vol. 1 (1908) 19:259–72, plus plates 21–28; vol. 2 (1909) 20:81–97, plus plates 2–13; vol. 3 (1910) 21:165–74, plus plates 19–28; vol. 4 (1911) 22:25–36, plus plates 1–17.

Griffiths, D., and R. F. Hare. 1907. The tuna as food for man. *USDA Bur. Plt. Indus. Bull.* 116. Washington, DC.

———. 1908. Prickly pear and other cacti as food for stock. *New Mex. A&M Bull.* 60. Las Cruces.

Hanselka, C. W., and J. C. Paschal. 1991. Prickly pear cactus: A Texas rangeland enigma. *Rangelands* 13:109–11.

Harborne, J. B. 1964. *Biochemistry of phenolic compounds*. New York: Academic Press.

Hardy, J. E. 1997. Flora and vegetation of the Solitario Dome, Brewster and Presidio counties, Texas. Master's thesis, Sul Ross State University, Alpine, Texas.

Harrington, H. A. 1980. The need for protection of our native cacti. *Cact. Succ. J. (U.S.)* 52:224–26, 232.

Hartmann, H. T., D. E. Kester, and F. T. Davies Jr. 1990. Plant propagation: Principles and practices. 5th ed. Englewood Cliffs, NJ: Prentice Hall.

Hatch, S. L., K. N. Gandhi, and L. E. Brown. 1990. Checklist of the vascular plants of Texas. College Station: Texas Agricultural Experiment Station.

Haustein, E. 1988. *The cactus handbook*. Secaucus, NJ: Chartwell Books.

Heil, K. D. 1984. Recovery plan for Sneed pincushion cactus and Lee pincushion cactus. Albuquerque, NM: U.S. Fish and Wildlife Service.

Heil, K. D., and E. F. Anderson. 1982. Status report, *Coryphantha hesteri*. Contract no. 14-16-0002-81-216. Albuquerque, NM: Office of Endangered Species, Fish and Wildlife Service.

Heil, K. D., and S. Brack. 1986. The cacti of Guadalupe Mountains National Park. *Cact. Succ. J. (U.S.)* 58:165–77.

———. 1988. The cacti of Big Bend National Park. *Cact. Succ. J. (U.S.)* 60:17–34.

Heil, K. D., and J. M. Porter. 1994. *Sclerocactus* (Cactaceae): A revision. *Haseltonia* 2:20–46.

Henrickson, J., and M. C. Johnston. 1986. Vegetation and community types of the Chihuahuan Desert. Pp. 20–39 in *Second Symposium on Resources of the Chihuahuan Desert Region, United States and Mexico,* edited by J. C. Barlow, A. M. Powell, and B. N. Timmermann. Alpine, TX: Chihuahuan Desert Research Institute.

Hernández, H. M., and R. T. Bárcenas. 1995. Endangered cacti in the Chihuahuan Desert. I. Distribution patterns. *Conservation Biology* 9:1176–88.

Hershkovitz, M. A., and E. A. Zimmer. 1997. On the evolutionary origins of cacti. *Taxon* 46:217–32.

Hester, J. P. 1939. New species of cacti. *Cact. Succ. J. (U.S.)* 10:179–82.

——. 1945. *Echinomastus mariposensis* sp. nov. *Desert Plant Life* 17:59.

Hewitt, T. 1993. *The complete book of cacti and succulents.* New York: Dorling Kindersley.

Hicks, D. J., and A. Mauchamp. 1996. Evolution and conservation biology of the Galápagos opuntias (Cactaceae). *Haseltonia* 4:89–102.

Hoffman, M. T. 1992. Functional dioecy in *Echinocereus coccineus* (Cactaceae): Breeding system, sex ratios, and geographic range of floral dimorphism. *Amer. J. Bot.* 79:1382–88.

Hunt, D. R. 1978. Amplification of the genus *Escobaria. Cact. Succ. J. (Gr. Brit.)* 40:13, 30.

——, ed. 1992. *CITES Cactaceae checklist.* Kew, England: Royal Botanic Gardens.

Hunt, D. R., and N. Taylor. 1986. The genera of Cactaceae: Towards a new consensus. *Bradleya* 4:65–78.

Hutto, R. L., J. R. McAuliffe, and L. Hogan. 1986. Distributional associates of the saguaro (*Carnegiea gigantea*). *Southwest. Nat.* 31:469–76.

Ideker, J. 1996. *Pereskia aculeata* (Cactaceae), in the lower Rio Grande valley of Texas. *Sida* 17:527.

Innes, C., and C. Glass. 1991. *Cacti.* New York: Portland House.

IOS Working Party. 1986. The genera of the Cactaceae: Towards a new consensus. *Bradleya* 4:65–78.

Johns, P. 1993. *Practical cacti growing.* Ramsbury, Marlborough, Wiltshire, England: Crowood Press.

Johnson, R. 1993. Thorny question: Will the prickly pear be kiwi of the '90's? *Wall Street Journal,* Feb., A1, A5.

Johnson, R. A. 1992. Pollination and reproductive ecology of acuña cactus, *Echinomastus erectrocentrus* var. *acunensis* (Cactaceae). *International J. of Plant Sciences* 153:400–8.

Johnston, M. C. 1977. Brief resume of botanical, including vegetational, features of the Chihuahuan Desert Region with special emphasis on their uniqueness. Pp. 335–59 in *Symposium on the Biological Resources of the Chihuahuan Desert Region,* edited by R. H. Wauer and D. H. Riskind. National Park Service Transactions and Proceedings Series, no. 3. Washington, DC: National Park Service.

Jones, J. G., and V. M. Bryant Jr. 1992. Phytolith taxonomy in selected species of Texas cacti. Pp. 215–38 in *Phytolith systematics,* edited by G. Rapp Jr. and S. C. Mulholland. New York: Plenum Press.

Jones, S. D., J. K. Wipff, and P. M. Montgomery. 1997. *Vascular plants of Texas, a comprehensive checklist including synonymy, bibliography, and index.* Austin: University of Texas Press.

Kartesz, J. T. 1994. A synonymized checklist of the vascular flora of the United States, Canada, and Greenland. 2nd ed. Vol. 1. Portland, OR: Timber Press.

Kennedy, K., and J. Poole. 1993. *Chisos Mountain hedgehog cactus* (Echinocereus chisoensis *var.* chisoensis), *recovery plan.* Albuquerque, NM: U.S. Fish and Wildlife Service.

Kinraide, T. B. 1978. The ecological distribution of cholla cactus (*Opuntia imbricata* [Haw.] DC.) in El Paso County, Colorado. *Southwest. Nat.* 23:117–34.

Kolle, D. O. 1978. A populational study of *Echinocereus chloranthus* (Cactaceae) in Trans-Pecos Texas. Master's thesis, Sul Ross State University, Alpine, Texas.

Kurlansky, M. J. 1980. On the trail of cactus rustlers. *International Herald Tribune*, 4 Nov.

Laras, A. 1999. Growing *Ariocarpus* from seed. *Cact. Succ. J. (U.S.)* 71:210–15.

Leding, A. R. 1934. A new spine character in cactus. *J. Hered.* 25:326–28.

Leuck II, E. E. 1980. Biosystematic studies in the *Echinocereus viridiflorus* complex. Ph.D. diss., University of Oklahoma, Norman.

Leuck II, E. E., and J. H. Miller, 1982. Pollination biology and chemotaxonomy of the *Echinocereus viridiflorus* complex (Cactaceae). *Amer. J. Bot.* 69:1669–72.

Leuenberger, B. E. 1986. *Pereskia* (Cactaceae). *Mem. N.Y. Bot. Gard.* 41:1–144.

Lichtenzveig, J., S. Abbo, A. Nerd, N. Tel-Zur, and Y. Mizrahi. 2000. Cytology and mating systems in the climbing cacti *Hylocereus* and *Selenicereus*. *Amer. J. Bot.* 87:1058–65.

Lindsay, G. S. 1955. Taxonomy and ecology of the genus *Ferocactus*. Ph.D. diss., Stanford University, California.

Lockwood, M. W. 1995. Notes on life history of *Ancistrocactus tobuschii* (Cactaceae) in Kinney County, Texas. *Southwest. Nat.* 40:428–30.

Lyons, G. 1979. The C.S.S.A. and conservation: A boom or a bust? *Cact. Succ. J. (U.S.)* 51:9–15.

———. 1999. Another view on conservation. *Cact. Succ. J. (U.S.)* 71:103–4.

Mabry, T. J., and A. S. Dreiding. 1968. The betalains. Pp.145–60 in *Recent advances in phytochemistry*, vol. 1, edited by T. J. Mabry, R. E. Alston, and V. C. Runeckles. New York: Appleton-Century-Crofts.

Mabry, T. J., K. R. Markham, and M. B. Thomas. 1970. *The systematic identification of the flavonoids.* New York: Springer.

Mandujano, M. del C., C. Montaña, and L. E. Eguiarte. 1996. Reproductive ecology and inbreeding depression in *Opuntia rastrera* (Cactaceae) in the Chihuahuan Desert: Why are sexually derived recruitments so rare? *Amer. J. Bot.* 83:63–70.

Mandujano, M. del C., J. Golubov, and C. Montaña. 1997. Dormancy and endozoochorous dispersal of *Opuntia rastrera* in the southern Chihuahuan Desert. *J. Arid Environ.* 36:259–66.

Mann, J. 1969. Cactus-feeding insects and mites. *U.S. National Museum Bull.* 256:1–158. Washington, DC: Smithsonian Institution Press.

Marshall, W. T. 1940. *Echinocereus chisoensis* sp. nov. *Cact. Succ. J. (U.S.)* 12:15.

———. 1952. A new and interesting cactus from Texas. *Saguaroland Bull.* 6:78–81.

Marshall, W. T., and T. M. Bock. 1941. *Cactaceae.* Pasadena, CA: Abbey Garden Press.

Martin, W. C., and C. R. Hutchins. 1981. *A flora of New Mexico.* Vol. 2. Vaduz, West Germany: J. Cramer.

Mauseth, J. D. 1977. Cytokinin- and gibberellic acid–induced effects on the determination and morphogenesis of leaf primordia in *Opuntia polyacantha* (Cactaceae). *Amer. J. Bot.* 64:337–46.

———. 1990. Continental drift, climate and the evolution of cacti. *Cact. Succ. J. (U.S.)* 62:302–8.

Mauseth, J. D., and W. Halperin. 1975. Hormonal control of organogenesis in *Opuntia polyacantha* (Cactaceae). *Amer. J. Bot.* 62:869–77.

Mauseth, J. D., R. Kiesling, and C. Ostolaza. 2002. A cactus odyssey: Journeys in the wilds of Bolivia, Peru, and Argentina. Portland, OR: Timber Press.

Mayer, M. S., L. M. Williams, and J. P. Rebman. 2000. Molecular evidence for the hybrid origin of *Opuntia prolifera* (Cactaceae). *Madroño* 47:109–15.

Mayhew, D. E., and A. L. Wiens. 1994. Viruses in cacti and other succulents. *Cact. Succ. J. (U.S.)* 66:117–21.

McCarten, N. F. 1981. Fossil cacti and other succulents from the late Pleistocene. *Cact. Succ. J. (U.S.)* 53:122–23.

McDonald, B. 1993. Endangered species protection could pose prickly problem. *San Angelo Standard-Times,* 11 Jun., 5D.

McFarland, J. D., P. G. Kevan, and M. A. Lane. 1989. Pollination biology of *Opuntia imbricata* (Cactaceae) in southern Colorado. *Canadian J. Bot.* 67:24–28.

McIntosh, N. D. P. 1994. Observations on the Rivas method of cactus culture. *Cact. Succ. J. (U.S.)* 66:132–35.

McLeod, M. G. 1975. A new hybrid fleshy-fruited prickly-pear in California. *Madroño* 23:96–98.

Mellen, G. 1991. The *Echinocereus fendleri* controversy. *Cact. Succ. J. (U.S.)* 63:208–12.

Meyer, B. N., and J. L. McLaughlin. 1981. Economic uses of *Opuntia Cact. Succ. J. (U.S.)* 53:107–12.

———. 1982. A note on the phytochemistry of *Opuntia* (Cactaceae). *Cact. Succ. J. (U.S.)* 54:226–28.

Miller, D. J. 1995. The distribution and descriptions of the fishhook cacti: *Ancistrocactus* at the Devils River State Natural Area, Val Verde County, Texas. Report to Texas Parks and Wildlife Department, Austin, Texas.

Miller, J. M., and B. A. Bohm. 1982. Flavonol and dihydroflavonol glycosides of *Echinocereus triglochidiatus* var. *gurneyi. Phytochemistry* 21:951–52.

Mizrahi, Y., A. Nerd, and P. S. Nobel. 1997. Cacti as crops. *Horticultural Reviews* 18:291–320.

Moore, W. O. 1967. The *Echinocereus enneacanthus-dubius-stramineus* complex (Cactaceae). *Brittonia* 19:77–94.

Morgan, G. R. 1976. Man, plant, and religion: Peyote trade on the mustang plains of Texas. Ph.D. diss., University of Colorado.

Morgan, G. R., and O. C. Stewart. 1984. Peyote trade in South Texas. *Southwest Hist. Quart.* 87:269–96.

Nassar, J. M., N. Ramirez, and O. Linares. 1997. Comparative pollination biology of Venezuelan columnar cacti and the role of nectar-feeding bats in their sexual reproduction. *Amer. J. Bot.* 84:918–27.

Newcomb, W. W. 1956. The peyote cult of the Delaware Indians. *Texas J. Sci.* 8:202–11.

Nobel, P. S. 1988. *Environmental biology of agaves and cacti.* New York: Cambridge University Press.

———. 1994. *Remarkable agaves and cacti.* New York: Oxford University Press.

Nobel, P. S., U. Lüttge, S. Heuer, and E. Ball. 1984. Influence of applied NaCl on Crassulacean acid metabolism and ionic levels in a cactus, *Cereus validus*. *Plant Physiology* 75:799–803.

Nowicke, J. W. 1996. Pollen morphology, exine structure and the relationships of Basellaceae and Didiereaceae to Portulacaceae. *Syst. Bot.* 21:187–208.

Osborne, M. M., P. G. Kevan, and M. A. Lane. 1988. Pollination biology of *Opuntia polyacantha* and *Opuntia phaeacantha* (Cactaceae) in southern Colorado. *Pl. Syst. Evol.* 159:85–94.

Parfitt, B. D. 1985. Dioecy in North American Cactaceae: A review. *Sida* 11:200–206.

———. 1991. Biosystematics of the *Opuntia polyacantha* complex (Cactaceae) of western North America. Ph.D. diss., Arizona State University, Tempe.

———. 1998. New nomenclatural combinations in the *Opuntia polyacantha* complex. *Cact. Succ. J. (U.S.)* 70:188.

Parfitt, B. D., and C. H. Pickett. 1980. Insect pollination of prickly-pears (*Opuntia*: Cactaceae). *Southwest. Nat.* 25:104–7.

Parfitt, B. D., and D. J. Pinkava. 1988. Nomenclatural and systematic reassessment of *Opuntia engelmannii* and *O. lindheimeri* (Cactaceae). *Madroño* 35:342–49.

Pearcy, R. W. O. Björkman, M. M. Caldwell, J. E. Keeley, R. K. Monson, and B. R. Strain. 1987. Carbon gain by plants in natural environments. *BioScience* 37:21–29.

Pendley, G. K. 2001. Seed germination experiments in *Opuntia* (Cactaceae) of the northern Chihuahuan Desert. *Haseltonia* 8:42–50.

Pickett, C. H., and W. D. Clark. 1979. The function of extrafloral nectaries in *Opuntia acanthocarpa* (Cactaceae). *Amer. J. Bot.* 66:618–25.

Pilbeam, J. 1981. *Mammillaria,* a collector's guide. New York: Universe Books.

Pimienta-Barrios, E., G. Barbera, and P. Inglese. 1993. Cactus pear (*Opuntia* spp., Cactaceae) international network: An effort for productivity and environmental conservation for arid and semiarid lands. *Cact. Succ. J. (U.S.)* 65:225–29.

Pinkava, D. J., M. A. Baker, B. D. Parfitt, M. W. Mohlenbrock, and R. D. Worthington. 1985. Chromosome numbers in some cacti of Western North America—V. *System. Bot.* 10:471–83.

Pinkava, D. J., L. A. McGill, and T. Reeves. 1977. Chromosome numbers in some cacti of western North America—III. *Bull. Torr. Bot. Club* 104:105–10.

Pinkava, D. J., and M. G. McLeod. 1971. Chromosome numbers in some cacti of western North America. *Brittonia* 23:171–76.

Pinkava, D. J., M. G. McLeod, L. A. McGill, and R. C. Brown. 1973. Chromosome numbers in some cacti of western North America. II. *Brittonia* 25:2–9.

Pinkava, D. J., and B. D. Parfitt. 1982. Chromosome numbers in some cacti of western North America. IV. *Bull. Torr. Bot. Club* 109:121–28.

———. 1988. Nomenclatural changes in Chihuahuan Desert *Opuntia* (Cactaceae). *Sida* 13:125–30.

Pinkava, D. J., B. D. Parfitt, M. A. Baker, and R. D. Worthington. 1992. Chromosome numbers in some cacti of western North America—VI, with nomenclatural changes. *Madroño* 39:98–113.

Pinkava, D. J., B. D. Parfitt, J. P. Rebman, and M. A. Baker. 1998. Chromosome numbers in some cacti of western North America. VII. *Haseltonia* 6:32–41.

Pinkava, D. J., B. D. Parfitt, and J. P. Rebman. 2001. Nomenclatural changes in *Cylindropuntia* and *Opuntia* (Cactaceae) and notes on interspecific hybridization. *J. Arizona-Nevada Acad. Sci.* 33:162–63.

Ponsinet, G., G. Ourisson, and A. C. Oehlschlager. 1968. Systematic aspects of the distribution of di- and triterpenes. In *Recent Advances in Phytochemistry*, vol. 1, edited by T. J. Mabry, R. E. Alston, and V. C. Runeckles. New York: Appleton-Century-Crofts.

Poole, J. M., and G. K. Janssen. 1995. Report on searches, surveys and monitoring for Tobusch fishhook cactus in 1995. Conducted for Endangered Resources Branch, Texas Parks and Wildlife Department, Austin, Texas.

Poole, J. M., and D. H. Riskind. 1987. Endangered, threatened or protected native plants of Texas. Austin: Texas Parks and Wildlife Department.

Porter, J. M., M. S. Kinney, and K. D. Heil. 2000. Relationships between *Sclerocactus* and *Toumeya* (Cactaceae) based on chloroplast trnL-trnF sequences. *Haseltonia* 7:8–17.

Potter, R. L., J. H. Petersen, and D. N. Veckert. 1984. Germination responses of *Opuntia* spp. to temperature, scarification, and other seed treatments. *Weed Science* 32:106–10.

Powell, A. M. 1995. Second generation experimental hybridizations in the *Echinocereus* × *lloydii* complex (Cactaceae), and further documentation of dioecy in *E. coccineus*. *Pl. Syst. Evol.* 196:63–74.

———. 1998a. *Trees and shrubs of the Trans-Pecos and adjacent areas*. Austin: University of Texas Press.

———. 1998b. Plant communities of the Chihuahuan Desert Region. *Wildflower* 14:38–42.

———. 1998c. Third generation experimental hybrids in the *Echinocereus* × *lloydii* complex (Cactaceae). *Haseltonia* 6:91–95.

———. 2000. Grasses of the Trans-Pecos and adjacent areas. Marathon, TX: Iron Mountain Press.

———. 2002. Experimental hybridization between *Echinomastus intertextus* and *E. warnockii* (Cactaceae). *Haseltonia* 9:80–85.

Powell, A. M., and B. L. Turner. 1977. Aspects of the plant biology of the gypsum outcrops of the Chihuahuan Desert. Pp. 315–25 in *Symposium on the Biological Resources of the Chihuahuan Desert Region,* edited by R. H. Wauer and D. H. Riskind. National Park Service Transactions and Proceedings Series no. 3, Washington, DC: National Park Service.

Powell, A. M., and J. F. Weedin. 2001. Chromosome numbers in Chihuahuan Desert Cactaceae. III. Trans-Pecos Texas. *Amer. J. Bot.* 88:481–85.

Powell, A. M., A. D. Zimmerman, and R. A. Hilsenbeck. 1991. Experimental documentation of natural hybridization in Cactaceae: Origin of Lloyd's hedgehog cactus, *Echinocereus* × *lloydii. Pl. Syst. Evol.* 178:107–22.

Professional Association for Cactus Development. 1996. *Journal of the Prof. Assoc. for Cactus Development* 1:1–119.

Ralston, B. E. 1987. A biosystematic study of the *Opuntia schottii* complex (Cactaceae) in Texas. Master's thesis, Sul Ross State University, Alpine, Texas.

Ralston, B. E., and R. A. Hilsenbeck. 1989. Taxonomy of the *Opuntia schottii* complex (Cactaceae) in Texas. *Madroño* 36:221–31.

———. 1992. *Opuntia densispina* (Cactaceae): A new club cholla from the Big Bend region of Texas. *Madroño* 39:281–84.

Ramaley, F. 1940. Control of prickly pear in Australia. *Science* 92:528–29.

Raun, G. G. 1996. Distribution and population density of *Coryphantha hesteri* in the Big Bend region of Texas. *Cact. Succ. J. (U.S.)* 68:115–18.

———. 1997. *Echinomastus* in the Trans-Pecos region of Texas. *Cact. Succ. J. (U.S.)* 69:122–26.

———. 1998. Unsalvageable law. *San Angelo Standard-Times,* 27 Feb., 6A.

Reeves, B. 1994. *Toumeya papyracantha*—What and where next. *Cact. Succ. J. (U.S.)* 66:184–88.

Rivera, E. R., and B. N. Smith. 1979. Crystal morphology and ^{13}carbon/^{12}carbon composition of solid oxalate in cacti. *Plant Physiol.* 64:966–70.

Robinson, H. 1973. New combinations in the Cactaceae subfamily Opuntioideae. *Phytologia* 26:175–76.

———. 1974. Scanning electron microscope studies of the spines and glochids of the Opuntioideae (Cactaceae). *Amer. J. Bot.* 61:278–83.

Rojas-Aréchiga, M., and C. Vásquez-Yanes. 2000. Cactus seed germination: A review. *J. Arid Environ.* 44:85–104.

Rösler, H., V. Rösler, T. J. Mabry, and J. Kagan. 1966. Flavonoid pigments of *Opuntia lindheimeri. Phytochemistry* 5:189–92.

Ross, R. 1981. Chromosome counts, cytology, and reproduction in the Cactaceae. *Amer. J. Bot.* 68:463–70.

Rowley, G. 1978. The illustrated encyclopedia of succulents. New York: Salamander Book, Crown.

———. 1980. Pollination syndromes and cactus taxonomy. *Cact. Succ. J. (Gr. Brit.)* 42:95–98.

Ruffner, G. A., and W. D. Clark. 1986. Extrafloral nectar of *Ferocactus acanthodes* (Cactaceae): Composition and its importance to ants. *Amer. J. Bot.* 73:185–89.

Sajeva, M., and J. D. Mauseth. 1991. Leaf-like structure in the photosynthetic, succulent stems of cacti. *Annals of Botany* 68:405–11.

Sánchez-Mejorada, R. H. 1982. *Some prehispanic uses of cacti among the Indians of Mexico.* Toluca: State of Mexico, Developmental Agriculture, Natural Resources.

Schmidly, D. J. 1977. *The mammals of Trans-Pecos Texas.* College Station: Texas A&M University Press.

Schmidt Jr., R. H. 1986. Chihuahuan climate. Pp. 40–63 in *Second Symposium on Resources of the Chihuahuan Desert Region,* edited by J. C. Barlow, A. M. Powell, and B. N. Timmermann. Alpine, TX: Chihuahuan Desert Research Institute.

Schulz, E. D. 1932. *Cactus culture.* New York: Orange Judd.

Schulz, E. D., and R. Runyon. 1930. Texas cacti. *Proceedings Texas Acad. Sci.* 14:1–181.

Schumann, K. 1898–1902. *Gesamtbeschreibung der Kakteen.* J. Neumann, Neudamm, Germany.

Scobell, S. A. 1999. Pollination ecology of claret cup cactus along an elevation gradient. Master's thesis, Indiana State University, Terre Haute.

Scogin, R. 1985. Nectar constituents of the Cactaceae. *Southwest. Nat.* 30:77–82.

Smith, D. W. 1973. A taxonomic and distributional study of the Cactaceae in Brewster County, Texas. Master's thesis, Sul Ross State University, Alpine, Texas.

Smith, G. F., H. D. Ihlenfeldt, J. Thiede, U. Eggli, and D. Metzing. 1999. The International Organization for Succulent Plant Society (IOS): Its role and potential services to the international scientific community. *Taxon* 48:715–20.

Spears Jr., E. E. 1987. Island and mainland pollination ecology of *Centrosema virginianum* and *Opuntia stricta.* *J. Ecology* 75:351–62.

Speirs, D. C. 1978. The evolution of cacti. *Cact. Succ. J. (U.S.)* 50:179.

———. 1989. The opuntias of Alberta, Canada (both of them). *Cact. Succ. J. (U.S.)* 61:235–36.

Spira, T. P. 1981. Nectar-sugars and pollinator types in California *Trichostema* (Labiatae). *Madroño* 28:44.

Stewart, O. C. 1989. *The peyote religion: A history.* Norman: University of Oklahoma Press.

Stuppy, W. 2002. Seed characters and generic classification of the Opuntioideae (Cactaceae). *Succulent Plant Research* 6:25–58. Sherborne, England: David Hunt.

Suzán, H., G. P. Nabhan, and D. T. Patten. 1994. Nurse plant and floral biology of a rare night-blooming cereus, *Peniocereus striatus* (Brandegee) F. Buxbaum. *Conservation Biology* 8:461–70.

Sutton, K., J. T. Baccus, and M. S. Traweek Jr. 1997. Habitat of *Ancistrocactus tobuschii* (Tobusch fishhook cactus, Cactaceae) on the Edwards Plateau of Central Texas. *Southwest. Nat.* 42:441–45.

Tate, J. L., ed. 1972. *Cactus cookbook.* 2nd ed. Cact. Succ. Soc. (U.S.). Riverside, CA: Riverside Color Press.

Taylor, N. P. 1978. Review of the genus *Escobaria*. *Cact. Succ. J. (Gr. Brit.)* 40:30–37.

———. 1979. Further notes on *Escobaria*. *Cact. Succ. J. (Gr. Brit.)* 41:17–20.

———. 1984. A review of *Ferocactus* Britton & Rose. *Bradleya* 2:19–38.

———. 1985. *The genus* Echinocereus. Portland, OR: Timber Press.

———. 1986. The identification of Escobarias (Cactaceae). Botley, Oxford, England: British Cactus and Succulent Society.

———. 1987. Additional notes on some *Ferocactus* species. *Bradleya* 5:95–96.

———. 1988. Supplementary notes on Mexican *Echinocereus*. *Bradleya* 6:65–84.

———. 1998. *Coryphantha robustispina* (Engelm.) Britton & Rose: The correct name for the taxon variously known as *C. scheeri* Lemaire and *C. muehlenpfordtii* Britton & Rose *(nom. illeg.)*. *Cactaceae Consensus Initiatives* 6:17–20.

Taylor, N. P., and J. Y. Clark. 1983. Seed-morphology and classification in *Ferocactus* subg. *Ferocactus*. *Bradleya* 1:3–16.

Toumey, J. W. 1899. Sensitive stamens in genus *Opuntia*. *Bot. Gaz*. 30:356–61.

Tufenkian, D. 1999. *Ariocarpus*: Easy to grow. *Cact. Succ. J. (U.S.)* 71:201–5.

Turner, B. L., and A. M. Powell. 1979. Deserts, gypsum, and endemism. Pp. 96–116 in *Arid Land Plant Resources,* edited by J. R. Goodin and D. K. Northington. Lubbock: International Center for Arid and Semi-Arid Land Studies, Texas Tech University.

Turpin, S. A. 1997. Cradles, cribs, and mattresses: Prehistoric sleeping accommodations in the Chihuahuan Desert. *J. of Big Bend Studies* 9:1–18.

Tuttle, M. D. 1991. Bats, the cactus connection. *National Geographic* 179:131–40.

U.S. Fish and Wildlife Service. 1987. *Tobusch fishhook cactus* (Ancistrocactus tobuschii) *recovery plan.* Albuquerque, NM: U.S. Fish and Wildlife Service.

Valiente-Banuet, A., A. Rojas-Martínez, M. del C. Arizmendi, and P. Dávila. 1997. Pollination biology of two columnar cacti (*Neobuxbaumia mezcalaensis* and *Neobuxbaumia macrocephala*) in the Tehuacan Valley, central Mexico. *Amer. J. Bot.* 84:452–55.

Van Devender, T. R. 1986. Pleistocene climates and endemism in the Chihuahuan Desert flora. Pp. 1–19 in *Second Symposium on Resources of the Chihuahuan Desert Region, United States and Mexico,* edited by J. C. Barlow, A. M. Powell, and B. N. Timmermann. Alpine, TX: Chihuahuan Desert Research Institute.

———. 1990. Late Quaternary vegetation and climate of the Chihuahuan Desert, United States and Mexico. Pp. 104–33 in *Packrat middens: The last 40,000 years of biotic change,* edited by J. Betancourt, T. R. Van Devender, and P. Martin. Tucson: University of Arizona Press.

Van Devender, T. R., and T. Burgess. 1985. Late Pleistocene woodlands in the Bolsón de Mapimí: A refugium for the Chihuahuan Desert biota? *Quaternary Research* 24:346–53.

Vargas-Mendoza, M. C., and M. Gonzalez-Espinosa. 1992. Habitat heterogeneity and seed dispersal of *Opuntia streptacantha* (Cactaceae) in nopaleras of central Mexico. *Southwest. Nat.* 37:379–85.

Vega-Villasante, F., H. Nolasco, C. Montaño, H. L. Romero-Schmidt, J. Llinas, and M. Jimenez. 1994. Infestation of peninsular cylindropuntias by the cochineal insect (*Dactylopius* sp.): First report. *Cact. Succ. J. (U.S.)* 66:114–16.

Walkington, D. L. 1966. Morphological and chemical evidence for hybridization of some species of *Opuntia* occurring in southern California. Master's thesis, Claremont Graduate School, Claremont, California.

Wallace R. S. 1986. Biochemical taxonomy and the Cactaceae: An introduction and review. *Cact. Succ. J. (U.S.)* 58:35–38.

———. 1995. Molecular systematic study of the Cactaceae: Using chloroplast DNA variation to elucidate cactus phylogeny. *Bradleya* 13:1–12.

Wallace, R. S., and S. L. Dickie. 2002. Systematic implications of chloroplast DNA sequence variation in subfam. Opuntioideae (Cactaceae). *Succulent Plant Research* 6:9–24. Sherborne, England: David Hunt.

Wang, X., P. Felker, A. Paterson, Y. Mizrahi, A. Nerd, and C. Mondragon-Jacobo. 1996. Cross-hybridization and seed germination in *Opuntia* species. *J. of the Prof. Assoc. for Cactus Development* 1:49–60.

Warnock, B. H. 1970. *Wildflowers of the Big Bend Country Texas*. Alpine, TX: Sul Ross State University.

———. 1977. *Wildflowers of the Davis Mountains and Marathon Basin Texas*. Alpine, TX: Sul Ross State University.

Wauer, R. H. 1980. *Naturalist's Big Bend*. Rev. ed. College Station: Texas A&M University Press.

———. 1997. *For all seasons: A Big Bend journal*. Austin: University of Texas Press.

Wauer, R. H., and D. H. Riskind. 1977. Transactions of the symposium on the biological resources of the Chihuahuan Desert Region. U.S. Department of the Interior, National Park Service Transactions and Proceeding Series, no. 3. Washington, DC: U.S. Department of the Interior.

Weedin, J. F., and A. M. Powell. 1978. Chromosome numbers in Chihuahuan Desert Cactaceae: Trans-Pecos Texas. *Amer. J. Bot.* 65:531–37.

———. 1980. IOPB chromosome number reports LXIX. Edited by A. Löve. *Taxon* 29:703–30.

Weedin, J. F., A. M. Powell, and D. O. Kolle. 1989. Chromosome numbers in Chihuahuan Desert Cactaceae. II. Trans-Pecos Texas. *Southwest. Nat.* 34:160–64.

Wells, P. V. 1966. Late Pleistocene vegetation and degree of pluvial climatic change in the Chihuahuan Desert. *Science* 153:970–75.

Weniger, D. 1969. The small-flowered echinocerei of Texas and New Mexico. *Cact. Succ. J. (U.S.)* 41:34–43.

———. 1970. *Cacti of the Southwest*. Austin: University of Texas Press.

———. 1984. *Cacti of Texas and neighboring states*. Austin: University of Texas Press.

Westlund, B. L. 1991. Cactus trade and collection impact study. Report prepared for Texas Natural Heritage Program, Texas Parks and Wildlife Department, Austin, Texas.

Wittler, G. H., and J. D. Mauseth. 1984. The ultrastructure of developing latex ducts in *Mammillaria heyderi* (Cactaceae). *Amer. J. Bot.* 71:100–10.

Worthington, R. D. 1986. Observations on the flowering cacti from the vicinity of El Paso, Texas. *Cact. Succ. J. (U.S.)* 58:213–17.

———. 1989. *An annotated checklist of the native and naturalized flora of El Paso County, Texas*. El Paso Southwest Botanical Miscellany no. 1. El Paso, TX: Richard D. Worthington, Floristic Inventories of the Southwest Program.

———. 1995. *Biota of the Franklin Mountains*. Pt. 2. El Paso, TX: Floristic Inventories of the Southwest Program.

Wright, Y. 1932. *Coryphantha hesteri,* sp. nov. *Cact. Succ. J. (U.S.)* 4:274.

Yuasa, H. H., H. Shimizer, S. Kashiwai, and N. Kondo. 1973. Chromosome numbers and their bearing on the geographic distribution in the subfamily Opuntioideae (Cactaceae). *Rep. Inst. Breed. Res., Tokyo Univ. Agric.* 4:1–10.

Zimmerman, A. D. 1985. Systematics of the genus *Coryphantha* (Cactaceae). Ph.D. diss., University of Texas at Austin.

———. 1993. Systematics of *Echinocereus* × *roetteri* (Cactaceae), including Lloyd's hedgehog-cactus. Pp. 270–88 in *Proceedings of the Southwestern Rare and Endangered Plant Conference,* edited by R. Sivinski and K. Lightfoot. Santa Fe: New Mexico Forestry and Resources Conservation Division.

Zimmerman, A. D., C. Glass, R. Foster, and D. Pinkava. Forthcoming. Cactaceae. In *Chihuahuan Desert Flora,* edited by J. Henrickson and M. C. Johnston. Los Angeles: J. Henrickson.

Zimmerman, A. D., and D. A. Zimmerman. 1977. A revision of the United States taxa of the *Mammillaria wrightii* complex with remarks upon the northern Mexican populations. *Cact. Succ. J. (U.S.)* 49:23–34, 51–62.

Zimmerman, D. A. 1972. The fruits of *Coryphantha hesteri. Cact. Succ. J. (U.S.)* 44:85.

Index

Boldface type indicates scientific names of species, main common names, and the page numbers where the discussion of each taxon begins.

abrojo, 99, 102, 107
Acacia, 34, 194, 313
Acanthocereus, 41, 45, 52, 66, **462**
  pentagonus, 462
  tetragonus, **462**
Acharagma, 9
Acleisanthes, 43, 44
  longiflora, 43
Agave, 4
  lechuguilla, 34, 303, 313, 443
aguijilla, 113
alfilerillo, 113
alicoche, 234, 241, **277**
  berlandier's, **277**
  yellow-flowered, **278**
alkali sacaton, 344, 346
alkaloids, 26, 28
Amanita, 26
Amaranthaceae, 29
Ancistrocactus, xiii, 9, 13, 19, 36, 40, 70, 249, 280, 298, 331, **332**, 333–36, 338, 339, 342
  brevihamatus, 328, 332, **333**, 334, 337–39, 343, 450
    brevihamatus, 333, **334**, 335–38, 340–43
    pallidus, 333–35, **339**, 340–42, 346
  scheeri, 45, 300, 332–39, 341, 342

tobuschii, 17, 60, 70, 332–38, 341, 342
uncinatus, 331
  wrightii, 331
anhalinine, 314
anhalonine, 314
Anhalonium, 309, 317
  engelmannii, 321
  fissuratum, 321
  jourdanianum, 316
  lewinii, 316
  williamsii, 309
anthocyanins, 26
anthoxanthins, 26
Anthoporidae, 42
Aquilegia, 43
  chrysantha, 43
Archilochus, 43
  alexandri, 43
Ariocarpus, 9, 15, 16, 35, 48, 50, 70, 72, 309, 311, 313, **317**, 318–21
  fissuratus, 5, 50, 57, 311–13, **317**, 318–21, 378, 431
    fissuratus, **317**, 318–20
    lloydii, 320
  kotschoubeyanus, 51
  williamsii, 315
Arizona queen of the night, 196
Asclepias, 268
Ashmeadiella, 42
  meliloti, 42
Asteraceae, 114
Astrophytum, xiii, 13, 66, 267, 280, **463**
  asterias, **463**
  myriostigma, 463

Austrocylindropuntia, 65
  exaltata, 65
Aztekium, 313

Barbados gooseberry, 464
barberry fig, 46
barrel, 284
  blue, 284
  candy, 292
  compass, 292
  fishhook, 292
  fish-hook, 292
barrel cactus, 5, 51, 52, **288**, 300, 328
  Arizona, **289**, 290
  fishhook, 292
  lower Rio Grande Valley, 289, **296**
  southwest, 292
  southwestern, 292
  whiskered, 296
Bartschella, 371
Basellaceae, 36
Basidiomycetes, 26
bee, 408
  plasterer, 408
beehive, 451
  needle "mulee," 451
  New Mexico, 416
betacyanins, 26
betalains, 26
betaxanthins, 26
Big Bend eggs, 399
Bisnaga, 295
bisnaga, 284
  costillona, 296
  de chilitos, 383
  de dulce, 284
  meloncillo, 284
bisnagre, 284

497

biznaga, 292
   de agua, 292
   de chilitos, 386
   de limilla, 296
   de tuna, 296
   espinosa, 296
   ganchuda, 296
   limilla, 296
biznagita, 360
   Warnock, 360
   white, 354
Bouteloua, 330
   gracilis, 344
   ramosa, 330
Brasiliopuntia, 66
Bromeliaceae, 1
Browningeae, 66

Cacalia, 114
   kleinia, 114
Cactaceae, 1, 8, 16, 18, 19, 26–29, 33, 34, 36–40, 57, 65, 202, 300, 368, 438, 469, 473, 475
Cacteae, 9, 12, 13, 16, 37, 66, 69
Cactoblastis, 51
   cactorum, 51
Cactoideae, 5, 6, 9, 12, 14, 16, 18, 25, 37, 38, 54, 55, 58–60, 65–67, 69
Cactus, 96–; cactus, 5–
   aggregate, 208
   Arizona barrel, **289**
   barbwire, 462
   barrel, 285, **288**, 292, 294, 295, 300, 328
   beautiful lace, 247
   beavertail, 189
   beehive, **392**, 409
   bicolor, 297
   bicolor, 306
   big-needle pincushion, **439**
   birdfoot, 406
   bisquit, 401
   black-spine claret cup hedgehog, 208
   bleo, 107
   Boke's, 370
   Boke button, 370
   Boke's button-, 366, **368**
   Boquillas button, 370
   bristlehair prickly pear, 186
   brown-flowered, 267
   bunch-ball, 208
   bunched cory, 446
   button, 364, 368
   calcaratus, 461
   candelabrum, 107
   candle, 118
   candy, 287
   cane, 107
   Castetter pricklypear, 129
   catclaw, 331
   cat-claw, 331
   Chaffey's pincushion, 396, **399**
   chain-link, 107
   chaparral, 196
   Chisos hedgehog, 237, **241**
   Chisos Mountain hedgehog, 243, 244
   christmas, 113
   claret-cup, **200**, **203**, 216
   claret-cup hedgehog, 205
   Classen's, 247
   cob, 428, **431**, **432**, 439
   cob cory, 436
   common button, **365**, 366
   cone, **322**
   Correll's green-flowered hedgehog, 251, **258**
   cory, **392**, 446
   cream, 387
   cylindricus, 107
   dahlia, **278**
   dasyacanthus, 399
   Davis' dwarf hedgehog, 250
   Davis hedge, 250
   deer-horn, 196
   desert christmas, 113
   desert pincushion, **396**
   devil, 88
   devil's head, 57, 287
   discus, 177
   divine, 316
   dog-turd, 88
   dumpling, 316
   dwarf hedgehog, 244, **247**
   Duncan's, 404
   Duncan's pincushion, **402**
   dwarf cory, 406
   dwarf hedgehog, **247**
   eagle-claw, **281**, **284**, **327**, 328, 331
   eaglesclaw, 284
   eastern beehive, 409, **410**, 413
   emoryi, 96
   Fendler's hedgehog, **225**, 228
   ficus-indica, 191
   finger, 461
   fishhook, 300, **332**, 338, 339, 375
   fish-hook, 292
   fishhook nipple, 375
   foxtail, 267, 381
   fragrant, 416
   giant fishhook-, 289, **292**
   goldspine hedgehog, 270
   golf ball, **377**
   golfball, 363
   Graham dog, 91
   Graham dog-, 91
   Graham's fishhook, **373**
   grama grass, 347
   grama-grass, **344**, 345, 346
   grape, 390
   graybeard, 252, **273**, 276
   green-flowered, 258
   green-flowered hedgehog, **250**, 258, 276
   green-flowered torch, 258, 263
   grooved nipple, 452, **458**
   Guadalupe pncushion, **416**, 420, 422
   gummifera, 386
   gummiferus, 386
   hair-covered, **387**, 388
   hedgehog, **197**, 300
   hedgehog-, 221
   hedgehog cory-, 455
   hemisphericus, 386
   Hester cory, 409
   Hester's dwarf, 409
   Hester's pincushion, **406**
   heyderi, 386
      hemisphericus, 386
   Heyder nipple-, 383
   Heyder's pincushion, 383, **384**
   hunger, 186
   imbricatus, 107
   junior Tom Thumb, 431
   Kuenzler's hedgehog, 276
   lace, 244
   lace-spine, 379
   lacespine, 379
   lacyspine, 379
   lady-finger, 277
   Langtry claret-cup, 208
   Langtry rainbow, **221**, 222
   least cory, 406
   Lee pincushion, 424
   Lee's pincushion, **416**, 422
   living rock, 5, **317**, 318
   Lloyd hedgehog, 214
   Lloyd's hedgehog, 209, **211**, 214
   Lloyd's hybrid hedgehog, 211
   Lloyd's Mariposa, 363
   longcentral woven-spine pineapple, **349**, 350, 354

mariposa, 358, **360**
Mariposa Lloyd's, 363
meiacanthus, 383
mountain cob, 439
mountain fishhook, 331
multistemmed sea-urchin, 452, **455**, 458
Nellie cory, 406
Nellie's cory, 406
Nellie's pincushion, 402, **404**
neomexicanus, 415
New Mexico beehive, 409, **413**
New Mexico rainbow, 258
nipple, **381**, 382, 386, **461**
nylon, 258
organ pipe, 28
pancake, 386
paperspine, 347
paper-spine plains, 347
paper-spined, 347
pencil, 113, **279**
pentagonus, 462
pereskia, 463
pincushion, **371**, 375, **392**
pineapple, 348, 451, 461
pitaya hedgehog, 234
pottsii, 381
Potts' nipple, 381
Potts pricklypear, 150
prolifera, 387
pudding, 316
radiosus, 415
  neomexicanus, 415
rainbow, 217, 221
rat-tail, 381
rattail, 381
red-flowered hedgehog, 208
red-goblet, 205
rhinoceros, 455
Roetter's hybrid hedgehog, **208**, 209, **210**, 211
Runyon's pincushion, **428**
rusty hedgehog, 251, **263**
Scheer cory, 449
Scheer's fishhook, 338
scheeri, 449
scolymoides, 461
  sulcatus 461
sea-urchin, **451**, **452**, **463**
short-spined fishhook, **333**, 334
silver column, 363
silverlace, **416**, **424**
sitting, 228
slender stem, 113
small-flowered hedgehog, 251, **255**

Sneed cory, 420
Sneed pincushion, 420
Sneed's pincushion, **416**, **418**
snipe, **333**, **339**, 343
snowcone nipple, 428
sour, 416
sphaericus, 390
spiny-star, 416
stanly-dog, 96
star, 318, **463**
starvation, 186
strawberry, 54, 56, 208, **228**, **229**, **234**, **237**, **241**
sulcatus, 461
sunset, 375
sunset fishhook, 375
sweet-potato, 196
Texas, 326
Texas claret-cup, **200**, **205**, 208
Texas cone, **323**, **324**, 326
Texas rainbow, **216**
texensis, 386
Tobusch fishhook, 333
torch, 228
tree, 107
triangle, **462**
triploid hybrid hedgehog, **214**, **215**, 216
tunicatus, 96
turk's head, 208, 284
turnip, 316
twisted-rib, **297**, 298
varicolor, 439
varicolor cob, 428, **431**, **436**
varicolor cory, 439
viviparus, 409
Warnock's, **357**, 358
Weedin's small-flowered hedgehog, 252, **270**
Weniger hedgehog, 225
Weniger's small-flowered hedgehog, 252, **267**
western green-flowered hedgehog, 251, 260
western Heyder pincushion, **383**, **384**, **387**
whiskerbrush pincushion, **439**, **443**
whisky, 316
white-flowered, 360
white-flowered fishhook, 343
white-spine cob, 428
woven-spine pineapple, **349**, **350**, 354
Wright's fishhook, **373**, 375

Yellow-flowered, 273
calcium oxalate, 29
Calvin-Benson cycle, 22
Calymmantheae, 66
candle, 247
  purple, 247
cardenche, 107
cardon, 5, 41, 107
cardoncillo, 118
Carnegiea, 3, 47, 50, 52
  gigantea, 3, 28, 33, 41, 66, 313
carotenoids, 26
Carpophilus, 40
Caryophyllales, 26, 29, 36, 466
catalinaria, 113
Centrospermae, 29
century plant, 4, 62
Cephalocereus, 52, 66
  senilis, 269
Cereeae, 66, 193
Cereoideae, 65
Cereus; cereus, 66, 193, 196, 197
  berlandieri, 277
  blanckii, 278
  chloranthus, 260
  coccineus, 204
    melanacanthus, 204
  desert nigtblooming, 196
  desert night-blooming, **193**, 194, 196
  dasyacanthus, 211, 221
    neomexicanus, 211
  dubius, 234
  enneacanthus, 237
  fendleri, 228
    pauperculus, 228
  greggii, 193
    cismontanus, 196
  imbricatus, 102, 107
  night blooming, 196
  night-blooming, 40, 196, 462
  paucispinus, 205
  pentagonus, 462
  pentalophus, 277
  poselgeri, 279
  pottsii, 196
  roemeri, 208
  roetteri, 208, 210
  spinulosus, 464
  stramineus, 234, 237, 241
  tetragonus, 462
  Texas night-blooming, 196
  viridiflorus, 255
    cylindricus, 255
challote, 316
chaute, 321

Index 499

chautl, 321
chautle, 321
Chenopodiaceae, 29
Chilita, 371
   pottsii, 381
   wrightii, 376
cholla, 5, 6, 12–14, 31, 32, 45, 48, 50–52, 55, 60, 62, 64, 69, **73**, 74–76, 105, 107, 112, 113, 478
   Big Bend, 103, **107**
   Big Bend cane, 109
   Big Bend devil, 91, 92
   candle, **113**, 114
   cane, 107
   Christmas, **109**
   club, 7, **73**, 76, 81–83, 85, 93
   clumped dog, **81**, 85
   common devil, 92, **93**
   common dog, 85
   creeping, 96
   cursed, 96
   Davis, 102
   Davis', 96, **99**
   devil, 81, 88, 93, 96
   devil's, 96
   dog, 34, 54, 64, 76, 81, 84, 85, 88–91, 93–95, 187, 241, 243
   Graham's dog, 86, **88**
   icicle, **96**, 99
   Jeffdavis, 102
   Klein, 118
   Klein pencil, 118
   Klein's, 118
   Klein's pencil, 118
   mounded dwarf, 91
   pencil, 113
   Schott's dog, **86**
   Schott's dwarf, 88
   sheathed, 99
   silver, 109
   silver-spine cane, 109
   silver tree, 109
   Stanly club, 96
   Stanly's, 96
   teddy bear, 48
   thistle, 99, 102
   tree, **102**, 103
   walkingstick, 107
claret-cup, 56, 57, 67, 208, 213
   echinocereus, 208
   Langtry, 208
   little, 208
   Mexican, 205
   southwest, 205
   Texas, 208
Clavatae, 85

clavelina, 99
clavellina, 88, 99
cob cactus, 431
   varicolor, 436
Coccidae, 51
Cochemiea, 371
cochineal, 50, 51
Cochiseia, 392
Colletidae, 42, 408
columbine, 43
Consolea, 20, 66
Corynopuntia, 7, 76, 81
   grahamii, 91
   schottii, 88
   stanlyi, 96
Coryphantha; coryphantha, xiii, 7–9, 13, 17, 35, 36, 47, 48, 70–72, 311, 324, 371, **392**, 393, 394, 397, 412, 419–21, 433–35, 440, 441, 461
   albicolumnaria, 428
   alversonii, 412, 419
   calcarata, 461
   chaffeyi, 394, 396–98, **399**, 400, 401, 403, 404, 415, 430
   chlorantha, 412, 419
   columnaris, 413
   conoidea, 326
   cornifera, 454
      echinus, 455
   dasyacantha, 362, 396, 397–404, 426, 430, 431, 434, 435, 437,
      dasyacantha, 399, 428
      varicolor, 399, 439
   delaetiana, 445, 460
   duncanii, 6, 17, 21, 33, 359, 394, 397–99, 401, **402**, 403, 404
   echinus, 13, 57, 352–54, 360–94, 395, 448, 449, **451**, 452, 454, 455, 457–60
      echinus, 451, **452**, 453, 454, 456, 457, 459, 460
      robusta, 452, 454, **455**, 456–58
   El Paso long-tubercled, 446, **450**
   emskoetteriana, 431
   engelmannii, 449
   fragrans, 415
   henricksonii, 434
   hesteri, 24, 55, 349, 267, 393, 405, **406**, 407–09, 413, 414, 419
   long-tubercled, **446**, 447

   macromeris, 9, 15, 324, 395, 411, 412, **439**, 440–45, 447, 449,
      macromeris, 395, 439, 440
      runyonii, 439, 440, 442
   minima, 3, 33, 55, 57, 249, 393, 402, **404**, 405, 406, 438
   missouriensis, xiii, **461**, 462
      caespitosa, **461**, 462
      robustior, 461
   muehlbaueriana, 431
   muehlenpfordtii, 449
   nellieae, 406
   neomexicana, 415
   neoscheeri, 449
   nickelsae, 445, 460
   obscura, 458
   oklahomensis, 413
   pectinata, 454, 455
   piercei, 431
   pottsiana, xiii, 57, 394, 395, 415, **428**, 429–31, 435
   pottsii, 381
   pygmaea, 420
   radiosa, 415
      neomexicana, 415
   ramillosa, 15, 33, 324, 439–42, **443**, 444, 445, 448, 460
   robertii, 429, 431
   robustispina, 57
      scheeri, 449
      uncinata, 449
   roederiana, 458
   runyonii, 431
   Runyon's, 439, 442
   scheeri, 54, 57, 395, **446**, 447, 449–51
      robustispina, 447, 449, 451
      uncinata, 451
      scheeri, 446, **447**, 448, 449
      uncinata, 70, 394, 446, 447, 449, **450**
      valida, 447, 449–51
   scolymoides, 455
   sneedii, 16, 29, 33, 395, 398, 412, 414, **416**, 417–27, 430–36
      albicolumnaria, 274, 362, 380, 399, 418, **424**, 425–28, 435–37
      guadalupensis, 418, **420**, 421–24

leei, 417, 420, 421,
   422, 423, 424, 427
 sandbergii, 426
 sneedii, 418, 419–23,
   425, 427
 villardii, 426
strobiliformis, 420, 431,
   434
   albicolumnaria, 435
   durispina, 424, 427,
     428, 431
   varicolor, 431
sulcata, 395, 448, 452,
   458, 459, 460
   nickelsae, 460
tuberculosa, 17, 380, 394,
   398, 399, 404, 408,
   417–21, 424, 426–28,
   430, **431**, 432–36, 438
   durispina, 420, 435
   orcuttii, 420, 435
   tuberculosa, 33, 362,
     398, 403, 428, 430,
     432, 433, 436–38
   varicolor, 362, 394,
     398, 400, 414, 428,
     431, 432, **436**, 437–39
varicolor, 436
vivipara, 16, 34, 40, 320,
   366, 394, 399, 406, 408,
   409, 410–13, 415, 417,
   419, 420, 425, 426, 431,
   435
   arizonica, 412
   deserti, 412
   neomexicana, 409,
     410, 412, **413**, 414,
     415, 421, 425, 426
   radiosa, 409, 410,
     412–15
   vivipara, 394, 395,
     409, **410**, 411–13, 415
   zilziana, 404
coyonostle, 107
coyonostli, 107
coyonostole, 107
coyonoxtle, 99,
coyote candles, 107
crack star, 321
Crassulacean acid metabo-
   lism, 22, 45, 59
creosotebush, 34
Cumarinia, 392
Cumulopuntia, 66
cylinder bells, 258, 263
Cylindropuntia, 7, 47, 55, 66,
   74–76, 113, 116–18
 caerulescens, 118
 davisii, 102
 imbricata, 107

 kleiniae, 118
 leptocaulis, 113
 recondita, 118
   perrita, 118
 tunicata, 99

Dactylopius, 50, 51
Dasylirion, 4
devil's claw, 287
devil's head, 284, 287
devilshead, 284
devils pincushion, 287
devil's root, 316
diabolic root, 316
Diadasia, 42
 afflicta, 42
 rinconis, 42
Dialictus, 42
Didiereaceae, 36
divine herb, 316
dog cholla, 34, 64
Dolichothele, 371, 391
 sphaerica, 390
druse, 29
dry whisky, 316, 321
drywhisky, 316
Duncan snowball, 404

eagle claw, 284
eagle claws, 284
eagle's-claw, 284
early bloomer, 354
Ebnerella, 371
 wrightii, 376
Echinocactus; echinocactus,
   45, 47, 51, 52, 69–71,
   280, 288, 301, 309, 313,
   322, 331, 348, 463
 asterias, 463
 bicolor, 301
   schottii, 306
   tricolor, 306
 brevihamatus, 333
 catsclaw, 331
 conoideus, 326
 erectocentrus, 360
   pallidus, 360
 flavidispinus, 308
 hamatacanthus, 292
 horizontalonius, 281
 horizonthalonius, 57, 280,
   281, 282–87, 291, 292,
   312
   centrispinus, 284
   curvispinus, 283, 284
   horizonthalonius, 284
   moelleri, 283, 284
   nicholii, 284
   intertextus, 349
     dasyacanthus, 354

   intertextus, 349
 jourdanianus, 316
 lewinii, 316
 longihamatus, 296
   crassispinus, 296
 macromeris, 443
 mariposensis, 363
 muehlenpfordtii, 449
 parryi, 284
 polycephalus, 287
 pottsianus, 428
 rapa, 316
 reichenbachii, 244
 scheeri, 339
   brevihamatus, 339
 setispinus, 299, 300
   hamatus, 300
   setaceus, 299, 300
 sinuatus, 296
 texensis, 56, 57, 281, 283,
   284, 285–87
 turkshead, 296
 uncinatus, 327
   wrightii, 327
 viviparus, 412
 warnockii, 360
 williamsii, 315
   anhaloninica, 316
   lutea, 316
 wislizeni, 289
 wrightii, 331
Echinocereeae, 66, 67, 69
Echinocereus; echinocereus,
   xiii, 5, 7, 11, 17, 27, 35,
   36, 38, 39, 41–43, 47,
   48, 50, 60, 66, 67, 69,
   197, 203, 215, 227, 241,
   245, 248–50, 264, 269,
   270, 276, 279, 352
 albiflorus, 228
 arizonicus, 200
 berlandieri, 277, 278
 blanckii, 278
 brevispinus, 234
   brevispinus, 235
   enneacanthus, 235
 bristolii, 216
 caespitosus, 246, 247
   perbellus, 247
 carmenensis, 253, 268,
   277
 carnosus, 237
 chisoensis, 25, 32, 34,
   198, 223, 237, **241**,
   242–44, 458
   chisoensis, 198, 237,
     241, 242, 243
   fobeanus, 241
 chloranthus, 252, 253,
   255, 263, 264, 268

Continued
  chloranthus, 252, 253, 263
  cylindricus, 252, 253, 255, 258
  neocapillus, 252, 260, 267, 268
  rhyolithensis, 253, 260, 263, 270, 272, **277**
  russanthus, 252, 260, 263, 267
  weedinii, 260
 claretcup, 208
 coccineus, 5, 31, 32, 36, 39, 40, 41, 43, 100, 199, **200**, 201–09, 211–15, 217–19, 221, 226–28, 231, 239, 250, 321
  aggregatus, 202, 208
  coccineus, 202
  conoideus, 204
  gurneyi, 201, 202, 205, 207–09
  melanacanthus, 201, 204
  neomexicanus, 201
  octacanthus, 205, 208
  paucispinus, 200–204, **205**, 206–09, 211, 213
  roemeri, 202, 208
  rosei, 200, 202, **203**, 204, 205, 209, 210, 215
 conglomeratus, 232, 240, 241
 ctenoides, 219–22, 224
 dasyacanthus, 26, 36, 43, 198, 201, 202, 207, 208, 210–13, 215, **216**, 217–24, 243, 250, 255, 256, 265, 266, 274, 360
  ctenoides, 220, 221
  dasyacanthus, 219
  hildmannii, 214, 219, 220, 221
  minor, 211, 219
  steereae, 221
 davisii, 3, 33, 55, 60, 61, 198, 244, **247**, 248–50, 252, 253, 256, 258, 260, 275, 405, 438
 degandii, 221
 delaetii, 269
  delaetii, 269
 dubius, 228, 230, 232–34, 237, 241
 ennaecanthus, 5, 27, 33, 42, 199, 204, 212, 227, **228**, 229–34, 237–41

brevispinus, 228–33, **234**, 235–39, 278
 carnosus, 234–37
 dubius, 228, 230, 233, 234, 238, 239
 enneacanthus, **229**, 230–40
 intermedius, 235, 236
 stramineus, 241
Fendler, 228
fendleri, xiii, 56, 198–201, 204, 207, 210, 212, **225**, 226–28, 243, 276, 278
 fendleri, 199, **225**, 226–28, 276
 kuenzleri, 226, 227, **276**
 rectispinus, 225, 227, 228
fobeanus, 241
hempelii, 226, 276
hildmannii, 214, 219, 221
kuenzleri, 276
lloydii, 211, 214
×lloydii, 211
longisetus, 253, 267
 delaetii, 269
 longisetus, 266
merkeri, 232, 234
multicostatus, 214
neocapillus, 253, 268, 270
neomexicanus, 214
×neomexicanus, 198, 199, 201, 211, **214**, 215, 216, 227
octacanthus, 208
pacificus, 200
papillosus, **278**
 angusticeps, 278
 papillosus, 278
pectinatus, 11, 198, 218–20, **221**, 222, 224, 246, 266
 ctenoides, 219, 221
 dasyacanthus, 219, 221
 minor, 211, 219
 neomexicanus, 219, 221
 pectinatus, 219, 220, 222–24
 texensis, 221
 wenigeri, 11, 198, 218, 219, **221**, 222–24, 246
pentalophus, 277, 278
 leonensis, 277
 pentalophus, 277
 procumbens, 277
perbellus, 247
pink-flowered, 228

pleiogonus, 214
polyacanthus, 200, 201
 neomexicanus, 204
 rosei, 204
poselgeri, 197, **278**, **279**
purple, 234
rayonesensis, 253, 266
reichenbachii, x, 67, 198, 218, 223, 224, 242, 243, **244**, 245–47, 250
 albertii, 67
 albispinus, 67
 baileyi, 67, 246
 caespitosus, 246
 chisoensis, 244
 chisosensis, 244
 fitchii, 246
 perbellus, 67, 245–47
 reichenbachii, 246, 247
roemeri, 67, 208
roetteri, 214
 lloydii, 214
×**roetteri**, 27, 36, 43, 199, 200, 202, 205–07, **208**, 209, 210, 213, 219, 221, 227
 neomexicanus, 27, 32, 43, 206, 209, **211**, 212–14
 roetteri, 209, **210**, 211
rosei, 203
rubescens, 214, 221
russanthus, 253, 263, 264
 fiehnii, 253, 265, **277**
 russanthus, 253
 weedinii, 253, 270
santaritensis, 200, 202
sarissophorus, 232, 234
spicemound, 241
spinosissimus, 221
standleyi, 257, 258
steereae, 221
stramineus, 5, 33, 42, 54, 56, 198, 199, 227, 228, 230–36, **237**, 238–41, 250, 253
 occidentalis, 240
 stramineus, 199, **237**, 240
triglochidiatus, 27, 200–202, 208, 209, 228
 coccineus, 205
 gurneyi, 27, 201, 209, 214
 melanacanthus, 201, 205
 mojavensis, 201
 neomexicanus, 201, 205, 216

octacanthus, 202, 208
paucispinus, 201, 208
rosei, 204
triglochidiatus, 201, 207, 208, 276
viridflorus, xiii, xvi, 27, 31, 36, 43, 67, 198–201, 210, 212, 214, 215, 217–19, 227, 248, **250**, 251–53, 255, 262, 264, 266, 269, 270, 274–77, 352, 408
    chloranthus, 215, 251, 253, 254, 257, 258, 260, 261–64, 266, 268, 270, 272, 273
    correllii, 248, 250–54, 256, 257, **258**, 259–61, 268, 269, 275, 277
    canus, 56, 250, 252–54, 257, 264, 268–70, **273**, 274–76
    cylindricus, 54, 199, 215, 218, 228, 250–54, **255**, 256–64, 266, 269–73
    davisii, 250, 253
    neocapillus, 56, 250, 252, 253, 255, 257, 260, 264–66, **267**, 268, 269, 274, 275, 277
    russanthus, xiii, 199, 218, 243, 251–53, 255, 257, 260–62, **263**, 264–75
    standleyi, 252, 257, 258, 260
    viridiflorus, 248, 250, 252, 253, 256–258, 260, 274, 275, **276**
    weedinii, 252, 253, 255, 257, 260, 263–66, 268, **270**, 271–74, 276, 277
Echinomastus, 5, 9, 35, 36, 38, 48, 71, 72, 280, 322, 324, **348**, 349, 351–53, 356, 357, 359, 360, 393
    dasyacanthus, 357
    intertextus, 348, **349**, 350–54, 356, 357, 359, 360
        dasyacanthus, 350, 351, 353, **354**, 355–57
        intertextus, 350, 351–57
    mariposensis, 33, 34, 348, 349, 352, 358, 359, **360**, 361–63, 380
    pallidus, 360
    uncinatus, 331
        wrightii, 331
    warnockii, 33, 348, 349, 351–54, 356, **357**, 358–60, 362, 363
Echinopsis, 27, 40, 50, 313
Ecobesseya, 392
empacho, 56
entrena, 107
Epiphyllum, 27, 60, 196
Epithelanta, xiii, 9, 39, 50, 71, **364**, 365–68, 378, 379
    bokei, 365–67, **368**, 369, 378
    greggii, 370
    micromeris, xiv, 321, 364, 365, 366–69, 378
        bokei, 370
        greggii, 365, 370
        micromeris, 365, 366, 368
Escobaria; escobaria, 9, 72, 392–94, 396, 397, 406, 409, 430, 434
    albicolumnaria, 424
    bella, 431
    chaffeyi, 399
    corncob, 436
    dasyacantha, 399, 401
        chaffeyi, 401
        duncanii, 404
        varicolor, 439
    duncanii, 404
    durispina, 428
    emskoetterana, 431
    emskoetteriana, 431
    guadalupensis, 420, 421
    hesteri, 409
    leei, 422
    Lee's carpet, 424
    minima, 406
    muehlbaueriana, 431
    nellieae, 406
    neomexicana, 415
    oklahomensis, 413
    orcuttii, 399
    runyonii, 431
    Runyon's, 431
    sneedii, 416, 418, 434
        leei, 424
    Sneed's carpet, 420
    strobiliformis, 428, 434
        durispina, 428
    tuberculosa, 427, 436
        durispina, 427, 428
        tuberculosa, 427
    varicolor, 439
    vivipara, 399, 413
        neomexicana, 415
escobarias, 396, 430
Escobesseya, 399
    albicolumnaria, 428
    albocolumnaria, 428
    dasyacantha, 399
    duncanii, 402
estria del tarde, 416
Euphorbia, 424
    antisyphilitica, 424
Euphorbiaceae, 29

Ferocactus, 5, 9, 10, 13, 17, 35, 45, 47, 48, 51, 52, 70, 71, 280, **288**, 295, 300, 331, 336, 393
    bicolor, 300, 306, 308
        flavidispinus, 308
    hamatacanthus, 5, 32, 70, 288, 289, 291, **292**, 293–96, 298, 299, 328–30, 335
        crassispinus, 296
        hamatacanthus, 289, **292**, 293–96
        sinuatus, 289, 292, 294, **296**
    setispinus, 288, 300
    wislizeni, xiii, 5, 56, 57, 70, 283, 288, **289**, 290–95, 450, 465
        herrerae, 292
        tiburonensis, 292
Fishhook, 375
    Graham, 375
    tangled, 375
    Wright, 376
    Wright's, 376
flavonoids, 26, 27
Fouquieria, 4
    splendens, 4

garambullo, 113
Glandulicactus, 70, 280, 294, 298–300, **327**, 330, 331, 335
    crassihamatus, 327, 331
    uncinatus, 34, 54, 283, 284, 291, 298–300, **327**, 329, 331, 333, 334
        uncinatus, 331
        wrightii, 294, **327**, 328–31, 335, 341
    wrightii, 331
glory of Texas, **301**, 302, **303**,
grama, 330, 344, 346, 351
    blue, 344, 346
    chino, 330

Grusonia, 55, 66, 75, 76, 81, 85, 91, 95
   aggeria, 85
   emoryi, 96
   grahamii, 91
   schottii, 88
   stanlyi, 96
Gymnocactus, 322
Gymnocalycium, 50, 313

Haagea, 371
Halictidae, 42
Hamatocactus, 66, 70, 280, 288, **297**, 299–301, 331, 336, 393
   bicolor, 67, 288, **297**, 298–301, 331, 335
      setaceus, 67, 299
   hamatacanthus, 296
   setispinus, 300
   uncinatus, 300, 331
      wrightii, 331
   wrightii, 331
Harrisia, 196
Hatch-Slack pathway, 22
hawkmoth, 43
heart twister, 208
Hechtia, 313
   texensis, 424
hegdehog, 197
   brown-flowered, 331
   brown-spine, 258, 263
   Chisos, 244
   claretcup, 208
   comb, 225
   Correll, 260
   dwarf, 247
   Fendler, 228
   Fendler's, 228
   golden rainbow, 221
   green-flowered, 276
   lace, 247
   Livermore golden spine, 273
   Lloyd's, 214
   Mexican claret-cup, 205
   New Mexico rainbow, 221
   purple, 228
   rusty, 267
   strawberry, 234
   straw-colored, 241
   strawpile, 241
   Texas, 331
   Texas rainbow, 221
   triploid-hybrid, 214
   warty, 234
   woolly, 270
Hevea, 29
Homalocephala, 280, 287
   texensis, 287

honeybee, 43
   European, 43
hordenine, 314, 315
horse crippler, 287
horse-crippler, 56, **284**, 285, 287
horse maimer, 284
huevo de venado, 196
hummingbird, 43
   black-chinned, 43
Hylocereeae, 66
Hylocereus, 39, 196
Hymenoptera, 42

Indian dope, 316
Indian fig, 46, **191**

Jatropha, 424
   dioica, 424
joconostli, 107

Krainzia, 371

Laetilia, 51
   coccidivora, 51
Larrea, 34, 153, 194, 303, 313, 394, 396, 446, 450
   tridendata, 34
Lasioglossum, 42
   forbesii, 42
latex, 28
lechuguilla, 34, 243, 317, 377, 379, 396, 422, 432
lemonvine, **463**
lengua de vaca, 191
Lepidocoryphantha, 392, 440, 441
   macromeris, 443
Leptocereeae, 66
Leptocladia, 371
Leptocladodia, 371
   leona, 381
little chilis, 383, 386
living rock, 321
lizard-catcher, 375
Lloyd's Mariposa, 363
long mamma, 443
Lophocereus, 47
Lophophora, 28, 35, 39, 50, 70, **309**, 311, 313
   diffusa, 309, 313, 314
   echinata, 316
      lutea, 316
   fricii, 316
   jourdaniana, 316
   lewinii, 316
   williamsii, 6, 21, 28, 57, 283, **309**, 310–16, 319, 321, 463
      anhaloninica, 316

      decipiens, 316
      echinata, 310, 316
      lewinii, 316
      lutea, 316
      pentagonia, 316
      pluricostata, 316
      williamsii, 310, 316
Luthurge, 42
   bruesi, 42

Manduca, 43
Maihuenia, 22, 40, 65
Maihuenioideae, 65
Maihueniopsis, 66
Mammillaria; mammillaria, 6–9, 15, 17, 27–29, 35, 36, 38, 47, 48, 50, 70, 71, 309, 311, 344, 365, **371** 376, 381, 390–93, 471
   albicolumnaria, 428
   applanata, 386
   Big Bend, 446
   borealis, 415
   buchheimeana, 386
   calcarata, 460, 461
   conoidea, 323
   dasyacantha, 396
   declivis, 386
   denudata, 379
   dense, 399
   duncanii, 404
   echinus, 451
   emskoetteriana, 431
   engelmannii, 449
   escobaria, 431
   fishhook, 375
   fissurata, 317
   fragrans, 413, 415
   fuzzy, 379
   gaumeri, 383
   grahamii, 70, 371, **373**, 374–76
      grahamii, 372, **373**
      oliviae, 373
   gummifera, 57, 386
      applanata, 386
      hemispherica, 386
      meiacantha, 383
   hemispherica, 386
      waltheri, 386
   hesteri, 409
   heteromorpha, 443
   Heyder, 386
   heyderi, 15, 29, 57, 321, 371, 373, 381, **383**, 385–87
      applanata, 386
      bullingtoniana, 384, 386, **387**

hemispherica, 386
heyderi, 384, 385–87
gummifera, 386
meiacantha, 383
hirschtiana, 413
lasiacantha, 62, 71, 359, 366, 367, 371, 372, **377,** 378, 379, 414
  denudata, 379
  lasiacantha, 379
  minor, 379
leona, 381
lewinii, 316
longimamma, 390, 391
  sphaerica, 390
macdougalii, 387
macromeris, 439
meiacantha, 15, 17, 371, 373, **381,** 382, 383, 385, 386
melanocentra, 383
  meiacantha, 383
  runyonii, 383
microcarpa, 375
micromeris, 365, 368
milleri, 375
montana, 413
muehlenpfordtii, 449
multiceps, 387, 390
  grisea, 390
  texana, 390
nellieae, 406
neomexicana, 415
pale, **390**
papyracantha, 344,
pectinata, 455
Potts', 377, **379**
pottsii, 71, 274, 329, 362, 371, 372, 377, **379,** 380, 381, 426, 429
prolifera, 56, 471, **387,** 388–90, 430
  multiceps, 387, 390
  prolifera, 390
  texana, 372, **387,** 388, 390
pusilla, 387
  texana, 387, 390
radians, 461
  sulcata, 461
radiosa, 412
  borealis, 415
  neomexicana, 415
  vivipara, 412
ramillosa, 446
robustispina, 449
runyonii, 383, 440
scheeri, 446
  valida, 451
scolymoides, 455
similis, 461
  caespitosa, 461
sneedii, 420
sphaerica, 371, **390,** 391
strobiliformis, 436, 460
sulcata, 458
texana, 390
texensis, 386
tuberculosa, 431, 436
varicolor, 431
vivipara, 412
  borealis, 415
  neomexicana, 413, 415
  radiosa, 413, 415
  vera, 412
waltheri, 386
williamsii, 315
wrightii, 70, 328, 371, 374, **375,** 376, 411, 412
  wilcoxii, 375
  wrightii, 372, 373, **375,** 376
Mammillopsis, 371
Mammilloydia, 371
manca caballo, 287
manca mula, 284
Manduca, 43
medicine of God, 316
Megachilidae, 42
Melocactus, 45, 57, 66
  communis, 284
mescal, 316
mescal button(s), 316
mescaline, 28, 313, 314, 321
Miqueliopuntia, 66
moon "p," 316
mucilage, 28
mulato, 368
Myrtillocactus, 3, 18, 45
  geometrizans, 3, 113

needle mulee, 449
Neobesseya, 392
  muehlbaueriana, 431
Neogomesia, 317
Neolloydia, 9, 55, 72, 280, **322,** 326, 348, 393
  conoidea, 9, 322, **323,** 324, 325, 352, 362, 363, 430
    conoidea, 322, **323,** 324–26
    texensis, 326
  gautii, 326
  intertexta, 354
    dasyacantha, 357
    intertexta, 354
  mariposensis, 363
  matehualensis, 326
  texensis, 326
  warnockii, 357
Neomammillaria, 371
  applanata, 386
  gummifera, 386
  hemispherica, 386
  heyderi, 386
  meiacantha, 383
  pottsii, 381
  runyonii, 383
  vivipara, 413
  wrightii, 376
night-blooming cereus, 40, 41
nopal, 120, 129, 147, 165, 177, 181, 186, 474
nopal cegador, 118, 120
Nopalea, 20, 66
nopales, 74
nopalitos, 45, 49
Normanbokea, 322
Notocacteae, 37, 66

Obregonia, 313
ocotillo, 4
Oehmia, 371
Opuntia; opuntia, x, xii, xiii, 5–7, 10, 12, 13, 25–27, 31, 32, 34–36, 38, 40–42, 44–51, 55, 58, 59, 62, 65–67, 69, **73,** 74–76, 99, 119, 127, 151, 155, 159, 172, 180, 282, 313, 466, 470, 477
  aciculata, 123
  aggeria, 34, 55, 67, 77, **81,** 82–85, 87–93, 95, 457, 469
  alta, 75, 180
  arborescens, 103, 104
  arbuscula, 113
  arenaria, 183, 187
  arizonica, 177
  atrispina, 79, 121, 122, **123,** 124–26
  aurea, 181
  aureispina, 139, 158, 161
  azurea, 55, 127, 129, **130,** 132, 133, 135, 137, 138, 140, 154, 155, 158–60, 163
    aureispina, 79, 80, 130–33, 135, **136,** 137–40, 184
    azurea, 130, 131, 133, 134, 136, 138
    diplopurpurea, 79, 127, **130,** 131–41, 156
    discolor, 79, 130, 131, 135, **139,** 140
    parva, 79, 130–33, **134,** 135–38

Index 505

Continued
  ballii, 149, 150
  basilaris, xiii, **189**
  basilaris, **189**
  bigelovii, 45, 48, 49
  bradtiana, 73
  cacanapa, 180, 190
  caerulescens, 117
  camanchica, 140, 153, 155, 156, 158–61, 165, **166**, 167–72, 174, 176, 178
    albispina, 169
    gigantea, 169
    longispina, 169
    major, 169
    minor, 169
    orbicularis, 169
    pallida, 169
    rubra, 169
    salmonea, 169
  Camanchican, 169
  charlestonensis, 144
  chihuahuensis, 133, 169
  chisosensis, 31, 79, 80, 131, **156**, 157, 158, 179, 184
  chlorotica, 127
    santa-rita, 134
  ×*columbiana*, 181
  compressa, 75, 143, 150
    macrorhiza, 147
  cyclodes, 178
  cylindrica, 107
  cymochila, 140, 143–47
  dark-spined, 125
  davisii, xiii, 55, 76, 77, 96, 97, **99**, 100–102
  Davis's, 102
  decipiens, 107
  delicata, 150
  densely spined, 93
  densispina, 33, 77, 82, 84, **91**, 92–95
  dillei, 177, 192
  discata, 177, 180, 181, 192
  discoid, 177
  drummondii, 75
  dulcis, 54, 80, 156, 161, 165, 167, **169**, 170–72
  echinocarpa, 109
  edwardsii, 151, 153
  Edwards Plateau, 153
  ellisiana, 40, 48, 171, 180, **189**, 190
  emoryi, 73, 76, 82, 83, 88, 92, **93**, 94, 95

Emory's 96
engelmannii, 5, 56, 75, 123, 152, 153, 155, 156, 158, 163, 169, **172**, 173, 176, 177, 178, 180, 181, 192
  cacanapa, 158
  cuija, 177, 180
  cyclodes, 158
  discata, 177
  dulcis, 171, 172
  engelmannii, 54, 55, 80, 158, **172**, 174–76, 178–81
  flavispina, 177
  flexospina, 177
  lindheimeri, 42, 55, 80, 153, 158, 172–74, 176, **177**, 178–81, 190, 191
  linguiformis, 55, 73, 172, 178, 180, 181, **190**, 191
  wootonii, 177
Engelmann's, 177
eocarpa, 172
expansa, 172
exuviata, 107
ferruginispina, 181
ficus-indica, 46, 49, 177, 189, 190, **191**, 192, 472
filipendula, 150
fragilis, 34, 181, 183, 188
frutescens, 113
  brevispina, 113
fulgida, 45, 49, 50, 98
furiosa, 99
fuscoatra, 147
fusiformis, 147
gilvo-alba, 75
golden-spined, 139
gosseliniana, 126
grahamii, 85, 88, 90
grandiflora, 147
gregoriana, 177
griffithsiana, 181
haematocarpa, 181
hemifusa, 75, 147
herrfeldtii, 120
hystrix, 99
imbricata, 5, 11, 13, 17, 31, 47, 48, 51, 55, 73, 77, **102**, 103–05, 107, 113–17
  arborescens, 55, **103**, 104–09
  argentea, 55, 103–04, 106, **107**, 108, 109

  imbricata, 106, 107, 116
  lloydii, 107
  vexans, 107
inermis, 49
kleiniae, 13, 78, 104, 110, 112, **113**, 114–17
  Davis Mountains, 117
  kleiniae, 117
  Rio Grande, 117
  tetracantha, 113, 116
leptocaulis, x, 13, 31, 47, 48, 56, 78, 104, **109**, 110–14, 116, 117, 119, 279
  brevispina, 112
  longispina, 112
lindheimeri, 133, 152, 158, 172, 176–78, 180, 190
  chisosensis, 156
  dulcis, 172
  ellisiana, 189
  lehmannii, 180
  lindheimeri, 158, 181
  linguiformis, 190
  tricolor, 190
linguiformis, 190
lloydii, 107
longiclada, 181
loomsii, 150
mackensenii, 78, 79, **150**, 151, 155, 165
  mackensenii, 127, 150, **151**, 152, 153
  minor, 78, 150, 151, **153**, 154–56, 160
macrocalyx, 120
macrocentra, xiii, 26, 55, 79, **125**, 126–29, 132, 134, 135, 136, 139–41, 154, 155, 160, 163
  aureispina, 136
  macrocentra, 128
  minor, 125–27, 134, 153, 156
  castetteri, 128, 129, 134
macrorhiza, 6, 78, 140–43, **144**, 145–47, 149, 150, 151, 153, 163
  macrorhiza, 143, 146, 151
  pottsii, 146, 150
magna, 107
megacantha, 49, 192
mesacantha, 147
  greenei, 147

macrorhiza, 147
microdasys, 118, 120, 189
   rufida, 120
missouriensis, 182
   rufispina, 186
   trichophora, 182
moelleri, 81, 85
mound-forming, 85
perrita, 117
parishii, 96
phaeacantha, 78, 80, 140, 141, 144, 153, 159, 161, **162**, 163–66, 168, 169, 171, 172, 176, 180, 189
   brunnea, 169
   camanchica, 166, 169
   discata, 158, 172, 174, 176, 177
   major, 151, 153, 158, 161, 165, 166, 168, 169, 176
   nigricans, 129, 165
   phaeacantha, 151, 153, **162**, 163–68, 171, 172, 176
   piercei, 165
   spinosibacca, 162
   wootonii, 177
pinkavae, 181
plumbea, 150
polyacantha, 6, 31, 34, 49, 55, 78, 141, 144, 147, 163, **181**, 183, 184, 186–89
   arenaria, 6, 33, 76, 181–83, 186, **187**, 188, 189
   erinacea, 181
   hystricina, 181, 186
   nicholii, 181
   polyacantha, 181, 183, 184–86, 188
   rufispina, 183, 186
   trichophora, 141, **182**, 183–87
pottsii, 6, 55, 78, 141, 143, 144, **147**, 148–50, 153, 163, 188, 379, 429
procumbens, 177
pusilla, 75
recondita, 118
reflexa, 181
×*rooneyi*, 139
rosea, 107
rubiflora, 165
rufida, 12, 13, 55, 78, **118**, 119, 120, 189, 193
   tortiflora, 120

sand-loving, 189
santa-rita, 126, 134
schottii, 81–84, **85**, 90, 94, 95
   grahamii, 77, 82, 84–87, **88**, 89–91
   schottii, 77, 82, 83–85, 86, 87–91, 93
sinclairii, 181
spinosibacca, 80, **158**, 159–62, 165
×*spinosibacca*, 162
spinosior, 104, 106, 107, 113, 116, 117
spinotecta, 107
stanlyi, 94, 96
   kunzei, 96
   peeblesiana, 96
   stanlyi, 96
stapeliae, 99
stenochila, 150
streptacantha, 46, 49
stricta, 75
strigil, 79, **120**, 121–23, 155
   flexospina, 122
subarmata, 178, 181, 190, 192
sweet, 172
tardospina, 152, 181, 192
tenuispina, 143, 144
tesajo, 113
texana, 181
tortispina, 79, **140**, 141–44, 146, 147, 149, 153, 155, 163
trichophora, 186
tricolor, 190
tunicata, 11, 75–77, **96**, 87–100, 102
   davisii, 102
   tunicata, 99
vaginata, 113
valida, 176
versicolor, 45
vexans, 107
violacea, 126, 128, 129, 133
   castetteri, 126, 128, 129, 134
   macrocentra, 128, 129, 133
   santa-rita, 126, 134
   violacea, 126, 128, 129
vulgaris, 49, 52
whipplei, 75, 98, 99, 102
winteriana, 181

wootonii, 177
wrightii, 117
xanthoglochia, 147
Opuntiae, 112, 153
Opuntioideae, 5, 6, 8, 9, 12–14, 16, 18, 19, 37, 38, 40, 54, 65, 66, 69, 76
organo, 241, 462
organ pipe, 41
Osmia, 42
   subfasciata, 42

Pachycereeae, 16, 66, 67, 69
Pachycereus, 3, 45, 47, 50, 52, 313
   marginatus, 3
   pecten-aboriginum, 52
   pringlei, 41, 45, 66
Papaver, 29
   somniferum, 29
Papaveraceae, 29
Paralictus, 42
Parthenium, 339
   argentatum, 339
pear, 120
   blind, 120
   cinnamon, 120
Pediocactus, 322, 344, 347
   papyracanthus, 347
pejote, 316
Pelecyphora, 50, 313
pellote, 316
Peniocereus, 41, 43, 66, 69, **193**, 196
   greggii, 6, 19, 43, 44, 56, **193**, 194–96
      greggii, **193**, 194, 196
      transmontanus, 196
   johnstonii, 196
   marianus, 196
   striatus, 44, 45, 196, 279
peote, 213
Pereskia, 4, 6, 18, 22, 27, 38, 52, 60, 65, **463**, 472
   aculeata, xiii, **463**
Pereskioideae, 9, 16, 18, 37, 38, 40, 65
Pereskiopsis, 20, 27, 65, 66
pest pear, 75
peyori, 316
peyot, 316
peyote, 21, 28, 50, 57, **309**, 310, 313–15, 321, 463
peyote cimarrón, 321
peyotl, 309, 316
pezote, 316
Phellosperma, 371

photosynthesis, C3, 22, 23
photosynthesis, C4, 22
Phycitinae, 51
Pilosocereus, 66
pincushion, 375
    desert, 399
    Duncan's, 404
    fishhook, 375
    flat cream, 387
    golf ball, 379
    pancake, 386
    small-spined cream, 383
Pinus, 96
    remota, 96, 317
piote, 316
piotl, 316
pitahaya, 205, 234, 241, 462
pitalla, 237
pitaya, 221, 234, 237, 241
    ashy-white, 225
    brown-flowered, 258
    Chisos, 244
    Davis' green, 250
    Fendler's 228
    green-flowered, 258, 263
    mound, 237
    purple, 234
    slender-spined, 221
    Texas rainbow, 221
    yellow-flowered, 221
piule, 316
Polaskia, 50, 313
Porfiria, 371
Portulacaceae, 36, 37, 40
prickly-pear, 120
    blind, 120
prickly pear, 5, 6, 12, 14, 17, 30–32, 39, 44–53, 55, 60, 64, 69, **73**, 74–76, 120, 129, 147, 156, 162, 165, 168, 169, 175–77, 179, 184, 186, 187, 189–92, 410, 472, 473
    bearded, 123
    Big Bend, 162
    Big Bend purplish, 131, **134**
    Big Hill, 131, **139**
    black-and-yellow-spined, 121, **123**
    black-spine, 129
    black-spined, 129
    blackspine purple, 129
    blind, **118**, 120
    bristlehair, 186
    brown-spine, 165
    brownspine, 165
    brown-spined, **162**, 165
    Castetter purple, 129
    chain, 147
    Chisos, 131, **156**, 158
    Comanche, **166**
    cow-tongue, 190
    crow-foot, 75
    desert, 177
    diploid purple, **130**, 131
    eastern, 75, 147
    Ellis', **189**
    El Paso, 189
    Englemann, 177
    Englemann's, **172**, 173
    flaming, 181
    golden-spined, 131, **136**
    Graham, 91
    grassland, 147
    hair-spined, 186
    little, 75
    longspine, 129
    long-spined, 127, 129
    long-spined purplish, **125**, 126
    low, 75
    Mackensen's, **150**, **151**
    marble-fruit, **120**, 121
    marblefruit, 123
    New Mexico, 165
    plains, **144**, 145, 147, **181**, 186
    Potts, 150
    Potts', **147**, 148
    purple, 127, 129, 140, 155, 156
    purple-fruited, 165, 177
    purple-tinged, 129
    redeye, 129
    red-spined, 186
    sand, 182, **187**
    sand-loving, 189
    short-spined purplish, **150**, **153**
    southern plains, **182**
    spinyfruit, 162
    spiny-fruited, **159**
    starvation, 186
    sweet, **169**, 170
    Texas, **172**, 173, **177**, 191
    tuberous-rooted, 147
    tulip, 165, 177
    twisted spine plains, **140**, 141, 144
Prosopis, 107, 113, 153, 194, 303, 313, 394, 396
Pseudocoryphantha, 412, 421
Pseudomammillaria, 371
Pterocactus, 65
Pyralid, 51

phycitine, 51

queen of the night, 196, **464**
queen-of-the-night, 196
Quiabentia, 27, 65
    raiz diabolica, 316

Rapicactus, 322
Recurvus, 295
Rhipsalideae, 66
Rhipsalis, 1, 7, 66
    baccifera, 1, 66
Roseocactus, 317
    fissuratus, 321

sacasil, **279**
saguaro, 3, 5, 33, 41, 60, 467
Schlumbergera, 7, 39
    truncata, 7, 113
Sclerocactus, 280, 322, 331, 344, 347, 348
    intertextus, 354
        dasyacanthus, 357
    mariposensis, 363
    papyracanthus, 347
    uncinatus, 331
        wrightii, 331
    warnockii, 360
Selaginella, 34, 247, 249, 268, 273, 399, 404
Selenicereus, 39, 196, **464**
    pseudospinulosus, 464
    spinulosus, **464**
silicon dioxide, 29
Solisia, 371
sotol, 4
Sporobolus, 344
    airoides, 344
star cactus, 321
star peyote, 463
star rock, 321
Stenocereus, 3, 29, 45, 47, 50, 313
    griseus, 3
    thurberi, 28, 41, 66
Stetsonia, 50, 313
strawberry, 237
    Mexican, 237
Strombocactus, 313
sunami, 321

Tacinga, 65
tapon, 368
tasajilla, 113, 118
tasajillo, 113, 118, 241
tasajo, 107, 118
tencholote, 99
tepenexcomitl, 284

Tephrocactus, 65
tesajillo, 113
tesajo, 113
Texas longhorn, 296
Texas pride, 306
the bad seed, 316
Thelocactus; thelocactus, xiii, 13, 35, 71, 280, 300, **301**, 306, 313, 322, 331, 352
   bicolor, 17, 297–99, **301**, 303, **303**, 304, 305, 328
      bicolor, 301, 302, **303**, 304–07
      bolaensis, 303
      dasyacanthus, 357
      flavidispinus, 302, 303, 305, **306**, 307, 308
      schottii, 306
      schwarzii, 302
   flatspine, 308
   flavidispinus, 306

Marathon Basin, **301**, 302, **306**
pottsii, 306
setispinus, 300
Toumeya, 71, **344**, 347
   papyracantha, 33, **344**, 345–47
Trichocereeae, 66
Trichocereus, 50, 313
Triglochidiatus, 41
triterpenes, 28, 52, 56
Tubinicarpus, 313, 322
tuna, 181
tuna de tierra, 316
tunas, 46
Tunilla, 66
Turbinicarpus, 313
turks head, 296
turk's head, 284, 296, 331

vaca, 331
vela de coyote, 107

visnaga, 287, 292, 296
visnagita, 354
   white-flowered, 354
viszagita, 363
   mariposa, 363
viznaga, 287
viznagita, 354
   blanco, 354

whiskerbrush, 446
white column, 428
white mule, 316
Wilcoxia, 197, 279
   poselgeri, 279

xoconostle, 107

Yucca, 4
yucca, 4

The authors and Texas Tech University Press are deeply grateful to the following contributors, without whose generous and timely support this work would not have been possible.

The Edwill Fund
Elizabeth Winston Mize
Grover and Sally Murray
Colorado Cactus and Succulent Society
Austin Cactus and Succulent Society
Lee J. Miller
National Capital Cactus and Succulent Society

Plate 1. Chihuahuan desertscrub, looking SE to the igneous Chisos Mts. Habitat of *Opuntia camanchica*, *O. azurea*, and *Echinocactus horizonthalonius*.

Plate 2. Grassland in N Brewster Co., TX, with the limestone Glass Mts in the background. Habitat of *Opuntia davisii*.

Plate 3. Alluvial basin supporting grassland in N Brewster Co., TX, in contact with igneous mountains. Habitat of *Opuntia tortispina*, *O. pottsii*, and *Echinomastus intertextus*.

Plate 4. Oak-juniper-pinyon woodland above basin grassland, Davis Mts in central Jeff Davis Co., TX. Habitat of *Opuntia phaeacantha* var. *phaeacantha*, and *Echinocereus viridiflorus* var. *cylindricus*.

Plate 5. Oak-juniper-pinyon woodland in the Chisos Mts, southern Brewster Co., TX. Habitat of *Opuntia chisosensis, O. engelmannii* var. *engelmannii*, *Mammillaria meiacantha*, and *Echinocereus coccineus*.

Plate 6. Montane woodland in upper Madera Canyon, Mt Livermore, Jeff Davis Co., TX. Habitat of *Opuntia polyacantha* and *Echinocereus viridiflorus* var. *weedinii* (altitude ca. 8,000 ft).

Plate 7. Emory Peak, Chisos Mts, southern Brewster Co., TX. Habitat of *Opuntia chisosensis*.

Plate 8. Chihuahuan desertscrub, Hen Egg Mt and Agua Fria in the background, southern Brewster Co., TX.

Plate 9. Chihuahuan desertscrub in igneous-derived substrate, Big Bend Ranch State Park, sedimentary exterior of Solitario Dome in background, southern Presidio Co., TX.

Plate 10. *Agave lechuguilla* and *Euphorbia antisyphilitica* on limestone of Reed Plateau, looking N across sedimentary substrate and alluvial soils of the desert floor to igneous desert mountains, southern Brewster Co., TX.

Plate 11. Metamorphic rock, the Caballos Novaculite formation in Marathon Basin, N Brewster Co., TX. Habitat of the "novaculite endemic" cacti and many others.

Plate 12. *Selaginella* mat on novaculite substrate, Marathon Basin. Microhabitat of diminutive cactus *Echinocereus davisii*.

Plate 13. *Yucca torreyi*, SW foothills of the igneous Chisos Mts, Brewster Co., TX. In the W, *Y. torreyi* is an "indicator species" of the Chihuahuan Desert, but in the E it extends far into the subtropical Tamaulipan scrub, where it is called *Y. treculiana*.

Plate 14. *Agave lechuguilla* in the limestone Dead Horse Mts; Chisos Mts in the background; Brewster Co., TX.

Plate 15. *Dasylirion heteracanthum*; Black Gap Wildlife Management Area, southern Brewster Co., TX.

Plate 16.
Fouquieria splendens, with its slender, spreading, spiny stems; *Yucca torreyi* in left foreground; Big Bend National Park, Brewster Co., TX.

Plate 17.
Pereskia grandifolia, a leafy cactus species from Brazil, South America.

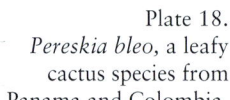

Plate 18.
Pereskia bleo, a leafy cactus species from Panama and Colombia.

Plate 19.
Representative cactus flower with yellow inner (petaloid) tepals (*Echinocereus dasyacanthus*).

Plate 20. Individual plants of *Opuntia imbricata* (tree cholla) are the largest native, wild cacti in the Trans-Pecos.

Plate 21. Low, mounded habit of *Opuntia densispina* (Big Bend devil cholla).

Plate 22. Medium-size specimen of *Ferocactus wislizeni* (Arizona barrel cactus) from the Franklin Mts, El Paso Co., TX.

Plate 23. Multistemmed (branched) specimen of *Echinocereus coccineus* (claret-cup cactus) from Pecos Co., TX.

Plate 24. Some cryptic Trans-Pecos cacti. Find *Epithelantha bokei* (Boke's button-cactus), left middle; *Echinomastus mariposensis* (Mariposa cactus), upper right in flower; and *Ariocarpus fissuratus* (living rock cactus), lower right.

Plate 25. Apical mat of hair pulled away from fruit in *Echinocactus horizonthalonius* (eagle-claw cactus).

Plate 26. Woody skeleton of *Opuntia imbricata* (tree cholla).

Plate 27. Fasciated (crested or cristate) individual of *Thelocactus bicolor* var. *flavidispinus* (Marathon Basin thelocactus).

Plate 28. Representative cactus flowers showing pericarpel, receptacular tube, sepaloid tepals, and petaloid tepals (*Echinocereus* × *roetteri* var. *neomexicanus*, Lloyd's hedgehog cactus).

Plate 29. Betacyanin-pigmented stems of *Opuntia azurea*, a purple prickly pear.

Plate 30. *Echinocereus chisoensis* (Chisos hedgehog cactus) in "nurse plant" association with *Larrea tridentata* (creosotebush) and *Opuntia aggeria* (clumped dog cholla).

Plate 31. Harvested cacti awaiting sale in southern Brewster Co., TX.

Plate 32. *Opuntia aggeria* (clumped dog cholla) near Study Butte, Brewster Co., TX.

Plate 33. *Opuntia aggeria* (clumped dog cholla); maturing fruits.

Plate 34. *Opuntia schottii* var. *schottii* (Schott's dog cholla); Langtry, Val Verde Co., TX.

Plate 35. *Opuntia schottii* var. *schottii* (Schott's dog cholla); maturing fruits.

Plate 36. *Opuntia schottii* var. *grahamii* (Graham's dog cholla) from Solitario uplift; cultivated; Presidio Co., TX.

Plate 37. *Opuntia schottii* var. *grahamii* (Graham's dog cholla); maturing fruits.

Plate 38. *Opuntia densispina* (Big Bend devil cholla); cultivated; Solis, Big Bend National Park, Brewster Co., TX.

Plate 39. *Opuntia densispina* (Big Bend devil cholla); maturing fruits.

Plate 40. *Opuntia emoryi* (common devil cholla); near Candelaria, Presidio Co., TX.

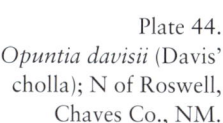

Plate 41. *Opuntia emoryi* (common devil cholla); mature and maturing fruits; near Candelaria, Presidio Co., TX.

Plate 42. *Opuntia tunicata* (icicle cholla); Glass Mts, Pecos Co., TX (photo by Marcos A. Rodriguez).

Plate 43. *Opuntia tunicata* (icicle cholla); maturing fruits; Glass Mts, Brewster Co., TX.

Plate 44. *Opuntia davisii* (Davis' cholla); N of Roswell, Chaves Co., NM.

Plate 45.
Opuntia davisii (Davis' cholla); mature fruit; near Marfa, Presidio Co., TX.

Plate 46.
Opuntia imbricata var. *arborescens* (tree cholla); N Brewster Co., TX.

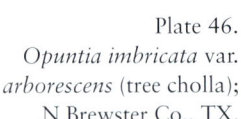

Plate 47.
Opuntia imbricata var. *arborescens* (tree cholla); mature fruit; near Blue Mt, Jeff Davis Co., TX.

Plate 48.
Opuntia imbricata var. *argentea* (Big Bend cholla); cultivated; Mariscal Mt, Big Bend National Park.

Plate 49.
Opuntia imbricata var. *argentea* (Big Bend cholla); mature and immature fruits; cultivated; Mariscal Mt, Big Bend National Park.

Plate 50.
Opuntia leptocaulis
(Christmas cholla); short-spine form; near Dryden, Terrell Co., TX.

Plate 51.
Opuntia leptocaulis
(Christmas cholla); mature fruits; S Big Bend National Park, Brewster Co., TX.

Plate 52.
Opuntia leptocaulis
(Christmas cholla); long-spine form; S Brewster Co., TX.

Plate 53.
Opuntia leptocaulis
(Christmas cholla); mature fruits; River Road, Big Bend National Park, Brewster Co., TX.

Plate 54.
Opuntia kleiniae (candle cholla); Musquiz Canyon, Jeff Davis Co., TX.

Plate 55.
Opuntia kleiniae (candle cholla); mature fruits, Musquiz Canyon, Jeff Davis Co., TX.

Plate 56.
Opuntia kleiniae (candle cholla); putative backcross; near Mitre Peak, Jeff Davis Co., TX.

Plate 57.
Opuntia kleiniae (candle cholla); putative backcross; same population as Plate 56; near Mitre Peak, Jeff Davis Co., TX.

Plate 58.
Opuntia kleiniae (candle cholla); the "Rio Grande *kleiniae*"; near Castolon, Big Bend National Park, Brewster Co., TX.

Plate 59.
Opuntia kleiniae (candle cholla); mature fruit of the "Rio Grande *kleiniae*"; near Castolon, Big Bend National Park, Brewster Co., TX.

Plate 60. *Opuntia rufida* (blind prickly pear); S Big Bend National Park, Brewster Co., TX.

Plate 61. *Opuntia rufida* (blind prickly pear); maturing fruits; cultivated; S Big Bend National Park, Brewster Co., TX.

Plate 62. *Opuntia strigil* (marble-fruit prickly pear); near Fort Stockton, Pecos Co., TX.

Plate 63. *Opuntia strigil* (marble-fruit prickly pear); immature and mature fruit; S of Fort Stockton, Pecos Co., TX.

Plate 64. *Opuntia atrispina* (black-and-yellow-spined prickly pear); cultivated; Pecos River high bridge, Val Verde Co., TX.

Plate 65. *Opuntia atrispina* (black-and-yellow-spined prickly pear); nearly mature fruit; cultivated; Pecos River high bridge, Val Verde Co., TX.

Plate 66. *Opuntia macrocentra* (long-spined purplish prickly pear); Hueco Mts, El Paso Co., TX (photo by Richard D. Worthington).

Plate 67. *Opuntia macrocentra* (long-spined purplish prickly pear); mature fruits; 35 mi N of Sierra Blanca, Hudspeth Co., TX.

Plate 68. *Opuntia azurea* var. *diplopurpurea* (diploid purple prickly pear); between Presidio and Ruidosa, Presidio Co., TX (possibly introgressed from var. *parva*).

Plate 69. *Opuntia azurea* var. *diplopurpurea* (diploid purple prickly pear); mature fruits; 10 mi SE of Ruidosa, Presidio Co., TX.

Plate 70. *Opuntia azurea* var. *parva* (Big Bend purplish prickly pear); River Road, Big Bend National Park, Brewster Co., TX.

Plate 71. *Opuntia azurea* var. *parva* (Big Bend purplish prickly pear); mature fruits; Big Bend National Park, Brewster Co., TX.

Plate 72. *Opuntia azurea* var. *aureispina* (golden-spined prickly pear); near Mariscal Mt, type locality, Big Bend National Park, Brewster Co., TX.

Plate 73. *Opuntia azurea* var. *aureispina* (golden-spined prickly pear); mature fruits; near Mariscal Mt, type locality, Big Bend National Park, Brewster Co., TX.

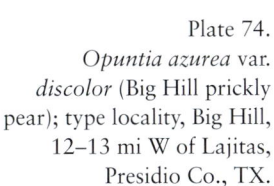

Plate 74. *Opuntia azurea* var. *discolor* (Big Hill prickly pear); type locality, Big Hill, 12–13 mi W of Lajitas, Presidio Co., TX.

Plate 75.
Opuntia azurea var. *discolor* (Big Hill prickly pear); mature fruits; type locality, Big Hill, 12–13 mi W of Lajitas, Presidio Co., TX.

Plate 76.
Opuntia tortispina (twisted spine plains prickly pear); near Marfa, Presidio Co., TX.

Plate 77.
Opuntia tortispina (twisted spine plains prickly pear); mature fruit; near Fort Davis, Jeff Davis Co., TX.

Plate 78.
Opuntia macrorhiza (plains prickly pear); cultivated; near Monahans, Ward Co., TX.

Plate 79.
Opuntia macrorhiza (plains prickly pear); nearly mature fruits; cultivated; near Monahans, Ward Co., TX.

Plate 80. *Opuntia pottsii* (Potts' prickly pear); near Marfa, Presidio Co., TX.

Plate 81. *Opuntia pottsii* (Potts' prickly pear); maturing fruits; cultivated; near Alpine, Brewster Co., TX.

Plate 82. *Opuntia mackensenii* var. *mackensenii* (Mackensen's prickly pear); cultivated; Terrell Co., TX.

Plate 83. *Opuntia mackensenii* var. *mackensenii* (Mackensen's prickly pear); mature fruit; cultivated; near Longfellow, Terrell Co., TX.

Plate 84. *Opuntia mackensenii* var. *minor* (short-spined purplish prickly pear); type locality, near Ruidosa, Presidio Co., TX.

Plate 85. *O. mackensenii* var. *minor* (short-spined purplish prickly pear); mature fruits; near Dryden, Terrell Co., TX.

Plate 86. *Opuntia chisosensis* (Chisos prickly pear); cultivated; Chisos Mts, Big Bend National Park, Brewster Co., TX.

Plate 87. *Opuntia chisosensis* (Chisos prickly pear); mature fruits; cultivated; Chisos Mts, Big Bend National Park, Brewster Co., TX.

Plate 88. *Opuntia spinosibacca* (spiny-fruited prickly pear); near tunnel, Big Bend National Park, Brewster Co., TX.

Plate 89. *Opuntia spinosibacca* (spiny-fruited prickly pear); mature fruits; cultivated; Big Bend National Park, Brewster Co., TX.

Plate 90. *Opuntia phaeacantha* var. *phaeacantha* (brown-spined prickly pear); near Alpine, Brewster Co., TX.

Plate 91. *Opuntia phaeacantha* var. *phaeacantha* (brown-spined prickly pear); fruits; Wildrose Pass, Jeff Davis Co., TX.

Plate 92. *Opuntia camanchica* (Comanche prickly pear); cultivated; Brewster Co., TX.

Plate 93. *Opuntia camanchica* (Comanche prickly pear); cultivated; Brewster Co., TX.

Plate 94. *Opuntia dulcis* (sweet prickly pear); near Boquillas, Big Bend National Park, Brewster Co., TX.

Plate 95. *Opuntia dulcis* (sweet prickly pear); cultivated; near Boquillas, Big Bend National Park, Brewster Co., TX.

Plate 96. *Opuntia dulcis* (sweet prickly pear); mature fruits; cultivated; near Boquillas, Big Bend National Park, Brewster Co., TX.

Plate 97. *Opuntia engelmannii* var. *engelmannii* (Engelmann's prickly pear); near Terlingua, Brewster Co., TX.

Plate 98. *Opuntia engelmannii* var. *engelmannii* (Engelmann's prickly pear); Alpine, Brewster Co., TX.

Plate 99. *Opuntia engelmannii* var. *lindheimeri* (Texas prickly pear); cultivated; from Jim Hogg Co., TX.

Plate 100. *Opuntia engelmannii* var. *lindheimeri* (Texas prickly pear); mature fruits; cultivated; near Dryden, Terrell Co., TX.

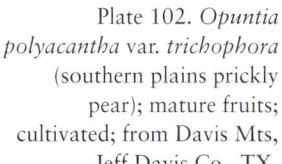

Plate 101. *Opuntia polyacantha* var. *trichophora* (southern plains prickly pear); Hueco Tanks, El Paso Co., TX (photo by Richard D. Worthington).

Plate 102. *Opuntia polyacantha* var. *trichophora* (southern plains prickly pear); mature fruits; cultivated; from Davis Mts, Jeff Davis Co., TX.

Plate 103. *Opuntia polyacantha* var. *arenaria* (sand prickly pear); El Paso Co., TX (photo by Richard D. Worthington).

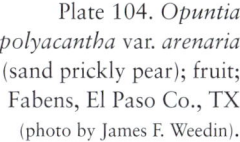

Plate 104. *Opuntia polyacantha* var. *arenaria* (sand prickly pear); fruit; Fabens, El Paso Co., TX (photo by James F. Weedin).

Plate 105. *Opuntia basilaris* var. *basilaris* (beavertail cactus); in cultivation; Desert Botanical Garden, Tempe, AZ (photo by David J. Ferguson).

Plate 106. *Opuntia basilaris* var. *basilaris* (beavertail cactus); fruits; in cultivation; Tucson Botanic Garden (photo by James F. Weedin).

Plate 107. *Opuntia ellisiana* (Ellis' prickly pear); cultivated; Alpine, Brewster Co., TX.

Plate 108. *Opuntia ellisiana* (Ellis' prickly pear); immature fruits.

Plate 109. *Opuntia engelmannii* var. *linguiformis* (cow-tongue prickly pear); cultivated; Alpine, Brewster Co., TX.

Plate 110. *Opuntia engelmannii* var. *linguiformis* (cow-tongue prickly pear); fruits; cultivated; Alpine, Brewster Co., TX.

Plate 111. *Opuntia ficus-indica* (Indian fig); in cultivation; Tucson Botanic Garden (photo by James F. Weedin).

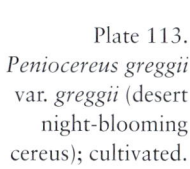

Plate 112. *Opuntia ficus-indica* (Indian fig); fruits; in cultivation; Tucson Botanic Garden (photo by James F. Weedin).

Plate 113. *Peniocereus greggii* var. *greggii* (desert night-blooming cereus); cultivated.

Plate 114. *Peniocereus greggii* var. *greggii* (desert night-blooming cereus); mature fruits; cultivated.

Plate 115. *Echinocereus coccineus* var. *rosei* (claret-cup cactus); N-central Hudspeth Co., TX.

Plate 116. *Echinocereus coccineus* var. *rosei* (claret-cup cactus); mature fruit; cultivated; from Hudspeth Co., TX.

Plate 117. *Echinocereus coccineus* var. *paucispinus* (Texas claret-cup cactus); N Brewster Co., TX.

Plate 118. *Echinocereus coccineus* var. *paucispinus* (Texas claret-cup cactus); mature fruits; cultivated; from N Brewster Co., TX.

Plate 119. *Echinocereus coccineus* var. *paucispinus* (Texas claret-cup cactus); male flower.

Plate 120. *Echinocereus coccineus* var. *paucispinus* (Texas claret-cup cactus); female flowers.

Plate 121. *Echinocereus roetteri* var. × *roetteri* (Roetter's hybrid hedgehog cactus); NW Culberson Co., TX.

Plate 122. *Echinocereus roetteri* var. × *roetteri* (Roetter's hybrid hedgehog cactus); Jarilla Mts, Otero Co., NM.

Plate 123. *Echinocereus roetteri* var. × *roetteri* (Roetter's hybrid hedgehog cactus); Jarilla Mts, Otero Co., NM.

Plate 124. *Echinocereus roetteri* var. × *neomexicanus* (Lloyd's hedgehog cactus); orange-flowered, putative F_1 generation; E Pecos Co., TX.

Plate 125. *Echinocereus roetteri* var. × *neomexicanus* (Lloyd's hedgehog cactus); mature fruit; cultivated; from E Pecos Co., TX.

Plate 126. *Echinocereus roetteri* var. × *neomexicanus* (Lloyd's hedgehog cactus); red-orange-flowered, putative F$_2$ generation; E Pecos Co., TX.

Plate 127. *Echinocereus roetteri* var. × *neomexicanus* (Lloyd's hedgehog cactus); pinkish-flowered, putative F$_2$ generation, E Pecos Co., TX.

Plate 128. *Echinocereus roetteri* var. × *neomexicanus* (Lloyd's hedgehog cactus); yellow-flowered, putative backcross to *E. coccineus*; Marathon Basin, Brewster Co., TX.

Plate 129. *Echinocereus* × *neomexicanus* (triploid hybrid hedgehog cactus); tentatively identified; Jarilla Mts, Doña Ana Co., NM (photo by Allan D. Zimmerman).

Plate 130. *Echinocereus dasyacanthus* (Texas rainbow cactus); S Brewster Co., TX (photo by Shirley A. Powell).

Plate 131. *Echinocereus dasyacanthus* (Texas rainbow cactus); mature fruit; in cultivation.

Plate 132. *Echinocereus dasyacanthus* (Texas rainbow cactus) introgressed from *E.* × *roetteri* (second- or later-generation backcross; = *E.* × *roetteri* var. *neomexicanus* nothospecies; E Pecos Co., TX.

Plate 133. *Echinocereus pectinatus* var. *wenigeri* (Langtry rainbow cactus); cultivated; from Terrell Co., TX.

Plate 134. *Echinocereus pectinatus* var. *wenigeri* (Langtry rainbow cactus); mature fruit; cultivated; from Terrell Co., TX.

Plate 135. *Echinocereus pectinatus* var. *wenigeri* (Langtry rainbow cactus); the white cross-bands and green bases of the inner tepals are characteristic.

Plate 136. *Echinocereus fendleri* var. *fendleri* (Fendler's hedgehog cactus); in cultivation.

Plate 137. *Echinocereus fendleri* var. *fendleri* (Fendler's hedgehog cactus); mature fruit; cultivated; from Presidio Co., TX.

Plate 138. *Echinocereus enneacanthus* var. *enneacanthus* (strawberry cactus); S Brewster Co., TX.

Plate 139. *Echinocereus enneacanthus* var. *enneacanthus* (strawberry cactus); mature fruit; in cultivation.

Plate 140. *Echinocereus enneacanthus* var. *brevispinus* (strawberry cactus); cultivated; from W Val Verde Co., TX.

Plate 141. *Echinocereus enneacanthus* var. *brevispinus* (strawberry cactus); mature fruits; cultivated; from W Val Verde Co., TX.

Plate 142. *Echinocereus stramineus* var. *stramineus* (strawberry cactus); S Brewster Co., TX.

Plate 143. *Echinocereus stramineus* var. *stramineus* (strawberry cactus); nearly mature and mature fruits; S Brewster Co., TX.

Plate 144. *Echinocereus chisoensis* var. *chisoensis* (Chisos hedgehog cactus); Big Bend National Park, TX.

Plate 145. *Echinocereus chisoensis* var. *chisoensis* (Chisos hedgehog cactus); immature fruits; Big Bend National Park, TX.

Plate 146. *Echinocereus reichenbachii* cf. var. *reichenbachii* (lace cactus); cultivated.

Plate 147. *Echinocereus reichenbachii* cf. var. *perbellus* (lace cactus); mature, dehiscent fruit; cultivated.

Plate 148. *Echinocereus reichenbachii* var. cf. *fitchii*; cultivated.

Plate 149. *Echinocereus reichenbachii* var. cf. *baileyi*; cultivated.

Plate 150. *Echinocereus davisii* (dwarf hedgehog cactus); three plants in flower, in a *Selaginella* mat, along with an immature plant of *Thelocactus bicolor* var. *flavidispinus;* Marathon Basin, Brewster Co., TX.

Plate 151. *Echinocereus davisii* (dwarf hedgehog cactus); one multistemmed plant in flower; Marathon Basin, Brewster Co., TX.

Plate 152. *Echinocereus davisii* (dwarf hedgehog cactus); nearly mature fruit; Marathon Basin, Brewster Co., TX.

Plate 153. *Echinocereus viridiflorus* var. *viridiflorus* (green-flowered hedgehog cactus); cultivated; from N NM.

Plate 154. *Echinocereus viridiflorus* var. *cylindricus* (small-flowered hedgehog cactus); cultivated; from Davis Mts, Jeff Davis Co., TX.

Plate 155. *Echinocereus viridiflorus* var. *cylindricus* (small-flowered hedgehog cactus); mature fruits; 9 mi SE of Alpine, Brewster Co., TX.

Plate 156. *Echinocereus viridiflorus* var. *correllii* (Correll's green-flowered hedgehog cactus); Marathon Basin, Brewster Co., TX.

Plate 157. *Echinocereus viridiflorus* var. *correllii* (Correll's green-flowered hedgehog cactus); mature fruits; Marathon Basin, Brewster Co., TX.

Plate 158. *Echinocereus viridiflorus* var. *chloranthus* (western green-flowered hedgehog cactus); putative hybrid with var. *cylindricus*; Jeff Davis Co., TX.

Plate 159. *Echinocereus viridiflorus* var. *chloranthus* (western green-flowered hedgehog cactus); mature fruit; Hudspeth Co., TX.

Plate 160. *Echinocereus viridiflorus* var. *russanthus* (rusty hedgehog cactus); Christmas Mts, Brewster Co., TX.

Plate 161. *Echinocereus viridiflorus* var. *russanthus* (rusty hedgehog cactus); W Chisos Mts, Brewster Co., TX.

Plate 162. *Echinocereus viridiflorus* var. *russanthus* (rusty hedgehog cactus); mature fruits; cultivated; S Brewster Co., TX.

Plate 163. *Echinocereus viridiflorus* cf. var. *russanthus* (rusty hedgehog cactus); Rosillos Mts, Brewster Co., TX.

Plate 164. *Echinocereus viridiflorus* var. *neocapillus* (Weniger's small-flowered hedgehog cactus); cultivated; from Marathon Basin, Brewster Co., TX.

Plate 165. *Echinocereus viridiflorus* var. *neocapillus* (Weniger's small-flowered hedgehog cactus); mature fruits; cultivated; from Marathon Basin, Brewster Co., TX.

Plate 166. *Echinocereus viridiflorus* var. *neocapillus* (Weniger's small-flowered hedgehog cactus); "hairy" seedling; Marathon Basin, Brewster Co., TX.

Plate 167. *Echinocereus viridiflorus* var. *neocapillus* (Weniger's small-flowered hedgehog cactus); "hairy" juvenile; Marathon Basin, Brewster Co., TX.

Plate 168. *Echinocereus viridiflorus* var. *neocapillus* (Weniger's small-flowered hedgehog cactus); young adult; Marathon Basin, Brewster Co., TX.

Plate 169. *Echinocereus viridiflorus* var. *weedinii* (Weedin's small-flowered hedgehog cactus); cultivated; from Davis Mts, Jeff Davis Co., TX.

Plate 170. *Echinocereus viridiflorus* var. *weedinii* (Weedin's small-flowered hedgehog cactus); mature fruits; cultivated; from Davis Mts, Jeff Davis Co., TX.

Plate 171. *Echinocereus viridiflorus* var. *canus* (graybeard cactus); cultivated; Solitario, Presidio Co., TX.

Plate 172. *Echinocereus viridiflorus* var. *canus* (graybeard cactus); mature fruits; Solitario uplift, Presidio Co., TX.

Plate 173. *Echinocereus viridiflorus* var. *canus* (graybeard cactus); seedlings; cultivated; from Solitario uplift, Presidio Co., TX.

Plate 174. *Echinocereus pentalophus* (alicoche); cultivated.

Plate 175. *Echinocereus pentalophus* (alicoche); nearly mature fruit; cultivated.

Plate 176. *Echinocereus berlandieri* (Berlandier's alicoche); cultivated; from S TX.

Plate 177. *Echinocereus berlandieri* (Berlandier's alicoche); nearly mature fruit; cultivated; from S TX.

Plate 178. *Echinocereus papillosus* var. *papillosus* (yellow-flowered alicoche); cultivated; from Jim Hogg Co., TX.

Plate 179. *Echinocereus papillosus* var. *angusticeps* (yellow-flowered alicoche); S TX (photo by James H. Everitt).

Plate 180. *Echinocereus papillosus* var. *angusticeps* (yellow-flowered alicoche); S. TX.

Plate 181. *Echinocereus poselgeri* (dahlia cactus); cultivated.

Plate 182. *Echinocereus poselgeri* (dahlia cactus); nearly mature fruits; cultivated.

Plate 183. *Echinocactus horizonthalonius* var. *horizonthalonius* (eagle-claw cactus); S Brewster Co., TX.

Plate 184. *Echinocactus horizonthalonius* var. *horizonthalonius* (eagle-claw cactus); mature fruit; S Brewster Co., TX.

Plate 185. *Echinocactus texensis* (horse-crippler); cultivated.

Plate 186. *Echinocactus texensis* (horse-crippler); mature fruits; cultivated.

Plate 187. *Ferocactus wislizeni* (Arizona barrel cactus); cultivated; from Franklin Mts, El Paso Co., TX.

Plate 188. *Ferocactus wislizeni* (Arizona barrel cactus); cultivated; from Franklin Mts, El Paso Co., TX.

Plate 189. *Ferocactus wislizeni* (Arizona barrel cactus); mature fruits; cultivated; from Franklin Mts, El Paso Co., TX.

Plate 190. *Ferocactus hamatacanthus* var. *hamatacanthus* (giant fishhook-cactus); cultivated; from S Brewster Co., TX.

Plate 191. *Ferocactus hamatacanthus* var. *hamatacanthus* (giant fishhook-cactus); cultivated; from S Brewster Co., TX.

Plate 192. *Ferocactus hamatacanthus* var. *hamatacanthus* (giant fishhook-cactus); mature fruits; cultivated; from S Brewster Co., TX.

Plate 193. *Ferocactus sinuatus* var. *sinuatus* (lower Rio Grande Valley barrel cactus); S TX (photo by James H. Everitt).

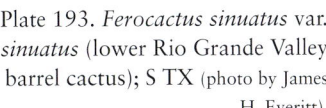

Plate 194. *Hamatocactus bicolor* (twisted-rib cactus); cultivated; from S TX.

Plate 195. *Hamatocactus bicolor* (twisted-rib cactus); mature fruits; cultivated; from S TX.

Plate 196. *Thelocactus bicolor* var. *bicolor* (glory of Texas); cultivated; from S Brewster Co., TX.

Plate 197. *Thelocactus bicolor* var. *bicolor* (glory of Texas); mature fruits; cultivated; from S Brewster Co., TX.

Plate 198. *Thelocactus bicolor* var. *flavidispinus* (Marathon Basin thelocactus); cultivated; from Marathon Basin, Brewster Co., TX.

Plate 199. *Thelocactus bicolor* var. *flavidispinus* (Marathon Basin thelocactus); mature fruit; cultivated; from Marathon Basin, Brewster Co., TX.

Plate 200. *Lophophora williamsii* (peyote); Presidio Co., TX.

Plate 201. *Lophophora williamsii* (peyote); cultivated; from Presidio Co., TX.

Plate 202. *Lophophora williamsii* (peyote); mature fruits; cultivated; from Presidio Co., TX.

Plate 203. *Ariocarpus fissuratus* var. *fissuratus* (living rock cactus); Black Gap Wildlife Management Area, Brewster Co., TX.

Plate 204. *Ariocarpus fissuratus* var. *fissuratus* (living rock cactus); Rosillos Mts, Brewster Co., TX.

Plate 205. *Ariocarpus fissuratus* var. *fissuratus* (living rock cactus); mature fruits; Big Bend National Park, TX.

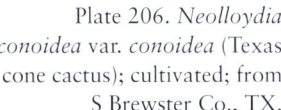

Plate 206. *Neolloydia conoidea* var. *conoidea* (Texas cone cactus); cultivated; from S Brewster Co., TX.

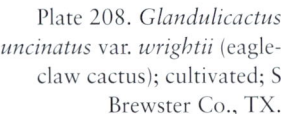

Plate 207. *Neolloydia conoidea* var. *conoidea* (Texas cone cactus); cultivated; from near Sanderson, Terrell Co., TX.

Plate 208. *Glandulicactus uncinatus* var. *wrightii* (eagle-claw cactus); cultivated; S Brewster Co., TX.

Plate 209. *Glandulicactus uncinatus* var. *wrightii* (eagle-claw cactus); cultivated; from S Brewster Co., TX.

Plate 210. *Glandulicactus uncinatus* var. *wrightii* (eagle-claw cactus); mature fruits; Rosillos Mts, Brewster Co., TX.

Plate 211. *Ancistrocactus brevihamatus* var. *brevihamatus* (short-spined fishhook cactus); cultivated; from Val Verde Co., TX.

Plate 212. *Ancistrocactus brevihamatus* var. *brevihamatus* (short-spined fishhook cactus); cultivated; from Val Verde Co., TX.

Plate 213. *Ancistrocactus tobuschii* (Tobusch fishhook cactus); cultivated; from McKinney Co., TX.

Plate 214. *Ancistrocactus tobuschii* (Tobusch fishhook cactus); mature fruits; Edwards Co., TX (photo by Mark Lockwood).

Plate 215. *Ancistrocactus scheeri* (Scheer's fishhook cactus); cultivated; from S TX.

Plate 216. *Ancistrocactus scheeri* (Scheer's fishhook cactus); mature fruit; cultivated; from S TX.

Plate 217. *Ancistrocactus brevihamatus* var. *pallidus* (snipe cactus); NE Brewster Co., TX (photo by Shirley A. Powell).

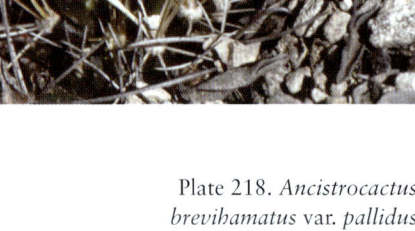

Plate 218. *Ancistrocactus brevihamatus* var. *pallidus* (snipe cactus); cultivated; Terrell Co., TX.

Plate 219. *Ancistrocactus brevihamatus* var. *pallidus* (snipe cactus); immature fruit; cultivated; NE Brewster Co., TX.

Plate 220. *Toumeya papyracantha* (grama-grass cactus); near Los Alamos, Sandoval Co., NM (photo by Dale and Marian Zimmerman).

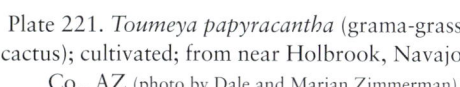

Plate 221. *Toumeya papyracantha* (grama-grass cactus); cultivated; from near Holbrook, Navajo Co., AZ (photo by Dale and Marian Zimmerman).

Plate 222. *Echinomastus intertextus* var. *intertextus* (woven-spine pineapple cactus); N Brewster Co., TX.

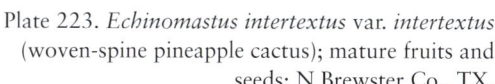

Plate 223. *Echinomastus intertextus* var. *intertextus* (woven-spine pineapple cactus); mature fruits and seeds; N Brewster Co., TX.

Plate 224. *Echinomastus intertextus* var. *dasyacanthus* (longcentral woven-spine pineapple cactus); Franklin Mts, El Paso Co., TX (photo by Richard D. Worthington).

Plate 225. *Echinomastus intertextus* var. *dasyacanthus* (longcentral woven-spine pineapple cactus); Doña Ana Co., NM.

Plate 226. *Echinomastus intertextus*; putative hybrid between *E. intertextus* var. *intertextus* and *E. warnockii*; Chisos Mts, Brewster Co., TX.

Plate 227. *Echinomastus warnockii* (Warnock's cactus); cultivated; from S Brewster Co., TX.

Plate 228. *Echinomastus warnockii* (Warnock's cactus); mature fruits; cultivated; from Mariscal Mt, S Brewster Co., TX.

Plate 229. *Echinomastus mariposensis* (Mariposa cactus); S Brewster Co., TX.

Plate 230. *Echinomastus mariposensis* (Mariposa cactus); mature fruits; S Brewster Co., TX.

Plate 231. *Epithelantha micromeris* var. *micromeris* (common button cactus); cultivated; from S Brewster Co., TX.

Plate 232. *Epithelantha micromeris* var. *micromeris* (common button cactus); mature fruits; cultivated; from S Brewster Co., TX.

Plate 233. *Epithelantha bokei* (Boke's button-cactus); cultivated; from S Brewster Co., TX.

Plate 234. *Epithelantha bokei* (Boke's button-cactus); cultivated; from S Brewster Co., TX.

Plate 235. *Epithelantha bokei* (Boke's button-cactus); mature fruits; Solitario uplift, Presidio Co., TX.

Plate 236. *Mammillaria grahamii* (Graham's fishhook cactus); W Presidio Co., TX.

Plate 237. *Mammillaria grahamii* (Graham's fishhook cactus); mature fruits; cultivated; plants from S Presidio Co., TX.

Plate 238. *Mammillaria wrightii* var. *wrightii* (Wright's fishhook cactus); W slope, Organ Mts, Doña Ana Co., NM (photo by Dale and Marian Zimmerman).

Plate 239. *Mammillaria wrightii* var. *wrightii* (Wright's fishhook cactus); near Santa Rita, Grant Co., NM (photo by Dale and Marian Zimmerman).

Plate 240. *Mammillaria lasiacantha* (golf ball cactus); pinkish midregions; Fresno Creek, Presidio Co., TX.

Plate 241. *Mammillaria lasiacantha* (golf ball cactus); brownish midregions; Rosillos Mts, S Brewster Co., TX.

Plate 242. *Mammillaria lasiacantha* (golf ball cactus); mature fruits; Candelaria, Presidio Co., TX.

Plate 243. *Mammillaria pottsii* (Potts' mammillaria); cultivated; from S Brewster Co., TX.

Plate 244. *Mammillaria pottsii* (Potts' mammillaria); cultivated; S Brewster Co., TX.

Plate 245. *Mammillaria meiacantha* (nipple cactus); cultivated; from Davis Mts, Jeff Davis Co., TX.

Plate 246. *Mammillaria meiacantha* (nipple cactus); mature fruit; cultivated; from Davis Mts, Brewster Co., TX.

Plate 247. *Mammillaria heyderi* var. *heyderi* (Heyder's pincushion cactus); S Brewster Co., TX.

Plate 248. *Mammillaria heyderi* var. *heyderi* (Heyder's pincushion cactus); mature fruits; cultivated; from S Brewster Co., TX.

Plate 249. *Mammillaria heyderi* var. *bullingtoniana* (western Heyder pincushion cactus); flowers and mature fruits; Peloncillo Mts, Hidalgo Co., NM (photo by Dale and Marian Zimmerman).

Plate 250. *Mammillaria prolifera* var. *texana* (hair-covered cactus); cultivated; from Terrell Co., TX.

Plate 251. *Mammillaria prolifera* var. *texana* (hair-covered cactus); mature fruits; cultivated; from Terrell Co., TX.

Plate 252. *Mammillaria sphaerica* (pale mammillaria); cultivated; from McMullen Co., TX.

Plate 253. *Coryphantha dasyacantha* (desert pincushion cactus); cultivated; from S Presidio Co., TX.

Plate 254. *Coryphantha dasyacantha* (desert pincushion cactus); mature fruits; cultivated; from S Presidio Co., TX.

Plate 255. *Coryphantha dasyacantha* (desert pincushion cactus); Jeff Davis Co., TX.

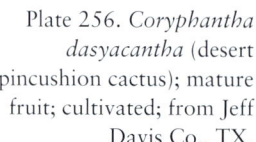

Plate 256. *Coryphantha dasyacantha* (desert pincushion cactus); mature fruit; cultivated; from Jeff Davis Co., TX.

Plate 257. *Coryphantha chaffeyi* (Chaffey's pincushion cactus); Chisos Mts, S Brewster Co., TX.

Plate 258. *Coryphantha chaffeyi* (Chaffey's pincushion cactus); mature fruits; cultivated; from S Brewster Co., TX.

Plate 259. *Coryphantha duncanii* (Duncan's pincushion cactus); Mariscal Mt, S Brewster Co., TX.

Plate 260. *Coryphantha duncanii* (Duncan's pincushion cactus); mature fruits; cultivated; from Mariscal Mt, S Brewster Co., TX.

Plate 261. *Coryphantha minima* (Nellie's pincushion cactus); Marathon Basin, Brewster Co., TX.

Plate 262. *Coryphantha minima* (Nellie's pincushion cactus); mature fruits, barely visible; cultivated; from Marathon Basin, Brewster Co., TX.

Plate 263. *Coryphantha hesteri* (Hester's pincushion cactus); 9 mi SE of Alpine, Brewster Co., TX.

Plate 264. *Coryphantha hesteri* (Hester's pincushion cactus); Marathon Basin, Brewster Co., TX.

Plate 265. *Coryphantha hesteri* (Hester's pincushion cactus); mature fruits, some spines cut away; cultivated; from N Brewster Co., TX.

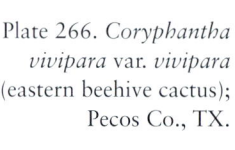

Plate 266. *Coryphantha vivipara* var. *vivipara* (eastern beehive cactus); Pecos Co., TX.

Plate 267. *Coryphantha vivipara* var. *vivipara* (eastern beehive cactus); cultivated; from Marathon Basin, Brewster Co., TX.

Plate 268. *Coryphantha vivipara* var. *radiosa*; Pecos Co., TX.

Plate 269. *Coryphantha vivipara* var. *radiosa*; Brewster Co., TX.

Plate 270. *Coryphantha vivipara* var. *neomexicana* (New Mexico beehive cactus); cultivated; from Jeff Davis Co., TX.

Plate 271. *Coryphantha vivipara* var. *neomexicana* (New Mexico beehive cactus); Sutton Co., TX.

Plate 272. *Coryphantha sneedii* var. *sneedii* (Sneed's pincushion cactus); Franklin Mts, El Paso Co., TX (photo by Richard D. Worthington).

Plate 273. *Coryphantha sneedii* var. *sneedii* (Sneed's pincushion cactus); Franklin Mts, El Paso Co., TX (photo by Richard D. Worthington).

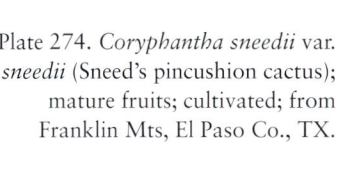

Plate 274. *Coryphantha sneedii* var. *sneedii* (Sneed's pincushion cactus); mature fruits; cultivated; from Franklin Mts, El Paso Co., TX.

Plate 275. *Coryphantha sneedii* var. *guadalupensis* (Guadalupe pincushion cactus); Guadalupe Mts, Culberson Co., TX (photo by Brent Wauer).

Plate 276. *Coryphantha sneedii* var. *guadalupensis* (Guadalupe pincushion cactus); fruit; cultivated; Mesa Garden, NM (photo by James F. Weedin).

Plate 277. *Coryphantha sneedii* var. *leei* (Lee's pincushion cactus); cultivated; from S NM.

Plate 278. *Coryphantha sneedii* var. *leei* (Lee's pincushion cactus); fruits; cultivated; plants from S NM.

Plate 279. *Coryphantha sneedii* var. *albicolumnaria* (silverlace cactus); near Terlingua, Brewster Co., TX.

Plate 280. *Coryphantha sneedii* var. *albicolumnaria* (silverlace cactus); cultivated; from S Brewster Co., TX.

Plate 281. *Coryphantha sneedii* var. *albicolumnaria* (silverlace cactus); white-flowered form; cultivated; from S. Brewster Co., TX.

Plate 282. *Coryphantha sneedii* var. *albicolumnaria* (silverlace cactus); mature fruits, red form; cultivated; S Brewster Co., TX.

Plate 283. *Coryphantha sneedii* var. *albicolumnaria* (silverlace cactus); mature fruits, green form; cultivated; from Solitario, Presidio Co., TX.

Plate 284. *Coryphantha pottsiana* (Runyon's pincushion cactus); cultivated; from Val Verde Co., TX.

Plate 285. *Coryphantha pottsiana* (Runyon's pincushion cactus); mature fruits; cultivated; from Val Verde Co., TX.

Plate 286. *Coryphantha tuberculosa* var. *tuberculosa* (cob cactus); Rosillos Mts, S Brewster Co., TX.

Plate 287. *Coryphantha tuberculosa* var. *tuberculosa* (cob cactus); mature fruits; cultivated; from S Brewster Co., TX.

Plate 288. *Coryphantha tuberculosa* var. *varicolor* (varicolor cob cactus); cultivated; from Jeff Davis Co., TX.

Plate 289. *Coryphantha tuberculosa* var. *varicolor* (varicolor cob cactus); mature fruits; cultivated; from Jeff Davis Co., TX.

Plate 290. *Coryphantha macromeris* var. *macromeris* (big-needle pincushion cactus); Study Butte, S Brewster Co., TX.

Plate 291. *Coryphantha macromeris* var. *macromeris* (big-needle pincushion cactus); mature fruits; cultivated; from S Brewster Co., TX.

Plate 292. *Coryphantha macromeris* var. *runyonii* (Runyon's coryphantha); cultivated; from Starr Co., TX.

Plate 293. *Coryphantha macromeris* var. *runyonii* (Runyon's coryphantha); mature fruit; cultivated; from Starr Co., TX.

Plate 294. *Coryphantha ramillosa* (whiskerbrush pincushion cactus); cultivated; from S Brewster Co., TX.

Plate 295. *Coryphantha ramillosa* (whiskerbrush pincushion cactus); mature fruits; cultivated; from S Brewster Co., TX.

Plate 296. *Coryphantha scheeri* var. *scheeri* (long-tubercled coryphantha); cultivated; from S Brewster Co., TX.

Plate 297. *Coryphantha scheeri* var. *scheeri* (long-tubercled coryphantha); mature fruits; cultivated; from Brewster Co., TX.

Plate 298. *Coryphantha scheeri* var. *uncinata* (El Paso long-tubercled coryphantha); El Paso Co., TX (photo by Richard D. Worthington).

Plate 299. *Coryphantha scheeri* var. *uncinata* (El Paso long-tubercled coryphantha); Fabens, El Paso, TX (photo by Dale & Marian Zimmerman).

Plate 300. *Coryphantha echinus* var. *echinus* (sea-urchin cactus); Pecos Co., TX.

Plate 301. *Coryphantha echinus* var. *echinus* (sea-urchin cactus); mature fruits; cultivated; from Pecos Co., TX.

Plate 302. *Coryphantha echinus* var. *robusta* (multistemmed sea-urchin cactus); habit; S Tornillo Creek, Big Bend National Park, Brewster Co., TX.

Plate 303. *Coryphantha echinus* var. *robusta* (multistemmed sea-urchin cactus); S Tornillo Creek, Big Bend National Park, Brewster Co., TX.

Plate 304. *Coryphantha echinus* var. *robusta* (multistemmed sea-urchin cactus); mature fruits; cultivated; from S of Chisos Mts, Brewster Co., TX.

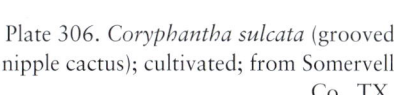

Plate 305. *Coryphantha echinus* var. *robusta* (multistemmed sea-urchin cactus); cultivated; from near Ruidosa, Presidio Co., TX.

Plate 306. *Coryphantha sulcata* (grooved nipple cactus); cultivated; from Somervell Co., TX.

Plate 307. *Coryphantha sulcata* (grooved nipple cactus); mature fruit; Somervell Co., TX.

Plate 308. *Coryphantha missourensis* var. *caespitosa* (nipple cactus); cultivated; from Parker Co., TX, S of Weatherford.

Plate 309. *Acanthocereus tetragonus* (triangle cactus) (photo by Scooter Cheatham).

Plate 310. *Acanthocereus tetragonus* (triangle cactus); mature fruit; cultivated.

Plate 311. *Astrophytum asterias* (sea-urchin cactus or star cactus); Starr Co., TX (photo by Jackie M. Poole)

Plate 312. *Pereskia aculeata* (lemonvine); cultivated.

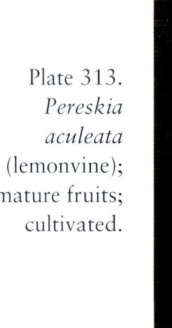

Plate 313. *Pereskia aculeata* (lemonvine); mature fruits; cultivated.